QM Library
KU-996-863
23 1375394 8

Human Haptic Perception: Basics and Applications

Martin Grunwald
Editor

Birkhäuser
Basel · Boston · Berlin

Martin Grunwald
University of Leipzig
Paul Flechsig Institute for Brain Research
Haptic and EEG-Research Laboratory
Johannisallee 34
04103 Leipzig
Germany

Library of Congress Control Number: 2008936532

Bibliographic information published by Die Deutsche Bibliothek
Die Deutsche Bibliothek lists this publication in the Deutsche Nationalbibliografie;
detailed bibliographic data is available in the internet at <http://dnb.ddb.de>.

ISBN 978-3-7643-7611-6 Birkhäuser Verlag, Basel – Boston – Berlin

Part of Springer Science+Business Media
Printed on acid-free paper produced from chlorine-free pulp. TCF ∞
Cover design: Micha Lotrovsky, 4106 Therwil, Switzerland
Back cover illustrations (from top to bottom):
- See page 89. With friendly permission of the authors.
- See page 305. With friendly permission of the authors.
- Photo taken by Martin Grunwald, with friendly permission.
- See page 150. With friendly permission of the authors
- See page 434. Photo taken by Dirk Göger, with friendly permission.
- Detail from "The Incredulity of Saint Thomas", painting by Michelangelo Merisi da Caravaggio
- Erich Kissing at work (www.erich-kissing.de), photo taken by Martin Grunwald, with friendly permission.

Printed in Germany
ISBN 978-3-7643-7611-6

e-ISBN 978-3-7643-7612-3

9 8 7 6 5 4 3 2 1

www.birkhauser.ch

Contents

If touch is not a single perception, but many instead, then its purposes are also manifold

Aristotle (384–322 BC) *De Anima*

Preface

The quotation by Aristotle aptly describes the complexity of content and structure embodied in the sense of touch. No other sense exhibits properties so variable in scope or remains so puzzling even today – understood only in terms of its principle features. Viewed from phylogenetic and ontogenetic perspectives, the sense of touch plays a central role relative to the other senses. Its fundamental significance to humans derives from its epistemiological function, making possible an awareness of surroundings and the consciousness of self. In this way, the sense of touch is sine qua non for thought, action, and consciousness.

Since the beginnings of scientific research into touch, the most varied of scientific disciplines have investigated one aspect or the other of this sense. The questions posed and the methods used to conduct such research are just as varied as the disciplines devoted to it. The particular results of these efforts, however, have not yet led to a comprehensive theory of the sense of touch. Too many questions remain unanswered. As of today, the wish expressed by Max Dessoire, to integrate the various areas of research pertaining to the sense of touch into one scientific doctrine – described by him as the doctrine of haptic perception – remains illusive. This state of affairs stands in contrast to the fact that there exists a great, almost unmanageable quantity of findings that are oriented toward elucidating basic principles and related to applications that, it must be admitted, clarify many facets of the sense of touch and are of technical benefit. Manifold are the findings in the realm of tactile perception where the investigated subject doing the perceiving behaves passively with respect to the stimulus. But just as fundamental is scientific clarification of the haptic perception process that derives from a subject who is actively engaged – either aware or unaware. This state of perception-cognition for tactile perception requires the entire scope of properties inherent to the sense of touch and is an everyday, universal process in our lives. Long before our birth, this active process constructs the initial, flexible neuro-sensorial matrix to which all other senses are obliged to relate, a set of circumstances that persists as a life-long requirement for life. And for this reason, ever more scientists worldwide are researching the biological, psychological, neurochemical, and social mechanisms of human haptic perception and its interactions with the other senses. Additionally, new areas of application are continually developing, e.g. in the field of rehabilitation, virtual interfaces, robotics, and in haptic design, where principles of human haptic perception are converted and then implemented in practice.

In view of this fascinating, dynamic background, this volume, which subdivides into six sections, compiles contributions from 46 international authors on the most varied topics of human haptic perception. In the first section, philosophical and historical aspects of the sense of touch are introduced. Here, authors from four different countries analyze the beginnings of scientific research into the sense of touch, from the start of the 19th to the middle of the 20th century, a time when psychological, terminological, and methodological foundations were laid for today's research. These contributions are intended to clarify the essential sources of the branches of research that exist today and should be helpful in placing current research into the required historical context. It will not hereby remain unstated that we are amazed by our own recognition that – during recourse to the progenitors of the science of touch – a number of concepts considered new today are, in fact, more than 100-years old.

The second section of this volume presents fundamental aspects of the anatomical, physiological and neurophysiological conditions in our bodies that provide the basis for the realization of human haptic perceptions. These biological aspects are essential to an understanding of the various psychological and clinicopathological processes of human haptic perception. Beyond that, they represent a link, in terms of function and content, between the human model and areas of virtual-technical application.

As nature would have it, haptic perception fulfills multifaceted psychological functions in all realms and stages of life. Several of these functions, as well as various psychological and psycho-physiological aspects of human haptic perception, are covered in section three of the book. Although such a presentation can never be exhaustive, the contributed topics range from prenatal mechanisms of haptic perception to learning, memory, illusions, synaesthesia – all the way to questions of haptic perception in space travel.

Section four continues with a presentation of various clinico-neuropsychological topics. Even if this subject area is not yet a part of the mainstream of clinicopsychological and neurological diagnostics and intervention, new and exciting perspectives have emerged in recent years that benefit the pathology of haptic and tactile perception both therapeutically and in clinical diagnosis. Of particular significance in this regard is the universal interconnection between haptic perception and body schema representation in relation to different mental disorders.

In parallel with rapid technical developments in recent decades, an innovative and, in part, spectacular field of research and applications has been established, having the goal of implementing the principles of human haptic perception in virtual scenes, different electro-mechanical interfaces, and in robotic systems. In this way, engineers, psychologists and neuroscientists are making great strides into the field of haptic simulation in the context of various technical systems. As a result, not only are new and beneficial applications being discovered and applied, but, by these means, new perspectives are emerging in the field of research methodology. The fundamental principles of this field of research and the areas of application are described in section five of the book.

Research into our senses has always been associated with the search for practical as well as industrial applications. The search for knowledge has thus never been far removed from the goal of practical utility. In part, such goals are, in fact, the motivation for the research. Even as we see this trend emerge more evidently in other realms of the senses and in our everyday lives, practical/technical applications as far as research pertaining to the sense of touch is concerned often still go unnoticed at large. The spectrum of these developments – so-called haptic design – ranges from new and improved surface properties for devices and products of all types, to changes in complex haptic properties in the operation of machinery or vehicles. Equally broadly diversified are practical applications in the fields of rehabilitation or assistance to help orient the blind and individuals with poor sight. A selection of such types of applications are illustrated in section six by way of examples.

This present volume is tied to the hope that the broadly diversified illustrations of the most varied aspects of human haptic perception will provide a useful tool to those unfamiliar with the field as well as to students and to scientist from various disciplines. Not least, the book should be a stimulus and a support for all those who are currently, or will be in the future, concerned with new perspectives on research and application in human haptic perception. The fact that not all of the planned aspects of human haptic perception could be taken into consideration in this volume is attributable to the natural limitations of such a project. The publisher and the authors sincerely hope that editions to follow will expand the spectrum of depiction.

The publication of this textbook has only been possible because two powerful and dedicated forces were active in equal measure – for which I would like to express my deepest thanks at this time. First of all, we had the many authors who believed in this project and who, by means of their contributions, created the inherent substance of this book. Equally, I thank Dr. Hans Detlef Klüber, of Birkhäuser Publishers, for his proposal to bring this book into being and for his patient support and optimism in all phases of this project. I would like to give special thanks for the trust placed, and the dedication contributed, by all of those who participated in this book project, as well as for the personal support offered by my colleagues, F. Krause and I. Thomas. I conclude this editorial effort on this volume with the sincere hope that the basic, interdisciplinary research and applications pertaining to the sense of touch will come to assume a central role within the life sciences in the future.

Leipzig, April 2008 Martin Grunwald

I.
Epistemological and historical aspects

1

Haptic perception: an historical approach

Robert Jütte

Traditional perceptions of the sense of touch

The idea that perception or sensation may be localised in certain physical organs (e.g., skin) has a long tradition. It pervades many cultures. The system of sensory physiology (of which touch is one important element) is shaped by the influence of both medical thought and the philosophy of nature. Let us turn first to ancient Indian medicine or natural philosophy, as it appears in the Vedas. The Vedas are the most ancient Indian religious texts and consist for the most part of hymns, liturgical chants, sacrificial formulas and magic spells. The Rgveda, the oldest of the vedic texts, has not yet a verb for 'touch' or 'feel' and no expression for the corresponding sensation which – in a later text entitled Atharvaveda – is called *sam-sparsa* (feeling) [1]. In the Ayurveda, which forms an appendix to the Atharvaveda, the primeval matter (*sattva*) acts upon the fives senses of knowledge or *buddhīndrīya* (hearing, touch, sight, taste, smell – Fig. 1). The sense of touch is associated with the wind, one of the five elements in ancient Indian philosophy. The skin, as one of the sense organs, is envisaged simply as the meeting point of the qualities or object assigned to this sense: skin – finger – grasping – feeling.

In ancient China, too, the human organism was perceived as a miniature copy of the universe. The doctrine of the five elements or the five phases of transformation is the basis of the idea that there are many numerical correspondences between nature and the human body (Fig. 2).

The sense of touch thus plays an important role in Chinese pulse diagnostics, e.g., in a clas-

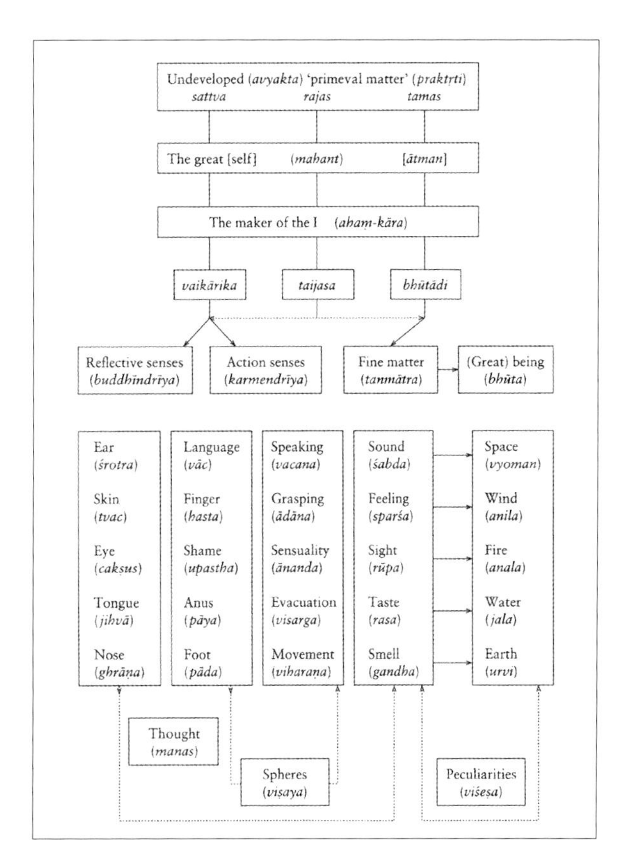

FIGURE 1. THE SYSTEM OF THE FIVES SENSES IN ANCIENT MEDICINE
Source: RFG Müller (1951) Grundsätze altindischer Medizin. *Munksgaard, Copenhagen, 83*

sical text entitled '*Seven sorts of Pulses which indicate danger of Death*' (dating back to the 3rd century AD). The metaphors used in describing these pulses concern tactile perception, for example: "*If the Motion of the Pulse resembles the hasty pecking of the Beak of a Bird, there is a failure*

Elements *hsing*	Direction *fang*	Sense organ *kuan*	Tastes *wei*	Smells *chou*	*Yin-yang*	Internal organ *tsang*	Part of body *thi*
Wood	East	Eye	Sour	Goatish	Yin in yang or lesser yang	Spleen	Muscles
Fire	South	Tongue	Bitter	Burning	Yang or greater yang	Lung	Pulse(blood)
Earth	Centre	Mouth	Sweet	Fragrant	Equal balance	Heart	Flesh
Metal	West	Nose	Acrid	Rank	Yang in yin or lesser yin	Kidneys	Skin and hair
Water	North	Ear	Salt	Rotten	Yin or greater yang	Liver	Bones (marrow)

Figure 2. Symbolic correlations of the sense organs in the Chinese tradition
Source: J Needham (1978) The shorter science and civilisation in China. *CUP, Cambridge, Table 9 (selection)*

of Spirits in the Stomach: one may also conclude that the Heart performs its Functions but ill, and that the Blood is in no good condition" [2]. Other descriptions of dangerous pulses do not refer to a primarily tactile perception, although some figurative comparisons may be explained by a tactile experience of sensing distinctive 'pulses'. According to Elisabeth Hsu these descriptions do not solely express the physicians' tactile experiences but are part and parcel of a more general familiarity with tactile perception.

The Greek philosopher Empedocles uses the word *pagamai* (flat of the hand or gripper) to denote the senses in general [3]. This means that his descriptions of sensory perception in general refer to the sense of touch. In his *Timaeus*, Plato deals systematically with the senses. Unlike the other senses, he does not attach the sense of touch to a specific physical organ. In his opinion sensations of pleasure and pain and other qualities perceptible to the senses, such as warm and cold, feature as "*disturbances that affect the whole body in a common way*" (Timaeus, 65c). Aristotle (384–322 BC) not only expanded the hitherto merely inchoate physiology of the senses, but also advanced them to a state of completion that retained its authority well after the Christian Middle Ages [4]. In the Aristotelian view, each function is determined by its object. Applied to the senses, this means that each sense organ is assigned to a specific object of perception. Aristotle's *De anima* deals with the senses one by one in the order of sight, hear, smell, taste and touch, placing special emphasis on each case on the object of the perception. The organ of the sense of touch is not the skin, but the heart. The corresponding medium (the flesh) is thus in the body itself, and not outside it. Aristotle describes the object of the sense of touch as palpable. The distinction between the palpable and the visible or the resonant lies in the fact that, while the latter are perceived through the agency of the medium, here "*it is as if a man were struck through his shield, where the shock is not first given through the shield and passed onto the man, but the concussion of both is simultaneous.*" (De anima, 423b, 15ff). For this reason, Aristotle considers the sense of touch to be more closely related than the other senses to the four elements, since the properties of the elements (e.g., dry and wet) are palpable (Fig. 3).

The Aristotelian doctrine of the fully unified and independent nature of the sense of touch was scarcely ever questioned in the subsequent centuries. The *De anima* of Albertus Magnus (c. 1197–1280) follows him in similarly classifying qualities such as hard and soft and rough and smooth as derivatives of the qualities primarily registered by the sense of touch (e.g., warm or cold) [5]. As we know today, these tactile qualities are, in fact, detected by sensors in the skin that pass on the corresponding stimuli to the brain *via* the peripheral nerves and the spinal cord. But until the 19th century, by which tome experimental physiology had made substantial progress, it was impossible to form any definite,

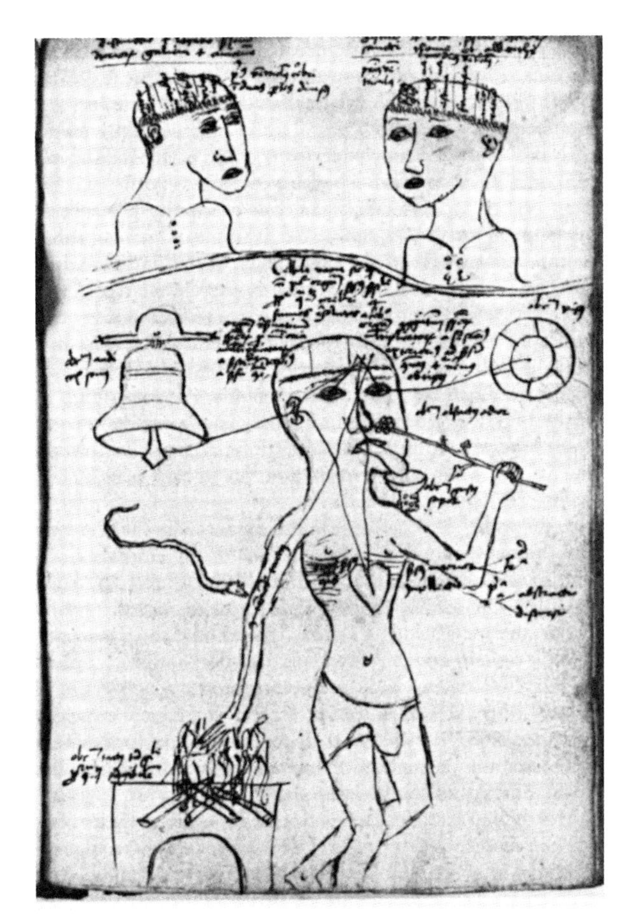

FIGURE 3. MEDIEVAL REPRESENTATION OF ARISTOTLE'S DOCTRINE OF THE SENSES (1496)
Source: R Jütte (2005) A history of the senses. From Antiquity to Cyberspace, *translated by James Lynn. Polity Press, Cambridge, 37*

let alone correct, idea of the way these stimuli were relayed, even though Albertus Magnus had already drawn attention to the central role of the nerves in the sense of touch.

The sense of touch held on to the special position in the hierarchy of the senses granted to it by Aristotle well into the Middle Ages, and even to the modern era. Much of this was due to the pivotal medieval philosopher and theologian St Thomas Aquinas (1224/25–1274). In his own *De anima*, Aquinas endorsed Aristotle's view that without the sense of touch there could be no other senses [6]. In his pithy formulation, touch is "*the first sense, the root and ground, as it were, of the other senses*" [7]. Aquinas's ranking of the senses is based on the doctrine of immutatio spiritualis or mental modification, an incorporeal yet material transcription of sensory stimuli. Thus, a beam of light striking the eye does not produce a physical change. In the case of hearing, smell and taste, on the other hand, a hybrid form of mental and physical change is already present, while in the case of touch a material transcription takes place.

The status of touch

The sense of touch is the extremist among the senses, for it has frequently been ranked both at the bottom and at the top of the scale of esteem. This apparent contradiction goes back to its variable status in Aristotle, for, while ranking it fifth in order of merit (after sight, hearing, smell and taste) the treatise on the soul also describes it as a sense that reaches its highest form of development in man (*De anima*, 412a, 22). The Arab scholar Avicenna (980–1037) provides one explanation of Aristotle's conflicting statements. As he understood it, what the Greek philosopher meant was that with respect to honour the primacy of the sense of sight applied, but that from a point of view of natural aptitude the sense of touch merited priority. This resolution of the contradiction met with the approval of many medieval scholars. Aquinas developed a complex theory based on Aristotle's doctrine of the soul, in which touch and sight are granted more or less equal rights. In common with Avicenna the thought that, in addition to the traditional hierarchy dominated by vision, there was a second hierarchical order in which the sense of touch played the major perceptual role. With his exhaustive and conclusive arguments for the alternative primacy of sensation, Aquinas shows himself to be a decidedly original thinker.

Aquinas opens his case for the superior status, and systematic primacy, of the sense of touch by noting that sensitive life forms define themselves by means of their sense of touch: "*touch [is] the first sense, the root and ground, as it were, of the other senses, the one which entitles a living thing to be called sensitive*" [8]. The haptic comprehension of the world is also of central importance for the survival of the individual and the species, since it is the means by which we distinguish between the edible and the inedible – an argument already advanced by Avicenna [9]. Another important argument for the priority of touch, according to Aquinas, is that it is the root (radix fontalis) of all sentient activity. It follows from this that the other senses are all derived from it: "*In the first place touch is the basis of sensitivity as a whole; for obviously the organ of touch pervades the whole body, so that the organ of each of the other senses is also an organ of touch, and the sense of touch by itself constitutes a being as sensitive*" [10]. Thus, the operations of the other senses are seen as subordinate to tactus. Aquinas's third argument for the precedence of touch rest upon its optimal performance in the process of gathering knowledge: "*Therefore the finer one's sense of touch, the better, strictly speaking, is one's sensitive nature as a whole, and consequently the higher one's intellectual capacity. For a fine sensitivity is a disposition to a fine intelligence*" [11]. Following Aristotle's idea of the flesh as the medium of touch, Aquinas argues that sensorial being with 'hard flesh' would not perceive things as well as those with soft flesh (e.g., man) and would therefore be less receptive to perception of any kind.

This remarkable reappraisal of touch may be encountered in the works of other Christian, Muslim and Jewish scholars of the Middle Ages, although their approach was sometimes slightly different and their distinctions less subtle. After the 13th century, the Jewish tradition, for example, underwent a change that also may be seen later in the allegorical representations of the five senses in the Renaissance or in the Age of baroque: the sense of touch falls increasingly into disrepute. The culprit, once more, is Aristotle, or, more precisely, his *Nicomachean Ethics* (III, 8b), where the sense of taste is associated with "*the pleasures of love*" and accused of "*disorderliness*". No less a figure than the Jewish medieval philosopher and learned physician Maimonides (1135–1204), who refers to this passage in his *Guide for the Perplexed* (II, 36), decided to approve it. It was, however, his later translators and commentators who forged a connection between this approving quotation of Aristotle and various places in the Bible (e.g., Deut. 4: 28), thereby helping to ensure that the mental association of touch with sinful behaviour (voluptuousness and unrestrained sex drive) became widespread. Thus, both Abraham ben Schemtov Bibago, who was a doctor at the Court of King Juan II of Aragon towards the end of the 15th century, and the celebrated Talmudist Moses Isserles (c. 1525–1572) refer explicitly to the sense of touch as shame (hebr. *cherpah*) [12]. The association of touch with the sexual urge in the language and visual imagery of the Middle Ages and above all, the modern era has other roots besides these. The Christian tradition should be mentioned here. It was Eve who first touched the apple when she seduced Adam. In this way, the haptical perception became a symbol of eroticism as such, to which poets and painters of not only the early modern age returned time and time again.

Tactile imagery

Aristotle had assigned no specific organ to the sense of touch and insisted that haptic perception was distributed all over the body. Nevertheless, if it was to be represented at all it had to be positioned somewhere in the body. The obvious organ was the hand, with which the human being feels, holds and 'grasps' in the metaphorical sense [13]. In the biblical scenes which were used to adorn 16th century allegories of touch the fifth and last sense is represented by various forms of hand-touching. The biblical passages used in these allegories symbolised the ambivalence of the tactile sense: on the one hand it is the only source of salvation and on the other the cause of doom (Fig. 4).

FIGURE 4. ALLEGORY OF TACTUS BY MARTEN DE VOS (1532–1603)
Source: S Ferino-Pagden (ed) (1996) Immagini del sentire. I cinque sensi nell'arte. *Leonardo Arte, Cremona, 115*

Following Aristotle's praise of the relative reliability of touch in situations where the other senses may be deceived, it is not surprising that, in the Bible, for example, touching and feeling are the most effective ways of convincing ourselves of the real existence of a thing or phenomenon (cf. Luke 24: 38–39, John 20: 27). Touching consequently becomes the simplest and most basic form of communion with the sacred. This plastic idea was not least a factor in the formation of the medieval cult of relics, in which a large role is played by the touching of the bodily remains of saints and items of their clothing.

According to the American medical and cultural historian Sander L. Gilman, many medieval pictorial representations of sensory perception refer to pleasure, and particularly to sexual lust [14]. Indeed, there can be little doubt that tactile experience had sexual connotations well into the modern era. When the characters of the decidedly earthy Shrovetide plays spoke of liking to 'feel a woman' they had more in mind than the fondling of breast and the various other forms of sexual harassment. Most of the pictorial representations that show a man reaching into a woman's décolletage are unambiguous iconographic symbols of sexual intercourse (Fig. 5).

On the other hand, the emblematic art of the 16th and 17th centuries is also concerned with the relation between bodily sensation and pain [15]. The Dutch art of this period is full of allegorical representations of the sense of touch centred on the theme of pain during sickness and treatment. The interpretation of the tactile sense as a medium of pain and other unpleasant physical experiences was largely influence by Cesare Ripa's *Iconologia* (1593), a standard work on iconography for artists in which the sense of touch is almost exclusively associated with pain.

Numbness

Before the advent of modern diagnostic techniques (e.g., computer tomography) the loss of physical sensation (*tactus imminutus*) was

Figure 5 The five senses (sight) by Hendrick Goltzius (1558–1617)

Source: S Ferino-Pagden (ed) (1996) Immagini del sentire. I cinque sensi nell'arte. *Leonardo Arte, Cremona, 117*

interpreted either as a localised injury to the part of the body in question, or as a symptom or side effect of a serious illness (e.g., leprosy). When examining people for leprosy in the Middle Ages and in the early modern period, physicians looked not only for tell-tale changes to the surface of the skin but also tested its sensitivity by inserting a needle into the calf or middle finger. Any numbness in the suspected patient was regarded as *signa univoca*, a clear symptom of the dreaded disease. Although local anaesthetics were not introduced into surgery until the end of the 19th century, the pain-numbing effects of certain substances had been known since antiquity. These substances had narcotic as well as anaesthetising effects, however, and were rarely used to make the body completely insensitive to pain. It was difficult to calculate the correct dosage, and many experienced surgeons considered the risk that the patient might not wake up from this artfully induced unconsciousness to be too high. For a long period in time, loss of sensation was considered treatable by some and incurable by others. Before the late 19th century, however, the art of medicine was usually at a loss in cases of 'blocked nerves', or damaged neural tissue.

The taming of touch

18th century pedagogues, for instance, regarded the tactile sensitivity of the blind as particularly worthy of emulation. Rousseau advised against children to become too used to heavy manual work, since their hands would become calloused and lose their "*fineness of touch*" [16]. He thought it essential to preserve this sensitivity so they would be able, for example, to identify objects by their feel in conditions of darkness. The German pedagogue Johann Christoph Gutsmuth (1759–1839), for example, promoted in the first half of the 19th century 'exercises in feeling" for school children. Among his contemporaries were some authors who insisted that the hands should be meticulously clean in order to maintain the sense of touch.

Medicine has long abided by certain protocols in respect of the touching and grasping of another body. By the early 19th century, however, traditional pulse-diagnostics was being supplemented increasingly by palpation (examination by touch) [17]. Nevertheless, in the 1870s, the renowned physiologist Hermann von Helmholtz (1821–1894) still regarded pulse-taking as the doctor's most important diagnostic tool (Fig. 6).

Since the 17th century, the sense of touch had gradually been excluded from the act of eating. Only bread continued to be eaten with the fingers in polite company. All other food required the use of a fork. The real world looked,

Figure 6. A doctor's house call by Daniel Chodowiecki (1726–1801)
Source: I Kleimenhagen ((1969) Chodowiecki und die Medizin. *Michael Tritsch, Düsseldorf, 110*

however, different from books on table manners. It was not until the late 19th century that the fork became the basic equipment of a lower-middle class kitchen, while working-class families largely continued to eat with knives and spoons. They probably used their hands as well [18].

The deterioration of tactility

The transformation of the sense of touch as a result of the industrial revolution is still uncharted territory. There is no historical research on the subject. Karl Marx already noted that the lathe had largely replaced the human hand in the production of tools and machines. Nevertheless, machines could not replace much heavy and hazardous manual work, so that calloused hands undoubtedly continued to be a characteristic physical feature of the industrial and manual worker as the 19th century wore one. But touch is not simply localised in the hand. Did the long-term changes in the world of life and work (technologisation, industrialisation) also lead to a deadening of the tactile sense in other sensitive areas of the body? The medical literature of the late 19th century offers a preliminary purchase on the question. The statistics for skin diseases which are available for some countries report no abnormal increases in loss of dermatological sensitivity that might indicate a rise or reduction of normal threshold of tactility. According to a statistical report by the American Dermatological Association, there were only 56 cases of anaesthesia (total loss of sensation) in the USA in the period between 1878 and 1911. 90 of the cases reported were diagnosed as hyperaesthesia, the opposite symptom: acute skin sensitivity. These two dermatological and neurological conditions represent a bare 0.09% and 0.014%, respectively, of all 'skin diseases' reported in the period [19].

Discovering the physiology of haptic perception

The first physiologist to undertake a systematic experimental investigation of the sense of touch was Ernst Heinrich Weber (1795–1878) (Fig. 7). He was aware from the outset of the difficulties of experimental work in this field, realising at an

FIGURE 7. PORTRAIT OF ERNST HEINRICH WEBER
Source: Institut für Geschichte der Medizin der Robert Bosch Stiftung

early stage that "*pure sensation tells us nothing about where the nerves that produce the sensations are being stimulated*" and that "*all sensations are originally only conditions that stimulated our consciousness*" [20].

Knowledge of the fine structure of the skin was sketchy at the time Weber began his experiments. Although the vital importance for the perception of vibrations of the large, lamellae-like corpuscles at the end of the nerve fibres of the subcutis (Pacinian corpuscles) was already recognised, their functions as receptors was still a matter of debate. So it was not surprising that Weber should continue to assume that the skin's senses of heat and pressure were one and the same. It was not until the 1880s that Magnus Gustaf Blix (1849–1904) in Sweden and Alfred Goldscheider (1858–1935) in Germany demonstrated that the skin contained both temperature and pressure points by subjecting tiny areas of skin to electrical stimulation [21].

'Sensorial circles' was Weber's term for highly sensitive areas of the skin. He thought that their anatomical substratum was the skin nerves and assigned a particular nerve to one or more of these circles. He also noted that, when applied the points of a pair of compasses to the skin, the pricks could not be experienced as two distinct sensations beyond a certain distance. This marked the discovery of a phenomenon described towards the end of the 19th century as the 'simultaneous space threshold' by Max von Frey (1852–1932), who was a pupil of the celebrated Leipzig experimental physiologist Carl Ludwig (1816–1894). As late as the 1960s, physiologists were still using Weber's compass to measure the sensitivity of various areas of the hand.

Weber's pioneering exploration of the sense of pressure led to the law of just-noticeable differences being named after him (Weber's law). Weber rejected the idea of a specific epidermal temperature sensor, but this did not deter him from looking for areas of the skin that were more sensitive than others to heat and cold. The role of the so-called pain sense, which can also be localised on the skin, was relatively minor, for he lumped pain together with the 'common sense', about which it was impossible to obtain scientifically exact information.

Weber's achievement in the physiology of skin sensation has been acknowledged by historians of science. His famous treatise *Tastsinn und Gemeingefühl* (*Sense of Touch and Common Sense*, 1846) not only included important research results in the field of physiology of the skin but also gave a boost to sensory physiology in general. Terms still used today in physiological research, such as 'threshold of stimulation', 'temperature sense' or 'simultaneous spatial threshold', were either first coined by Weber or named after his experiments.

Towards the end of the 19th century many of the questions which had either been left open or simply not considered in Weber's theories of the skin senses were finally answered by the Würzburg physiologist Max von Frey. In 1894, he proved the existence of pain points. In doing so, he added a fourth sense to the senses of pressure, warmth and coldness. A year later, he discovered that each of these four forms of sensation or modalities possessed its own organ. The so-called 'Krause-Endkolben' (Krause's corpuscles), for example, were responsible for the sensation of coldness. These form a sensitive receptor, consisting of a round or oval body with built-in nerve threads, which is situated in the epidermis.

For his experiments Frey constructed a simple device that enabled him to apply tiny stimuli to the skin. It consisted of a series of brushes of different degrees of stiffness which were fixed with sealing wax to a movable rod (Fig. 8). This instrument, known universally in physiological research as the 'Frey brush', is used to locate pain points and to determine their threshold values. The experiments performed on prisoners of both sexes without their consent by Cesare Lombroso (1836–1909) in the 1880s show how such pain-measuring instruments could also be used for inhumane and scientifically dubious investigations. The Italian psychiatrist and criminologist attempted to substantiate his ideologically loaded theory of the inferior physical sensitivities of professional criminals by means of a series of physiological experiments. To test the reflex responses of his human guinea pigs, he stuck needles into them and treated them with electric shocks, and must certainly have left them with more than merely unpleasant sensory impressions [22].

The pioneering experimental sensory physiology of the 19th century is characterised by the definition of the senses according to sensory modalities. In contrast to earlier times, it is now no longer the individual senses that are enumerated and classified, but sensory perceptions of all kinds. The interpretation of the sense of touch as a skin sensation consisting of a number of discrete aspects (pressure, heat, cold and pain) may be regarded as typical. The methods and instruments developed by leading 19th century sensory physiologists produced results that are still valid today.

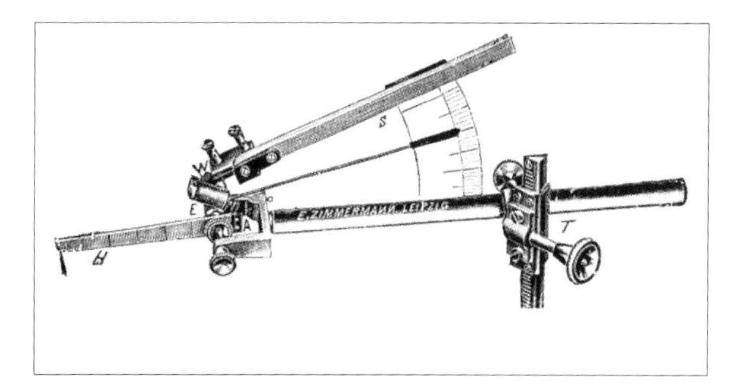

FIGURE 8. ELECTRO-MAGNETIC INSTRUMENT BY FREY TO MEASURE THE TACTILE SENSE (1923)
Source: http://vlp.mpiwg-berlin.mpg.de/technology/search?-max=10&img.exportcaption=Weber

Towards a 'haptic age'?

According to an article in the German newspaper *Stuttgarter Nachrichten* of 18 October 1999, we have now entered a 'haptic age', and the writer was not just referring to a trend in underwear. The piece also noted other signs of changing times: for instance, the stress on tenderness in sexual therapy, as well as the increasing demand for healers with 'magic hands' (magnetopaths, masseurs and chiropractors) and the supposedly growing numbers of adults in need of 'loving sex'. For years before this, the German weekly Die Zeit (11 April 1997) published a '*Manifesto for the Emancipation of the Sense of Touch*' which appeared to reflect a 'new pleasure in touching'. Historians, whose task it is to take stock of a variety of different sources, will obviously be sceptical, pointing out that similar trends can be observed in previous times. In the first half of the 20th century, for instance, some thinkers were maintaining that the sense of touch was losing its significance in sensory knowledge, while others were insisting on the priority of haptic experience. Hermann Friedmann (1873–

1957), an adherent of the then up-and-coming Gestalt theory of the psychology of perception, declared in 1930 that the objects of tactile experience were eking out "*a brief and wretched existence in a stunted tactile memory and the aesthetically impoverished realm of tactile fantasy*" [23]. The Hungarian psychologist Geza Révész (1878–1955) took an opposite view. In the early 1940s he stressed that knowledge acquired by means of the sense of touch was more convincing and persuasive [24]. Moving forward to the philosophy of the present, we encounter Jean Baudrillard's theory that the hand is "*no longer the prehensile organ that focuses effort: rather, nothing more than the abstract sign of manipulability, to which buttons, handles, and so on are all the better suited*" [25]. The invention of touch screens, interactive monitors which dispense with the keyboard and mouse, making computers easier to use than ever, would appear to bear out Baudrillard's theory that the increase of tactile experience in the media age has not necessarily produced an enrichment of sensory perception. In the face of the increasing lack of affectionate and intimate bodily contact in the 20th century, the American futurologist John Naisbitt called for a new form of social behaviour, for which he coined the term 'high touch'. It addresses the growing need for a kind of closeness and togetherness that might compensate for the negative aspects of high-tech society. He sees the rapid growth of self-help groups and the worldwide boom in new systems of learning and therapy (e.g., feeling therapy and other forms of body therapy) as early signs of a development in this direction [26]. On the other hand, the late 1960s witnessed the beginning of a remarkable rediscovery of the sense of touch. For it was around this time that a generation which had grown up with television began to discover that, in addition to the ever more dominant sense of sight, there was also a sense of touch. Besides the sit-in, the flower-power generation and the hippie movement also invented the 'touch-in', at which people who had never met before would kiss and embrace and seek to offer each other tenderness.

At almost the same time, towards the end of the 1960s, the American media theorist Marshall McLuhan took many critics of the television age by surprise with his thesis that, contrary to what its name appeared to suggest, television was a tactile rather than a visual medium, since the cathode rays which produced television images actually 'stroked' the retina of the eye. Hence the title of this famous book *The Medium is the Massage* (1967) [27]. Today, in the multi-media age, McLuhan's dictum has acquired new associations: our continuous zapping between dozens of channels provides plenty of exercise for the sense of touch.

The new generation of industrial robots is also now equipped with tactile sensors, which enable their grasping hands to be controlled with greater precision. Nevertheless, the sense of touch continues to be a headache for scientists working in virtual reality and robotics, since this sense is not located in a particular organ but spread over the whole body. The invention of the 'sensing glove' by Scott Fisher in 1985 was an important milestone on the way to cyberspace. With it, hand movements, or rather the feeling, grasping and moving of objects, could be simulated as virtual reality for the first time (Fig. 9). A simple and inexpensive version of this glove is now available as a computer toy [28]. The 'data suit', with its armoury of pressure- and temperature-sensitive sensors, represents a further development of the glove. Although this suit has so far not been put to any practical use, it is occasionally worn by cyber artists.

Look into the future

What will the future look like? Haptics are not only to be put to recreational (e.g., computer games) and artistic uses. Scientists are working on sensitive prosthetics allowing for example, the restoration of grip function in patients with spinal cord injuries. Thus, according to James Geary, "haptics research will certainly extend the human hand's reach, across biological frontiers and into virtual worlds" [29].

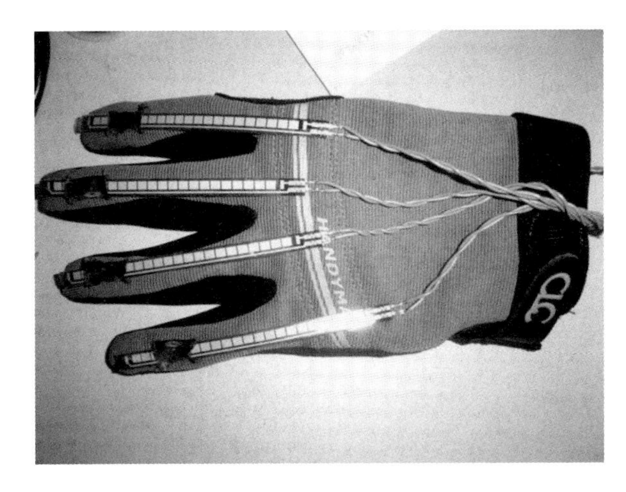

FIGURE 9. TECHNO GLOVES BY JEFF HAMALAINEN (2004)
Source: www.tufts.edu/.../ hamalainen1.jpg

Summary

The idea that perception or sensation may be localised in certain physical organs (e.g., skin) has a long tradition. It pervades many cultures. The system of sensory physiology (of which touch is one important element) is shaped by the influence of both medical thought and the philosophy of nature. In Indian and Chinese tradition the human organism was perceived as a miniature copy of the universe. For the western tradition the Greek philosopher Aristotle (384–322 BC) is of utmost importance. He not only expanded the hitherto merely inchoate physiology of the senses, but also advanced them to a state of completion that retained its authority well after the Christian Middle Ages. The Aristotelian doctrine of the fully unified and independent nature of the sense of touch was scarcely ever questioned in the subsequent centuries.

The first physiologist to undertake a systematic experimental investigation of the sense of touch was Ernst Heinrich Weber (1795–1878). The pioneering experimental sensory physiology of the 19th century is characterised by the definition of the senses according to sensory modalities. In contrast to earlier times, it is now no longer the individual senses that are enumerated and classified, but sensory perceptions of all kinds. The interpretation of the sense of touch as a skin sensation consisting of a number of discrete aspects (pressure, heat, cold and pain) may be regarded as typical. The methods and instruments developed by leading 19th century sensory physiologists produced results that are still valid today.

Selected readings

Böhme H (n.d.) Der Tastsinn im Gefüge der Sinne. Anthropologische und historische Ansichten vorsprachlicher Aisthesis. www.culture.hu-berlin.de/HB/texte/tasten/html

Classen C (1993) *Worlds of Sense. Exploring the senses in history and across cultures*. Routledge, London, New York

Kalkmann HW (ed) (2005) *La main dans la main. Ausstellungskatalog*. Kunstverein Bad Salzdetfurth, Bad Salzdetfurth

Kunst- und Ausstellungshalle der Bundesrepublik Deutschland (ed) (1996) *Tasten*. Steidl, Göttingen

Loenhoff J (2001) *Die kommunikative Funktion der Sinne: theoretische Studien zum Verhältnis von Kommunikation, Wahrnehmung und Bewegung*. UVK-Verl.-Ges., Konstanz

Michler M (1999) *Die Hand als Werkzeug des Arztes: eine kurze Geschichte der Palpation von den Anfängen bis zur Gegenwart*. Steiner, Wiesbaden

Wehr M, Weinmann M (ed) (1999) *Die Hand: Werkzeug des Geistes*. Spektrum Akad. Verl, Heidelberg, Berlin

Switzer A, Cirincione D, Karle W (1977) Die Rolle des Körpers in der Feeling Therapy. In: HG Petzold (ed): *Die neuen Körpertherapien*. Junfermann, Paderborn, 376–387

Wilson FR (2002) *Die Hand – Geniestreich der Evolution: ihr Einfluss auf Gehirn, Sprache und Kultur des Menschen*. Rowohlt-Taschenbuch-Verl., Reinbek

2

German pioneers of research into human haptic perception

Martin Grunwald and Matthias John

Introduction

The science of the human sense of touch, also known as 'haptic', had a long and rich tradition in Germany before World War II. Philosophers and physiologists, medical doctors and psychologists approached the subject of touch from different points of view and each of them developed an unmistakable, and sometimes enduring, part of the knowledge we have today. Among these scientists are some whose works about research into the sense of touch were internationally known and influential while others, though, were only received in the German speaking world. Regardless of this criteria, several of the most important representatives of German touch research before World War II and their scientific contributions are outlined below. The accounts are ordered chronologically according to the respective year of birth (E.H. Weber, M. von Frey, M. Dessoir, G. Révész, D. Katz, and E. von Skramlik). In creating this list, an exemplary selection was made which can only include a part of the range of German scientists who have researched the human sense of touch in the last 150 years. A comprehensive presentation is impossible in a work such as this and must be left for a historical-psychological study. The fact, however, that such a work does not exist (even in a most basic form) shows that the history of this research is not exactly the focal point of today's scientific interest. Several authors are neglected in this 'sketch' – partially, because their effects are sufficiently known and acknowledged and partially because, to date, too little information is to be found about them. In any case, other researchers have delivered works to us about the human sense of touch and deserve, at least, to be mentioned here: G. T. Fechner, W. M. Wundt, J.N. Czermak, L. A. H. von Strümpell, E. Mach, O. Funke, H. Lotze, E. Hering, G. Meissner, T. Hausmann, P. Mahner, A. Goldscheider, G.A. Brecher, O.F. Scheuer, R. Hippius.

E. H. Weber

The scientific and systematic examination of the human sense of touch began with an anatomist and physiologist from Leipzig, Germany, Ernst Heinrich Weber (1795–1878). He studied in Wittenberg and Leipzig and became professor of comparative anatomy in 1818 in Leipzig and, in 1840, professor of physiology. In 1840, he laid out first findings about the systematic and experimental testing of sensory thresholds in the human sense of touch in his dissertation paper, '*De pulsu, resortione, auditu et tactu annotatines anatomicae et physiologicae*' [1]. As early as this paper, his findings about two-point discrimination were developed and he could prove that the ability to discriminate between two tactile stimuli, applied at the same time on two points of the body, was different in different places on the skin. To show this, E.H. Weber used a compass with dulled points. The sensibility thresholds valid here are used still today as a diagnostic tool in, e.g., neurology. E.H. Weber continued to explore the subject of the research of the human sense of touch even after his dissertation. The central results of his work regarding the sense of touch were published in a short, German-language abstract in 1835 [2]. In 1851 [3], a larger, more comprehensive work with the title '*Die*

FIGURE 1. ERNST HEINRICH WEBER (1795–1878)

Lehre vom Tastsinn und Gemeingefühl' ('*The Science of the Sense of Touch and General Sensation*') followed. Within, he presented several newer and older studies as well as a discussion about the few national and international publications about anatomy and the physiology of the sense of touch that had been published at the time (e.g., C. Bell, J. Müller).

E.H. Weber did not pursue his research into the human sense of touch because of philosophical considerations but, rather, as he briefly stated in his publications from 1835 and 1851, this 'sensory organ' offered him the practical possibility to "*conduct the most varied of experiments and to collect data from various points of view*" without damaging the human body while doing so. In addition, Weber was hoping to collect information which he could "*later use [with respect to] the visual and other senses*".

As a result, practical considerations and the hope of discovering basic principles of other perceptual modalities were the main aim of his experimental studies. He didn't use only experimental findings for the foundations of his studies of the sense of touch but, as was usual in his times, phenomenological and generally formulated arguments found their way into his presentations and analyses. In this way, he confronted the basic dimensions of human perception while, at the same time assuming that, for "*inexplicable reasons, the soul*" is forced to process all sensations in relation to the categories of space, time and quantity. With these categories E.H. Weber laid the foundation for the interpretation of his own findings about spatial two point stimuli discrimination. The time-space structure of our perception is, for Weber, a natural axiom although Kant's philosophy may have contributed to this line of thought. At the same time, however, he postulated a 'sense of locality' (or 'feel for locality') which appears to be all the more differentiated the more numerously nerve fibres are present in the various sensory organs. In order for the sense of locality to come into existence, E.H. Weber hypothesises that the sensory organs are separated into "*small, departments located next to each other*" and that they are connected to the brain with individual nerves. E.H. Weber depicted these areas of the skin as 'feeling circles' [3] or 'sensation circles' [3]. The ability to differentiate between two spatially different stimuli is, according to him, based upon the fact that the two stimuli lie within two different sensation circles. He also suspected that the nerve fibres running from each 'department' were connected to a topographically- ordered, corresponding area in the brain. As support for his hypothesis he presents results from observations made during tests conducted on patients paralysed on one side of the body (hemiplegia) [3]. At the same time he points out that his hypothesises are based on founded, yet non-proven assumptions as the basic relationships between the paths of the nerves related to the sense of touch and the brain had only been insufficiently examined thus far. E.H. Weber also remained uncertain about the question of where the seat of the human soul might be – a question which had occupied science for hundreds of years and which the newer science, in its global form, was gradually getting away from. Weber presented indications which suggest that the brain is the base of the soul's activities but stated, at the same time, that he doubted whether the information available thus far could conclusively answer the question [3].

The theoretic and practical autonomy of the 'sense of locality' within the sense of touch postulated by E.H. Weber and its connection to sensory activities of the skin become clear in the following quote: "*The sense of locality helps us to better know the movements of our limbs and with the help of the movements of our limbs – dependent upon our own free will – we get to know our skin and orient ourselves on one and the same (the skin). Both abilities, from the beginning extremely limited, compliment and complete each other*" [3].

The sense of locality's ability to differentiate spatially is, according to Weber, dependent on the number of nerves in each sensory organ. With this in mind he concludes that the ability to differentiate is much better formed in the visual organs, with their high density of nerves, than in the sense of touch. With precise methodology and academic system, he examined the spatial, aligning discrimination of the skin and every possible part of the body with the help of his experimental paradigm of two-point discrimination. The results, which he presented in a table with values, showed him that the spatial aligning discrimination for pressure stimuli were differently precise from bodily part to bodily part. In E.H. Weber's terminology, the sense of locality is, as a result, unevenly precise in its development.

In his papers he did not only describe the studies of the ability to discriminate with two-point-cues, but, in separate papers and essays, outlines new phenomena and issues. These studies, at descriptive-analytic and experimental levels, show him as a sensitive, alert and systematic observer and, concurrently, as a pioneer of a new, experimental science. In different chapters, he analysed the different abilities of the sense of touch in recognising objects and object properties. During his studies he realised that passive finger contact with objects of various sizes resulted in an object recognition that was worse than when the objects were actively explored [3]. If he had, to this point, concentrated more on the aspects of passive stimuli discrimination, then he now turned his attentions more and more to the active touch processes and, as a consequence, had to deal with the sensory functions of the muscles. This led him to the question of how much influence the muscular activities – and the accompanying pressure stimuli which affect the skin – have on the perception of touch in situations such as weight estimation. The methodological difficulties that E.H. Weber faced during these examinations were described as clearly as the results which were gained by them.

Regarding the general function of the senses, E.H. Weber wrote that all senses are created for the perception of the world around us and that they are also, in principle, capable of triggering qualitatively different sensations. The lightest pressure sensation is as perceivable as stronger ones which can result in pain. The slightest change in temperature is as noticeable as the smallest change in weights. We are able to determine the position of our limbs precisely at any time without the help of visual contact. The quality and aligning discrimination of individual sensory systems is, however, dependent upon how finely we are able to perceive the properties our environment. The smallest sensation thresholds and the greatest stimuli-amplitudes – to the point of painfulness – determine the range and sensory dimensions of our perception. As a scientist, it was his concern to determine these dimensions – for instance, in the form of variation thresholds – with systematic and experimental precision and to record them in size and figures. "*Our sensory perception of the world is dependent upon the smallest measures which we possess and with which we can judge time and space*" [3].

Nevertheless, in his studies he constantly referred to findings regarding other modalities – providing that he was aware of them. He sovereignly and competently connects the absolute thresholds of visual and acoustic perception with his findings regarding the sense of touch and presents relationships in a sober, empirical way. While doing so, he sticks strictly to the available data and facts – he was not interested in philosophical interludes or historical feuds.

With the coolness of a scientific experiment, he conducted comprehensive examinations into the ability of various body parts (e.g., finger, mouth, lips) to perceive temperature. He noticed, among other things, that areas that were scarred as a result of burning were less able, if at all, to adequately judge temperatures and attributed this to the destruction of the touch organs. He also pursued the question of whether inner organs and mucous membranes are capable of registering temperature sensations. He tested, in experiments on himself and others, whether or not hot and cold water led to sensations in the stomach or nasal cavities and he was not afraid of administering clysters with different (hot or cold) water temperatures. He observed that the left hand was much more capable of determining differences in temperature than the right and attributed this effect to the difference in the thickness of the skin. He also described numerous illusionary effects which appeared to be dependent upon whether the tested area was a single finger or an entire hand. His explanation for such effects was a "*summation of the impressions*" in the brain [3]. That human skin shows variance in the quality and accuracy of warmth and cold perception is, according to Weber, a sign of corresponding 'organs' in the skin – organs which are sensitive to the related stimuli. However, the anatomical research at the time was unable to provide information about this and, as a result, E.H. Weber decidedly left this question to future researchers so that they could answer these questions more definitively [3]. This open and honest reference to the need for future research is to be found – firmly formulated – in many places in the texts written by Weber. Despite the poor quality of the data available to him, he attempted to explain the observed effects as best he could with the existing knowledge. In the case of temperature perception, it was at least plausible for him, that the volume of the touch organs responsible for the sensation of pressure change as a result of warmth or cold stimuli and that these volume changes could be responsible for the perception of warmth and coldness. He formulated this, however, strictly and expressly as a hypothesis.

He named the body's ability to perceive temperature the 'sense of temperature'. This, for him, existed alongside the 'sense of locality' and the 'sense of pressure' and, together, these three senses were the abilities of the sense of touch. These terms do not, however, cover the entire, complex reality of the sense of touch and, for this entirety, he created another term: general sensation. This word creation existed alongside the, at the time, widespread term 'common sense' (meaning 'coenaesthesis') for the integrative and apperceptive performance of the sensory apparatus. These sensations, which Weber called 'Gemeingefühle' ('general bodily feelings' or 'common sensation') include the "*...consciousness of the state of sensation that is carried to us from all of the body parts which are connected by sensory nerves – with the exception of the specific sensory feelings from the nameable senses themselves.*" [3]. He later writes, "*General sensation is finest in the organs of touch and in the muscles which belong to those body parts richest in nerves*" [3]. A central quality of general sensation is, according to this conception, pain and, as a result, the perception of pain became a direct component of the research E.H. Weber conducted into the subject.

Weber was obviously aware of the ambiguity of the term though and it comes then as no surprise that he attempted to describe – with examples from clinical practice – the ways that sensations of the sense of touch and those of common sensation are related to each other. One example thereof is his description of an 18 year old shoemaker who suffered from hypochondria. He was fully insensitive to strong pain stimuli (being stuck by a needle or hot iron being placed on his skin) but was, however, able to feel the lightest touch of a feather. In another case, a medical doctor from Geneva was described who was paralysed on the right side of his body. He showed a limitation, rather a complete insensitivity to the 'common sensation' although he remained sensitive to touch stimuli on the skin of the affected area.

Weber does not only use clinical cases to describe the changes in and limitations of 'common sensation' with limited pain perception but, rather, examines (with his own system) the development of pain sensations when various types of stimuli are applied. He, for instance, gradually raised the temperature of water in which he submerged his own hands – or the hands of test subjects – until it became painful. He recorded and presented these results as painstakingly as those of a test in which he held a key which had been cooled in quicksilver on various body parts for seconds or even minutes. Even the various types of pain that resulted from these tests (stabbing or burning pain) were recorded in the protocol.

He examined the fact that hair roots were capable of producing pain sensations when pulled upon or pressed. These tests were conducted using various weights which were tied to the ends of the hairs.

Alongside inner organs, mucous membranes and vessels, Weber was of the opinion that especially the muscles were responsible for providing us with precise information about different qualities of common sensation. In this, he includes the performance that can be achieved when subjects judge the difference in weights of two objects when the pressure stimuli are considerably removed. Also, we thank common sensation of the muscles for the fact that we are always aware of the (active or passive) position and location of our limbs. According to Weber, the fact that we can do this with such precision and accuracy without visual aid was reason enough to name this 'Kraftsinn' or 'sense of power'. This suggestion was, however, made with great caution [3] as he did not wish to expand the existing terminological framework within the study of the sense of touch. He could not have been aware of the terminological uncertainty that this word would later cause. Even if Weber did not expand upon this 'sense-term', later authors used it without it helping to make the terminology of haptic studies clearer.

On the one hand, E.H. Weber generated, with scientific creativity, systematic and experimental findings about the abilities of the human sense of touch such as which had not been seen before. On the other hand, some of his trials remained unproven and some of his unproven hypotheses would later be assumed.

Several of the questions he posed would not be taken up again for many decades. He described, for example, the appearance of sensory illusions after the amputation of limbs and several effects that illnesses in the brain and spine have upon the sense of touch – especially the emergence of hallucinations. He devoted himself at length to the question of in which ways the attention processes changed the perception of touch stimuli [3] and that elementary memory capacities are necessary and possible in order to even compare two stimuli which come one after the other [3]. In a similar way, he dealt with the question of which aspects self touch (the touching of one's own body) are dependent upon.

Experts in various disciplines, contemporaries and following generations of scientists researching the sense of touch and other sensory systems were influenced greatly by E.H. Weber. Wilhelm Wundt accredited him with the title 'The Father of Experimental Psychology' but other anatomists and physiologists also declared his works as milestones in their subjects [4]. That he was quite aware of the interdisciplinary and strong influence of his works can be seen in a programmatic quote from a paper published in 1835: "*The study of senses is a point in which, at some time in the future, the research of physiologists, psychologists and physicists must merge*" [2].

Max von Frey

Max von Frey (1852–1932) was born in Salzburg and studied medicine at several German universities. Later he worked with Ewald Hering and with the famous physiologist Carl Ludwig at the famous 'Physiologischen Anstalt' (opened 1869

FIGURE 2. MAX VON FREY (1852–1932)

in Leipzig) where he habilitated as a professor of psychology in 1882. In 1891 he received an associate professorship and in 1899 a full professorship of physiology at the University of Würzburg. Alongside physiological papers about the musculature [5] and the physiology of the circulatory system [6], the name von Frey is still closely associated with his comprehensive examinations of the human skin senses. In 1894 he began, at first with Friedrich Kiesow, to publish fundamental studies about the sensitivity of the skin and tactile sensors in the *Zeitschrift für Psychologie und Physiologie der Sinnesorgane* (*Journal of Psychology and Physiology of the Sensory Organs*). Numerous other essays, lectures and handbook articles followed until the end of the 1920s so that one might say he dedicated an entire lifetime of research to the sense of touch [7–9].

He developed the 'von Frey hairs' – an examinational instrument used still today to determine pressure points on the skin. To do this, von Frey used fibres of various diameters which first bend when a certain vertical pressure on the skin has been reached. With the help of this instrument, von Frey observed that the human skin has pressure points and that in their immediate vicinity no sensation of pressure can be created. von Frey referred to these touch-sensitive points as *key points*. In addition he was able to show that these key points are distributed unevenly on the skin's surface.

von Frey, who carried out a large number of individual studies about the function of the skin's sense of pressure, was a very influential person in his day. The Würzburger Physiological Institute, under his direction, was practically a synonym for physiological and psychological research into the sense of touch and was recognised for its strict guidelines of experimental methodology. The core of these studies was von Frey's notion that the performance of the touch sense could only be measured and explained by studying the perception of pressure. He did not believe in the 'deep-pressure sense' often discussed by other researchers. For von Frey, the qualities of the sense of touch existed, for the most part, to make pressure and contact perception possible and the perception organ existed exclusively in the skin. That contact and pressure could be perceived even after the skin had been frozen or anaesthetised was, for von Frey, explained by the fact that other peripherally located areas of the skin were stimulated (as a result of stretching). This attitude was confirmed in several experimental studies conducted by von Frey and his colleagues. As a result of his believing that no pressure sensitive areas existed under the skin – in contrast to the beliefs of Strümpell, Head and Sherrington – von Frey acrimoniously managed his line of argumentation. In doing so he took a central position in the German researchers' debate over the existence of a 'superficial' or 'deep' sense of pressure [10].

Especially impressive is von Frey's argumentation when he attributes the perception of passively led arm movements not to the effects of joint, tendon or muscle receptors but rather to the deformation of the dermal tissue under the affected joint. He noticed that the anaesthetising of the skin over the joint led to a 1° loss in perception of the passively led arm movements (in comparison to the same procedure with skin that had not been anaesthetised). For von Frey, changes in the tension of the skin were the key to understanding minimal positional changes of the limbs. Slightly more placably, he suggested that other sensory tools may very well be at work in movements of greater scale [11].

However, von Frey categorically rejected the idea that the joints or muscles contain receptors which also contribute to the perception of changes in position.

Max von Frey's essential contribution to touch sense research (alongside his experimental testing of key points) is his doggedness on the verification of his theses – an attribute which repeatedly enriched the contemporary discussion with critical findings. Without von Frey and his relentless adherence to his theses the debate and the scientific evidence regarding a 'deep sense of pressure' would certainly have been less fruitful. In this sense, his misunderstandings contributed greatly to the expansion of the theoretical and practical basis of the research into the sense of touch.

Figure 3. Max Dessoir (1867–1947)

Max Dessoir

Max Dessoir (1867–1947) is barely known among today's scientists for his experimental studies into the sense of touch. The Berliner philosopher and psychologist is, at best, remembered today as an art scholar, pioneer of psychological aesthetics or as the coiner of the phrase 'parapsychology'. His comprehensive paper '*About the Sense of the Skin*' ('*Über den Hautsinn*' 1892 [12] – first written as a medical professorial dissertation in Würzburg and then published in Du Bois-Reymond's *Archive for Anatomy and Physiology* – was much better known to his contemporaries. This is shown, for one, by his work's reception in the USA where it was discussed vigorously [13].

In this article, Dessoir sets himself the goal of systematically working through all available experimental findings regarding the research of the sense of touch. In addition, he presents several of his own experimental and phenomenological studies about the psychology of the sense of touch. These had been conducted, for the most part, in the Königlich tierärztlichen Hochschule (Royal Veterinary College) in Berlin.

Besides a distinct sensitivity for methodical-experimental questions, especially the analysis of the sources of error, Dessoir strove for a clarity in the definition of terminology. He defined the 'Skin Sense' as the conceptual designation for pressure and temperature perception [12] and, at the same time, made it clear that the classical terminology for the sense of touch is fundamentally insufficient. He was apparently the first to explicitly address this problem of terminology, the ever-present difficulty in integrating the large number of differentiable sensory impressions into an understandable context. He didn't know at the time that this would survive as his greatest scientific contribution to research of the sense of touch – Dessoir stumbled into the contemporary debate about 'Temperature Sense', 'Muscle Sense', the 'sense of pressure', etc. However, unlike his predecessors, he was not satisfied with a simple criticism of the different sub-senses but, instead, he pursued the intention of creating an encompassing generic term which included the different aspects of the sense of touch for scientific teaching of the sense of touch. Dessoir realised that despite the rich findings and ambitious activities of his times, such an academic teaching of the sense of touch appeared to be unbelievably far away. In light of this he stated, "*If the clarity of the presented circumstances is to be improved by a precise method of description, the introduction of several new expressions is not to be avoided*" [12].

Alluding to the terms optic and acoustic, Dessoir suggested that the teaching of the sense of touch therefore be called haptic. This word then subsumed not only the aspects of contact and pressure sensations but also the so called touch and muscle perceptions. Dessoir called the first two aspects of the sense of touch 'contact sense' and the others as 'Pselaphesie' (Ψηλαφάω – to feel). In doing so, Dessoir, without explicitly stating so, separates the perception of touch from the perceiving subjects point of view into an active and a passive area. The actively exploring subject's perception of touch which result from tactile and muscular sensations are, in his opinion, independent phenomena (*Pselaphesie*), which should be considered separately from the 'contact sense' also on the conceptual level. The necessity of this terminological determination was to be recognised only years later. Even if the terms *contact sense* and *Pselaphesie* did not take hold in the scientific community it is still to be credited to Dessoir that the teaching and researching of the largest and most comprehensive of human organs received the name haptic – a term that even today is of practical and theoretical importance.

His experimental contributions include replications of examinations into temperature perception, pain perception and that of electrical applications in which he was critical of the methodology used. Especially the studies into temperature perception make up a large part of his work – a result of the great interest triggered by Goldscheider's exploration of the warm and cold points of the human skin. With experiments on himself and others he studied various human mucosa (throat, nasal septum, stomach, etc.) and their degrees of temperature sensitivity. The, sometimes very painful, experiments were conducted only on himself and he even tested whether an exposed dental nerve triggered not only pressure but also temperature sensations. In such a way he also tested whether the end (head) of a man's penis is sensitive to temperature. Experiments which he could not conduct on volunteers or, as was standard at his time, named test persons were carried out on dogs. He precisely described the test procedures and the observed behavioural reactions of his test animals.

Dessoir's disbelieving and committed spirit did not stop short of using his own body and allowed itself a critical relationship to existing knowledge. It is, then, not surprising that Dessoir also doubted the existence of specific hot or cold receptors in the skin as a result of a lack of anatomical-histological findings regarding the matter.

After publishing his professorial dissertation, Dessoir turned to other areas of psychology and, particularly, aesthetics and philosophy. Among others he wrote a comprehensive work about the history of psychology [14] as an expansion and re-working of his philosophical dissertation. This book is, even today, considered authoritative for the history of 18th century psychology. His reputation suffered in experimental psychology circles as he turned to the scientific study of occult phenomena (parapsychology) [15] although he approached this area with self-criticism and used psychological-experimental methodology [16].

Géza Révész

Révész was born on 9 December 1878 in Siófok, Hungary, and died in 1955 in Amsterdam. He studied and received his post doctorate degree in law in Budapest and then he joined Georg Elias Müller in Göttingen, Germany to work in the field of experimental psychology. Révész taught psychology in Budapest and emigrated to The Netherlands in 1920 or, according to some accounts, in 1921. He became a professor at the University of Amsterdam in 1931 and published, for the most part, in German and in the German language publications and was named an honorary member of the Deutschen Gesellschaft für Psychologie (German Society for Psychology).

Géza Révész left many and various marks within psychology. His comprehensive work includes numerous articles about music psychology [17, 18], the psychology of language [19], aptitude [20] and thought and he is considered the father of psychology for the blind. Several of his papers were translated into English so that he became known outside of German speaking countries [21–23]. Besides music and language psychology, Révész also intensively studied and researched the human sense of touch. In 1938 he published a two volume work with the title *Die Formenwelt des Tastsinnes – Grundlegung der Haptik und der Blindenpsychologie* [24] – (*Fundamentals of Haptic and the Psychology of the Blind*). With this work, following the definition of *haptic* as the study of the sense of touch, Révész attempted to integrate the findings of his time into the closed concept of haptic perception. In doing so he supports his perceptual-psychological work with experimental and observational studies, especially ones in which he conducted with patients who had been born blind and ones which had become blind later in life. Noticeably, and in contrast to his psycho-physiological colleagues, he combines clinical presentations with profound philosophical and epistemic arguments. He intensively devoted himself to the Kant thesis that all people – even those born blind – are equipped *a priori* with a spatial perception. The psychological and physiological experts of his time doubted the ability of those born blind to possess this ability as they represented the opinion that only the visual sense/sight could make this possible. This position seemed to be supported by operations performed on people born blind which were unable to detect determine object or spatial information. However, Révész was methodically brilliant and proved with his analysis that the findings (from those born blind) speak for the acceptance of a haptic spatial perception which exists completely independently of visual perception. The

Figure 4. Géza Révész (1878–1955)

work which Révész conducted regarding haptic deception is very important and supportive in this regard. In opposition to the dominant thesis of the time from Gelb and Goldstein, 1919 [25], which stated that spatial perception is only possible with the help of the visual sense, Révész proved that haptic illusions appear in the same fashion regardless of whether the subject is blind or not. For Révész, these findings are a clear indication of the fact that the 'optical theory' of visual perception is invalid. In fact, he insisted that spatial perception is possible by means of haptic and visual perception – the principles of formation of this spatial perception working independently from one another. Révész poses the question of whether haptic and visual perceptions underlie the same or completely independent organisational principles. Supported by his findings, he comes to the conclusion that human haptic underlies an organisation which is formed wholly independent of the principles of visual perception. This made Révész, without a doubt, one of the most radical and decisive representatives of experimental touch research in Germany. With courage and entelechy, he not only demanded an independent methodology for haptic research but, rather, postulated that the entire dimensions of human haptic be a totally separate field of research. In doing so he takes place in the general discussion regarding the position of the sense of touch in relation to others in the hierarchy of the senses – a discussion taking place since Aristotle – and takes a stance against the Platonian position in which vision represents the dominant sense. He doesn't doubt the importance of the sense of vision but doesn't agree with the general degradation of the sense of touch as a lower sense. Rather, he showed, with his studies of the blind, which incredibly complex accomplishments the visually-independent sense of touch is capable of.

As a result, the influence of Révész's research of blindness on the conception of haptic is undeniable. It is only logical that Révész, in his attempt to organise haptic perception, speaks of 'optical haptic' (optohaptic) when referring to people with normal sight and of an 'autonomous haptic' when referring to those people born blind. He decidedly refers to the visual influences which haptic perception is subject to in seeing people. In contrast, haptic perception in those born blind can be considered uninfluenced by visual experience and, therefore, autonomous. According to Révész, the principles which underlie both forms of haptic could only be explained when comparative studies were carried out on people with the visual sense and on people who had been born blind. He repeatedly states that the findings of the time were by all means insufficient in order to form a closed haptic theory. He does not only criticise the numerous deficits in his colleagues working methods but, rather, emphatically insisted that haptic research must get away from the paradigms of optics.

"*In haptic, the questioning and methodology, the hypotheses and teaching and even the interpretation of the empirical data were, as mentioned, oriented on visual perception. Under these circumstances, haptic could not emancipate itself from optic. As a result, the entire research has stagnated*" [24]. He also argued that the phenomenological, descriptive approach within psychology can represent a research instrument which is capable of producing statements about haptic perception.

The systematic of the touching process as delivered by Révész is based upon the work of Hippius [26] which examined the active and passive exploration of shapes and forms. Like David Katz, he emphasises the essential importance of movements in the touch process. This process itself is understood by Révész as a successive-analytical process with the goal of object recognition – in opposition to visual perception – to which he attributes a simultaneous, synthetic process with the same goal.

Révész postulates another difference to visual perception in the so called 'stereo-plastic principle' which he himself refers to as the "*fundamental principle of haptic perception*" and which is non-existent in visual discernment. With this principle, Révész describes the human desire to use the sense of touch in exploring all touchable,

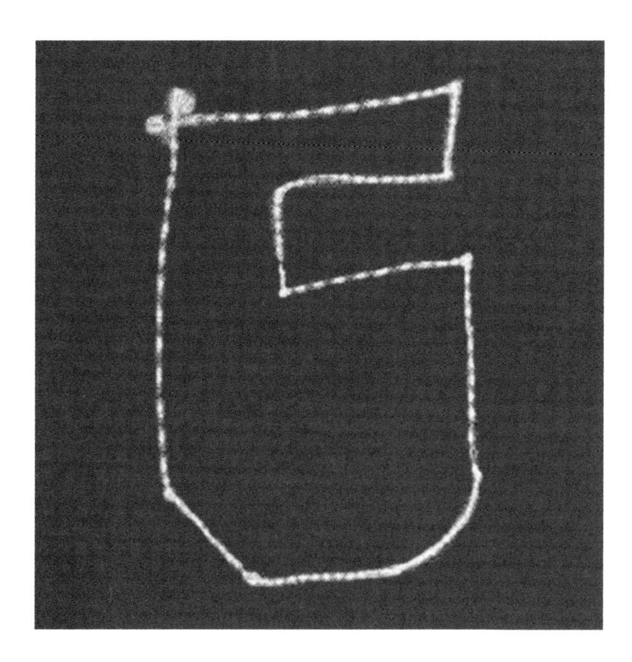

FIGURE 5. PHOTOGRAM OF A ONE-FINGER OBJECT EXPLORATION.
Each point equals 0.04 seconds. (Révész [24])

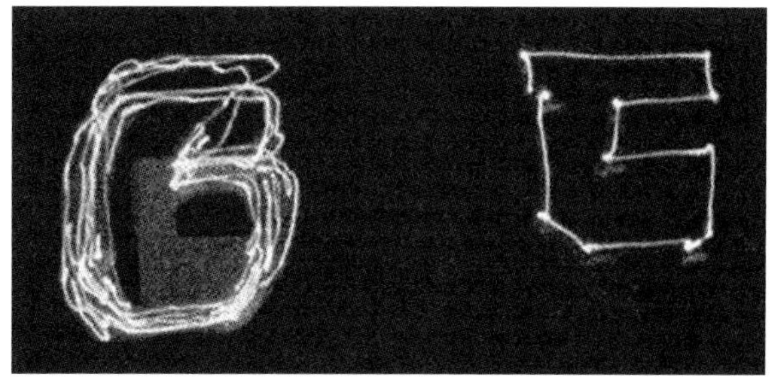

FIGURE 6. PHOTOGRAM OF TWO-HANDED EXPLORATION
(Révész [24])

tangible objects. The need to explore objects haptically is, according to Révész, an exploratory principle which we can suppress only with effort. It also leads us to collect information about every perceivable characteristic within the realm of haptic perception. He was unable to see a comparable curiosity – need to perceive – in visual object exploration.

Révész also describes the influences of internal attitude during the haptic exploration process as essential, and again, as a major difference to visual perception. He differentiates between 'receptive' (passive, contemplative, without the aim of recognition) and 'intentional attitudes' (haptic exploration with the explicit goal of object recognition) [24]. Révész called the results of a receptive haptic exploration process 'haptomorphe Gestalten' and those of intentional attitudes 'optomorphe Gestalten'.

Here and at other places in Révész's typology, you can find numerous allusions to the terminology and postulated principles of the psychology of gestalt. As the psychology of gestalt is, however, for the most part, limited to the areas of visual perception, Révész constantly attempted to define the particularities of haptic gestalt principles as compared to visual ones. His interest was especially drawn to form recognition by means of touch which, according to his observations, developed successively. In order to test his views, he had subjects feel unknown bottle forms with both hands and interrupted the touch process at various time intervals. He recorded the results of the active form recognition up to the point of interruption (see Fig. 5). With the help of co called 'photograms' Révész analysed the work movement of individual fingers during the scanning process. To do so, a light bulb was attached to the top of the middle finger. Photographic recordings were made of the light which was emitted at a high-frequency and constant intervals from a constant distance. The recorded light-point recordings allowed Révész to observe that, during the exploration process, certain aspects of a form (corners and edges) were explored longer and more frequently than other elements (Fig. 6). With this and other studies using film technology, Révész examined the 'successive nature' of haptic object exploration. At the same time he observed various exploratory phases from comprehensive to specific (Fig. 7). Within the framework of these trials, he also examined the question of how haptic exploration is determined by symmetrical (or asymmetrical) aspects

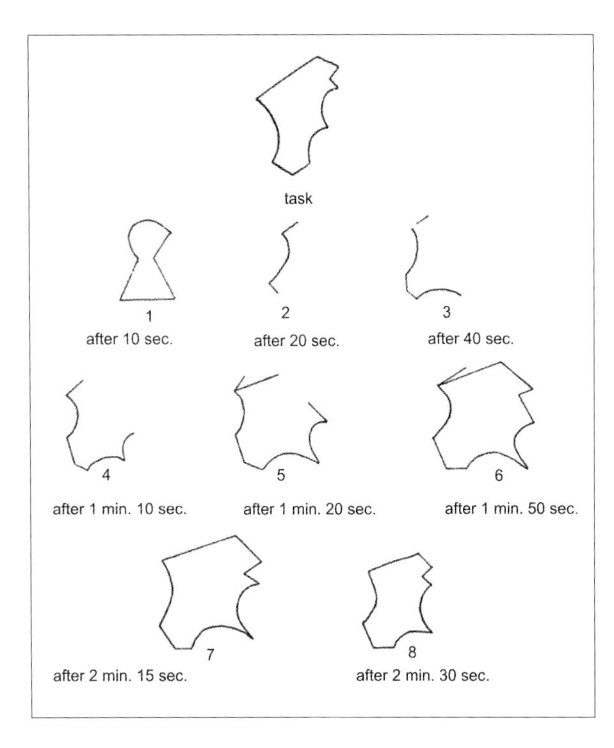

Figure 7. Graphic reproductions of flat objects (tasks) after different exploration intervals *(Révész [24])*

of the object's design. Révész posits a 'symmetrical movement-gestalt' in which we tend to perceive asymmetrical objects as symmetrical. Alongside the descriptive-measurement analysis of the exploratory process, Révész analysed the ability of seeing people and those born blind to haptically gauge an object's proportions and translate them into an active structuring process. An important result of his descriptive-analytical studies is the realisation that haptic perception is always accompanied by an 'optification' and that visual perception is accompanied by a 'haptification'. This means that, according to Révész, the haptic exploration process in seeing people is accompanied by a tendency to create a visual imagination of the object being felt. In a parallel fashion, visual perception is checked and verified by haptic experience ('haptification'). In this way, Révész formulated a structural principle in which haptic perception takes up a central position within the human sensory systems.

David Katz

The German psychologist David Katz (1884–1953) studied under Georg Elis Müller and received his doctorate in Göttingen, Germany. Together with Géza Révész, he conducted perceptual and memory experiments on chickens. In 1919 he became a professor of psychology and education at the Institute for Psychology at the university in Rostock, Germany. Together with his wife Rosa Katz he worked on the issues of child psychology, the phenomena of hunger and appetite as well as the senses of touch and 'vibration'.

In 1933 he fled Nazi Germany, first going to England and then, in 1937, to Stockholm where he became a professor for education and psychology and where he died in 1953.

Alongside individual studies and papers, he published, in 1925, the monograph *The World of Touch*[1] – a work that is still well known and influential today [27] . With this work he decidedly challenged the miserable position of touch sense research within psychology at the time and also the treatment of this sensory system as a 'lower sense'. In addition to this, Katz hoped that a higher regard for the sense of touch within psychology would result in application-oriented contributions for the fields of education and industrial applied psychology. Furthermore, he pointed out that the prevailing positions in the cognitive sciences would alter greatly if the human sense of touch were granted its proper and appropriate position. Katz was convinced that a higher regard for the occurrences taking place in the sense of touch would result in its showing a fundamental character in, and practical uses for, the fields of psychology and education. That the hand takes an exceptional position here as a special 'touch-sense organ' is, for Katz, as self-evident as the important con-

1 The work was translated into English in 1989 by L.E. Krueger [27].

FIGURE 8. DAVID KATZ (1884–1953)

nection between hand activities and brain maturation already recognised in developmental psychological circles. Katz looked into all of the important findings from research into the sense of touch which were available at the time. In as much, his work from 1925 is not only a collection of the arguments based on his own studies but, also, an attempt to present the entire field of the sense of touch as a self-contained subject from a psychological perspective. Hence, it becomes understandable that Katz deals with the physiological basics or the anatomical discussions very superficially if at all. Influenced by Gestalt psychology and with knowledge of the already prevalent focus on visual perception (and the resulting limited and marginalised position of the sense of touch in psychology) he pushed for a general reassessment of this sensory area with a very broad and comprehensive report about the touch sense phenomena which had been discussed up to that time. He not only discussed findings and phenomenological descriptions of perceptions of the sense of touch, but also possible capacities for memory and hallucinations.

Katz's content and methodology are obviously influenced by E.H. Weber, even though he didn't work with the meticulous strictness (nor did he wish to as a result of his more material and practical orientation) that was prescribed and demanded by the physiological and psychophysiological predecessor and contemporaries. He referred to various aspects of E.H. Weber's works with reverence and honest acknowledgement, but categorically criticised the strict psychophysiological elementary analyses of his predecessor and contemporaries. Katz considered the separate analysis of individual stimuli and singular phenomena in an artificial laboratory environment to be 'atomism' and he didn't believe that it could lead to a comprehensive understanding of the human sense of touch. Katz stressed that the sensory-physiological approach to sensory research often merely examined 'artificial products' and rarely the real world and its conditions. Katz acknowledged the basic successes of sensory-physiology but, for the sense of touch, this fragmentation of the realm of one's sensory experience was an unacceptable methodology. He commented ironically on the single-stimuli studies of his colleagues: "*Most people will probably die without ever having experienced the irritation of an isolated pressure or warmth point. The same can be said about a real spatial threshold*" [27].

Consequently, Katz uses a phenomenological-experimental methodology in his own trials. This method contains measured values as well as differentiated verbal judgements of the test persons – these described aspects of his touch-sense research taking on a considerable importance. Only a few of his studies show the unconditional methodological hardness and precision and the limitation to only a few variables that was common, standard practice in the psychophysiological research of his day. Katz did not want, however, to present results based merely in systematic studies as so many of his colleagues had. He was much more interested in redefining the position of the sense of touch within the hierarchy of the senses and within psychology. As a result, his 1925 work contains manifold and diverse references to relationships between the various senses. With comparisons and analogies of broad and specific quality, he draws connections between the sensory modalities and refers, whenever possible, to the exceptional qualities of the possible touch processes. Katz was not afraid of a direct, but well thought out, comparison of the sense of touch and its specific capabilities with those of the other senses. That significant differences between vision and touch become apparent was made very clear by Katz – especially with the help of examples regarding the ability to differentiate between various surface properties.

His dealings with the research methods of his day and those of his predecessors led Katz to the opinion that, until then, very little or no attention had been paid to "*movement as a formative force of the sense of touch*" [27]. He complained that the atomistic methodology of sensory-physiology – which was significantly influenced by research into the optical system – had blocked the scientists' view of the exploratory movements of the touching test person despite the fact that, according to Katz, the actual performance of the human sense of touch is considerably determined by the fact that these active movements in the fingers and hands are responsible for touch processes coming about at all. Again, Katz referred to E.H. Weber here and, in order to caricaturise the one-dimensional research approaches of his contemporaries, he resorted to biting analogies which several of his colleagues surely did not find humorous: "*Examining the sense of touch in a state of rest is almost like wanting to investigate the musculature of the legs after having put them in a plaster cast*" [27].

For Katz, the central importance of the movements for the performance of the sense of touch is mainly a result of observations made while subjects haptically explored surface properties and qualities. Like E.H. Weber before him, he correctly recognised that the very fine spectrum of object and surface properties can only be determined when the subject actively moves his exploratory extremities over the object structure in question. In order to examine this central function in an experimental situation, Katz chose a paradigm supported by the recognition the surface qualities of various paper types. In one such experiment he had subjects compare two sheets of paper chosen from a selection of 14 industrially prepared paper types and then order them in an attempt to judge the test person's differentiation thresholds. Katz is aware of the fact that the test person's tendency to unsystematically order the paper according to roughness represented a methodological problem. Katz used this test arrangement – with variations – as a basis for several different experiments into different questions. He, for example, reduced the size of the stimuli (paper pieces) to 2 mm in diameter and, in turn, determined the differentiation thresholds regarding the provided papers. In other experiments, he limited the subjects' ability to move their fingers sideways while exploring the paper's surface and observed that the test person's differentiation abilities were almost completely forfeited. He came to the same conclusion when the paper's surface was presented to a finger at rest (passive). He varied the experiments by isolating the subject's finger tips with different materials (e.g., 0.1 mm thick layer of collodion, 0.5 mm bandage, rubber coatings) and presenting them with the same tasks. To his surprise, the test persons were very well able to differentiate between the various types of paper. He also gave his subjects the task of differentiating between these papers by means of feeling them with a wooden peg. Again, they were capable of differentiating between them. Katz recognised that vibrations which appear during the touch process represent the crucial factor in stimulus recognition. He eliminated the influence acoustic information, as far as possible, by closing the auditory canal of his subjects during the touch process.

Katz did not, however, limit himself to testing the subjects' finger tips in these experiments. They were also required to differentiate between

Figure 9. Schumann Tachistoscope
This device made it possible to lead pieces of paper along the fingertips of a hand at rest at a constant defined speed. The paper samples were attached to a rotating disk. With the adjustable motor it was possible to achieve constant, steady speeds (Katz [27])

the stimuli using their toes, lips and teeth. He also attempted to determine which factors and measures were capable of impeding the perception process. To this purpose, the fingertips were rubbed extensively with a Turkish towel or cooled with snowmelt. With this methodology in mind, it comes as no surprise that Katz also included amputees in his experiments. He had 19 forearm amputees and 16 above-elbow amputees feel various materials (e.g., cardboard, leather, cloth) with the stump of their amputated limbs and requested that they recognise and name the material. Both groups showed substantial recognition capabilities – on average, nine of 12 materials were correctly identified.

To examine the influence of the duration of stimulation during the passive presentation of stimuli, Katz used the Schumann Tachistoscope (Fig. 9) that allowed him to vary the exploration times. Katz observed maximum distinguishability between paper types at a medium stimuli speed of 15 cm/s. At this speed, a stimulation time of 1.25 s for each sheet of paper was carried out. Although Katz did not record (in writing) a systematic variation of stimulation times, he claimed that at a speed of 60 cm/s (ca. 0.3 s) the ability of the subject to differentiate was nullified. All paper types appeared to be equally smooth. These, and other, studies led Katz to the conclusion that the vibrations which accompany the touch process are crucial to the perceptual results when assessing object surfaces.

With a similarly elaborate apparatus Katz examined the questions of how quickly test subjects were able to correctly identify different types of paper and of which influence the number of fingers used had on this. The recognition times with the use of one finger were ca. 5 s and when using five fingers the test subjects were notably faster. Katz explained the superiority of using five fingers with reference to Helmholtz [28] and the fact that processes of the central nervous system lead to a qualitatively better interpretation of the individual pieces of information.

Katz repeatedly described the (sensory) movements test subjects made during his various trials and experiments. He illustrated the complexity of the finger's work movements and stressed that they are apparently never in a state of rest. This phenomenological approach to the explorative movements seems to not have satisfied Katz very well so he undertook an attempt to suitably record and document the work movements of the finger graphically. To do

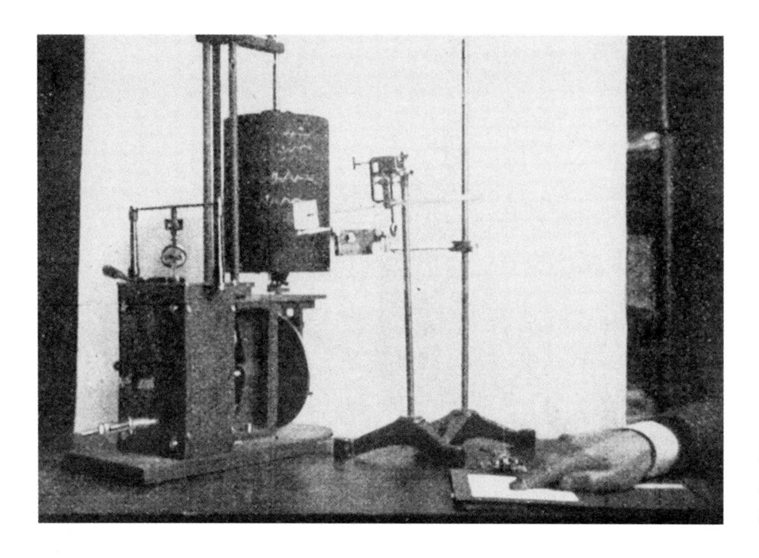

Figure 10. Device for recording finger movements during the finger exploration of various types of paper *(Katz [27])*

so, he used a very simple device to which the index finger is attached by a silk cord during the touch process. This cord is then attached *via* a roller system to a recording device. In this way, exploratory movements of the finger were depicted and plotted on a curve (Figs 10 and 11). With quite a low number of test subjects, and the resulting uncertainty in the quality of the data, Katz ascertained that finger movements towards the body were made more quickly than corresponding movements made in the direction away from the body. Additionally, he states that the movements made towards the body were also performed with more pressure. Katz was very well aware of the methodological limitations of his trials but he stressed the necessity and legitimacy of such pilot studies. As he can 'observe' only one mistake in each of these trial designs, he wanted to make another attempt which would offer the possibility to examine the movements of the finger 'in its natural dressing'. This is a goal which is to be admired – especially when one considers the humble technological possibilities that Katz had at his disposal. To accomplish this, he used a sooted, glazed paper on which the finger left traces as it moved across the surface. Alongside the numerous 'touch methods' which he was able to document, Katz stressed the observation that the object's surface was not touched continuously but, rather, with interruptions.

Alongside this explicit stressing of the movements in active touching processes, the importance of the fact that he dedicated himself to the experimental analysis of the exploratory movements must be recognised – even if he did so with simple methods. His work marked the beginning of a search for answers to these questions within haptic – questions which are, even today, extremely important and methodically difficult to research.

He also observed the fact that temperature perception plays an important part in the judgement of materials and surface qualities. Katz systematised the question and had his test subjects, in various experiments, judge the impression of warmth or cold made by different materials with the same temperatures. He then compared the relationships of these judgements to the specific thermal conduction coefficients of each material. Appropriately, he asked himself why different materials, although they have the same temperature, appear to be warmer or colder. The observed test persons' classification of the materials (e.g., aluminium = cold, wood = warm) remains constant independent of whether the

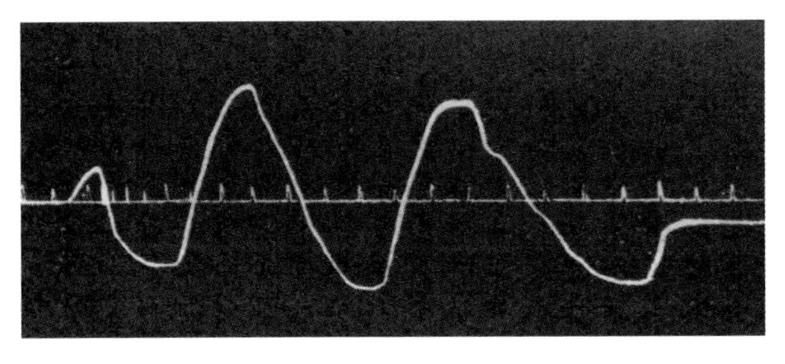
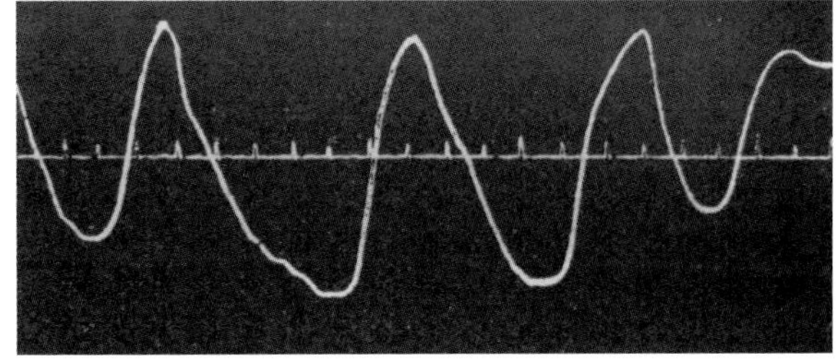

FIGURE 11. GRAPHIC DEPICTION OF THE WORK MOVEMENTS OF A SINGLE FINGER – FROM TWO SUBJECTS DURING THE EXPLORATORY TOUCHING OF PAPER SURFACES *(Katz [27])*

materials had been cooled to subzero temperatures or heated to temperatures of up to 26°C. He reached the same results when the subjects were asked to use their lips rather than fingers. In a similar fashion Katz examined the ability to judge this at very high temperatures and concludes that specific and typical perceptual patterns (which he named '*Temperaturgestalten*' or 'temperature patterns') exist for certain classes of materials.

In this section of his monograph of 1925, Katz correctly emphasises, that his most important contribution to haptic research was the recognition of the involvement of vibration sensations in active object and surface exploration. He was able to show that, in active exploration, these vibration sensations always worked alongside pressure sensations and played a considerable role in the recognition of objects and surfaces. Additionally, Katz clearly depicted the clinical-diagnostic importance that would later be granted vibration perception and the importance that this sensation has in the daily lives of the blind and deaf. Considering that he had worked so very fruitfully in other sensory areas, it is not surprising that he discovered so many connections between the 'sense of vibration' and the visual or acoustic systems. Katz was not strict in his use of the term 'sense of vibration'. He did not assume an individual sense but suspected, like Max von Frey, that the pressure receptors were able to receive pressure as well as vibrational stimuli at the same time.

Nevertheless, he emphasized the individual qualities which can be collected by the 'sense of vibration' and which can not be interpreted as a specific form of pressure qualities. Katz's statement that the 'sense of vibration' is actually not a 'proximal sense' is also interesting. Vibrations can be transported over long distances and then be detected by our bodies as such sensations. Also, vibrations can be perceived – under certain conditions – by many different people at once – in strong contrast to the individually experienced pressure sensations. Again here, the 'sense of vibration' shows its special place among the various sensory areas. For Katz it was self-evident that we live not only in an optical and acoustic world but, rather, in a vibrational one as well.

Emil von Skramlik

In the list of German scientists which have intensively researched the human sense of touch, Emil Ritter von Skramlik (1886–1970) takes a special place. First of all, he was one of the last to systematically devote himself to experimental haptic research in Germany before World War II and, not less importantly, he wrote the most comprehensive monograph about haptic and tactile perception to be published to date. The two volume work titled *Psychophysiologie der Tastsinne* [29] (*The Psychophysiology of the Senses of Touch*) – published in 1937 – contains almost

Figure 12. Emil von Skramlik (1886–1970)
(Photo: Myriam von Skramlik)

900 pages! Within it, the physiology professor from Jena presents nearly all of the findings of all international studies regarding human tactile and haptic perception from physiological, anatomical and psychophysiological research that had been done. At the same time, von Skramlik gives a detailed overview of the numerous experiments which he himself did or sponsored, and which delve into a wide variety of aspects of haptic perception. A notable characteristic of these (mainly) psychophysiological experiments is the strict and systematic methodology which distinguishes him as an excellent physiologist. These things are not only characteristic of his research into the sense of touch, but also his numerous papers about the physiology of the heart or the physical effects of nicotine on the human organism. He wrote two text books about physiology [30, 31].

This strict methodology is a valuable part of the von Skramlik scientific legacy which resulted from his experimental investigations into the sense of touch. Another part is his originality, marked by the wide spectrum of questions asked and his topics regarding the psychophysiology of touch. von Skramlik's first papers are oriented on the psychophysiological line of questioning based on Weber and Fechner. In such a way he examines threshold values in weight estimation, temperature perception and that of movement and pain. His meticulous studies into these questions were accompanied by a critical opinion of his contemporary physiologists which, to his mind, commonly worked without the proper consideration of the anatomical relationships involved. This was also a reason that he, in his book from 1937, extensively discussed the anatomical characteristics of the skin, and that he analysed the touch receptors known at the time.

In another part of the book he stresses the value of research into the sense of touch for clinical diagnostics. An increasing number of publications were coming out of neurological and psychiatric practical test procedures concerning the testing of the ability of the sense of touch in the case of various illnesses and diseases. However, according to von Skramlik, these testing procedures were too often used in a clinical context without the underlying processing mechanism being understood in a healthy human. He understood the good sense and importance of studying haptic perception disturbances in the fields of psychiatry and neurology but decidedly pointed out that he felt the underlying processes in a healthy human being must first be examined before such a diagnostic application may be put into practice.

Fascinated by Aristotle's well known finger illusion and motivated by a lack of systematic examinations into this phenomenon, he developed and depicted numerous other haptic deceptions which he tested on himself or others (Fig. 13). von Skramlik also expanded his examinations from the deceivability of touch

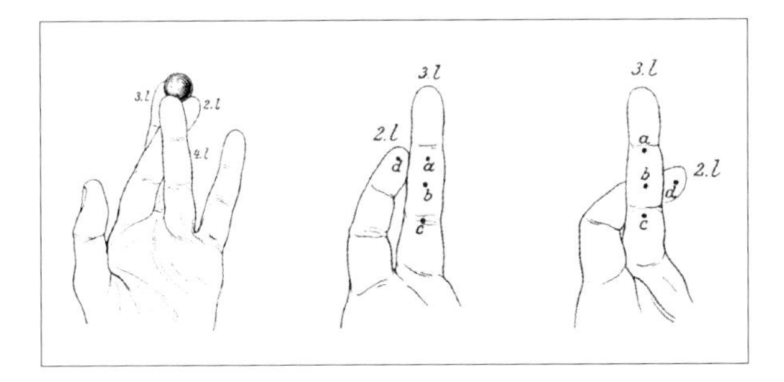

FIGURE 13. EXAMPLE OF A HAPTIC ILLUSION AFTER VON SKRAMLIK [29]

into the area of the perception of haptic apparent movements. He objectively let non-movable or non-flexible objects be manipulated by force and defined the appearance of apparent movements or illusions thereof (Fig. 14).

In order to analyse movement illusions, von Skramlik designed a series of test arrangements. He mainly described the observed perception effects in a phenomenological way. Besides the development and examination of new illusionary phenomena he also systematically examined the previously discovered optical illusions (Poggendorf, Delboeuf) and their effect on the sense of touch. In his results he states that the speed of the movements represents a decisive factor for the appearance of haptic illusions (Fig. 15). He suggested that one's work and daily habits may explain the perceptive illusions which he observed.

The basis for many studies which von Skramlik conducted (either himself or those which he had conducted by postgraduates and colleagues as Director of the Physiological Institute in Jena at the Friedrich-Schiller-University) was a critical examination of the methodology behind existing experimental findings. As a rule, he derived from such examinations (or the fact that for specific questions, insufficient or no information at all existed) the need to conduct his own investigations.

As E.H. Weber and D. Katz had observed before him, von Skramlik realised that the quality of haptic performance was also dependent on whether the exploratory movements of the subject were carried out actively or passively. Haptic, i.e., active touch, processes are always connected to the motor function of the perceiving subject. von Skramlik's predecessors and contemporaries recognised that active touch perception could not only be a function of the already known pressure receptors in the skin, but rather receptive systems within the muscles and joints were being considered more and more in order to explain the phenomena related to the sense of touch. There was a hefty debate at the time – some spoke for a self-contained *Kraftsinn* (sense of force) (E.H. Weber) or *Muskelsinn* (muscle sense) (C. Bell) as well as 'stress' or 'position' senses. Others attempted to explain the phenomena by postulating a 'deep-pressure

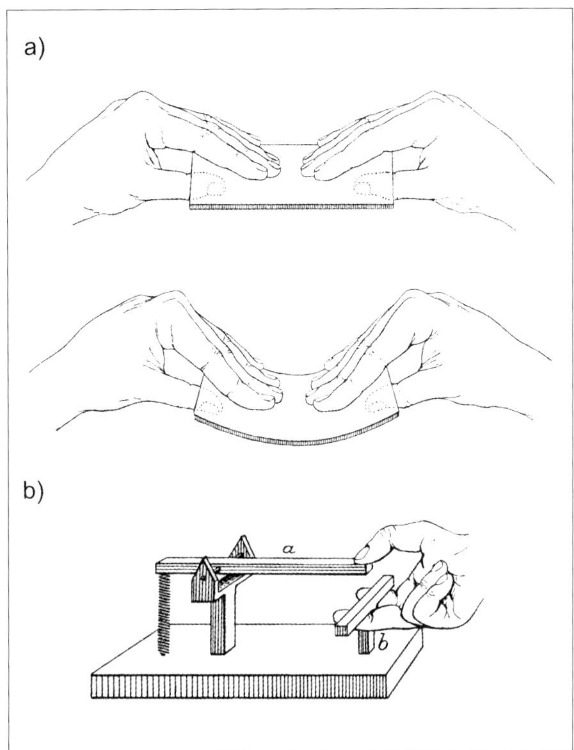

FIGURE 14. TEST ARRANGEMENT FOR THE ANALYSIS OF APPARENT MOVEMENTS AFTER VON SKRAMLIK [29]

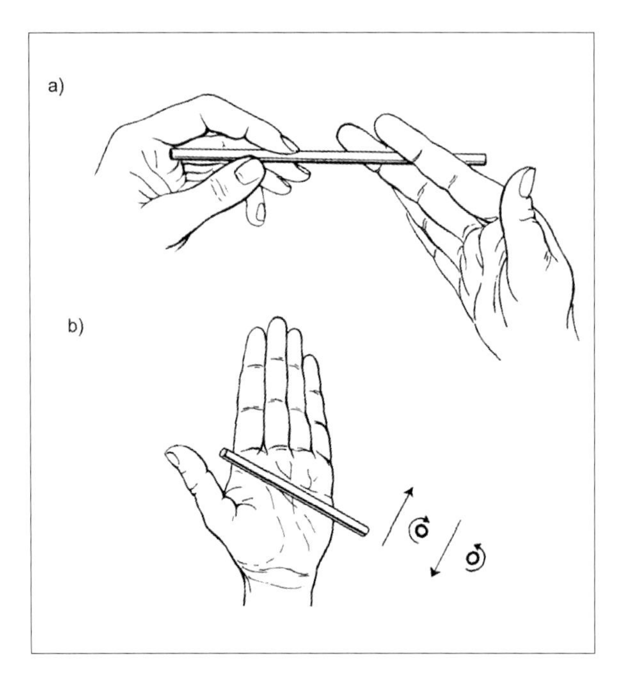

FIGURE 15. ILLUSIONS OF MOVEMENT AFTER VON SKRAMLIK [29]]

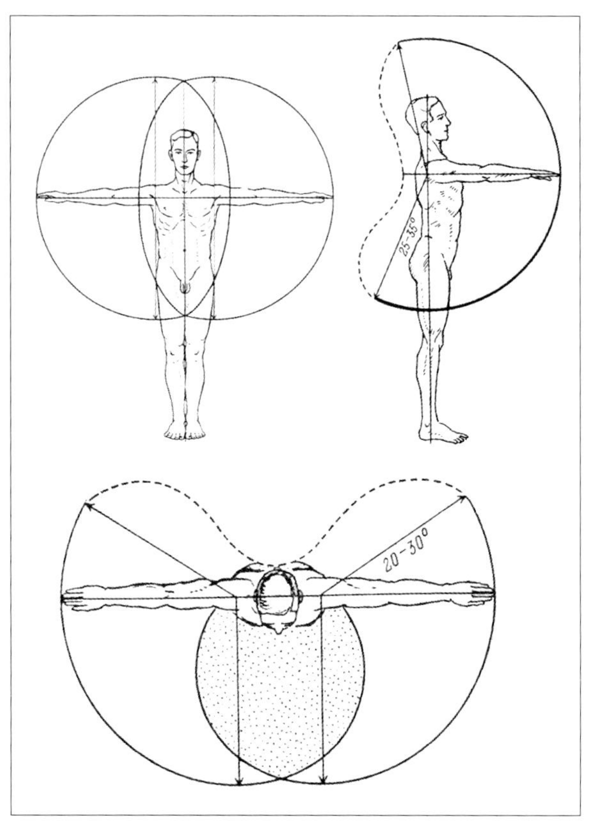

FIGURE 16. HAPTIC SPACE AFTER VON SKRAMLIK [29]

sense' and a 'surface pressure sense'. Influenced by the work of E.H. Weber and von Frey, von Skramlik attempted to solve the definition dilemma by assuming that all haptic performance is created by cooperation between the senses of pressure and power. Building upon this bipolar understanding of the foundation of the touch system, von Skramlik spent less time with the simple physiological one-point stimuli settings and turned his attention to the question of which importance the movements of human extremities and the entire body have on the human perception of touch. Without explicitly formulating it so, von Skramlik understood the entire human being as a moving organ of measurement which is constantly acting in three dimensional space. According to von Skramlik, power and pressure stimuli are thus, constantly being processed without our consciously being aware of it. This enormous everyday achievement – which von Skramlik discovered in numerous and various areas of life and which he recognised as an independent psychological property – required, according to him, a special scientific analysis.

It wasn't the object related touch performance that interested him as a scientist but, rather, experimental examinations regarding spatially related haptic activities of the human. With this in mind, he performed many experiments and trials about how precisely a human being can arrange his body and limbs in three dimensional spaces. He was interested in the degree of variance that appeared when the tests were repeated and the same instructions regarding room and position were given to the subjects and also in the tendencies and trends which could be recognised in the subjects' variations

Following Uexküll's [32] 'active area' (*Wirkraum*) concept and the 'tactile space' (*Tastraum*)

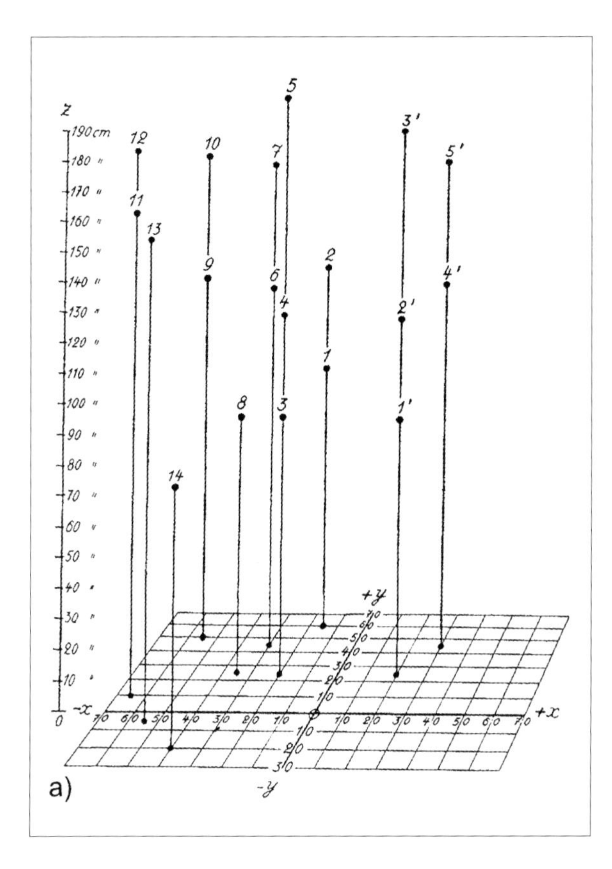

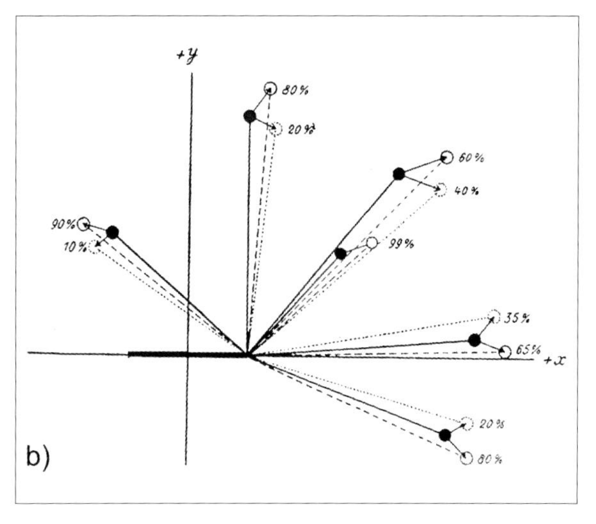

FIGURE 17.
a) Examples of spatial points presented in a x,y,z coordinate system, b) Examples of spatial points reproduced (black spots) by subjects with closed eyes and the average discrepancy in percentages (white spots). (von Skramlik [29])

concept of Turro [33], von Skramlik developed the concept of 'haptic space' (Fig. 16). This is characterised by the fact that all movements of the human body, especially the extremities, occur with a specific relationship to various bodily axes and the position of the head. In classifying the different tactile spaces, he differentiated between one and two handed tactile spaces as well as a four-extremeties tactile space.

In order to examine movements in haptic space more precisely, von Skramlik had his subjects reconstruct coded, three dimensional coordinates (x,y,z) within the space. This occurred both with one or both hands and with closed eyes. While this was being done the target points were arranged along the body's longitudinal axis in various positions and in various distances from the body. The variations between the target points and the actual spatial point recreated showed systematic variance. The shown points in the median plane were reproduced, as a rule, too far away from the body. It is also true that noticeable variations occurred systematically between the target and actual points identified when the tasks were carried out with the left and then the right hands (Fig. 17a and b). Furthermore, von Skramlik ascertained that, with closed eyes, the pointing to basic bodily axes (e.g., median bodily axis) is always connected to a systematic variance (Fig. 18).

The objective positioning of one's own bodily axes is subjectively – when visual perception is removed – distorted by an internal 'haptic coordinate system'. But not only one's own bodily

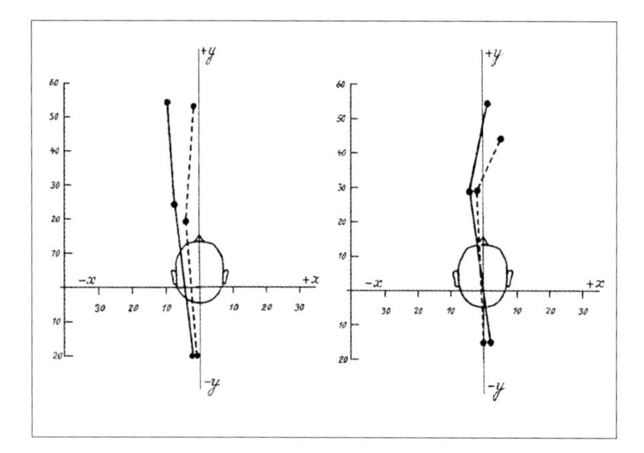

FIGURE 18.
Examples of variances in the median bodily axis with variable distances from the body with closed eyes for the right hand (solid line) and for the left hand (dashed line). (von Skramlik [29])

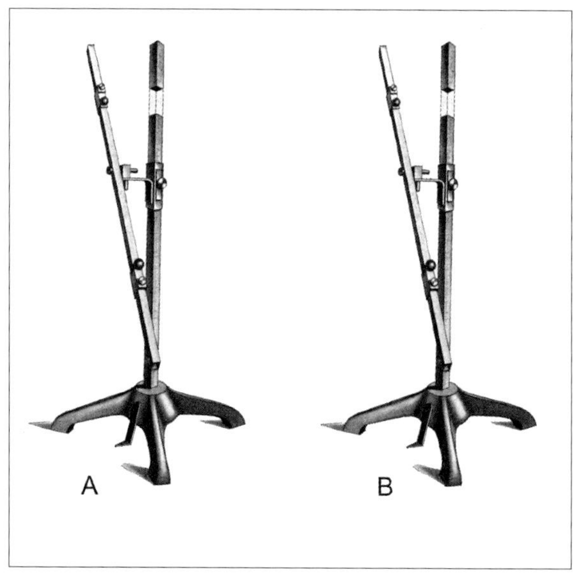

FIGURE 19.
A standard stimuli (line) with defined x,y,z coordinates are to be haptically reproduced on apparatus B with the same spatial coordinates without the use of vision. (von Skramlik [29])

axes and the reproduction of related spatial points are subject to the idiosyncratic assessment of the 'haptic coordinate system' as von Skramlik called it. For example, major variances occurred between the spatial coordinates of two objects when the subject was to reproduce their arrangement in space – e.g., parallel arrangement near the body – see Figure 19.

According to von Skramlik, it appears that the 'haptic coordinate system' varies from Euclidean geometry on all levels. Even in the arrangement of two rods in a right angle results in an objective variation of 3–4° (Fig. 20) and the vertical arrangement of a line is apparently never positioned perfectly – although in this case the variances are dependent upon whether the stimulus is in the median plane or to the right or left of the bodily centre.

The precise methodology and system of his experiments shows that von Skramlik was attempting to understand the laws and principles of this specific perceptual system by analysing single performance components with gradual variations in the conditions. It is therefore only natural that von Skramlik examined the influence of the head's posture on basic haptic performance. He observed that the subjective perception of the body's medial axis was objectively changed when the position of the head was altered. He also examined the influence of the head's position on the perception of various lengths presented with various spatial relationships to the subject's body (Fig. 21).

The search for the human 'haptic coordinate system' led von Skramlik to the conclusion that haptic space was significantly defined by the influence of the senses of pressure and power and not, as other researchers believed, by the vestibular system. Following this line of thought, changes in position of the entire body should be understood as haptic performance which is negotiated by the senses of pressure and power. Consequently von Skramlik used refined devices to examine the perception of changes in space and position with relation to the entire body (Fig. 22).

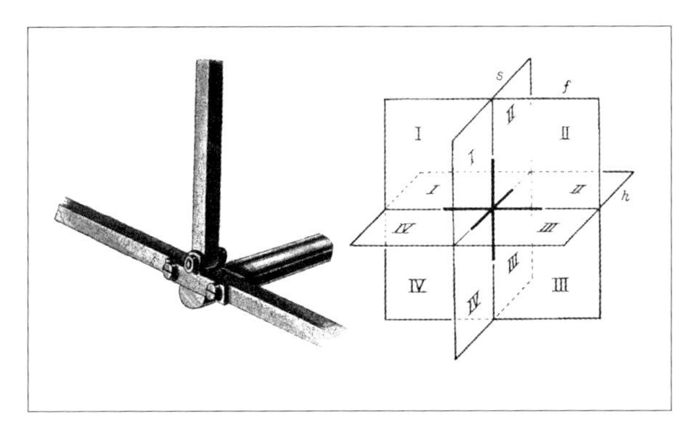

Figure 20.
Apparatus for the determination of the 'haptic right angle' after von Skramlik [29]

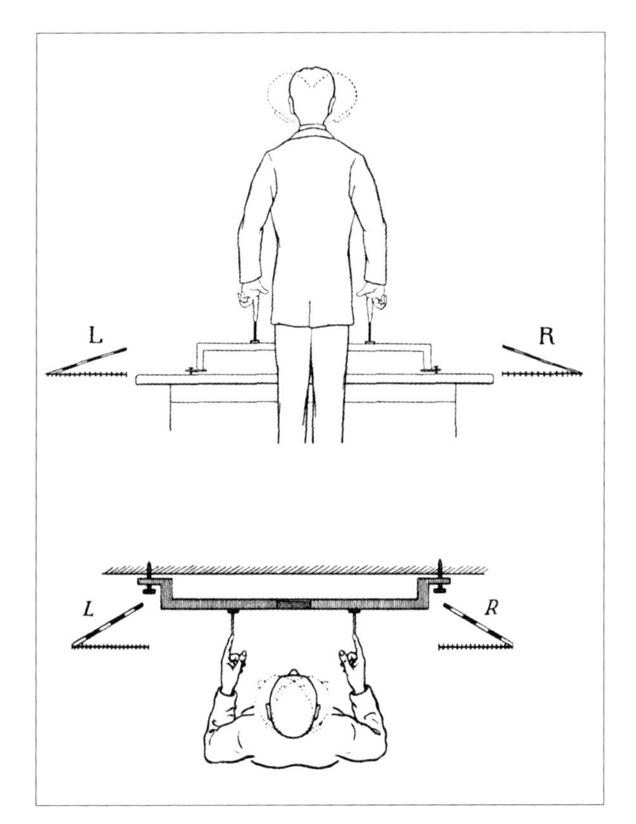

Figure 21.
Haptical presentation of distances with varying head postures and varying positions in relation to the body (von Skramlik [29])

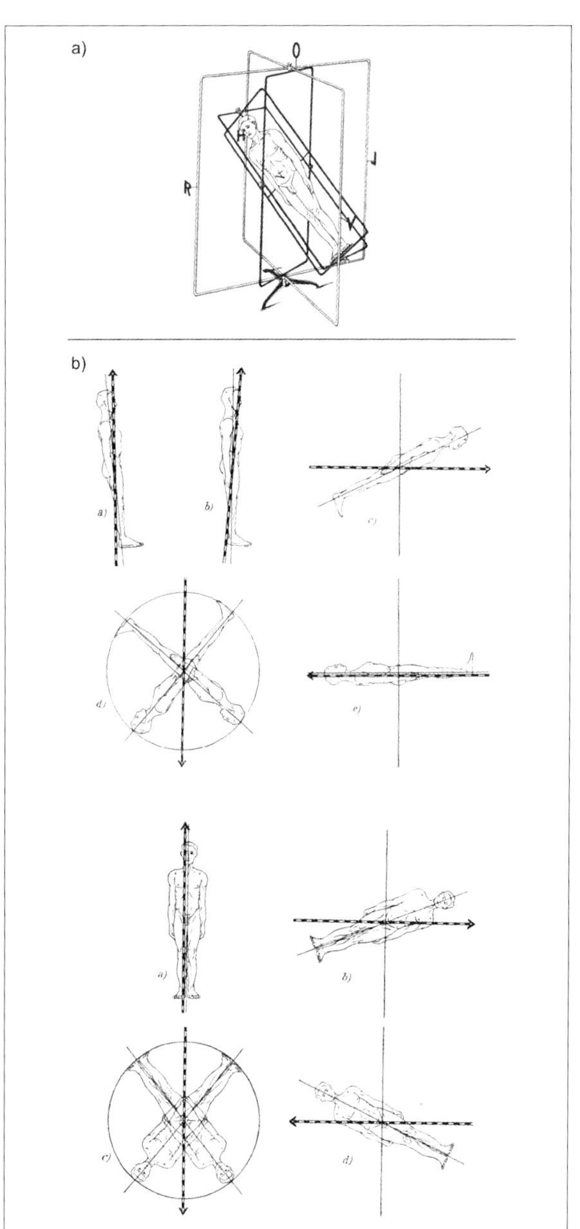

Figure 22.
a) Schematic depiction of the experimental apparatus which was used to affix the test subjects to a movable, full body framework. b) Subjectively perceived position of the body (thick, solid line) and the objective position of the body (thin line through the body's axis) (von Skramlik [29])

The subjects – normally untrained – were placed in a laboratory with closed eyes and set into precarious positions. These were then changed and the subject was to say when they believed they had reached, for instance, a vertical or horizontal position. Besides the expected variations between subjective and objective positions, von Skramlik again documented great individual variations in the results of these tests. von Skramlik attributes these variations between the individual subjects to the various living and educational situations of the individuals themselves. In opposition to the otherwise purely academic nature of his experiments, von Skramlik argued that the orientation of the body in space should be examined further because of the potential that it offered for submarine and air travel. Prompted also by the potential industrial use of the results, von Skramlik initiated examinations into the viscosity of fluids and the ability to haptically differentiate between powder particles.

von Skramlik observed and described the fact that, despite carefully controlled and planned experiments, the individual performance of the test subjects could vary greatly in many tests. In opposition to his predecessors, von Skramlik admitted that, in the area of haptic perception, the psychophysiological principles of Wundt and Fechner could only be partially valid but in individual cases did not allow for adequately precise predictions. These effects become visible in the 'tactile measurement' tests that von Skramlik performed. He attempted to answer the question of which variables had an influence on the haptic/tactile sizing of objects and lengths. The results indicate that the speed of the motions is the dominant factor in one's ability to estimate the size of an object. Consequently, von Skramlik looked for the influence of the motion speed on perception. He analysed the point at which two haptically presented lengths were perceived as the same when the touch speeds were different. In these tests, a finger of one hand was run along a predefined length using a special device (see Fig. 23).

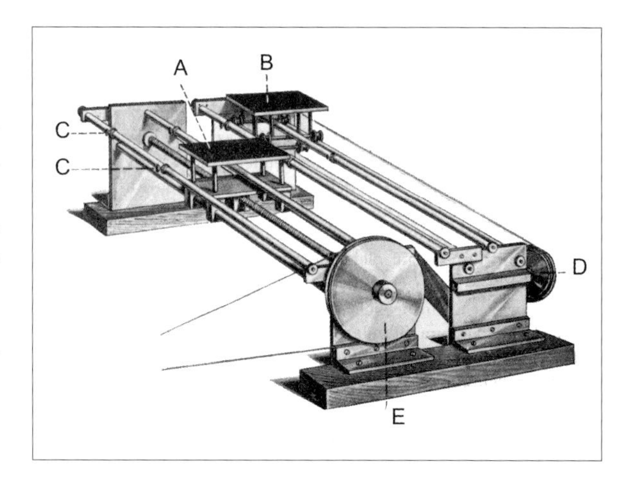

Figure 23. Device for the presentation and reproduction of lengths.
One hand rested on the movable cart for the registration of the standard length. (B) The speed of the sampling was 60 cm/s. During the movement the subjects' fingertip was in contact with the length's target ring (C) on the rods in front of the finger. After this, the subject's hand was moved over the same distance at varying speeds and he/she was to say when the control length hand been covered (von Skramlik [29]).

A standard length with a moderate tempo served as a control and the following tests occurred with variations in speed. Again the degree of variation between the individual subjects' results were surprising to von Skramlik. However, he noted one stable finding. Namely that, in general, the subjectively estimated length was decidedly overestimated when the speed was reduced.

The range and number of haptic perception experiments carried out by von Skramlik are so extremely manifold that they cannot be sufficiently encompassed by a historical overview such as this. Without a doubt, von Skramlik was one of the most productive and creative German scientists in the field of haptic perception before World War II. His work and, especially his book from 1937, marked the end of a long lasting and intensive research tradition with the goal of investigating the psychophysiological principles of the human sense of touch – touch in all of its various dimensions. It is to be hoped that, in the future, not only the work of von Skramlik but that of other lesser known – or even completely unrecognised – scientists will be appreciated by historical research and that their work and ideas will be made accessible to the current field of colleagues that are looking into these phenomena. In this way, hopefully, we will not only find a measure of respect for the past scientific efforts made, but rather will be able to use this reflection of the past to help ourselves better understand our own position in the search for the laws and functional principles of the sense of touch.

Summary

Of the many institutional and private activities in the research of the sense of touch in Germany before the Second World War only a few could be mentioned in the previous pages: H. Weber, M. v. Frey, M. Dessoir, G. Révész, D. Katz, E. v. Skramlik. What they all have in common is the fact that they enriched the research in this field on a national and international level with original contributions. E.H. Weber's first experimental studies of skin sensitivity and two-point discrimination were a fundamental step for psycho-physiological research. Von Frey enriched expert discussions with scores of experimental data about skin sensitivity – using his "Frey Hairs" – and the question of the existence of a "deep-pressure sense". In 1892, Max Dessoire introduced the term "haptic" as the "science of the human touch" and, in doing so, gave this field of perception research its name. Géza Révész and David Katz worked and fought for an independent research methodology in the research into the sense of touch and for a better position for this sense within psychological research in general. They analysed, in various ways, the importance of the active vs. passive exploration of objects and surfaces and they attempted to work out how stimulus exploration functions.

In contrast, the importance of Emil von Skramlik's works for touch-sense research is still largely unknown. His two-volume, 1937 work about the psycho-physiology of the sense of touch was only partially received in Germany and remains unknown abroad. However, within its pages, von Skramlik presents a number of his own studies ranging from illusionary experiments to body-position analysis. The spectrum of his scientific work is extremely wide and the future will show how greatly his results impact the modern research into the touch of sense. This small selection of scientific works and people is meant to motivate the reader to more actively analyse, learn from and appropriately recognize the historic and current value of these and other works which have not been considered closely enough to date.

Acknowledgements

I would like to warmly thank Professor Rand Evans for his help in checking these texts for any linguistic imperfections and also Ms. Tröbs and the staff of the University of Leipzig's library for their help and support.

Selected readings

Weber EH (1978) *The Sense of Touch*. London; New York: Academic Press for Experimental Psychology Society, 1978

Révész G (1958) *The Human Hand; a Psychological Study*. London: Routledge & Paul

Katz D (1989) *The World of Touch*. Edited and translated by Lester E. Krueger. Lawrence Erlbaum Associates Inc

Ribot T (1886) *German Psychology Today: The empirical school*. (2nd ed.). [J.M. Baldwin, Trans.). New York: Scribners

3

British pioneers of research into human haptic perception

Jonathan Cole

The history of early research on haptics in the United Kingdom is, to a large extent, the history of several great men, reflecting in part the relative small numbers engaged in research. There was a golden age for British neurology and neurophysiology at the turn of the 19th and 20th Centuries, before and during the First World War. After this the focus moved to a smaller scale with the work of Adrian and Matthews; for instance, being largely on the mechanism of the nervous impulse and on biophysics rather than on the functioning of larger systems. Such were the successes of such an approach, with the accruing of Nobel Prizes for Adrian and Sherrington, Huxley and Hodgkin, and Katz, that whole system approaches only became in vogue once more within the UK in the latter decades of the last century.

Bell's handbook

Research on sensorimotor integration in the UK, and arguably elsewhere too, began with Bell. Sir Charles Bell (1774–1842) was a Scottish anatomist, artist, surgeon and physiologist. He is remembered for his eponymous facial palsy, for his long thoracic nerve and for the debate with Magendie over the sensory and motor natures of the dorsal and ventral nerve roots. But it is in his book, *The Hand; Its mechanism and vital endowments as evincing design*, that his enormous contribution to the subject of haptics was made [1]. The work considers many aspects of the hand, from comparative anatomy to even substitution of other organs for it. At one point he even discusses how animals were suited to the progressive changes in the earth and the elements, without taking any further steps towards Darwin. But the heart of the book is on what today we might call neuroscience, both experimental and theoretical.[1]

Throughout this rich and beautifully illustrated work (Bell was an accomplished artist), he also reveals the depth of his thought and his insights many of which appear to have been based on deduction.[2] He drew attention to the complexities of even a simple movement, which appear to have largely been taken for granted, "*we use our limbs without being conscious, or at least, without any conception of the thousand parts which must conform to a single act… by an effort of the cultivated mind we must rouse ourselves to observe things and actions of which sense has been lost by long familiarity.*"

Though, of course, it was Sherrington who coined the word proprioception, or self knowledge[3], it was Bell who first described movement and position sense. "*We awake with a knowledge of the position of our limbs; This cannot be from a recollection of the action which placed them where they are; it must therefore, be a consciousness of their present condition.*" Bell, op cit.

Interestingly Ian Waterman, a man I have studied who has lost cutaneous touch and movement and position sense below the neck, describes exactly this problem when he wakes [2]. He has no idea where his legs are in bed under the sheets, and so moves them to find out.[4] "*When a blind man, or a man with his eyes shut, stands upright… by what means is it that he maintains an erect position? How is it that a man inclines in due degree towards the winds? It is obvious that he has a sense by which he knows the inclination of his body and that he has a ready aptitude to adjust it.*

It can only be by the adjustment of muscles that the limbs are stiffened. There is no source of

FIGURE 1. SIR CHARLES BELL (1774–1842)

knowledge but a sense of the degree of exertion in his muscular frame…

In truth we stand by so fine an exercise of this power, and the muscles are, from habit, directed with so much precision and with an effort so slight, that we do not know how we stand. But if we attempt to walk on a narrow ledge, or stand in a situation where we are in danger of falling we become subject to apprehension; the actions of the muscles are magnified and demonstrative to the degree in which they are excited.

We are sensible of the position of our limbs… although we touch or see nothing. It must be a property internal to the frame… what can it be but a consciousness of the degree of action…" Bell, op cit.

He thus clearly describes a sense of movement and position, and also immediately brings our attention to two other aspects. Though describing them as a sense, he realises that they lie between consciousness and what is unattended to and how even similar actions can be either automatic or filling our attention according to context, for instance when we walk or walk across a narrow ledge. In the phrase, "from habit", he implies that learning movements may be dealt with differently to habitual ones. Ian Waterman, without proprioception, describes how he does not know from one day to the next whether he will be able to make a movement; he is nearly devoid of habits, or automatic movements, all requiring mental concentration of a various degree.

"*When we stand, we cannot raise or extend the arm without a new position of the body, and a poising of it, through the action of a hundred muscles.*

Nothing appears simpler to use than raising an arm… yet in that single act, not only are innumerable muscles put into activity, but as many are thrown out of action, under the same act of volition." Bell, op cit.

Thus he describes the multiple synergies of action between muscles for even a simple single act. In the next passage he clearly showed he was thinking of active touch, of touch being through a combination of cutaneous sensation and manipulation of the tactile organ,

"*Accompanying the exercise of touch, there is a desire of obtaining knowledge; in other words, a determination of the will towards the organ of the sense. Bichat says it is active when others are passive.*

… in the use of the hand there is a double sense exercised; we must not only feel the contact of the object, but we must be sensible to the muscular effort which is made to reach it, or to grasp it with the fingers." Bell, op cit.

"*When treating the senses, and showing how one organ profits by exercise of the other and how each is indebted to that of touch, I was led to observe that the sensibility of the skin is most dependent of all on the exercise of another quality. Without a sense of muscular action or consciousness of the degree of effort made, the proper sense of touch could hardly be an inlet to knowledge at all… the motion of the hand and fingers, and the sense or consciousness of their action must be combined with the sense of touch…*

Great authorities made no account of the knowledge derived from the motions of our own frame. I called it a sixth sense.

...the sense of touch is exercised by means of a complex apparatus – by a combination of the consciousness of the action of the muscles with the sensibility of the proper nerves of touch...

The motion of the fingers is especially necessary to the sense of touch. These bend extend, or expand, moving in all directions like palpa, embracing the object and feeling it on all surfaces, sensible to its solidity..." Bell, op cit.

It is not clear that in using the word "sensible" he meant aware, though at this time the concept of non-conscious processing may well not have been current. But elsewhere in the book he did consider such automatic or reflex movements, and in a slightly unexpected context perhaps,

"*everyone changes their position and shifts the weight of his body; were you constrained to retain one position during the whole hour you would rise stiff and lame. The sensibility of the skin is guiding you, which if neglected, would be followed by the death of the part. ... A patient with paralysis of the lower part of the body, we must give especial directions to the nurses and attendants that the position of his limbs be changed at short intervals... If this is neglected, you know the consequence to be inflammation of the parts that press upon the bed... local irritation, then fever and mortification and death.*" Bell, op cit.

At that time, of course, before the work of Head and Riddoch and others during the First World War and then Guttmann after the Second, those with spinal injury soon perished.

He was also aware, anticipating the work of Adrian among many others, that the nervous system is most interested in changing stimuli rather than static ones,

"*For the nervous system it holds universally that variety of contrast is necessary for sensation, the finest organ of sense losing its property by the continuance of the same impression.*" Bell, op cit.

He even considered whether this sixth sense was due to peripheral feedback or to internal corollary discharge,

"*At one time I entertained a doubt whether this proceeded from a knowledge of the condition of the muscles or from a consciousness of the degree of effort directed to them in volition...*"[5] Bell, op cit.

Then towards the end of his book Bell turned to an aspect of sensation we still shy from today; its affective nature (but see Olausson et al. [3]).

"*The exercise of the muscular frame is the source of some of our chief enjoyments. This activity is followed by weariness and a desire for rest; and although unattended with any describable pleasure of local sensation, there is diffused through every part of the frame a feeling almost voluptuous.*" Bell, op cit.

Lastly he has a short sentence on the possible interactions of brain and hand. Though not entirely clear what he meant at this distance, it is clear that he understood something of the complexities involved in organising and controlling movements of the hand, and that the brain might be divided functionally and possibly anatomically into, say, a motor hand area.

"*The hand is not a thing appended... there must be an original part of composition of the brain which shall have relation to these parts before they can be put into activity.*" Bell, op cit.

Bell's book on the hand is so rich in insight that one can but wonder at how much he induces from observation and deduction. This is not to suggest that he was not a great experimentalist, though others have suggested so.[1] After Bell, however, insights have been forged mainly through experiments and empiricism.

Sherrington and deafferentation

In the late 19th and early 20th century Sir Charles Sherrington (1857–1952), worked uncovering the secrets of organisation of the nervous system, concentrating on reflex activity. But he also made important observations on the effects of deafferentation on functional movement in non-human primates. In 1895, he and Mott sectioned the dorsal, sensory, roots of monkeys, so depriving them of feedback from the limb and followed the animals for up to 3 months [4].

FIGURE 2. SIR CHARLES SHERRINGTON (1857–1952)

"*After section of the dorsal roots from either whole cervical or lumbar roots, the monkey no longer used its arm or leg; movements of the hand and foot are practically abolished; the movement of grasping both with foot and hand never occurred. But movements of the elbow and knees and especially those of the shoulder and hip are much less affected.*

When the animal runs about it does not attempt to use the leg and the forelimb swings helplessly."

They observed that when the animal was allowed to climb a rope or side of a cage, it did not use its deafferented limb, either fore- or hindlimb. If feeding time was deferred and then the animal was offered food with the intact arm tied behind its back, then it still did not use its affected limb, choosing instead to thrust its neck forward. If the food was placed in the de-afferented arm then it did not lift it, preferring to use the foot. When it tried to use its deafferented foot, then it would miss its object widely. This profound sensory ataxia, was seen in patients due to syphilitic tabes dorsalis.

"*The defect increases from base to the free apex of the limb… it is successively greater at the knee and ankle and greatest (amounting as regards volition to absolute loss) in the digits.*

In this respect it curiously closely simulates the impairment of motility ensuing upon ablation of the limb region of the cortex cerebri; but it is, in the monkey, more severe than the impairment following cortical ablation."

During forced movements, e.g., when held, the terminal digits were moved, but without clear grasping movements. They concluded that the delicately-adjusted movements of the fingers and smaller hand muscles were affected most, just those movements represented in the cortex.

They then move onto an extraordinary suggestion.

"*In the case of certain movements, e.g., grasping movements of the hand and foot, opposition of thumb and first finger, the animal is rendered absolutely powerless to perform them… Although we are aware of the danger of introducing terms relating to consciousness into descriptions based solely on motor re-actions, we believe that we cannot more lucidly state the condition of the animals than by saying that the volitional power for grasping with the hand had been absolutely abolished by the local loss of … sensibility. Further this volitional power was lost immediately from the time of operation.*"

By volitional power they meant the will to move the limb. In other words deafferentation not only prevents controlled movement of the limb, but also abolishes the *will* to move it. Such observations, made on monkeys at the beginning of the 20th century, had to wait for deafferentation in man to be confirmed or refuted.

There are rare examples where lack of the possibility of action precludes intention towards that action at a perceptual level.[6] This was described by Sacks in his book *A Leg to Stand On* [5]. More recently Cole [6] discussed three examples of subjects describing volitional impairment towards action, where subjects had "forgotten what to do". One was a man, like Sacks, who had suffered a painful peripheral injury to the ankle; when the injury had recovered he did not know how to move the foot. A second, when suffering from viral meningitis, tried to move his neck, rendered rigid by the infection. Not only could he not move it, but he had the percept that he no longer knew how to. Lastly subject IW, with a total deafferentation below the neck and no movement or position sense, described a similar lack of volition towards movement at the beginning of his impairment, which passed once he could see and move his body and limbs. These observations offer some support for the idea that sensory loss, or impairment, can affect volitional aspects of movement as well as coordination of those movements. Sherrington and Mott were extraordinary to consider this after their short experiments on non-human primates.

They subsequently checked whether the deafferented cortex was still able to produce movements, by using absinthe to induce epilepsy, and found that, if anything, movements were more easily elicited than in controls and in distal muscles as well as proximal ones. Lack of movement was not due to a cortical degeneration or loss of functioning. Mott and Sherrington also went onto to make single dorsal root sections and found minimal losses of movement, suggesting considerable overlap in sensory afferents between roots, certainly in relation to function. They considered the contributions of cutaneous and muscle sensory afferents to these deficits, but because of the overlap of these in dorsal roots were unable to reach firm conclusions, though they thought that sensory afferents from the skin were important for hand coordination.

Subsequently Taub and Berman [7], repeating the experiments with single limb deafferentation, confirmed their results. But they went on to make bilateral root lesions, after which, in the macaque, deafferented limbs were used with some accuracy in a free environment. The ability to climb, walk and appose thumb and finger were preserved, results confirmed by Bossom and Ommaya [8]. They presumed that a single limb deafferentation led to favouring of the remaining intact one. Perhaps if both are deafferented the animal has to relearn movement, to an extent, without peripheral sensory feedback.[7] These experiments, however, lead to complete loss of sensation from a limb, with absence of pain and temperature sensation too, and this leads to problems, to say the least, with pressure ulcers and the animal's behaviour towards the unfelt limbs. More recently study of the acute sensory neuronopathy syndrome, in which movement and position sense and cutaneous touch are lost but small fibre function including pain and temperature are retained, are both more pertinent and acceptable, allowing study of the effects of sensory loss over months and years.

Head and sensory testing

Sherrington worked experimentally on animals, presaging methods which dominated British physiology for much of the 20th century. But the turn of the century was also a time of rich clinical research. One of its major figures was Sir Henry Head (1861–1940), a British neurologist. One of his enduring works was to test peripheral sensation methodically and to then use these tests to relate the site of neurological loss to its sensory deficit.

His tests of sensation are fascinating because of their thoroughness, though they could not be used in a busy outpatient clinic [9]. He stated that they had to be measurable and objective, and should "*Not [be] hampered by time or opportunity… not compelled to consider the wishes of the patient*". Though thorough he did not want to "*exhaust the patient's attention*". He started above threshold and was keen to "*determine not threshold but difference between normal and abnormal areas*". He also used terms for what is measured rather than terms for any underpinning mechanism, like stereognosis etc. He talked of measured prick rather than analgesia and was concerned not only with the area of reduced or lost sensation but with what is left.[8]

FIGURE 3. SIR HENRY HEAD (1861–1940)

He also wrote of the necessity of avoiding a priori hypotheses. Between a physical stimulus and act of sensation, he stressed, lay a multitude of transformations. "*In man sensation is a highly differentiated reaction to physiological processes transformed from peripheral end-organ to highest receptive centres.*" The role of sensory testing is to try, as best one can, to unpick the simplest levels of deficit. In order to do that he suggested that one needs a quiet room, the right weather and no distractions, one example being a full bladder.

Spontaneous and reduced sensation

He differentiated spontaneous from reduced sensation. In this he was not simply thinking of parasthesiae. Even numbness was thought of as being not only reduced sensation, or "*loss of feeling*", but also "*a positive abnormal sensation*". This is often forgotten, but clearly revealed by those who live describing their complete sensory loss after spinal cord injury,

"*This is not the same sort of sensation somehow [as after dental anaesthesia]... That numbness is sort of a void, as though something has been taken away and is no longer there. My loss of sensation is not quite like that; it is a different feeling...*"

"*It is a sensation of nothing. I immediately compared with before. To me it was a sensation. You're saying that if you can't feel you can have a sensation. But it wasn't numbness. It was nothingness. It was a sensation because you can sense nothing. It was a definite sensation.*" [10]

Wittgenstein asked "*Is the feeling of nothing a feeling?*" [11]. It seems the answer is that it can, as Head understood.

Though ostensibly trying to test peripheral sensation, he was more than aware of the central components in any such judgement and of the relations between movement and positions sense and higher perceptions.

"*Inquire if the patient has noticed that he is unaware of the position in which the affected limb lies, and if he preserves an idea or mental picture of the limb. Many with cerebral lesions complain that when they wake at night they do not know where the arm is lying, and it sometimes seems as if part of the limb had disappeared.*"

Loss of sensation

The next part of his examination will be familiar to contemporary students of neurology and medicine. Touch was tested using cotton wool, though with "*extreme caution, some brands are so stiff that pressure may result*". He was aware, however, of the need to use stimuli appropriate to the subject. "*Finer cotton wool may fail to act as a stimulus to the horny palm of a workman*". He was also aware of the way in which even light touch could alter with certain central lesions, "*In cases of thalamic overreaction cotton wool may produce a sensation which is affective, pleasurable, tickling or itching.*"

It is interesting also that even then he was thinking of peripheral sensation not simply as touch sensation but as having an affective component, something still not widely accepted.[9]

He used Von Frey hairs extensively, and not just for their force but also in relation to their cross sectional area, giving tables of their pressure per area and tension (0.4–3.5 g pressure per unit area 8–100 g/mm and tension 1–32 g/mm). A camel hair was used but with reluctance, since when moved slowly it was not felt but when fast it was. Pressure touch, or deep touch, was tested with a blunt object of the same temperature as the skin. There were various pressure aesthesiometer which were designed to be only able to apply a certain pressure.

Pain and temperature

He tested superficial pain with a pin or needle prick, being careful not to include touch sensation in the perception. A spring algesimeter needle enabled a limited application force. Pressure pain was tested with a larger, bigger ended algesimeter. He stipulated three of more readings from any point.

Temperature was tested with flat bottomed silver tubes diameter 1.25 cm filled with water or ice and larger copper tubes 4 cm diameter. Heat spots were tested with a "*soldering iron with an apex 1 mm square*", turned down one presumes.

Thus far his methods might not be surprising, thought the length of time taken would be difficult to achieve now. But he also tested other aspects of sensation some of which have largely discarded now.

Other cutaneous perceptions

For roughness he used a 'Graham-Brown aesthesiometer' which was a mass of brass with a polished surface from which a tooth projected using a graduated screw. The tooth raked the skin. More prosaically he also used emery or glass paper drawn over the skin. Tickling and scraping were tested by drawing either his pulps of the fingers over the soles of feet or cotton wool over the pinna. For vibration a tuning fork, 128 Hz, was placed on a body part firmly supported on bed or pillow. With the eyes shut the subject was asked to say when vibration no longer felt. At this point tuning fork was transferred to the homologous area opposite side, when vibration could usually be felt. Head was concerned about adaptation to the stimulus. Localisation was tested using Henri's method, in which a patient was asked to mark on a life sized diagram the exact spot on the patient's body which had been stimulated. He also allowed them to point to the place on another person. Two point discrimination was tested using the simultaneous application of two points of a blunt compass, with the discrimination one or two touches, and the successive application of two points to obtain a distance discrimination.

Joint and position sense was tested by placing, passively, one part of the body in a given position and asking the patient to place the other limb in the same position. Apparently Horsley had a glass plate with 3 d graduations to quantify limb positions over 10 trials were used. Patients were distracted during the passive movements. To test passive movement itself he changed the position of the limb or part of it and asked patients to indicate when they could appreciate it. Gordon Holmes's protractor was used to quantify this. Head noted that this was influenced by rate of movement and the touch of the clinician.

Appreciation of weight was assessed using circular discs of lead skin covered with chamois leather, with the percept measured either when the hand supported, when force over area was being tested, and unsupported when receptors involved in the perception of joint position and movement would be used. He asked which of two weights successively was successively heavier or lighter. Further testing was performed of the appreciation of size, using leather circles on the palms, two dimensional shape, with a circle, square and triangle in stiff leather applied to the palm, "*evenly and firmly*". Form was tested in three dimensions, using active manipulation in the hands, with both common objects and those made up. Texture was assessed using manipulation of silk, cloth, velvet, etc.

Head's analysis of the patterns of sensory deficit in relation to the level of lesion moved from peripheral nerve injury to spinal cord to sub-cortical and cortical losses, through a number of papers [9]. His enthusiasm to collect patients runs through them, with him describing one man, injured by a gunshot wound to the head in World War I, as a "*superb example*" of cortical damage, reflecting the age.

He found, however, that to test patients under imperfect clinical conditions and over a short time was insufficient. So he and Rivers embarked on one of the great and epic neurological experiments [9, 12]. He arranged for his own superficial radial nerve to be cut surgically in the forearm and then Rivers tested its recovery over nearly two years of painstaking study. This was performed in Rivers' rooms in Cambridge at weekends since Head worked in London.

They found, unsurprisingly, that initially all sensation to testing of touch, pain and temperature were lost and that at the junction between the analgesic area and normal there was an area of hyperaesthesia. Recovery began around day 32 post surgery, with different rates for light touch and for pin prick. As the latter returned earlier Head experienced mislocalisation of sensation and confusion at the extremes of testing between hot and cold. For these and other reasons they argued for two forms of peripheral afferent; the epicritic, well localised, low threshold sensation like light touch, and the protopathic seen in areas with temperature and pain afferents without light touch ones, in which there was greater radiation of sensation and confusion between hot and cold sensation at the extremes. This distinction has not entered widespread use.

Head and cortex

In his work on patients with cortical lesions suffered often from low velocity rifle injuries in Word War I he was also led to one of his most famous theories and one not dominated by sensation. Motor schemata are internal models of sensorimotor function, first suggested in relation to dissociation of localisation of sensation on the body in patients with preserved touch but absent proprioception,

"*Recognition of the locality of a stimulated spot demands reference to another 'schema'. For a patient may be able to name correctly, and indicate on a diagram... the exact position of the spot touched and yet be ignorant of the position in space of the limb...This faculty of localisation is evidently associated with ... another schema or model of the surface of our bodies.*"

"*It is to the existence of these schemata that we owe the power of projecting our recognition of posture, movement and locality beyond the limits of our own bodies to the end of some instrument held in the hand.*"

He was quite clear that these schema were not accessible to consciousness, were not modality specific to vision or motor systems, and were dynamic,

"*We have been able to show that the standard against which a change in posture is estimated is not an image either visual or motor; it lies outside consciousness. Every recognisable change in posture enters consciousness already charged with its relation to something which has gone before, and the final product is directly perceived as a measured postural change, For this combined standard, against which all subsequent changes in posture are measured, before they enter consciousness, we have proposed the word 'Schema'. Man perpetually builds up a model of himself, which constantly changes. Every new posture or movement is recorded on this plastic schema and the activity of the cortex brings very fresh group of sensations evoked by altered posture into relation with it.*

"*Before the afferent processes caused by movement of a joint can evoke a change in consciousness, they have already been integrated and brought into relation with the previous physiological dispositions.... Just as on a taximeter the measured distance is presented to us already translated into shillings and pence, so the final product on special changes rises into consciousness as a measured postural appreciation.*

For this standard we have proposed the word 'schema'. By means of perpetual alterations in position we are always building up a model of ourselves, which constantly changes. Every new posture or movement is registered on this plastic schema and the activity of the cortex brings every fresh group of sensations evoked by altered posture into relation with it.

Recognition of posture and movement is obviously a conscious process. But the activity on which depend the existence and normal character of the schemata lie for ever outside consciousness; they are physiological processes with no direct psychical equivalent. The conduct and habiliments of the actor who appears before us on stage are the result of activities behind the scenes of which we must remain ignorant, so long as we are only spectators of the play." [9]

In these famous passages Head developed ideas about sensorimotor activity which was outside, though influencing, consciousness, ideas about non-conscious body schema and conscious body image which remain current to this day [13]. Ironically therefore, much of the rest of Head's work has an overwhelmingly sensory feel, as he attempts to convince that there are specific sensory losses associated with various parts of the nervous system and that the sensory cortex is the seat of sensory memories and comparisons. Even when discussing astereognosis, using wooden figures and knives and coins, he stressed the cortical deficits in recognition rather than in coordination and sensation.

"*The sensory cortex is the organ to which we owe the faculty of relating one sensation to another...Not uncommonly a patient with a cortical lesion can recognise that some movement has taken place, but is unable to discover its direction or amplitude.*"

Later he considers the flavour of cortical loss compared with lesions lower in the nervous system and suggests that,

"*sensory changes, produced by a lesion of the cortex, must be considered in terms of the psychical act and not of the physical stimulus. The function of the cortex is to endow it with spatial relationships, with the power of responding in a graduated manner and with those qualities by which we recognise the similarity or difference of objects brought into contact with the body...*" p. 728.

By this he may mean that the cortex allows judgement, certainly in relation to his tests of sensation, but also and beyond that, the cortex is the seat of psychical activity, by which he may mean conscious interpretations of sensory inputs.

Adrian and interrogation of the nervous system

ED, later Lord, Adrian (1889–1977) was one of the great experimental physiologists of the 20th century. Working in Cambridge, he showed the all or none principle of single sensory nerve impulses, invented the concentric EMG needle and investigated the nature of the EEG after it was first described by Berger. A winner of the Nobel Prize in 1932 with Sherrington, he also made fundamental observations on the nature of sensation.

He showed that peripheral sensory receptors often had slower adaptation rates than their afferent nerve fibres and the differences between rapid and slowly adapting receptors. He plotted tuning curves of the response of muscle spindles to stretch. In short, he began to move beyond a descriptive form of physiology and began to interrogate and impose ideas on the nervous system, seeking out its responsiveness. Much of this rigorous work isolated peripheral receptors and nerves and so epitomises the reductive nature – and excellence – of post World War I physiology in the UK.

In delving down into receptor response curves, however, he was not unaware of their physiological roles. In discussing end organ responsiveness in terms of rapidly and slowly adapting, he suggests that we have left out the most important factor of all,

"*that the receptors form part of an organism which has the power of movement. ...If the organ-*

FIGURE 4. LORD ADRIAN (1889–1977)

ism were motionless the phasic receptors would discharge impulses whenever the environment changed, but would cease as soon as it had settled down... we take advantage of this when going to sleep... consciousness of our body fades rapidly [as] skin receptors become very rapidly adapted to a constant environment." [14]

Later he wrote,

"*The importance of ... exploring movements is shown by the fact that the verb to feel is used as often in an active as in a passive sense. We feel the hardness of a surface by pressing our finger against it, just as we feel the shape of an object by moving it about in the hand.*"

"*The fact that the receptors can be moved about in relation to the external word enlarges their scope enormously... a motile animal can explore a stationary world by changing the relation of the receptors to the environment... [this] enables us the extract information about the external world, not only from the exteroreceptors on the surface of the body, but also from the proprioceptors.*" [14]

This echoes Bell, but with rather more data. By recording for the first time from single afferents he was able to uncover something of their responsiveness to their adequate stimulus, though he was well aware of how these inputs required huge, then unknown, central elaboration. At the time also, the division between different peripheral nerve classes was unknown.

"*Do slow impulses produce pain reactions because they can excite particular neurone systems or because the fibres which transmit them are distributed in a particular way?*

The idea of specific fibre reactions is attractive."

Yet he was also aware of the problems in explaining what happened between activation of a sensory end organ and perception. Writing in 1928 he wrote that there was,

"*an unsatisfactory gap between two such events as the sticking of a pin into my finger and the appearance of a sensation of pain.*" [15]

The search for explanation and the neuroscience of the felt quality of sensation and perceptions, the 'hard problem' of qualia, largely eludes us to this day.[10]

Conclusions

This short survey of work in relation to haptics within the UK in the 19th and early 20th century has been from the perspective of neurological and neurophysiological work. It has therefore focussed on the work of a few men, influential though they were. Their results show two themes, the relation and occasional gap between data and ideas and the progressive movement from whole body work to a reductionist, empirical approach.

Though Bell performed experiments and made important observations as a result, some of his most important insights came from deduction and theorising. To suggest, for instance, that perception of movement might come from either peripheral feedback or from awareness of central effort was extraordinary. Similarly Sherrington's huge contribution was experimental, but he was also able to suggest still in the 19th century that deafferentation might lead to a loss of volition. Head's great work came from careful clinical observation; his theoretical work on body schema has also been hugely influential but some of his other ideas, like protopathic sensation has been less so. Adrian's work went far beyond the demonstration of the sensory nerve impulse; he began to ask the nervous system questions, while at the same time being aware of how far his experiments were from perception.

These men, bar Bell, may not have recognised that their work was relevant to haptics research. But, by laying out the relation between movement and sensation, determining methods for testing sensation in man and determining the nature of clinical sensory deficits, and by determining the nature of sensory impulses and the responsiveness of cutaneous sensory organs, they all helped lay the ground for later workers.

Summary

In reviewing the development of work in the United Kingdom on haptics over the last two centuries I have focussed on the contributions of a few great men. (Sir) Charles Bell, in the early 19th century is not as well known as he deserves in this field since he first described movement and position sense, affective sensation and asked whether knowledge of position was due to central corollary discharge or peripheral feedback. (Sir) Charles Sherrington made many contributions, not least in coining the word proprioception. But within this field he made experimental observations on the effects of deafferentation on nonhuman primates. As a result of these experiments he also asked whether deafferentation might not only prevent controlled movement of the limb, but also abolishes the *will* to move it. (Sir) Henry Head moved clinical medicine towards physiology and made detailed observations on sensory loss and its testing, as well as – in his famous experiment of 1907, severing his own radial sensory nerve to observe its recovery. In an apparent side he also described the body schema for the first time. Lastly Lord (Edgar) Adrian moved physiology towards mechanisms and the cellular, neuronal level, with brilliant results. But he too made salutary observations about sensory mechanisms at a whole organism level. These three knights and a lord, including two Nobel Prize winners, were not the only within Britain to consider haptics research, but their contributions alone are sufficient to give the UK some significance within the field.

Selected readings

Adrian ED (1928) *The Basis of Sensation* (reprinted: New York and London: Hafner, 1964)

Bell C (1833) *The Hand; Its mechanism and vital endowments as evincing design*. London: William Pickering (reprinted by Pilgrims' Press, Brentwood, England, 1979)

Head H (1920) *Studies in Neurology, Volumes 1 and 2*. Oxford; OUP and London: Hodder and Stoughton

Rivers WH, Head Sir H (1908) A human experiment in nerve division. Brain 31: 323-450

Sherrington Sir C (1906) *Integrative action of the Nervous System*. Yale: Scribner (reprinted 1947, Cambridge, CUP)

Notes

1 Bell was a reluctant experimentalist out of temperament and humanitarian instinct. In the days before anaesthesia, experiments on animals were done while the animal was still conscious and without analgesia, (conditions similar to those under which surgeons operated on patients). Bell wrote in a letter,
"*I should be writing a third paper on the nerves but I cannot proceed without making some experiments, which are so unpleasant to make that I defer them. You may think me silly, but I cannot perfectly convince myself that I am authorised in nature, or reli-*

gion, to do these cruelties – for what? – for anything else than a little egotism and self-aggrandisement; and yet, what are my experiments in comparison with those which are daily done? And daily done for nothing." [16]. His biographers commented,
"*The particular period of his surgical activity before the advent of anaesthesia prevented one of his gentle humanity from equalling the exploits of surgical competitors who were cast in a tougher mould.*"

2 In the 20th century the philosopher Wittgenstein's work contains many insights relevant to neuroscience with a similar deductive approach [17].

3 Sherrington defined it as "*reactions on sense arising in motor organs and their accessories.*" [18]

4 When writing his biography I often imagined it as a dialogue between Bell and Waterman.

5 In this passage he also considered phantom limb sensations, though did not call them this (before the work of Silas Weir Mitchell in the 1860s), and used these as further support for a muscular sense (as Horatio Nelson had used his phantom limb sensation as evidence for the immortality of the soul).
"*After a limb has been removed by the surgeon the person still feels pain, and heat and cold. Urging a patient to move who has lost his limb I have seen him catch at the limb to guard it... and long after his loss, he experiences a sensation not only as if the limb remained, but as if it were placed or hanging in a particular position or posture.., [which] seems to change with the change in posture if the body. These are additional proofs of a muscular sense.*"

6 The relation between will and action was considered by the philosopher Wittgenstein, in the 1940s.
"*Willing is not the name of an action.*
Willing, if it is not to be a sort of wishing, must be the action itself. It cannot be allowed to stop anywhere short of the action. If it is the action, then it is so in the ordinary sense of the word; so it is speaking, writing, etc. But it is also trying, attempting, making an effort – to speak, to imagine" [11].
He found it difficult to separate the act from the will to make the action. Much more recently Haggard and colleagues, among others, have used empirical methods to try to tease apart these two. By measuring the timing of perception of an intention to an action and of its sensory consequences, and by disturbing these with transcranial magnetic stimulation, Haggard and colleagues showed how these two percepts were bound together [19, 20]. Haggard and Clark [21] showed that the timing of this awareness of action depends on binding between the intention and its sensory effect. Conscious experience of action may be related to the process of action execution in a predictive manner [22], rather than be based on reconstruction of intentional action based on retrospective inference. Awareness of voluntary action or a sense of agency, that sense of intending and executing an action, appears to derive from a stage later than intention but earlier than movement itself, see Marcel [23], at least for short duration intentional action in an experimental situation.

7 There is another area in which Sherrington's conclusions have subsequently been found wanting. He considered the effect of cutting the sensory nerves from the hoof of a horse, a widespread procedure of economic importance known as foundering. With metalled roads horses developed soreness and lameness from pounding the hard surfaces. This operation prevented them from feeling the pain, but also deafferented the foot.
Sherrington wrote that, "*the horse trots normally after neurotomy has rendered the whole foot insensitive.*" Charles Bell had considered the same operation some 70 years before and had come to a different conclusion. After the operation, he wrote, "*the horse puts out its feet freely and the lameness is cured... It is obvious however, that there is certain defect; the horse has lost its natural protection, and now must be indebted to the care of his rider. It has lost not only the pain which should guard against over exertion, but the feeling of the ground, which is necessary to his being perfectly safe as a roadster*" [1].

8 It is tempting here to compare the neurology of loss, which much of clinical neurology has been focused on documenting and the neurology of what is left, of living with impairment, which has been more the province of rehabilitation and the writing of, say, Oliver Sacks among others.

9 Sherrington realised this well. "*Mind rarely, probably never, perceives any object with absolute*

indifference, that is without 'feeling.' All are linked closely to emotion" [24].

10 In his 1906 work *Integrative action of the Nervous System*' [25] Sherrington makes the distinction between sense organs which act a distance receptors, "*in the leading segments [the head] we find... receptors which acting as sense organs initiate sensations having the physical quality termed 'projicience'. Organs of odours, light and sound produce sensations projected into the world outside the 'material me'. Projicience refers them, without elaboration by any reasoned mental process, to directions and distances in the environment corresponding with the real directions and distances.*" He then compares these with "*non-projicient receptors of intero and proprioception*". These, he suggests, "*underpin affective tone (physical pain and pleasure), while the distance receptors guide the conative [will (into action)] tone.*" This may be an oversimplification which Bell was aware. Merleau-Ponty among others realised the relation between distance perception and movement ease and ability.

4

Early psychological studies on touch in France

Yvette Hatwell and Edouard Gentaz

In France, early psychological studies on touch belong to two domains which developed quite independently in the 18th, 19th and the first part of the 20th century. The first one concerned speculative (philosophical) and/or informal considerations made by practitioners and observers on the functioning of touch in early blind people, and more incidentally in cognitively impaired children. In the middle of the 19th century, studies on blindness led to the major invention by Louis Braille (1829 and 1837) of the raised-dot alphabet, universally used today, which allows early blind persons to have tactually access to written culture. The second trend is scientific and experimental. In the 19th century, this research concerned mainly the anatomy and physiology of somesthetic sensory receptors, neural pathways and cortical projections. Simultaneously, the methodological bases for experimental sciences were established. Further, and especially during the first half of the 20th century, studies labelled 'physiological psychology' were introduced in the academic institutions in Paris. Finally, thanks to the French psychophysiologist and psychologist Henri Piéron (1881–1964), true experimental psychology appeared in France at the beginning of the 20th century. Regarding touch, these experimental works were mainly psychophysical and concerned the cutaneous and kinaesthetic sensibility of sighted adults, but higher levels of touch functioning were also studied. After the Second World War, the two trends mentioned above were associated and the experimental psychology of touch extended to blindness and sensory compensations [1, 2].

Early analyses of touch and touch-vision relations in blindness and education

Although the nature of the sensory experiences and the mental representations of complete congenitally blind people were already questioned in the Greek and Roman literature, thinking about blindness became central in a number of eminent philosophers' writings during the 18th century (the 'Enlightenment'). Initially, the reasons for this interest were not to provide practical help to the blind, but to obtain philosophical arguments challenging the dominant inneist conception of the origin of knowledge proposed by Descartes and others. Later, this focus on the blind and the observation of the capabilities of some of them led to the search of how the use of touch could allow them to receive instruction and train for a job.

Diderot and the 'Letter on the Blind for the use of those who see' (1749)

The first main French contribution to these questions is the *Letter on the blind for the use of those who see* published in 1749 by Denis Diderot [3]. This philosopher belonged to the empiricist trend developed by Locke, Berkeley or Condillac, and was very interested by the discussion raised by the Irish philosopher William Molyneux. In his well-known letter (1693) to Locke, Molyneux asked whether a born-blind man whose sight would be suddenly restored would immediately identify visually a cube and a sphere without first touching them. An affirmative answer would show the innate nature of perceptual knowledge whereas a negative answer would demonstrate that experience is necessary and that no cross-modal transfer is possible without previous visual-tactual association. Of course, Locke's [4] empiricist answer was that there would be no immediate shape identification by the blind and that learning was necessary to acquire knowledge by vision. By this time, there was not yet successful cataract surgery and this question was purely theoretical. It is noteworthy that, since 1728 and further, when some blind adolescents and adults recovered sight after surgery, their visual behaviour confirmed Locke's answer: Except for figure-ground segregation, no shape, size, orientation or number identifications were possible [5, 6].[1]

Diderot agreed with Locke's answer to Molyneux but, in his *Letter on the blind*, his theoretical considerations were based not on speculations but on the direct observation of a born-blind man living in a small city (Puiseaux) in the centre of France. Diderot was fascinated by the capabilities of this person who was married, had a son and made a living by selling the liquors he prepared. Diderot described very genuinely how this blind man could thread a needle and explore tactually volumetric lines and figures, with a preference for symmetrical ones. He noted his high ability to discriminate weights and textures, and how he could detect the presence of an obstacle near him by using the air movements sensed on the face. This was the first mention of the 'sense of obstacles' attributed to the blind in literature and named 'facial vision'. When Diderot asked the blind man if he would be glad to recover vision, the surprising answer was that he would prefer having longer arms to explore better around him by touch. In the same *Letter on the blind* Diderot also described how Saunderson, another early blind man living in England (who had died 10 years before) became a well-known mathematician, taught geometry and published a number of scientific articles. Diderot was particularly interested by the cubes and small tactile materials used by Saunderson to carry out complex calculus.

This set of observations led Diderot to comment on the role of vision and touch and more generally of the different senses. He insisted on the fact that the intensive use of touch by blind people enhances in them its discriminative capacities and therefore demonstrates the crucial role of experience. This 'sensory compensation' supported the empiricist and sensualist views of the philosopher who stated that "*the assistance given by the senses to each other prevents them to improve*" (1749). Based on these observations, in his *Letter on the Blind*, Diderot also proposed some general philosophical considerations on the origin of knowledge, and on the reasons to reject the assumption of the intervention of God as Creator of the world. After this publication, Diderot was arrested and spent four months in prison because, according to the police, he wrote things going against morality and religion.

Valentin Haüy and the first school for blind children (1785)

Valentin Haüy (1745–1822) was very impressed by the work of the abbey Charles-Michel de l'Epée (1712–1789). L'Abbé de l'Epée imagined a method, using manual signs and vibratory stimulations, in order to teach deaf-mute children to speak and write and he opened the first school for them in Paris. Encouraged by this success and very influenced by Diderot's (1749) *Letter on the Blind*, Haüy tried to find methods allowing blind children to read and write tactually. He conceived volumetric displays accessible to touch and potentially adapted for the blind. Mainly, he used enlarged wooden roman letters of our alphabet to emboss hard paper, thus obtaining characters in relief on the back of the sheet, and he designed tablets and cubes to support arithmetic operations.

In the early 1780s, Valentin Haüy met a young early blind boy (François Le Sueur), aged 17 years, who begged near the church. V. Haüy offered to take care of him if he agreed to learn to read and write with these new displays. Fortunately, this boy was very bright and highly motivated. In 1784 in Paris, V. Haüy explained to an assembly of academic scientists and politicians the feasibility and social utility of the creation of a school where blind children could learn to read and receive a professional training. Then, François Le Sueur gave a demonstration of his capacity to read the embossed letters tactually, write them, operate calculus, correctly use raised-dot geographical maps, etc. As a result, Haüy obtained the permission and the funds to open a school for the blind in Paris (1785). This school, the first in France and in the world, was named *Institute for Blind Children*. It later became (1818) the *Royal Institute for Young Blind* and is today the *National Institute for Young Blind*.

FIGURE 1. LOUIS BRAILLE (1809–1851)

Louis Braille and the raised-dot alphabet

Louis Braille (1809–1851) (Fig. 1) was three-years-old when he completely lost sight after an accidental wound in one eye. His father, who was literate and lived in a small village near Paris, nevertheless sent his son to the local school and helped this very gifted child to acquire some basic knowledge. When he was aged 10 years, Louis Braille was accepted as a pupil at the *Royal Institute for Young Blind* created by Valentin Haüy (but this school had now another director). He learned very rapidly to read and write tactually with the raised roman letters created by Haüy. He became a lecturer when he was 15-years-old and a professor some years later, and he was greatly appreciated by his blind pupils. However, very rapidly too, he objected that the embossed raised roman letters used in the school were difficult to discriminate and were not adapted to touch. Actually, very few pupils in the Institute achieved correct reading and writing, and even those who could master these abilities could not have further reading practice when they left school because no books were available.

When he entered as a pupil at the *Royal Institute for Young Blind*, Louis Braille experienced a system designed by an army officer, Charles Barbier, to allow soldiers to read tactually orders and instructions by night (for details, see [7]). In it, different raised-dot patterns represented a sound each. Each pattern was composed of a combination of raised dots taken from a 12-dot matrix (two columns and six rows), and the association of the sounds represented by a series of patterns led to the formation of a word. Barbier had the intuition that such a system may also be useful for the blind and he proposed to the director of the school to test it with his pupils. Although sceptical, the director accepted.

Louis Braille immediately observed that it was much easier to discriminate tactually raised-dot patterns than the Haüy's linear and curvilinear in relief roman letters used in the school. But he noted too the disadvantages of Barbier's system. First, because the number of dots was too large, exploratory movements of the index fingertip were necessary to decode each pattern. Secondly, each pattern represented a sound and not a letter. Therefore, using a trial and error method, Louis Braille replaced progressively the 12 raised-dot matrix by a smaller matrix contain-

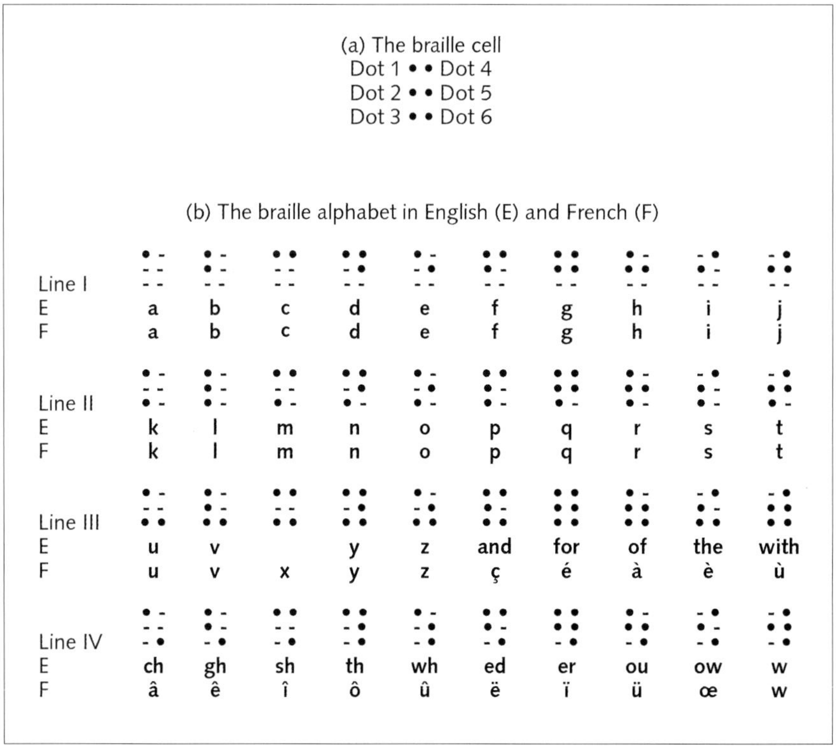

(a) The braille cell
Dot 1 • • Dot 4
Dot 2 • • Dot 5
Dot 3 • • Dot 6

(b) The braille alphabet in English (E) and French (F)

Line I										
E	a	b	c	d	e	f	g	h	i	j
F	a	b	c	d	e	f	g	h	i	j

Line II										
E	k	l	m	n	o	p	q	r	s	t
F	k	l	m	n	o	p	q	r	s	t

Line III										
E	u	v		y	z	and	for	of	the	with
F	u	v	x	y	z	ç	é	à	è	ù

Line IV										
E	ch	gh	sh	th	wh	ed	er	ou	ow	w
F	â	ê	î	ô	û	ë	ï	ü	œ	w

FIGURE 2. THE BRAILLE ALPHABET IN ENGLISH AND FRENCH

ing six dots (two columns and three rows) and he decided that each pattern would now represent a letter of the alphabet and not a phoneme. After a number of empirical adjustments on himself and on his friends at school, Braille published a book (using the embossed linear characters of V. Haüy) in which he presented his new method of writing words and music for the use of blind people in 1829 (at the age of 18). In this first version of his alphabet, Braille maintained some continuous lines linking two points but later, in the second version (1837), these lines were suppressed and the whole alphabet was made of dots (Fig. 2). The patterns of dots were logically constructed, with the first 10 letters using only the combination of the four upper dots and the next series adding a dot, and so on. The number of possible combinations of the six dots of the matrix is 64 (with no dot representing a space between words), and therefore Braille could represent the whole alphabet, accentuated letters, punctuation and, with the addition of a coded sign, numbers and musical notation.

Tactually, the alphabet proposed by Braille in 1837 is perfectly adapted to the discriminative abilities of this modality. Although based solely on empirical adjustments, it meets the fundamental requirements for maximal perceptual efficiency, as it will later be demonstrated by psychophysical experimental studies. As stated earlier, the discontinuous raised-dot patterns are easier to differentiate by touch than the in relief lines and curves of roman letters. In addition, Braille found empirically that the lowest cutaneous two-point limen in the fingertip was 2 or 2.5 mm (after training) and this was further confirmed by modern psychophysical measures. Therefore, in the braille alphabet, the distance between two dots belonging to the same letter is approximately 2.2 mm, and the distance between two dots belonging to different letters is slightly higher. As a result, a letter comprising all the six dots (such as the French 'é' or the contracted sign representing the English word 'for') may be perceived in a single fixation of the pulp of the index (without vertical or lateral movement). Finally, because each pattern represents a letter and not a phoneme, this system has the status of an alphabet in which any language may be transcribed.

This braille method was rapidly adopted by the pupils of the *Royal Institute for Young Blind* in spite of the opposition of the teachers who argued that it rendered impossible any written communication between blind and sighted people. After some years of hesitation, it was accepted in Paris in 1847. Gradually, other countries adopted it too (for example, the US in 1869). Today, it is the most widely used system and has been adapted all over the world to each local language. It has allowed blind people to leave their previous status of beggars and opened to them the highest levels of instruction and musical achievements. Incidentally, it should be noted that Louis Braille's six-dot matrix must necessarily be transformed into an eight-dot matrix (two columns and four rows) in order to write computer programs.

Informal psychological observations on touch: Pierre Villey

Pierre Villey (1879–1933), a blind man who was first pupil and then professor at the *National Institute for Young Blind* in Paris, published a number of comprehensive books [8–10] on blindness. By observing himself and his blind pupils, and because he was well acquainted with the psychological, educational and sociological literature on the blind, he insisted on the education of touch in the blind [10]. He recommended that young blind children receive the Montessori sensory tactile exercises (see below) to develop their sensory abilities; he insisted on the role of manual exploratory movements, showed that what is called 'sensory compensation' did not refer to finer discrimination acuity in the blind but rather to higher attention devoted to some sensory cues neglected by the sighted [8]. In addition, he analysed the braille alphabet and, by reference to contemporary psychophysical studies, he justified why the distance between two dots in this alphabet was 2 or 2.5 mm. Finally, he described what is called the 'sense of obstacles' in the blind and denied that it was based on facial sensations, as suggested by Diderot (1749). Instead, he stated that obstacle perception derived entirely from low level auditive stimulations, as experimentally observed by Dolanski [11] and further by Supa, Cotzin and Dallenbach [12] (see below).

This survey shows that during this period, the interest devoted to touch stemmed from the philosophical, practical and educational problems raised by early blindness in children and adults. This interest extended to cognitively handicapped children, as it will now be described.

Touch in early educational techniques: Jean Itard and Victor, a 'wild child'

The relationship between the development of perceptual abilities and the development of the cognitive abilities was particularly studied by Jean Itard in the education of a 'wild' child, Victor de l'Aveyron [13]. In December 1800, Jean Itard, a doctor, became the director of the institution for deaf-mute children created by the abbey Charles-Michel de L'Epée. At the same period (January 1800), a wild child aged about 10–11 years was discovered living alone in a forest in Aveyron (a region in the south of France). This child was afterwards examined by the famous French psychiatrist, Pinel. The results of his examination were clear: this child was a severely retarded person and was not educable. In contrast, Itard assumed that the impairment of this child (now named Victor) had no biological origin but was rather due to cultural and educational deprivation. Consequently, Itard suggested that Victor could benefit from the new teaching method based on the sensualist theories of Condillac and Locke. These theories assigned a crucial role to the education of senses in cognitive development.

At the beginning, Victor actually showed some features in common with a severely retarded person. His perceptual and cognitive abilities were very weak. The sense organs did not seem to be damaged but they were not functional. Itard [13] wrote: "*the eyes see but they do not look, (…) the hands are used only to grasp objects but not to perceive their shapes*". The touch abilities were surprising because, for example, Victor was not able to differentiate the warm and cold. Thus, Itard proposed to Victor numerous warm and cold baths and, after one year, the child became able to discriminate different temperatures. However, Victor could still not match an object explored with one hand (without vision) with an identical object simultaneously explored both with the other hand and vision.

Therefore, Itard proposed several intramodal tactual exercises to train Victor to discriminate different objects with his hands. These objects varied first according to both their shapes and textures (chestnut *versus* acorn; walnut *versus* stone) and then according to their shapes only (e.g., metal alphabetic letters). After six years of education [14], Victor became able to tactually discriminate a 'C' and a 'G' or an 'I' and a 'J'. Moreover, Itard observed that these tactual exercises increased Victor's concentration and attention. "*Never before had I seen him so serious, quiet and meditative*", wrote Itard.

At the same time, Victor was trained to discriminate different types of sounds (from sounds made by objects to vowel sounds). Thus, Victor learned to recognise some words written on a blackboard by following each letter with his index finger. Victor also learned to handwrite through several types of exercises: imitation of simple and complex gestures, follow-up of patterns and of letters with a rod. After six years of training, Itard concluded that Victor could no longer be considered as a severely retarded person: he was able to read and write some words and to express his basic needs. According to Itard, these results showed that "*the perfection of sight and touch (…) powerfully contributed to the development of Victor's intellectual faculties*".

After this study, Itard became very famous in all European countries, but he stopped his educational work with Victor.[2] Séguin (a doctor and educator) modified somewhat Itard's techniques for the education of the senses and applied them for ten years in his school for mentally handicapped children in Paris. Then, he moved to the USA where he created institutes for deficient children [15]. In 1898, Maria Montessori discovered Itard's and Séguin's publications (forgotten at the time). She wrote in 1926 that "*the meticulous descriptions of Itard were the first attempts at experimental pedagogy. Inspired by his tests, I designed a great many educational material*" [16]. These tentative studies in the educational field thus revealed the potential abilities of touch and suggested that training touch was beneficial for development.

Physiological and experimental psychology of touch

General context

During the 19th century, academic psychology in France was completely integrated to philosophy and was also linked to medicine (for details, see [17, 18]). But the interest for experimental psychology developed progressively, especially with the influential work of Ribot (1839–1916). Ribot was a philosopher who wanted to separate psychology from philosophy and introduce in it the experimental methods developed in Germany by Weber and Wundt. He published books describing English [19] and German [20] psychological research and, although he did not practice himself experiments (instead, he was interested by mental pathology), he greatly influenced the beginnings of experimental psychology in France. In 1885, he was allowed to teach experimental psychology at the University of Paris (la Sorbonne) and, in 1888, he obtained an official position of professor of 'experimental and comparative psychology' at the Collège de France, an academic structure in Paris parallel to the Sorbonne. He created there the first *Laboratory of Physiological Psychology* which was initially directed by Beaunis and then (1894) by Binet. Janet (1859–1947) succeeded to Ribot as professor of experimental psychology. He manifested too an interest for personality dissociations and considered that the comparison between normal and mentally troubled people was a kind of natural experimentation. On the other hand, Binet [21] published an *Introduction to Experimental Psychology* where he presented arguments in favour of scientific psychology and considered that the methodological laws established by Bernard [22, 23] for experimental medicine and sciences should apply too to psychology. Finally, the main founder of classical experimental psychology in France was Henri Piéron (1881–1964) (Fig. 3). Piéron worked mainly on sensation and he taught experimental psychology in Paris from 1907 to 1954. He directed the *Laboratory of Physiological Psychology* (1912) which became further the *Laboratory of Experimental and Comparative Psychology* of the Sorbonne, then the *Laboratory of Experimental Psychology* of the University René Descartes-Paris 5. In 2006, this research centre has been divided in two: the *Laboratory of Psychology of Perception,* and the *Laboratory of Psychology and Cognitive Neurosciences.*

In 1894, Alfred Binet and Henri Beaunis founded *L'Année Psychologique*, one of the first journals in the world devoted exclusively to the publication of national and international experimental articles and comprehensive reviews of psychological research. This journal, directed by Piéron from 1912 until his death in 1964, is still today a major reference in experimental psychology.

During the time where experimental psychology tended to exist independently from the dominant academic philosophical psychology, experimental anatomy and physiology developed and a number of discoveries improved the knowledge on the nervous system and on touch functioning.

Figure 3. Henri Piéron (1881–1964)

Anatomical and physiological research on somesthetic sensibility

In 1822, Magendie [24] published experimental observations showing that localised lesions of the posterior roots of the spinal chord resulted in sensory troubles whereas motor functions were not impaired. Conversely, localised lesions of the anterior roots generated motor troubles whereas no sensory deficits were observed. Some months later, Magendie [25] fairly reported that this major discovery of the distinct spinal nervous pathways of sensory inputs and motor outputs had actually been suggested in London by Bell 13 years ago, in a monograph privately published in only 100 exemplars.

Concerning touch, Skramlik [26] reports in his major book a number of contributions of French researchers to the study of this sense. For example, the anatomical and physiological characteristics of the cutaneous and kinaesthetic nerves were studied by several authors [27–29] and [30]. Beaunis [27], Chapentier [31] and Pallot [32] worked on the 'muscular sense' and on weight discriminations whereas Delage [33] examined the role of the semi-circular channels of the internal ear in the perception of directions, especially by reference to the vertical orientation. Bizet [34] and Pitres [35] described the sensation of 'phantom limb' in amputee persons and respectively called this phenomenon a 'hallucination' or an 'illusory sensation'.

Experimental psychology of touch

A number of more psychological experimental studies were published by Henri Piéron and his associates during the first decades of the 20th century. Piéron was interested by the factors acting on the sensory process from its beginning (the neural excitation of sensory receptors) until its ending, i.e., when a subjective sensation is experienced. He worked mainly on vision and audition, but also on cutaneous, kinaesthetic, thermal and pain sensibility. His approach was essentially psychophysical: He was concerned by the mechanisms underlying sensory knowledge, and studied the relations between the properties of the physical stimulus (intensity, frequency, localisation, etc.), the characteristics of the nervous impulses generated by the stimulus and the nature of the sensation reported by the subject. Impressed by the work of Johannes Müller who suggested [36] that, actually, we do not know the state of the objects but only the state of our nerves, Piéron insisted on the differences between the physical properties of the stimulus and the nervous and subjective responses of the subject. He concluded one of his main books (entitled *Sensation, a guide for life*) [37] in this way: "*Sensations are biological symbols of the external forces acting on the organism, but they cannot have more similarity with these forces than the similarity existing between these sensations and the words representing them in the symbolic language system used in interhuman social relations*" (p. 413) (translated by author).

Concerning more specifically the study of touch (we will not consider here the research on thermal and pain sensibility), Piéron and his associates (including his wife) conducted a series of studies on the multiple characteristics of this modality. Some of them concerned the design of new displays allowing technical improvements in the measure of the intensity and the nature of the stimulation: An esthesiometer [38] enhancing the stability and reliability of the cutaneous stimulation, a gravimeter for weight discriminations [39], a display evaluating the tactile sensitivity to thickness differences between metal blades [40]. Incidentally, it should be noted that Piéron's wife published all her articles under the name of 'Madame (Mrs) Henri Piéron'.[3]

The experimental design is generally correctly planned and described in these experimental studies, but the number of subjects is not always given and the results are sometimes presented as simple statements not accompanied by numerical data (as it was usually done by the German gestaltist psychologists). Other articles give detailed quantitative data but, of course, without statistical analyses. We will evoke only some of these studies to illustrate their diversity and their limits, and to show how they anticipated the more sophisticated research on this domain which will take place after World War II.

Concerning the basic sensory cutaneous functioning, Piéron published comprehensive book chapters [41, 42] summarising the current knowledge on touch which therefore became available for students and researchers. Fesssard and Piéron [43], Piéron [44], and Piéron and Segal [45] described the characteristics of cutaneous excitability to electric stimulation and response latency, and Piéron [46] tested the cutaneous sensibility to vibratory stimulations. Toltchinsky [47] compared the cutaneous discriminations in different parts of the hand and the body. Gault [48] examined the role of these stimulations in the language communication with deaf people. The role of attention was underlined by Binet [38, 49] who was impressed by the high intraindividual variability of the two-point limen measured by the esthesiometer. He showed that this variability was strongly dependant both on the variations of attention and on the degree of uncertainty of the responses accepted by each subject. As a result, he questioned whether a reliable measure of cutaneous acuity may actually be obtained.

Higher level studies of touch functioning have also been published during the first half of the 20th century. Laureys [50] compared what we call today the intramodal efficiency of vision and touch in a task of perceptual estimation of the volume of a cube (5 cm or 4 cm side). The subject was asked to select from comparison objects the cube which volume was 1/8th of the standard cube. There were three conditions of presentation of the cubes: visually, or tactually while the blindfolded subject grasped the object or tactually while the subject grasped the cubes with his hand placed behind his back. The author concluded that the highest accuracy was obtained with vision. In her comments on the early French research on visual-tactual relations, Streri [51] carried out an analysis of variance on the raw data given by Laureys [50]. This analysis confirmed the superiority of the visual condition over the two other ones but revealed too an interaction between modalities and object size, with less errors on the smaller cube than on the larger, especially with normal hand grasping.

Philippe [52] tried to obtain a 'pure tactile sensation', that is a cutaneous sensation not modified by previous visual perception and representation of the display (visualisation). Blindfolded subjects received no visual or verbal information about the display used (which was partly different from the classical Weber's esthesiometer), and were asked to draw and verbalise their 'pure' sensory impressions. The author observed that the subjects who were previously acquainted with the classical esthesiometer had a lower discrimination limen and visualised very well the two pins used as stimuli. Those completely naive had higher limen and could not identify the nature of the stimulus.

The role of visualisation was also evidenced by Tastevin [53, 54] who conducted a very complete examination of the tactile Aristotle illusion. In this illusion, when a small ball is placed between two crossed fingers, the blindfolded subject perceives two balls. Tastevin tested on himself and on some blind and sighted adults the conditions affecting this systematic error. He rejected the classical explanation assuming that the illusion resulted from the contacts of the ball with the unusual sides of the crossed fingers. Instead, he argued that actually this illusion derived from errors in the localisation of the cutaneous stimulations on the crossed fingers. Tastevin stimulated simultaneously the palmar faces of crossed fingers with a compass having one arm ended by a very small ball and the other by a point. During these stimulations, the subject could either see or was asked to close his eyes. Unfortunately, no numerical data of results were provided, but the author reported that without seeing, the sensations on the crossed fingers were erroneously localised in their normal position and not inverted. This means that the sensation generated by the arm of the compass ended by the point was located where the ball actually was, and the sensation generated by the ball was located where the point was. But, as soon as vision was available, the correct localisation of the stimulations on the crossed fingers reappeared. Tastevin labelled 'visual catching' (in French, *captation visuelle*) the fact that vision modified tactual perception when the two modalities provided conflicting data. In the contemporary studies of perceptual conflicts, this phenomenon corresponds very exactly to what is called 'visual capture' (for reviews, [55] or [56].

An extensive study of intersensory transfer between vision and kinaesthesis was conducted by Madame Henri Piéron [57]. In this a-theoretical article (its introduction has only nine lines) her aim was to know the conditions for a maximal efficiency of this transfer. The experimental design was very complex since there were six conditions of presentation and reproduction of lines which lengths varied from 5 to 9 cm. The duration of visual presentations were either long (2 s) or short (0.05 s), and the kinaesthetic presentations, called 'muscular' by the author, were either passive or active, and either with the dominant hand or with the non-dominant hand. After the presentation of the stimulus, two tests were performed: the recognition of the presented length and the reproduction of this length.

The results were presented as means of responses in each condition so that Streri [51], when commenting on this study, could not carry on these data an analysis of variance as she did on Laureys's [50] ones. But some of the conclusions proposed by Mrs Piéron are interesting because they have been further supported experimentally in modern psychology (for a review, see [58]). Mrs. Piéron observed that the intramodal transfer, i.e., the transfer of data inside the same modality, was more accurate than the intermodal transfer between two different modalities, that the kinaesthetic intramodal transfer was more accurate in the active condition than in the passive one, and that the long visual presentation was more efficient than the short one. Another interesting finding is that the transfer from vision to kinaesthesis was better performed than the transfer in the reverse direction. Asymmetries in cross-modal transfer have often been observed in modern research in children and adults [56, 59, 60].

Finally, the work of Dolanski [11] on the obstacle sense of the blind deserves some comments. As stated in the first part of this chapter, Diderot (1749) attributed to tactile perceptions located on the face and due to air movements near obstacles the capacity of totally blind persons to detect them before contact. Dolanski (who was blind since the age of 5 or 6 years) tested on himself and on some other blind and blindfolded adults different experimental conditions. The subjects were either equipped with fabrics on their face preventing any tactual stimulation due to air movements, or they were equipped with auditory masks preventing any auditive stimulation. Dolanski did not provide quantitative data of the results but he concluded from his observations that the obstacle sense was not affected by the suppression of tactual stimulations on the face, whereas the suppression of auditive stimulations resulted in a great number of errors. Therefore, he rejected Diderot's notion of 'facial vision' and explained the obstacle sense by a higher sensibility of blind people to auditive cues not taken into account by sighted persons. This is exactly the conclusion of the statistically based studies published later by Supa et al. [12] and more recently by Strelow and Brabyn [61] or Ashmead et al. [62].

General conclusions

In this chapter, we have reviewed the main French contributions to the study of touch functioning before World War II. To conclude, we will show how these early studies were original and how well they prepared the way for modern research in this field.

The informal observations due to philosophers and practitioners working on blindness had a major achievement, the invention by Louis Braille of the raised-dot alphabet allowing blind children and adults to now have access tactually to written culture. On the other hand, the positive role of the education of touch in cognitively disabled children led to a number of contemporary remediation techniques. For example, during the Second World War, Fernald [63] employed a 'multisensory' technique (largely based on the principles described by Itard, Séguin and Montessori) with children exhibiting reading difficulties. In this technique, known as the 'multisensory trace', the children trace a written word with their index finger while pronouncing it and looking at it. These multisensory techniques, which activates different channels (vision, audition and haptics) seem to bring a suitable solution because it could made explicit connections between the different activities involved in reading and spelling [64]. They allow the child to retain at the same time the visual image of the letter and the movement necessary for writing it. Some contemporary studies showed that multisensory methods are particularly suitable for the remediation of reading difficulties [64–66].

According to Bryant and Bradley [64], backward readers, just as beginning readers, tend to treat as two separate things what they learn about the sound of a written word and what they learn about its visual appearance. Multisensory methods permit to help them connecting these two different kinds of learning. Recently, Bara et al. and Gentaz et al. [67–69] successfully used a multisensory training with pre-reading kindergarten children in showing the positive effects, on the understanding and use of the alphabetic principle, of adding visual-haptic and haptic exploration of letters in a training designed to develop phonemic awareness, letter knowledge and letter/sound correspondences.

The second set of early French contributions concerns scientific and experimental research. During the last decades of the 19th century and the first ones of the 20th, experimental methods applied progressively to a trend of the French psychological research which became independent from philosophy. The basic properties of cutaneous and kinaesthetic nervous excitability were studied both in their neurophysiological and psychological components. The interest for the basic neural processes following the stimulation of sensory receptors inevitably led the researchers to study their end product, that is sensation itself and its integration in the higher levels of perception. It is striking to note how rich, diverse and pertinent were the psychological questions posed in the first half of the 20th century and how well they constitute the roots of modern psychology. The concept of visualisation evoked by Philippe [52] anticipated the pilot publication of Worchel [70] showing experimentally the role of prior visual representation in haptic spatial perception. Similarly, the modern 'visual capture' observed in visual-tactual perceptual conflicts [55, 56, 71] was already described by Tastevin [53, 54]. Madame Henri Piéron [57] raised the major questions studied today about the intersensory transfer between vision and haptics, and Dolanski [11] showed the role of auditive cues in the obstacle perception of the blind. Of course, the weakness of these studies relies on their poor quantitative and statistical analyses but it is noteworthy that, in spite of these limitations, the conclusions of their observations were often validated further.

After World War II, the informal study of touch in blindness by practitioners and the experimental psychology of touch in sighted adults became associated and from then on, systematic experimental research on early blindness developed in France (for example [1, 2, 72, 73].

Summary

Early psychological studies of touch in France belong to two domains which developed quite independently from the middle of the 18th century until the early 20th. The first set concerns informal observations due to philosophers and practitioners working on blindness and it had a major achievement, the invention by Louis Braille of the raised-dot alphabet allowing blind children and adults to have access tactually to written culture. The second set of contributions is scientific and experimental. During the last decades of the 19th century and the first ones of the 20th, experimental methods applied progressively to French psychological research. The basic properties of cutaneous and kinaesthetic perceptions were studied both in their neurophysiological and psychological components (for example, role of visualisation, or crossmodal transfer). After World War II, the informal study of touch in blindness by practitioners and the experimental psychology of touch in sighted adults became associated and from then on, systematic experimental research of touch in early blindness developed in France.

Acknowledgements

We thank Martin Grunwald for having sent us some main pages of the pilot book of Skramlik (1937) where references to French contributions were quoted. We are also grateful to Mrs Michèle Coullet-Maurin who kindly translated these pages in French for our use.

Selected readings

Diderot D (1749/1972) *Lettre sur les aveugles à l'usage de ceux qui voient*. Paris: Garnier-Flammarion

Weygand Z (2003) *Vivre sans voir. Les aveugles dans la société française du Moyen-Age au siècle de Louis Braille*. Paris: CREAPHIS

Villey P (1914) *Le Monde des aveugles*. Paris: Flammarion (2nd edition: Paris: Corti, 1954)

Bloch H (2006) *La psychologie scientifique en France*. Paris: Armand Colin

Skramlik von E (1937) *Psychophysiologie der Tastsinne*. Archiv für die Gesamte Psychologie; vierter Ergänzungsband: Akademische Verlagsgesellschaft

Piéron H (1932) Le toucher. In: G Dumas (ed): *Nouveau Traité de Psychologie*. Paris: Masson, 1055–1219

Streri A (1994) L'étude des relations entre les modalités sensorielles au début du siècle. In: P Fraisse, J Segui (eds): *Les origines de la psychologie scientifique française: le centenaire de l'Année Psychologique*. Paris: Presses Universitaires de France, 205–222

Notes

1 However, we know today that the observation of visual abilities immediately after the recovery of sight in adulthood is not a pertinent method because, when the ocular system has not been stimulated by light during the first years of life, the retina and other ocular structures suffer from atrophy. Therefore, the recovered visual acuity is very poor.

2 No exact information is available concerning the further evolution of Victor. After his separation with Itard, a nurse took care of him until his death some 30 years later. Although his cognitive abilities improved during the years of training with Itard, ceiling effects appeared and Victor stayed cognitively handicapped. Therefore, he required assistance during all his life.

3 By this time, the feminist movement was not strong enough to preserve the personal identity of married women! Please note that in the list of references, the publications of Mrs. Piéron appear under the name of: Piéron HM.

Haptics in the United States before 1940

5

Rand B. Evans

In the United States, the history of research in the skin senses begins not in a research laboratory but in an asylum. It was only in the late 19th century that research universities based on the German model were established and departments of psychology emerged.

Beginnings: Samuel Gridley Howe and Laura Bridgman

Our starting point is in 1837 at the New England Asylum for the Education of the Blind in Boston, later known as the Perkins School for the Blind, when seven-year-old Laura Bridgman was brought there. Both blind and deaf due to scarlet fever contracted when she was 26 months old, she had only a limited ability to smell or taste. Her parents had communicated with her entirely through rudimentary touch signals, "*A pat on the head expressed approval, on the back disapproval*" [1].

Samuel Gridley Howe, director of the Asylum, had received some medical training at the Harvard Medical School [2]. Howe was also an avid follower of phrenology, the movement that held that the brain contains the faculties that give us our abilities and capacities as human beings [3]. A major tenet of phrenology was that through the exercise of the faculties of the mind, which has a physical manifestation in the matter of the brain, a person could realize their maximum potential. It was a matter of finding the best channel to stimulate those faculties. What may have interested Howe in particular was that the phrenologists held that the intellectual faculty of language "*gives a facility of acquiring knowledge of arbitrary signs*..." [4].

Howe began teaching Laura by pasting on everyday objects their names printed on tags in raised letters. After she had associated the tactual impression with the object, the tags were taken off and Laura was taught to select the object when given the tactual impression of the name or the name tag when encountering the object. Then he transferred her to touching printed metal letters containing the words and finally to the 'deaf and dumb' alphabet using finger signs pressed into the palm to allow people to communicate with Laura and she with them [5]. Over time, Laura was able to learn the names and meanings of concrete objects but also to be able to communicate her thoughts in writing. The case attracted a great amount of interest [1]. It was the first account of a deaf-blind person learning language.

The techniques employed by Howe at the Perkins School were taught to others, such as Anne Sullivan, who as the teacher of Helen Keller, the 20th century's equivalent of Laura Bridgman, used many of the same techniques in her teaching [6].

G. Stanley Hall and the skin senses at Johns Hopkins University

One young American psychologist, G. Stanley Hall, became particularly interested in the case of Laura Bridgman. He gained access to her records and published an exhaustive article in the journal *Mind* about her and her abilities [1].

Figure 1. Drawing of Laura Bridgman and Samuel Gridley Howe
Courtesy of the Perkins School

Figure 2. G. Stanley Hall
Source: Am J Psychol *Vol. 35 (1924), Facing page 313.*

It may be partly because of this interest that, when Hall set up his laboratory of psychology at Johns Hopkins University shortly after arriving there in 1883, he began several experiments on touch. The University had been founded as a research university on the German model in 1875.

While Hall had attended the lectures of Wilhelm Wundt in psychology and Karl Ludwig in physiology, when he studied in Germany in 1879 his doctorate was nominally with William James at Harvard. Hall's actual doctoral research was done however with Henry Bowditch at Harvard's School of Medicine, Harvard's first Professor of Physiology.

One of Hall's first research collaborators at Johns Hopkins was H. H. Donaldson. Donaldson was an advanced student in nervous physiology of the Johns Hopkins biologist, H. Newall Martin. Donaldson was interested in working with Hall and chose as his topic a study of the skin senses [7]. Specifically, he was interested in how movement of a stimulus point across the skin was perceived. In doing this research he used a small diameter brass stimulus of about 1.5 mm drawn gradually across the skin. Donaldson reported that when the stimulus point was slowly moved:

"*...it often happened that it seemed to stand still for a time or even be lost, when suddenly a sharp sensation of cold, distinctly localized, would recall its presence and position*" [8].

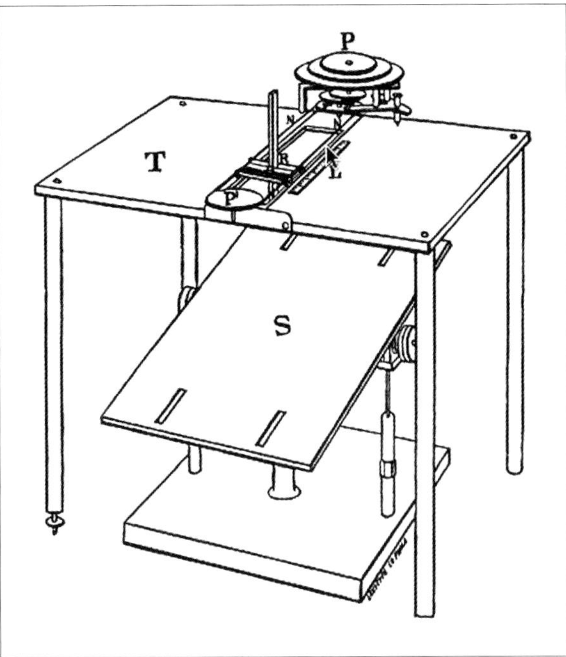

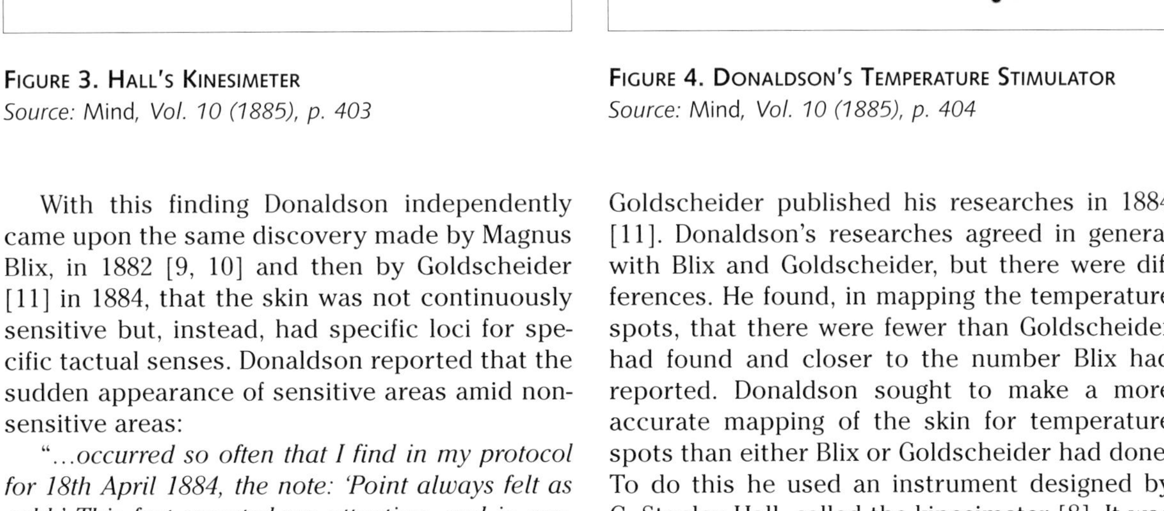

Figure 3. Hall's Kinesimeter
Source: Mind, *Vol. 10 (1885), p. 403*

Figure 4. Donaldson's Temperature Stimulator
Source: Mind, *Vol. 10 (1885), p. 404*

With this finding Donaldson independently came upon the same discovery made by Magnus Blix, in 1882 [9, 10] and then by Goldscheider [11] in 1884, that the skin was not continuously sensitive but, instead, had specific loci for specific tactual senses. Donaldson reported that the sudden appearance of sensitive areas amid non-sensitive areas:

"*...occurred so often that I find in my protocol for 18th April 1884, the note: 'Point always felt as cold.' This fact arrested my attention, and in connexion with the other work I made several maps of these cold-spots on different parts of the body. When the experiments had reached this point, an important paper by Magnus Blix came into my hands*" [8].

Blix's researches were not well known, however, until they appeared in German translation. In 1884 [12, 13] Goldscheider had similarly been unaware of Blix's research until after his own observations on sensory spots were made and Goldscheider published his researches in 1884 [11]. Donaldson's researches agreed in general with Blix and Goldscheider, but there were differences. He found, in mapping the temperature spots, that there were fewer than Goldscheider had found and closer to the number Blix had reported. Donaldson sought to make a more accurate mapping of the skin for temperature spots than either Blix or Goldscheider had done. To do this he used an instrument designed by G. Stanley Hall, called the kinesimeter [8]. It was the instrument Donaldson had used in his earlier research but altered for use as a mapping device. It is pictured in Figure 3.

The kinesimeter allowed differing temperatures to be delivered to specific and repeatable sites on the skin by means of a metal stimulator similar to that used by Blix with a rounded stimulus point of .09 mm in diameter.

With this Donaldson calculated that he was able stimulate every part of the 2 cm by 3 cm patch of skin under study. He marked the skin with a series of dots in indelible ink so that the same procedure could be repeated over the same points on the skin over a period of days. The instrument itself was indexed so that the spots that produced temperature sensations could be recorded as well as the nature of the report (hot, cold, etc.). Donaldson published examples of the maps for hot and for cold responses on the same patch of skin. He found that the temperature for which cold was reported was about 15°C and for 'hot' about 50°C. Donaldson used the terms 'cold spot' and 'hot spot' to relate to the sensitive areas on the skin for these sensations [14], although Altruz suggested the term 'warm' spots be used, instead, because of his theory that the experience of 'hot' comes from the simultaneous stimulation of 'warm' and 'cold' spots [15].

Following Johannes' Müller's doctrine of specific nerve energies, Donaldson surmised that there must be receptors for these distinct sensory spots. To find these had the skin under some of his cold spots and 'hot spots' excised. Looking at them under a microscope, however, he found no special receptors but only free nerve endings. He reported that "*no difference could be made out between the spot at which cold had been felt and that at which heat was observed. There were numerous nerves beneath these spots, but they were almost as numerous in neighbouring parts*" [8].

In another research, Donaldson found that the cornea, after being rendered insensitive to pain was still sensitive to heat and cold [16], demonstrating the independence of pain as a sensory quality.

Hall and Donaldson then returned to the original study of the perception of movement on the skin that Donaldson was working on before he stumbled across the temperature spots. This time they used a larger stimulus, 1.5 cm that would stimulate pressure receptors without stimulating individual temperature spots. Using Hall's kinesimeter they found, in terms of continuous movements on the skin, that there was no indication of a 'motor sense' in the skin but that there were a variety of sources of information concerning localization of movement. They concluded that:

"*'local signs' are quite heterogeneous, and that, in the strong tendency we have to move the touching dermal surface over objects in contact with it, we are seeking not merely to multiply but to diversify our sensuous data for judging the nature of the impressions and to fill up the dermal 'blind spots' between which impressions are sifted in to us.*" [17].

An important point suggested by this research is that passive reception of an unmoving stimulus and active reception of it are very different things.

Donaldson, after leaving Johns Hopkins, did not do other tactual work, but G. Stanley Hall continued his interest in the subject. In 1886, Hall supervised Joseph Jastrow's dissertation that compared spatial perception in vision and in touch [18] and, in 1887, Yuzero Motora's thesis on the sensitivity of the skin to gradual pressure changes [19].

Jastrow went on to organize the psychology displays at the Columbian Exhibition in Chicago, the 1892–1893 World's Fair. Volunteers among the fair visitors were tested in several sensory areas, among which included the skin senses [20]. Tactual an kinesthetic measures taken there included the judgment of weights in which a series of weights had to be arranged in order of weight, the perception of roughness of different surfaces, the measurement of sensitivity to pain from pressure and the two-point threshold on the tip of the forefinger. Tests also were given on the judgments of relative length by touch as well as an eye-hand coordination test for accuracy of touching a target [20].

In 1888 Hall left Johns Hopkins for the presidency of the newly founded Clark University. Hall did little research personally after that, but his students at Clark were often active in research on the skin senses with Hall's direction or that of his colleague and former student from Johns Hopkins, Edmund C. Sanford.

There were several active psychological laboratories that produced tactual research now and then. One was at Harvard. The demonstrational laboratory had been founded by William James for his classes in 1875 although the research laboratory came along somewhat later. By 1892 the laboratory had reopened in Dane Hall. Hugo Münsterberg, who had come to Harvard to take over the laboratory from William James, could list eight researches on touch having been done by 1893 [21].

Titchener and the skin senses at Cornell University

The most active psychological laboratory in the United States prior to 1940 in terms of tactual research was the Cornell University laboratories of E. B. Titchener.

Titchener arrived at Cornell University in Ithaca New York in the autumn of 1892, with a new Ph.D. in psychology from Wundt's laboratory at Leipzig. He set about improving and expanding on the laboratory established the year earlier at Cornell by another of Wundt's students, Frank Angell. In 1898 and 1900, Titchener published articles on his Cornell laboratory and his concept of an ideal laboratory for experimental psychological research and its equipment. Describing his own laboratory, he listed in both articles, by name, one room devoted to 'Hapics and Organic Senses' [22, 23]. Titchener was perhaps the first psychologist in America to adopt Dessoir's suggestion to use the term 'haptic' as the parallel term in the study of touch to 'optic' in vision [24]. It was Titchener (1901) who contributed the definition of 'Haptics' to James Mark Baldwin's influential *Dictionary of Psychology*. He defined Haptics as:

"*The doctrine of touch with concomitant sensations and perceptions – as optics is the doctrine of sight, and acoustics that of hearing... It may cover (and this is probably its best use) the whole range of function of skin, muscle, tendon, and joint, and even of the static sense – thus including the senses of temperature and pain, and the perceptions of position, movement, &c.; or it may be restricted to cutaneous sensations and perceptions in the narrower sense*" [25].

FIGURE 5. EDWARD B. TITCHENER
Source: Author's collection

Neither Dessoir nor Titchener's use of the term has the current meaning of haptics, of course. That appears to have been suggested by Revesz in 1950 [26]. Titchener made use of the broad meaning of this term in his laboratory at Cornell. Except for labeling the research area of his laboratory as 'haptics', however, Titchener usually made use of the more typical terminology to describe the skin senses. In his *An Outline of Psychology* [27] published in 1896, he listed all of the classes of sensations under two major headings, "*Sensations of the Special Senses (external stimulus)*" and "*Organic Sensations (internal stimulus)*" [27]. Under "*Sensations of the Special Senses*" he listed Aristotle's five classic sense departments. The last of these is that of 'cutaneous sensations'. He divided cutaneous sensations between sensations connected with mechanical stimulation of the skin and those connected with thermal stimulation [27]. Under experiences connected with mechanical stimuli he argued that the only genuinely simple cutaneous quality of sensation is pressure. What we think of as simple sensations are not simple qualities of sensation at all "*but mental processes which are really of a complex nature, and arise from the excitation of two or more senses*" [27]. He listed and described the bases of the experiences that called contact, hardness and softness, sharpness, roughness and smoothness, wetness and dryness, resistance, touch and impact. While we think of these as coming from the skin itself, Titchener argues that in reality these processes are all mixtures of cutaneous and organic sensations [27].

The second source of skin sensations listed by Titchener is that connected with thermal stimulation. He recognized hot and cold spots served by free nerve endings located near blood vessels as their receptors [27].

In 1896 Titchener did not consider pain to be a special sense but a 'common sensation' produced by the overstimulation of sensory receptors [27]. However, by 1899, when the second edition of his *Outlines of Psychology* was published, he rejected the notion of the 'common sense' and recognized pain as a separate special sense along with pressure and temperature [28].

The second major classification besides the "*Sensations of the Special Senses*" was the 'organic sensations'. These have an internal stimulus source. Under organic sensations, Titchener listed muscular, tendinous, and articular sensations, sensations of the alimentary canal, and qualities of the circulatory, respiratory and sexual sensations. His consideration of the complex nature of apparently simple cutaneous experiences laid the groundwork for Madison Bentley's 'synthetic experiment' in 1900 [29].

When his *Textbook of Psychology* appeared in 1909–1910 to replace the *Outlines*, Titchener gave additional space for detail of his classification and changed some of his terminology.

Touch, for Titchener, involved, excluding the parts of the skin involved in smell and taste, "*all the sensations aroused by contact with the bodily surface with objects from the material world.*" [30]. His classifications were basically those of the *Outlines of Psychology* but there were some differences. He classified the cutaneous senses as being made up of the pressure sense, temperature senses, and the pain sense. The other class of senses was the subcutaneous senses or organic senses. These included the kinesthetic senses. Titchener adopted the term kinesthesis from H.C. Bastian [31]. Titchener's use of 'kinesthetic sense' was that of the sense of movement including 'muscular sense', the 'tendinous sense', the 'articular sense', the 'ampular sense', the 'vestibular sense' and others. Aside from these, Titchener recognized other organic sensations as belonging to the overall sense of touch. He discussed the sensitivity of the abdominal organs, the digestive and urinary systems, sensations of the circulatory and respiratory systems and sensations of the genital system. Titchener's mention of the sensations of the glans penis in the male and of the clitoris in the female caused his *Textbook of Psychology* to be banned at some universities in the United States.

Virtually all of these aspects of touch experiences were systematically investigated during Titchener's 35 years at Cornell. Titchener's students, such as Karl M. Dallenbach, Michael Zigler and John Paul Nafe and, in turn, their students continued to influence the field of touch in the United States into the 1980s.

Thus, with G. Stanley Hall and his former student, E. C. Sanford at Clark University and particularly with E. B. Titchener at Cornell and his students, we find the basis of the 'familial line' of most psychologists researching in the field of touch in the United States. The remainder of this paper explores some of the areas of research that formed this body of research from this line up to 1940.

Localization of touch

The earliest researches coming out of the Cornell laboratory were based on the localization of pressure sensations on the skin. They were, in one way or another, investigations on local signs in tactual localization.

Walter Bowers Pillsbury was the first of Titchener's graduate students. His doctoral thesis research began during the 1893 academic year. He used E.H. Weber's 'second method' of localizing pressure stimulations to the skin. In this method the experimenter stimulates a point on the skin with a thin charcoal point and then the subject not having visual contact with the stimulated point on the skin, attempts to touch it with another charcoal pen. Czermak's criticism against Weber had been that using as his statistic, the average error, the calculated resulting error of localization was less than the two-point threshold itself [32]. Pillsbury sought a methodological technique to correct for this. He did so by recording the direction and extent of each error rather than using the average error. More importantly, though, Pillsbury introduced the notion of visual imagery as a component in the tactual localization, a finding counter to Lotze's local signs theory [33].

Titchener's first woman graduate student at Cornell, Margaret Washburn, became the first woman Ph.D. in psychology in America. The title of her doctoral thesis under Titchener was, "*Concerning the influence of visual association on the space perception of the skin*". Titchener had the article that was based on her thesis published in *Wundt's Philosophische Studien* in 1895 [34].

Washburn expanded on Pillsbury's observations on the use of visual imagery in tactual localization. Her hypothesis was that the localization of tactual sensations in normally sighted people is not unmediated but takes place by means of visual associations. So, if the skin is stimulated by a pressure stimulus, the individual stimulated uses something like a visual map of the body to locate the site of stimulation. Washburn had five subjects, two of whom had good visual imagery, two whose imagery was only moderate and one, a woman of 50 years old, who had been blind since she was 5 years old. Using Weber's 'second method' as Pillsbury had done, her experimental results supported her hypothesis: the better the ability to visualize, the better the ability to tactually localize. The blind subject was the least accurate in localization, the person with superior visualization skills, the most accurate. Washburn's results supported an empirical view of skin localization against the nativistic view of local signs.

Celeste Parrish's research took the argument one more step [35]. She had the subjects identify perceived the point of stimulation without touching it. She found much larger errors of localization than when the subject could touch the skin or move the response pen around until the site was found. This was parse out the difference between pure visualization and memory for a prior sensation.

Two-point threshold

Another topic Titchener assigned had to do with the two-point threshold. The issue of what constitutes two points and one point was a troublesome issue [36]. In 1918 Cora Friedline, researched the problem in a systematic way using Titchener's method of analytical introspection. She found that between a distinct two points and a distinct one point experience there was a range of other experiences that were neither. She also found that the discrimination of the difference between one point and two points required a high degree of attention and a low amount of fatigue. She thus demonstrated that, beyond the question of the distribution of sensory 'spots' there was the issue of attitude and judgmental factors. Her findings led to the practice of observers reporting 'two points' or 'not two points' rather than 'one point' or 'two points' in psychophysical experiments on the two point threshold [37].

The synthetic experiment

Perhaps the most significant line of research from Titchener's laboratory and the one most relevant to modern haptics was that begun with

Madison Bentley's doctoral thesis, '*The synthetic experiment*' [29]. Most of the earlier studies from American laboratories had been involved in simple psychophysical studies of the relationship between stimuli and sensations or between sensations and their underlying physiology. With the synthetic experiments, Titchener was pushing more directly into more complicated states. He was working within the framework of experimental introspective psychology in which one analyzes every-day complex experiences into their simplest components, the elements of experience. Bentley's thesis added another component, however, synthesis. If an analytic result shows the true elemental components, it should be possible, "*by adding the elements, one by one, to rebuild the original experience. Surely, no better test of the accuracy of an analysis is possible than the reinstatement of the whole through synthesis of the products of dissection*" [29]. This is a restatement of one of Titchener's beliefs gained from Wilhelm Wundt, that "*the test of analysis is synthesis*" [27].

The demonstration of synthetically produced experiences was not new with Bentley's research, however. Altruz was doing much the same thing as early as 1896 on the experience of 'heat' [14, 15]. Also, Thunberg devised a thermal grill in 1896 that formed the prototype of many other 'heat grills' for demonstrating synthetic 'heat', an experience of 'heat' produced by stimulating warm and cold receptors simultaneously [38]. Ironically, of all the touch blends to be reproduced synthetically, 'heat' would be one of the most difficult to obtain reliably [36]. It has only been recently, with advances in neurophysiology, that the reason for this difficulty is better understood [40].

The touch complex or 'touch blend' Bentley chose was 'wet'. Introspectively 'wet' renders down to the simple sensations of pressure plus cool. If, this is so, then it should be possible to stimulate the skin with stimuli that produce coolness and pressure but which are not 'wet'. The perception should still be that of wet even though there is no moisture involved. Bentley's subjects perceived a 'wet' experience when the cool liquid was contacted through a thin membrane. The membrane was dry and protected the finger from the liquid but it communicated the pressure and coolness of the water through the membrane. The experience produced was that of 'wet'. In the same way cool flour in a funnel-shaped receptacle was perceived as being a fluid when the finger was inserted [29].

After Bentley's experiment, the synthetic experiment project languished for some years. There was Elsie Murray's analyses of tickle [41] in 1908 and her introspective study of organic sensations such as 'dull' and 'sharp' in 1909 [42]. There was also Cutolo's thesis on 'heat' in 1918 that showed how difficult synthesis can be [43]. It was only in 1922 when the synthetic experiments began showing up in print again. E. G. Boring surmised that it was the pressure of the developments of Gestalt psychology that made Titchener decide to begin these synthetic exploration of perception [36]. Evans, however, has suggested that it was developments in Titchener's own thinking in the period after 1910 that was responsible and which led Titchener eventually to abandon elements and recast his elementistic psychology into one of dimensions of consciousness [44]. In this system, a given experience appears at the juncture of several quantitative and qualitative dimensions. The indication of Titchener's new line of thinking is evident in 1920, when he devised a theoretical model for touch experience based on Bentley's findings and that of other introspective studies available up to that time [45]. Titchener's model for touch, his 'touch pyramid' is shown in Figure 6. He admitted that many of the representations were tentative but he expressed a desire to publish the figure:

"*because I think it marks the direction of recent work; because it may perhaps serve as the starting-point of new experiments; and also for the more personal reason that I have shown the model in my lectures of the past two years, and that the observers in certain investigations show to be published from the Cornell laboratory have therefore been familiar with its terms and series*" [45].

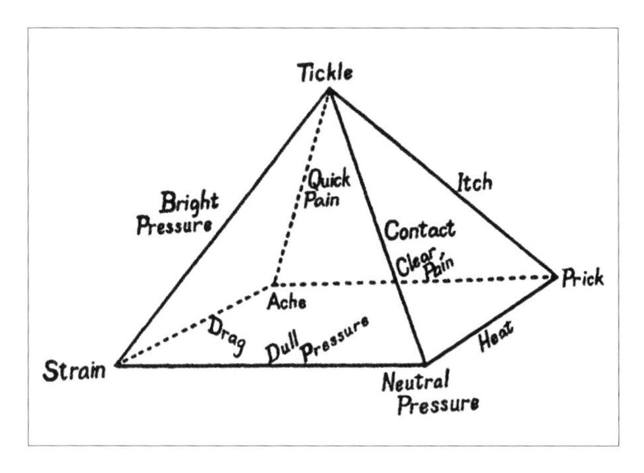

FIGURE 6. TITCHENER'S TOUCH PYRAMID
Source: Am J Psychol *Vol. 32 (1920), p. 213*

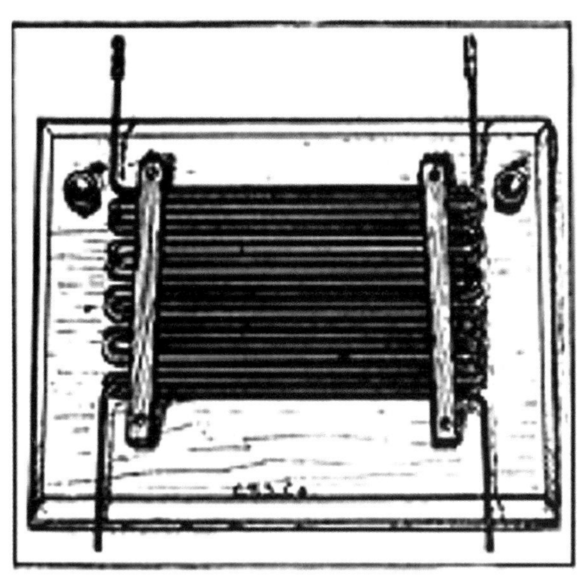

FIGURE 7. DALLENBACH'S HEAT GRILL
Source: C.H. Stoelting Catalogue, 1930

The model proved to be the guide for the vocabulary and the theory in much of the literature for a long time. Titchener's laboratory buzzed with the new synthetic experiments. There would be added to the studies already noted, synthetic experiments on the perception of 'oily' [46], stickiness [47], rough and smooth [48], 'clamminess' [49], 'liquidity', 'semi-liquidity' and 'hardness' [50], 'tickle and itch' [51], 'hardness' and 'softness' [52], among others. In each case the analysis was reproduced, more or less, by a synthesis of artificial stimuli.

The study of tactual complexes continued into the 1930s at Cornell, particularly by Karl M. Dallenbach and his students. A particular problem that Dallenbach attacked is that of 'hot' or 'heat'. Cutolo [43] in 1918 had found that the fusion of cold and warm to produce hot can be difficult, particularly using punctate stimuli. He did not find a total fusion into hot but instead experiences like a 'sting' or 'smack'.

Cutolo's study had been followed up in 1920 at Harvard by J.H. Alston, a student of Titchener's former student, E. G. Boring. Alston also used a thermaestheiometer, to deliver punctate temperature stimuli to different receptors. He found that the experience of 'hot' can be obtained by stimulating a cold spot and an adjacent warm spot, even when the distance between the spots is relatively large (10–15 cm). The resulting 'hot' is difficult to localize introspectively, however. He found the imperfectly fused 'hot' is more likely to fuse with the cold spot rather than the warm spot or be perceived as a sequence of warm and cold. He concluded that "*both psychologically and psychophysically heat is more intimately related to cold than to warmth*" [53].

In 1930, Dallenbach and his student, Ferrall used a heat grill of Dallenbach's own design (see Burnett and Dallenbach [54]) in which warm and cold water ran in close parallel tracks. The study agreed with Cutolo that the experience is not a total fusion to 'heat' but is something like pressure and prick [55]. That is where Titchener had placed heat in his touch pyramid of the tactual experiences, somewhere between pressure and prick.

The matter appears to depend on the completeness of the fusion of cold and warm and the temperatures chosen. The use of a heat grid is more likely to give fusion than punctate points. The demonstration using Dallenbach's heat grill was common in the 30s with most untrained subject perceiving 'heat' when cold and warm grids are placed in close proximity to each other. Dallenbach and his students continued to try to find the relationship between warm, pressure, cold and pain in 'heat' and in other temperature complexes all through the 1930s [55–58].

Eccentric projection

Perhaps the study in Titchener's laboratory concerning the synthetic experiment that is most relevant to present-day haptics is that of 'eccentric projection'.

George Trumbull Ladd had made use of that term in 1887 in his *Elements of Physiological Psychology*.

"*Sensations, as the elements of so-called 'presentations of sense' are psychical states whose place – so far as they can be said to have one – is the mind. The transference of these sensations from mere mental states to physical processes located in the periphery of the body, or to qualities of things projected in space external to the body, is a mental act. ... Certain compound sensations – as, of light and color, of smoothness and hardness – are not regarded merely as psychical acts whose cause lies in the extended object; they are regarded as qualities of its surface and appear to belong to it as forms of its objective being. Such a result, however, must be regarded as brought about by the action of mind, in forms and according to laws of its own*" [59].

Ladd proposed that there are two stages in this 'achievement' of mind. These are 'localization', or the transference of the composite sensations from mere states of the mind to processes or conditions recognized as taking place at more or less definitely fixed points or areas of the body; and the 'eccentric projection' (sometimes called 'eccentric perception') or the giving to these sensations an objective existence in the fullest sense of the word 'objective' as qualities of objects situated within a field of space and in contact with, or more or less remotely distant from, the body. The law of eccentric projection is generally stated thus: Objects are perceived in space as situated in a right line off the ends of the nerve-fibers they irritate [59].

William James in his *Principles of Psychology* accepted the idea of eccentric projection but rejected Ladd's belief that the mind could, somehow, project itself out into empty space and into a distant object [60]. In his *Principles of Psychology*, James places the seat of eccentric projection in the 'joint-feelings'. To James, "*The localization of the joint-feeling in a space simultaneously known otherwise (i.e., to eye or skin), is what is commonly called the extradition or eccentric projection of the feeling*" [60]. James goes on:

"*The content of the joint-feeling, to begin with, is an object, and is in itself a place. For it to be placed, say, in the elbow, the elbow as seen or handled must already have become another object for the mind, and with its place as thus known, the place which the joint-feeling fills must coalesce. That the latter should be felt 'in the elbow' is therefore a 'projection' of it into the place of another object as much as is being felt in the finger-tip or at the end of a cane can be. ...*

But for the distance between the 'here' and the 'there' to be felt, the entire intervening space must be itself an object of perception. ...When it is filled by our own bodily tissues (as where the projection only goes as far as the elbow or finger-tip) we are sensible of its extent alike by our eye, by our exploring movement, and by the resident sensations which fill its length. When it reaches beyond the limits of our body, the resident sensations are lacking, but limbs and hand and eye suffice to make it known. Let me, for example, locate a feeling of motion coming from my elbow-joint in the point of my cane, and the seen end of it then absorbs my sensation just as my seen elbow might absorb it, or I am blind and imagine the cane as an object continuing my arm, either because I have explored both arm and cane with the other hand, or because I have pressed them both along my body and leg." [60]

It was only one additional step from this to say that it should be possible to create the experience with only the sensations whether the object is there or not. It was, after all, a restatement of George Berkeley's theory of perceptual object.

Titchener gave the problem to Louis B. Hoisington for his dissertation project. His thesis was published under the title "*On the Non-Visual Perception of the Length of Lifted Rods*" [61]. True to the principles of the synthetic experiment, Hoisington not only analyzed the experience of the eccentric projection with the rods but also was able to reproduce it synthetically with pressure stimuli without any physical rods at all.

He had his subjects grasp a series of rods, each by their wooden handle, which was identical for each rod. There was a standard rod and several other rods that differed in terms of length. By holding the handle lightly in the hand the subjects lifted the rods to a horizontal position. Their task was to estimate the length of the rods by the sensations only by lifting them. In doing this Hoisington was able to determine the direct stimuli that produced the impressions of differing lengths of the rods. The subject's introspective results were that the perception of length was determined by a ratio of two pressures in the hand. One of the pressures was referred to as 'kick' which was the "*pressure-sensation at the base of the thumb stimulated by the up-thrust of the back end of the handle*." The other pressure was simply called 'pressure' and was the "*sensation in the forefinger stimulated by the down-pull of the end of the rod*" [61]. Hoisington summarized:

"*If we hold a rod in the vertical, there is nothing but the weight; no notion of length arises. It is only as the instrument becomes inclined that the perception of length comes in; weight without an extended center of mass never gives rise to the perception of length; though we do not at all deny that, given a center of mass 'out there', variations in weight may set off a kind of derived judgment of difference in length.*" [61]

Hoisington pointed out that:

"*The absolute intensity of the two impressions can not be the basis for the judgment of length; for we judge a long light rod, which produces relatively weak pressures, as longer than a short heavy bar, which produces much more intense impressions. The answer to the question: What is the experimental foundation for the perception? Thus becomes clear; our process of elimination and our introspective analyses lead us to the same result: the ratio between the 'kick' and the 'pressure'. In a short rod, the amount of 'kick' relatively to the 'pressure' is small; in long rods, the two pressures more nearly approach each other in intensity; with every change in the center of mass there takes place a change in the ratio of 'kick' to 'pressure'. A sensation-complex of two opposed pressures, more or less fused in experience, differing in intensity within certain ratio-limits, is the sensory pattern fundamental for our perception of length as given cutaneously...*" [61]

When Hoisington masked the pressure and kick sensations by the use of spoon-like coverings on the hand, he reported that the experience of length disappeared.

At this point Hoisington initiated a 'synthetic series', in which he produced the length experience with no physical length being present. This was done by an ingenious 'feedback' device in which the handle was attached to an elaborate pulley and weight arrangement which produced, when hefted by blindfolded subjects, differing ratios of 'kick' and 'pressure' in the hand. The result was an immediate illusion of length [61].

This study was followed up by studies involving the non-visual perception of the length of vertically whipped rods [62] and of horizontally whipped rods [63], each finding similar results, though also including some kinesthetic components due to the more active nature of the situation.

Facial vision

In a way, the solution to the question of 'facial vision' in the blind came about through a version of the synthetic experiment. The ability of the blind to somehow perceive and avoid objects in their path was documented at least as far back as Diderot in the 18th century (see Supa, Cotzin and Dallenbach, for a review of the early literature [64]). In 1872, in England, W. H. Levy, himself blind, called the ability 'facial perception'. He reported that

"*I seem to perceive objects through the skin of my face, and to have the impressions immediately transmitted to the brain. The only part of my body possessing this power is my face*" [65].

There were certainly no lack of theories of how the blind manage to avoid obstacles (See Hayes [66]). William James reviewed much of the literature on the subject in his *Principles of Psychology* [60] and concluded that the ability to perceive objects was due to the objects producing non-auditory pressure sensations on the tympanic membrane. In 1893, Dresslar in G. Stanley Hall's laboratory at Clark University conducted a series of experiments and determined that the ability was due to sound and that, when the ears were tightly stopped, the ability disappeared [67]. It was only in the 1940s and 1950s, however, that Dallenbach and his students at Cornell, in an exhaustive series of experiments, demonstrated that the facial experience was illusory and that the awareness the blind have of objects before them is due to their hearing the echoes reflected off the objects around them. In the several experiments, Dallenbach and his students performed, they eliminated everything except echoes and the ability remained. If the echoes in the room were eliminated, then the ability disappeared. In one synthetic experiment the blind subject put on earphones in a sound proof room and listened to a person walking in another room, holding a microphone. The subject was able to tell the walking person when an object was first 'perceived' and then when he was about to walk into it, completely from the sounds heard on the earphones. It was the sounds of footsteps or the tapping of a cane reflecting from objects that the blind were responding to, not facial vision [68–70].

The influence of Gestalt psychology

Wertheimer's introduction of the concept of phi and the demonstration of the phi phenomenon, in which discrete visual stimuli turned on and off in sequence across a space appear as a single stimulus moving across that space, stimulated research on both sides of the Atlantic [71]. Vittorio Benussi, a student of Alexius Meinong at Graz, demonstrated that the phenomenon was also reproducible on the skin [72–74]. H. E. Burt at Harvard demonstrated that it was also demonstrable in audition [75] and both Burt [76] and Anna Whitchurch [77] at Cornell carried out experiments exploring this topic in the realm of touch.

Burt had a ten-point kinohapt in which any two of the ten rods could be released by solenoids with a well controlled time differential between impressions. This allowed differing space and time relationships to be studied in the sequencing of the pressure stimuli. A tactual 'phi phenomenon' effect was noted, but there was more. One of Burt's observers reported that even though there were only two points placed on the skin, they perceived "*A walk in which there are a number of discrete points in succession.*" Another subject reported movement in which the illusory points "*skip along like playing four or five notes on a piano in succession*" [76]. These observations are remarkably like the experience observed and recorded almost 60 years later by Frank Geldard [78] and called it the 'cutaneous rabbit'. Burt did nothing more with the observation other than report it. When Geldard independently came on the same phenomenon, he recognized it as sensory saltation [79].

Another Gestalt influence on the psychology of touch came with the publication of David Katz's monograph *Der Aufbau der Tastwelt* in 1925 [80]. There was an extensive review of Katz's work in the *Psychological Bulletin* by Michael Zigler the following year [81]. Katz criticized the work of Blix, von Frey, Donaldson, Titchener, Thunberg and others for being too 'atomic' in their study of touch and thus distracting psychologists from the complex experiences E.H. Weber dealt with in his work on touch. Zigler, a student of Titchener, disagreed with Katz and responded that the researches from Titchener's laboratory often involved the more complex experiences of 'everyday life'. He pointed out that the large number of experiments done using the synthetic experiment in Titchener's laboratory and elsewhere demonstrated that this was so. Katz showed no familiarity with this research in his historical introduction to his monograph. In fact, much of the research in Titchener's laboratory in the 1920s had a distinctly phenomenological tenor to them. Titchener, no longer dealing with elements, viewed sensory experience in terms of the intersection of dimensions of quality, intensity, extensity (space), protensity (time) and attensity (clearness) [44]. Zigler gave a thorough explication of Katz's ideas on the psychology of touch. Even after his criticisms, Zigler predicted that Katz's book would "*stimulate psychologists to rework the field of touch from the phenomenological point of view, and will thus serve as groundwork and point of departure for numerous studies of detail*" [81]. Zigler's review in the *American Journal of Psychology* was, for many American psychologists, their first glimpse of Katz's work. Katz's *Aufbau* was not translated into English until 1989 [80], so Zigler's review remained a major source of information in English for many years.

An important point of Katz's book was his emphasis on active touching. He was correct to the degree that many, perhaps most of the experiments on touch, even the touch of form, had used passive touch. As James J. Gibson expressed it many years later:

"*Active touch refers to what is ordinarily called touching. This ought to be distinguished from passive touch, or being touched. In one case the impression on the skin is brought about by the perceiver himself and in the other case by some outside agency… Active touch is an exploratory rather than a merely receptive sense*" (Gibson [82]).

For Titchener, touch experiences played an important systematic role in the final developments of his systematic psychology in the 1920s. Doing away with sensory elements in favor of dimensions was one thing, but the experiences of simple affect, pleasantness and unpleasantness was something else. These simple feelings or affects were not considered in Titchener's earlier system to be sensory but were subjective experiences that accrued to sensations, images and perceptions. This changed for Titchener with one pivotal experiment conducted by John Paul Nafe in 1924. The purpose of the study was to attempt to make 'feelings' palpable. The result of the experiment harkened back to Elsie Murray's work in 1908 on somatic senses [42] and demonstrated, at least to Titchener and his students' satisfaction, that the feeling of pleasantness is a 'bright pressure' and the feeling of unpleasantness is a 'dull pressure' (Nafe [83]). Nafe expanded on this notion after he left Cornell for Clark University, particularly after Titchener's death in 1927. In fact Nafe attempted to reduce all the tactual senses down to different simple analyzed states of 'brightness' (Nafe [84]).

The vibration sense

Katz's *Aufbau* brought to the United States a controversy that had been ongoing in Europe for some time, whether vibration is a separate sense or if it is merely mediated by the pressure sense (see Geldard [85]). At the time Katz' *Aufbau* appeared, he was already involved in a controversy with von Kries on the matter. Katz held that vibration must be considered a separate sense from pressure.

Katz's book gave support to researchers in the United States who, since the 1920s had been using the Tadoma method, developed at the Perkins School for the Blind, to allow deaf-blind individuals to interpret speech directly with touch [86, 87]. In the Tadoma method, a deaf-blind person places their hands on the speaker's face and learns to interpret their speech from the vibrations experienced through the hands.

Two of the champions of the notion of communicating speech by means of touch were Robert H. Gault of Northwestern University and his student Louis Goodfellow. Even before the appearance of Katz's book, Gault had been interested in the possibility of the deaf being able to perceive speech through the skin senses. He and his students published a large number of researches on tactual communication in the 1920s and 1930s [88, 89].

In 1923 Gault conducted a series of studies at Northwestern University in which he reported being able to teach the ability to perceive vowels, words and even short sentences with reasonable accuracy only through touch.

The experiments were conducted in various ways using sound tubes or an electromagnetic device that delivered vibrations directly to the hand. Gault concluded "*it is possible to learn to interpret oral speech by tactual impressions*" [90]. Katz gave additional impetus to this movement with his lectures in 1929 at the University of Maine, published as *The Vibratory Sense and other Lectures* [91]. Between 1924 and 1934 alone, Gault alone or with one of his students published at least 27 articles on the topic. His various students published, independently, at least as many [88].

In 1932 Gault produced his own device called the teletactor, which was described as being designed to allow the deaf to hear through their fingers. It was described as:

"*very much like a combination of the telephone and the radio. The speaker talks into a microphone. By wires the voice is transmitted to the receiver, much more powerful and sensitive than the telephone receiver, topped by an aluminum plate, which vibrates with each tone of the voice. On this plate the deaf person places his fingers, and feels the sound of the voice*" [92].

Gault's initial research showed promise for the tactual senses to become, at least, a complement to lip reading, but, the skin simply does not have the temporal resolving ability necessary to process the high frequencies of speech [93]. While Gault's attempts eventually proved impractical, his work is considered the beginning of speech communication by vibrotactile devices [93].

The argument that there is a separate vibrational sense apart from pressure was largely put to rest by a series of papers by Geldard in 1940 [85, 94–96]. The last article in the series asked the question: "*is there a separate 'vibratory sense?'*" The answer came down as a resounding "*no*". Geldard found no evidence that would require postulating a separate sense of vibration. The driving of the pressure sense was sufficient to explain the degree to which vibration can be experienced [96].

When W. L. Jenkins surveyed the literature on touch for in S. S. Steven's influential *Handbook of Experimental Psychology* in 1951, the topic of a vibration sense was relegated to a paragraph and the arguments against its existence were those of Geldard's articles [97]. Calne and Pallis in their review of the literature on the vibration sense in 1967 concluded, on the basis of neurophysiological evidence developed after Geldard's article, that the vibratory sense is a 'temporal modulation' of the tactile sense, something analogous to the relationship between flicker fusion and vision [98].

While Katz's view of a vibration sense may have lost favor, his general philosophy of touch and particularly his view on active touch, has remained influential [82, 99].

The tau effect

Another influence of Gestalt psychology on tactual research was the phi phenomenon. In 1924 Harry Helson received his Ph.D. in Psychology from Edwin G. Boring at Harvard. His thesis topic was on Gestalt psychology and he published two long articles on the topic in the *American Journal of Psychology* in 1925 [100, 101]. Along with Kurt Koffka's article in the *Psychological Bulletin* in 1922 [102], these article were the best available source of information on Gestalt psychology for American readers.

One of the studies that came out of Helson's contact with Gestalt psychology, was his work on the tactual 'tau effect'. The immediate stimulus for Helson's research was from Gelb's research [103] and Benussi's [74] criticism of Wertheimer's concept of 'pure phi' or the pure experience of movement. Helson related the criticism in his article on Gestalt:

"*Although Koffka and Koehler are inclined to accept the fact of a pure phi, objections to it have been raised. Benussi repeated Wertheimer's experiments both in touch and vision and found that 'something moves' even though the Os could not tell what moved. Benussi says that 'experience gives no pure movement in the sense of a movement without an object*" [101].

Helson asked the question relating to the phi phenomenon, "if two [Müller-Lyer illusion] figures are exposed in succession, the one of which seems larger or smaller than the other, will there be a difference in the kind of movement seen as compared with that produced by two stimuli which appear to be equal in size?" [102]

Helson and King tested this question on touch by using differing tactual extents as their 'figures'. The stimuli were produced by a version of the Dallenbach Kinohapt. This device had solenoids bearing hard rubber stimulus rods which could be dropped and raised on the skin at precise locations and times and durations. The three stimuli were presented in sequence with the time between the first and second kept constant at .5 s and that between the second and third stimuli were varied between .3 inches and .7 inches. The observers reported if the second extent was longer than, equal to or shorter than the first. What Helson and King found was that there was a relationship concerning the perceived distance between stimulated points and the rate of the stimulation [105]. Helson gave the following example based on a situation in which three spots are touched on the back of the hand or arm in quick succession:

"*If one is asked to judge whether the second spatial interval is equal to, greater than or less than the first interval, it will be found that the observer's report depends more upon the time interval between stimulations than upon the actual distances between places touched… Thus, if we stimulate three spots on the skin so that the first distance is 20 mm and the second 10 mm, but the time interval between the second and third stimulations is twice as great as that between the first and second, then the distance between the second and third spots will be judged as nearly twice as great as the first. The conditions may be reversed to give the opposite effect*" [104].

Helson proposed that the phenomenon "*be called the tau effect because it obeys definite laws, can be measured and is not due to 'imagination', 'attention', 'suggestion' or any other peculiarly mentalistic mechanism*" [104].

Helson's emphasis on avoiding 'mentalistic' descriptions demonstrates one of the tensions in American psychology that had been growing all through the 1920s and would become even more manifest in the 1930s. The introspectionism of Titchener and his followers was being pushed aside for the 'objective' psychologies of behaviorism and neurophysiology. While research continued to be done in the 1930s in perception, sensation became more and more in the province of neurophysiology or of physiological psychology. Perception became more the province of Gestalt psychology. Until the 'cognitive revolution' of the 1960s, research using introspective data was carried out with increasing trepidation.

Haptics as it is ordinarily understood today arose from the juncture of work in psychology, physiology, physics and engineering. That would come after World War II, but the psychological fundamentals on which it was based was well established beforehand.

Summary

The history of haptics in America begins with the use of tactual cues in the training of the deaf mute Laura Bridgman in the 1830s. The use of touch for communication in this case and the introduction of psychophysics to America led psychologists in the 1880s and afterwards to explore the largely neglected study of touch. The major contributors to these lines of research drew upon the European literature and expanded upon it. The laboratory of E. B. Titchener at Cornell University was a major source of this research. This work of Titchener's students and their students there and at other locations is emphasized up to 1940.

Acknowledgement

Much of the material used in this paper was gathered while the author was a guest researcher at the Max Planck Institute for the History of Science in Berlin, Germany, in 2005. Special thanks are due to Dr. Hans-Jörg Rheinberger, Director of Department 3, Dr. Henning Schmidgen for their many kindnesses and to the library staff for their excellent assistance.

Selected readings

Schiff W, Foulke E (1982) *Tactual perception: A sourcebook*. Cambridge University Press

II.
Neurophysiological and physiological aspects

Anatomy of receptors

Zdenek Halata and Klaus I. Baumann

The human hand is a complex organ serving the functions of grip and touch. The mechanoreceptors of the hand can be categorised into those located within skin and subcutaneous tissues and those associated with joints and muscles providing the central nervous system with information about position of movement of hand and fingers. In addition to mechanoreceptors there are numerous free nerve endings reacting to thermal and/or painful stimuli generally referred to as polymodal nociceptors. They are found in the connective tissue of the locomotion apparatus as well as the skin and even enter the epidermis. Morphologically these are terminal branches of afferent nerve fibres without any specific structures around these 'free' nerve endings in marked contrast to the different types of mechanoreceptors.

Mechanoreceptors of joints

Joints are surrounded by mechanoreceptors in the connective tissue forming the joint capsules. The first type – Ruffini corpuscle – is found in the outer fibrous layer (membrana fibrosa) of the joint capsules. A Ruffini corpuscle consists of one or several cylinders formed by flat perineural cells and is supplied by one myelinated axon (5–10 μm diameter) loosing its myelin sheath on entering the cylinder and branching several times (Fig. 1). Each branch is covered incompletely by terminal Schwann cells anchoring between the fibrils with differently shaped nerve terminals. The nerve terminals contain accumulations of mitochondria and empty vesicles. Free surfaces are coved by basal lamina of the termi-

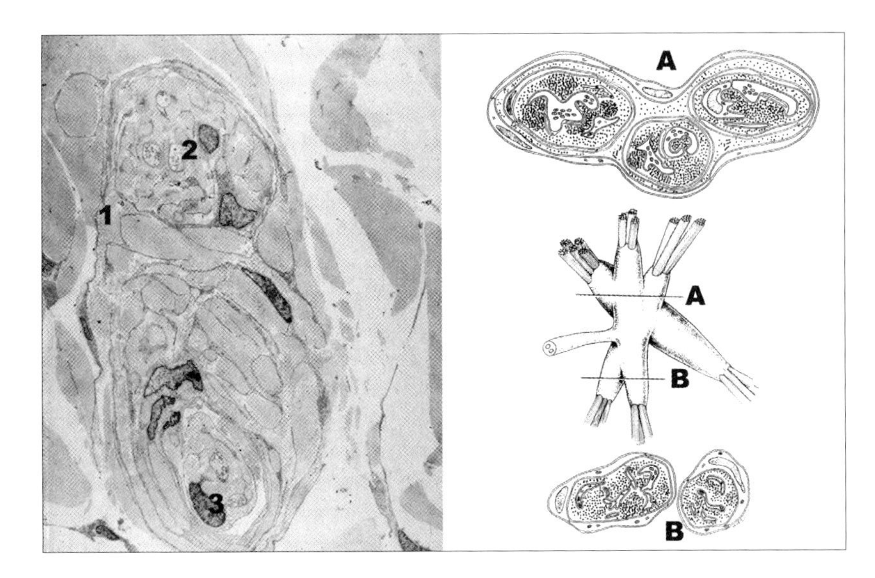

Figure 1. Ruffini corpuscles
Left: Electronmicrograph of a cross section through a Ruffini corpuscle. (1) perineural capsule, (2) nerve terminal. (3) nucleus of terminal Schwann cell. Magnification × 6,000.
Right: Diagrammatic representation of a Ruffini corpuscle from a joint capsule. The longitudinal axis of the cylinders run parallel to the collagen fibrils of the fibrous layer. A and B represent cross sectional images through the planes as indicated.

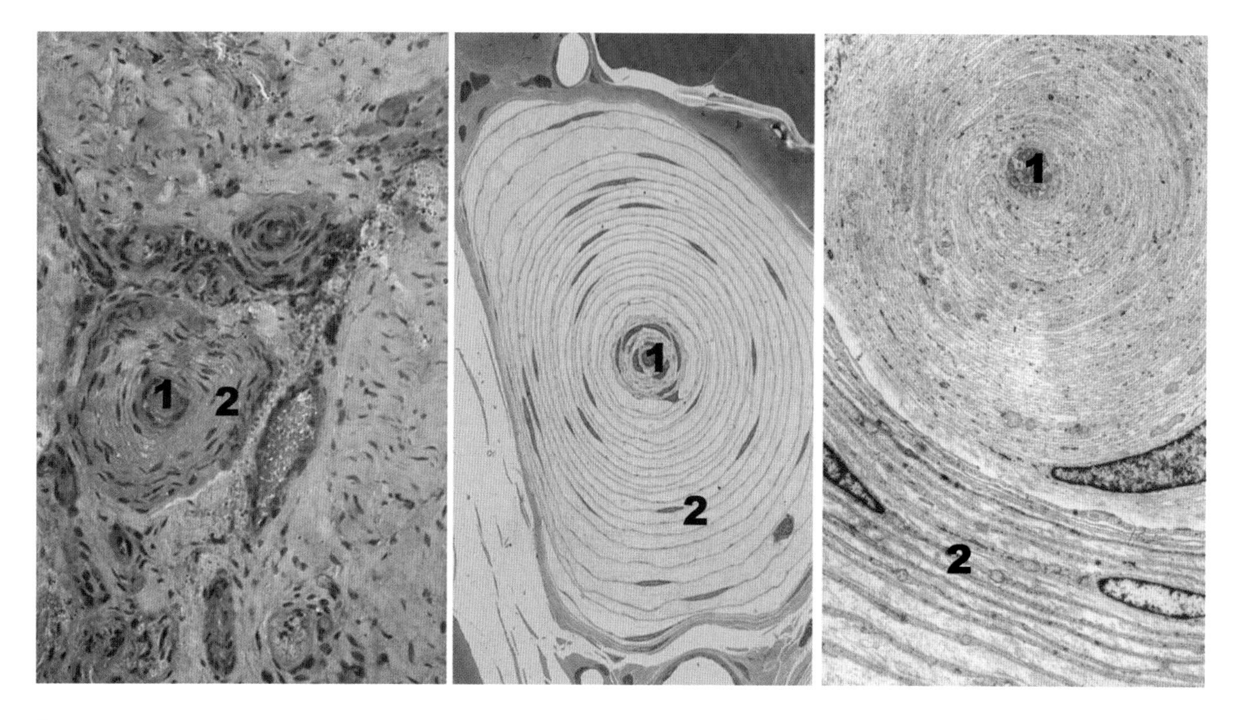

Figure 2. Vater-Pacini corpuscles: Cross sections through Vater-Pacini corpuscles
Left: Paraffin section of the fibrous layer of a joint capsule showing a Vater-Pacini corpuscle. (1) inner core, (2) perineural capsule. Magnification × 800.
Middle: Vater-Pacini corpuscle from the epimysium of hand muscles. (1) inner core with nerve terminal, (2) perineural capsule. Magnification × 1,200.
Right: Electron micrograph showing the central part of a Vater-Pacini corpuscle. (1) axon with inner core, (2) perineural capsule. Magnification × 6,000

nal Schwann cells. The longitudinal axis of the cylinders measures approximately 200–300 μm and follows the direction of the collagen fibrils in the fibrous layer of the joint capsule. Structurally the Ruffini corpuscles of the locomotion apparatus are identical with those of the skin.

Functionally Ruffini corpuscles were found to respond with high sensitivity to stretching of the collagen fibres. The discharge pattern of the action potentials is slowly adapting with very regular interspike intervals during maintained stimulation [1].

The second type of joint mechanoreceptor is the so called Vater-Pacini corpuscle, also referred to as Pacinian corpuscle. They are the largest type of mechanoreceptors found in mammals with a length of up to 2 mm and a diameter of up to 1 mm (Fig. 2). The myelinated axon of 6–10 μm diameter looses the myelin sheath on entering the inner part or the corpuscle, but in older people the myelin sheath may extend up to one third of the inner core of the corpuscle. The axon ends in the centre of the corpuscle with a ball shaped thickening.

Cytoplasmic spines extend from the non-myelinated part of the axon anchoring between the lamellae of the inner core of the corpuscle. The axoplasm of the 'ball' and near the origins of the cytoplasmic processes contains accumulations of mitochondria and empty vesicles. The inner core of the corpuscle is formed by symmetrically arranged lamellae of terminal glial cells. In a longitudinal section these lamellae are shaped

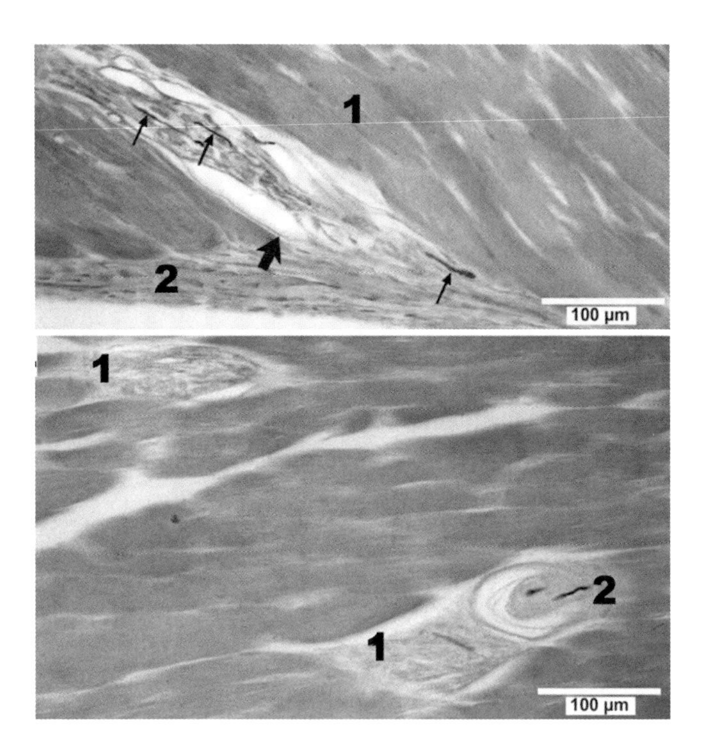

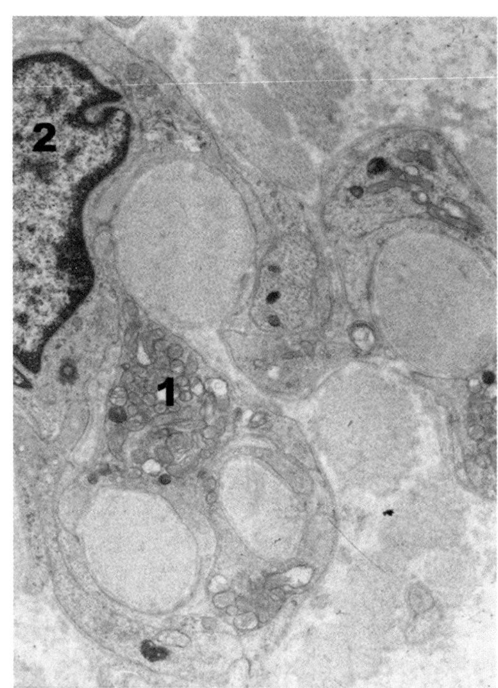

FIGURE 3. GOLGI TENDON ORGANS

Left: Silver stained longitudinal section (above) and oblique section (below) of Golgi tendon organs (1). Small Pacini corpuscles (2) are also seen. The nerve terminals are indicated by thin arrows, the thick arrow marks the perineural capsule.

Right: Electronmicrograph of a cross section through a Golgi tendon organ. (1) nerve terminal with accumulation of mitochondria, (2) nucleus of a terminal Schwann cell. Magnification × 12,000.

like half moons on either side of the axon resembling a 'hot dog'. Two symmetric longitudinal clefts separate the two lamellar systems. Thin collagen fibrils are running as spirals through the longitudinal cleft and between the lamellae. The number of lamellae varies with the size of the corpuscle and can reach up to 60 layers. The perineural capsule consists of layers of thin perineural cells extending from the perineurium of the nerve. The cells are flat and show micropinocytotic vesicles. The outer side is surrounded by a basal lamina. Collagen fibrils also run through the clefts between these flat perineural cells. The number of layers of perineural cells forming the capsule depends on the size of the corpuscle. Large corpuscles may have up to 70 layers.

Functionally Pacinian corpuscles respond to vibration stimuli with an optimum of sensitivity in the frequency range of about 200 Hz and vibration amplitudes below 0.1 µm [2].

Mechanoreceptors of the musculature

Two types of mechanoreceptors are found in the musculature: muscle spindles and Golgi tendon organs.

Golgi tenton organs (GTO) were first described by Golgi in 1880 and are found at the juncture between skeletal muscle and tendon, rarely also between muscles and their fascia (Fig. 3). GTOs have the shape of a spindle with a maximum

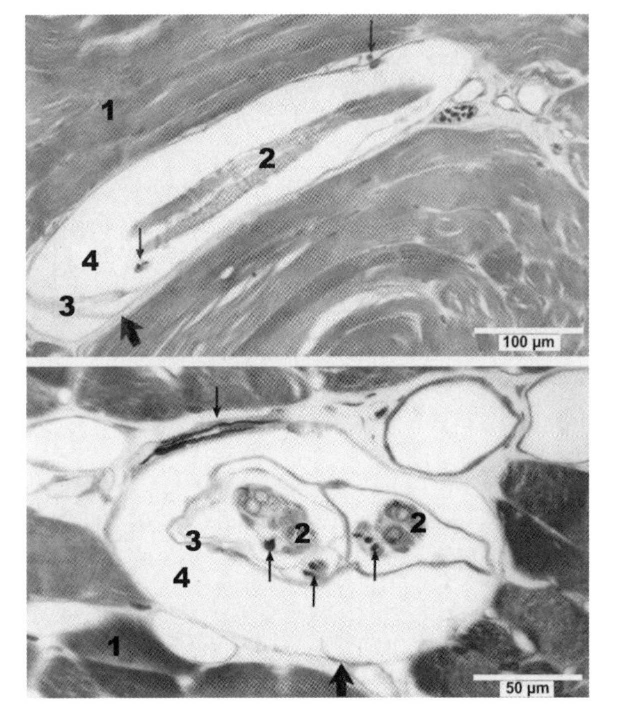

FIGURE 4. MUSCLE SPINDLES IN SILVER STAINING
Above: Longitudinal section of a muscle spindle.
Below: Cross section through a muscle spindle. (1) Extrafusal muscle fibres, (2) intrafusal fibre, (3) inner spindle sheath, (4) subcapsular space. Thin arrows incidate nerve fibres, thick arrow the perineural capsule.

length of about 1,600 µm and a width of up to 120 µm [3, 4]. The spindles are pointed on both sides towards the muscle and towards the tendon, but the end facing the muscle is usually about 25% thicker than the end facing the tendon. Most GTOs are simple spindles, but sometimes one or the other side can branch into 2 or 3 lappets. Occasionally small Pacinian corpuscles can be found within GTOs.

GTOs are usually supplied by several (3 to 6) myelinated afferent axons (5–15 µm diameter) entering the GTO on the long side where the perineurium verges into the capsule of the GTO formed by flat perineural cells surrounded by a basal lamina. The capsule is open at the pointed ends of the GTO. Within the GTO, the afferent nerve looses its myelin sheath and branches several times. Lamellae of terminal glial cells anchor the afferent nerve terminals between collagen fibres. Blood vessels enter the GTO together with the afferent nerve while the venous outlet is usually at the pointed ends.

Functionally, GTOs monitor the tension developed by the muscle and are able to elicit protective spinal reflexes *via* inhibitory synapses with the corresponding motor neurons [5, 6].

Muscle spindles (Figs 4 and 10) belong together with the GTOs to the group of encapsulated mechanoreceptors of the locomotion apparatus. They are 2–10 mm long and about 0.2 mm thick. In the lumbrical muscles, the longest muscle spindles are found in the middle of the muscles while those closer to the muscle ends are shorter.

According to Robertson [7], the fusiform capsule of muscle spindles consists of an inner spindle sheath, an outer spindle sheath. Both layers are formed of flat perineural cells. Thin collagen fibres are running in spirals through the periaxial space between the two layers. The outer layer is surrounded by endomysium of the extrafusal muscle fibres. The perineurium of the supplying nerve is a continuation of the arachnoid matter and extends into the capsule of the muscle spindle, explaining the assumption that cerebro spinal fluid can be found within the muscle spindles [8]. Thin muscle fibres are running through the muscle spindles and referred to as intrafusal fibres. They are encased by connective tissue of the endomysium. Two types of intrafusal fibres can be distinguished depending on the arrangement of their nuclei in the middle (equator) of the muscle spindle: thin nuclear chain fibres, where the nuclei form a row, and thick nuclear bag fibres, where the nuclei are accumulated like in a bag. There are more nuclear chain than nuclear bag fibres and one human muscle spindle contains 1 to 5 nuclear bag fibres and 2 to 11 nuclear chain fibres [9]. The nuclear bag fibres insert into collagen fibres of the endomysium of the extrafusal muscle fibres outside the capsule of the the muscle spindle. In

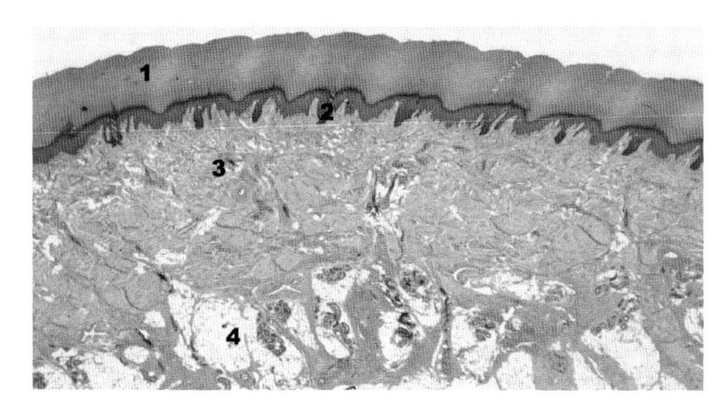

FIGURE 5. RIDGED SKIN FROM HUMAN FINGER SKIN.
(1) stratum corneum, (2) glandular ridge, (3) reticular layer of the dermis, (4) subcutaneous adipose tissue. Paraffin section, HE, magnification × 400.

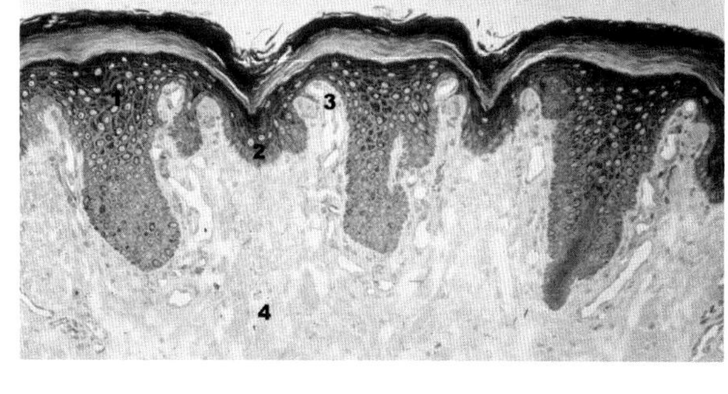

FIGURE 6. CROSS SECTION THROUGH THE SKIN OF A MONKEY FINGER TIP WITH GLANDULAR AND ADHESIVE RIDGES.
(1) glandular ridge, (2) adhesive ridge, (3) papilla of the papillary layer of the dermis, (4) reticular layer of the dermis. Semithin section, magnification × 1,500.

contrast, the nuclear chain fibres insert on the inner side of the capsule [10]. Intrafusal muscle fibres have more satellite cells than extrafusal fibres. Similar to the extrafusal muscle fibres, three different types of intrafusal fibres can be distinguished: type A – fast twitch fibres, type B – slow twitch fibres and type C – fast twitch fibres with small diameter.

Afferent axons entering the periaxial space near the equator of the spindle with diameters of 6–15 μm (Ia-fibres) or about 5 μm (group II-fibres) branch several times before forming characteristic nerve endings around the intrafusal fibres. Ia-fibres form primary nerve endings (also called anulospiral endings) near the equator of nuclear bag and nuclear chain fibres. In contrast, group II nerve fibres form secondary nerve endings above and below the equator of the muscle spindles mainly of nuclear chain fibres. These can be of anulospiral or flower spray type. In the polar regions of muscle spindles thin myelinated nerve fibres (group III) and non-myelinated nerve fibres (group IV) form free nerve endings. In addition to sensory innervation, muscle fibres are supplied by axons of motor neurons of Aβ- and Aγ-type.

Muscle spindles monitor the length or changes in length of muscles. For a detailed review see [11].

Mechanoreceptors of the connective tissue between skin and muscle fascia

Within the subcutaneous tissue between skin and muscles large Pacinian corpuscles can be found often arranged in groups of up to three corpuscles. Their structure is identical to those described in the joint capsules. In addition, Ruffini corpuscles are found in the reticular layer of the dermis. They have the same structure as described above for those in the fibrous layer of the joint capsules.

Cutaneous mechanoreceptors

Among the cutaneous mechanoreceptors, estimated to reach a total number of 17,000 per hand [12], one type – Merkel cell nerve endings – is found close to the surface of the skin within the epithelium, while the other types of mechanoreceptors are found in the dermis – either in the papillary layer close to the epidermis (Meissner corpuscles) or lower down in the reticular layer (Ruffini corpuscles). All types are adapted to the typical structure of the skin in fingers and palms which is ideally suited to deal with mechanical stimuli. A schematic drawing showing the loca-

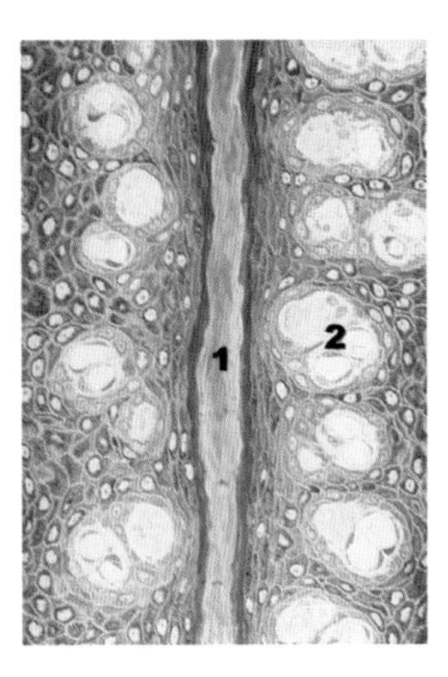

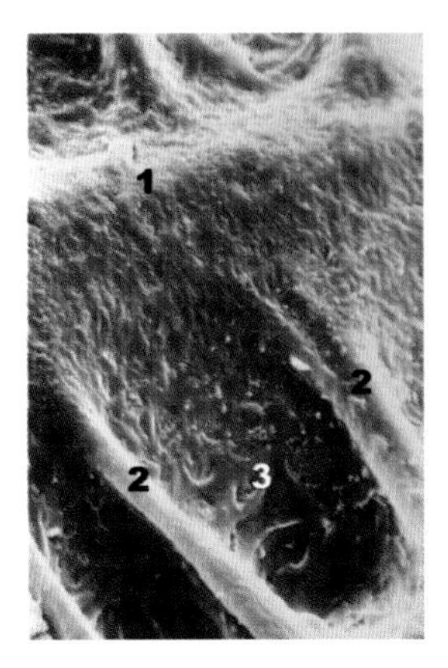

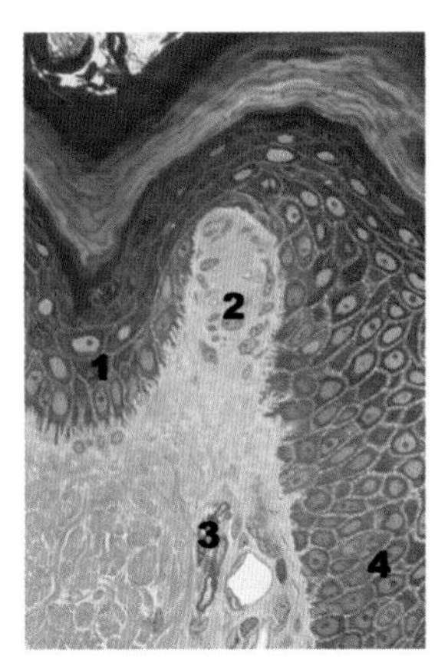

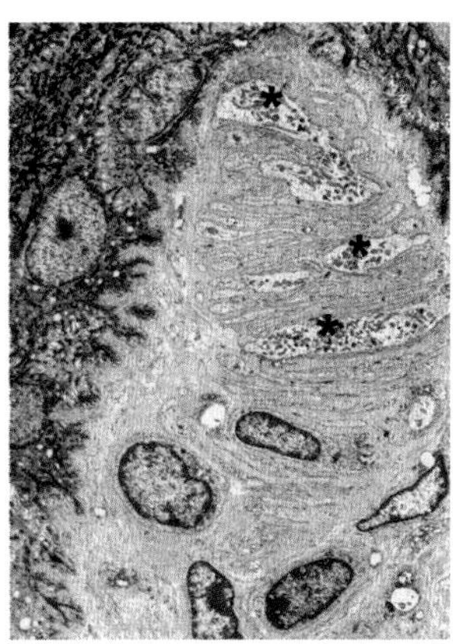

Figure 7

Left: Horizontal section through the skin of a monkey finger tip parallel to the surface. (1) superficial furrow between two glandular ridges. (2) dermal papillae of the papillary layer. Semithin section, magnification × 1,500.
Right: scanning electronmicrograph of the basal layer of the epidermis from below. (1) glandular ridge, (2) adhesive ridge, (3) epithelial crypt. Magnification × 1,600.

Figure 8 Cross sections through the skin of a monkey finger tip. Meissner corpuscles

Left: Light micrograph of a Meissner corpuscle. (1) adhesive ridge, (2) Meissner corpuscle within the papillary layer of the dermis, (3) myelinated axons of the Meissner corpuscle, (4) glandular ridge. Semithin section, magnification × 3,000.
*Right: Electron micrograph of a Meissner corpuscle. *nerve terminals with accumulation of mitochondria. Magnification × 5,000.*

tion and characteristic features of all types of cutaneous mechanoreceptors is shown in Figure 10.

The surface has skin lines visible with the naked eye. The basal side of the epidermis shows below the epithelial ridges glandular ridges through which sweat ducts pass to the surface (Figs 5, 6 , 7 and 10). Below the groves on the epithelial surface are adhesive ridges. They run parallel to the glandular ridges but are thinner and less deep. Perpendicular to both are cross ridges. In this way, the lower surface of the epidermis forms crypts between glandular and adhesive ridges and perpendicular running cross ridges (Fig. 7). These crypts contain the papillary layer of the dermis and Meissner corpuscles (Figs 8 and 10).

Thus, *Meissner corpuscles* are positioned just below the basement membrane in dermal papillae adjacent to adhesive ridges. On the other side, between Meissner corpuscles and the glandular ridges are capillary loops. In relation to the skin surface, Meissner corpuscles are found below the epidermal grooves and are thus more superficially located than the intraepidermal Merkel cell receptors at the bottom of the glandular ridges (Fig. 10). One square millimetre of skin can contain up to 24 Meissner corpuscles [13]. However, this number decreases with age to about 6 per mm^2 in 70–84 year old human subjects [14].

Meissner corpuscles are oval in shape. The longitudinal axis measures 100–150 µm and is perpendicular to the skin surface. The width of the corpuscles is in the range of 40–70 µm [15]. The corpuscle consists of terminals from myelinated nerve fibres separated by thin lamellae of terminal glial cells also called lamellar cells [16]. The proximal part of the corpuscle has a cup shaped perineural capsule. Collagen fibres of the dermal papilla run between the lamellae of the distal part without capsule and anchor in semi-

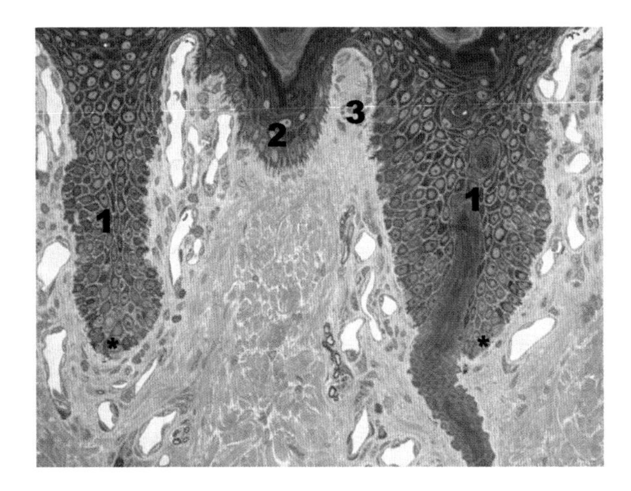

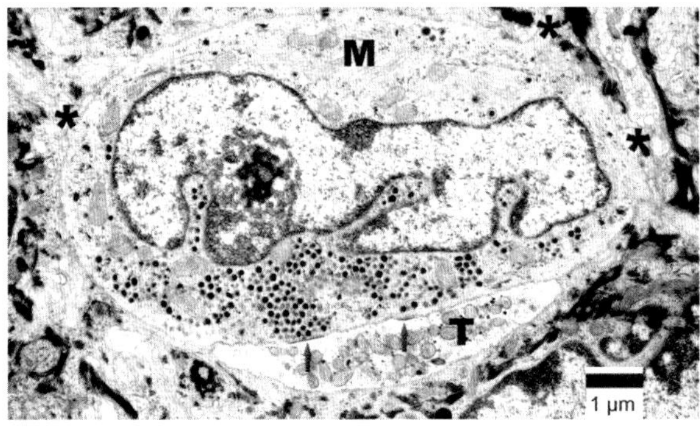

FIGURE 9. CROSS SECTIONS THROUGH THE SKIN OF A MONKEY FINGER TIP. MERKEL CELL NERVE ENDINGS
Above: Light micrograph of two glandular ridges (1) with Merkel cells (*) in the basal layer of the epithelium. (3) Meissner corpuscle. Semithin section, magnification × 1,500.
Below: Electron micrograph of a Merkel cell (M) with associated nerve terminal (T). Characteristic finger-like cytoplasmic processes are marked by asterix, synapse-like contacts between Merkel cell and nerve terminal are marked by arrows.

desmosomes of the epidermis. One corpuscle can be supplied by several myelinated axons of 3–5 μm. The axons loose their myelin sheath on entering the corpuscle and branch into discoid terminals containing large numbers of mitochondria and vesicles of about 80 nm in diameter.

Meissner corpuscles are known to respond to small changes (tens of μm) of indentation of the skin with short, rapidly adapting spike trains. In contrast to Pacinian corpuscles their optimum frequency is in the range of 20–30 Hz [17, 18].

Merkel nerve endings are the only mechanoreceptors found within the epithelium. Groups of up to 19 Merkel cells are positioned in the basal layer of the glandular ridges of the epidermis near the ducts of sweat glands (Figs 9 and 10). Merkel nerve endings consist of Merkel cells and discoid nerve terminals. In contrast to keratinocytes, Merkel cells are oval in shape and have spiny cytoplasmic protrusions anchoring between keratinocytes (Fig. 9). The nucleus is characteristically lobulated. In the cytoplasm facing the nerve terminal are typical osmiophilic (dense cored) granules of 80–120 nm in diameter (for review see [19]). Each group of Merkel cells is supplied by one myelinated nerve fibre (3–5 μm) from the superficial nerve plexus of the dermis. The axon loses its myelin sheath within the papillary layer of the dermis. Covered by a Schwann cell without forming a myelin sheath the axon approaches the basement membrane. Dichotome branchings of the axon are seen at the last Ranvier node and at the epidermo-dermal border. Within the epidermis the axon is 'naked' and branches further finally forming discoid terminals with accumulations of mitochondria and empty vesicles (20–40 nm) at the basal side of Merkel cells. Synapse-like connections can be seen between Merkel cells and nerve terminals.

Merkel nerve endings respond to indentation of the skin with long lasting, slowly adapting trains of action potentials [20]. There has been a long controversy whether the Merkel cell is involved in the mechano-electric transduction process or not (for review see [19]). Recent experimental data suggest a glutamatergic synaptic transmission between Merkel cell and nerve terminal [21–24]. Thus, it appears likely, that a dual transduction mechanism may exist with the Merkel cell responsible for the slowly adapting responses [19, 25, 26].

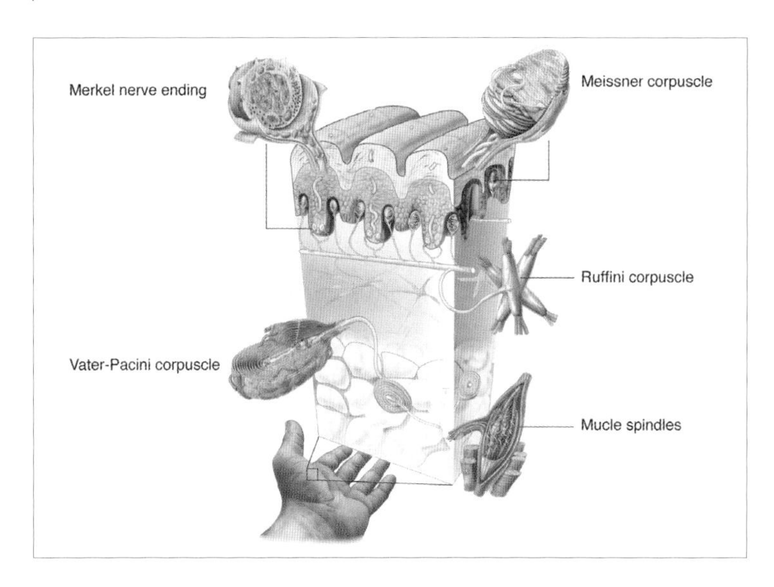

Figure 10. Schematic drawing of mechanoreceptors in the skin and muscles of the hand
In the epidermis, Merkel nerve endings are located in the basal layer of glandular ridges. Meissner corpuscles can be found in the papillary layer of the dermis, while in the reticular layer Ruffini corpuscles are seen. Deep in the subcutis are large Vater-Pacini corpuscles and in the muscles (especially lumbrical muscles) muscle spindles are found between 'extrafusal'

Summary

The human hand serves important functions of grip and touch. Numerous specialised nerve endings are found in the skin, subcutaneous tissues, muscles, ligaments and joint capsules. Some of them form corpuscular mechanoreceptors specialised in monitoring pressure or changes in pressure exerted on the skin surface (Merkel and Meissner corpuscles), stretching of the dermis or ligaments and joint capsules (Ruffini corpuscles) and vibrations transmitted from tools to the hand (Pacinian corpuscles). In addition the length and tension of muscles is controlled *via* information provided by muscle spindles and Golgi tendon organs. A vast number of free nerve endings are distributed throughout the skin and locomotion apparatus serving as polymodal nociceptors responding to various potentially harmful stimuli and partly also to changes in temperature.

Selected readings

Gescheider GA, Bolanowski SJ, Hall KL, Hoffman KE, Verrillo RT (1994) The effects of aging on information-processing channels in the sense of touch. 1. Absolute sensitivity. *Somatosens Mot Res* 11: 345–357

Gescheider GA, Thorpe JM, Goodarz J, Bolanowski SJ (1997) The effects of skin temperature on the detection and discrimination of tactile stimulation. *Somatosens Motor Res* 14: 181–188

Hämäläinen H, Järvilehto T (1981) Peripheral neural basis of tactile sensations in man: I. effect of frequency and probe area on sensations elicited by single mechanical pulses on hairy and glabrous skin of the hand. *Brain Res* 219: 1–12

Johansson RS, Vallbo AB (1979) Tactile sensibility in the human hand: relative and absolute densities of four types of mechanoreceptive units in glabrous skin. *J Physiol* 286: 283–300

Johansson RS, Vallbo AB, Westling G (1980) Thresholds of mechanosensitive afferents in the human hand as measured with von Frey hairs. *Brain Res* 184: 343–351

Johansson RS, Vallbo AB (1980) Spatial properties of the population of mechanoreceptive units in the glabrous skin of the human hand. *Brain Res* 184: 353–366

Johansson RS, Vallbo ÅB (1983) Tactile sensory coding in the glabrous skin of the human hand. *TiNS* 6: 27–32

Vallbo AB, Johansson RS (1984) Properties of cutaneous mechanoreceptors in the human hand related to touch sensation. *Hum Neurobiol* 3: 3–14

Physiological mechanisms of the receptor system

7

Antony W. Goodwin and Heather E. Wheat

Introduction

When we manipulate objects in our environment, a vast array of receptors in the skin, joints and muscles is activated. This information is relayed to the central nervous system and underlies two distinct but complementary aspects of hand function. Most obviously, these neural signals lead to haptic perception. We may sense how rough or smooth a surface is, or how curved an object is, whether it is soft or hard, whether the surface is slippery or sticky, how heavy it is and so on. Less obvious, but equally important, is the use the motor control system makes of these sensory signals in order to ensure appropriate hand movements resulting in stable grasps and effective complex manipulations. Some examples of common manipulations in our daily lives are: lifting a cup of coffee, opening a door, getting dressed, typing a manuscript, threading a needle.

Key features of the hand

The skeleton of the hand is made up of many bones (Fig. 1). Tendons inserting in each of the phalanges allow movements with many degrees of freedom, which is essential for the wide range of hand movements underlying the myriad complex manipulations we make. Some of the muscles from which the tendons originate are small muscles located in the hand, but most of the muscles are larger and are located some distance away in the forearm. Glabrous or hairless skin covers the front surface of the hand. This has a ridged appearance which is particularly striking on the fingertips giving rise to the fingerprint patterns which are unique to each individual. The back surface of the hand is covered in hairy skin.

During hand movements, many muscles are activated, angles change at many joints, contact with objects occurs at multiple points, and skin stretches or relaxes over the hand. Thus, sensory signals are relayed by mechanoreceptors in the muscles, in the joints, in the hairy skin, and in the glabrous skin. These mechanoreceptors are the terminal points of fibres, principally large myelinated axons that form the sensory components of the peripheral nerves innervating the hand, which are the median, ulnar and radial nerves. The cell bodies of these axons are located in the dorsal root ganglia and the central branches enter the spinal cord *via* the dorsal roots.

Glabrous skin is particularly relevant because the fingertips are the regions that most frequently contact objects during manipulation [1]. The skin mechanics of the fingertip is highly complex. The central part of the fingerpad is relatively flat, but the sides and end are highly curved. When objects are grasped, deformation of the skin and subcutaneous tissue is constrained by the bone of the distal phalanx.

Innervation of glabrous skin

Afferent types

Glabrous skin is densely innervated by axons of the Aβ group; these are myelinated axons with diameters in the approximate range of 7–14 micrometers with conduction velocities in the range of 15–60 m s^{-1}. Single fibre responses have been recorded in humans, using the technique

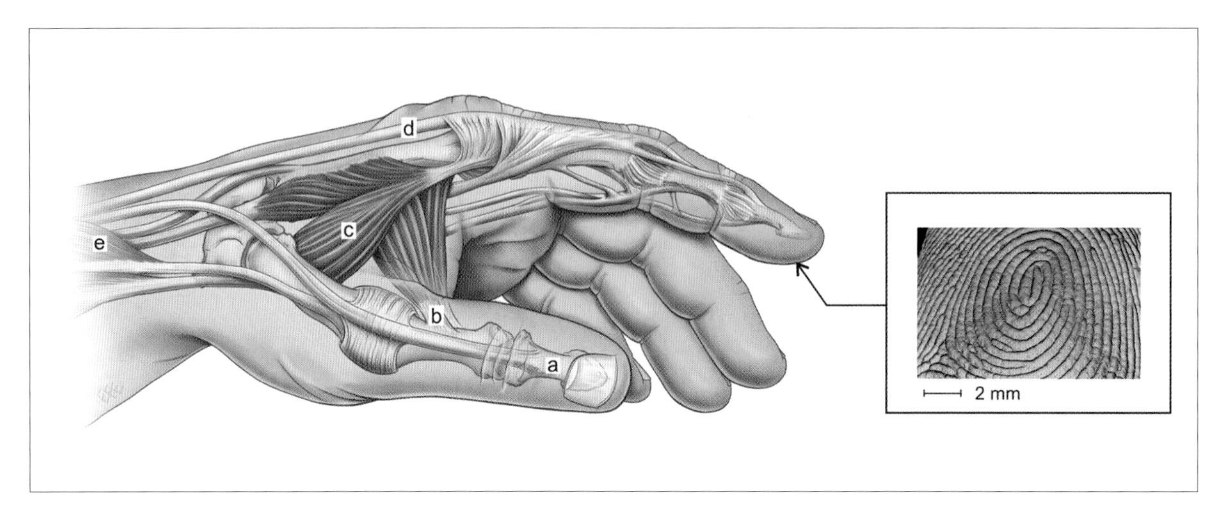

FIGURE 1. KEY FEATURES OF THE HAND
Phalanges (a) are moved by tendons (b) inserted in them and originating from muscles in the hand (c), and by tendons (d) from muscles in the forearm (e). Inset shows scanning electron micrograph of glabrous skin with its characteristic fingerprint patterns. Hand drawn by Levent Efe, CMI.

of microneurography, and in monkeys and other experimental animals, using fibre splitting (see Box). These primary afferent fibres form four distinct groups which are readily classified by their responses to indenting probes [2]. Two classes respond during the dynamic phase, when the probe is advancing into the skin or when it is retracting from the skin, and also during the static phase, when a steady indentation of the probe is maintained. They are termed slowly-adapting (SA) afferents (Fig. 2). The remaining two classes respond only during the dynamic phase and are termed fast-adapting (FA) afferents.

The afferents also differ in the spatial characteristics of their receptive fields (the region of skin from which the afferent can be activated). For type I afferents the receptive fields are confined to a small region of skin with sharp borders of rapidly increasing threshold. There are multiple zones of maximum sensitivity within the field. In contrast, the type II afferents have larger receptive fields with less well-defined borders of gradually increasing threshold and devoid of multiple zones of maximum sensitivity. The four afferent classes correspond to the four types of mechanoreceptors present in glabrous skin (see Chapter by Halata). SAI and FAI afferents innervate Merkel complexes and Meissner corpuscles respectively and the multiple zones of maximum sensitivity probably result from a single fibre terminating on more than one receptor. SAII and FAII afferents terminate on Ruffini organs and Pacinian corpuscles respectively.

FAI and FAII afferents have different adaptation characteristics evident from their different responses to probes vibrating sinusoidally into and out of the skin. FAI afferents are most sensitive, or have lowest thresholds, around 30 Hz whereas FAII afferents are most sensitive around 300 Hz [3]. Adaptation results from mechanical filtering introduced by the highly structured Meissner and Pacinian corpuscles. There may also be a component of adaptation in the transduction process occurring at the axon terminals [4].

Innervation density

SAI and FAI afferents are not uniformly distributed over the glabrous skin of the hand [5].

Box 1. Recording from single afferent fibres

Responses of single primary afferent fibres from the hand have been recorded in humans and monkeys. In humans, the technique of microneurography is used in awake subjects. A sharp microelectrode, usually tungsten insulated with varnish, is inserted through the skin into the median, ulnar or radial nerve. Action potentials from single axons can be recorded while receptors from the hand are activated by the experimenter. By passing a small current through the electrode, a single or small number of axons can be stimulated and the subject can report the resulting sensation. Anaesthetised monkeys are the experimental animal model of choice because only primates have hands. A small bundle of fibres is detached from the peripheral nerve and is successively split, or teased, until a micro-bundle with a single active fibre remains and can be placed on a hook electrode to record the action potentials. One advantage of this form of recording is the high stability, enabling single axons to be recorded from routinely for many hours. A second advantage of anaesthetised monkey experiments is that a completely immobilised finger allows precisely controlled stimuli to be applied.

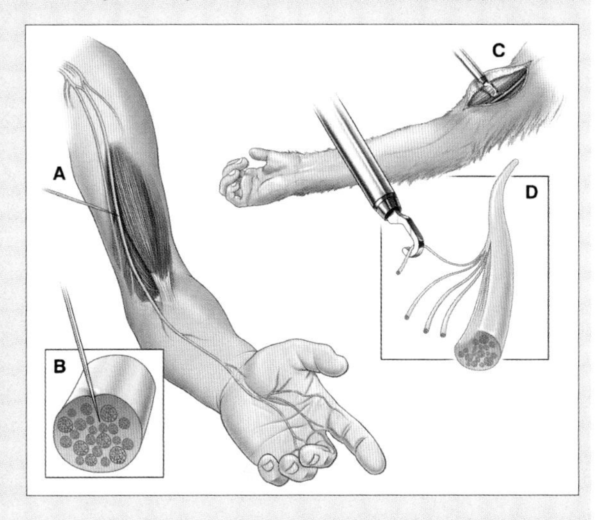

Figure Box 1. Recording from single sensory fibres in peripheral nerves.
A, B. Microneurography in an awake human. A microelectrode is inserted through the skin into the peripheral nerve. C, D. Fibre splitting in an anaesthetised monkey. A small bundle of fibres, separated from the nerve, is successively split. The final micro-bundle is draped over a hook electrode. Illustration by Levent Efre, CMI.

Innervation density increases in a distal direction, reaching a peak at the fingertips of 0.7 and 1.4 mm^{-2} for SAI and FAI afferents, respectively. This high innervation density enables the type I afferents to relay high resolution spatial information about handled objects. For type II afferents, innervation density is relatively uniform over the glabrous skin with a density of 0.09 and 0.22 mm^{-2} for the SAII and FAII afferents, respectively. This relatively low density precludes the type II afferents from playing a significant role in high spatial resolution processing.

Information relayed by afferents

During manipulation, a wealth of information about the object being handled and about the parameters of the hand movements is relayed by the mechanoreceptive afferents from glabrous skin. Psychophysics experiments in which objects are applied to the immobilised fingers of human subjects show that the glabrous skin on the fingertips provides information about the shape of the object, its position on the skin, the texture of the surface, the soft-

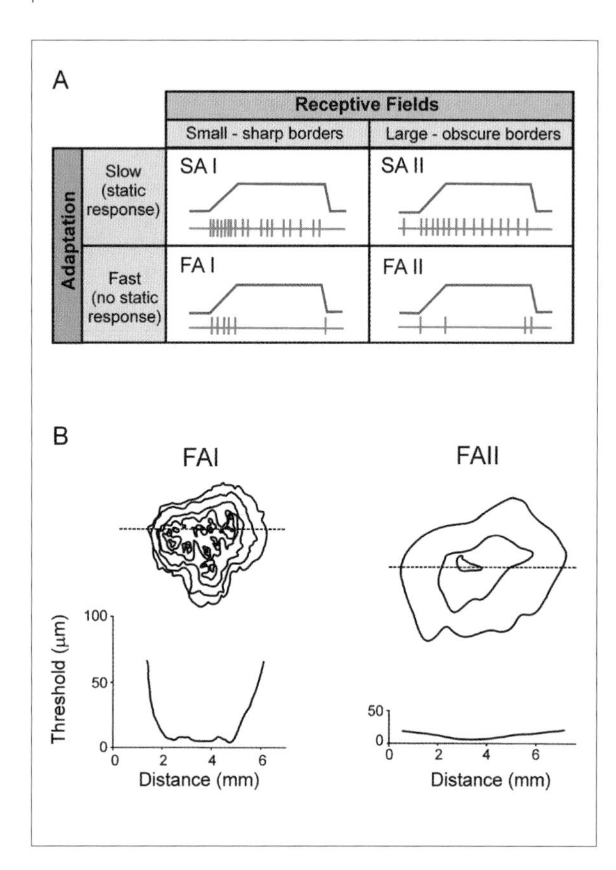

FIGURE 2. CLASSIFICATION OF AFFERENTS FROM GLABROUS SKIN

A. Table showing the four classes grouped by their temporal adaptation properties and by their spatial receptive field characteristics. Green lines show temporal profiles of the indenting probe and orange ticks indicate action potentials. B. Lower curves contrast the sharp and gentle borders of a type I and type II afferent respectively. Upper contour plots contrast the multiple zones of sensitivity of type I afferents with the uniform field of type II afferents. Modified from [2].

ness of the object, its degree of slipperiness, and the three-dimensional forces between the fingers and the object [6–8]. When fibres innervating the fingertips are blocked by injection of local anaesthetic, the person looses sensation and becomes incapable of the manipulations we take for granted [9].

Single fibre responses

Handling an object results in stresses and strains at the mechanoreceptors in the skin. These lead to receptor potentials, which in turn cause action potentials to propagate along the primary afferent axons. Information about the objects handled and about the hand movements is contained in the characteristics of the streams of action potentials reaching the spinal cord. Therefore, the first step in understanding tactile neural processing is an appreciation of the effect of each of the parameters of the object and the manipulation on the responses of individual afferents.

Force

When an object is grasped, three-dimensional forces are exerted on the skin. These may be resolved into a component at right angles to the skin, sometimes termed the grip force, and a component tangential to the skin, termed the load force. The effect of grip force can be ascertained by indenting an object into the skin, at the receptive field centre, which results in either a slowly or a fast adapting response (Fig. 2). Increasing the force of indentation results in an increase in the fibre's response. For the SAI afferent illustrated in Figure 3A, both the static and the dynamic components of response increase. A similar result holds for increasing the depth of indentation. Obviously increasing indentation depth results in an increase in force, but the relationship between depth and force is non-linear [10]. It has been appreciated for some time that fibres with receptive fields directly under the contacted object have force-dependent responses. Less well appreciated is the fact that when a fingerpad contacts an object, fibres with receptive fields all over the glabrous skin of the terminal segment respond with force-dependent responses, even though their receptive fields are remote from the stimulus (Fig. 3B).

Changes in load force, tangential to the skin, also affect afferent responses [11]. Tangential force may be applied in any direction in the plane tangential to the skin. In Figure 3C, the responses of an SAII afferent to changing directions of

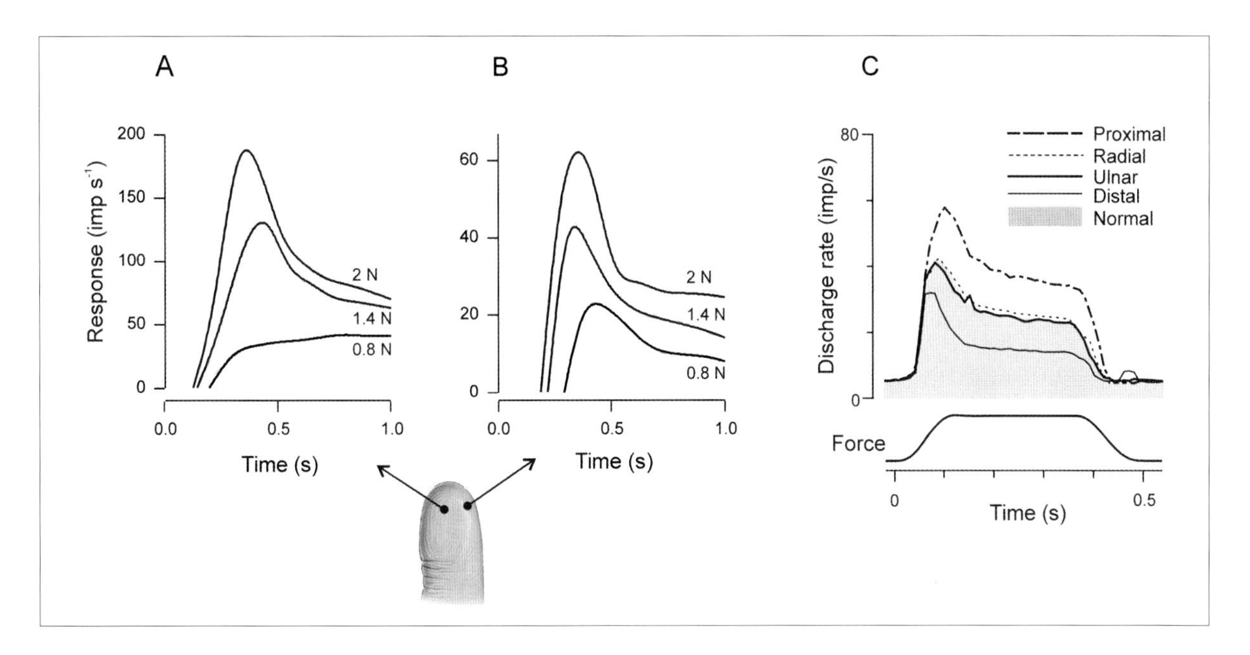

Figure 3. Effect of force on single afferent responses

A. Increasing grip force results in an increased response for this SAI afferent with a receptive field on the centre of the fingerpad. B. The response of an SAI afferent, with a receptive field on the side of the digit, increases with increasing grip force. C. Changing the direction of the three-dimensional contact force changes the response of the SAII afferent. A & B authors' data. C modified from [11].

force are illustrated. For this unit, the response is greatest when there is a normal force plus a tangential force in the proximal direction and is least when there is a normal force plus a tangential force in the distal direction.

Shape

The objects we handle vary in shape and size. Cutaneous primary afferent responses are affected by the local shape of the object in contact with the skin. Afferent responses have been characterised for a range of shapes either scanned over the receptive field or indented into the receptive field. Local shape is conveniently expressed in terms of the local curvature of the object; the curvature at any point is the reciprocal of the radius of curvature at that point. In Figure 4A, a wavy surface of increasing curvature scanned across the receptive field of an SAI afferent results in increasingly modulated responses of the afferent [12]. For a sphere of increasing curvature (decreasing radius) indented into the skin (Fig. 4B) there is a monotonic relationship between response and curvature [13]. For scanned surfaces, curvature affects both SAI and FAI responses. For indented surfaces, the effect of curvature on SAI responses is marked but is minor for FAI afferents. An extreme manifestation of curvature sensitivity is the observation that SAI afferents respond vigorously to edges and much less so to flat surfaces [14]. Edge enhancement is much less marked in FAI afferent responses.

Position

Sometimes the term cutaneous receptive field is used in a misleading way, with the implication

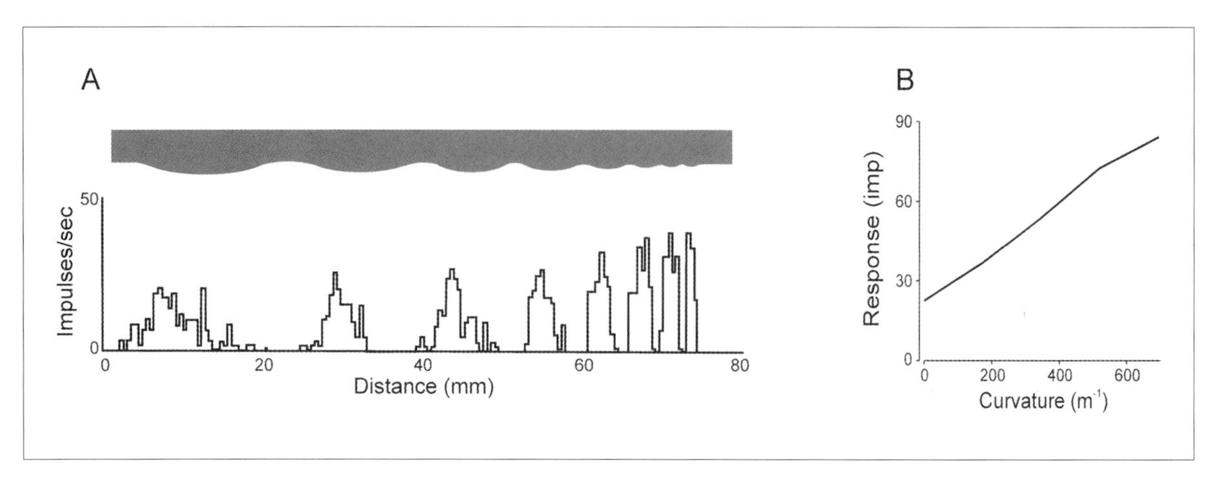

FIGURE 4. SINGLE SAI AFFERENT RESPONSES INCREASE AS THE CURVATURE OF THE OBJECT INCREASES
A. Wavy surface of increasing curvature scanned over the fingerpad. B. Spheres of increasing curvature (decreasing radius) indented into the fingerpad. A. modified from [12]. B. modified from [13].

that the afferent responds equally to any stimulus within the receptive field and does not respond to stimuli outside the receptive field. However, receptive fields are not uniform in structure and the afferent's response varies with the position of the object in the receptive field. In Figure 5A a sphere of curvature 521 m^{-1} (radius 1.9 mm) was stepped along a line through the centre of the receptive field of an FAI afferent. Response is greatest when the initial contact point is at the centre of the receptive field, and the response decreases symmetrically in a graded fashion as the contact point moves away from the receptive field centre. SAI afferents have a similarly shaped receptive field profile (Fig. 5B). For the type II afferents (not illustrated in this figure), the receptive fields are broader and less well defined, which is consistent with a role other than high spatial resolution tactile processing.

Slip

When we grasp and lift an object, the grip forces and load forces used have to be appropriate for the weight of the object and for the frictional conditions between the skin and the object. In particular, if the grip force is insufficient, the object will slip. Any slips or micro-slips during manipulation, including at the commencement of the manipulation, can be detected by the cutaneous mechanoreceptors, allowing the motor control system to adjust forces and attain a successful manipulation. SAI, FAI and FAII afferents respond to slip [15]. For smooth surfaces, or surfaces with limited features or texture, the FAII afferents are the principal detectors of slip or micro-slip [16].

Elementary population responses

During manipulations, a wide range of tactile information must be relayed to the central nervous system, and yet there are only four classes of cutaneous mechanoreceptors in glabrous skin. An alternative view of the same problem is that responses of individual mechanoreceptive afferents are affected by many different stimulus parameters. Thus, responses of single afferent fibres are ambiguous. An example is shown in Figure 5B. Changing either the curvature of the object or its position in the receptive field chang-

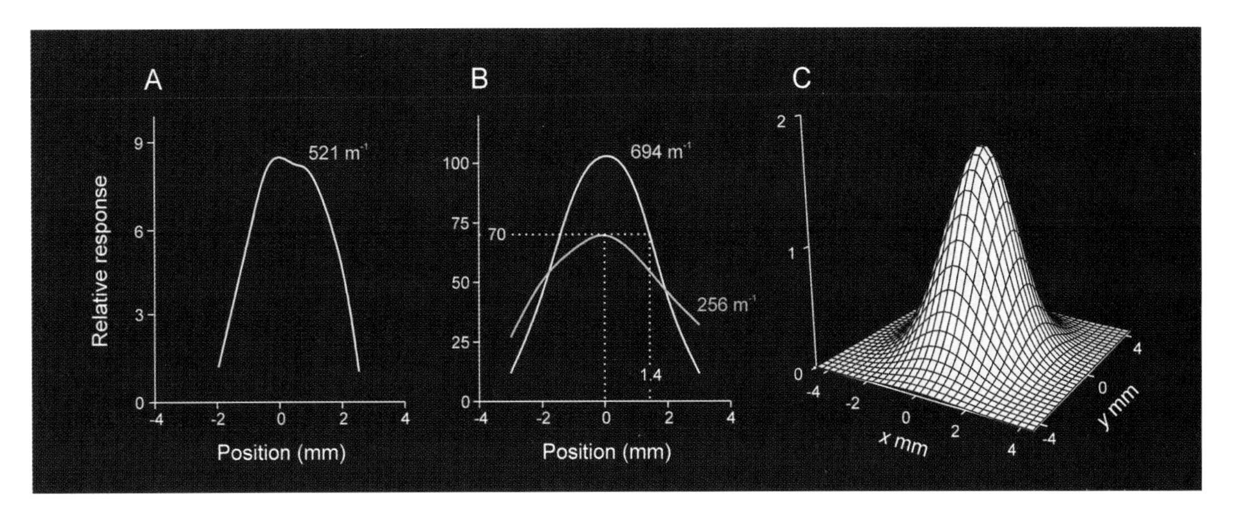

FIGURE 5. RECEPTIVE FIELD PROFILES OF SINGLE AFFERENTS TO SPHERES INDENTING THE SKIN
Response is plotted as a function of the position of the sphere in the receptive field. A. Response of an FAI afferent to a sphere of curvature 521 m^{-1}. B. Responses of an SAI afferent to two different spheres with curvature 256 and 694 m^{-1} respectively. C. Two-dimensional receptive field profile for a sphere of curvature 694 m^{-1}, which can also be interpreted as a simplified population response. Modified from [13].

es the response of the afferent. A response of 70 could have resulted from a sphere of curvature 256 m^{-1} at position 0 mm in the receptive field or from a sphere of curvature 694 m^{-1} at position 1.4 mm or –1.4 mm.

This ambiguity is readily resolved by considering, together, the responses of all the active afferents in the population. An elementary population response follows directly from the receptive field profile of an individual afferent, which depicts the afferent's response to the stimulus located at different positions in the receptive field. An alternative interpretation of the curves in Figure 5B is that they show the profile of responses of a population of SAI afferents, with the afferents' receptive field centres located at increasing distances from the point of contact of the stimulus on the skin.

The yellow curve in Figure 5B is the receptive field profile, for a sphere of curvature 694 m^{-1}, along one axis through the receptive field, but obviously the picture is similar in all directions through the two-dimensional receptive field. The complete two-dimensional receptive field profile is shown in Figure 5C. This can be interpreted as the two-dimensional population response to a sphere of curvature 694 m^{-1} contacting the skin at position $x = 0$, $y = 0$. Responses are depicted for afferents with receptive field centres distributed in both x and y dimensions on the skin in the region of contact with the sphere.

Inspection of Figure 5B and C suggests how the ambiguity about individual stimulus parameters is eliminated when responses of the population are considered. Changing the curvature (shape) of the sphere will change the shape of the population response in a corresponding way; more curved spheres will result in higher and narrower (more highly curved) population responses. Shifting the position of contact on the skin will result in a corresponding shift of the whole response. An increase in contact force will result in a corresponding upward scaling of the entire response. Such changes in shape, position or scaling of the population response can be determined independently of each other,

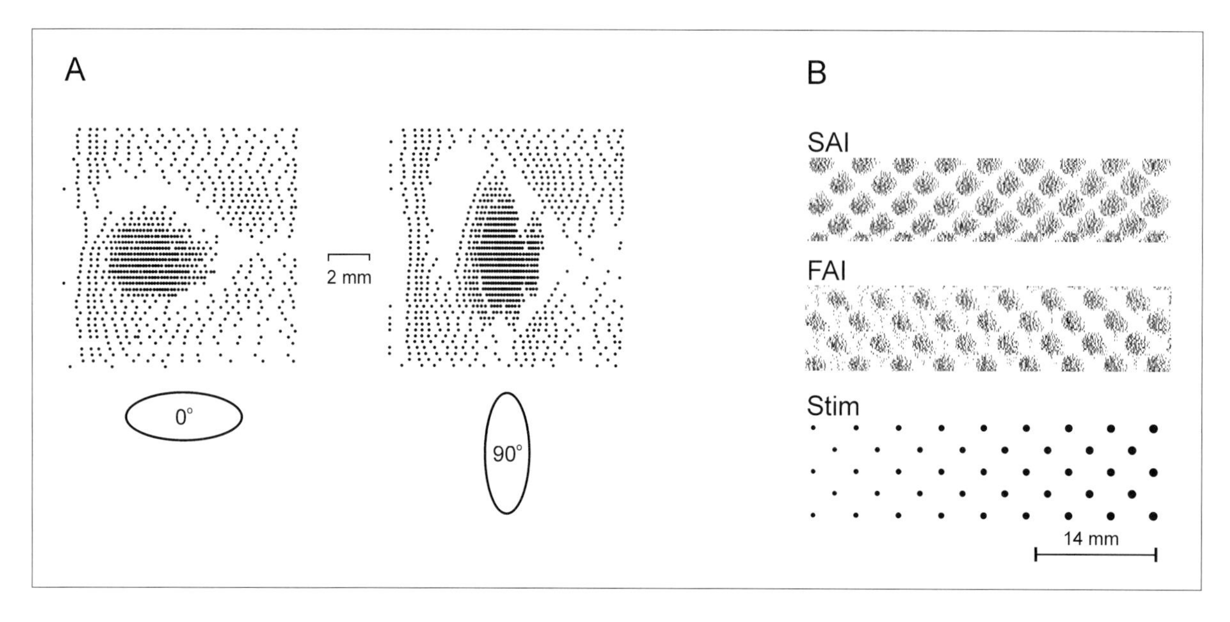

Figure 6. Responses of single fibres to stimuli scanned across the receptive field displayed as spatial event plots.

A. An SAI afferent responding to a 1×5 mm torroid scanned at an orientation of either 0° or 90°. B. Responses in a single SAI afferent (top) and a single FAI afferent (middle) are isomorphic with the pattern of raised dots in the stimulus (bottom). A. modified from [18]. B. Modified from [19].

resulting in independent signalling about these parameter values to the central nervous system. More formal mathematical analysis verifies these observations [17].

Moving stimuli

In some tasks, like lifting a cup of coffee, the object is grasped and indents the skin without lateral or tangential motion between the skin and the object. In other tasks, such as identifying an object by touch, the hand is moved over the object and there is tangential motion across the skin. To quantify the responses of primary afferents under these conditions, stimulators are used to scan objects over the fingertips, usually at constant velocity.

In a common approach in such experiments, the object is scanned over the skin in a straight line, then shifted a small amount in a direction at right angles to the scanning direction, and then again scanned over the skin. By repeating this procedure a number of times, the occurrence of action potentials in the afferent can be displayed as a function of the two-dimensional position of the object in the receptive field. These plots, usually called spatial event plots, are equivalent to the two-dimensional receptive field profiles described above for an afferent responding to an indenting stimulus. Alternatively, they can be interpreted as population responses depicting the responses of a two-dimensional population of afferents innervating the skin over which the object is scanned. At a qualitative level, it can be 'seen' how moving stimuli are represented in such elementary population responses. For example, when a torroid is scanned over the skin (Fig. 6A) changing its orientation will change the orientation evident in the plot and changing the shape of the ellipsoid will change the shape of the spatial event plot [18].

Textured surfaces

Determination of surface texture is a quintessential tactile task; commonly we rub our fingers over an object to determine how rough or smooth the surface is. There have been many studies of primary afferent responses to precisely manufactured textured surfaces, usually either gratings of alternating grooves and ridges or patterns of raised dots. In many cases, spatial features of the surface are evident in the spatial event plots of single afferents, which are interpreted as simplified population responses. In Figure 6B, the spatial event plots of an SAI afferent and an FAI afferent both reflect the pattern of dots in the surface scanned over the fingertip.

Human perception of the roughness of these experimental surfaces can be correlated with measures extracted from the spatial event plots. Roughness correlates closely with the spatial variation in SAI afferent responses and less well with FAI afferent responses [19].

Realistic population responses

The simplified population responses described above are useful for indicating how stimulus parameters might be represented in neural responses, but they are also, in some ways, misleading. In both Figures 5C and 6, there is an implicit assumption of infinite innervation density. In reality, those plots are sampled by a finite number of afferents, which have a spacing between receptive field centres of about 1.2 mm and 0.85 mm for thc SAI and FAI afferents, respectively. This separation is large compared to the scale of those plots. Thus, the representations are relatively sparsely sampled with a significant reduction in resolution. Moreover, it is not known to what extent innervation density over the fingertip is uniform. In addition, with age and with many common neuropathies there is a reduction in innervation and the remaining fibres may have a patchy distribution further compromising the stimulus representation.

A second inherent assumption in the ideal population responses is that all afferents have the same sensitivity. However, recordings show that the sensitivities of individual afferents within a class vary widely. For an indenting probe, the most sensitive SAI afferents are about ten times as responsive as the least sensitive SAI afferents [13]. Such variation distorts ideal plots like those shown in Figures 5 and 6.

The third factor that influences real population responses is the variability of neural responses or neural noise. This variability is superimposed on the ideal responses.

To form a realistic understanding of how the peripheral nerves send signals to the brain about multiple stimulus parameters, it is necessary to model or simulate real population responses that are noisy, distorted and sparsely sampled versions of the ideal population responses. Computer simulations are an effective way of generating more realistic population responses [6].

Innervation of hairy skin

Mechanoreceptors in the hairy skin covering the back of the hand fall into two groups: those associated with hair follicles and those located in skin between follicles. Merkel and Ruffini endings in the skin give rise to SAI and SAII afferents respectively, with properties similar to those in glabrous skin. Afferents from Pacinian corpuscles in the skin and subcutaneous tissue also have properties like their glabrous counterparts. In addition, there are rapidly adapting afferents from the skin, termed field units, and rapidly adapting afferents from the follicles.

When the hand moves during manipulation, there are distinct patterns of stretch and relaxation in the hairy skin, leading to activity of the mechanoreceptors. Movement of the joints activates rapidly adapting afferents which signal the occurrence of a joint movement. Slowly adapting afferents display both a static and a dynamic response with the potential to signal both joint position and movement. Recordings from single afferents, combined with psychophysics experiments, suggest that skin on the dorsum of the hand provides accurate proprioceptive information during hand movements [20, 21].

Muscle and joint receptors

Information about hand movements can potentially be provided by receptors in the many joints of the hand [22]. Most joint afferents only fire at the extremes of joint angle and the extent of their contribution to proprioception is uncertain. Both the extrinsic and intrinsic hand muscles (Fig. 1) contain muscle spindles which are sensitive to the length and velocity of the muscle and thus to the position and velocity of the joints. Afferent recordings combined with psychophysics have demonstrated the potential for muscle spindles to represent, with precision, hand movements [22].

How proprioceptive information from cutaneous afferents, muscle spindles, and joint afferents are combined is not known. Nor is it understood how proprioceptive information is combined with other cutaneous information (described above) about the properties of the object being manipulated and the manipulation itself.

Summary

When we manipulate an object, mechanoreceptors in the skin, joints and muscles are activated. Action potentials in the primary afferent fibres from these mechanoreceptors convey information about the object and about the manipulation itself to the brain. The glabrous skin of the fingertips is richly innervated. These afferents convey information about the shape and texture of the object, about the position of contact on the skin, about the forces between the digits and the object, and about the presence or absence of slip. Responses of individual afferents are affected by multiple parameters of the stimulus so that information relayed by them is ambiguous. However, when whole populations of cutaneous afferents are considered, it is clear how precise and unambiguous information about multiple parameters can be conveyed to the central nervous system. Mechanoreceptors in hairy skin, joints and muscles relay proprioceptive information to the brain. Ultimately, the range of sensory signals is integrated and forms the basis of tactile perception as well as contributing to the sensorimotor control of hand function.

Selected Readings

Darian-Smith I (1984) The sense of touch: performance and peripheral neural processes. In: JM Brookhart, VB Mountcastle, I Darian-Smith, SR Geiger (eds) *Handbook of Physiology – The Nervous System III.* American Physiological Society, Bethesda

Gandevia SC, McCloskey DI, Burke D (1992) Kinaesthetic signals and muscle contraction. T*rends Neurosci* 15: 62–65

Goodwin AW, Wheat HE (2004) Sensory signals in neural populations underlying tactile perception and manipulation. *Ann Rev Neurosci* 27: 53–77

Johnson KO, Hsiao SS (1992) Neural mechanisms of tactual form and texture perception. Ann *Rev Neurosci* 15: 227–250

Vallbo AB, Johansson RS (1984) Properties of cutaneous mechanoreceptors in the human hand related to touch sensation. *Hum Neurobiol* 3: 3–14

Neural basis of haptic perception

Steven Hsiao and Jeffrey Yau

Introduction

A major challenge in neuroscience is to understand the neural basis of behavior. The problem is multifaceted. First one must understand which afferent type(s) and cortical pathways are involved in the aspect of perception that you want to understand. Then one must understand how information is represented and coded in the neural responses. The earliest attempts at addressing the neural pathways underlying perception relied on lesion studies in which animals were trained to perform specific behavioral tasks and then were retested following the ablations. If the animal could not perform the task then the area that was ablated was deemed essential for the behavior. More recently researchers use functional imaging techniques to address these questions. Understanding the neural codes underlying behavior has been elusive. The intellectual breakthrough came from studies that combined psychophysical studies in humans with neurophysiological studies in monkeys [1, 2]. In these pioneering studies, Mountcastle and his colleagues showed that there is a tight correspondence between human vibratory detection and the neural activity recorded in the peripheral afferents of non-human primates. These findings not only demonstrated that the neural mechanisms used by the two species are similar but more importantly provided a scientific approach for studying the neural mechanisms of behavior.

In Goodwin and Wheat's chapter they describe the physiological basis of the peripheral receptor systems that underlie cutaneous perception. However, that is only part of the story. In addition to the four kinds of mechanoreceptors there are also other receptors that provide information about sensory inputs from the hand. These include receptors specialized for pain (two kinds), temperature (two kinds), itch (one kind) and, as noted by Goodwin, four kinds of afferents that are located in the muscles, tendons and joints that provide information about body position, movement and force. Together these afferents provide a rich multidimensional percept of the size, shape, texture, and temperature of objects that we hold and manipulate with our hands. In this chapter we describe the neural basis of size, shape, and texture perception. We first layout the basic architecture of the somatosensory system and briefly describe the regions of the brain that are involved in haptic perception. Next, we describe neural coding studies of texture and two- and three dimensional form and discuss how these aspects of haptic perception are represented in the somatosensory system. Finally we propose a working hypothesis of how we believe mechanoreceptive and proprioceptive information is integrated to form central representations of three-dimensional shape.

Anatomical basis of haptic perception

As described in the previous chapter, the neural basis for haptic perception begins with the activation of peripheral afferents in the skin, muscles, and joints that provides the initial representation of the external world. There are 13 different kinds of afferent fibers each with specialized receptor endings that allow them to encode information about different sensory inputs from the hand. Eight of the 13 provide information that is impor-

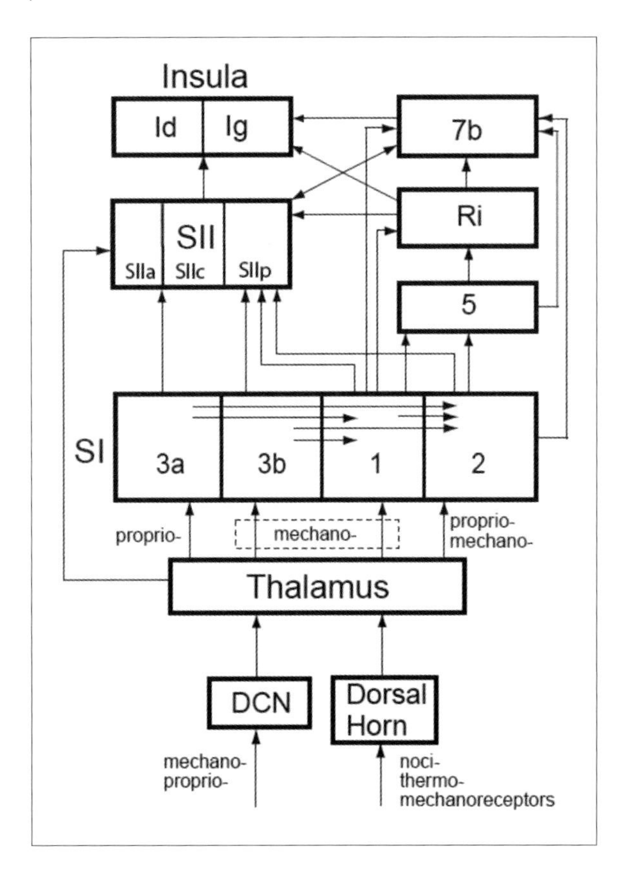

FIGURE 1. BLOCK DIAGRAM OF THE ANATOMICAL AREAS INVOLVED IN HAPTIC PERCEPTION

tant for haptic perception; four of these provide information about discriminative touch (slowly adapting type 1 – SA1, slowly adapting type 2 – SA2, rapidly adapting – RA, and Pacinian – PC) and four provide information about body position and movement (Muscle spindle types 1 and 2, golgi tendon organs, and joint receptors). During normal palpation, all afferent types (except perhaps the pain and itch receptors) are continuously active and provide a dynamic representation of the spatio-temporal profile of stimuli in contact with the skin along with a dynamic representation of the positions, movements, and forces of our limbs, digits and joints [3].

The large diameter A-beta axons initially carry sensory input from the mechanoreceptive and proprioceptive afferents that innervate the hand to the cortex. The ascending pathway from the hand to cortex comprises three synaptic stages (Fig. 1) that appear to function as relay stations. The cell bodies of the primary afferents, which reside in the dorsal root ganglia send their projections to neurons in the dorsal column nuclei (DCN). Projections from these second-order neurons in the DCN cross the midline at the medulla in a fiber tract called the medial lemniscus and synapse onto type I cells in the ventral posterior lateral (VPL) nucleus of the thalamus. Neurons in VPL then send their projections to neurons in Layers III and IV of primary somatosensory cortex (SI), located in the post-central gyrus on the parietal lobe and to neurons in second somatosensory cortex (SII), located on the upper bank of the lateral sulcus.

VPL is organized as a patchwork of rod like structures that are functionally segregated into a central core region containing neurons that are thought to have cutaneous SA1 like properties. Surrounding the central core is another cutaneous region with neurons that have larger receptive fields that are both slowly and rapidly adapting. Surrounding this region is an outer shell region that contains neurons that respond mainly to deep or proprioceptive input. VPL is the first processing stage where modality information becomes segregated along separate parallel pathways. There have been few neurophysiological studies of the coding properties of neurons in the DCN or VPL; however evidence suggests that there is little convergence or divergence of sensory input at these stages. For example, recently it has been shown that there appears to be a tight correspondence between the firing of neurons in the DCN and their cutaneous and proprioceptive afferent inputs [2, 4, 23]. Those studies show that there is a high efficacy of firing whereby single impulses evoked in peripheral afferent fibers appear to directly generate spikes in target DCN neurons. This suggests that afferent information is faithfully reproduced in the responses of DCN neurons. While similar findings have yet to be demonstrated in thalamic neurons, other studies suggest that neurons in the thalamus have small receptive fields, like peripheral afferent fibers [5], which suggest that there is also little convergence on neurons in VPL as well. This is not to say that nothing happens as information ascends from the periphery to the cortex. Other studies, particularly from the rodent have shown that neurons in these subcortical areas receive large feedback projections from cortex and receive inputs from other subcortical nuclei; these connections can potentially modulate the afferent inputs. Further studies are needed to fully understand how sensory inputs from the mechanoreceptive and proprioceptive afferents are processed in the DCN and VPL.

Neurons in VPL send their projections, albeit with varying strengths, to the four areas that comprise primary somatosensory cortex (3a, 3b, 1 and 2) and a small projection to SII cortex. Neurons in the core region of VPL send their main projections to area 3b, and to a lesser extent area 1, suggesting that these areas are involved in processing cutaneous input. Neurons in the shell region project to area 3a and area 2, which suggest that these areas play a prominent role in processing proprioceptive input. The four areas of SI are extensively connected. Neurons in area 3b send their projections to both area 1 and 2. Neurons in area 1 also send their projections to area 2. Neurons in area 3a send a large projection to area 2 and a small projection to area 1. Several observations about cortical function can be inferred from these connectivity patterns. First, SI does not simply process sensory input in parallel, but also shows an internal hierarchy with neurons in areas 3a and 3b having responses that are more closely linked to the proprioceptive and mechanoreceptive inputs, respectively, and with neurons in areas 1 and 2 being further down the pathway playing an integrative role. Of particular importance to haptic perception is the observation that neurons in area 2 receive both cutaneous and proprioceptive input, which supports the notion that area 2 plays an important role in stereognosis, or three-dimensional (3-D) form perception. Evidence that these areas play different roles in perception comes mainly from ablation studies. Animals with area 3b ablated have profound deficits in practically all aspects of haptic perception. This includes the ability to recognize two- and three-dimensional shape, texture, and motion. Animals with ablations of area 1 appear to show selective deficits in texture perception and ablations of area 2 result in specific deficits in their ability to recognize large three-dimensional shapes. The functional role of these areas has not been reproduced using modern fMRI techniques.

Haptic sensory information is processed beyond SI along two cortical pathways. The dorsal stream follows projections from SI to Brodmann's areas 5 and 7, situated in the posterior parietal cortex. Neurons in these areas appear to play a role in multisensory integration as well as sensory initiation and attention. In addition, these parietal regions interact with motor areas and are implicated in guiding motor tasks and providing information of how to perform different tasks. Recent imaging studies in humans suggest that an area in the posteromedial parietal sulcus (pIPS) is important for tactile spatial acuity. The ventral stream flows from the four areas of SI cortex to the secondary somatosensory cortex, which is located in the superior bank of the lateral sulcus (Silvian fissure). The composition and organization of SII has been contentious, with traditional descriptions positing the existence of one or two body maps [6]; more recent studies suggest that SII is composed of at least three distinct areas [7]. Two of these areas have responses that appear similar to responses seen in area 2 and are believed to be the next stage of processing after area 2 for integrating proprioceptive and cutaneous inputs. These areas flank a central field that appears to be concerned with cutaneous inputs. The representation of touch in the lateral sulcus in humans is even more elaborate with researchers identifying at least four distinct areas [8].

When SII is ablated animals have severe deficits in processing cutaneous and proprioceptive information – similar to the deficits seen when area 3b is ablated, suggesting that SII plays an important role in processing haptic information [9]. The areas that are important for processing haptic information beyond SII are poorly under-

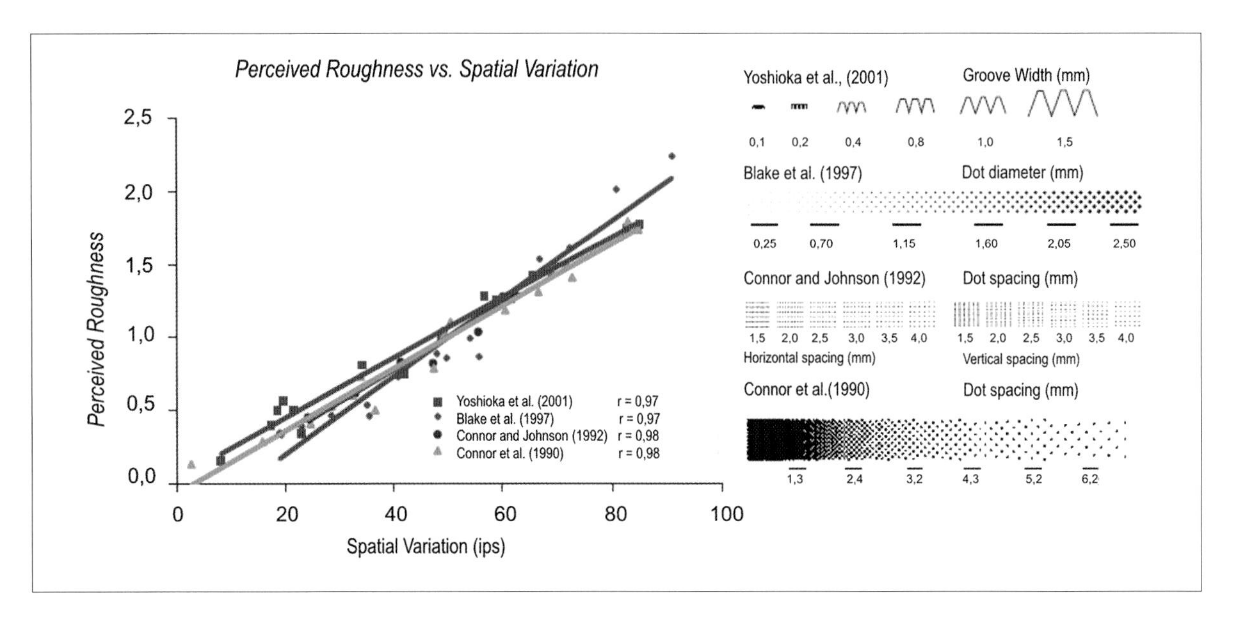

FIGURE 2. NEURAL CODING OF TACTILE ROUGHNESS

Plot on the left shows consistency of the hypothesis that roughness is coded by the spatial variation in firing rates among SA1 afferents. Plots on the right show examples from the four studies that correlated SA1 spatial variation with roughness perception in humans.

stood. Reciprical connections between SII and the frontal lobe exist suggesting a role for SII and medial premotor cortex in haptic decision making [10]. Furthermore, neurons in SII have been shown to project to regions in the insular cortex, which may be important for higher level form processing and may also have a role in the integration of discriminative touch with the affective component of haptics. The elaborate connection of the insula with regions in the temporal lobe most likely underlies the complex interactions between haptic perception and tactile memory. Finally recent studies have also shown visual cortical areas play a role in tactile shape processing but not in roughness perception [11, 12].

Neural basis of 2-D form and texture perception

Two-dimensional form perception (e.g., Braille reading) is invariant over a wide range of stimulus conditions. It is relatively unaffected by scanning speeds (up to about 80 mm/sec), contact force (in the range of 0.2–1.0 N), and whether patterns are passively scanned across the finger or if the finger actively scans the pattern. The threshold of spatial acuity on the fingertip is about 1.0 mm. This threshold has consistently been found using a wide range of tactile stimuli and corresponds to the ability of humans to distinguish two closely positioned point probes (called the two point limen), detect gaps in an otherwise continuous surface, discriminate the orientation of gratings, and identify features of complex spatial patterns such as embossed letters of the alphabet [13–15]. Remarkably, subjects show similar performance between touch and vision on a letter identification task when the embossed letters are scaled in size to match the density of SA1 afferent fibers and cones in the visual task [16]; these results suggest that the mechanisms of form processing in the two systems are similar [17]. The spatial system is

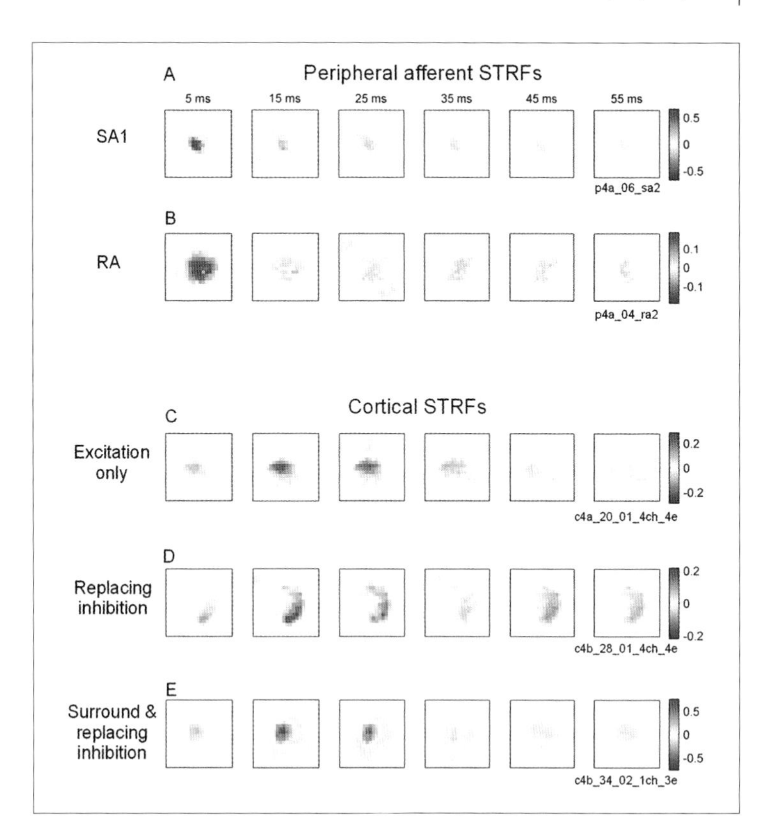

FIGURE 3. REPRESENTATIVE EXAMPLES OF SPATIOTEMPORAL RECEPTIVE FIELDS RECORDED FROM THE PERIPHERAL (SA1 AND RA) AND CORTICAL NEURONS IN AREA 3B *Red is excitation and blue is inhibition. Color code represents the relative intensity of firing.*

also extremely rapid, exemplified by the abilities of Braille readers who can read at rates up to 100 words per minute, which translates to reading rates faster than a character every 100 ms.

There is an accumulation of evidence to suggest that 2-D form is subserved by the SAI afferent system. As described in the previous chapter, these afferents are closely associated with the Merkel neurite complexes, which densely innervate the skin of the hand. As discussed in the previous chapter, the SA1 afferents send to the central nervous system, in the population response, an isomorphic representation of the 2-D spatial profiles of patterns on the skin. A comparison between the spatial acuity of subjects in behavioral tasks and the spatial acuity of the different mechanoreceptive afferents demonstrates that the perception of fine spatial form can only be conveyed by SA1 fibers. As the dimensions of the spatial patterns increase, the functional division between the SA1 and RA (also called fastly adapting or FA1) afferents becomes less obvious. The threshold for spatial acuity of the RA system is about five times greater than the corresponding threshold found for SAI afferents.

SAI afferents are also responsible for texture perception. In a series of psychophysical experiments of human subjects and neurophysiological experiments on monkeys, Johnson and Hsiao and their colleagues tested a wide range of hypothetical neural codes that could account for the psychophysical behavior [18]. These included codes based on mean firing rate, spatial and temporal variations in firing rate and modal codes that combined input from different afferent populations. The only code that consistently accounted for the human behavior was one based on the spatial variation in firing rates among SAI afferent fibers separated by about 2.0 mm. In the next section we will describe cortical mechanisms that compute spatial variation. Although SAI afferents have also been implicated in coding for hardness and softness, this hypothesis has not been explicitly tested in a combined psychophysical and neurophysiological study. The neural coding of stickiness is not known.

Cortical mechanisms of form and texture processing

As stated earlier, the first place where form and texture are processed in cortex is area 3b of SI cortex. In the periphery neurons have small receptive fields (RF), which underlie the isomorphic neural representation. In contrast neurons in cortex have larger and more complex RFs. Figure 2 shows a comparison of SA1 and RA afferent receptive fields, as well as sample receptive fields of a typical neuron in area 3b. While the receptive fields of the peripheral afferents are for the most part circular and simple, the receptive fields of neurons in area 3b of SI cortex are larger (2–3 times larger than first-order afferents, but still mostly restricted to a single fingerpad) and consist of excitatory and one or more inhibitory subregions. The change in RF structure is indicative of the transformation that the information is undergoing as it flows centrally. Area 3b neurons show a broad range of receptive field structures demonstrating that the representation of spatial form has, in this first cortical processing stage, shifted from activation of an array of point receptors to the activation of a set of feature selective neurons. For example, while none of the peripheral neurons showed orientation-tuned responses, a majority (about 70%) of area 3b neurons are selective for orientation [19]. Area 3b neurons typically have a center excitatory region flanked on one or more sides by an inhibitory field. This receptive field structure enables neurons to encode features like the orientation of bars (Fig. 3). The mechanisms then appear to be similar to the ones used by simple cells of the primary visual cortex (V1). In addition to the spatial component of the response, the RFs also have a temporal component which lags the spatial response. This delayed or replacing inhibition plays several functions. First the inhibition defines a period of integration which allows features to be integrated. Second it functions to prevent temporal smearing of the response and allows the neurons to show invariance to changes in scanning velocity.

What then is the relationship between these cortical receptive fields and form and texture processing? The answer to the second question is easy. Although the cortical neurons have a variety of RF structures, a subset of the neurons have spatial structures exactly like those predicted from the roughness coding study discussed earlier. Thus the mean firing rate of a subset of area 3b neurons accounts perfectly for human roughness perception. The answer to the first question is not as straightforward. While the RFs of area 3b neurons are sensitive to spatial gradients and show feature selective responses, it is still not known how complex spatial shapes are represented.

The receptive field properties of neurons in area 1 and area 2 are larger, more complex, and even less well understood. The classical receptive fields of area 1 neurons typically span one digit, but have been observed to span multiple fingers as well. These neurons function less like point detectors as compared to area 3b neurons and demonstrate more nonlinear processing of form. In this manner neurons in area 1 appear to convey a higher degree of feature selectivity for 2-D spatial form. Many neurons in area 1 also show sensitivity to motion direction suggesting that area 1 plays a role in motion processing. At the next cortical processing stage, area 2 neurons have even larger receptive fields, often spanning the entire hand. These neurons, which receive both cutaneous and proprioceptive input, show highly nonlinear responses to shape stimuli and show selective responses to large objects that vary in curvature. Whether these neurons are processing information related to motion, form and texture is not known. They appear to be selective for curvature and angles. Their responses to 2-D form are not well explained by the arrangement of their receptive field structures, as was the case in areas 3b and 1. Instead, these neurons may be representing spatial patterns in a more abstract, higher-order feature space involving the representation of large objects.

In the dorsal stream, neurons in areas 5 and 7 show complex responses that appear to be driven less actively by purely somatosensory stimuli. Area 5 neurons have large receptive fields that are often bilateral and are affected by the animal's behavior and may be important for guiding motor movements during grasping tasks. Neurons in area 7 also have complex response properties, often showing interactions with the visual system. These neurons do not underlie haptic perception, *per se*, but instead are related to mechanisms of focusing attention.

In the ventral stream, neurons in SII show complex responses to cutaneous and proprioceptive input. These neurons receive input from each of the areas in SI as well as thalamic input (from the ventroposterior inferior nucleus – VPI). They have large receptive fields that cover much of the contralateral hand and a large majority of SII neurons (as many as 90%) exhibit bilateral responses [7]. SII neurons are also selective for complex spatial patterns. More impressively, many SII neurons show responses that are spatially invariant on a single fingerpad (stimulus selectivity is consistent for patterns presented to multiple locations on the fingerpad) [20, 21]. Stimulus selectivity is also consistent across fingers and even across hands, in cases of neurons with bilateral receptive fields. Given the convergence of input from thalamus and all of the areas comprising SI cortex, it is highly likely that SII neurons plays a prominent role in processing 3-D shape information. These results in addition to other studies showing that neurons in SII cortex are affected by selective attention support the notion that SII cortex lies further along the processing pathway leading to haptic perception [22]. Little is known about how tactile shape information is processed in insular cortex and beyond.

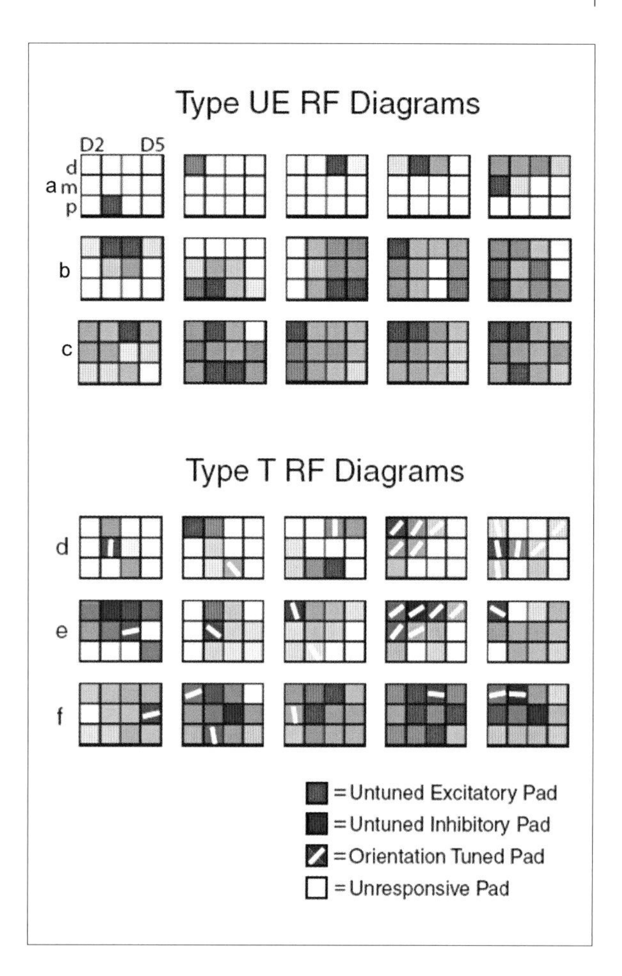

FIGURE 4. EXAMPLES OF RECEPTIVE FIELDS OF SII NEURONS
Each block represents the response from the proximal, middle and distal finger pads of digits 2,3,4 and 5. Red is a pad that was excitatory, blue – inhibitory. Pads with lines through them were tuned to the orientation of a bar indented into the center of the pad. Type UE were neurons with only untuned excitatory pads. Type T were neurons with one or more tuned pads.

Neural basis of body position and 3-D form

Humans have a large capacity to recognize objects without visual input. Blind subjects perform with an accuracy of 96%, with a mean response time of less than 5 s when asked to identify 100 common objects using their hands. Furthermore, accuracy increases as the number of digits that contact the object increases and is dependent on how the objects are explored. The manner in which we explore objects is very purposeful; we use stereotyped movements depending on the object features we are trying to perceive. There are eight different exploratory procedures that are used. Two of these procedures are used for 3-D shape perception. These include enclosure, where the hand is wrapped around objects with the fingers molded around its contours, and contour following, where the edges of the object are traced with the fingerpads. Both procedures require the integration of proprioceptive with mechanoreceptive input.

Much of the previous section described the mechanisms underlying the perception of cutaneous inputs from the mechanoreceptors; far less is known about the proprioceptive inputs and how they represent body position. As mentioned in the previous chapter, proprioception is a complicated sensory modality, the investigation of which has been rich and tumultuous. Most of our understanding of proprioception is limited to studies of the peripheral mechanisms that convey limb position. Historically, it was thought that the neural mechanism underlying the perception of joint angle and limb position relied on joint receptors. This logical candidate was called into question with the demonstration that the vast majority of joint receptors responded optimally at the extreme ranges of extension and flexion and that they responded minimally at intermediate angles. A subsequent study showed that anesthetizing joint receptors in the knee joint capsule did not impact subjects' abilities to judge and discriminate joint angle and patients who have had joint replacements have a clear sense of where there limbs are in space. Clearly another candidate receptor(s) was required to explain the human ability to perceive joint position.

Muscle spindles, embedded in muscle fibers, became a leading candidate when it was shown that the application of a vibratory stimulus to tendons of the biceps or triceps, thereby activating muscle spindles, altered estimates of perceived elbow angle by as much as 40º. Other studies have shown that vibratory stimulation of muscle spindles can also produce illusions of limb movement. For example, the application of the vibratory stimulus to the tendons of the triceps induces an illusion in which your arm is perceived to extend. This effect can have dramatic sensory results when combined with cutaneous input, as illustrated in the Pinocchio effect. In the Pinocchio effect, if you apply the vibratory stimulus to the tendons of your left arm while contacting your nose with a finger of your left hand, the resulting illusion is one in which you perceive that your nose grows.

Muscle spindles alone could not account for proprioception, however, as Gandevia and McCloskey cleverly demonstrated using the entrapment maneuver. This posture, achieved by extending all the fingers while flexing only the middle digit, serves to functionally detach the extensor and flexor muscles from the distal phalanx. Subjects who maintain their hands in this position retain their ability to judge static joint position. This suggested a role for cutaneous mechanoreceptors in the skin. All classes of mechanoreceptors are responsive to skin stretch produced during joint movement. SAII afferents have been proposed as the best candidate receptor for encoding joint position from skin stretch. Studies have shown that it is possible to induce an illusion of movement simply by stretching the skin near finger joints.

It has long been known that the proprioceptive sense is not determined purely by signals from peripheral receptors. Instead, position sense can be derived, or at least influenced, by a centrally generated motor signal. In one experiment, subjects were first asked to fatigue their arm muscles with exercise and then to judge the arm position. The subjects made systematic errors in reporting arm position that directly related to the force (or effort) required to maintain the position of the fatigued arm. Another study advocating the role of motor commands in judgments of limb position, found that subjects perceived a change in wrist position of up to 20º when they attempted to flex and extend their wrists despite

BOX 1. EXPERIMENTAL METHODS

The data shown in Figures 2, 3, and 4 were derived from combined psychophysical and neurophysiological experiments. In the psychophysical experiments (Fig. 2), humans scanned their fingers across embossed spatial patterns, like the ones shown in the figure and reported a number that was proportional to the subjective magnitude of the sensation. If one surface felt twice as rough as another then they were asked to give a number for that surface that was twice as large. The subjective reports were then normalized within subjects and then averaged across subjects to give the psychophysical responses shown in Figure 2.

All of the neurophysiological experiments were performed on macaque monkeys. In the peripheral experiments (Fig. 2) the animals were anesthetized and recordings were made from single afferent fibers in the arm. In the experiments shown in Figure 2, recordings were made while the same roughness patterns that were presented in the psychophysical experiment were scanned across the animal's hand.

In the cortical experiments (Figs 3 and 4), the animals were awake and performing a visual distraction task. In these experiments, fine microelectrode wires (80 microns in diameter) were driven through the dura mater into either primary or secondary somatosensory cortex. After single neurons were isolated, recordings were made while tactile stimuli were presented to the animal's hand. The stimuli in use in Figure 3 were random probes indented into the finger pad. The spatial-temporal receptive fields were computed by computing spike triggered averages of the neural response. In the experiments shown in Figure 4, the stimuli consisted of oriented bars that were indented into the distal pads of digits 2, 3, 4 and 5.

the fact that their arms were paralyzed and anesthetized with an ischemic nerve block (pressure cuff on the arm). These studies demonstrate that there are complex interactions between sensory and motor processes.

While the cortical mechanisms underlying proprioception are not understood, it is clear the brain must integrate information from joint receptors, muscle spindles, cutaneous mechanoreceptors, as well as centrally generated motor commands, to generate a precise representation of the body's position in space. This information is then combined with signals conveying discriminative aspects of touch. These signals are presumably combined in the brain regions described above to form a unified perception of 3-D form.

Our present working hypothesis of haptic object perception is that the cutaneous inputs provide information about the local features at each point of contact with objects. The SA1 afferents provide information about local form and texture, the RA afferents provide information about whether the object is static or moving, the PC afferents provide information about local vibrations and the thermoreceptors provide information about the temperature. The cutaneous information is then integrated with proprioceptive inputs (perhaps the SA2 system) which determine where each of the points of contact lie in three-dimensional space. When grasping objects, different populations of neurons acquire distinct local 3-D views depending on the hand conformation. These populations define the size and shape of specific objects and are matched against stored representations.

Summary

The neural basis for haptic perception begins with the activation of peripheral receptors that innervate the skin, muscles and joints of the hand. The signals in the peripheral afferents convey information regarding the various aspects of discriminative touch. These afferents ascend the medial-lemniscal tract *via* the dorsal column nucleus and relay neurons in VPL/VPM of thalamus. These projections are organized in a somatotopic map and preserve the place and mode of the sensory signals. Neurons in the thalamus differentially innervate the areas comprising the primary somatosensory cortex. Neural signals conveying cutaneous input are sent to areas 3a and 1 while those conveying proprioceptive input are sent to areas 3a and 2. Sensory information is processed in parallel and serially. Receptive fields become larger through the convergence of sensory input. Form processing begins as an isomorphic representation, but becomes more nonlinear and feature selective at intermediate and higher cortical processing stages. The projections diverge into two streams beyond SI cortex. The dorsal stream projects to areas 5 and 7 where neurons are involved in sensorimotor processes, multisensory integration, and attention. The ventral stream projects to the second somatosensory cortex where neural responses become more complex. Ultimately the neural signals interact with regions in the temporal lobe to be used in memory, and frontal areas for higher order cognitive functions.

Selected readings

Haggard P (2006) Sensory neuroscience: from skin to object in the somatosensory cortex. *Curr Biol* 16: R884–R886

Mountcastle VB (2005) *The Sensory Hand. Neural Mechanisms in Somatic Sensation*. Cambridge, MA: Harvard Uni Press.

Hsiao SS, Vega-Bermudez F (2002) Attention in the somatosensory system. In: RJ Nelson (ed): *The Somatosensory System: Deciphering the Brain's Own Body Image*, 197–217. Boca Raton: CRC Press

Johnson KO (2002) Neural basis of haptic perception. In: S Yantis (ed): *Stevens Handbook of Experimental Psychology*, 3rd Ed., Vol. 1. *Sensation and Perception*, 537–583. New York: Wiley

Johnson KO, Hsiao SS, Yoshioka T (2002) Neural coding and the basic law of psychophysics. *Neuroscientist* 8: 111–121

The neural bases of haptic working memory

9

Amanda L. Kaas, M. Cornelia Stoeckel and Rainer Goebel

Introduction

When deciding which kiwi fruit or pear needs eating first or which drink has the right temperature to be consumed on a warm day, we are likely to explore and compare hardness or temperature using our hands. The process that enables us to keep the relevant information active for task performance over a short period of time is called 'working memory' (WM) [1]. WM allows us to hold stimulus characteristics on-line to guide behaviour in the absence of external cues or prompts [2]. Without active WM, initial percepts decay quickly with different time constants for different input modalities (Box 1).

The neural basis of *haptic WM* is usually studied indirectly in sequential discrimination paradigms, when information from one stimulus has to be retained for comparison with a second stimulus. In an attempt to avoid confounds due to hand and finger movements, passive stimulus presentation prevails over the more natural active exploration of tactile object features such as shape and texture.

Haptic perception can be decomposed into *tactile* and *kinaesthetic perception* [3]. Functionally, the tactile (cutaneous) sense provides awareness of stimulation of the outer surface of the body, whereas the kinaesthetic sense provides us with an awareness of static and dynamic body posture. The term *tactual perception* was coined to refer to all perceptions mediated either by cutaneous sensibility and/or kinaesthesis. Haptic perception (and therefore haptic memory) is more than the sum of kinaesthetic and passive tactile processing. While passive tactile stimulation and isolated *kinaesthesia* produce tactual sensations, only active exploration allows for the perception of objects in space [4,

Box 1. Sensory memory

After a stimulus has ceased, some sensory information is retained independently of active rehearsal or interference. This 'ultra' short-term form of memory has been explored most intensively for visual (*iconic memory*) and auditory stimuli (*echoic memory*). In both modalities, sensory memory is considered to be very short lived, in the order of a few hundreds of ms. However, a study by Melzack and Eisenberg [71] indicates that somatosensory 'afterglows' can persist over minutes when the lip was stimulated with nylon monofilaments. Nevertheless, it might be prudent not to generalise these results to other body sites and kinds of tactile stimuli. Results from Gilson and Baddeley [8], Millar [19] and Sinclair et al. [13] indicate that some information that is robust against interference persists for 5–10 s for the location of touch, three-dimensional shapes and vibrotactile frequency. Results from Harris et al. [14, 15] point to two different processes within the initial seconds following vibrotactile stimulation, the first one lasting up to 1 s. This would be in line with other modalities. It is safe to assume that the exact characteristics of tactile sensory memory are not yet thoroughly explored and might differ for the wide variety of tactual features.

Box 2. A cognitive model of working memory

In 1974, Baddeley and Hitch [72] introduced a multicomponent model of WM that has remained influential until today. The theory proposes two *slave systems* (the *phonological loop* for verbal material and the *visuo-spatial scratch-pad* for visual information) that are responsible for short-term maintenance of information. Although Baddeley [73] acknowledged the possibility of additional slave systems for other kinds of information (e.g. tactile) no further slave systems were added in later elaborations (Fig. 1).
A *central executive* is responsible for directing attention to relevant information, suppressing irrelevant information and inappropriate actions, and for coordinating cognitive processes when more than one task must be done at the same time.
WM allows for temporal storage and manipulation of a limited amount of information and thereby supports human cognitive processes by providing an interface between perception, long-term memory and behavioural response [6, 70]. Only task relevant features of the initial memory trace (Box 1) are retained, and maintenance of this information is susceptible to interference (Box 3). The upper capacity limit of WM has been estimated at four so-called 'chunks', but capacity is reduced when information is not merely maintained but also manipulated [36] (Box 3).

How does working memory relate to other memory systems?

A well-know taxonomy of human memory systems makes a distinction between explicit memory which is consciously accessible and implicit memory which is not [74]. Implicit (nondeclarative) memory is related to perceptual and motor skills. Implicit memory has been related to structures in the neocortex, striatum, amygdala, cerebellum and reflex pathways. Explicit (declarative) memory contains factual knowledge of people, places and things as well as their meaning. It has been subdivided in episodic memory, for events and personal experiences, and semantic memory, for facts. The initial stages of explicit long-term memory are thought to be mediated by structures in the medial temporal lobe, including the hippocampus, perirhinal, enthorhinal and parahippocampal cortex. The association areas in parietal, frontal and temporal cortex are believed to be the ultimate repositories of distributed explicit memory traces. WM is required for encoding and recall of explicit knowledge, and perhaps for some types of implicit knowledge as well. However, not all information processed in WM enters long-term memory.

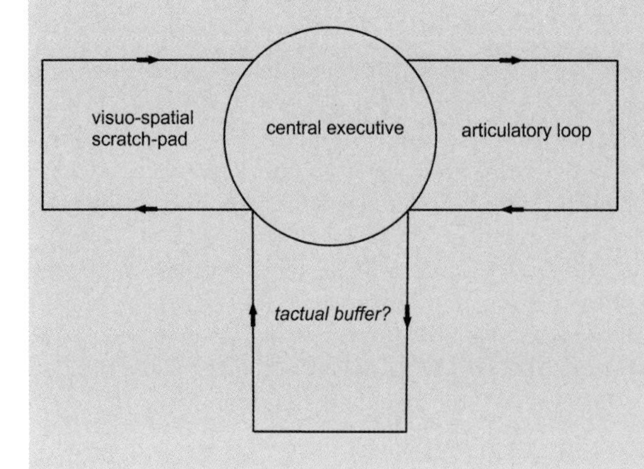

Figure 1. WM
A WM model, adapted from Baddeley [70]. Is there enough evidence to assume a discrete buffer for tactual information along with its own neural network?

5]. Our review will include literature on tactual processing more generally, aiming to provide a fruitful basis for further investigations of haptic WM.

The concept of working memory

Maintenance ('short-term memory') and manipulation of information for further processing depends on mechanisms of active control generally ascribed to the 'central executive' module of WM in cognitive psychology (Box 2). Active control is crucial for the manipulation of stored information as in mental rotation or synchronous translation between languages. While researchers working with non-human primates seem to focus on pure maintenance, psychologists often emphasise questions related to the WM capacity limit (Box 1), number and characteristics of separate memory buffers and the manipulation of stored information. There is a vast dataset on the on-line manipulation of information in the visuo-spatial and in the verbal domain [6], but there is no convincing empirical example for the manipulation of tactual information within WM so far.

WM is a higher-order cognitive process which might be affected by the sensory input modality in which information was originally acquired, but possibly even more so by the computational demands posed by the task 'at hand'. In a natural setting, information on outside events is usually picked up by multiple senses simultaneously. Therefore, many WM tasks with tactual input might draw on more general supramodal spatial or verbal working memory processes, either primarily or along with tactual coding. This is especially relevant if a task can be solved on the basis of higher-order 'non-tactual' information rather than based on the original tactual characteristics. Verbal labels are most likely applied whenever the task permits. Visuo-spatial imagery will be involved in the storage of spatial information or shapes that cannot easily be named. In fact, tactile encoding is often paired with visual recognition in experimental settings [7].

Tactual working memory from a behavioural perspective

Tactile working memory

Early studies dedicated to tactile short-term memory used a task where the location of touch had to be recalled after a variable delay [8–11] or after different interference tasks (arithmetic *versus* tactile [12]; tactile, arithmetic, and verbal [11]) (Box 3). It was found that active rehearsal improved and interference impaired performance for delays between 5 and 30 s. A more recent study [13] used delayed discrimination of vibrotactile stimuli of 12 different frequencies and also found that performance was better for unfilled delays as compared to trials with interference (counting backwards) for delays over 5 s. Initially, performance decayed quickly even without interference.

Neither vibration nor location of touch deal directly with the characterisation or identification of external objects. However, together with early studies using haptic shape discrimination (see below), these studies investigated potential analogies between the tactual, verbal and the visuo-spatial domain with respect to memory decay, with and without interference.

Harris et al. [14] used a different approach to further investigate the time course of tactile memory traces. Vibrotactile stimuli were delivered to the same as compared to different fingers (either ipsilateral or contralateral). For short interstimulus intervals (< 1 s), performance on a frequency delayed matching-to-sample test was better when both stimuli were delivered to the same finger. With longer delays (1–2 s) performance was equally good for corresponding fingers of both hands but worse when the test pulse was delivered to a distant finger. After a delay of 5 s performance was at chance level. During the first second, poststimulus interference was most effective when delivered to the same finger. Transcranial magnetic stimulation (TMS) of the contralateral primary somatosensory cortex (S1) disrupted memory performance when delivered

Box 3. Typical working memory paradigms

WM is involved in many everyday activities such as listening to complex sentences, briefly remembering telephone numbers, or mental arithmetic. For the standardised investigation of WM in the lab, variations of the same paradigm which allow separation of different WM stages are used. Typically a stimulus that has to be remembered is presented (*encoding*). The stimulus is then removed and is retained during a *delay* interval of variable length. After the delay, a *probe* is presented, that is usually compared to the initially encoded information (*recall*, *retrieval*). In its simplest form this paradigm is know as the *delayed matching-to-sample task* requiring a response to a (non-)matching probe only, or requiring a *two-alternative forced-choice*,[1] in which the subject gives a yes/no response (e.g. 'yes' if features are identical, 'no' if features are different).

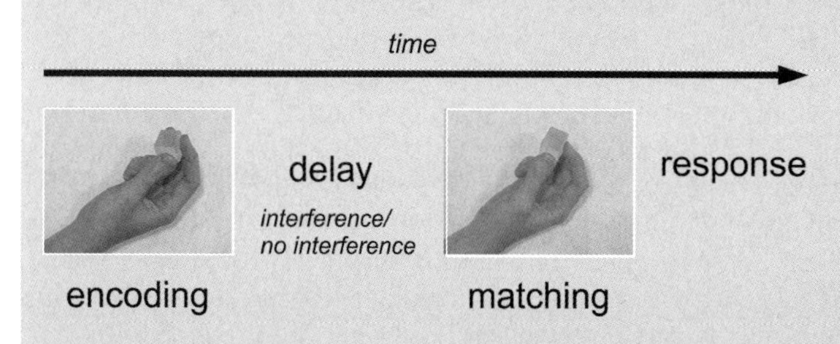

Figure 2. Delayed matching-to-sample *Schematic outline of a delayed matching-to-sample task. Subjects are required to decide whether probe and sample are identical or whether the probe differs from the sample in a certain way (is longer/shorter/rougher etc.).*

The *Brown-Peterson-Paradigm* [75, 76] provides the subject with a distractor task during the delay, known to induce *interference*. Interference can be a function of the length of the delay. Different distractor tasks can also affect task performance to different degrees allowing for conclusions about the processes involved in information retention during the delay.

WM Capacity constraints can be investigated by manipulating *memory load*. Memory load can vary depending on the number or the complexity of items that must be maintained in WM for a given task. Grunwald et al. [45, 46] varied memory load by the use of shapes of different complexity. In the *Sternberg paradigm*, multiple items are presented sequentially, and after presentation of a probe stimulus, subjects have to indicate whether the probe item was part of the list of items. Reaction time increases with the number of items in the list. Adapting this paradigm for the tactual modality, Fiehler et al. [16] had subjects explore up to three different engraved line patterns with a stylus. After tracing a probe, subjects had to indicate whether this matched any of the previous patterns. Another way to vary memory load is by using the *n-back task* [77]. Subjects monitor a series of stimuli and respond whenever a stimulus is presented that is the same with regard to a certain feature as the one presented n trials previously, where n usually varies between 1 and 3.

A way to engage information manipulation in a controlled setting is the *mental rotation paradigm* [78]. It requires the subject to match two- or three-dimensional objects that are presented in different orientations. Visual imagery is thought to subserve spatial operations necessary for matching the objects. Interestingly, reaction times increase linearly with the amount of rotation necessary to match the objects.

---

1 Note that the way the term 'two-alternative forced-choice' (2AFC) is used here differs from how it was defined in classical signal detection theory. There, the 2AFC design is used to circumvent response biases, by forcing the subject to choose which of two (simultaneously) presented stimulus displays contain the target signal.

Box 3. (continued)

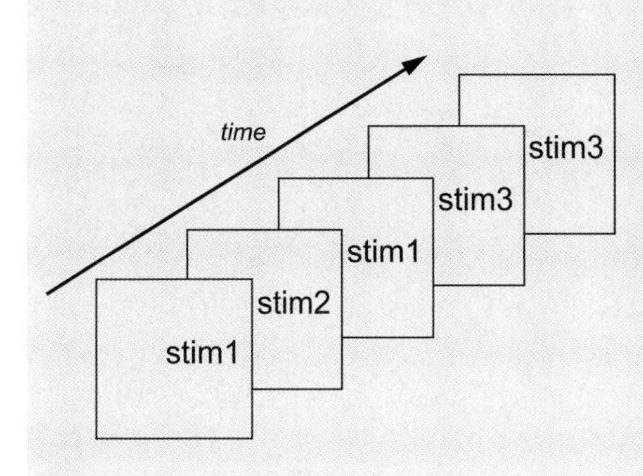

Figure 3. N-back
Schematic outline of the n-back task. In the 0-back condition subjects would have to respond to a predefined stimulus, e.g., Stim2. In the 1-back condition subjects are required to respond to the second occurrence of stim3 and in the 2-back condition to the second occurrence of stim1.

early (300 or 600 ms) but not when it was delivered late (900 and 1200 ms) in the delay period [15]. The authors conclude that WM resides in S1 and is somatotopically organised. We would suggest that the early processes investigated by Harris et al. [14, 15] most likely correspond to the initial percept (Box 1) that is qualitatively different from the later processes that are usually investigated in the context of WM. Their results indicate that the initial percept of vibrotactile stimuli is accessible up to 1 s.

Kinaesthetic working memory

Kinaesthetic processing is at least two-fold. It can serve perception (in the context of haptic object processing) or locomotion and action. Note that kinaesthetic memory for arm movements (as in a pointing task) is more relevant in the latter context and studies of this kind of 'kinaesthetic motor memory' will be disregarded here.

We know of only one study that investigates kinaesthesia in the context of shape and length discrimination, using a delay between stimulus and probe [16]. Although it is a neuroimaging study, its design and behavioural results shall be presented in some detail here.

Fiehler et al. [16] used engraved multisegment meaningless line patterns that were explored with a stylus. Memory load varied between one and three line patterns. Clearly, the pressure perceived *via* the stylus contributed to the perception of the engraved pattern, underlining the difficulty of experimentally isolating kinaesthesia. In the most difficult condition, where three sets had to be memorised, 69% of the responses were still correct after an average delay of 10 s. However, participants were trained for 30–60 min prior to the experiment using the same 28 stimuli that were used later. It cannot be excluded that excessive training encouraged the development of an appropriate verbal or visual strategy.

Haptic working memory

In early studies on haptic WM, subjects actively explored dot-patterns [17, 18] or three-dimensional shapes mounted on a surface [19–21] in a matching-to-sample paradigm (Box 3). The shapes did not correspond to commonly known real objects or shapes. According to Millar [19] retention became susceptible to haptic interference after 10 s. Bowers et al. [21] used three-dimensional objects of which shape, texture and presentation duration were to be remembered. Despite interference, performance was close to perfect for shape and texture. However, both stimulus types could be easily named and the fact that recall was verbal further imposed the usage of a verbal coding strategy. The fact that performances in [20] and [21] were close to perfect even with delays up to 80 s and different interference tasks (verbal or tactual) strongly suggests that the tasks used were too easy to learn something useful about tactual WM.

The results from Millar [19] recall the findings from Gilson and Baddeley [8] on the location of touch. Both suggest the involvement of two processes in short-term storage of tactual information. For haptic information, the first would be effective during the initial 10 s and robust against interference. After 10 s memory traces would become susceptible to interference from either the same – haptic – domain [19] or a different domain [8].

Clearly, parallel findings are no proof that the same processes or networks are involved in the storage of different kinds of information such as the location of touch and the shape of three-dimensional objects. Even studies using highly comparable (vibrotactile) stimuli are rather contradictory at this stage as to the duration of the sensory trace. While findings from Sinclair et al. [13] indicate the first process lasts as long as 5 s, results from Harris et al. [14, 15] strongly indicate that higher level-coding starts as early as 1 s after stimulus presentation. This would be in line with current knowledge about sensory traces in the visual and auditory domain. The divergent findings in the tactual domain might be explained by assuming that sensory decay has a different time constant for different types of stimuli. In addition, the onset of higher-level coding processes might vary depending on stimulus material and task strategy.

The initial robustness against interference led Gilson and Baddeley to assume that the first process corresponds to a sensory after image lasting for up to 10 s. Active rehearsal and therefore WM seemed to be more important beyond the initial 10 s, although Gilson and Baddeley [8] stated themselves that "*the nature of tactile 'rehearsal' presents an intriguing problem*".

Interestingly in this context, haptic orientation matching *improved* when a 10 s delay was imposed between encoding and matching [22]. This finding suggests that haptic (allocentric or world-based) spatial representations only develop with time and rely on a visuo-spatial rehearsal strategy. This is supported by the finding that congenitally blind subjects, lacking visual experience, did not show improved performance with the introduction of a delay [23].

In a similar vein, Millar [19] noted that memory performance for actively explored three-dimensional shapes became more stable with ongoing trials, when subjects started to employ verbal labelling as a longer-term process. This was supported by a later study, where she found that memory for unfamiliar dot-patterns depended only on the length of the delay while memory for familiar shapes (which were coded verbally) was disturbed by other attentional demands during the delay [18].

However, even when the use of verbal strategies was explicitly discouraged (by using vibrotactile stimuli of 12 different frequencies), Sinclair et al. [13] found that performance was better for unfilled delays as compared to trials with interference (counting backwards), while performance decayed quickly in both conditions during the initial 5 s. This suggests that, although phonological and visuo-spatial buffers play an important role in tactual WM, a specifically tactile WM buffer might also exist.

To compare tactual and visual WM capacity, Bliss et al. [24] used the more complex n-back paradigm (Box 3). In the tactual condition plastic letters were explored with the index finger. Performance was worse for tactually as compared to visually presented letters. The authors take this as evidence for the limited capacity of tactual WM as compared to visual WM. However, as the material used was clearly verbal, we would argue that the difference is due to faster visual *versus* relatively slower tactual encoding. Indirect evidence for this conjecture are the results from blind subjects, who outperformed sighted subjects in the tactual n-back task with plastic letters. With Braille letters they showed the same level of performance as sighted people with visually presented stimuli [25].

Investigating the manipulation of information in WM, Marmor and Zaback [26] and Carpenter and Eisenberg [27] adopted the mental rotation paradigm (Box 3) of tactually perceived objects. Analogous to findings in the visual domain, reaction times increased with the angular difference between objects [26]. Although this effect was seen in blind subjects as well [27], it still seems likely that sighted subjects use visual imagery as the predominant strategy for mental rotation.

Taken together, conclusions from the studies reported here are at least two-fold. Firstly, the duration of the initial sensory trace might differ for different tactual stimuli and features (vibration, location of touch, shapes, etc.). If the aim is to investigate tactual WM, delays beyond the duration of the initial sensory memory should be used. Secondly, to emphasise processes that are specific to tactual WM the use of verbal or visual coding strategies have to be discouraged by choosing appropriate stimuli.

The neural bases of working memory

The cognitive model of WM inspired neuroscientists to search for specialised systems acting as buffers for the storage and manipulation of information. For a long time, prefrontal cortex (PFC) was the favourite candidate for this function. The first demonstrations of the involvement of PFC in (visual) WM came from animal studies done by Fuster and Alexander in 1971 [28]. Prefrontal cells in the macaque monkey cortex showed sustained activation when a certain target location had to be kept in mind for future actions.

A distinction has been made between ventro- and dorsolateral PFC. Some researchers have suggested that the dorsolateral (dlPFC) and ventrolateral PFC (vlPFC) are involved in WM for spatial *versus* detailed object-related information (shape, colour), extending the differentiation between a dorsal occipitoparietal and ventral occipitotemporal processing stream to PFC [2]. Other groups provided evidence for a subdivision based on processing demands. Whereas the vlPFC has been suggested to be involved in updating and maintenance, dlPFC might contribute to executive control processes in WM, such as selecting, monitoring and active cognitive manipulation of information [29, 30]. Anterior PFC (aPFC) might subserve the selection of processes and subgoals.

In recent years, the ideas on the contribution of the PFC to WM have changed: there is evidence that its involvement is not so much WM specific, but rather supports sustained attention, coding of prospective targets and stimulus-response contingencies. Hence it might be that the PFC serves attentional functions, such as monitoring and action selection during delay-periods rather than memory storage itself [31]. Neurons in the dlPFC of the monkey have been found to adapt flexibly, in a domain-independent manner, to represent whatever information is critical for task performance [32, 33].

After the introduction of whole-brain imaging techniques such as PET and fMRI, an increasing number of studies also reported involvement of other, more posterior regions during the WM delay. Apart from prefrontal cortex, short-term maintenance and manipulation of information for upcoming tasks activated regions related to sensory processing and attention. This led to reconsideration of the view that the PFC serves as a memory buffer. Instead, it appears that WM for a particular kind of information is associated with the coordinated recruitment of posterior regions involved in initial sensory processing, as well as frontal regions involved in the mental processing related to the upcoming task [33–35].

Burton and Sinclair [13] suggested a model for tactile WM based on connectivity profiles between S1 and S2 (secondary somatosensory

Box 4. The neural substrates of working memory as seen by different methods

Cell recordings

(Single-)cell recording is an invasive method, requiring the implantation of micro-electrodes. Sustained spiking activity of the recorded neuron(s) during the delay period is usually interpreted as a contribution to WM maintenance. While providing detailed data, cell recordings cover only a tiny region of the brain, which leads to the problem of selection of the right sites for neuronal recording. WM involvement can only be shown for a (small) chosen area. Furthermore, due to the invasiveness of the method, most cell recordings stem from animal experiments. While it is safe to assume a close correspondence between processing networks in non-human primates and humans for basic sensory processing this is definitely not the case as soon as language-related processes come into play.

Electroencephalography (EEG) and Magnetoencephalography (MEG)

Non-invasive techniques with high temporal resolution in the range of ms such as whole-brain EEG and MEG can potentially provide important information on the timing of different WM processes in humans. EEG might be especially suited for use with active tactual exploration, since it suffers less from movement-related artefacts than the other imaging techniques. However, the spatial resolution of EEG is very limited, and good temporal resolution is only advantageous when stimulus events can be controlled on the same time-scale, as is usually the case for visual and auditory experiments. This is problematic in tactual exploration where cognitive processes are less time-locked. By integrating spectral power density over larger time windows this problem can be partly overcome. A functional relationship between the theta band and cognitive control functions and WM has been postulated. Grunwald et al. [45] were able to show that spectral power in the EEG theta frequency band – functionally related to WM load – showed a linear correlation with the complexity of haptic reliefs near the end but not at the beginning of self-paced haptic exploration.

Functional magnetic resonance imaging (fMRI) and positron emission tomography (PET)

With a temporal resolution less than a second and a spatial resolution in the range of a few millimetres, whole-brain fMRI is ideal to study haptic WM non-invasively in humans. PET is less sensitive to movement-related artefacts but provides less spatial resolution and data have to be integrated over larger time-windows.

To analyse both PET and fMRI neuroimaging data, a procedure known as cognitive subtraction has been commonly applied. By using two similar tasks only one of which (task 1) requires WM, subtraction of brain activation during task 2 from task 1 will reveal brain areas related to WM. Cognitive subtraction is a powerful method when tasks are matched appropriately.

Event-related designs (fMRI only) can be used to separate brain activity during different stages of a task, e.g. during encoding, delay and retrieval. This is not trivial because activation related to the different task components can be blurred [79]. The challenge is to separate delay-related activity from perceptual and encoding processes, retrieval, decision making, motor preparation and responses.

cortex), parietal, insular and frontal cortical regions. Areas within this processing hierarchy are expected to differ in the level of integration and transformation of sensory input as necessary for memory formation and response preparation [36]. In addition, there is evidence that the activation patterns in each region are not static, but are likely to change during the WM delay.

Below, we will discuss the neural bases of tactual sensory and WM, based on evidence from 'cell recordings' (Box 4) in non-human primates and results obtained by non-invasive whole-brain imaging techniques in humans.

The neural bases of tactual working memory in non-human primates

Two groups have investigated tactual WM in non-human primates using cell recordings (Box 4). One group used passively applied vibrotactile flutter [37] while the other group used active exploration of surface textures [32, 38, 39].

The set of experiments reviewed by Romo and Salinas ([37]; Fig. 4) illustrates the distributed network involved in WM for vibrotactile frequencies (Box 5). Macaque monkeys were trained to memorise a vibrotactile stimulus (around 20 Hz) applied to the fingertip of their restrained hand, while the other hand was placed on an immovable key. After a short delay (3 s in most cases), a second frequency was applied to the finger tip, and the monkeys had to decide whether this was equal or different from the first, pressing a button with their free hand. In a set of studies data were recorded from posterior sensory and frontal areas.

Neurons in the postcentral gyrus, functionally known as S1, showed increased firing during application of stimulus and probe, which was correlated to the frequency applied (Fig. 4) but not during the delay. S2, located ventrally from S1 in the parietal operculum, receives its projections principally from S1. A subset of S2 neurons showed frequency dependent firing rate modulations, which lasted for a few hundred milliseconds after stimulus offset [40]. During application of the probe stimulus, neuronal responses in S2 combined information from the past and the current stimulus and were correlated with the final behavioural outcome of the decision process [41].

Neurons in the inferior PFC convexity showed a variety of response profiles during the delay period [42]. Some showed sustained firing throughout the delay ('persistent' neurons); others had increased firing rates only at the beginning or near the end of the delay ('early' or 'late' neurons). When delay duration was increased from 3 s to 6 s, response profiles for late neurons shifted proportionally. These late neurons reflected memory of the encoded stimulus, and not response preparation: their graded response pattern depended on the frequency of the encoded stimulus and their firing pattern was independent of the probability of a certain response. Subsets of neurons in medial premotor cortex (MPC, comprising supplementary and pre-supplementary motor cortex) started firing in a stimulus-specific manner near the end of the delay or started firing during presentation of the probe. MPC neurons that were already firing during the delay later reflected the difference between stimulus and probe in a graded fashion. However, MPC responses were mainly related to the intended motor response.

Active haptic stimulus exploration and longer delay intervals were used in a series of studies by Zhou and Fuster [32, 38, 39]. Monkeys explored the surface texture of a vertical rod with one hand. After a delay of 10–20 s, they decided which surface of two probe rods was identical to the sample. In separate experiments, neuronal spikes were recorded from S1 and S2, primary motor cortex (M1), superior parietal lobule (SPL), corresponding to Brodmann area (BA) 5a and IPL, corresponding to BA 7b. Responses were mainly recorded from cortex contralateral to the operant hand.

In contrast to the findings of Romo and colleagues who worked with vibrotactile stimuli, haptic delayed matching of surface textures led to sustained firing frequency changes (excitation or inhibition) in single-unit recordings from the S1 hand area at least across parts of the memory retention period [38]. For some recordings, the increase in firing rate was stimulus-specific. Significant correlations between the firing rate in S1 during sample presentation and firing rate during the delay were found for neurons in S1, S2, M1 and SPL, but not in IPL or in a small set

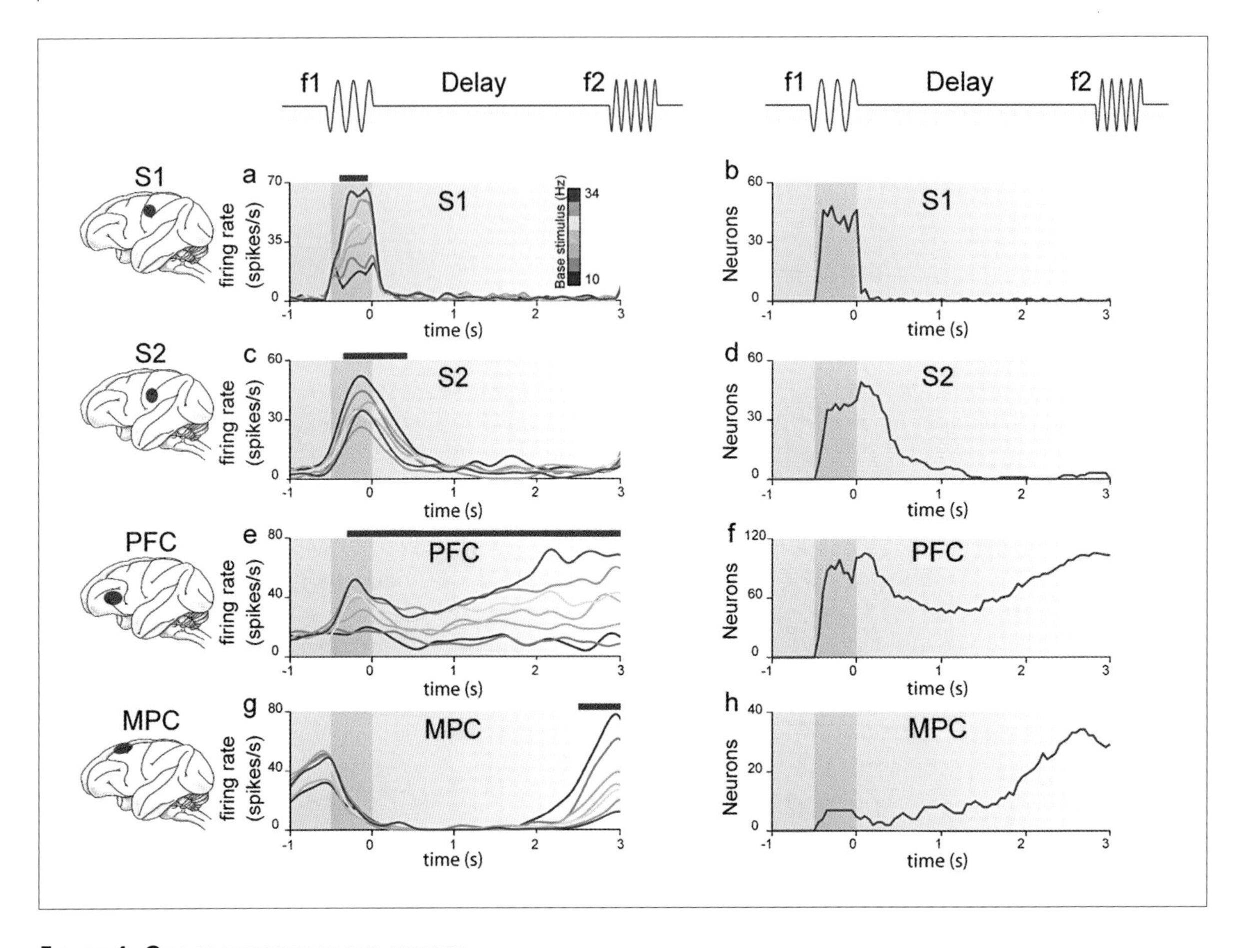

FIGURE 4. CELL RECORDINGS IN THE MONKEY

(This figure was adapted with permission from Nature Reviews Neuroscience, *Romo et al. [37] Nature Publishing Group)*

The figure shows neuronal responses evoked by the base stimulus in the flutter discrimination task (top panel). Recordings are shown for S1, S2, PFC, and MPC. a, c, e, g *Spike density functions for four single neurons. The brown bars indicate times during which the neuron's firing rate carried a significant monotonic signal about the base stimulus. Colours indicate responses corresponding to different base frequencies, as indicated by the scale gradient in* a. *E.g.,* a *and* e *show neurons that fired most strongly with high stimulus frequencies, while* c *and* g *show neurons that fired most strongly with low frequencies. Parts* b, d, f, h *show the number of recorded neurons carrying a significant signal about the base stimulus, as a function of time relative to the beginning of the delay period for the four different areas. Data is shown for a fixed delay of 3 s. The grey area indicates base stimulation period.*

of neurons recorded from the ipsilateral S1 [32]. Such correlations indicate the maintenance of content-specific information. For incorrect trials the correlation with the encoding frequency in S1 was either absent or negative (in SPL) during the delay. This suggests that S1, M1 as well as IPL were involved in the maintenance of haptic information. Activation in S1 was also observed in a visuo-haptic version of the task, in which monkeys were shown a visual icon represent-

Box 5. Experimental challenges

To ensure that observed brain activation is related to the perceptual and mnemonic processes of interest, exploration schemes need to be kept as similar as possible between conditions. Otherwise, differences in reaction times or brain activation might just be related to different finger movements. On the other hand, distinct features, such as shape or surface texture, are best extracted by different manual exploration strategies [80]. It is already quite a challenge to have participants explore naturalistic objects features in a controlled manner in the context of a psychophysical study. Hand movements have to be controlled and restricted even further in MEG and fMRI experiments, because they can lead to measurement artefacts.

Both problems are overcome by passive tactual presentation or the use of vibrotactile stimuli. The latter are well-controlled in that they are not only time-locked but also objectively quantifiable. Vibrotactile flutter (5–50 Hz) stimulates a known subset of somatosensory afferents, most prominently fibres originating from rapidly adapting cutaneous mechanoreceptors (Meissner corpuscules). On the other hand, generalisation of results obtained with these optimally controlled stimuli to active touch as it occurs in more naturalistic settings is obviously limited.

ing a specific surface texture, and after a delay matched this to a haptic probe [32]. In this study, sample specific firing in S1, MI, and IPL as well as performance increased with training, demonstrating how S1 activity evolves during cross-modal association learning.

Zhou et al. [32] hypothesise that haptic WM extends into the prefrontal cortex, which integrates the sensory information for WM maintenance. They proposed that the PFC is not the seat of WM, but a participant in the network subserving this function. PFC guides the transition from representation to operation by activating the relevant cortical network as well as subcortical components to achieve a behavioural goal. In this process, re-entrant feedback is of crucial importance for PFC functioning, especially the continuous flow of sensory information about environmental changes produced by actions, and the recurrent and reciprocal re-entry maintaining the link between sensory inputs and executive outputs until the goal is attained [32].

Cell recording studies in monkeys have provided us with a detailed description of the neuronal basis of tactual WM. However, information is limited to the areas and neurons that were chosen for recording, i.e., mainly contralateral parietal and prefrontal areas. Furthermore, tactual WM processes in humans might differ from those seen in monkeys. Tactual WM in humans might involve additional higher-order processes in general and language in particular. Human neuroimaging studies now provide the means to study whole-brain activity changes non-invasively in a rich set of tasks.

Human neuroimaging studies of tactual working memory

Most of these studies have used functional magnetic resonance imaging (fMRI) to investigate the neural bases of tactual WM in humans with millimetre resolution. One early study by Klingberg and colleagues [43] used Positron Emission Tomography (PET), which has a more limited spatial and temporal resolution than fMRI. Very few studies used magnetoencephalography (MEG; [44]) or electroencephalography (EEG; [45, 46]). A short description of the strengths and weaknesses of the different methods is given

Box 6. The interpretation of neuroimaging data

Working memory load and brain activation

One common strategy to find out more about the capacity limit of WM is to systematically manipulate memory load (see *memory load*). Some would state that any brain area showing load-dependent activation increases is involved in WM maintenance. Others would argue that linear increases with increasing memory demands are the consequence of control processes (e.g., the strategic organisation of memoranda) [81]. Attention has been implicated as limiting factor of WM capacity.

Working memory or response selection?

Some studies experimentally control the time point of response selection in the evolution from perception to action. This is done in order to separate retrospective, mnemonic coding (stimulus related) from prospective coding (response related). Onset of a region which plays a role in WM per se should occur before response selection takes place.

in Box 4. Challenges related to the interpretation of neuroimaging results are discussed in Box 6. Apart from the different methods used, the studies reviewed here also show substantial differences concerning delay length, the focus of analysis, hand used for stimulus encoding and the type of stimuli presented.

The delay interval varies from approximately 1 s [44, 47, 48] to 17 s [49, 50]. It became evident from behavioural studies that depending on the stimuli used active rehearsal mechanisms might not be recruited to bridge delays below 5 s. Results from studies using short delays might therefore reflect initial sensory memory rather than WM.

With respect to data analysis, some researchers have analysed encoding, delay, and matching as separate units, or even focused on one of these processes exclusively. Others have taken a whole trial, or blocks of similar trials as the unit of analysis. We will focus on cortical regions activated during the WM delay, although it is difficult to fully separate delay-related activity from encoding and response related processes (Box 6).

The body of studies on tactual WM does not yet allow for reliable conclusions on lateralisation effects. Stimulus presentation was not systematically varied between the left and the right hand. Instead, presentation was mostly to the right hand of right-handed subjects. In one study, the left hand was used for exploration, while the right hand was used for active matching [51]. Occasionally, stimuli were presented to both hands at the same time [45, 46, 52, 53], without controlling for differential exploration strategies of each hand. Only Kostopoulos et al. [54] varied the hand to which the stimuli were presented (albeit across subjects). However, in this study the retention interval was not analysed separately.

Keeping these considerations in mind, we will firstly describe studies using passive tactile stimuli, then move to the results from the two studies using kinaesthetic stimulation, and finally we will present the findings from research using truly haptic input.

Tactile working memory

Inspired by the impressive groundwork in non-human primates laid out by a.o. Mountcastle and Romo, many researchers opted for passive delayed discrimination tasks with flutter [54–56] or pressure pulses [48], which are quantifiable and easily controlled (Box 5). Interestingly, the majority of studies using delayed vibrotacile discrimination also chose relatively short delays (~1 s to 5s) and compared stimulus discrimination (involving WM) with mere stimulus detection [43, 48, 55]. One study contrasted WM trials with long (8 s) *versus* short (2 s) delays [56].

These studies revealed a host of parietal and frontal regions (Fig. 5a). In the frontal cortex, Numminen et al. [48] and Preuschhof et al. [55] report delay related activation in the ventral premotor cortex (PMC, corresponding to BA 44/45) and the medial frontal gyrus (probably corresponding to an area anterior to the functionally defined supplementary motor area (SMA) and known as preSMA). Numminen et al. [48] also report activation of dlPFC (corresponding to BA 9). Others report delay related activation for the cingulate gyrus [56] and the anterior insula [43, 56]. In the parietal cortex, activity was focused in S1, more specifically in BA 2 [48] and the IPL [48, 55, 56].

Only Klingberg and colleagues [43] report activation in the temporal lobe, i.e. in the fusiform, parahippocampal and middle temporal gyrus. These regions have been associated with processes related e.g. to object and space perception. It is not clear how this is to be interpreted in the context of vibrotactile WM.

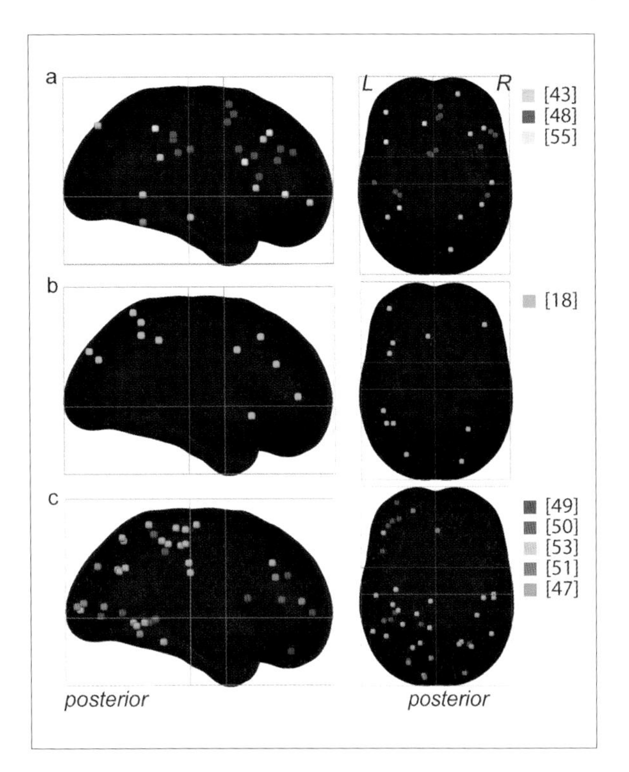

Figure 5. Summary of the results from human fMRI and PET tactual WM studies

A glass brain containing the centres of gravity of cortical activation clusters related to WM for vibrotactile (a), kinaesthetic (b) or haptic (c) stimuli reported by the fMRI and PET studies reviewed in this chapter.

The figure only displays studies contrasting the (delays of) WM trials versus *a control condition. MNI coordinates (reported by [16, 50, 55]) were converted to Talairach space using the algorithm written by Matthew Brett (http://imaging.mrccbu.cam.ac.uk/downloads/MNI2tal/mni2tal.m). For sake of clarity, subcortical clusters were excluded. The number of clusters reported by Ricciardi et al. [53] was reduced by including only clusters with a z-value over 5.*

Kinaesthetic working memory

We only know of one study investigating the neural correlates of kinaesthetic WM, which uses active exploration of engraved line patterns with a stylus [16].

Subjects sequentially traced up to three different complex 2-D shapes (three = high workload condition) whereas they traced three circles in the baseline condition. Encoding, delay and matching phase of trials with a differing kinaesthetic workload were contrasted to equivalent intervals in a baseline condition. Similar to WM for vibrotactile stimuli, Fiehler et al. [16] found a parieto-frontal network for the maintenance of kinaesthetic information (Fib. 5b). When contrasting delays with high workload with the baseline condition, parietal activation was observed including the anterior and caudal intraparietal sulcus (aIPS, cIPS), the supramarginal gyrus (SMG), and SPL. In the frontal lobe, activity was centred in ventral PMC, preSMA, vlPFC, and the anterior medial frontal cortex. In aIPS/SPL and aIPS/SMG activation correlated with workload (Box 6).

Interestingly, this study implies an important role of SPL in kinaesthetic WM, in contrast to the WM network for vibrotactile information, which seems to involve the IPL.

Fiehler and colleagues found that encoding of increasingly more complicated hand movements was correlated with graded activity in S1. However, as sensorimotor processing demands increased accordingly, this does not unambiguously speak for a role of S1 in kinaesthetic encoding *per se*.

Haptic working memory

Human neuroimaging studies on haptic discrimination and WM used either 2-D [45, 46] and 3-D spatial patterns [53] or spatial object features such as length [49, 50] orientation [51], and shape [47]. Peltier et al. [47] contrasted shape discrimination with the discrimination of textured fabrics, interpreted as a non-spatial object-characteristic. In contrast, Pietrini et al. [52] had their subjects perform a within-category discrimination of face masks and bottles.

During haptic WM, activity was again found in parietal and frontal areas but with a clear left lateralisation (Fig. 5c). In parietal cortex, activated regions included bilateral postcentral cortex [47, 53], aIPS (left aIPS: [49]; bilateral aIPS: [47]), bilateral ventral and posterior IPS [47], left SPL [50, 53]), and left inferior parietal lobule (IPL: [53]).

The frontal activations reported in the above studies were restricted to the left hemisphere, and most commonly involved the more anterior or prefrontal portion of the frontal lobe. A few foci were observed in left precentral gyrus [53] and left posterior inferior frontal gyrus [49]. These regions might correspond to regions involved in motor processing. The regions found in left PFC included the aPFC [49, 51], middle frontal cortex [49, 53], dlPFC [49], ventral PFC [49], and medial frontal cortex [51]. It is an open question whether the left lateralisation in frontal cortex is due to the specific processing demands of haptic WM, or related to right-hand stimulus exploration and matching.

Other regions engaged in haptic WM were located in the occipital cortex [51, 53]; and in (occipito)temporal cortex: in the fusiform gyrus [49] and the inferior temporal cortex, a region potentially corresponding to the lateral occipital complex (LOC: [47, 53]).

Is there a common network for tactual working memory?

S1

The role of S1 in tactual WM is not yet clear. Sustained firing in S1 was found in the macaque during maintenance of surface texture [32]. The only neuroimaging studies showing activity in S1 specific to WM demands used passive tactile stimulation (pressure pulses) and very short delays of up to 1.25 s [48]. It cannot be ruled out that enhanced S1 activity was due to selective attention in WM trials.

Based on behavioural data derived for a delayed flutter discrimination task, Harris et al. [14, 15] suggested that S1 might sustain the memory trace during the very first second of WM only. Indeed, Salinas et al. [40] did not observe sustained delay activity in S1 throughout a 1.5–3 s delay using single cell recordings. The importance of S1 for information maintenance might therefore change dynamically with the transition from passive sensory memory to actively sustained WM.

S2

Activation of S2 was mainly reported for stimulus encoding. This was true for passive tactile [54], active kinaesthetic [16] and haptic stimuli [49, 50]. S2 does not seem to play a major role in the maintenance of tactual information.

IPS, SPL, and IPL

The anterior portion of the IPS might be important in kinaesthetic WM. It was found active during encoding and short-term maintenance of hand movements [16] but also during the memory delay in a haptic shape discrimination task [49, 50]. Thus besides processing dynamic movements aIPS might also code for hand configuration. Interestingly, the corresponding area in the monkey, known as AIP, responds to the visual and motor components of grasp and is tuned to specific shapes to be grasped [57]. In humans aIPS is also involved in the (cross-modal) transformation of visual object information for hand preshaping in grasping, and during object holding and manipulation [58–62].

The anterior SPL was found to support maintenance and discrimination of object length [50] and kinaesthetic patterns [16]. The SPL contains Brodmann area 5, which receives input originating from deep and cutaneous receptors mainly from BA 2 in non-human primates. Lesions to anterior SPL can produce tactile apraxia, an impairment of exploratory hand movements in the absence of sensory or motor deficits, associated with tactile agnosia [63].

Passive tactile discrimination of vibrotactile flutter [55, 56] and pressure stimuli [48] seems to involve IPL rather than SPL. However, delay periods in the cited studies were mostly short and not analysed separately so that transient rather than sustained activation of IPL cannot be ruled out.

Frontal areas

Just as WM for other stimulus modalities tactual WM involves a set of prefrontal areas, probably *via* the same anatomical connections as described in the monkey [64–66].

As mentioned above, stimulus-specific *versus* process-specific division of labour between dlPFC and vlPFC have been suggested. Data on tactual WM does not seem to follow either of these distinctions. As can be seen from Figure 5, although the kind of processing demands were similar for all studies (information maintenance without manipulation), frontal activations are located in vlPFC and dlPFC, aPFC and medial frontal areas, without any evident pattern for purely tactile *versus* kinaesthetic *versus* haptic stimuli. The dlPFC was involved in tactual discrimination of 3-D patterns [53], spatiotemporal patterns of pressure pulses [48] and vibrotactile discrimination [43]. However, vibrotactile discrimination also involved vlPFC [42, 43], as did the active length discrimination of 3-D shapes [49, 50]. The latter task involved aPFC as well, an area also showing sustained activation for the maintenance of haptically explored orientations [51].

A region of the inferior frontal gyrus (BA 44, sometimes including BA 45) often known as Broca's area, has been reported in several tactual WM studies. Some researchers have suggested a link between (imagery of) complex hand movements and speech, referring to the importance of action recognition for communication [67]. Other interpretations associate Broca's area with hierarchical selection and nesting of action segments, with the anterior-most regions corresponding to higher-order plans, and the posterior-most regions to single motor acts [68]. Activation of Broca's region in the context of tactual WM might also indicate the use of verbal coding strategies.

Other areas

Some studies report activation of visual association areas (e.g., fusiform gyrus, LOC) during haptic WM tasks employing bar orientation [51], real-life objects [52] and spatial patterns [53]. Similar results were found for blind people without visual experience [52, 53], suggesting that activation in 'visual' areas, particularly the lateral occipito-temporal complex, might reflect meta-modal processing of macro-spatial object characteristics [69]. However, the use of a visual imagery strategy cannot be ruled out and might in fact be a cognitive substrate of meta-modal processing in a visuo-spatial store.

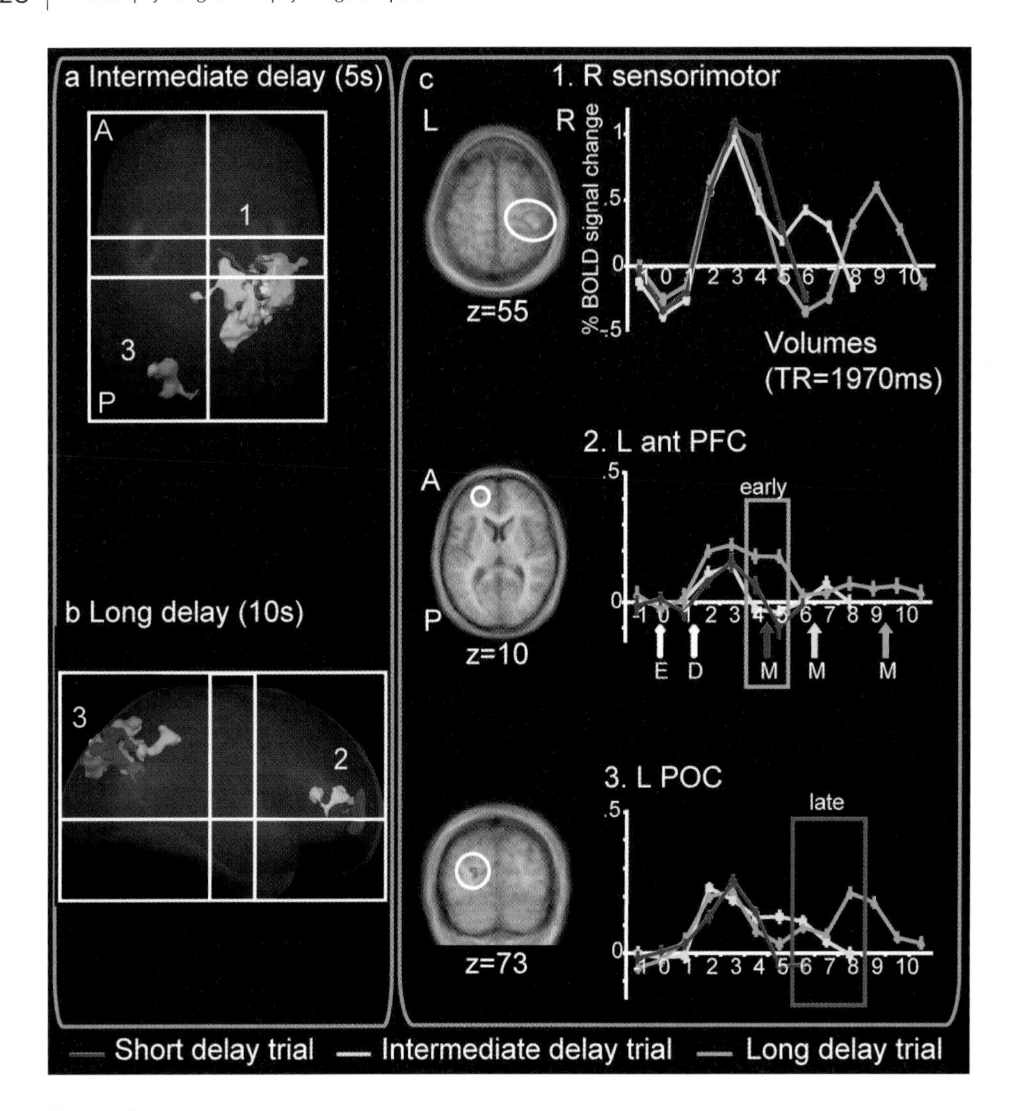

FIGURE 6. ACTIVATION DURING RETENTION INTERVALS OF DIFFERENT LENGTH

This figure was modified with permission from Cerebral Cortex, *Kaas et al. [51] Oxford University Press.*

Selected areas which showed delay-length dependent activity in an fMRI study using a delayed haptic orientation matching task. (a+b): Activation clusters from individual subjects (indicated by different colours) projected in a brain model (right and top views) (a): Regions in the right sensorimotor cortex (1) and left parieto-occipital cortex (POC, 3) showed higher activation for trials with retention intervals of 5 s versus *0.5 s. (b): Left aPFC (2), and left POC (3) showed higher activation when haptically explored orientations had to be maintained across a 10 s interval, compared to a 5 s interval. (c): Equivalent delay-length dependent clusters from a group analysis. The graphs show the % signal change across the delayed match-to-sample trials of different delay length (red = 0.5 s; yellow = 5 s and blue = 10 s) for right sensorimotor cortex (5 s trials–0.5 s trials), left aPFC (10 s trials–5 s trials) and left POC (in both 5 s and 10 s trials; displayed is the result from the 5 s–0.5 s contrast). The arrows indicate onset of Encoding (E), Delay (D), and Matching (M). A square window marks the time points of interest for the left aPFC and left POC. The group results were projected on the group average of the Talairach-normalised brains.*

The relative importance and coding schemes of the areas described above are not static across the delay, but change dynamically. For example, spectral power in the EEG theta frequency band – functionally related to memory load – showed a linear correlation with the complexity of haptic reliefs near the end of a 10 s retention interval, but not at the beginning [45]. Kaas et al. [51] showed how fMRI activation hotspots shifted from S1 during the first few seconds, to PFC and parietal occipital regions in the later stages (Fig. 6) of a 10 s delay interval in a haptic orientation matching task.

Summary

Research addressing working memory (WM), a process maintaining and manipulating information for upcoming tasks, has largely neglected the haptic domain. The few available tactual studies mostly used stimulus material allowing for a verbal or a visual coding strategy. Evidence from these behavioural, animal neurophysiological and human neuroimaging studies suggests that WM for tactual information is supported by similar mechanisms and cortical areas observed in other sensory modalities.

Initial sensory memories are thought to reside in the primary areas related to perception itself. Depending on the stimuli used, initial sensory memories for tactual stimuli might persist longer than auditory or visual sensory memories. There is discordance about if and when actively controlled WM processes start playing a role and which kind of tasks interfere with tactual WM. Different findings are again most likely explained by the kind of stimuli used in different experiments. As in other encoding domains subjects are likely to employ multi-modal rather than purely tactual coding strategies involving verbal coding and visual imagery whenever the task permits.

This is partly reflected by the findings from neuroimaging studies in humans and non-human primates. These studies revealed that WM for tactual information relies on an interplay between multi-modal prefrontal areas and parietal association cortex. The involvement of additional sensory areas in the anterior parietal, occipital, and temporal cortex most likely depends on the characteristics of the information that has to be kept online. Which particular sensory and meta-modal areas are recruited is probably determined by the level of abstraction and the type of manipulations required for successful task performance.

Prospective coding of haptic object characteristics is of great importance for accurate grasping and skilful object manipulation. Therefore it appears likely that these features have a WM representation. More studies are needed focusing on truly haptic stimulus features, such as roughness and hardness, using a design that discourages the use of verbal or visual coding strategies. Given the limited amount of studies dedicated to the topic, a more definite characterisation of the neuronal correlates of haptic WM as opposed to networks supporting WM for other types of information is yet to be explored.

Acknowledgements

We would like to thank Joel Reithler, Christoph Bledowski and Jutta Mayer for their helpful comments on the manuscript.

Suggested reading

Myake A, Shah P (eds) (1999) *Models of working memory: mechanisms of active maintenance and executive control.* Cambridge University Press, Cambridge

Neuronal plasticity of the haptic system

10

Christoph Braun

Introduction

To survive in a continuously changing environment either as individual or species, organisms need to adapt. Since environmental changes occur on different time scales ranging from milliseconds to hundreds of centuries, nature provides an arsenal of different mechanisms and strategies of adaptation. Immediate adaptations are necessary in life-threatening situations. Alternatively, changes in climate, food availability, and the appearance of competitors and predators alter an individual's behaviour, and might, on a longer timescale, even shape organisms across generations. Adaptation of the latter type occurs mainly on an evolutionary basis, which acts on genetic information. In contrast, adaptation referring to changes in individual's behaviour is realised by learning which is mediated on a neuronal level.

In this chapter, we will focus on learning as a specific capability of adaptation and also on its underlying neuronal mechanisms. In psychology and ethology, *learning* is described as a change of individuals' cognition and behaviour due to a previous experience. Learning is reflected in the acquisition of new abilities and is associated with alterations in neuronal connectivity by forming new neuronal networks. The brain's capacity of rewiring the nervous system according to its history of activation is commonly referred to as neuroplasticity. Neuroplastic changes of connectivity are assumed to persist at least for some period of time, even if inducing factors like effects of training or an altered environment are no longer present.

Unfortunately, the concept of neuroplasticity is still rather vague. So far, there exists no clear definition of a minimum time duration for which the altered connectivity has to persist in order for it to be referred to as being plastic. Usually, changes in neuronal processing that are acquired during at least several weeks of motor skill training or during the experience of an altered environment are referred to as long-term plasticity. By contrast, short-term plasticity relates to immediate but volatile changes in neural connectivity that is induced within hours or even seconds due to different experimental conditions. Yet, there is no clear-cut distinction between short-term plasticity on the one hand and modulation of neuronal activity due to task-switching on the other.

Studying the neurobiological basis of *learning* requires the documentation of changes in neuronal network organisation. In humans and animals with highly developed brain functions, only alterations in local networks with a well-structured organisation can be examined. According to their distributed nature, learning-induced changes in more complex networks are hard to detect. Therefore, the neuronal basis of learning and plasticity has been most extensively studied in primary sensory (SI) and motor cortex (MI). Both areas reveal a systematic topographical organisation of functions that is even accessible on a macroscopic level.

In SI, skin regions are represented somatotopically, i.e., neighbouring skin areas are represented in adjacent cortical zones except for some discontinuities due to the mapping from the two-dimensional skin surface to the one-dimensional strip of SI [1]. It has been shown

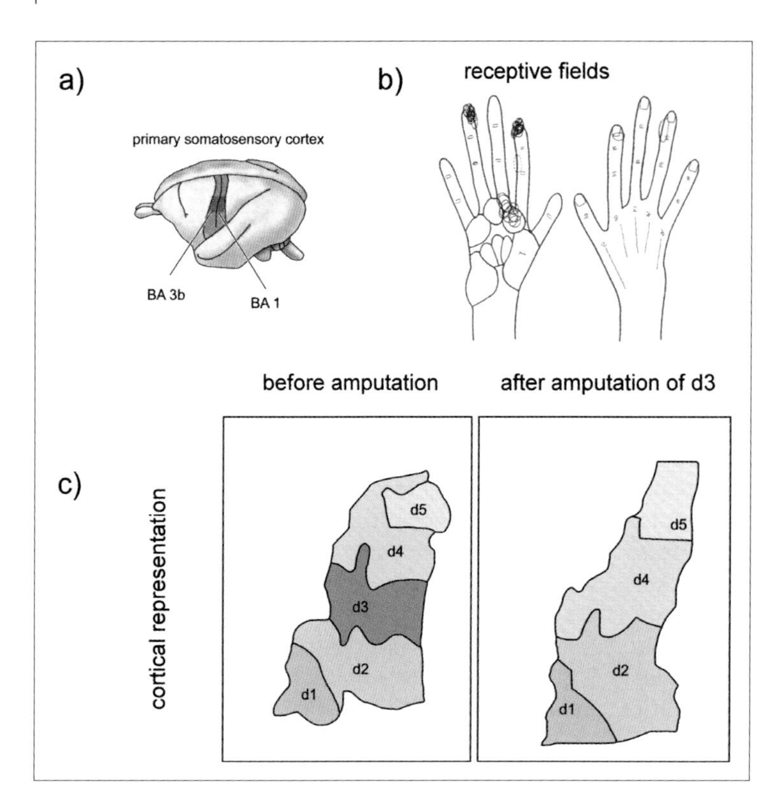

Figure 1. Cortical reorganisation after finger amputation in owl monkey

a) Localisation of the cortical hand area in primary somatosensory cortex. b) Representations of individual digits d1 to d5 before (left column) and after amputation of digit d3 (right column). c) Receptive field size for neurons in the area representing digit d3 prior to amputation. After amputation (right column), receptive fields spread to digit d2 and d4.

that the extent of the representational zones is related to the density of sensory receptors in the corresponding skin area and thus proportional to its spatial resolution [2]. Depending on the site of tactile stimulation, more or less neurons become activated. Whereas primates' hand and mouth areas are largely represented, the representation of whiskers in barrel cortex in rats is strongly magnified. Concurrently, with the magnification of skin regions in SI, the receptive field size of neurons in primary somatosensory cortex varies across the body surface. A receptive field refers to the skin area that activates a certain cortical neuron. In general, the larger the cortical magnification is expressed, the smaller the receptive field size. In experiments studying neuronal plasticity in SI, changes in cortical magnification and receptive field size, or both, are assessed and serve as an index of cortical reorganisation (Fig. 1).

Experimental studies

First reports demonstrating plastic changes of the central nervous system in vertebrates date back to the 1980s, when Michael Merzenich and colleagues [3] studied the functional organisation of somatosensory cortex in owl monkeys of whom the third digit (d3) had been amputated on both hands. Mapping the functional organisation of SI before amputation, a lateral to medial 'homuncular' representation of digit one to five (thumb to little finger: d1 to d5) was obtained in the cortical hand area with distinct zones for each finger. Remapping cortical representations and receptive fields of the animals' fingers 2–6 months after the amputation, the authors discovered the previous cortical representation of d3 had been occupied by input from d2 and d4. In addition, cortical neurons, which formerly responded only to d3 input, were found to respond to stimulation of d2 and d4. These findings suggest that reduced sensory input due to deafferentation after amputation changes the functional organisation of SI (Fig. 1).

With the availability of non-invasive, functional brain imaging methods providing high spatial resolution; such as functional magnetic resonance imaging (fMRI), electro- and magnetoencephalography (EEG/MEG), and near-infrared spectroscopy (NIRS), the findings of Merzenich have also been confirmed in humans [4]. Comparing somatosensory representations between the intact and the affected hemisphere in patients with hand amputation, Yang and colleagues [4] demonstrated that stimulation of the lower lip on the side of the amputated hand yielded activation of contralateral SI at a location that corresponds to the former hand area.

In further experiments in amputees, Flor et al. [5] observed that the amount of cortical reorganisation is positively correlated with the strength of phantom pain. Phantom pain is felt in about 80% of patients with amputations [6] in a part of the body that no longer exists due to amputation or afferent nerve damage. From studies on reorganisation and phantom pain, it was concluded that phantom pain is the result of unbalanced and dysfunctional activities in the cortical representation of the deafferented limb [5] and might be seen "*as maladaptive failure of the neuromatrix to maintain global bodily constructs*" [7]. In addition to the reorganisation of SI, alterations of MI organisation have also been observed in those patients. Interestingly, the amount of reorganisation and the strength of phantom pain varied with the employment of the prostheses replacing the lost limb. On average, patients using a functional prosthesis revealed less reorganisation and phantom pain than patients using a cosmetic prosthesis, and these patients were better off than those not using any prosthesis [8].

Plastic changes in the functional organisation of SI have also been observed in other chronic pain diseases. Maihöfer et al. [9] and Pleger et al. [10] reported plastic changes in SI in patients suffering from complex regional pain syndrome (CRPS). CRPS is a chronic pain disease that can even be elicited by only minor injuries; however, during its aetiopathology, a variety of symptoms such as burning pain, muscle spasms, restricted or painful movements, and changes in nail and hair growth may emerge. In these patients, decreased distances between the cortical representations of thumb and little finger of the affected hand were observed indicating a dedifferentiated hand representation. Until today, it is unclear whether the distorted hand representation in CRPS patients is the primary cause for the symptoms or if the hand representation is distorted due to pathological peripheral sensory input.

Besides the experience of pain, relocation and mislocalisation of haptic sensation has been described as a perceptual consequence of limb amputation [11, 12] and some chronic pain diseases [13]. Applying faint tactile stimuli to the face of a patient with hand amputation, Ramachandran [11] could elicit missensations that were felt in the amputated hand. The effect was explained by afferent information from the face relaying to the SI area assigned to the hand before amputation. According to these findings, the perception of tactile stimuli at locations not being stimulated – referred to as mislocalisation – was taken as psychophysical index of the altered functional organisation of SI. Interestingly, specific mislocalisation profiles can also be found in healthy subjects, however, to a lower extent. Thus, there are convincing reasons to assume that mislocalisation profiles reflect the organisation of SI on a perceptual level (Fig. 2).

In contrast to cortical reorganisation induced by persistently reduced sensory input after cortical deafferentation, plastic changes resulting from augmented sensory input have also been reported in animals and humans [14–17]. In these studies, increased input was due to exhaustive stimulation and intensive sensory training. In a study by Jenkins et al. [14], monkeys were trained for 10 days to touch a rotating disk with weak pressure such that fingers were not carried along with the disk. Changes in somatosensory circuitry induced by the training comprised

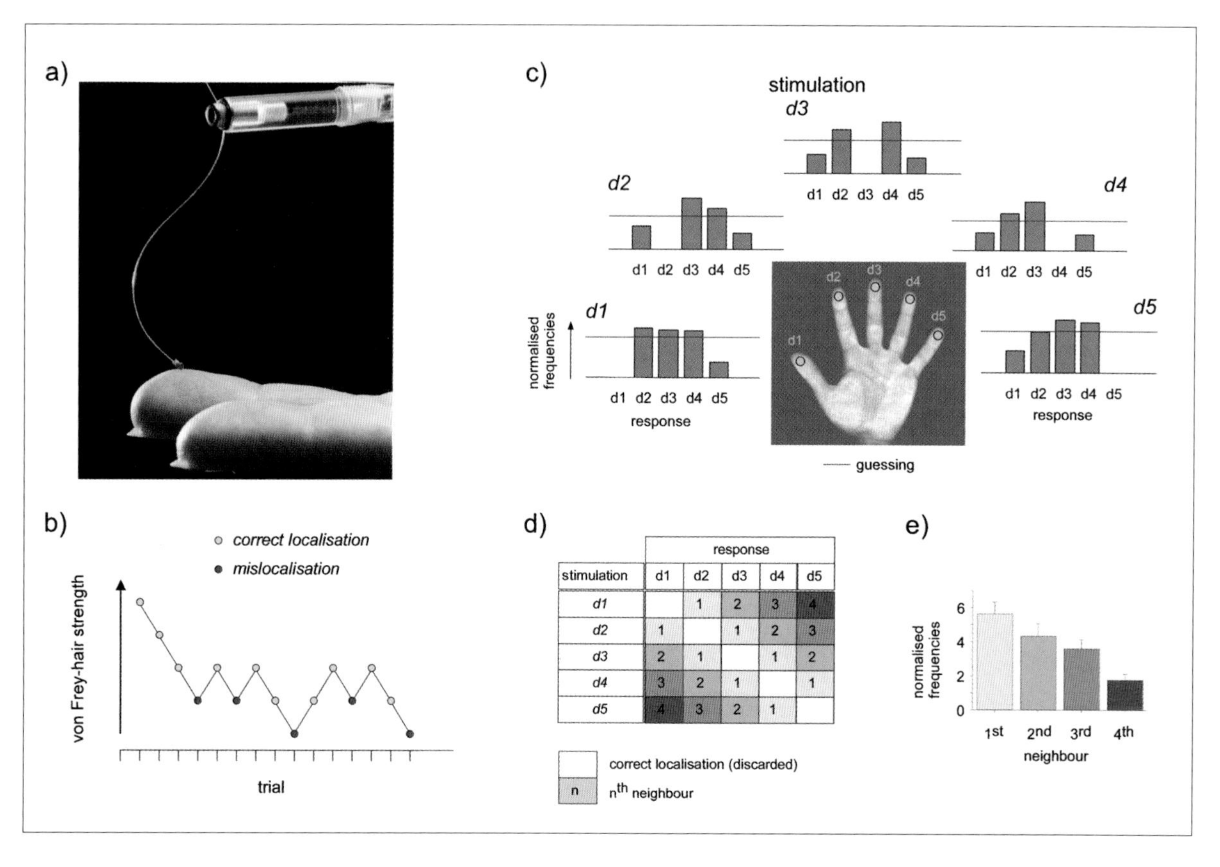

FIGURE 2. MISLOCALISATION

a) Mislocalisation is studied using von Frey-hairs different strength to stimulate fingers of one hand. The stimulation site is chosen randomly. b) In order to stimulate at the threshold a stronger von Frey-hair is used whenever the stimulus has not been localised correctly. Conversely a thinner hair was used when the stimulation site has been localised correctly. From the mislocalisation profiles obtained for each finger c) the response bias with respect to the neighbourhood of fingers can be calculated. Incorrectly localised stimuli are predominantly attributed to neighbouring finger and less often to more distant fingers d).

an increase of cortical representations of the stimulated finger and a concomitant decrease of receptive field size of the neurons in the corresponding SI region. Similar results were obtained by Recanzone et al. who trained monkeys extensively in frequency discrimination [18–22]. In humans, increased cortical representations due to exhaustive sensory input have been reported for professional musicians [23]. Intensive playing of a stringed instrument provides strong and highly relevant haptic input to the grip hand. Using neuromagnetic imaging, Elbert and colleagues demonstrated a larger hand representation for the left grip hand as compared to that of the right hand for professional players but not for amateurs. In professionals, the amount of reorganisation depended highly on the age of the inception of the musical training. The earlier they started, the larger was the cortical hand representation [23]. Although details of these results conflict with findings by Hashimoto et al. [24], evidence for plastic changes of cortical finger representations in professional musicians have been corroborated.

In the late 1940s, Donald Hebb formulated his self-named rule describing the strengthening of the "*efficiency of one cell to fire another cell*" after repeated or persistent excitation [25]. In short, Hebb's rule is often paraphrased as "*neurons that fire together wire together.*" Accordingly, synchronous stimulation of different neighbouring skin regions should result in overlapping cortical representations, while asynchronous stimulation should yield segregated representational zones [26–29]. In accordance with this rule, changes in the functional organisation of somatosensory cortex have been reported in monkeys after surgical fusion of the third and fourth digits of a monkey's hand. Due to the artificial syndactyly, monkeys experienced a stronger amount of synchronised sensory input at the fused fingers than before surgery. Consecutively fusion of the fingers induced strongly overlapping representations in the third and fourth finger [30]. Following the same rationale, Wang et al. [31] performed an experiment applying synchronous stimulation using two narrow bars to the distal and proximal phalanges of digits 2 to 4 of one hand, respectively. Distal and proximal phalanges of an individual finger never received sensory input at the same time. After 6 weeks of training, cortical representations of proximal and distal regions of each finger were clearly segregated showing no overlap. However, cortical representations of distal phalanges of the different fingers shared a common region in SI and thus were largely overlapping. The same was true for proximal phalanges. In summary, synchronous haptic input caused integration and asynchronous stimulation segregation of the corresponding representations in SI.

In humans suffering from congenital syndactyly, similar effects of cortical reorganisation in SI have been found [32]. A chaotic arrangement of cortical finger representations has been discovered in these patients when mapping the organisation of SI with MEG. However, after surgical separation of the fingers the functional organisation of SI changed from an abnormal to a normal latero-medial arrangement of finger representations. Another example for neuroplastic effects for synchronous stimulation comes from blind Braille readers using three fingers in contrast to those using only one finger. Overlapping representations for index, middle, and ring finger – the reading fingers – were found in the group using three fingers. In contrast, the group using only one finger revealed clearly segregated finger representations [33, 34]. Interestingly, changes in the functional organisation of SI were associated with perceptual changes: the three-finger readers showed less accurate localisation performance than one-finger readers. It has been concluded that the lack of clearly segregated finger representations in three-finger readers is beneficial for more fluent Braille reading, because it provides a wider haptic sensing span. In contrast, overlapping representations are disadvantageous for the precise localisation of tactile stimuli. Like in amputees, mislocalisation performance appears to be directly related to the functional organisation of SI.

Time course of plasticity

Following the fundamental studies showing changes in the functional organisation of SI, a vast amount of publications have appeared replicating and extending previous findings and knowledge [35]. In particular, it has been shown that in addition to plastic changes occurring after a couple of weeks of modified sensory input and training, there are also short-term changes appearing almost instantaneously after alterations in stimulus input [36, 37]. Birbaumer and colleagues [38] have demonstrated that the functional organisation of SI can be modified in phantom pain patients within a few minutes after anaesthetic intervention. During anaesthesia of the brachial plexus a latero-inferior shift of the lip representation in SI on the hemisphere contralateral to the amputated hand was observed, however, only among patients who reported relief from pain. In these patients, the localisation of the representation of the lower lip became similar to the localisation obtained by mirroring the lip representation of the unaffected side at the mid-sagittal plane. This finding not only supports the relationship between the experience of phantom pain and the amount of cortical reorganisation in SI, but also indicates that the altered functional organisation of somatosensory cortex in amputees cannot be explained by missing afferent input alone. Since anaesthesia is expected to reduce peripheral input that should already be diminished due to the amputation, no further reorganisation is expected; however, the opposite was found. Obviously, reorganisation and phantom pain interact and cause an imbalance of cortical activation that might even be affected by afferents from the stump. Most importantly the experiment demonstrates that reorganisation can be induced on a short-term timescale.

In general, there seems to be a certain correlation between the duration of training or exposure to an altered environment and the stability of the cortical representations. Short-term changes of cortex reorganisation seem to be less stable than long-term effects. In an experiment by Godde et al. [26], two fingers were stimulated simultaneously for two hours a day for three days. Already after the first training session an improvement of two-point discrimination was observed that was fully reversible after 8 h without training. Interestingly, repeated synchronous stimulation resulted in a prolonged recovery indicating stabilisation of the perceptual improvements.

Changes in cortical organisation within the range of minutes have been demonstrated in a study by Braun et al. [39]. Tactile stimuli were delivered to the thumb, middle, and little fingers of both hands with a stimulus onset asynchrony (SOA) of 250 ms. In the control condition, stimuli were applied randomly to individual fingers. In the experimental condition, fingers were stimulated in a fixed finger sequence starting at d1, continuing at d3, and ending at d5 on one hand referred to as the experimental hand. The three fingers of the other hand were stimulated in random sequence.

Magnetic source imaging revealed that switching from the control to the experimental condition did not change the functional organisation of somatosensory cortex for the control hand. For the experimental hand, however, the cortical distance between the representations of d1 and d3 was smaller during the fixed stimulation than during the random stimulation.

Accordingly, like during synchronous stimulation, stimulation in a fixed sequence order is expected to yield overlapping finger representations (Fig. 3).

These experiments suggest that there is ongoing adaptation of cortex organisation to sensory input patterns presumably to optimise stimulus processing. With increased frequencies of occurrence of certain input constellations, corresponding neuronal connection patterns are strengthened and finally become the default circuit. On the other hand, there are also findings suggesting that changes in the functional organisation of SI endure only for a certain period of time and return to baseline, despite the persistence of the altered sensory input [40]. Accordingly, changes in the functional organisation of cortex reflect dynamic processes of rewiring rather than the static pattern of neuronal connectivity induced by the persistence of a new environment.

Bottom-up *versus* top-down induced plasticity

In the previous sections, we have seen that cortical reorganisation was mainly driven by sensory input. Depending on the statistical properties of sensory input, integration or segregation of representational zone has been observed. It became obvious, however, that also factors such as (1) motor activity, (2) task specificity, (3) direction of attention, and (4) proprioception could also modulate the functional organisation of SI.

Effect of motor activity

When we grasp a fragile object like a crystal ball, grip force is continuously adjusted under

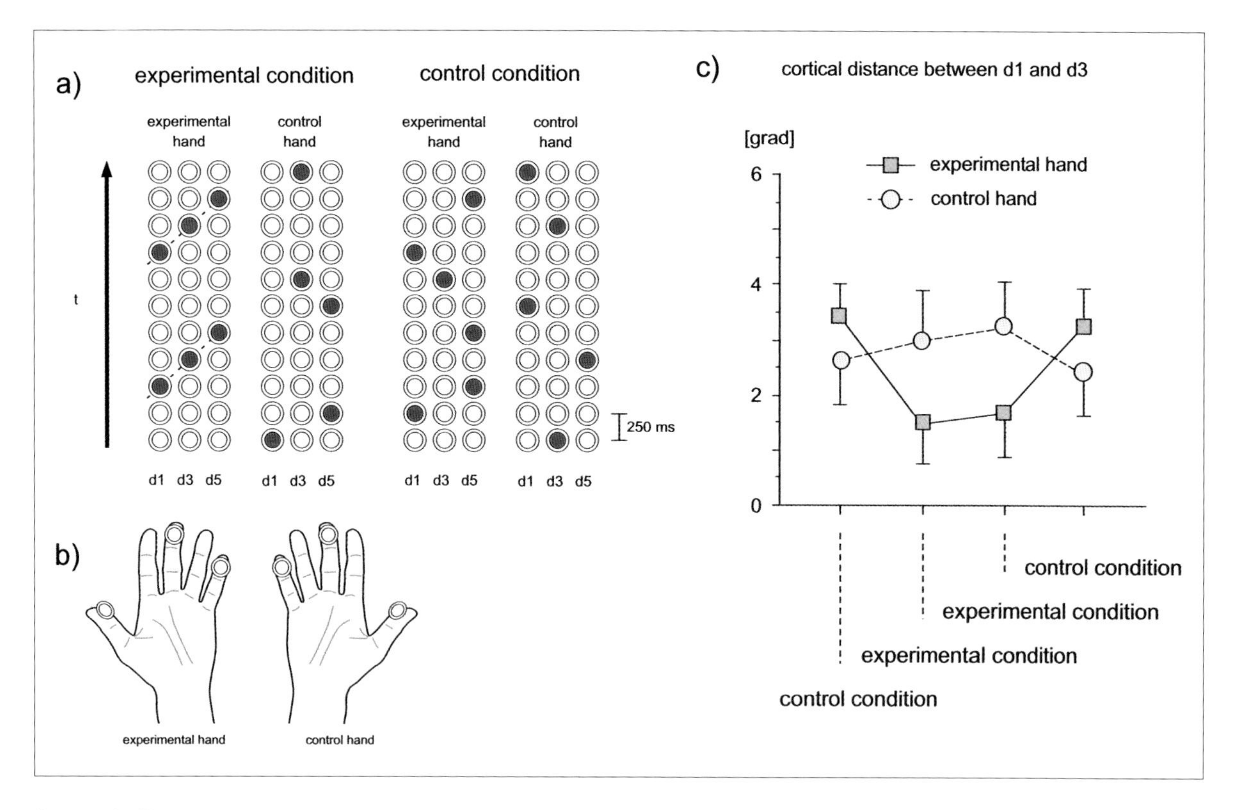

FIGURE 3. EFFECTS OF STIMULUS SEQUENCE ON THE ORGANISATION OF SI

In the experimental condition digit 1, 3 and 5 (b) were successively stimulated in this order with an SOA of 250 ms (a). The stimulated finger is indicated in red. The site of stimulation at the control hand was chosen randomly. In the control condition index, middle and little finger of both hands were stimulated randomly. The experiment started and ended with the control condition. In between the experimental condition was repeated twice. c) The cortical distances between the representations of d1 and d3 diminished during the fixed sequence stimulation during the experimental condition indicating short-term plastic changes of SI. The localisation of finger representations is specified in polar coordinates. Distances between representation of d1 and d3 are specified in polar angle.

careful sensory control. Here, a different mode of sensory processing is employed as compared to a power-grip when we hold a firm, solid object with the whole hand. Processing of sensory information from individual fingers is therefore only required for tasks involving particularly fine adjusted pinch-grip. Comparing the hand representations for the pinch-grip task during which the human subjects were asked either to generate a constant force or track a predefined force level, remarkable differences were found [41]. Using MEG for functional localisation, it was concluded that finger representations were less segregated during rest, than those during the more demanding pinch-grip tasks. Most segregated finger representations were obtained for the tracking task which was more demanding than the control task. Thus, sensory maps appear to adapt dynamically to task requirements: the more sensory precision is required for motor control, the more segregated are the corresponding cortical representational zones. Similar results were obtained for rest, handwriting, and brushing like movements, motor tasks that require various degrees of sensory control. Again, overlapping finger representations were found for the rest condition and segregated ones were obtained for movement tasks [42].

Dysfunctional interaction between the organisation of somatosensory cortex and motor control has been suggested to be the pathogenetic cause of certain motor disorders. Several studies have revealed that the functional organisation of primary somatosensory cortex is abnormal in focal dystonia [43–46], a motor disorder accompanied with task-specific cramping due to involuntary co-contractions of agonistic and antagonistic muscles [47, 48]. Often focal dystonia occurs as a vocational disease in people who rely extensively on fine motor skills, like musicians [49, 50], writers, and surgeons. Accordingly, such impairments are labelled musicians' or writer's cramp. Among others, world-famous pianist and composer Robert Schumann is supposed to be one of the victims of focal dystonia [51]. Interestingly, examinations of the functional organisation of SI revealed degraded hand representation with large overlaps of the different finger areas in these patients [45, 52]. The relationship between symptoms of focal dystonia and impaired organisation of SI could even be verified in an animal model [46, 53]. Byl et al. [53] trained a couple of owl monkeys to perform a well-defined grasping movement. She found that extensive repetition of the task led to an impairment of the performance accuracy and to symptoms typical for focal dystonia. Similar to the findings in focal dystonia patients, the functional organisation of SI was heavily degraded in those monkeys. From these findings it has been hypothesised that the reorganisation of SI is the causal source for the cramps in focal dystonia [54, 55]. Yet, the opposite could also be true: the altered hand representation in somatosensory cortex might be the consequence of a dysfunctional motor system. This argument is supported by findings in writer's cramp patients showing task dependent changes of SI organisation during handwriting, which were in the same range as in healthy controls [56]. Thus, although the functional organisation of SI differs between writer's cramp patients and healthy controls, adaptations of somatosensory hand representation in SI during writing and brushing are similar in patients as in controls. Since studies investigating the role of the organisation of somatosensory cortex in the genesis of focal dystonia have been only correlative in nature, it is still an open question whether degraded representations of SI in focal dystonia represent the basis or the consequence of focal dystonia.

Effects of task and context

Other factors that are supposed to influence the functional organisation of SI are context and tasks associated with the presentation of tactile stimuli. Our amazing capacity to adapt the functional organisation of SI to tasks requirements has been described in an experiment during which thumb and the little finger were stimulated synchronously for 1 h a day for 4 weeks [57]. In order to increase the attention to the stimulation a tactile discrimination task was introduced. The experiment was originally designed to induce fusion of the cortical representations of simultaneously stimulated fingers. During training subjects were asked to identify the stimulus patterns (see Fig. 4d) in each trial. Between training sessions subjects' mislocalisation profile was investigated by stimulating individual fingers with near-threshold stimuli. The mislocalisation profile summarises to which finger incorrectly localised stimuli were assigned. Usually, the distribution of mislocalisations to other than the stimulated finger is not random but shows a maximum for fingers neighboured to the stimulated one and lower to more distant fingers [58]. Before and after the training the functional organisation of SI was examined by stimulating subjects passively. As expected from experiments using correlated sensory input, the representation of thumb and little finger in SI showed stronger fusion of representational zones in the post-training as compared to the pre-training session. Most interestingly, these changes in the functional organisation of SI were also reflected in an altered mislocalisation profile. At the beginning of the training, stimuli

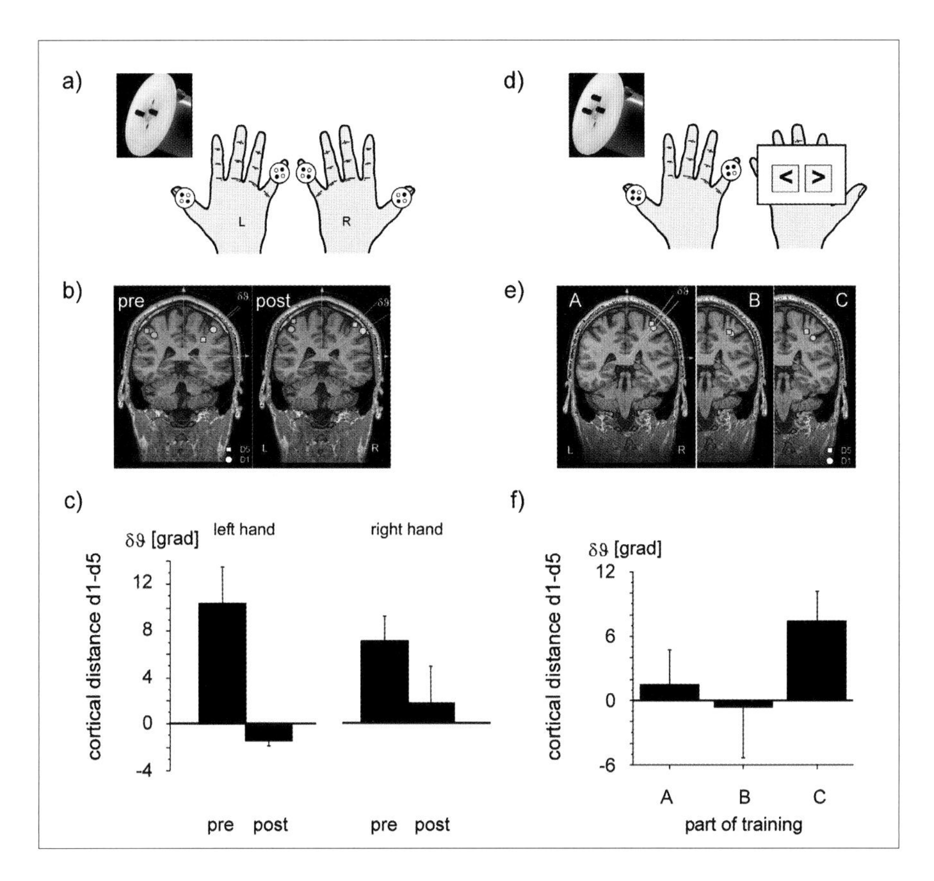

FIGURE 4. SYNCHRONOUS STIMULATION OF THUMB AND LITTLE FINGER IN COMBINATION WITH A TACTILE DISCRIMINATION TASK

During the pre- and post-training session a)–c) stimuli were perceived passively. The stimulus consisted of two pins that indented the skin surface. Pins were arranged along the finger axis a). Synchronous stimulation induced overlapping representations of d1 and d5 which was indicated by a shorter distance between source locations b) and c). During training subjects had to discriminate the direction of the stimulus pattern d). Across training an enlargement of the distance between the cortical representation of thumb and little finger was observed indicating increasing segregation of finger representations in SI e)–f). Training was subdivided in three blocks.

applied to the thumb and little finger were preferentially mislocalised at the index and ring finger, respectively. After training, incorrectly localised stimuli that were delivered to the thumb were more often felt at the little finger, and stimuli applied to the little finger were mostly attributed to the thumb. Thus, the results of the examination of the functional organisation of SI, were completely in accordance with the psychophysically assessed mislocalisation profile revealing also overlapping representations of the simultaneously stimulated fingers.

Examining the functional organisation of SI during the sensory training, i.e., when subjects performed the discrimination task, unveiled an ostensibly contradictory result with respect to the pre-post training comparison. Across training, dipole source localisations modelling the centre of cortical representation activation of thumb and little finger moved apart indicating a segregation of individual finger representations (Fig. 4d–f). Thus, pre- and post-training comparison of the SI organisation revealed integration and the analysis of the SI organisation along the training revealed segregation of finger representations. These puzzling findings can only be explained if it is assumed that the functional organisation of SI adapts specifically to task and condition requirements and representational maps that emerge during training can be switched immediately. Since SI has to be regarded as a neuronal network, it is feasible that adaptation to different stimulus properties can occur at the same time. In the above described experiment, it appears as if SI adapted simultaneously to the coincident stimulation of thumb and little finger as well as to the demanding task of tactile pattern discrimination. Depending on stimulation conditions and task requirements in the pre- and post-training experiment and in the discrimination task during training either one or the other mode of network operation was employed and dominated the functional organisation of SI.

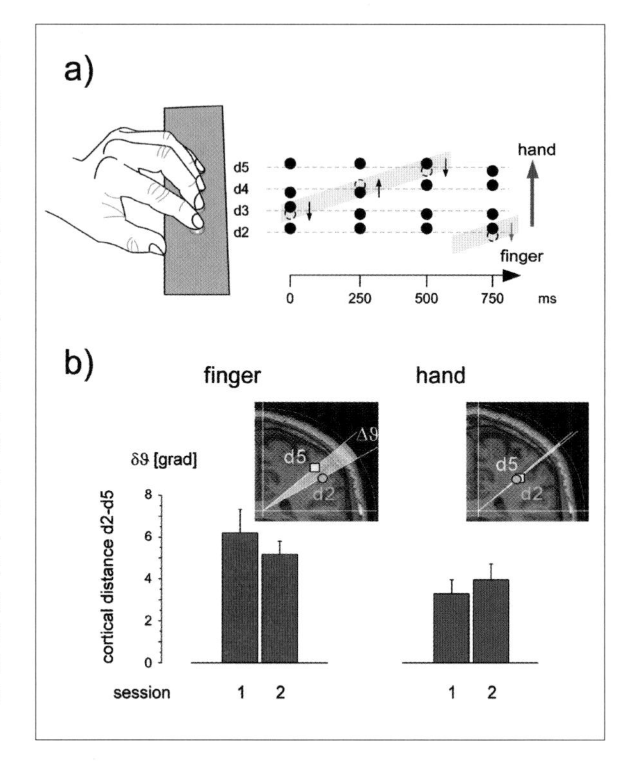

FIGURE 5. EFFECTS OF ATTENTION ON THE FUNCTIONAL ORGANISATION OF SI

a) digit 2 to 5 of one hand were stimulated with stimuli moving across the finger tip. Successive stimuli were applied to adjacent fingers resulting in a hapticly perceived apparent motion. In the finger *condition subjects had to report the stimulation direction at the index finger. In the* hand *condition they had to identify the apparent motion across the hand. The experiment was replicated in a second session. b) The distance between the representation of d2 and d5 was smaller during the* hand *condition (right) than for the* finger *condition (left) suggesting stronger overlap of finger representations in the* hand *condition. In contrast to the finger condition, subjects had to integrate the input from multiple fingers in the* hand *condition in order to succeed in the discrimination task.*

Effects of attention

Closely related to stimulation context and task requirements, attention is assumed to be a potent factor underlying the dynamic modification of functional organisation of SI. Hints for the impact of attention arise from studies by Buchner et al. [59, 60] and Iguchi et al. [59, 60]. In a recent study [61], manipulation of the focus of attention without changing physical properties of the stimuli was done by applying stimuli which were moving either from left to right or *vice versa* to all finger tips of one hand except for the thumb. In one sweep, successive stimuli were applied to the neighbouring fingers. Therefore an apparent stimulus motion that moved from left to right or *vice versa* across the hand could also be perceived. Movement direction of stimuli at individual fingers and movement direction across the hand was uncorrelated. In two different tasks subjects had to report the direction of the perceived stimulus motion either at the index finger (finger condition) or the across fingers (hand condition). While in the hand condition, the focus of attention had to cover multiple fingers in order to solve the task, the focus of attention had to be directed to the index finger alone in the finger condition. Accordingly, overlapping finger representations were found for the hand condition and segregated representations for the finger condition. It is thus concluded that directed attention to the tip of a single finger causes stronger separation of the corresponding cortical representation from the representations of other fingers.

Effects of attention and task load on the functional organisation of SI have also been shown by Schaefer et al. [62]. They found larger hand representations for a demanding problem solving task than during the control task that comprised passive stimulation only. As a supra-ordinate structure controlling the representation in SI, a prefrontal–cortical gating system was suggested.

Effects of proprioception and body schema

The usage of hand held tools requires that the sensory input is merged with information about the shape of the tool in order to perform appropriate motor actions. From previous fMRI studies body schema is assigned to multimodal associative cortical areas like posterior parietal cortex [63]. Since the person using a tool usually acquires a 'feeling' of the tool extensions it has been hypothesised that the shape of the tool is incorporated in the body schema and might possibly affect the processing even in SI. Examples for both short- and long-term effects of tool use have been provided by Schäfer et al. [64]. Since tool use is always associated with motor activity and requires elevated concentration at least during unskilled performance, it remains questionable whether changes in SI organisation during tool use reflect rather effects of motor activity and attentional demands than an altered body schema.

Limits of neuroplasticity

From the findings of neuroplastic changes of the nervous system one might extrapolate that changes in neuronal connectivity due to illnesses might be recovered as long as the correct training is found and intensively pursued. Indeed, studies in blind people revealed that large-scale reorganisation may take place and visual cortex takes over somatosensory function [65–69]. However, the finding of cortical activation during tactile stimulation alone, does not tell whether visual cortex is functionally involved in somatosensory processing. In order to verify that the activation in visual cortex is indeed related to the processing of somatosensory information sensory processing in visual cortex must be disturbed. To this aim stimulation of visual cortex by transcranial magnetic stimulation (TMS) was applied in blind people during a tactile discrimination task [70]. The study revealed that the performance of tactile discrimination was worse under verum than under sham stimulation of visual cortex. Moreover, it could be shown that TMS stimulation of visual cortex could even induce haptic sensations in blind people [71]. Results of both experiments indicate that haptic perception in blind people employs visual cortex. It might even be concluded that visual cortex has taken over somatosensory function in these group of patients.

Despite large-scale reorganisation in blind patients, there are also findings that do not reveal noteworthy reorganisation, even though it could have been expected because of functional reasons. In a study involving adolescents that had received a unilateral subcortical periventricular lesion during the last trimenon of gestation, it was demonstrated that motor functions were completely reorganised from the affected to the intact hemisphere [72]. Thus, movements of both the intact and the paretic hand were controlled by the contralesional side. Rather surprisingly, transhemispheric reorganisation of SI was not observed in these patients, although sensory and motor activities interact closely during everyday life activities. Using diffusion tensor imaging MRI techniques, Staudt et al. [73] nicely demonstrated that thalamo-cortical pathways had grown around the lesioned area and had reached their target zone in ipsilesional SI. Tactile stimulation of the paretic hand revealed contralateral evoked activities in the lesioned hemisphere without any clue for contralesional activation. Summarising, cortical reorganisation has been shown in a variety of studies to take place. There are conditions where changes in the organisation of SI are beneficial for the processing of sensory information. Neuroplastic effects seem to be important for the fine adjustment of cortical networks and optimised stimulus processing. In other cases of cortical reorganisation such as amputation, changes do not appear to be advantageous and predominantly reflect intrinsic neuronal network properties [74] or imbalances of network activation that are not necessarily related to repair mechanisms. Furthermore, cortical reorganisation appears to be strongly bound to anatomical constraints. Even in the developing brain there are irrevocable anatomical constraints that limit the amount of cortical reorganisation.

Molecular and cellular mechanisms of neuronal plasticity

Our understanding of the mechanisms of neuroplasticity has dramatically improved since Eric Kandel and his group demonstrated elementary learning processes on a cellular and molecular level in the marine snail Aplysia [75, 76]. In Aplysia, gentle stimulation of the siphon causes a mild defensive reaction comprising the retraction of the gill and the shelf of the animal. In terms of classical conditioning, this stimulus is referred to as the conditioned stimulus (CS). When the siphon stimulation is paired with a noxious electrical shock to the tail (unconditioned stimulus: US) for a couple of repetitions, the withdrawal reaction elicited by siphon touching alone becomes strongly enhanced. Obviously synaptic transmission from sensory siphon neurons to motor neurons retracting the gill becomes more efficient due to paired stimulation. A training session of 30 repetitions yields conditioning effects that last for several days [75]. By recording cellular activity and by studying ion channels basic mechanisms of learning have been clarified. However, in more complex nervous systems and also with respect to short- and long-term plasticity, additional cellular and molecular mechanisms seem to be involved. While short-term changes in network behaviour are assumed to rely mainly on changes of functional connectivity, i.e., on variations of the sub-threshold activation of neurons and on the efficacy of synapses, in long-term changes structural changes comprising the new formation of spines, axon growth and the establishment new neuronal connections (Fig. 6). Neurogenesis, the proliferation of new cells from stem cells and their differentiation to nerve cells seems to be restricted to some brain regions and, at least to our current knowledge, do not play a role in haptic plasticity. In the following, general molecular principles and intrinsic mechanisms like sub-threshold activation, long-term potentiation and depression, disinhibition, N-Methyl-D-Aspartat (NMDA) receptors and Hebb's rule, and structural plasticity are mentioned that are involved in neuroplasticity. The list is by no means complete and the descriptions are rather superficial. Nevertheless, it may provide reference points to relate macroscopically observed plastic changes to a molecular or cellular level. Mechanisms are not mutual exclusive. In contrast one mechanism might be based upon another. Take the Hebb's principle as an example, the strengthening of synapse between two simultaneously firing nerve cells might be explained by NMDA receptor characteristics, by long-term potentiation or even by axonal sprouting.

Sub-threshold activation

Neuronal output is determined by the sum of dendritic activations. Besides phasic depolarisation due to specific input the postsynaptic potential might additionally be modulated by tonic sub-threshold activation. Depending on the tonic activation a defined input might generate more or less spiking activity of a nerve cell. Thus, the strength of the input-output coupling of a neuron

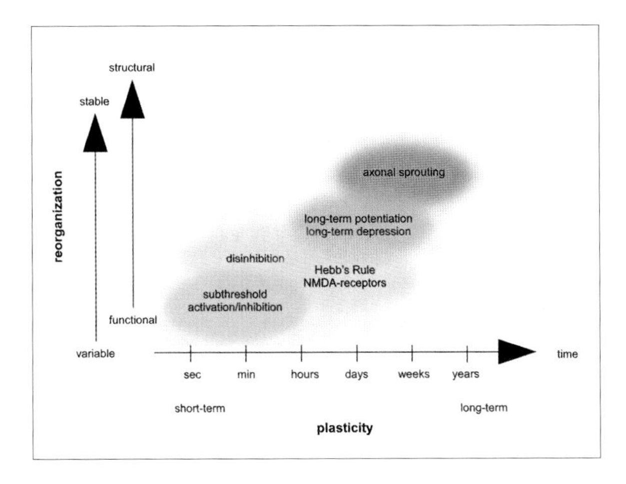

FIGURE 6. MOLECULAR AND CELLULAR MECHANISMS OF NEURONAL PLASTICITY

Mechanisms involved in short-term reorganisation processes are associated with rather volatile changes of functional connectivity. Long-term reorganisation is accompanied by stable structural modifications of neuronal connections.

is modified by variable sub-threshold activation. This mechanism might be effective in dynamically changing network connectivities in order to optimise stimulus processing or to maintain the balance between excitation and inhibition within a neuronal circuit. The implementation of subthreshold activation has been shown for rat barrel cortex where the output of the neuron whose primary whisker was bent and was affected by the activity of the surrounding non-primary whiskers *via* sub-threshold activation [77, 78]. Sub-threshold activation has been shown to be modulated by both thalamo-cortical and corticocortical connections [79].

Disinhibition

For a diversity of species it has been shown that the receptive field size of neurons SI is modulated by inhibitory neurons in SI [80]. An important role of inhibitory neurons in SI has been proposed by Jones [81, 82]. The inhibitory neurons are of GABA-ergic nature as has been shown by the administration of bicuculline, a GABA antagonist, causing an enlargement of receptive fields [83]. Recent findings show that the inhibition is tonic and is driven by a tonically active periphery [84].

Long-term potentiation and long-term depression

Physiological studies of the perforant pathway in rabbit hippocampus by Terje Lømo in 1966 [85] revealed that repeated stimulation of nerve cells with stimulation frequencies of about 100 Hz induces subsequent changes in the responsiveness of the neuron. The response to a single stimulation pulse could be increased by a stimulation burst preceding the single pulse. Long-term potentiation (LTP) is assumed to be a basic mechanism of learning and memory. Although different molecular mechanisms contribute to LTP in different brain regions and species, one major mechanism is the improvement of sensitivity of the postdentritic membrane to neurotransmitter signals released by the presynaptic cell. Increase in sensitivity is accomplished by higher efficacy of the receptors and by their increase in number. Invoking LTP new receptors are formed as a consequence of the repeated input signal. Moreover retrograde signalling by messenger molecules like NO might also be involved.

NMDA receptors and Hebb's rule

Excitatory cortical synapses are mainly glutamatergic. There are two types of receptors for this transmitter AMPA- and NMDA-receptors controlling membrane channels for cat ions like calcium, sodium and potassium. While AMPA receptors are fast and in a wide range insensitive to the membrane potential, NMDA-receptors are slow and their permeability for cations is voltage-dependent. For negative membrane potentials like the resting membrane potential the conductivity of NMDA-receptors for Na^+ ions is low. However, for positive membrane potentials the conductivity exceeds those of the AMPA receptors by 50 times. Thus the NMDA-receptors act like coincidence detectors and are assumed to play an important role in memory and learning by building up new functional and structural connections depending on sensory input.

Already in 1949 Donald Hebb [25] postulated that synaptic efficacy might be strengthened in a use-dependent manner. The more one cell activates another neuron the stronger will be their coupling. A wider interpretation of Hebb's rule postulates that the connection of two simultaneously active neurons is strengthened. The neuronal implementation of this rule might be accomplished by NMDA-receptors. Yet, there are certainly other molecular mechanisms that contribute to the use-dependent strengthening of neuronal connections.

Structural plasticity: axonal sprouting

Finally, it has also been shown that after long-term changes in sensory input new neuronal connections are established by axonal sprouting and growth. Furthermore, the emergence of dendritic spines on the post-synaptic membrane can also be taken as an index for the formation of new connections.

Summary

Research on the neuronal plasticity of SI, which has been presented here only fragmentarily and incompletely, has evidenced that the changes of the functional organisation of SI depend on bottom-up and top-down influences. While short-term changes in the functional organisation of SI seem to be rather labile and volatile, long-term effects appear to be more stable. We suppose, that depending on the persistence of an altered environment, labile effects become stabilised and finally form the default network. Changes of the functional organisation of SI are reflected also on the level of perception. Besides changes in discrimination performance, the profile of mislocalisation seems to be a sensitive correlate of cortical reorganisation in SI.

Despite the somatotopic organisation of SI suggesting a hard-wired entry of sensory information to the cortex, there is remarkable capacity of plastic changes of functional and structural connectivity even in adults where neurogenesis is complete. Although there is general agreement that the capacity for plastic changes is higher in younger than in old people, there are also limitations due to anatomy even in the developing brain. Considering patients who had received periventricular lesion during gestation, no taking over of haptic functions by the intact hemisphere could be observed despite clear occurrence of transhemispheric reorganisation of motor functions. Even if reorganisation of haptic information processing would have been beneficial for sensory-motor function, the development of somatosensory pathways and connections followed anatomical constraints. Summarising the studies, plastic changes in the haptic system appear as mechanism to react and adapt to changes in environment on different time scales. Concerning injuries and lesions of the sensory system, plastic changes of the nervous system appear not to represent a general repair mechanism, because there are various examples in patients who suffer from dysfunctional effects of neuronal plasticity such as pain. Injuries in the somatosensory system might cause an imbalance between activation and inhibition in the somatosensory system. Thus, spontaneous occurrence of plastic changes after lesions of the nervous system are assumed to reflect rather intrinsic network characteristics and the imbalance of neuronal activation than the restoration of functions. However, there is no doubt that intensive training of lost sensory and motor functions, strongly affects the functional organisation of cortex and forms the basis for recovery.

Acknowledgements

I am very grateful to Yiwen Li Hegner who assisted me writing the chapter on neuroplasticity. She inspired me with suggestions, comments, and came up with helpful additions and constructive corrections. Finally, she and Rachel E. Ott to whom I also owe many thanks edited the text with respect to readability. Furthermore, I thank my co-workers Renate Schweizer, Anne Wilms, Michaela Burkhardt, Monika Haug and Anja Wühle who contributed to this work. Last but not least, I want to thank the technicians in our laboratory Wolfgang Kern, Jürgen Dax, Gabi Walker und Maike Borutta for their valuable help. Research was supported by the German Research Council (SFB 550) and by the Volkswagenstiftung.

Selected readings

Feldman DE, Brecht M (2005) Map plasticity in somatosensory cortex. *Science* 310: 810

Flor H, Nikolajsen L, Staehelin Jensen T (2006) Phantom limb pain: a case of maladaptive CNS plasticity? *Nature Rev Neurosci* 7: 873

Kandel ER, Schwartz JH, Jessel TM (2000) *Principles of Neural Science* (4th ed). Appleton, New York

Schaechter JD, Moore CI, Connell BD, Rosen BR, Dijkhuizen RM (2006) Structural and functional plasticity in the somatosensory cortex of chronic stroke patients. *Brain* 129: 2722–2733

Stavrinou ML, Della Penna S, Pizzella V, Torquati K, Cianflone F, Franciotti R, Bezerianos A, Romani GL, Rossini PM (2007) Temporal dynamics of plastic changes in human primary somatosensory cortex after finger webbing. *Cereb Cortex* 17: 2134–2142

III. Psychological aspects

Haptic perception in the human foetus

11

Peter G. Hepper

Introduction

Recent years have seen increased exploration of the sensory development of the foetus [1]. All five senses, auditory, visual, cutaneous, olfactory and gustatory and been demonstrated to begin functioning in the prenatal period. It is the aim of this chapter to review the evidence of haptic perception in the human foetus. The development and function of other senses is briefly presented. However before reviewing this evidence it is necessary to pause to consider some general issues which must be borne in mind when evaluating the evidence presented by these studies.

Sensory and perceptual development in the foetus

First, it is important to distinguish between sensation and perception [1]. Often researchers reporting studies in this area use the two terms interchangeably but this is an error as the two are different. Sensation refers to the transduction of the physical signal by the sensory receptor and turning this physical signal into neural impulses. Thus a sound stimulates the hair cells of the inner ear and neural impulses are generated by the hair cell and transmitted to the cochlear nerve for onward neural transmission to the brain. Sensation refers simply to the act of transducing the physical stimulus into neural stimuli within the nervous system. Perception is the process which adds meaning to these neural impulses as they interact with the various centres and pathways in the brain. Thus if the sound stimulus were the voice of the infant's mother, the sensation experienced would be the various parameters of the sound, for example its frequency, duration, intensity. The perception experienced would include, for example, recognition of the voice as mother, the meaning of the words spoken by the mother, and potentially the urgency or otherwise of the information conveyed. Caution must thus be exercised in interpreting the results of studies examining foetal sensory and perceptual development. Although the foetus may respond to the sound of its mother's voice, whether the foetus perceives this stimulus as its mother's voice or is responding to a sensation due to the presence of a sound, needs careful experimentation and consideration.

A second issue concerns the onset of the development of sensory abilities. To determine whether the foetus detects a particular stimulus we have to rely on observation of a change in its behaviour, e.g., a sudden behavioural jump to the onset of a loud sound. Observation of the foetus's movement may be achieved through ultrasound [2, 3] or recording its heart rate [4, 5].

With appropriate methodology and controls the presence of a response following presentation of a stimulus can reliably be used to assume that the foetus has detected the stimulus presented. The inverse however is not true. In the absence of a response, great care must be exercised in drawing the conclusion the foetus cannot sense the stimulus. To respond the foetus requires both the neural apparatus to detect the stimulus and elicit a response (the one we are looking for) and the appropriate connections between sensory and motor pathways. Lack of any one may result in a lack of response. This is particularly important to bear in mind when con-

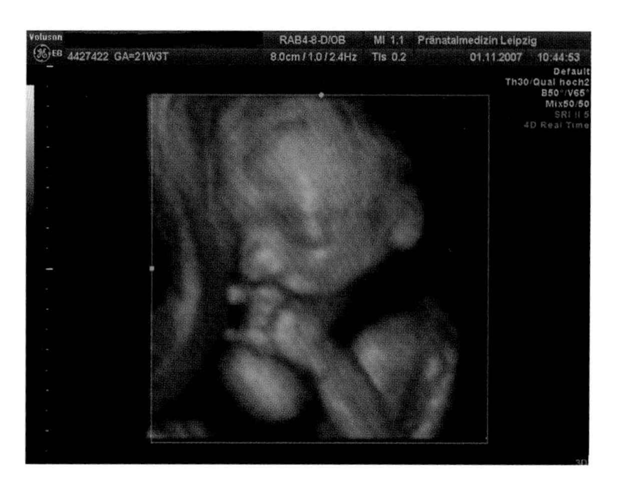

FIGURE 1. HUMAN FOETUS IN THE 21ST WEEK OF GESTATION.
Picture from Prof. Holger Stepan, Department of Obstetrics, University of Leipzig.

sidering the age of onset of sensory responding in the foetus.

The final issue concerns the natural foetal environment. It is very easy to elicit a response to sound in late gestation foetuses by playing sound through a loud speaker placed on the mother's abdomen. However this is not natural. If we are to consider the role of sensory stimulation for normal development it is important to consider whether, in the normal course of pregnancy, the foetus will be naturally exposed to these stimuli. It may be very easy in the experimental scenario to present a wide range of stimuli to the foetus, but these may not be experienced by the foetus in the normal course of pregnancy.

Sensory development in the foetus

Before considering the development of cutaneous sensation and perception in the foetus a brief overview of the sensory development of the other sensory systems will be provided.

By far the most studied sensory system in the foetus is that of audition [6]. This undoubtedly reflects the ease with which auditory stimuli can be presented to the foetus. Responses to auditory stimuli have observed using movement [7], heart rate [8] and most recently in late gestation using fMRI techniques [9]. Evidence indicates that the foetus begins to respond to sound around 24 weeks of gestation [10] but in a very restricted range of frequencies and at high intensity. As the foetus develops the sensitivity of its hearing increases and the range of frequencies responded to expands [11].

Studies which have played specific stimuli to the foetus (e.g., music, voices) have observed that the newborn responds to these sounds differently compared to other, unfamiliar sounds [12, 13]. This strongly suggests that these stimuli sound similar when 'heard' *in utero* to the way they sound outside of the womb, providing strong evidence for continuity of auditory experience before and after birth. The mother's skin attenuates sounds outside the womb but around the fundamental frequency of the human voice there is little attenuation and speech sounds will pass relatively unaltered into the foetus's environment [10, 14]. As well as the external environment providing a rich source of sounds the mother's physiological systems also provide noise, e.g., heart beat, digestive system [14]. Thus the foetus responds to sound and is naturally stimulated by sound in the womb.

The same cannot be said of vision. It is possible to elicit a motor response, observed on ultrasound, or a change in heart rate, by discharging a camera flash light over the mother's bare abdomen [15–17] from 26 weeks of gestation. However it is unlikely in the natural course of pregnancy that visual stimuli will penetrate the womb in such a distinct and discrete fashion. At best there may be a generalised glow if the women exposed her abdomen to sunlight but not a specific point source of stimulation. Thus while the foetus can respond to visual stimuli it may be unlikely that it will be naturally stimulated during pregnancy [1].

Studies have found that the foetus appears responsive to both smell and taste stimuli [18]. It is difficult to separate smell and taste *in utero* and both senses are usually considered together.

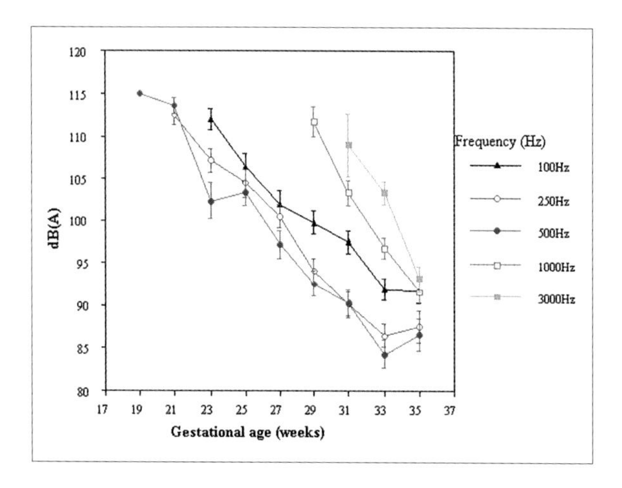

Figure 2. The mean sound intensity level required to elicit a response at each frequency for each gestational age [from 11]

Foetuses have been observed to increase swallowing when a sweet substance was injected into the amniotic fluid and decrease swallowing when a sour substance was introduced [19, 20]. Other studies have found that the foetus learns the taste of substances exposed by the mother's diet and responds differentially to this after birth compared to substances not experienced prenatally [21–23]. Because of the difficulties of presenting discrete chemosensory stimuli to the foetus it is unknown when the foetus first responds to such stimuli, but estimates place this at around 15 weeks' gestation [1]. The amniotic fluid provides a rich source of chemosensory stimuli [24] especially as substances found in the mother's diet will be present [25].

Cutaneous sensation

Studies of cutaneous sensation are extremely limited due to the difficulties of presenting these stimuli to the foetus. However there have been some experimental studies and these are reviewed here.

Of all the foetus's senses it appears that that of touch is the first to develop and respond to stimulation [26]. At 8 weeks the foetus responds to a touch on the lips, the first location to respond and receptivity then spreads to the cheeks, forehead, palms of the hands and upper arms. Only the back and top of the head do not elicit a response to touch by 14 weeks. However it must be noted that these studies were performed on exteriorised foetuses and thus their response may differ from foetuses *in utero*. There are many sources of tactile stimuli for the foetus, the walls of the uterus, the umbilical cord, its own body parts and potentially other foetuses in multiple pregnancies.

It is unlikely that the foetus will experience much by way of a change in temperature. The mother's physiology controls the uterine temperature carefully and ensures it remains around 0.5–1.5 °C above that of the mother with little variation. One experimental study squirted cold water (4 °C) at the foetus during labour and this elicited a change in heart rate [27]. Whether this response was due to temperature or pressure is unknown. Mothers often report that when taking a hot bath they feel an increase in the movements of their foetus but whether this reflects a response to temperature or abdominal muscle relaxation is unknown.

Little is known about proprioception in the foetus. Repeated observations of individual foetuses using ultrasound suggest that the individual has a preferred position to lie in. The foetus develops a preference for having its head turned to the right towards the end of pregnancy [28, 29]. The position of its arm shows a developmental trend with at 12 weeks its elbow flexed and wrists and fingers extended, at 16–28 weeks, the fingers are flexed and by 28 weeks the elbow, wrist and fingers are flexed [30]. At the end of pregnancy the foetus, usually, adopts a head down position in preparation for birth which could be indicative of a functioning vestibular sense. Moreover Hooker [31] reports that around 25 weeks of gestation the foetus starts to exhibit a righting reflex. The fact that the foetus moves during pregnancy from around 8 weeks' gestation suggests that stimulation arising from this movement may be present dependent upon the development of the required neural pathways. The foetus is susceptible to the influence of gravity [32] and also the motion of the mother (linear and angular acceleration) will be transmitted to the foetus [1]. Thus the vestibular sense, if functioning, will be actively stimulated.

The one sense that has received considerable attention is that of pain [33]. Increases in heart rate have been observed in foetuses following foetal scalp blood sampling and increases in movement have been observed after tactile stimulation during amniocentesis [34, 35]. A biochemical stress response has been observed during foetal blood transfusions [36]. The physiological and pharmacological systems responsible for the sensation of pain are present during late gestation [33]. However pain is a subjective phenomenon and the question of pain perception in the foetus has yet to be resolved.

Haptic perception

The individual components of the cutaneous sensory system combine to provide information for the individual's haptic perception. There are no direct studies that I have found that directly investigate haptic perception in the human foetus. However there are two areas of naturalistic observations that may be suggestive of a rudimentary haptic perceptual ability in the foetus.

One behaviour that appears early in gestation is hand-head contact [3]. The foetus first exhibits movements around 7–8 weeks of gestation, these being whole body movements and startles. Individual limbs begin to move in isolation with arm movements observed at 9 weeks of gestation and leg movements at 10 weeks of gestation [3]. At 10 weeks of gestation the foetus's hand comes into contact with the head [3]. Using 4-D ultrasound a greater visualisation of these movements has been enabled and seven different types of movements/contact have been reported: hand-head, hand-mouth, hand near mouth, hand-face, hand near face, hand to eye, and hand to ear [2, 37]. These movements continue throughout pregnancy with a major change occurring between 32–36 weeks. Until this point the majority of hand contacts with the head area were unimanual, i.e., involving only one hand. From 36 weeks there is a shift to both hands touching the head area together [38]. It is impossible to state whether there is any intentionality in the foetus's movements, especially during the early part of pregnancy. Moreover it is difficult to determine what feedback and information is obtained from the stimulation experienced by the foetus. With touch receptors in the hand and face operational to some extent by around 10–12 weeks of gestation, and certainly thereafter, contact of the head by the hands will result in simultaneous stimulation of hands and face and may begin the process of linking these, and other sensory stimulation involved in a single perceptual experience.

One interesting naturalistic experiment which can further our understanding of cutaneous and haptic perception before birth is that of multiple pregnancies. Here in addition to the cutaneous stimuli provided by the womb and the foetus itself, there is another individual in the womb, whose movements, actions and reactions may directly stimulate other occupants of the uterine environment. A number of studies have examined the behaviour of twin pregnancies. Twin foetuses are reported to touch each other *in utero* [39]. As pregnancy progresses the incidence of simultaneous movements increases, with reports that in the early half of pregnancy most movements are independent (95%), but towards the end of pregnancy there has been a significant increase in simultaneous movements (25% of all occurrences of movement) [40–43]. Detailed observations of the actions and reactions of twins *in utero* reveal significant facts regarding the development of cutaneous sensation [44]. There is little stimulation between foetuses before 11 weeks of gestation as the foetus's and uterine size is such that there is little contact between twins. However from 15 weeks' gestation onwards twin–twin stimulation is a constant feature of the individual's uterine environment. Stimulation and response occur

Table 1. The number of individuals observed sucking either their left or right thumb at three different gestational ages *[from 45]*

	Gestational age (weeks)			
	15–21	28–34	36-term	Total
Right thumb	71	88	93	252
Left thumb	10	4	8	22
				n=274

earlier in monochorionic twin pregnancies than in dichorionic twin pregnancies. The nature of the response to contact also changes. Early in pregnancy the response to touch is a whole body startle response but by 20 weeks of gestation this occurs in only 6% of contacts. At 12 weeks of gestation the most frequent response to contact is a general movement involving most of the body but by 22 weeks of gestation this has decreased from 77% of responses at 12 weeks gestation to 36% of all responses. From 12 weeks onwards there is an increasing trend for responses to become localised to the point on the body that was stimulated. For example, a contact on the leg results in a movement only in that leg. Such behaviour indicates that the foetus is able to localise the source of contact from at least half way through gestation and emits a specific localised response. Of course whether this reflects the development of a reflex arc or a more centrally determined response is unknown.

Another feature of early foetal movement is the appearance of lateralised behaviour or handedness. Foetuses show a preference for sucking their right thumb over their left thumb [45]. Moreover observation of individual arm movements indicates that approximately 90% of foetuses are right-handed, i.e., they move their right arm more than their left [46], and this is predictive of later handedness [47].

This preference is observed from the initial onset of individual arm movements at 10 weeks of gestation [48]. This may have implications for the development of the CNS. It has been postulated that there is a link between handedness and hemispheric specialisation of brain function [49] and that it is the hemispheric specialisation of the brain that leads to the development of lateralised motor behaviour. However the observation of motor asymmetries at 10 weeks of gestation is before these movements are thought to be under the control of the brain and suggests an intriguing possibility: asymmetric motor behaviour contributes to the development of hemispheric specialisation.

Crucial prenatal experience?

In general terms early sensory experience is important for the development of the sensory receptors and brain areas associated with them. Studies of the visual system have revealed the

Table 2. The number of individuals who were classified as either strongly or weakly left and right handed according to the thumb sucked prenatally *[from 47]*

		Handedness at 10–12 years			
		Strongly left	Weakly left	Weakly right	Strongly right
Thumb sucked prenatally	Right	0	0	19	41
	Left	0	10	5	0

importance of sensory experience for the development of feature detectors in the brain [50]. Removing sensory input from the whiskers during development results in the under development of those brain areas responsible for detecting their stimulation [50]. Sensory experience thus appears important for the normal development of sensory receptors and brain areas responsible for dealing with this sensory information. The role of prenatal sensory experience for the development of the senses and brain has yet to be explored but there is little reason to suggest that processes of sensory experiential neural development clearly demonstrated after birth would begin only after birth, but more likely, would also operate before birth. With regard to haptic perception: the cutaneous sensory receptors are functional before birth; all will be actively stimulated before birth by the foetus's environment and its movement. Thus the stimulation received by the foetus through its touch, movements, actions and reactions may contribute to the development of the sensory and neural systems responsible for haptic perception and indeed may be crucial for their normal development.

Summary

Rather than develop in an environment of sensory deprivation the human foetus exists in surroundings of naturally occurring stimulation. Moreover its sensory systems are sufficiently developed to detect and respond to stimulation occurring in this environment. Cutaneous stimulation is present, and forms a major part of the foetus's 'world' throughout pregnancy and cutaneous receptors are the earliest to begin function, providing over 30 weeks of stimulation *via* these receptors. There is a small amount of evidence (a result of lack of study rather than negative results) indicating potentially active haptic perceptual processes during pregnancy. The role of these processes in shaping the future development of the individual is unknown. However much evidence is accumulating that demonstrates the role of experiential factors in influencing the development of the nervous system and behaviour. It is likely that the sensory and perceptual processes involved in processing cutaneous and haptic stimuli contribute to, and perhaps are required for, the normal development of these senses in particular, and the development of the individual more generally.

Acknowledgements

The support of the Royal Jubilee Maternity Service is gratefully acknowledged.

Selected readings

Hepper PG (2005) Unravelling our beginnings: Fetal psychology – an embryonic science. *Psychol* 18: 474–477

Nijhuis JG (ed) (1992) *Fetal behaviour. Developmental and perinatal aspects*. Oxford University, Oxford

Lecanuet J-P, Fifer WP, Krasnegor NA, Smotherman WP (eds) (1995) *Fetal development. A psychobiological perspective*. LEA, Hillsdale, NJ

12

Haptic behavior in social interaction

Peter A. Andersen and Laura K. Guerrero

Touch "*is the core of sentience, the foundation for communication with the world around us, and probably the single sense that is as old as life itself.*" *Indeed, touch is "the most intimate of senses*" [1].

From infancy to adolescence and through all of life, interpersonal haptic behavior plays a vital role in our lives. Human haptic behavior extends far beyond the sensory world to every aspect of the social world. Interpersonal touch expresses warmth, affection, intimacy, immediacy, and love [2–4] but can also threaten and even injure. Haptic behavior also plays a central role in promoting health and happiness throughout the lifespan [1, 5–8]. Within social relationships, touch differs based on sex differences and relational stage. Cultural differences in touch also exist. Finally, sometimes touch is avoided, either because people have a predisposition that causes them to be touch avoidant, or because there is a taboo against touch. These issues are explored in this chapter, starting with the social significance of touch.

The social significance of touch

Experts believe that touch is the first sense to develop and the last sense to depart when we die [1]. From the time babies are in the womb, tactile stimulation plays a critical role in human development. Touch provides a channel for connecting to others and learning about the world. As Moszkowski and Stack [9] noted, "*touch is an important modality through which infants and mothers communicate; it is also a vital means through which infants self-regulate and explore their surroundings*" (p. 307). People who are deaf or blind are able to adapt to the loss of these senses and lead healthy, productive, and socially meaningful lives, but "*an existence devoid of tactile sensation is another matter; sustained physical contact with other humans is a prerequisite for healthy relationships and successful engagement with the rest of one's environment*" (p. 28) [1]. Children who are deprived of contact with others are disadvantaged socially, emotionally, cognitively, and physically.

Emotional, cognitive, and physical development

Considerable work has focused on the importance of touch for emotional and physical development in young human children and other primates. Harlow's classic work compared baby monkeys' preferences for nourishment *versus* contact comfort [10–12]. Harlow and his team raised baby monkeys in isolation from their mothers. They provided the baby monkeys with two types of 'surrogate' mothers, one had a hard wire body but contained milk while the other surrogate mother did not have any food, but was covered in soft terry-cloth that was warmed from a light bulb inside its body. Consistently, Harlow and his colleagues found that the baby monkeys preferred the warm and soft surrogate mothers and only went near the wire surrogates when they were hungry. The baby monkeys in this experiment were also unusually aggressive toward themselves and others, suggesting that being deprived of contact with real monkeys adversely affected their behavior.

Contact is just as important for humans. Montagu and Spitz summarized some of the earliest and most compelling research supporting the link between health and touch in young children [7, 13–14]. This research, which came from records of 19th and early 20th Century orphanages and children's hospitals, showed that around 30–40% of infants in these institutions died before their first birthday, with many other children dying sometime later in early childhood. Those who did survive tended to be plagued with psychological and physical problems the rest of their lives.

Lack of touch appears to be a proximal cause of these high mortality rates. Although most of these children received adequate food and shelter, they were seldom held by caregivers who were stretched thin trying to attend to the large number of infants in these institutions. This lack of tactile stimulation produced physical symptoms such as lethargy, non-responsiveness, self-aggression (e.g., biting self, hitting one's head against the crib), and repetitious or anxious behavior (e.g., constantly rocking back and forth; laying in a fetal position all day). These symptoms, plus depression, a lack of motivation to live, and crowded conditions, likely made these children more susceptible to diseases. Sometimes, however, there was no readily apparent cause for death, with children simply shutting down and dying. Montagu referred to this ailment as *marasmus*, which means that a person literally 'wastes away'.

Research on institutionalized children as well as feral children (i.e., children who are isolated or raised with animals rather than humans) also provides evidence that people's brains develop differently when they are deprived of human interaction. Several studies have shown that children are especially likely to suffer from decreased cognitive ability when they have spent long periods of time in neglectful environments. For example, neglected children fare better the sooner they are placed with a nurturing family [15–17]. Dennis found that the earlier children were adopted and taken away from a neglectful environment, the higher their IQ scores were in adolescence [18]. Children adopted before the age of two had an average IQ of above 100; those adopted between the ages of two and six had an average IQ of about 80; and those who remained in institutions had an average IQ around 50. Gerhart suggested that affectionate interaction with caregivers is critical for healthy brain development during the first 18 months of life, especially in terms of developing pathways for understanding social and emotional processes [19].

When researchers compare magnetic resonance images (MRIs) of brains of neglected children *versus* children raised in nurturing environments, they uncovered startling differences. The brains of the neglected children are smaller and not as well developed [20], with some studies suggesting that children who are rarely touched have brains that are about 20% smaller than children who receive frequent affectionate touch [21]. Children who grew up in isolation or lived with animals in the wild are likely to suffer especially significant cognitive problems, including difficulty with basic language skills. For example, one famous case of a feral child named Genie, who was locked up in a dark room alone for over 10 years, showed that it was impossible for a child to recover – both in terms of social competence and language development – after such a long period of isolation [22]. Feral children's brains also are especially underdeveloped in areas of the brain that process language and children raised in isolation also have severe problems adjusting to social interaction. Feral children who were raised with animal families (such as wolves or dogs) are more likely to acquire some of the social skills necessary to interact with humans, such as empathy and the ability to show affection.

Finally, touch has benefits for low-birth weight infants and other young children with health issues. Weiss, Wilson, Morrison, and Delmont videotaped mothers feeding their 3-month old low-birth weight infants and then checked back to see how the children were doing when they were one year old [23]. They found that children had better visual-motor skills and gross motor development if their mothers had used more stimulating touch when feeding them. Other research has shown that premature babies who are massaged by nurses gain more weight and are released earlier from the hospital than are babies who are not massaged [21]. Healthy babies also benefit from frequent and appropriate levels of tactile simulation [24].

Attachment

Secure attachments are another ingredient in the recipe for healthy social and physical development. Research on attachment originated with Bowlby's work [25–26]. Based on his own observations of institutionalized children, Bowlby concluded that deprivation of maternal contact has life-long consequences for humans [27]. Children who were separated from their mothers often showed distress, followed by detachment and/or ambivalence. When the separation continued over time, children become increasingly aggressive and/or avoidant. Bowlby believed that humans have an innate and adaptive propensity for forming attachments with others. In his view, humans are hard-wired to engage in proximity-seeking behaviors that help them develop and maintain healthy attachments with other people from childhood through to old age. When attachment bonds do not develop properly in early childhood, children become insecure, have more unresolved nightmares, and have negative perceptions of themselves and/or others.

Children who form secure attachments with their caregivers learn to trust others, explore their environments freely, and develop multiple attachments to various people across the lifespan [28]. Those who are unable to form secure attachments are likely to avoid, fear, or obsessively search for intimacy and emotional closeness with potential attachment figures [29]. According to attachment theorists, interaction with caregivers provides a foundation for later attachments. Young children are most likely to develop a secure attachment style when the caregiver is attentive, responsive, and sensitive to their needs. An avoidant attachment type can emerge as a defense mechanism against neglect or over-stimulation, and anxious attachment is often the product of inconsistent parental communication (i.e., the caregiver is neglectful sometimes but loving other times). Research has shown that more than 85% of children who are abused or neglected have disorganized or insecure attachment styles even after they have been removed from the negative environment [30–31].

Although research suggests that attachment styles can be modified based on new experiences and social interactions, securely attached children are still more likely to develop secure attachments with others as adults than are insecurely attached children [32]. In romantic relationships, adults who have secure attachment styles are more likely to use a variety of affectionate behaviors, including touch [33–34]. Research also suggests that people who have positive recollections of being cuddled, hugged, and touched in other positive ways by caregivers when they were children tend to have happier relationships [35] and to be more self-confident as adults [36].

Touch in social and personal relationships

Touch is certainly a critical component of healthy child development; it is also highly consequential in adulthood. Haptic behavior conveys a myriad of messages, ranging from comfort, love, and sexual interest to violence and dominance [37–38]. Patterns of touch have also been shown to vary based on sex (i.e., men *versus* women) and type of relationship.

The bright side of social touch

Affectionate touch is an important immediacy behavior that reflects physical and psychological closeness [39], helps maintain relationships [40], and can directly and unambiguously communicate one's feelings [3]. Adults who give and receive affectionate communication, including

FIGURE 1. AS DEMONSTRATED BY THESE SISTERS, TOUCH IS A BONDING GESTURE THAT COMMUNICATES PHYSICAL AND PSYCHOLOGICAL CLOSENESS AS WELL AS INTERPERSONAL WARMTH.

touch, tend to report and exhibit more physical and mental health [41–42]. When people give or receive affection, stress-related adrenal hormones tend to decrease, while oxytocin tends to increase [3]. Oxytocin is a hormone associated with lactation, sexual satisfaction, and positive moods. Thus, heightened levels of oxytocin can occur in response to various types of touch, ranging from hugs, to breastfeeding, to sexual activity. In medical settings, touch by nurses appears to have a calming effect that reduces patient anxiety [43–44].

Haptic behavior also constitutes the primary way that people communicate comfort. In one study, Dolin and Booth-Butterfield had college students describe what they would do to comfort a roommate who was going through a distressing romantic breakup [45]. The vast majority of respondents mentioned touch as a way of comforting their roommate. The most commonly mentioned behavior was hugs, followed by pats on the arm or shoulder. Other forms of haptic behavior reported included holding the roommate's hand, letting the roommate cry on one's shoulder, and stroking the roommate's hair. Clearly, touch is an important means of communicating comfort.

The dark side of social touch

On the opposite end of the spectrum, violent touch can have a host of deleterious effects on people and relationships. Such touch is not uncommon. Estimates suggest that around 16% of married couples, 35% of cohabiting couples, and 30% of dating couples can remember at least one time in the past year when violent touch was used in their relationship [46], with behaviors such as shoving, pushing, grabbing, and shaking particularly prevalent [47].

Sometimes violent touch is used as a form of *intimate terrorism* that occurs when one partner (usually the male in a heterosexual relationship) intentionally and strategically uses violent touch (and threats of violent touch) to control the other partner [48–49]. Other times violent touch occurs less strategically as part of *common couple violence* where partners resort to violence as a way of trying to gain control of an argument (rather than control of the other person). Often, common couple violence is the result of an escalation of conflict and is less severe than intimate terrorism [38]. Violent touch is also associated with deficits in interpersonal skill, with people more likely to resort to violence when they do not have the communication skill necessary to manage the conflict [46].

Differences in touch based on sex and relational stage

Considerable research has investigated whether men and women differ in terms of how much touch they give and receive. Some studies have found that men initiate more touch in public settings and professional contexts [50–51], but reviews of literature have shown that, in general, women tend to give and receive touch more than men [52]. One reason behind this finding is that touch between male friends is less acceptable in US society than is touch between women or opposite-sex friends or partners. Interestingly, however, Floyd's work has shown that femininity and masculinity are not related to affectionate communication as one might expect [3]. Floyd's work showed that feminine individuals are likely to report using affectionate behavior, as one would expect, but so are masculine individuals (albeit to a lesser extent). Perhaps men who are comfortable with their masculinity feel free to show affection, including touch, without worrying about being stereotyped as unmanly.

In heterosexual romantic relationships, whether men or women touch more appears to be partially dependent on relational stage and age. Three observational studies of public touch showed that men are more likely to initiate touch in the beginning stages of relationships [53–55]. In Guerrero and Andersen's study, couples were unobtrusively observed as they stood in lines at movie theaters or the zoo [53]. A team of coders began observing the couples when they first started standing in the line. After touch patterns were recorded, the researchers approached the couple and asked them a few questions about their relationships. Men were likely to have initiated the first touch if the couple described themselves as in a new or casual dating relationship. Conversely, women were more likely to have initiated the first touch if the couple was married. Guerrero and Andersen suggested that social norms dictate that men have the prerogative to initiate touch in the beginning stages of a romantic relationship, but women, who often focus more on maintaining intimacy in their relationships, are more likely to touch once the relationship has become intimate and committed.

The Willis and Dodd study produced similar findings [55]. Men under the age of 20 who were in the early stages of a romantic relationship initiated the most touch. Women in their 40s were more likely to initiate touch than women in their 20s or 30s, especially if they were in a stable relationship. Another study by Hall and Veccia demonstrated that although men and women touched each other equally overall, age made a difference – men under 30 years of age were more likely to initiate touch than were older men [56].

Couples also display different levels of touch based on relational stage. Guerrero and Andersen's observational study showed that couples in serious dating relationships displayed twice as much touch as couples in casual dating or married relationships [57]. Yet spouses in married relationships were most likely to reciprocate (or match) one another's touch. In serious dating or escalating relationships, people may use touch to show their budding commitment to one another and to let others know that they are a couple. Thus, touch may be a means of escalating a relationship. For married couples, such touch may be superfluous; instead spouses may show intimacy through especially high levels of reciprocation. McDaniel and Andersen replicated these same findings of a curvilinear relationship between touch and relational stage with an international sample during airline departures [58]. Additionally, Emmers and Dindia also replicated these findings by investigating reports of private touch. In their study, couples reported the most private touch when they thought their relationships were moderate to moderately high in intimacy [59]. At very high levels of emotional intimacy, couples reported that private touch leveled off or dropped somewhat.

Cultural differences in haptic behavior

In addition to sex and relational stage, culture exerts a substantial influence on haptic behavior. As a result haptic interpersonal behavior varies considerably around the world from culture to culture [60–63] in terms of type of touch, location, total amount, and whether touch is manifested in public or private [58, 64–65]. The consensus of these studies is that the least haptically active region on earth is Asia, including Myanmar – formerly Burma – China, Hong Kong, Japan, South Korea, the Philippines, Taiwan, Thailand, and Vietnam [58, 65–67]. Northern

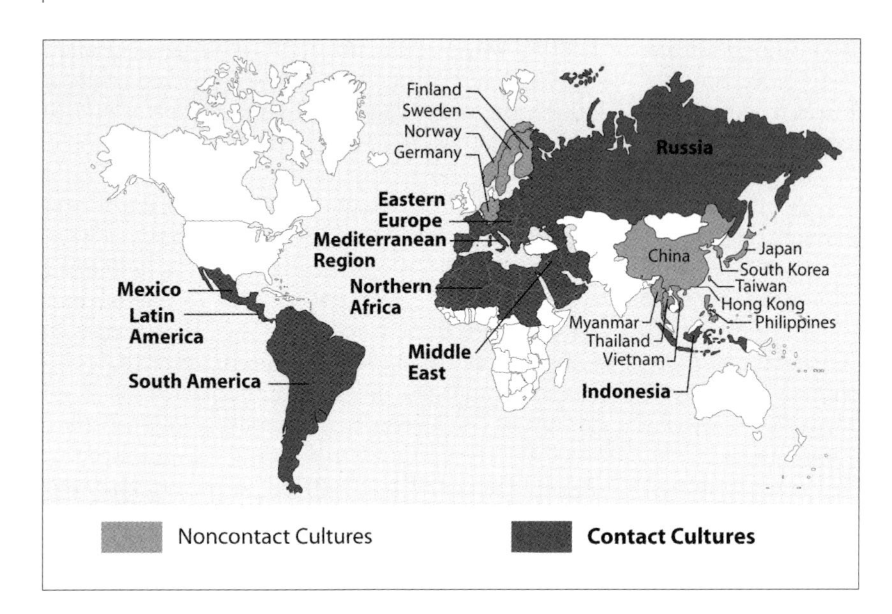

FIGURE 2. IMMEDIACY ORIENTATIONS OF SELECTED COUNTRIES AND REGIONS

European cultures, such as those from Finland, Germany, Great Britain, Norway, and Sweden are also relatively low in contact (albeit not as low as Asia). Anglo-Americans, the primary culture of the United States, and Canada are also among the least interpersonally haptic regions.

Countries where people display high levels of haptic activity and close interpersonal distances have been labeled contact cultures [61]. The Mediterranean region including France, Greece, Italy, Portugal, and Spain, along with most of the Latin America region, are the most interpersonally haptic active areas of the world. High levels of interpersonal haptic behavior have also been observed in most Arab countries (including those in northern Africa), Eastern Europe (including Russia) and Indonesia, where touch between members of the same sex (and especially between men) is expected and acceptable even though public touch between members of the opposite sex is sometimes taboo [65, 68–71]. Australia and the United States are moderate in their haptic contact level. Why is this the case?

As shown in Figure 1, with the exception of Eastern European countries, low-contact cultures are generally located in cool climates farther from the equator cluster whereas high-contact cultures tend to cluster in warmer locations. These latitudinal explanations for differences in haptic behavior include variations in energy level, climate, and metabolism [72–73]. In cooler climates the long harsh winters tend to create a more task-oriented culture whereas cultures in warmer climates tend to be more interpersonally oriented and interpersonally 'warm'. Warmer latitudes tend to host higher-contact cultures even within the United States [73], with students at sun-belt universities more touch-oriented than those in the frost belt. Similarly, Pennebaker, Rimé, and Blankenship [74] concluded:

"*Logically, climate must profoundly affect social processes. People living in cold climates devote more time to dressing, to providing warmth, to planning ahead for food provisions during the winter months. … Those living in warmer climates wear fewer clothes… and are more likely to see, hear and interact with their neighbors year 'round.*" (p. 329)

It is no coincidence that the most haptically active cultures are located nearer the equator where both skin and other people are more visible and available.

Touch avoidance

Just as cultural differences exist in touch, individual differences do as well. Due to both genetic and cultural factors some individuals are touch avoidant whereas others approach touch positively. If touch is as valuable and important to our social life as the previous discussion suggests, then avoiding touch would affect our self and our relationships with others in powerful ways. For several decades we have been exploring the mysteries of touch avoidance and have learned a lot about it.

The roots of touch avoidance research go back four decades to the work of Jourard who found men and women show a consistent trait of more or less 'touchability' that is an important communication predisposition and would affect the nature of their interpersonal relationships [75–76]. Investigating and measuring touch avoidance began by the mid 1970s [60, 77].

Touch avoidance is an attitudinal predisposition against most forms of interpersonal touch [60, 77] that has consistently been associated with less touch [53, 57], negative reactions to actual touch [78], greater interpersonal distances [79], reduced intimacy [80–83], less self disclosure [60, 84], and less relational satisfaction for oneself and one's partner [53, 85].

Two dimensions of touch avoidance have been revealed, a same sex dimension and an opposite sex dimension [60, 67, 77]. Relatively older individuals are more touch avoidant although it has not been established if this is a maturation effect or a cohort effect [60, 86]. Since it is primarily opposite sex touch avoidance that is associated with age, suggests a hormonal explanation; younger people in their prime reproductive years are more likely to approach opposite sex touch rather than avoid it [87]. But we may also be socialized to avoid touch as well, particularly same-sex touch. Same sex touch avoidance is, in part, a function of homophobia [88].

Touch avoidance is negatively related to self-esteem. Research suggests that children who have nurturing parents may be less touch avoidant, be more socially confident, and more socially skilled [60, 84]. A stereotype exists that individuals with high self-esteem are standoffish and aloof, but these results suggest the opposite; people who are comfortable with themselves are more comfortable with touch.

Across the touch avoidance literature and across the regions of the United States as well as European and Asian populations, males have consistently been found to be more avoidant of same-sex touch whereas females are more avoidant of opposite-sex touch [53, 57, 60, 67, 84]; an exception to this pattern may be found in Arab cultures, where male-male touch is frequent and acceptable. An alternative way of understanding this finding is that people from many different cultures avoid males. This cross cultural finding may be due to the fact that women throughout the world have greater consequences associated with sexual activity than men [89]. However, Crawford found that androgynous men had significantly lower same sex touch avoidance than very masculine men [90]. Likewise, Eman, Dierks-Stewart and Tucker found that androgynous and masculine individuals are less touch avoidant than feminine individuals, especially in opposite sex touch [91]. This suggests that more masculine individuals are socialized to engage in opposite sex touch but not same sex touch.

Touch taboos

One of the paradoxes of touch is that because it is such intimate, involving behavior, much of it is prohibited or constrained. In North America and Northern Europe, there are a number of haptic behaviors that constitute tactile transgressions and should be avoided [65, 92–93]. As we previously discussed, substantial cultural differences in tactile behavior make generalization to the entire world risky. However, the following nine principles, taken from Andersen and Jones, are helpful in understanding common touch taboos [65, 93].

The first principle is that touch must be relationally appropriate. Touch between supervisors and employees or among strangers may be excessively intimate and threatening. Too much touch on a first date can be relationally inappropriate and threatening. Nonfunctional touch should be avoided since it is perceived as excessively intimate and creates negative attributions.

Second, hurtful touches should always be avoided. This includes bone-crushing handshakes, play wrestling that gets too realistic, and aggressive nonconsensual sexual behavior. Even accidental touches such as bumping into someone or stepping on their toe are considered a tactile transgression that requires an apology.

Third, it is best to avoid startling another person with haptic behavior. Sometimes people try to startle friends as a joke, but most people do not think it's funny. Such inappropriate touches can cause relational problems and create undo anxiety for the recipient.

Fourth, touch should not be used to displace or relocate another person. Grabbing a spouse by the hand and dragging them along to speed them up or moving a person out of the way with a hand on the back in a crowd can produce surprisingly strong negative reactions. In most places a person has a right to the territory they presently occupy and relocating them is a tactile transgression.

Fifth, touches should not interrupt other people. Kissing your wife or husband during a phone call, hugging a child in the middle of his or her homework, or fondling your lover in the middle of a favorite television show are perceived as annoying rather than affectionate. Particularly annoying is the buttonholing technique where a person touches another person who is engaged in a conversation to get her or his attention.

Sixth, critical statements should not be accompanied by touch. Such a 'double whammy' is unsupportive and often perceived as unnecessarily aggressive or condescending. Imagine someone patting your arm while saying, "*You need to work harder if you hope to write a good report*". Even if the person used touch to try to soften the comment, it could be perceived as condescending or contemptuous.

Seventh, it is important to consider the situation where touch occurs. It may be fine to kiss one's spouse in the bedroom but maybe not in the boardroom or the classroom. Many people are uncomfortable with public displays of affection, so it is important to avoid certain types of touch in public settings unless you know how your partner will react.

Eighth, avoid touch that other people can interpret as unenthusiastic or insincere. This is particularly true of handshakes and hugs. A limp wristed handshake or an unenthusiastic hug is worse than not touching at all. Such half-hearted haptic displays send decidedly negative interpersonal and relational messages.

Finally, people should refrain from touches that other people perceive as inappropriate. Previously we documented the fact that various cultures, different families, and individuals have diverse tactile preferences. It is the touch initiator's responsibility to avoid bothering anyone haptically. People should refrain from any touch that the receiver dislikes, rebuffs, or asks to be stopped. Remember, "no" means no. Respect people's right not to be touched.

Summary

Haptics go beyond sensation or perception; touch is a fundamental part of human relationships and has the power to attract or repel, help or hurt, love or wound. Within social relationships, touch differs based on sex differences and relational stage. Various cultures perceive touch in a variety of disparate ways. While most people seek touch with loved ones and close friends, many people avoid touch, especially with strangers but even in close, intimate relationships. Beyond cultural rules about touching, touch avoidance, an interpersonal predisposition that causes some people to dislike and avoid touch, can have negative effects that undermine their closest relationships. In short, haptic behavior is the *sine qua non* of interpersonal interaction in all close relationships and perhaps the most basic and fundamental form of human communication.

Selected readings

Andersen PA (2008) Nonverbal communication: Forms and functions (Second edition). Waveland Press, Long Grove, Illinois

Field T (2002) Infant's need for touch. Human Development 45: 100–103

Guerrero LK, Andersen PA (1994) Patterns of matching and initiation: Touch behavior and avoidance across romantic relationship stages. J Nonverbal Behavior 18: 137–153

Hall JA, Veccia EM (1990) More "touching" observations: New insights on men, women, and interpersonal touch. J Personality and Social Psychol 59: 1155–1162

Jones S (1994) The right touch: Understanding and using the language of physical contact. Hampton Press, Cresskill, NJ

Learning effects in haptic perception

Hubert R. Dinse, Claudia Wilimzig and Tobias Kalisch

Introduction

Human haptic perception is not a constant, but subject to manifold modifications throughout lifespan. Major determinants are development and aging as well as alterations following injury and compensatory brain reorganization. While during early development haptic perception is refined due to maturation and experience, haptic perception during aging deteriorates due to many factors, not all of them fully understood. Besides these lifespan factors, it is common wisdom that haptic perception and skills in general improve through practice (Fig. 1), see also [1]. Perceptual learning involves relatively long-lasting changes to an organism's perceptual system that improve its ability to respond to its environment and are caused by this environment. In case of Blinds or in Musicians, both characterized by superior haptic perception, improvement is assumed to be due to use-dependent or experience-dependent neuroplasticity mechanisms. In any case, enhanced haptic perception is due to learning processes occurring in brain areas devoted to processing of haptic information. Modeling these processes not only contributes to a better understanding of perceptual learning, but also of cortical processing constraints present under baseline conditions.

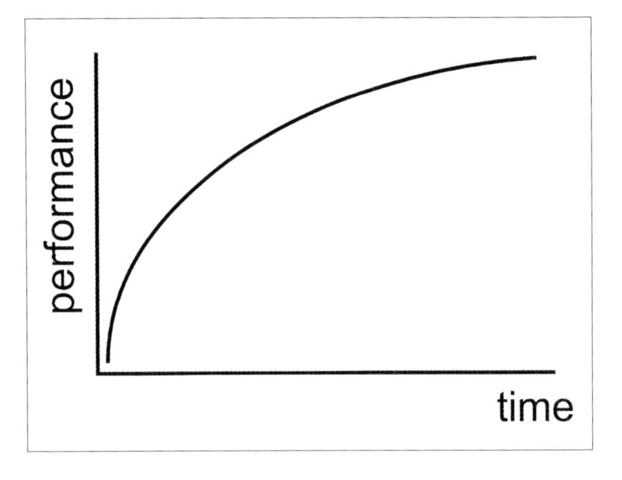

FIGURE 1. TIME COURSE OF IMPROVEMENT OF SKILLS OVER TIME.
Note rapid gain during the very early phase of practicing, while later additional improvement develops much slower indicating different mechanisms (see text).

Hierarchy of tactile-haptic performance

Gibson [2] defined the haptic system as "*The sensibility of the individual to the world adjacent to his body by use of his body*". The haptic perceptual system includes the peripheral receptors and is closely linked to the movement of the body. Many people have emphasized that haptic perception is closely linked to the concept of active touch. In our studies on haptic learning we refer to the idea of a hierarchy of tasks and task complexities that differentially involves in a graded way influences from the periphery, cortical processing with differing amount of cognitive demand, and tasks including motor aspect such as fine motor performance. Haptic learning can target specific tasks without necessarily altering the entire hierarchy. The investigation of either connected or unrelated aspects of tactile and haptic performance can help to unravel underlying mechanisms of haptic performance.

Table 1. Hierarchy of tactile – haptic-sensorimotor performance

Test	Peripheral	SI cortex Low cognitive	SI + higher areas High cognitive	sensorimotor action
	touch threshold Nerve cond. velocity	spatial discrimination two-point acuity hyperacuity gratings, dot pattern localization	recognition of known or arbitrary objects letters Braille characters	reaction times tapping peg board grasping grip strength
		everyday or arbitrary textures surfaces	symmetry shapes faces heights or widths	
		temporal discrimination frequency interval		

Dependence of haptic perception and learning on skin accessibility

The crucial role of skin coverage of the finger tips on tactile and haptic sensations and motor execution has been addressed in many studies. Figure 2 summarizes a comparison of different tasks performed in young adults and healthy elderly demonstrating that skin coverage severely impairs tactile function, however, in a highly task-dependent way. Moreover, the susceptibility is age-dependent, where some tasks such as two-point discrimination suffer little, while haptic object recognition suffers dramatically (Kalisch and Dinse, unpublished data, cf. also [3]). Remarkably, also motor performance depends on intact skin sensitivity. Recording of finger movement in expert touch typists under anesthesia of the right index fingertip increased typing errors of that finger seven-fold [4]. These studies emphasize that maximal tactile-haptic performance requires unrestricted skin access. However, little is known how prolonged skin coverage and training under coverage conditions as is the case in surgeons alter dependence of haptic performance from coverage.

Perceptual learning in haptic perception

Although training and practicing plays a crucial role mediating improvement of skills, perceptual learning is not achieved by a unitary process [5, 6]. Psychophysicists have distinguished between relatively peripheral, specific adaptations and more general, strategic ones [7–9], and between quick and slow perceptual learning processes [10]. Cognitive scientists have distinguished between training mechanisms driven by feedback (supervised training) and those that require no feedback, instead operating on the statistical structure inherent in the environmentally supplied stimuli (unsupervised training). In any case, perceptual learning exerts a profound influence on behavior because it occurs early during information processing and

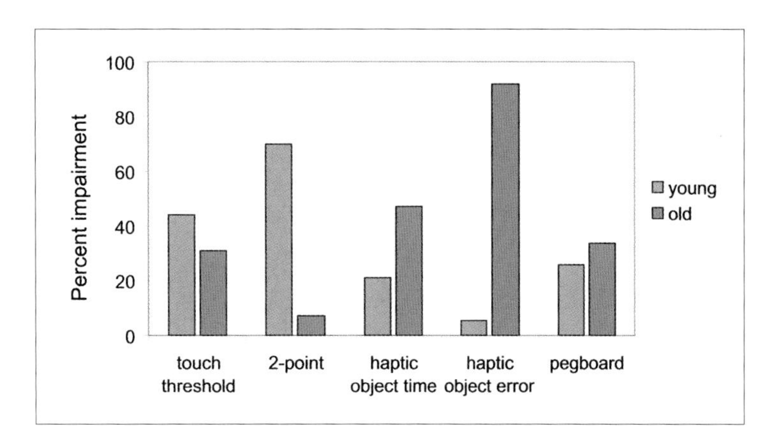

FIGURE 2. EFFECT OF SKIN COVERAGE ON HAPTIC PERFORMANCE
Shown is the task- and age-dependent impairment (percent difference compared to uncovered fingers) evoked by covering the fingers of young and elderly subjects with a rubber glove for various tactile, haptic and motor tasks.

thus shifts the foundation for all subsequent processes.

While there is a considerable amount of literature on perceptual learning of sensory and motor abilities, much less is known about training-based enhancement of haptic skills.

Time scales of perceptual learning

Practice-induced improvement can occur on many different time scales. For example, in some tasks, a few trials are sufficient to drive improvement. In other situations, performance improves only after weeks or months of practicing. What remains to be clarified are conditions that require either short- or long-term training. Also, under which conditions do improvements saturate. It has been suggested that one notable source of human inefficiency is that, unlike the ideal observer, human learning relies more heavily on previous decisions resulting in no learning on trials following a previous incorrect decision [11].

In studies of long-term training over weeks and months, one typically finds an early phase of rapid improvement followed by a second phase of much slower learning (cf. Fig. 1). It has been suggested that the early phase represents task understanding and the development of an appropriate strategy, while the second phase of improvement is due to selective and specific changes in cortical processing [12]. This view is consistent with the observation that gain obtained during the first phase is unspecific and transfers easily to other sites, whereas the gain accumulated during the second phase is highly specific to the trained condition. For an elaborated account of this assumption for visual perceptual learning see [13].

General rules in perceptual learning

An important question in evaluating studies of perceptual learning is how do we know what to learn? In other words, how does a neural system know which information is behaviorally relevant and which is not? There must be mechanisms that gate what is learned (i.e., to control what aspects are allowed and what aspects are restricted).

Mechanisms in perceptual learning (according to [5]):
- By *attention weighting* perception becomes adapted to tasks and environments by increasing the attention paid to important dimensions and features.
- By *imprinting*, processing mechanisms are developed that are specialized for stimuli or parts of stimuli.

Box 1.

How much daily training is needed for learning an auditory frequency and temporal-interval discrimination task was investigated by varying the number of training trials per day (360 or 900 trials per day for 6 days). Improvement on frequency discrimination required more than 360 trials of training while learning on temporal-interval discrimination occurred within the 360 training trials with additional daily practice having resulting no further improvement [14]. In another study, improvement in human localization (eight-alternative forced-choice with feedback) performance of a target randomly sampled from four elongated Gaussian targets with different orientations and polarities was observed within four trials lasting less than 1 min. Comparing human learning to an optimal Bayesian algorithm it was found that human learning was slower and incomplete with respect to the optimal algorithm [11]. Short-term practicing effects were also addressed in a haptic object recognition task, where subjects performed three consecutive sessions lasting between 10–30 min. Subjects improved for both time needed to complete the task and for numbers of incorrect recognitions by about 15% independent of age and gender [15].

- By *differentiation*, stimuli that were once indistinguishable become psychologically separated.
- By *unitization*, tasks that originally required detection of several parts are accomplished by detecting a single constructed unit representing a complex configuration.

Specificity

Perceptual learning is often characterized by a high specificity to stimulus parameters such as location or orientation of a stimulus with little generalization of what is learned to other locations or to other stimulus configurations. Selectivity and locality of this type implies that the underlying neural changes are most probably occurring within early cortical representations that contain well-ordered topographic maps to allow for this selectivity.

Transfer

In addition, a transfer of the newly acquired abilities is often considered an important marker of that processing level at which changes are most likely to occur: Limited generalization is taken as evidence for high locality of effects in early representations. In contrast, transfer of learned abilities is taken as evidence for the involvement of higher processing levels as is often observed in task and strategy learning [12, 16–19]. However, as summarized below, transfer and generalization is highly task-specific.

Feedback and reinforcement

Feedback provides valuable information about the correctness of performer's responses during learning. Correct feedback conditions in a Vernier discrimination task yielded a larger improvement of performance than manipulated and no feedback conditions. However, providing feedback that was uncorrelated to the performer's responses prevented learning [20]. On the other hand, studies of perceptual learning for hyperacuity found that even within-session learning occurs without feedback [21]. One explanation that perceptual learning can occur in the absence of external reinforcement is that internal reinforcement can serve as a learning signal. For instance, stimuli that are highly discernable can serve as a template that subjects can use to assess stimuli in more difficult conditions. It has been suggested that fast-learning studies often do not require feedback but that slow-learning protocols do [22]. According to recent fMRI (functional magnetic resonance imaging) data, the caudate nucleus responded to positive and negative feedback during perceptual learning in a manner analogous to its processing of extrinsic affective reinforcers indicating that this region may be a critical moderator of the influence of feedback on learning [23].

Attention and motivation

One way in which perception becomes adapted to tasks and environments is by increasing the attention paid to perceptual dimensions and features that are important [5, 24]. This view is well in accordance with recent models of visual attention that assume that efficient stimulus processing relies on an interaction between sensory (controlled) and executive (controlling) processes [25–27]. Within this framework, motivation might reflect an important factor for establishing robust perceptual representation even when conditions are sub-optimal or not efficiently controllable by executive functions [28–29]. For an unifying account about the role of modulators of perceptual learning see [30, 31].

External and internal noise

According to different theories practice improves discrimination by enhancing the signal, diminishing internal noise or both. This question has been studied mostly in the visual domain where signal strength and noise has been systematically varied. Data suggest that mechanisms of perceptual learning may rely on a combination of external noise exclusion and stimulus enhancement *via* additive internal noise reduction [32–33].

Magnitude of learning-induced changes

The amount of improvement of haptic performance depends on time (cf. Fig. 1). As discussed above, during the early phase subjects often improve dramatically, while the gain accumulated later in training saturates. Typical values of improvement are in the range of 10–25% compared to baseline conditions. It should be emphasized that the percent changes observed during haptic learning are several fold smaller than the typical alterations accompanying the decline of age-related haptic performance at high age.

Neural representation of learning

In search of a neurobiological basis of perceptual learning, recent studies have shown that skill acquisition is associated with selective plastic reorganizational changes in the cortical representations of those body parts that received stimuli during perceptual learning. Recanzone and co-workers were the first to demonstrate alterations of cortical sensory processing developing during perceptual learning [41]. They showed that tactile frequency discrimination training in monkeys over many sessions leads perceptually to a reduction of the discrimination threshold and cortically to an expansion of the maps in somatosensory cortex that represent the used finger. Most important, this study showed that there was a linear correlation between the individual amount of perceptual improvement induced by the training and the individual amount of cortical map expansion [41]. These data imply that cortical map size is a reliable predictor of the individual performance. This form of adult sensory plasticity has now been observed in all sensory systems including auditory, somatosensory and visual areas.

Since perceptual learning involves plasticity in the primary sensory cortices, the outcome of learning can be visualized not only in animal models, but also in human subjects using functional magnetic resonance imaging or high-resolution EEG. Pleger and co-workers [42, 43] and Dinse and co-workers [44] demonstrated that the amount of tactile discrimination improvement correlated with the amount of enlargement of the somatosensory hand representation (cf. Fig. 3). Thus, tactile learning paradigms provide a 'window' through which fundamental mechanisms of perceptual learning can be investigated. By this it becomes possible to evaluate the efficacy of perceptual learning independently on two levels: psychophysically and in terms of brain organiza-

Box 2.

In the tactile domain, Sathian and co-workers [34] provided evidence that practice-related improvement is highly specific for properties of the stimulus used in training. They reported substantial interdigital transfer of practice effects for discrimination of gratings varying in either spatial parameter and also for spatial acuity-dependent discrimination of grating orientation. Although tactile learning was task specific as shown in other sensory systems, there was generalization across fingers, unlike visual learning, which is highly location specific. In another, so-called tactile hyperacuity task, subjects had to discriminate a row of three dots in which the central dot was offset laterally from a row without such offset. Performance at the right index fingerpad improved with practice, which transferred completely to the left index finger demonstrating intermanual transfer [35]. To study tactile interval discrimination, subjects were trained for 900 trials per day for 10–15 days. Learning at the trained base interval generalized completely across untrained skin locations on the trained hand and to the corresponding untrained skin location in the contralateral hand. There was partial generalization to untrained base intervals similar to the trained one, but not to more distant base intervals. Interestingly, learning with somatosensory stimuli generalized to auditory stimuli presented at comparable base intervals [36]. These results demonstrate temporal specificity in somatosensory interval discrimination learning that generalizes across skin location, hemisphere, and modality.

The ability to detect the symmetrical or asymmetrical characteristics of individually presented plastic nonrepresentational shapes was tested over 4 days. This brief training led to improved performance as indicated by a decrease in scanning time and number of identification errors. However, learning depended on form parameters of symmetry and the type of scanning strategy used to explore a stimulus [37]. These data suggest that perceptual experience plays a major role in the efficient pick-up and utilization of haptic information.

Fast perceptual learning was investigated with respect to the perception of the heights or widths of wielded non visible rectangular objects. In this particular task, inertial differences are the basis for perceived size differences. Variables analyzed were attunement (attending to the task-relevant inertial variable), calibration (scaling spatial extent to the task-relevant inertial variable), and exploratory behavior (wielding so as to differentiate the task-relevant inertial variable). Subjects had to perform 25 trials with a set of practice objects; those trials were followed and preceded by 18 trials with a set of test objects. Practice, with knowledge of results, improved both attunement as measured by regression of perceived spatial extent on the inertial variables, and calibration, as measured by constant and variable error. In the absence of knowledge of results only variable error improved with practice. In both conditions, however, exploratory behavior decreased in duration and complexity, as measured by recurrence quantification analysis [38]. The data revealed a substantial complexity of mechanisms involved in fast perceptual haptic learning dependent on knowledge.

The role of knowledge in haptic exploration was specifically addressed in a haptic classification task. First, haptically available object properties were assessed by means of a questionnaire. Then the hand movements executed during haptic classification of manipulable common objects were examined. Manual exploration consisted of a two-stage sequence, an initial generalized 'grasp-and-lift' routine, followed by a series of more specialized hand-movement patterns. Interestingly, the specialized movements were largely driven by the knowledge of the properties of the specific objects [39].

Healthy subjects of different age-groups (covering 20–89 years) performed a cross-modal 3D shape-matching test based on artificial, cubic objects modified according to [40], which had to be explored haptically by hand movements and afterwards allocated to visually presented samples. There was a significant decline of cross-modal visuo-haptic performance with increasing age [15]. Both parameters of the visuo-haptic test, i.e., the time to complete the test and the number of errors increased significantly with age. Furthermore the age-related decay of haptic performance was stronger in female subjects. Interestingly, the individual level of sensory hand functions (touch threshold or two-point discrimination threshold) did neither predict the age-related decrease of haptic performance, nor the gender-specific differences in haptic performance. On the other side general cognitive abilities were found a reliable indicator of visuo-haptic performance in old age to a greater extent than sensory hand functions.

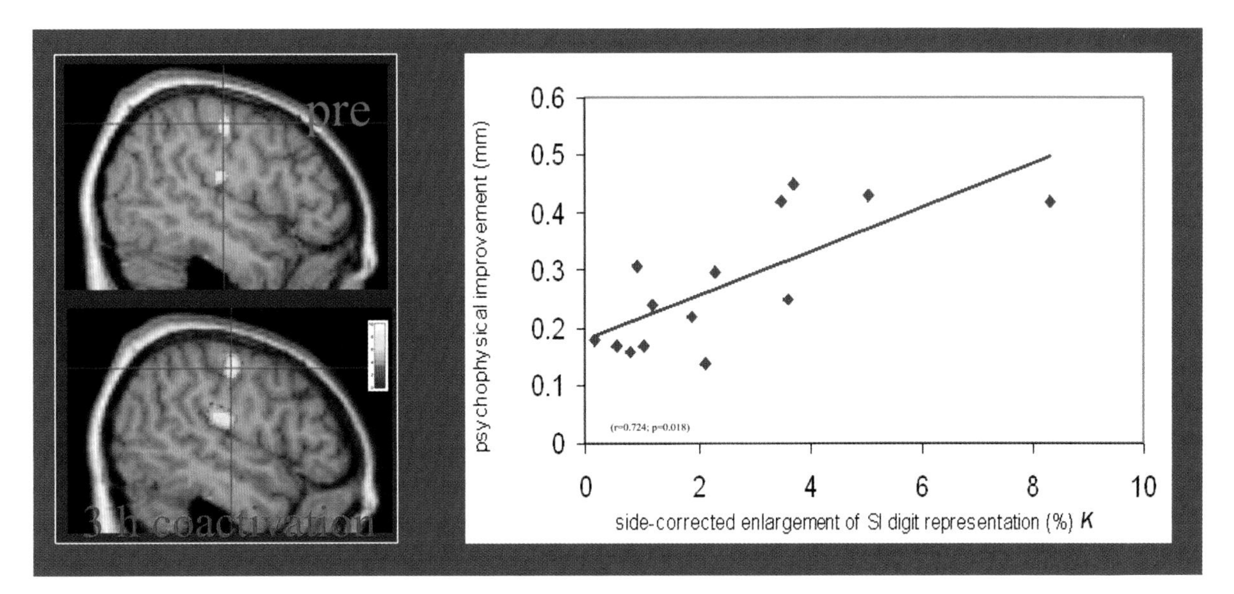

FIGURE 3. SINGLE SUBJECT COACTIVATION EFFECT ON CORTICAL MAPS

(A): BOLD (blood oxygenation level dependent) signals detected pre- and post coactivation in the contralateral SI in the postcentral gyrus, and in the contralateral SII in the parietal operculum above the Sylvian fissure. Comparing pre- with post-coactivation fMRI sessions revealed enlarged activation and increased BOLD signal intensity in SI and SII (secondary somatosensory cortex) representing the coactivated IF (index finger). These changes of BOLD signals recovered 24 h after coactivation (not shown). Linear correlation analysis between perceptual and cortical changes in SI (Pearson) revealed that psychophysical improvement correlated with the size of cortical map expansion indicating a causal relation (B). To express the coactivation effect on BOLD signals of the contralateral SI, the corresponding number of activated voxels per cluster were used to calculate $K=((right_{post}-right_{pre})-(left_{post}-left_{pre}))/right_{pre}$. *"K" was positively correlated with changes in psychophysical thresholds (r=0.744; p=0.002). Modified according to [43], with permission from* Science.

Box 3.

A recent neuroimaging study addressed the question of whether sensory cortical regions support retrieval of perceptual knowledge. Semantic decisions that indexed tactile, gustatory, auditory, and visual knowledge specifically activated brain regions associated with encoding these sensory experiences. Retrieval of tactile knowledge was specifically associated with increased activation in somatosensory, motor, and premotor cortical regions. In contrast, decisions involving flavor knowledge increased activation in an orbitofrontal region previously implicated in processing semantic comparisons among edible items. Perceptual knowledge retrieval that references visual and auditory experiences was associated with increased activity indistinct temporal brain regions involved in the respective sensory processing [45]. These results indicate that retrieval of perceptual knowledge relies on brain regions used to mediate sensory experiences with the referenced objects.
Individual differences in regional brain volumes modified by the effects of age and sex on the acquisition of a perceptual-motor skill in healthy adults were studied in subjects performing five five-trial blocks of a mirror-drawing task on three separate days. Index of performance was time to completion and number of errors committed. All participants improved with practice, but younger adults performed better than their older counterparts, and women performed better than men. Four brain regions – lateral prefrontal cortex, hippocampus, cerebellar hemispheres and the caudate nucleus – were analyzed. All regional volumes were negatively associated with age. Larger lateral prefrontal cortex was associated with better performance on the mirror drawing task, and this link was stronger in the older participants and was strengthened at the later stages of learning. Larger caudate was related to better performance, especially at later learning, among men, but among women the link was evident only during early learning. Thus, mirror-drawing represents a task that evokes activation which is highly dependent on age, gender and skill level [46].

tion offering a unique way to quantify and visualize learning processes.

Role of intensified or reduced use

Continuous and long-lasting practice of sensorimotor skills results in defined expansions of cortical representational areas as described for blind Braille readers and musicians [47–53]. As a consequence, their tactile and haptic performance becomes better.

While MEG and EEG studies demonstrated that musicians have altered cortical representations, there is still controversy about the specificity of these reorganizational changes. There are two possibilities: These changes are highly specific in the sense that they allow for improvement of the trained motor or perceptual skill only, i.e., the neural changes arising from skill training are assumed to have little consequences for information processing beyond that skill. In an alternative scenario, neural changes result in a widespread modification of the entire sensory processing. In this case, extensive consequences in terms of perceptual and behavioral abilities are to be expected that must generalize beyond the trained skill. There seem to be no simple answers as both has been found in sensory and motor domain suggesting some modality-specificity of this form of adaptation.

While there is agreement that use is a major factor driving plasticity of cortical processing and cortical maps thereby driving the development of outstanding sensorimotor skills, little is known how periods of disuse affect cortical representations and parallel perceptual performance in humans.

Box 4.

To address this question, the impact of about 3 weeks of immobilization of finger and hand due to cast wearing was investigated. Cast wearing significantly reduced the use of the immobilized limb, which was associated with a severe loss of tactile acuity and a reduced BOLD (blood oxygenation level dependent) response of cortical representational maps of the immobilized finger. In contrast to these maladaptations, compensatory effects on the left, healthy hand were observed: the frequency of use was substantially enhanced, and in case of right-hand immobilization, the perceptual performance of the left, healthy hand became even superior in comparison to healthy controls. About 2 weeks after cast removal, perceptual and cortical changes were fully reversible; however, enhanced tactile acuity of the healthy hand persisted [54]. The results demonstrate that disuse as occurring under the everyday-life situation of cast wearing impairs tactile performance and functional brain organization on a time scale of days. Accordingly, the concept of use-dependent plasticity can be extended into maladaptive regimes under conditions of disuse.

Role of passive stimulation

It is now clear that in addition to training, practice and experience, i.e., intensified or reduced use perceptual performance can also be improved by passive stimulation (i.e., through the statistics of the input) on a time scale of only a few hours or less. According to recent behavioral studies perceptual learning of visual motion can occur as a result of mere exposure to a subliminal stimulus, without external reinforcement, the subject actively attending to a task, or the motion-stimulus being a relevant feature of the particular task [55–58]. The coactivation paradigm developed by Dinse and co-workers induces improvements of tactile and haptic performance, and results in cortical reorganization thereby adding an important alternative for inducing perceptual changes without training [31, 42–44, 59–72].

In order to explain this effectiveness it has been suggested that all these types of learning occur through similar processes. Namely that the key to learning is that sensory stimulation needs to be sufficient to drive the neural system past the point of a learning threshold [31]. Factors such as reinforcement or motivation appear to play a permissive role, but optimizing sensory inputs by implementing specific high-frequency or burst-like stimulation, which are known to induce synaptic plasticity in brain-slice preparations, can also serve to boost responses that normally are insufficient to drive learning past this learning threshold.

Coactivation

Utilizing the knowledge about brain plasticity accumulated over the last years [61, 74], it has been suggested that specific stimulation protocols can be designed through which it becomes feasible to change purposefully brain organization and thus perception and behavior. Based on this concept, we have developed a so-called coactivation protocol, through which we can enforce localized activation pattern in the brain [42–44, 59, 60, 65]. Coactivation follows closely the idea of Hebbian learning [75]: Synchronous neural activity, which is regarded instrumental to drive plastic changes, is generated by the simultaneous tactile 'co-stimulation'. Coactivation leads to an improvement of tactile acuity and an expansion of corresponding cortical maps. The amount of perceptual gain resulting from this procedure linearly correlates with the amount of cortical reorganization and sug-

Box 5.

There have been many reports according to which prolonged and unattended stimulation is ineffective to drive plastic changes. Pairing of sensory stimulation with electrical stimulation of the Nucleus basalis (NB) has been shown to result in reorganization of cortical maps, however, simple sensory stimulation alone without NB stimulation was ineffective [73]. These apparent discrepancies can be settled in the light of the observation that simple (i.e., small field) prolonged passive stimulation had no effect on discrimination abilities, and that more massive, 'coactive' stimulation was required for plasticity [43, 65, 31]. Accordingly, in order to be effective, the passive stimulation has to include specific constraints such as high-frequency, burst-like stimulation described above.

gests a causal relation (Fig. 3), [31, 42–44, 62, 64, 65]. Coactivation-induced improvement of tactile acuity is quite large (15–20%), whereas gain after years of practice in musicians was in a range of 20–25% [65].

Technical aspects

To apply coactivation, a small device consisting of a small tactile stimulation device is taped to the tip of a finger. The device allows stimulation of the skin portions underneath thereby coactivating the receptive fields within this area of approximately 1 cm^2 (Fig. 4). In the standard version, coactivation is applied for 3 h using stimuli drawn from a Poisson process at inter-stimulus-intervals between 100 to 3,000 ms with an average stimulation frequency of 1 Hz and a pulse duration of 10 ms. Pulse trains required to drive the solenoid were stored on MP3-player, permitting unrestrained mobility of the subjects during coactivation.

In order to optimize coactivation to maximize effectiveness we have developed a number of protocols (Patents DE) such as employing electrical stimulation, multifinger coactivation [72], and high-frequency stimulation [69, 70], the latter allows to drive changes in 20 min only. Other options include combination with transcranial magnetic stimulation [67], drugs [44, 63, 71], or repeated applications (unpublished) that potentiate the effects of coactivation. In Figure 5 the effects of different coactivation protocols on magnitude and stability of evoked tactile improvement are summarized.

Other passive methods

Coactivation is not the only form of unattended activation-based learning. Under the assumption that simultaneously applied coactivation induces synchronous neural activity at a cortical site selected by the location of tactile stimulation, we went one step further and short-cut the sensory pathway. To this end we applied so-called high-frequency (5 Hz) transcranial magnetic stimulation (TMS) from outside the scull directly to selected brain areas in order to induce synchronous activity. Applying 20 min of 5 Hz TMS over the finger representation of SI lead to similar changes as observed after coactivation. These changes were paralleled by an expansion of the cortical finger representation and an increase of cortical excitability in SI (primary somatosensory cortex) and recovered after about 2–3 h [76, 77]. These findings demonstrate that meaningful improvement of perceptual performance can even be obtained by specific (high-frequency) stimulation of brain areas from outside the scull. Most notably, this intervention does not leave the cortical processing in a disorganized state, but on the

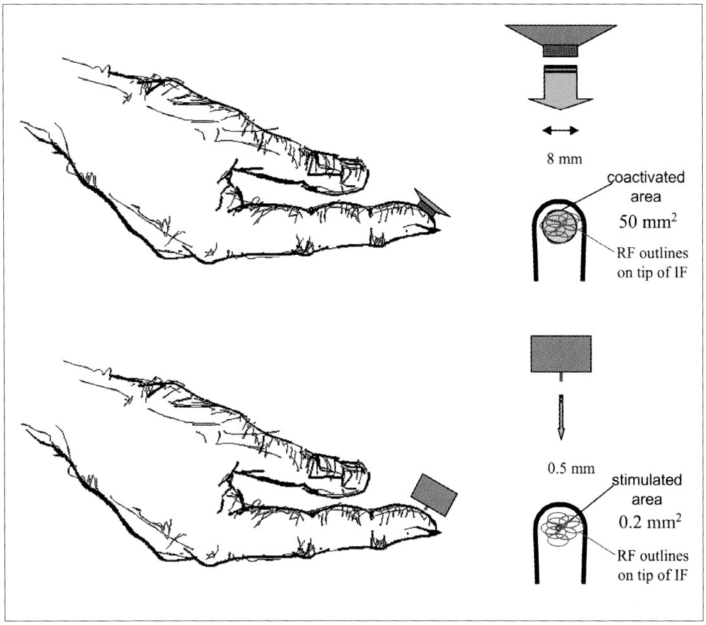

Figure 4. Application of coactivation
A. A small solenoid with a diameter of 8 mm is mounted on the tip of the right index finger (IF) to coactivate the receptive fields representing the skin portion under the solenoid (50 mm²). B. Control protocol to demonstrate the requirement for 'co-activation'. Application of a so-called single-site stimulation: A small device consisting of only one tiny stimulator (tip diameter 0.5 mm) is mounted on the tip of the right IF to stimulate a single 'point' (0.8 mm²) on the skin. Duration of stimulation is 3 h. Reprinted with permission from Neuron *[43].*

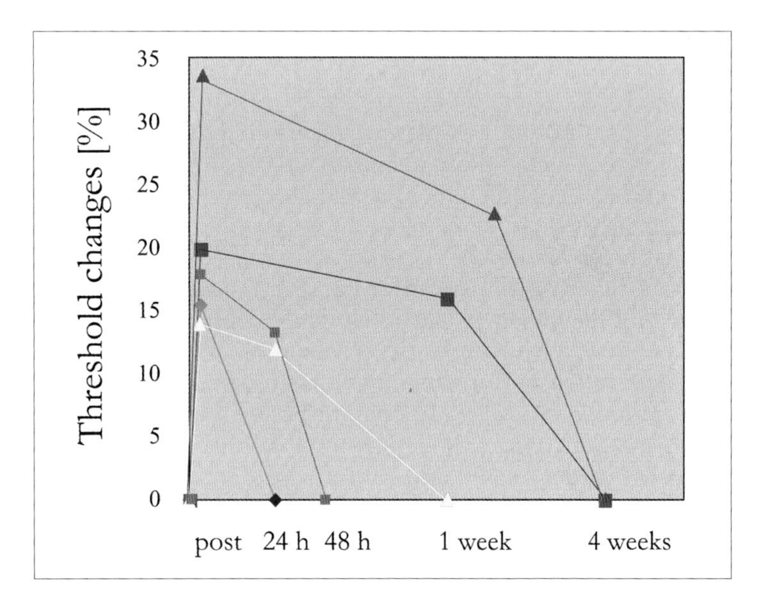

Figure 5. Time course of improvement (percent difference compared to baseline conditions) of tactile acuity for various coactivation protocols
Dependent on the refinement of the protocol used, induced changes are comparable in magnitude to changes observed in subjects undergoing heavy schedules of training. Repeated application of coactivation over many days stabilizes the improvement. Green: 1-finger coac [43], pink: 5-finger coac [72], yellow: 20 min high-frequency [69], blue: 5-finger 20 min high-frequency [unpublished], red: 5-finger 20 min high-frequency, repeated every 3 days [unpublished].

contrary leads to the emergence of a different, yet organized and meaningful behavior, as indicated by the improvement of discrimination performance.

Wide-range effectiveness of coactivation

Central to coactivation is that it does not require training of particular tasks, which makes coactivation-induced improvements task-independent. Therefore, testing for tactile acuity does not imply that acuity is in some way specific to coactivation. On the contrary, we assume that coactivation alters the entire way of neural processing related to tactile and haptic information. Consequently, one can expect that coactivation not only affects two-point discrimination thresholds, but basically all tasks related to tactile, haptic and sensorimotor processing. To test this assumption, we initiated a more comprehensive study in which we looked for a number of haptic tasks beyond two-point discrimination (Tab. 2).

In fact, consistent with these assumptions, coactivation leads to a remodeling of a wide range of tactile, haptic and even sensorimotor tasks (Tab. 2). Also consistent, both improvement and impairment is induced. The only parameter not affected by coactivation is touch threshold, which, according to Table 1, represents the most peripheral parameter. In contrast, of all parameters tested all have been affected by coactivation: tactile acuity (as measured by two-point or grating discrimination), frequency discrimination, localization, dot-pattern discrimination, haptic object recognition, decision making (multiple choice reaction times) and sensorimotor performance such as assessed by pegboard. Interestingly, all of the described parameters improved with the exception of localization tasks (within finger or across fingers). Similar results about a trade-off between discrimination and localization performance have been demonstrated under everyday-life conditions for blind Braille readers [50, 78], indicating that this is not coactivation-related, but rather represents a typical property.

Cellular studies suggest that there might be only a few, basic mechanisms that control synaptic transmission. In particular, the N-methyl-D-aspartate (NMDA) receptor has been implicated in synaptic plasticity [79–81]. In order to demonstrate that coactivation is mediated by established plasticity mechanisms, we tested its dependency on NMDA receptor activity. We found that memantine, a substance known to block selectively NMDA receptors [82], eliminates coactivation-induced learning, both psychophysically and cortically [44]. These data show that coactivation-induced perceptual learning and associated cortical changes are controlled by basic mechanisms known to mediate and modulate synaptic plasticity. More generally, these data demonstrate that by using specific drugs the outcome of coactivation can be further amplified.

Table 2. Joint changes of tactile-haptic-sensorimotor performance after coactivation

Task/parameter	effect of coactivation
touch threshold	no effect
two-point discrimination	+
Localization	-
finger mislocalization	-
frequency (flutter) discrimination	+
reaction times	+
Braille sign recognition	+
fine motor movements (finger–hand)	+
haptic object recognition	+
every day life performance	+

Musicians

The potential of coactivation is not limited to young adult subjects, but has been applied in musicians and elderly subjects, whose performance is already altered.

In professional pianists, tactile performance was significantly superior to that found in non-musician controls indicating that piano players benefit from their daily routine by developing significantly reduced tactile discrimination thresh-

Box 6.

Surprisingly, despite the better baseline performance coactivation resulted in an even higher gain of tactile acuity. While the baseline performance correlated well with the duration of daily piano practicing, the coactivation-induced improvement was correlated with the number of years of extensive piano playing [68]. These findings imply stronger capacities for plastic reorganization in pianists, and points to enhanced learning abilities. This kind of meta-plasticity suggests that extensive piano practicing alters somatosensory information processing and sensory perception beyond training specific constraints. The data also suggest that the potential of coactivation applies also in subjects characterized by high-level baseline performance. Support for enhanced learning capabilities comes from a recent study employing paired associative stimulation (PAS), which consists of an electric median nerve stimulus repeatedly paired (200 times at 0.25 Hz) with a TMS pulse over the hand motor area. Musicians showed a wider modification range of synaptic plasticity than no-musician controls, which suggests that Musicians regulate plasticity and excitability with a higher gain than normal [83].

olds, although piano playing is little related to tactile acuity abilities. Therefore, the question was whether there is room for further improvement.

Age

In contrast, elderly subjects are characterized by a significant decline of tactile, haptic and sensorimotor abilities. However, this decline is not irreversible, but treatable through various measures (Fig. 6).

The typical approach to ameliorate age-related changes is to subject elderly to intense schedules of training and practicing, and there is no doubt about the effectiveness of training-based intervention even at high age [84–88]. However, since many elderly suffer from restricted mobility, additional and alternative approaches are needed which supplement and enhance, or even replace conventional training procedures.

Taken together, these results show that age-related decline of perception and behavior can at least be ameliorated. This is important as it implies that age-related loss of mechanoreceptors appear to play a minor role in mediating and maintaining tactile and haptic function.

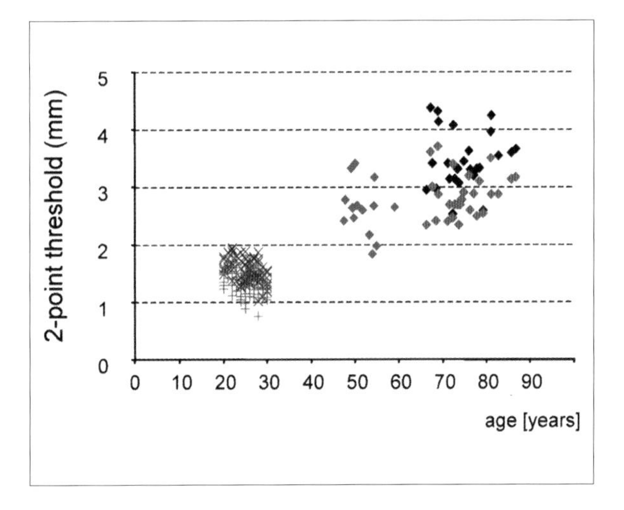

Figure 6. Tactile two-point discrimination thresholds of the tip of the right index finger as a function of age (total of 120 subjects)

After coactivation (pink symbols), thresholds of the coactivated subjects (young control group and group of elderly) were significantly reduced. Coactivation-induced improvement in the group aged 66–86 was several-fold stronger in magnitude as compared to the young subject. As a result, after coactivation thresholds of the elderly resembled those found in the subjects 47–55 years old. Reprinted with permission from Annals of Neurology *[66].*

Box 7.

To show that the age-related impairment of tactile acuity can be ameliorated by coactivation, we studied discrimination thresholds in healthy subjects aged 66–86 years. We found that 3 h of coactivation reduced two-point thresholds in aged subjects similar as described above for young adults. In this way, tactile acuity of a 80 year old subject comes to match that typically found at an age of 50 [66, 89]. However, the lowest thresholds in elderly after coactivation were still above thresholds typically observed in young subjects. Further experiments are therefore required to clarify if the performance observed after coactivation represents the lowest limit in acuity that can be reached by elderly given the anatomical and morphological changes accumulating over age, or if discrimination thresholds can be further reduced using more refined intervention methods. Recent data from a study on tactile acuity in elderly blind subjects revealed the maintenance of very low discrimination threshold even at high age [90], which supports the notion that tactile acuity in elderly can be preserved at a level typically found in young subjects under conditions of heavy regular practicing.

There have been alternative attempts to interfere with the age-related decline of sensory capacities. For example, the addition of noise can improve the ability to transfer reliably information, a phenomenon known as stochastic resonance. Electrical noise stimulation to the hand of elderly subjects lowered touch thresholds [91], while noise stimulation to the foot improved sway parameters in young and elderly subjects [92]. While stochastic resonance affects thresholds by enhancing inputs otherwise subthreshold, coactivation most likely alters the modes of neural processing due to specific changes of synaptic efficacy and synaptic connections [42–44]. Taking a pharmacological approach, a recent report showed enhanced encoding of motor memories in elderly adults, up to levels present in younger subjects, by restoring dopaminergic function in the elderly subjects through administration of a single oral dose of levodopa [93].

Trade-offs and interdependence

One of the hottest questions in learning is whether the enhancement of one ability occurs on the cost of others. In other words, is there a trade-off between abilities we gain while others are lost at the same time? A related question addresses the putative interdependence between various tasks. Both aspects are interesting because they allow insight into constraints of cortical processing (cf. Modeling).

In tactile perception the answer is clear: When subjects improve their tactile discrimination abilities (i.e., to perceive two closely spaced stimuli as two), this develops on the costs of localization performance, i.e., to accurately tell the position of a stimulus on the skin. Interestingly, both types of changes evolve without altering touch thresholds, which imply a trade-off between localization and discrimination on the one hand, but independence of both parameters from touch threshold on the other hand. Another example comes from haptic exploration, where haptic recognition is poorly associated with tactile acuity, but correlated with measures of non-verbal intelligence (Raven matrix [15]). In Figure 7 the joint dependences under baseline conditions are summarized for touch threshold, two-point discrimination and haptic object recognition revealing a complete lack of correlations. These findings imply that there exist cortical processing modes that cannot be optimized in parallel. In this view, the average performance encountered in an individual performer reflects most likely a sort of a balanced compromise trying to achieve optimal, but not maximal performance. Through perceptual learning it is possible to further enhance certain skills, but not all, resulting in a parallel improvement of some skills on the cost of impairment of others.

How that can be explained in terms of cortical processing is described in the next paragraph.

Modeling tactile learning

There is now compelling evidence that cortical population activity provides the most promising bridging between behavior on the one hand, and neural activity on the other hand [94–96]. To understand tactile spatial discrimination performance and its alterations during learning we provided a general framework using a mean field approach [97] (Fig. 8). In this approach, cortical population activity within cortical maps is modeled with a Mexican hat interaction of short-range excitation and longer-range inhibition [98, 99]. Within such a representation tactile stimulation evokes either single peaks of activation coding for the subjective experience of a single point on the skin while bimodal distributions are read out for two-points.

In the cortex the skin surface of the finger is represented as a topographically organized, two-dimensional map in which neighboring stimuli evoke neighboring peaks of activation of populations of neurons. In a simplified one-dimensional version, two neighboring inputs are represented along a one-dimensional cut through the two-dimensional topographic map.

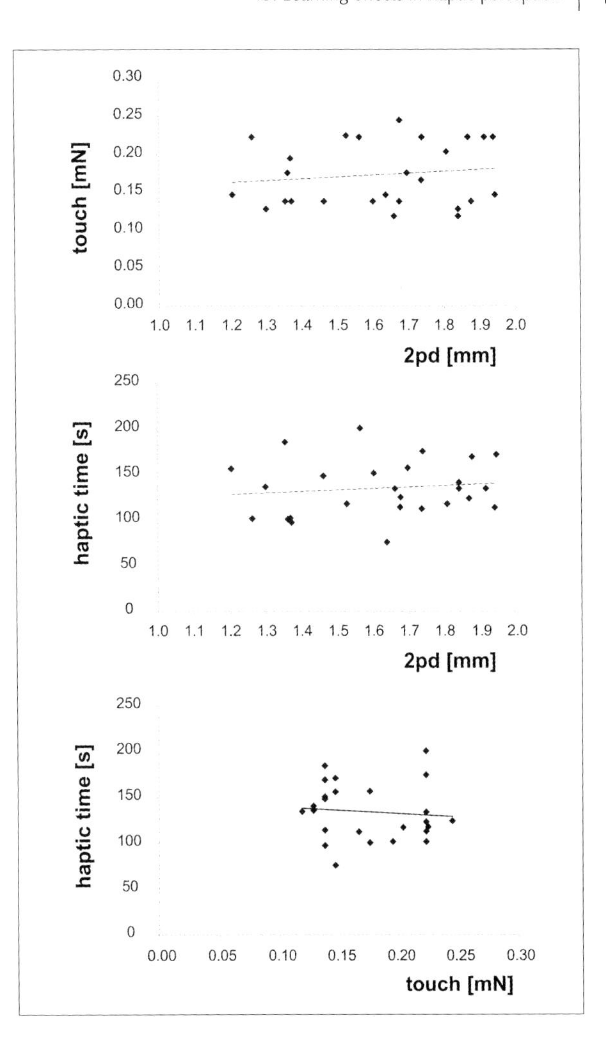

FIGURE 7. RELATION BETWEEN THREE MEASURES OF TACTILE PERCEPTION

Touch threshold (touch, the lowest force that allows detecting the presence of a stimulus on the skin of the fingertip), spatial two-point discrimination (2pd, the critical distance at which two simultaneously presented stimuli on the skin of the fingertip are perceived as two), and haptic object recognition (haptic time, time needed to correctly identify arbitrary three-dimensional objects made from Lego® bricks). There is an overall lack of correlation between measures: Top: touch vs. two-point (Pearson, $r=0.127$, $p=0.537$)*; middle: haptic recognition* versus *two-point (Pearson,* $r=0.118$, $p=0.566$)*, bottom: haptic recognition* versus *touch (Pearson,* $r=0.099$, $p=0.632$)*. Data from [15].*

Influence of lateral inhibition on discrimination and localization performance

In the model, stimulation at a single site evokes single peaks of activation coding for the subjective experience of a single point, while simultaneous stimulation at two sites (like in two-point discrimination experiments) evokes two peaks, which are read out for perceiving two points. It has been shown experimentally that for large distances two peaks of activation interact only weakly, while lateral inhibition leads to substantial suppression for shorter separations [100].

In contrast, localization refers to the ability to precisely read out the location of monomodal

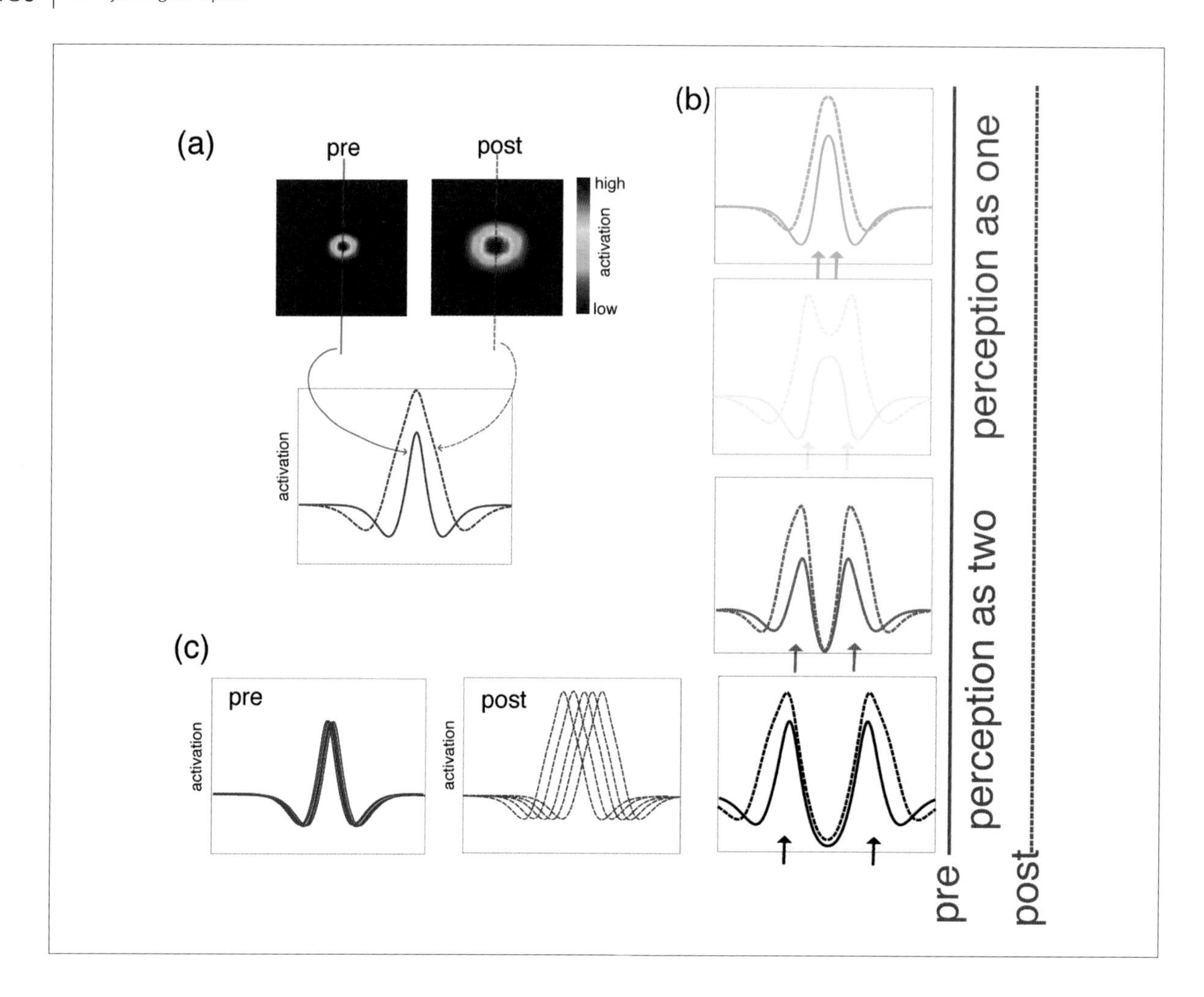

distributions of population activation. Several approaches have been formalized how single parameters may be estimated from population distributions of activation [101]. Spatial discrimination and localization are affected oppositely by lateral interaction: Lateral inhibition reduces coexisting, ongoing activation present at other locations of the cortical population. As coexisting activation is necessary for bimodal activation profiles coding for the perception of two points strong lateral inhibition deteriorates discrimination performance. On the other hand lateral inhibition suppresses the influence of noisy fluctuations of activation leading to a subsequent reduction in the variance of the read out peak position. Thus lateral inhibition improves localization performance but decreases discrimination performance.

Influence of learning on a computational level: suppression of inhibition

There is ample experimental evidence that learning processes are mediated through a decrease of inhibitory interaction. For example, coactiva-

Figure 8. Modeling learning induced changes of cortical population activity and its implications for tactile perception

a. Cortical population activity evoked by single site stimulation modeled as localized activation using a mean field approach with center-surround inhibition. Shown are a two-dimensional simulation (corresponding to a cortical activity distribution as measured by fMRI) for a pre and a post condition (a. top row) and a one-dimensional simulation (a. bottom row). For simplification we use in the following a one-dimensional model (i.e., a one-dimensional cut through cortical surface) but emphasize that all results are applicable for two-dimensional models accordingly. When the strength of inhibitory interaction is weakened through learning, the size of the cortical representation becomes larger, both for the width of the cortical area activated by the stimulus and the height of the activation (cf. pre-post).

b. For two-point stimulation (simultaneous stimulation at two sites) two spatially very close stimuli evoke a monomodal distribution of activation (green and yellow, solid lines; arrows depict location of incoming stimuli) which corresponds to the percept of one stimulus. For larger distances a bimodal distribution of activation develops, which is assumed to code for the sensation of perceiving two stimuli (magenta and black, solid lines). Note that for medium distances the two peaks of activation inhibit each other resulting in reduced maximum levels of activation (magenta), while for large distances between inputs this inhibition is reduced resulting in higher levels of activation (black). Reducing the inhibitory interaction through learning does not change the model performance for very close (green, dashed line) or very distant (black, dashed line) inputs. However, for intermediate distances the reduced inhibition allows for more co-existing activity (yellow, dashed line). As a result, activation profiles become bimodal shifting the discrimination of two separate stimuli towards smaller separation distances. The weakened inhibition is also evident in the smaller reduction in overall-level of activation for medium distances (magenta, dashed line).

c. Fluctuations in peak position, either from spontaneous noise within the neural population or from noise at the peripheral (input) level, shift the exact location of the activation peaks on a trial by trial basis (different lines refer to the model response in different trials. Strong inhibition prevents fluctuations enabling a precise read-out of the exact peak-location. After learning which reduces inhibition fluctuations increase. Using the same input and model parameters as used in a. and b., noise components integrated into the model (here: Gaussian white noise) lead to substantial scatter of the exact peak location. Due to the strong inhibitory component this scatter is weak before learning because the existing inhibition prevents influence from co-existing activation on other sites of the population of the population thus diminishing the influence of noise. After weakening the inhibitory component through learning the very same mechanisms that improve discrimination by permitting bimodal distributions (see b.) now deteriorate localization performance: Co-existing activation, in this case noisy fluctuations, lead to substantial scatter of the peak, thus making read-out of the exact peak location less precise.

tion has been shown to suppress paired-pulse inhibition in human subjects [102]. As a result, decreasing the strength of lateral inhibition in the model increases the size of the cortical representation as observed in fMRI and EEG studies [42–44, 64]. For discrimination (stimulation at two sites) the decrease in inhibitory interaction leads to bimodal activation profiles for distances that before learning evoked only single peaks, which provides a compelling explanation for the learning-induced improvement of discrimination [31, 42–44, 59–72]. In contrast, for localization reduced inhibition permits coexisting activation, which leads to a more powerful influence of noise thus deteriorating localization abilities. Combined, modeling lateral cortical interaction in population representations provides a unifying framework that explains task-specific requirements in tactile and haptic perception and their alterations evoked by perceptual learning.

Summary

Haptic perception in humans underlies a continuous remodeling resulting in altered performance skills. Most notably, intensive practicing improves skills, while little practicing impairs perception indicating that high-level skills cannot be maintained without ongoing training. Remodeling of haptic perception occurs not only under conditions of specific task training, but also under everyday-life conditions characterized by intensive execution of more complex schedules as in Braille reading blinds, or musicians. Time scales for improvement range from seconds to years suggesting the involvement of many different mechanisms. There is a controversial discussion about the role of top-down *versus* bottom-up processes, the role of feedback, and about conditions under which no further gain can be accumulated despite ongoing training. In addition, passive stimulation procedures that fulfill requirements for inducing synaptic plasticity are similarly affective in altering human haptic perception. Because of the passive nature these approaches provide alternative therapeutic means for intervention in impaired human subpopulations. Independent of how haptic learning is evoked there is now agreement that development of skills is based upon neuroplasticity mechanisms that alter cortical processing in sensorimotor areas.

Acknowledgements

Studies on tactile and haptic learning were supported by the Deutsche Forschungsgemeinschaft DFG and the Schering-Stiftung. We acknowledge valuable discussion on perceptual learning with many colleagues in the field, in particular Dr. M. Fahle and Dr. A. Seitz.

Suggested readings

Dinse HR, Kalisch T, Ragert P, Pleger B, Schwenkreis P, Tegenthoff M (2005) Improving human haptic performance in normal and impaired human populations through unattended activation-based learning. *Transaction Appl Perc* 2: 71–88

Dinse HR, Ragert P, Pleger B, Schwenkreis P, Tegenthoff M (2003) Pharmacological modulation of perceptual learning and associated cortical reorganization. *Science* 301: 91–94

Fahle M, Poggio T (2002) *Perceptual Learning*. MIT Press

Goldstone RL (1998) Perceptual learning. *Ann Rev Psychol* 49: 585–612

Karni A, Sagi D (1991) Where practice makes perfect in texture discrimination: evidence for primary visual cortex plasticity. *Proc Natl Acad Sci USA* 88: 4966–4970

Recanzone GH, Merzenich MM, Jenkins WM, Grajski K, Dinse HR (1992) Topographic reorganization of the hand representation in cortical area 3b of owl monkeys trained in a frequency discrimination task. *J Neurophysiol* 67: 1031–1056

Seitz A, Watanabe T (2005) A unified model for perceptual learning. *Trends Cogn Sci* 9: 329–334

Seitz AR, Dinse HR (2007) A common framework for perceptual learning. *Curr Opin Neurobiol* 17: 148–153

Important websites

The Haptics Community Web Site:
http://haptic.mech.northwestern.edu/

Haptics-L: The electronic mailing list for the international haptics community:
http://www.roblesdelatorre.com/gabriel/hapticsl/

The wiki of the international community of scholars studying perceptual learning:
http://www.perceptuallearning.org/

The Perceptual Learning mailing list:
http://cogmod.osu.edu/mailman/listinfo/plearn

14

Implicit and explicit memory effects in haptic perception

Soledad Ballesteros

Introduction

There is considerable evidence in the literature showing that memory is not a unitary entity. Old neuropsychological findings from different groups of patients and more recent laboratory studies are in agreement about the major memory systems in the human brain. A distinction is made between *declarative* or *explicit memory* and *nondeclarative* or *implicit memory*. Of special interest is the distinction in declarative memory between *episodic* and *semantic* memory [1]. *Episodic memory* includes personal experiences and conscious recollection of events in our past. This type of memory is defined by the capacity to retrieve voluntarily facts and events in the spatial-temporal context. On the other hand, *semantic memory* refers to our general knowledge, including the meaning of words and concepts.

Two decades ago, Graf and Schacter [2] used the terms *implicit* and *explicit* memory to refer to two different forms of memory as well as two ways of accessing previously encoded information. Explicit memory refers to conscious recollection of previous experience with stimuli (words, pictures, objects, etc.) while implicit memory is inferred when previous experiences with the stimuli do not require intentional or conscious recovery of previously perceived information. Most research on implicit and/or explicit memory has focused on verbal stimuli or pictures presented visually while the number of studies that have presented the stimuli tactually has been very limited.

Researchers on touch have called our attention to the historical lack of interest in the study of touch [3, 4]. For example, Heller [3] pointed out that psychologists had emphasised the study of visual shape perception. As he recognised, touch did not function as efficiently as vision in detecting outlines of shapes, because the touch modality is slower than vision and scans the stimulus sequentially. These different methods of operation made researchers think that touch is less important than vision. However, this situation has changed lately. During the last decade a number of laboratories around the world have dedicated a great deal of effort and research resources to studying how touch works [5]. It is true that human vision is an outstanding perceptual modality that allows sighted people to rapidly gather highly precise information from objects in space and their spatial relations. However, when human perceivers actively explore objects with their hands, a large number of sensory inputs and high quality of sensory information are extracted for further processing [6].

Active touch is a complex modality that extracts and encodes a wealth of information from cutaneous and kinaesthetic receptors. Haptic perception (the combination of the information gather by cutaneous and kinaesthetic receptors) is very efficient in object identification [7]. In conjunction with findings from behavioural studies, the cognitive neuroscience approach has incorporated new neuroimaging techniques that are causing great expectations in the field. They have been able to explore the activation of different parts of the cerebral cortex while performing different cognitive tasks. The study of the cerebral cortical areas involved in tactile and haptic perception and memory is an area of research that has rapidly grown during the last few years [8–12].

This chapter is organised into three major parts. The first part reviews empirical results from studies conducted with young adults who used raised-line shapes and three-dimensional (3-D) objects to investigate the possible dissociations between implicit and explicit forms of accessing previously encoded, tactually explored stimuli, without vision. The interest is in the recovery of the previously encoded information under implicit (unconscious) and explicit (conscious) retrieval. The second section is focused on whether the perceptual representations of visual and haptic objects that mediate perceptual priming as a measure of implicit memory are modality specific, and whether there are dissociations between implicit and explicit memory measures. A summary of the main behavioural and neuroimaging findings will be followed by a discussion of their theoretical implications. The final section considers a number of recent findings that indicate that haptic as well as cross-modal visual/haptic implicit facilitation is preserved with ageing. These behavioural results, in conjunction with other recent brain imaging findings, allow us to speculate on brain areas involved in haptic as well as cross-modal implicit and explicit memory in ageing. These findings are changing the understanding of visual and haptic memory in normal and pathological ageing.

Implicit and explicit memory for stimuli explored by touch

Despite the large number of studies conducted in memory, the literature is dominated by the use of verbal stimuli presented in the visual and the auditory modalities, while touch has been largely neglected. Although this section is devoted to long-term haptic memory, I will briefly mention a few findings on haptic short-term memory. Millar [4] reviewed a substantial body of research from blind and sighted children under the general framework of the information processing and memory-systems approach. She presented evidence showing the existence of short-term memory in the tactual modality. However, spans for touch are shorter than for vision and limited to two or three tactual items. According to Millar, the small memory spans (two to three items) for unfamiliar patterns explored by touch can be explained by the paucity of reference information for coding inputs spatially.

The small memory spans for haptically explored stimuli have also been observed in young adults. Moreover, the small haptic memory span (see Heller [13]) deteriorates in normal ageing and is very poor in the elderly with mild cognitive impairment (MCI). In our laboratory, we [14] investigated the interference caused by a secondary task (visual or haptic) in the performance of a primary (visual or haptic) task. In the study, we used a dual-task paradigm with two primary tasks (Visual Matrices and Haptic Corsi Blocks) combined with two secondary tasks (one visual and the other haptic) to assess the visual and the spatial (haptic) component spans in young adults, older healthy adults, and mild cognitive impairment (MCI) adults. A Corsi Haptic device was constructed to avoid the use of vision while performing the haptic tasks. The MCI group obtained the smallest span measures on both modalities, and their poorer performance was on the haptic tasks. The shortest spans obtained by the three groups occurred when the primary and the secondary tasks were conducted in touch (Haptic Corsi Blocks and Moving a Cube – to turn a cube once counter-clockwise around inside a box). The mean spans were 3.5, 2.5, and 1 for young adults, healthy older adults, and MCI, respectively. In contrast, when the primary and the secondary tasks were conducted in vision (Visual Matrices and the Arrows task – to press a key when two arrows pointed to the same direction and another one when the arrows pointed to opposite directions), the spans were larger 5.5, 3.5, and 3, for young adults, healthy older adults, and MCI, respectively. The largest spans were obtained when the primary task was visual and the secondary task was tactual. In this condition of less interference, spans were 6.5, 4.5, and 3.5 for young adults, older adults, and MCI, respectively).

An important issue in memory research is the nature of the internal representation created after the experience with some class of stimuli; that is, whether the representation is intrinsic to the modality or whether it is more general (e.g., spatial). Millar [4] suggested that when patterns can be organised within spatial reference frames, memory for touched patterns is further aided by spatial coding (see Chapter 16). Another issue is whether the representation resulting from touch is specific to the modality or is cross-modal, in the sense of being accessible by other modalities (see below).

A major distinction in long-term memory that has emerged in the past two decades is drawn between implicit and explicit memory. A phenomenon of great theoretical interest that has produced a wealth of experimental results is the dissociation between performance on implicit memory and explicit memory tests [15]. Explicit memory is usually assessed by free recall, recognition, and cued recall tests. Implicit memory, in contrast, is assessed incidentally using tests that do not require intentional, conscious retrieval of previously encoded information. In other words, implicit memory is unveiled when previous experiences with the stimuli do not require intentional recollection of previously presented information. Implicit memory is shown by the existence of repetition priming effects; that is, better performance in accuracy and/or response time for stimuli that have been previously encountered in comparison with performance with new stimuli not presented during the study phase.

Most studies on implicit memory have used verbal materials as stimuli presented either, visually or *via* audition. Other studies have focused on nonverbal materials presented mostly visually [16–19]. Compared to the large number of visual studies, haptic studies are scarce although active touch has proved to be very efficient in identifying familiar objects by active exploration [7]. Moreover, touch is also quite efficient in detecting a number of important spatial properties such as surface curvature [20, 21] and bilateral symmetry [22–26] see Chapter 16).

The haptic system is a complex perceptual system that encodes inputs from cutaneous as well as from kinesthesic receptors distributed over our body [27]. The human hand is an especially miraculous instrument that serves human beings in many ways [6]. When the active human explorer moves the hands to extract information about objects he/she is able to extract different types of information such as its surface texture, its symmetric or asymmetric character, its weight, its shape, its orientation, temperature, compliance, and so on. Millar [28] suggested that haptic perception depends on the complementary information from tactual acuity, active movement, and spatial cues.

Implicit and explicit memory for raised-line shapes

Several attempts in our laboratory since 1992 failed to show perceptual priming for raised-line novel patterns (2×2 cm) constructed by connecting five or six dots in a 3×3 dot matrix (see Chapter 16, Fig. 6). We tried several forms of encoding from structural to semantic encoding but we did not succeed in showing perceptual facilitation at the implicit test that followed the encoding phase [29]. The experimental procedure consisted of a study phase in which blindfolded haptic observers explored a series of shapes, one by one, with the forefinger of his/her preferred hand. They were allowed 10 seconds to explore each pattern. After a delay, implicit memory was assessed indirectly with a 'symmetry-asymmetry' detection task followed by an explicit recognition test. We attributed the failure to find haptic perceptual priming to the difficulty of encoding spatial information under reduced kinesthesic feedback. This was due, perhaps, to the small size and the complexity of the raised-line patterns as well as to the difficulty of referring the small nonsense shapes to a spatial reference frame under blindfolded conditions.

Srinivas, Greene, and Easton [30] succeeded in obtaining haptic priming using a set of easier novel raised-line patterns and highly intensive encoding in which participants had to produce an accurate verbal description of the shapes. Differences in the simplicity of the shapes and/or in the procedure used at encoding might account for the success in obtaining positive results. More specifically, Srinivas and colleagues adapted the Musen and Treisman's [31] visual paradigm to haptic implicit and explicit memory tasks. At encoding, Srinivas et al. [30] exposed undergraduate students to a set of novel three-line novel patterns presented haptically. Participants haptically identified 20 shapes during 10 seconds each, providing a verbal description of the position, orientation, and intersections of the lines in each raised-line pattern. Two were the encoding conditions. In the elaborative encoding condition, after describing the shape, participants generated a function for the stimuli. In the physically encoding condition, participants were asked to count the number of vertical and horizontal lines that contained the stimulus. Later, at test, haptic explorers were presented with old (studied) and new (non-studied) raised-line patterns and were asked to draw them. Episodic recognition was assessed by asking haptic explorers to discriminate 'old' *versus* 'new' raised-line patterns.

The results showed that haptic identification of raised-line novel patterns is unaffected by type of encoding, as elaborative processing at study produced similar haptic perceptual facilitation as physical or shallow encoding [32]. In contrast, explicit recognition was enhanced under semantic encoding. Srinivas et al. [30] concluded that a similar organisation of object information in memory occurs whether the patterns are perceived by vision or by touch. The results supported the modular architecture proposed by Tulving and Schacter [15], and that implicit and explicit memory tasks rely on presemantic and semantic representation, respectively, across vision and touch. An early study conducted by Hamann [33] reported haptic implicit memory in blind participants and a Braille stem-completion test to assess perceptual facilitation.

Implicit and explicit memory for three-dimensional (3-D) objects

Although active touch is efficient and fast in identifying familiar objects without vision, the question of how haptic objects are represented and retrieved implicitly has received little attention in the literature. Two decades ago, Wippich and Warner [34] conducted a pioneering study on implicit memory for objects and non-objects (unfamiliar objects) presented haptically. Implicit memory was assessed by subtracting the time needed to answer questions related to a haptic dimension between the study phase (the first presentation of the object) and the test-phase (the second presentation of the same object). In another study, Wippich [35] explored the components (motor and/or sensory) underlying performance in implicit and explicit tests of memory for haptic information. In the first experiment, the encoding tasks and the test conditions were varied. During the encoding-phase half of the participants in the active exploration condition were blindfolded and explored familiar objects haptically in order to answer questions about different object properties as fast as possible. The other half of the participants in the motor encoding condition performed the corresponding hand movements without sensory contact with the objects. At testing, a group of participants repeated the active exploration task with old and new objects. The other group encoded the objects without performing hand movements and based their judgments only on sensory impressions. The results showed significant effects of implicit memory for haptic information, as revealed by faster reaction times to old rather than to new objects. However, priming effects were only obtained in the active touching test condition. This result was observed for both encoding tasks, that is, even for participants who had only performed symbolic hand movements at encoding. Thus, the repetition of specific motor processes seems to underlie the effects of repetition priming in active touch. In the second experiment, participants took part in a passive touch paradigm, both at encoding and at test (sensory encoding). The priming effects observed under these conditions indicate that sensory processes can be relevant, too. Recognition performance was found to be influenced by other variables than measures of implicit memory.

More recently, other haptic studies explored possible dissociations between implicit and explicit memory tasks presenting familiar and unfamiliar objects to young [29] and older explorers [36]. However, a relatively small number of studies explored cross-modal facilitation and dissociations between vision and touch in young adults [37, 38] and older adults [39].

Ballesteros and colleagues ([29] Exp. 1) investigated whether active touch stores and retrieves information about familiar 3-D objects implicitly, and whether implicit memory could be dissociated from explicit, conscious retrieval of objects presented to touch without vision. As vision and touch are perceptual modalities well adapted to extract a wealth of objects' structural information, we hypothesised that haptic perceptual priming could be observed for objects presented to touch without vision at study and test. We also asked whether implicit memory for objects explored haptically would be sensitive to changes in the mode of exploration when conditions changed from study to test. That is, the question was whether haptic priming is specific, and changing the exploration conditions from study to test would diminish or even eliminate the perceptual facilitation. In other words, the question was whether the modification of the exploration condition would negatively affect implicit memory. Another question was whether this experimental manipulation would differentially affect implicit and explicit retrieval.

In the study, we used a *speeded object naming task* to assess implicit memory while explicit memory was assessed using a *recognition 'old–new'* test. At study, blindfolded young adults explored a series of familiar natural and artificial (man-made) objects with both hands, without gloves, one at a time while judging whether the object was heavy or light, round or sharp, large or small, and soft or rough. After a 5 min dis-

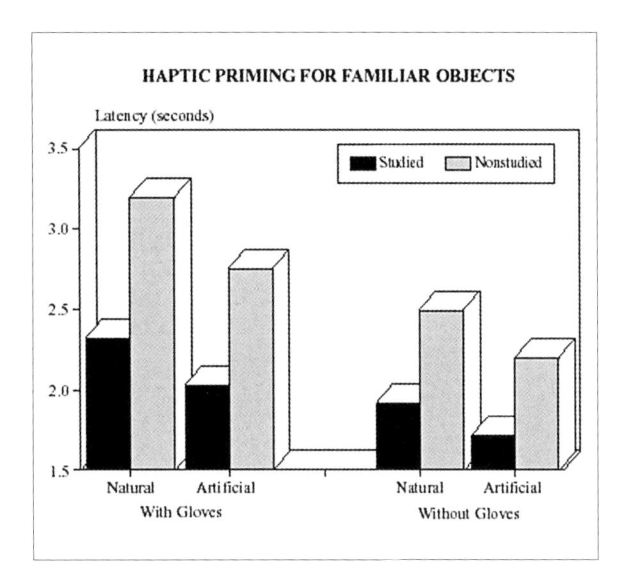

FIGURE 1.
Mean response time (in sec) corresponding to correct responses in the Speeded object naming task *for studied and non-studied objects as a function of type of object (natural and man-made) and mode of exploration (without gloves as in the study-phase and with gloves unlike the study-phase). From Ballesteros, Manga, Reales (1999) p. 792. Reprinted with permission of* Psicothema.

tracter task, half of the participants performed the implicit *speeded object identification task* consisting of naming old and new objects as quickly and accurately as possible. The other half of the participants performed the explicit recognition test. These two groups were further divided in two subgroups. One performed the implicit task and the other the explicit task as in the study phase (without gloves). The other two subgroups performed the implicit and the explicit memory tasks using latex gloves (the changed condition).

Figure 1 shows mean response times corresponding to the implicit memory results for studied and non-studied natural and artificial (man-made) objects as a function of the experimental manipulation. Performance on the test condition that matched the study condition is shown on the right side of Figure 1 while the left side displays results corresponding to the study-to-test changed condition. The main findings showed that in both experimental conditions (with gloves and without gloves) previously encoded objects were correctly named significantly faster than new objects; that is, substantial haptic priming was obtained for both types of objects (natural and artificial objects). Although participants required more time to name the objects using gloves, the sensory information was not critical as the haptic perceptual facilitation was similar for both exploration conditions.

The results of the recognition test showed that overall explicit memory was 90% correct. However, wearing gloves at test produced significantly lower performance compared to the unchanged (exploring the objects without gloves as at study) condition (88% *versus* 94% correct). The dissociation observed between performance in the implicit and the explicit haptic memory tasks suggests that both memory measures (implicit speeded object naming and 'new–old' object recognition) tap different object representations. The implicit test is specialised in encoding structural object information while explicit recognition tapped low-level cutaneous sensory information.

Is there implicit memory for novel object explored by touch? The next question we asked ([29] Exp. 2) was whether implicit memory would be shown for unfamiliar objects for which we did not have previous mental representations. As we were interested in the study of haptic representations of object shape and wished to avoid as much as possible verbal labelling and semantic encoding, in this study we used a set of three-dimensional wooden objects. This study used blindfolded participants. Implicit memory was assessed incidentally after study using a *speeded symmetry detection task* [29] while explicit memory was evaluated with an *'old–new' recognition task*. The stimuli were bilaterally symmetric and bilateral asymmetric 3-D objects (see Chapter 16, Figure 6, bottom). All the objects were made from a cubic piece of wood of approximately 7 cm on each side. There were two types of encoding conditions. One was the elaborative encoding condition while the other was the structural encoding condition. Participants in the elaborative encoding condition were asked to explore 20 objects, one by one, and to provide the name of a familiar object that each wooden object reminded them of. In contrast, participants in the structural encoding condition had to judge the complexity of 20 unfamiliar 3-D wooden structures on a five point scale. At test, participants in the implicit condition were presented with previously encoded objects and with new objects one by one in a different random order. They were asked to indicate as fast and accurately as possible whether the objects were symmetric or asymmetric. Participants in the recognition test were asked to specify whether each haptically explored object was presented at study or was a new object.

The results showed haptic priming for objects encoded structurally but not for those encoded semantically. In contrast, explicit haptic recognition was superior for objects semantically encoded at study than for those encoded structurally (hit–false alarms = 0.66 and 0.39, respectively). Priming for unfamiliar objects was observed under structural encoding, but not under semantic encoding while recognition was more precise under semantic encoding. These findings for studies with familiar and unfamiliar objects presented to touch, without vision, using implicit and explicit memory tests were dissociated experimentally and seem to tap different object representations. While the implicit task seems to rely on structural, shape-based representations of objects, the explicit recognition task appears to rely on episodic representations that include low-level cutaneous information.

Cross-modal priming between vision and touch: The structural descriptions hypothesis

Behavioural findings

A question of considerable interest is whether the perceptual representations of 3-D objects are shared between vision and touch. Implicit memory is a perceptual phenomenon and as such, repetition priming has been considered modality specific [40]. The first cross-modal implicit memory studies mostly used words as stimuli presented visually at study and in the auditory modality at test (or *vice versa*). The common findings were that priming diminished or disappeared when modality changed compared to conditions in which stimuli were presented to the same modality at study and test. The reduced cross-modal facilitation observed in those studies might be due to the lack of overlap between the sounds and the letters [38]. The question is whether cross-modal priming could be observed when objects are presented to touch and vision, two modalities that deal with objects.

Several behavioural studies [32, 37, 38] have reported cross-modal perceptual facilitation between vision and touch. For example, Easton, Srinivas, and Greene [32] used verbal stimuli and manipulated modality to find out whether the perceptual representations that mediate priming were specific to a particular processing modality. Implicit memory was assessed using a word-stem completion test.

The results showed that within-modal priming was larger than cross-modal priming. The authors attributed the results to the confounding of simultaneous and successive processing. When this variable was controlled by presenting words to both modalities sequentially, cross-modal priming was similar to within-modal priming. As visual-haptic priming with verbal stimuli could be mediated by lexical representations, Easton and colleagues [32] used novel two dimensional raised-shapes and familiar 3-D objects. Results showed robust although diminished cross-modal facilitation which lead them to conclude that it could be the case that the perceptual representations that underlie perceptual priming as a measure of implicit memory are not necessarily modality specific. Nevertheless, the robust cross-modal facilitation between modalities for nonverbal stimuli raises the possibility that the representations that support perceptual priming are not modality specific.

Box 1 presents the experimental paradigm used by Reales and Ballesteros [38] to investigate whether the perceptual representations of visual and haptic objects were modality specific and whether there were dissociations between implicit and explicit measures. We reasoned that the lack of priming or the reduced perceptual facilitation obtained in verbal studies might be caused by the lack of overlap between the perceived stimulation. In the case of vision and touch, we [38] speculated that visual and haptic object representations might be so similar that they could be shared. Cross-modal priming is a good method for investigating the extent to which the representations of vision and touch overlap.

We experimentally explored the hypothesis that the perceptual representations of visual and haptic familiar objects that mediate perceptual priming are not modality specific. To further this aim a study was conducted using a complete cross-modal priming and explicit memory design between these two perceptual modalities (vision and active touch). The idea was to show that to experience real objects at study in one modality (vision or touch) affected performance positively when the same objects were incidentally presented again to the other modality (vision or touch). In several studies, implicit memory for objects explored haptically showed the same facilitation when the same objects were haptically presented again at test (the same encoded modality) as when they are presented visually (the other modality). That is, within-modal priming was similar to cross-modal priming. The experiments also showed that manipulating levels of processing at study did not have any effect on either within-modal or cross-modal priming. The same perceptual facilitation was found under shallow and under deep encoding conditions (Fig. 4). Experiment 1 in the series showed that implicit memory for familiar objects assessed by a speeded naming task showed the same priming for visually or haptically studied objects and that cross-modal priming was equivalent to within-modal priming. As Figure 4 shows, the absence of levels-of-processing effects suggests that the

Box 1. Priming procedure used to study cross-modal implicit memory for touch and vision

Reales and Ballesteros [38] asked whether there is visual-haptic (and haptic-visual) cross-modal priming for familiar objects. The first experiment of a series investigated whether priming was maintained at the same level when at encoding objects were presented to a modality and then at test, incidentally, repetition priming was assessed at the other modality.

The hypothesis

The hypothesis that guided the study was that as vision and touch are two perceptual modalities well suited to deal with 3D object structure, both modalities might shared the same mental representations. So, we hypothesized that perceptual priming will be maintained in changed exploring conditions; that is the hypothesis under test was whether priming effects would be not reduced when modality is changed between the study phase of the experiment and the test phase. To test this hypothesis the following experiment was conducted.

Method

Participants

Forty-eight university students participated in the experiment for course credit. All have normal or corrected to normal vision and normal touch.

Materials and Equipment

The stimuli were 60 familiar stimuli as those shown in Figure 2. The size of the objects were appropriate for participants to be able to hold the object between their hands allowing enclosing, a very efficient hand movement procedure. The objects neither did make special noises nor did emit special odours that would facilitate identification.

Figure 2. Examples of familiar objects used in our experiments

To present the stimuli and register response times, a 3D real visual-haptic tachistoscope was used that allowed to present real objects to vision and touch. (see Fig. 3). The equipment was completed with a vocal key to stop the internal clock of the computer.

FIGURE 3. THE VISUAL-HAPTIC 3-D TACHISTOSCOPE USED FOR PRESENTING STIMULI HAPTICALLY AND VISUALLY
The apparatus was equipped with a liquid crystal window located in the methacrylate panel at the eye level. In the visual trials, this window allows the participant to see the object located at the presentation platform. The platform had a piezosensor board underneath the presentation platform. The apparatus was interfaced with a PC computer to run the experiment and register response times.

Experimental design

The design was a mixed factorial design that resulted from crossing 2 study modalities (vision and touch) with 3 study conditions (semantic encoded, physical encoded and nonstudied objects). The last variable was manipulated within subjects while the first one was the between-subject variable. The 48 participants were randomly assigned to the 4 experimental conditions (vision/vision; vision/touch; touch/touch; touch/vision). Moreover, the 60 familiar objects were divided randomly in two groups of 30 objects each that was further subdivided in two subsets of 15 stimuli. The two sets of 30 stimuli appeared equally often as studied and as not studied, half of which appeared equally often as semantically or physically encoded.

Procedure

Participants were tested in the laboratory individually. The session started with a study phase and after a short delay they performed incidentally the test-phase in which memory for previously encoded (semantically and physically encoded) and non-encoded (new) objects.

Encoding task. During the study-phase, participants were shown for 10 seconds (visually or haptically, according to the experimental condition) a series of objects, one by one. In the semantic encoding condition, they generated a sentence in which appeared the name of the object. In the physical condition, they have to rate the object´s volume in a 5-point scale.

Test task. After performing for 5 seconds a distracter task, observers performed a *speeded object naming task*. They incidentally were asked to name as soon as possible the objects presented one by one (30 old and 30 new objects) on the presentation platform of the 3D tachistoscope.

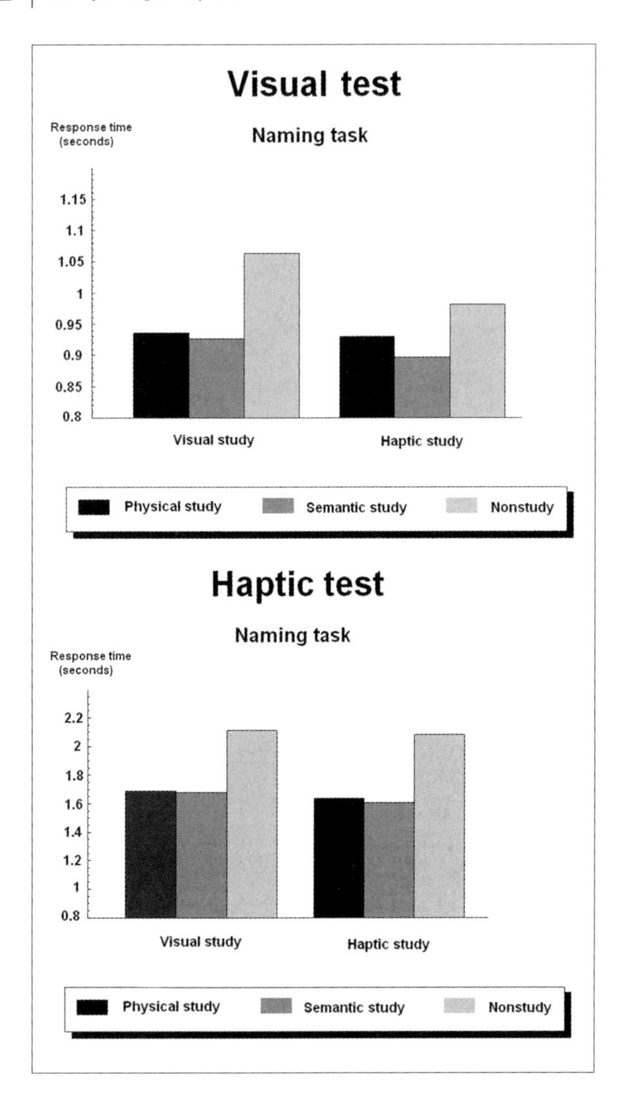

FIGURE 4.
Response time (in seconds) in the haptic and visual implicit object naming test as a function of level of study (physical study, semantic study, and nonstudied) and study modality (visual encoding or haptic encoding). Reprinted from Reales, Ballesteros (1999) Exp.1, p. 649, with permission of the American Psychological Association.

mental representations that support implicit memory are pre-semantic. We [38] used various implicit memory tests, including speeded object naming, picture-fragment completion (e.g., the level of completeness at which a fragmented picture could be identified), and speed of deciding whether a line drawing depicted a real object. All of these different implicit tests showed substantial cross-modal and intra-modal priming (faster responses for previously studied objects compared to new objects). In some cases, however, the magnitude of cross-modal and intra-modal priming effects was equivalent. Moreover, explicit and implicit memory tests were influenced by different variables. When pictures were used in an implicit test, priming was greater when pictures were presented at study than when real objects were presented, but an explicit test that used pictures benefited when real objects had been studied. The authors argued that the priming effect arises from an abstract structural description of objects that is accessible by vision and touch. The main results obtained in Experiment 1 were replicated after a delay of half an hour between study and test (Exp. 3). The equivalent facilitation obtained in Experiment 3 under shallow encoding replicated these results and further supported the idea that repetition cross-modal priming as well as within-modal priming in vision and touch is pre-semantic. Figure 5 shows the results from the implicit memory task when the implicit memory test followed encoding immediately and after a half an hour delay from study, both when study and test were conducted in the same modality and when they were performed in different modalities.

Explicit memory was also assessed in the two conditions, immediately and after a half hour delay from study using a free recall test. The results from the explicit test were consistent with other findings in the literature. Explicit memory benefits from repetition and is impaired as a function of the delay between study and test. Explicit memory for objects explored haptically was superior to that for objects explored visually. We speculated that this result could be explained by the wealth of attributes that perceivers are able to extract under haptic exploration (e.g., texture, size, shape, weight, and hardness).

The complete cross-modal priming between these two modalities suggests further that the

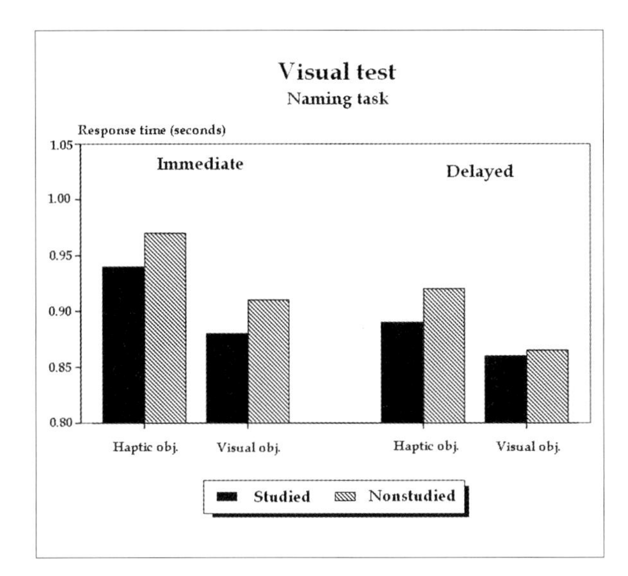

FIGURE 5.
Response time (in seconds) in the visual object naming test as a function of study modality (touch or vision) and delay conditions (immediate after encoding or half an hour delayed) and item type (studied or nonstudied). Reprinted from Reales, Ballesteros (1999) Exp. 3, p. 656. With permission of the American Psychological Association.

mental representations that support implicit perceptual memory might be shared between these modalities, vision and touch. Vision and touch are both well adapted to deal with an object's 3-D structure [38]. Lexical or semantic explanations can be discarded because, as it was mentioned, performing a deep encoding task at study did not produced large perceptual facilitation compared to shallow encoding. There is behavioural evidence that vision and haptics encode object structure similarly using a common representation. This conclusion is further supported by a number of recent neuroimaging studies that have shown overlapping activation in the same regions after visual and haptic processing. The activation occurs in extraestriate areas of the occipital cortex previously considered visual areas [8–10].

Object-based representations in the extrastriate visual areas: Neuroimaging results

An area of considerable current interest is the study of the relationship between brain functioning, neural connections and the perceptual modalities, especially vision and touch but also audition. Clear trends in neuroscience include a shift toward the idea that the different sense modalities are not separated, but interrelated. A number of researchers defend the idea that some parts of the human brain are not totally specialised and that there are some areas in the brain that do not respond only to one particular modality. In other words, some regions in our brain may share representations between modalities. The sharing of information between vision and touch has been confirmed in neuroimaging studies. The first report that extrastriate visual cortical areas are active during tactile perception came from Sathian's laboratory [41]. In this positron emission tomography (PET) study, a discrimination of grating orientation was used as the experimental task of interest, comparing it with a control task requiring discrimination of grating groove width. The results yielded activation at a focus in the extrastriate visual cortex, close to the parieto-occipital fissure.

During the last decade, neuroscientists have presented compelling evidence that shows common activation in the lateral and middle occipital areas (LOC and MOC) during haptic and visual object identification. These two areas in the extrastriate visual cortex associated previously with visual perception, are considered as the potential locus of the object structural system [10, 42]. Reales and Ballesteros argued [38] that cross-modal priming paradigms are a useful tool for studying the extent to which the representations of vision and touch overlap. James et al. [10] used this paradigm to investigate the neural substrate underlying the visual and haptic representations of objects. These investigators used a set of 48 3-D novel objects (Fig. 6) to avoid any possibility that verbal labelling mediated cross-modal facilitation.

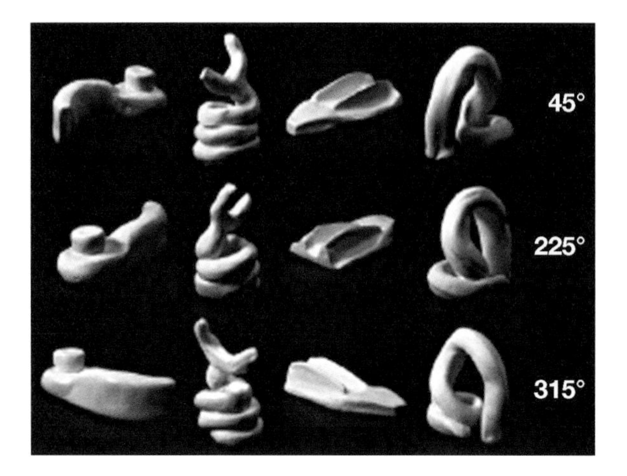

Figure 6. Examples of the 48 novel 3-D clay objects
(James et al. (2002). With permission of Elsevier Science Ltd)

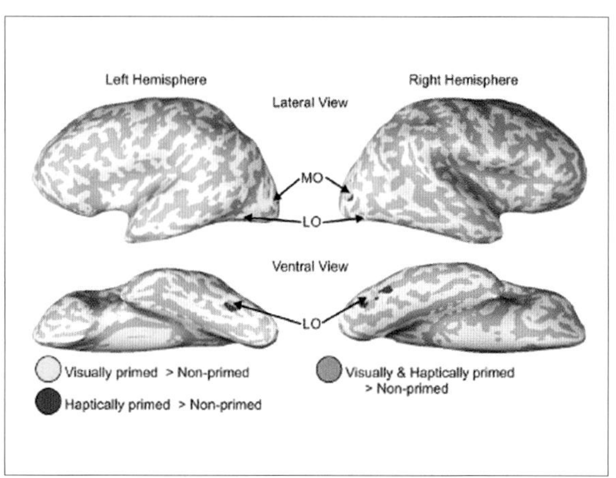

Figure 7. Maps of the brain activation obtained during the priming task
Activation was seen in the middle occipital area (MO) and the lateral occipital area (LO). Reprinted from James et al. (2002), with permission of Elsevier Science Ltd.

The effect of cross-modal haptic to visual priming in brain activation was assessed using functional magnetic resonance imaging (fMRI). Behavioural data were not collected. The authors argued that as the levels of activation were the same for within-modal as for cross-modal priming, these results parallel previous behavioural studies [32, 37, 38]. At study, participants explored a series of objects visually or haptically. Then, in the scanner they were tested visually on a computer screen and with stimuli haptically presented, or visually presented, and nonstudied objects. The results (Fig. 7) indicated that the object representation system of the ventral visual pathway were used for haptic object processing. Neuroimaging results showed that haptic and visual study of novel objects produced activation in several brain areas including the somatosensory cortex but a cross-modal perceptual priming effect was observed in a region previously considered involved in visual processing, the lateral occipitotemporal complex (LOC).

James and colleagues [10, 42] argued that cross-modal priming uses common haptic and visual representations and suggested that the neural substrate is MO, activated by both visual and haptic exploration. Although it could be argued that the activation found in ventral occipital areas is due to visual imagery, Amedi et al. [9] observed that LOC response during a visual imagery task produced considerably less activation in this area than did a haptic exploration task performed with the same objects. This and other imagery results suggest that the activation in LOC during haptic exploration reflects the activity of a multi-modal network.

Transcranial magnetic simulation (TMS) is a technique consisting of applying a magnetic pulse to the brain to interrupt processing. Studies in this area have shown that TMS applied to the occipital cortex contralateral to the hand used to explore gratings disrupts task performance. However, when TMS was applied to the ipsilateral occipital cortex, performance was not disrupted [43]. Sathian's lab has conducted new and very interesting work in cross-modal activation of visual cortical areas during 2-D form perception (see [12]). Pascual-Leone and collaborators have conducted pioneering work with blind participants. Using brain imaging, Pascual-

Leone's group has shown that the visual cortex of the blind person is activated when reading Braille [44, 45].

In short, a wealth of neuroimaging findings suggest that areas considered to be involved in visual processing are also normally involved in haptic tasks such as tactile grating orientation, cross-modal priming of novel objects as well as in reading Braille.

Implicit and explicit haptic memory for objects in normal ageing and dementia

Older adults have difficulty in remembering items, especially in free recall tests. Compared to the large number of studies on visual memory in ageing, it is surprising that there has been little research exploring differences in haptic memory in older adults. Numerous studies have shown that long-term memory, especially *episodic memory*, declines with age. For example, significant differences between young and older adults have been reported for words [46], faces [47], and even TV news [48]. Cross-sectional episodic data from the Betula project, a longitudinal, ongoing prospective study directed by Nilsson on ageing, memory, and dementia that started in 1988 in Umea, Sweden, showed a dramatic decrease in performance as a function of age [49]. The memory tasks used in the Betula study included free recall, cued recall, recognition of action and short sentences, free recall of words, recognition of names and faces, among other tests. Similar data have been reported by Park and collaborators (see [50]). Implicit memory tasks usually have shown age invariance. Repetition priming as a measure of implicit memory has shown that age differences, when they exist, are rather small compared to age differences in episodic (explicit) memory (for a review, see [51]). The stimuli used in ageing studies have been mostly words and pictures. Standtlander, Murdock, and Heiser [52] studied explicit recognition and showed that younger adults recalled more stimuli than older adults. Moreover, real objects were better recalled than a list of concrete, high imagery words. Interestingly, touching an object improved recall in young and older adults. However, exploring objects only haptically produced even higher recall in both ages. Why was the best performance found in the haptic modality? The answer perhaps is that movement benefited encoding [34, 53].

Ballesteros and Reales [36] investigated for the first time the status of implicit memory for familiar objects presented haptically in Alzheimer's disease (AD) patients and older controls. In addition, a group of young adults participated in the study that also looked at the dissociation between implicit and explicit haptic memory tests. At study, participants were presented individually with a series of familiar objects, one by one at the presentation platform of our apparatus. The encoding task consisted of actively exploring the object with both hands and generating a sentence including the object's name. After a 5 min break, the participants performed a speeded object naming test to assess implicit memory followed by a recognition 'new-old' test to assess explicit memory. The main results are displayed in Figure 8.

AD patients showed the same intact haptic priming in the speeded object naming task as both the older adult and the younger adult groups. Despite the intact priming, AD patients' explicit memory for haptically explored objects was highly impaired compared to healthy adults (young and older adults). These findings suggest that implicit memory for haptically explored objects is preserved in patients with mild AD. The impaired recognition performance of AD patients clearly contrasted with the intact repetition priming observed in the implicit memory task. AD patients can not rely on conscious retrieval to improve their performance on object naming as they have very poor recollection of previous experience with objects.

Researchers investigating visual priming have suggested that the extrastriate visual areas are crucial in perceptual priming [54]. The dissociation between these memory capacities supports the idea that different memory systems may be

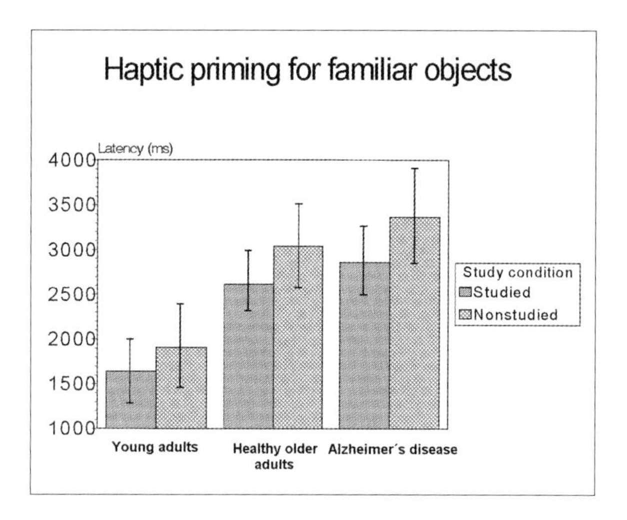

FIGURE 8.
Response time (in ms) in the speeded object naming task as a function of studied versus *nonstudied objects, and group. Reprinted from Ballesteros, Reales (2004) Intact haptic priming in normal aging and Alzheimer's disease: evidence for dissociable memory systems.* Neuropsychologia *42: 1063–1070. © Elsevier Science Ltd.*

implicated in implicit *versus* explicit memory tests. One of these brain systems may be the structural perceptual object system and the other relies on the medial-temporal lobe system that mediates explicit object recognition [15].

To explain this finding we speculated that visual and haptic object representations might be shared between the two modalities, since they are well suited to represent the structure of objects. As reviewed above, several studies have shown common activation in the lateral occipital area (LO) during haptic and visual object identification [9, 55]. Recent functional magnetic resonance imaging (fMRI) research with young adults suggested that the potential locus of the object structural description system is the extrastriate cortex [10, 42]. The results showed that haptic exploration of novel objects produced activation, not only in the somatosensory cortex, but in areas of the occipital cortex associated previously with visual perception as well. The areas MO (middle occipital) and LO (lateral occipital) were activated equally in haptic-to-visual priming and visual-to-visual priming. These neuroimaging findings are consistent with previous behavioural results [32, 37, 38] and pointed to the implications of these areas under haptic object exploration.

Moreover, other investigators using various types of tactile stimuli and brain imaging techniques confirmed the role of the occipital cortex in processing haptic information. For example, Sathian and his collaborators [12, 41] used positron emission tomography (PET) to localise activation in regional cerebral blood flow while young adults performed a grating orientation tactile task. *Post mortem* examinations of AD patients have shown little damage of primary visual, auditory and somatosensory cortices, the basal ganglia and the cerebellum, but substantial damage to association neocortices in the frontal, parietal and temporal lobes was found [55]. There is ample evidence implicating the medial-temporal lobe system (hippocampal complex) and diencephalic structures in explicit memory [56, 57]. Lesions to medial-temporal and diencephalic brain regions produce a selective deficit in declarative (episodic and semantic) memory called amnesia [58]. As patients with AD suffer severe damage to these areas [59] they showed major impairment of explicit object recognition. Our findings provided the first demonstration, to this researcher's knowledge, that AD subjects maintained intact haptic priming as assessed by a speeded object identification task, despite highly impaired recognition performance.

Summary

The present chapter has described the newly emerging literature that has shown: 1) Implicit memory assessed by the demonstration of perceptual priming for haptically explored objects as well as the dissociations between implicit (involuntary, unconscious) and explicit (voluntary, conscious) forms of memory. 2) Haptic perceptual priming, as a measure of implicit memory, is not hyperspecific; for example, modifying the cutaneous sensory information pick-up apparently did not diminish priming but impaired recognition. The dissociation between these two types of memory measures suggests that they tap different object representations. The implicit representations seem to rely on structural, shape-based representations of objects; in contrast, explicit memory appears to tap low-level cutaneous sensory information because it includes all types of distinctive information. 3) Haptic repetition priming is not modality specific. Several studies have shown that cross-modal facilitation (from vision to touch and from touch to vision) is similar to within-modal facilitation (vision–vision and touch–touch). The complete cross-modal transfer between vision and touch is not a transient phenomenon but is long-lasting. The representations that support priming are pre-semantic, since a level-of processing manipulation at encoding has no effect on the priming observed. In short, modality shifts from study to test do not eliminate priming. The observed cross-modal perceptual facilitation used a visual and haptic common representation. 4) Recent brain imaging studies have shown the neurological basis for objects explored by touch. For example, cross-modal priming effects were found in areas such as the lateral occipital cortex (LOC) and the middle occipital areas (MOC). These findings suggest that cross-modal priming between these two modalities is based on a common visual/haptic representation and that the neural substrate of this representation is the extrastriate visual cortex previously associated with visual processing. 5) Haptic as well as visual perceptual priming is preserved in normal ageing and Alzheimer's disease. In contrast, explicit memory for haptic objects was impaired in AD patients compared to older adult controls and young adults. It should be noted that areas in the extrastriate cortex that seem to play an important role in processing the structural description of objects are relatively well preserved at early stages of AD. In contrast, substantial brain damage occurred at early stages in areas involved in explicit memory such as the hippocampus and the medial-temporal lobe system.

Acknowledgements

The research reported in this chapter was supported by the *Dirección General de Investigación Científica y Técnica* (grants PB90-0003, PB94-0393, BSO2000-0108-C02-01 and SEJ2004-00752/ PSIC). I am very grateful to Dr. José Luis Dobato, neurologist from the *Hospital-Fundación Alcorcón* for his collaboration in locating and the careful evaluation of the Alzheimer's patients that participated in our studies. Finally, I thank to Morton Heller for his comments.

Further readings

Ballesteros S, Reales JM (2004) Intact haptic priming in normal aging and Alzheimer's disease: Evidence for dissociable memory systems. *Neuropsychologia* 44: 1063–1070

This article showed intact implicit memory for objects explored haptically without vision in normal older adults and Alzheimer's disease (AD) patients. In contrast, the explicit recognition of AD patients was highly impaired compared to healthy older adults.

Millar S (1999) Memory in touch. *Psicothema* 11: 747–767.

In this article, Millar reviewed a series of studies dealing with short-term memory for stimuli presented to touch. She noticed that haptic spans are shorter than visual spans and explained this result in terms of the paucity of reference information to organise inputs spatially. Movements are also important as they can be used in haptic rehearsal.

Three chapters of the book edited by Morton Heller and Soledad Ballesteros (2006) *Touch and blindness: psychology and neuroscience* published by Lawrence Erlbaum Associates provide complementary information from three different laboratories on the neural substrate that supports visual and haptic object representations (Chapters 7, 8, and 9):

- James, James, Humphrey, Goodale: Do visual and

15

Attention in sense of touch

Matthias M. Müller and Claire-Marie Giabbiconi

Attention in somatosensation

In everyday life the human brain is confronted with an enormous amount of sensory input at any given moment. To guarantee coherent and adaptive behaviour, selective attention is needed to focus the limited processing resources on the relevant part of the available information while ignoring the rest [1, 2]. This chapter provides an overview on some topics of current research in attention and the sense of touch. We will mainly focus on studies using mechanical stimuli rather than electrical stimulation. The physical characteristics of electrical stimuli (sharp and short duration of only a fraction of a millisecond) makes them closer to pain stimuli and, thus, suboptimal to mimic the complex interactions between different mechanoreceptors of the glabrous skin [3–5]. In contrast, mechanical stimuli have not such a sharp onset (mostly they are delivered in form of a sinusoid, see below) and make contact with the skin for much longer. Therefore, mechanical stimuli seem to mimic everyday experience of touch more closely as opposed to electrical stimuli. A second aim of this chapter is to discuss possible neuronal mechanisms that explain how to-be-attended tactile stimuli are processed preferentially in the brain.

Research on attention has a long history and definitions of attention varied from time to time. In 1890, William James [6] wrote: "*my experience is what I agree to attend to*". Along the same line, John Driver defined attention as "*a generic term for those mechanisms, which lead our experience to be dominated by one thing rather than another*" [7]. The main behavioural signature of attention is the improved accuracy in analysing and speeded detection or discrimination of attended stimuli.

But what makes somatosensation so different from other senses to look exclusively into the impact of attention onto that sensory system? Hsiao and Vega-Bermudez [8] nicely illustrated that point with the following example: "*if you switch your focus of attention to your foot, you immediately become conscious of sensations arising from receptors in your foot that were non-existent a moment earlier. This simple observation demonstrates the power of selective attention*". Based on that illustration the authors concluded that attention in somatosensation plays its role in the selection of specific sensory inputs at certain body locations. While this is not particularly different from other modalities when we focus on spatial accounts of attention, Forster and Eimer [9], among others, discussed an important difference compared to other sensory modalities. Contrary to vision for instance, in touch one interacts with proximal stimuli impinging on our body surface. In other words, while visual and auditory stimuli might be miles away from our body, somatosensory stimuli are not. They have an immediate impact onto our body surface, which makes somatosensory stimuli very different from auditory or visual ones. A further difference might lay in the fact that to a great extent primary somatosensory areas also appear to represent non-spatial attributes of tactile events [10, 11]. This feature not only enables a fast analysis, which is ecologically quite useful, it also makes non-spatial attentional selection accessible at a very early level of stimulus processing.

Somatosensory attention and transient stimuli

Most of the studies conducted so far focused on spatial somatosensory attention. Within this framework, attention is focused on a particular body location and stimuli presented at that location are processed faster compared to stimuli presented to any other body location. Only a few studies, which will be reported below, investigated feature-based attention. In feature-based attention subjects are required to discriminate between different features, such as texture or stimulation frequency, rather than a certain body location alone. Of course, spatial and feature-based attention in many cases is mixed, because to discriminate between different features, stimuli have to be presented at a certain body location. The crucial point here is the task subjects have to fulfil.

The effect of spatial attention on early perceptual processes in the visual [12] and auditory [13] modality has been illustrated by an amplitude modulation of early components of the respective sensory-specific event related potential (ERP). In somatosensation, a number of studies in monkeys and humans have also shown that the neuronal response to an attended tactile stimulus is enhanced compared to when this stimulus is unattended (for a further review see [14]). Single-unit recordings in monkeys showed an increase in firing rate to transient tactile stimuli in primary somatosensory cortex (SI) [15]. Imaging techniques, such as positron emission tomography (PET) and functional magnetic resonance imaging (fMRI) found attentional modulation of the blood oxygen level dependent (BOLD) response in primary and in secondary somatosensory cortex (SII) [14, 16–19].

In human non-invasive electrophysiological studies, transient tactile pulses elicit the somatosensory evoked potential (SEP) or the magnetic counterpart, the somatosensory evoked magnetic field (SEF). The SEP consists of the components P50, N70, P100, N140 and a positive late component [20–23]. Figure 1 depicts examples of SEPs and the attentional modulation thereof.

Mima and colleagues [24] reported of early SEF components with a latency of 38 and 68 ms.

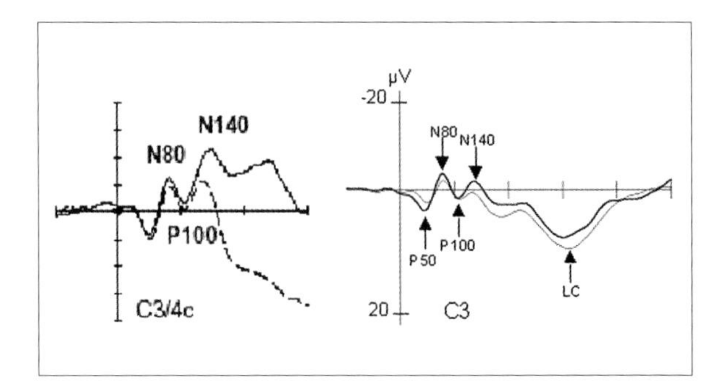

Figure 1. Examples of somatosensory evoked potentials (SEPs) to transient stimuli

The SEP consists of typical positive and negative going peaks, namely the P50, N80, P100, N140 and late component (LC). Numbers refer to the approximate peak latency after stimulus onset while P and N refer to a positive or negative going peak, respectively. Attending to a tactile stimulus resulted in marked increases of the N140 component and less positive going LC. Left panel is take from [9] and right panel from [23], with friendly permission.

These early responses were modulated with attention and had their sources in SI. However, they also reported an attentional modulation of later SEFs (125 and 138 ms), which had their sources in SII. These very early effects of spatial attention in SI might, however, be due to the nature of electrical median nerve stimulation used in that study (see remark above). With mechanical stimulation results are mixed, since early as well as late spatial attentional modulations of SEP components were reported. Eimer and colleagues [25] showed effects as early as overlapping with the N80 component. On the other hand attentional modulations no earlier than at the level of P100, and N140 [9, 23] have also been described (see Fig. 1). Together, these results suggest that the stage at which spatial attention alters the neural responses to transient stimuli is not a hard-wired mechanism but might be influenced by a number of variables.

One such variable might be the level to what extend stimuli have to be processed in order to perform a particular task. In humans [9, 25] as well as in animals [26] empirical evidence exists that extended in-depth stimulus processing alters to a greater extend responses in SII compared to rather superficial stimulus processing such as in a very easy detection task. In their animal study Chapman and Meftah [26] found that SII cells are highly sensitive to stimulus properties related to texture discrimination. Attention had the additional effect to increase their responsiveness. The consequence of extended in-depth processing and, thus greater neuronal modulation in SII, is a shift in the first component of the SEP that exhibits a pronounced attention effect such as a shift to the N140 rather than an earlier component.

A further finding in the field is the greater distractibility by competing stimuli in the somatosensory system compared to vision and audition. In other words, a rather small physical change at the to-be-ignored body location involuntarily pulls attention to that location. Of course such an involuntary shift is highly effective in detecting harmful stimuli even at to-be-ignored locations. Driver and co-workers [27] used mechanical oscillations driven at 300 Hz to produce fingertip vibrations simultaneously on both little (i.e., fourth) fingers. Participants performed a speeded two-choice discrimination task (discrimination between a long continuous *versus* two short bursts). Their results showed that incongruent simultaneously presented vibrations presented at the to-be-ignored hand delayed response times and increased error rates to the target substantially. The authors concluded that: "*...tactile attention cannot be perfectly restricted to one hand, to the complete exclusion of vibrotactile information on the other hand*". Kida and co-workers found additional empirical evidence that such to-be-ignored input will involuntarily pull attention to that body location when transient stimuli were applied [28–30]. These studies consistently showed that to-be-ignored sudden onset events are linked with a *greater* N140 amplitude, compared to the N140 evoked by deviants in an oddball stream [28–30]. This increase of N140 amplitude for the to-be-ignored stimulus is seen as evidence for the involuntary shift of attention.

From the visual domain it is known that such shifts of the attentional focus in space are time consuming [31–34]. Lakatos and colleagues extensively explored the time to detect the presence of a tactile stimulus as a function of its distance from the body location to which attention was oriented in the first place [35]. Their results showed that it took subjects a bit more than 200 ms to shift their attention from their right to left wrist. From these findings one can conclude that spatial separation of stimulation sites as well as inter stimulus intervals [36] play an important role to determine which component of the SEP is modulated with attention.

Somatosensory sustained attention and vibrating stimuli

Transient stimuli presented with long inter-stimulus intervals allow the underlying generators to recover after each stimulus. Thus, transient stimulation reveals a detailed time course of post-stimulus processing (as manifested in the evoked potential), but it is very difficult to maintain a focused state of attention upon stimuli that only occur transiently [37]. As mentioned above, the somatosensory system in particular seems to be organised in a way to automatically shift attention to that body location where something happens. Furthermore, in everyday life we make the experience that sustained attention to a certain body location is very often required. Most strikingly, when we search for an object in a place where our visual system might face some limits (for instance searching for our key-ring in a dark bag). Spatial and non-spatial information gathered from our sense of touch highly determines the success of this search. To mimic such sustained attentional demands, a highly successful technique is to present vibrotactile sinusoidal stimuli with a certain vibration frequency for a longer time period, often up to several seconds.

In humans mechanical vibrotactile stimulation elicits a sinusoidal electrophysiological brain response called the somatosensory steady-state evoked potential (SSSEP), which has the same temporal frequency as the driving stimulus and may include higher harmonics [38–42]. The largest SSSEP amplitudes have been found for stimulation frequencies in the 20 Hz range [40]. An example of SSSEPs, elicited with vibrotactile stimuli at 20 and 26 Hz is shown in Figure 2. Mechanical stimulation of the glabrous skin in this frequency range is related to the sense of flutter [43–45]. In animals, it was found that rapidly adapting afferent units with projections to areas 3b and 1 of the primary somatosensory cortex are coding low frequency mechanical stimuli [46–49]. From SI further projections into the secondary somatosensory cortex of the hemisphere contralateral to the stimulated side have been reported [50]. A further finding, by means of optical intrinsic signal processing, showed that the absorbance in cortical areas 3b and 1 of the squirrel monkey remained very stable for up to 30 s of 25 Hz stimulation [50]. The same was true for SI neuron spike discharge activity for stimulus durations of up to 7–8 s [50]. Such findings indicate a relatively slow adaptation, i.e., very limited habituation of the flutter system to continuous stimulation.

So far, most of the reports on human SSSEPs focused mainly on its physiological properties, such as its 'temporal resonance' in order to compare it to similar brain responses obtained in vision and audition [40]. Just recently, it was shown that the amplitude of the SSSEP is significantly increased when subjects attended to the vibrating stimulus delivered to one finger as compared to when this finger was ignored [51]. In this experiment, vibrotactile stimuli were delivered to the left and right index finger simultaneously for several seconds. Each finger was stimulated with a different vibration frequency in the flutter frequency range (20 Hz at left, 26 Hz at right index finger). Subjects had to attend to one finger and to detect small decreases in amplitude in the stimulus stream of the to-be-attended finger. Attending to one finger resulted in a significant increase in SSSEP amplitude compared to when this finger was unattended (see Fig. 2). It is worth mentioning that similar to what has been found in the above mentioned animal studies,

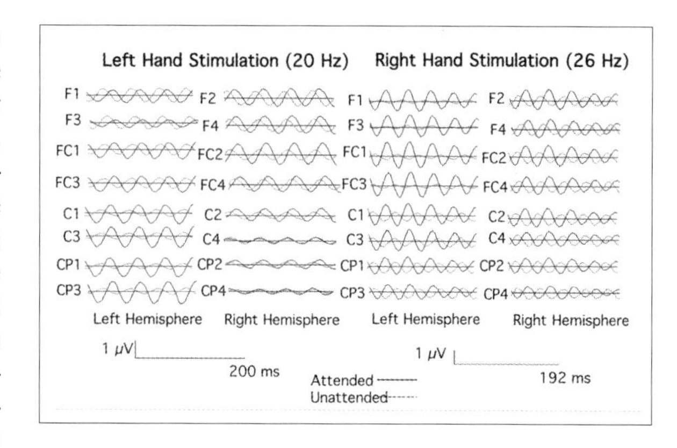

FIGURE 2.
Example of steady state somatosensory evoked potentials (SSSEPs) elicited by vibrotactile stimuli with 20 Hz (at left hand) and 26 Hz (at right hand). SSSEP amplitude was significantly increased with attention. Figure adapted from [51].

SSSEP amplitude exhibited no signs of habituation when we compared the first and last second of stimulation [51].

In line with the anatomical connections of flutter described above activation of the primary and secondary somatosensory cortex was found but the SSSEP *attention effect* was almost exclusively located in SI (see Fig. 3). This is a highly interesting finding, which allows one to speculate upon the possible neuronal mechanisms that generate this effect. We will turn to that point in the following section.

Neuronal mechanisms of selective attention in touch

Different models are described in the literature to explain the neuronal basis of preferential stimulus processing of the attended stimulus. In somatosensation two models play an important role. The first model relies on a sensory gain control mechanism that has been described in the visual modality [12]. Basically, preferential

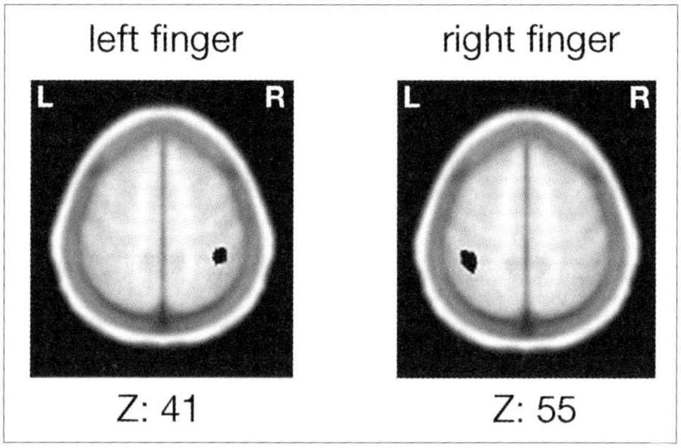

FIGURE 3.
Cortical sources of the attention effect (attended minus unattended) for left and right index finger stimulation with vibrotactile stimuli (see Fig. 2). The attention effect was very pronounced in primary somatosensory cortex (SI) at the hemisphere contralateral to the stimulated index finger (Statistical Parametric Maps; $p<0.01$*; Z=location of axial slice). Figure adapted from [56].*

stimulus processing is managed by increasing the neuronal gain (amplification) of the to-be-attended stimulus and/or decreasing the neuronal gain (suppression) of the to-be-ignored stimulus. The second model counts on a change in the temporal pattern of action potentials. Here, neurons that code the to-be-attended stimulus synchronise their action potentials resulting in a separation of this neuronal population from the surrounding background noise [52–55]. In the following, these models and their link to somatosensation will be outlined in more detail.

Sensory gain control as a neuronal mechanism of attention

A prominent model in the visual modality to explain how the brain processes the attended stimulus preferentially is the sensory gain model [12]. In this model a gain control affects the overall neural response in a particular brain region without changing the time course or patterning of the neural activity [12]. Based on ERPs, sensory gain is reflected by (a) no change in the waveform of the ERP, (b) no change in the latency of the ERP component of interest, (c) no change in scalp voltage topography of this component between attended and unattended stimuli, and (d) no change in cortical sources within a certain time window between attended and unattended stimuli, which is to some extend related to (c).

As depicted in Figure 1, there is no difference in the latency of any of the SEP components between the attended and unattended condition. In Figure 4, the topographical distribution of the P50, P100, N140 and late positive component (LC) of the SEP for attended and unattended tactile stimuli is depicted. These topographical distributions resulted from the SEP depicted in Figure 1 (left panel).

Obviously there is almost no change in the topography between the two conditions. The only difference, as depicted in Figure 1, is a change in amplitude. Thus, results already fulfil three important requirements (a–c) of the sensory gain control model. The forth and last requirement is the invariance of cortical sources between attended and unattended stimuli. Imaging studies cited above provided supportive evidence for requirement (d). Importantly, sensory gain control is not only present in transient stimuli but explains amplitude modulations of the SSSEP to vibrating stimuli as well [56].

Results so far are supportive for sensory gain control as a plausible neuronal mechanism for selective stimulus processing in the somatosensory cortex. But what is unanswered up to now is the exact neuronal interplay of amplification and suppression. While studies that show a signal difference between an attended and unattended stimulus cannot unambiguously judge whether the attended stimulus was amplified or the unattended stimulus was suppressed because both would result in an amplitude difference, some studies in animals and humans focused on that question.

Hsiao and colleagues [15] used a cross-modal task in which monkeys had to match tactile and visual stimuli. Monkeys learned to discriminate

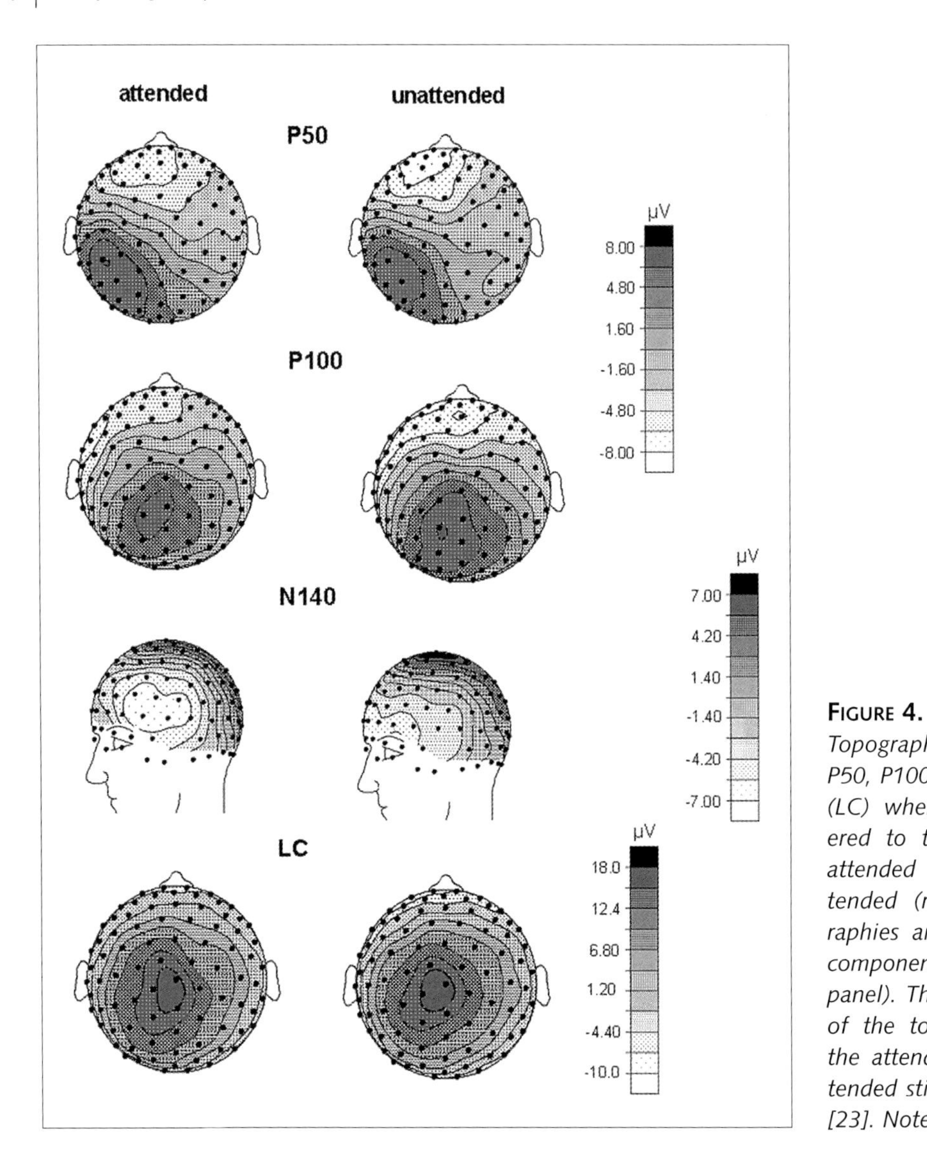

FIGURE 4.
Topographical distributions of the P50, P100, N140 and late component (LC) when a tactile stimulus delivered to the right index finger was attended (left column) versus *unattended (right column). The topographies are calculated from the SEP components depicted in Figure 1 (right panel). There is no marked difference of the topographical distribution of the attended compared to the unattended stimulus. Figure adapted from [23]. Note: Different scaling.*

paired associations in this cross-modal task. They also performed a detection task in the visual domain. The authors reported that their attentional manipulations correlated with a combination of *increase* and *decrease* in neuronal firing rates. In fact, more than 80% of the neurons in SII exhibited attentional modulation and these neurons showed a mixed response of increase (~60%) and decrease (~20%) of their firing rates. A related pattern was also found in SI. Such a mixed response of amplification and suppression was confirmed by a subsequent study [57]. Responses of SI neurons were modulated by the animal's focus of attention. 50% of neurons in SI showed both enhanced and suppressed responses in trials where the animal was cued to respond and attend to the hand contralateral to the recording site in SI. Similar to animal studies, in humans it was reported that fMRI responses in SI showed a mixed pattern, which is depicted in Figure 5: (1)

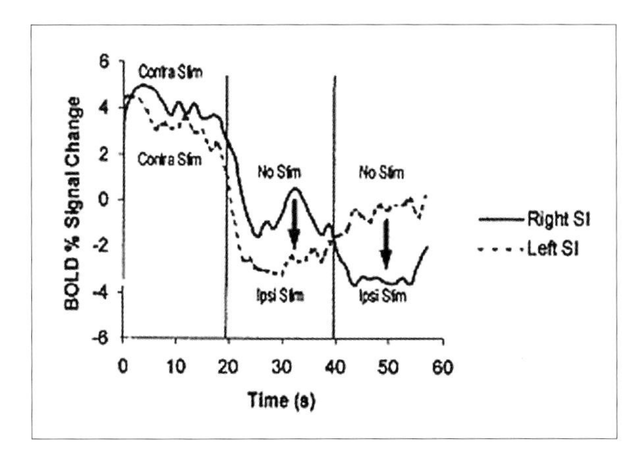

FIGURE 5.

Blood oxygen level dependent (BOLD) response of voxels in right (solid line) and left (dashed line) SI. Vertical lines indicate onset of blocks with different stimulation. In the first block, task-relevant stimulation was delivered to left and right fingers. In the second block, stimulation was delivered to the left index finger and no stimulation at the right finger. In the third block, the conditions were reversed: stimulation was delivered to the right index finger with no stimulation of the left finger. Results show suppressed activation in ipsilateral SI compared to rest under task-relevant stimulation conditions (arrows). Figure taken from [58] with friendly permission.

facilitation (amplification) in contralateral SI and (2) suppression in ipsilateral SI [58].

From these findings it seems quite obvious that sensory gain control introduces an effective and fine tuned mixture of amplification and suppression that seems to be highly dependent on the task at hand. A very interesting finding in animals was that suppressive responses occurred mainly during the early and amplification during the late phase of a trial [59].

Synchronisation of neuronal responses as a neuronal mechanism of attention

Imagine a random dot display in which all dots move arbitrarily. Such a display produces visual noise that does not allow to extract or to identify any meaningful object or feature within the display. Then all of a sudden for a short period in time a subset of dots move coherently in one direction. Most certainly this pattern produces a pop-out effect in such that every observer will recognise this pattern of coherent motion. If these dots form an object, say a geometric figure, then coherent motion will produce a pop-out of that object from the random background noise. Importantly, no change in luminance or colour or grey scale is needed to produce such a pop-out.

On the neuronal level this pop-out can be produced by synchronised neuronal activity, without changing the gain. This idea of synchronisation of neuronal responses as a mechanism of selective stimulus processing was put forward by Niebur and colleagues [54, 55] based on theoretical computational models, which were influenced by von der Malsburg [60]. The general principle is depicted in Figure 6. As schematically depicted in the lower traces, action potentials of two neurons do not occur at the same time point when stimuli are unattended. However, when the monkey attended a certain stimulus, then action potentials of the two neurons become synchronised. In humans an increase of synchronised neuronal activity in the EEG recordings was shown in the visual modality [52, 61]. Recently the neuronal synchronisation hypothesis of attention was also strongly supported by intracortical recordings in the visual cortex of the behaving monkey [62]. The authors stated: "*...these localised changes in synchronisation may serve to amplify behaviourally relevant signals in the cortex*".

In somatosensation, Steinmetz and colleagues [63] analysed the correlation between pairs of neurons recorded in SII as a function of attention. Animals were cued to perform blocks of trials of either a tactile or a visual task. When attention was cued to the tactile task a dramatic increase in synchrony between cells in SII resulted compared to when animals performed the visual task. In humans, increased neuronal synchronisation with attention was reported in a tactile delayed-match-to-sample task [64]. Subjects had to discriminate dot patterns similar to Braille deliv-

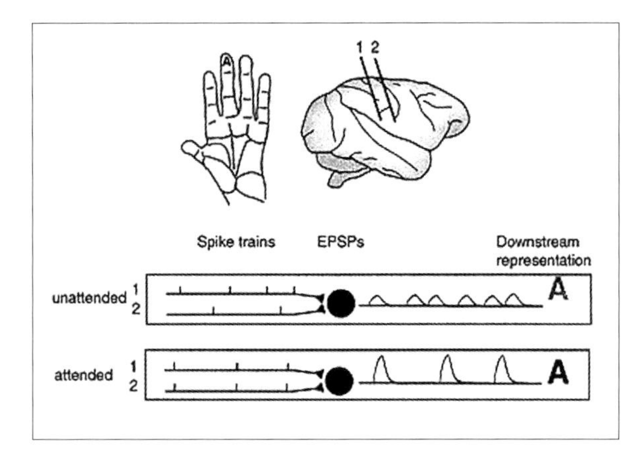

FIGURE 6.
Schematic representation of the neuronal synchronisation hypothesis of attention. Monkeys were required to either identify a certain letter or to perform a non-tactile task while the finger rested onto an embossed letter (A in this case). Spike activity from two cells in somatosensory cortex was recorded (1 and 2). When monkeys performed the non-tactile task action potentials were not synchronised in time (unattended). This was different when monkeys attended to the letter. As depicted in the lower panel, neuronal discharge rates between the two neurons were highly synchronised. These synchronous inputs produce larger excitatory postsynaptic potentials (EPSPs) in the receiving neuron that resulted in higher postsynaptic firing rates, leading to an enhanced representation of the stimulus letter. Figure taken from [55] with friendly permission.

ered to their index fingers. Attention was cued to one finger (either left or right) and subjects were asked to report the reoccurrence of a sample pattern in delayed presented test patterns. This manipulation resulted in synchronised responses in a higher frequency range, with its generators presumably in SI, as inferred from scalp topographies. This result not only shows the possibility that selective stimulus processing can be performed by synchronised neuronal activity in the human brain, but also that SI represent indeed non-spatial attributes of tactile stimuli as argued in the introduction of this chapter.

Summary

Our sense of touch differs from vision or audition that stimuli have a direct immediate impact onto the body. This requires fast behavioural responses. In line with this requirement, physical changes of stimuli at body locations that had to-be-ignored automatically pull attention to that body location to analyse the significance of that change. Attention to easy to detect stimuli alters neuronal responses in SI. Deeper processing, such as in a stimulus discrimination task is more related to neuronal changes in SII. Two cortical neuronal mechanisms that allow preferential stimulus processing of the to-be-attended stimulus were described. On the one hand a sensory gain control mechanism introduces a fine tuned mixture of signal amplification and suppression. On the other hand temporal synchronisation of action potentials of neurons that code the to-be-attended stimulus produces a 'pop-out' from the background noise. Whether both mechanisms interact or whether the one or the other is predominantly active for certain tasks are subject to future research.

Selected readings

Chapman CE, Meftah EM (2005) Independent controls of attentional influences in primary and secondary somatosensory cortex. *Journal of Neurophysiology* 94: 4094–4107

Forster B, Eimer M (2004) The attentional selection of spatial and non-spatial attributes in touch: ERP evidence for parallel and independent processes. *Biological Psychology* 66: 1–20

Giabbiconi CM, Trujillo-Barreto NJ, Gruber T, Müller MM (2007) Sustained spatial attention to vibration is mediated in primary somatosensory cortex. *NeuroImage* 35: 255–262

Hsiao SS, Vega-Bermudez F (2002) Attention in the somatosensory system. In: RJ Nelson (ed.) *The somatosensory system: Deciphering the brain's own body image*. Boca Raton: CRC Press, 197–217

Haptic object identification

Soledad Ballesteros and Morton A. Heller

16

Introduction

The study of touch has interested psychologists and neuroscientists for a long time but interestingly, the world of touch is flourishing nowadays. Important developments are taking place in psychology of touch, neurosciences, virtual reality and robotics. Today, haptics is more than ever a multidisciplinary field as well as a world research effort. In 2005, after a number of interesting and successful conferences organised independently in the United States and Europe, the First World Haptics Conference took place in Pisa (Italy). The Second World Haptics Conference was celebrated in Tsukuba (Japan) in 2007. The papers presented at World Haptics cover three main areas; the Science of Haptic Perception, the Technology of Haptic Interfaces, and the Applications of Haptic Interfaces and Teleoperation Systems. The great success of these international worldwide scientific meetings made us feel that the world of touch is more vibrant than ever.

The startling growth experienced by the field of haptics during the last decade represents a realisation that fundamental perceptual problems may be solved in this modality [1]. Recent approaches to the study of touch from the experimental psychological methodology and sophisticated neuroscience techniques are likely to provide interesting advances in the field.

In this chapter we first consider the 'hot' issue of active *versus* passive touch. Tactile passive perception refers to stimulating the stationary finger or hand with a moving or static external stimulus whereas the term haptic perception is reserved for referring to the active exploration and manipulation of surfaces and objects with our hands. This mode of exploration allows one to extract a wealth of sensory information for further processing. The central core of the chapter is devoted to the perception of a series of important dimensions experienced in surfaces and to the recognition of two-dimensional raised-outline stimuli and the identification of three-dimensional objects by active touch.

Active *versus* passive touch

Gibson [2, 3] drew a distinction between active and passive touch. For Gibson, haptics did not consist of the mere addition of kinaesthesis and cutaneous information as one experiences objects or patterns. We do not simply act as receptor surfaces, with an existence that merely serves the needs of experimenters in perception. People are actors and perceivers in a real world. This means that we often act in a manner that maximises the quantity and quality of information that we obtain from the environment. Gibson was critical of much of the older research that assumed that we can generalise from passive cutaneous stimulation to other, more normal circumstances in which we engage in active movement to obtain information from the world. Active touch differs from passive touch in the intentionality of our exploratory behaviours. The distinction is somewhat different from the older reafferance/exafference dichotomy. Gibson thought that active touch was more likely to yield objective percepts and veridical perception, while passive touch tended towards subjectivity.

There are certainly circumstances where stimulation is imposed on a stationary and passive perceiver. The question is whether these are atypical, or whether or not we can generalise from the study of passive touch to more natural, active situations.

The distinction that Gibson drew between active and passive touch has certainly been controversial [4]. In his original experimental report, Gibson [2] pressed cookie-cutter shapes on the palms of subjects in a passive condition, and allowed subjects the option of actively feeling the outlines of the shapes, in an active condition. In all cases, the subjects made visual matches to the haptically experienced stimuli. While there were a number of different passive conditions, active touch yielded clearly superior recognition performance. Unfortunately, there were some serious methodological difficulties with the experiment. Confirmation of these sorts of findings can be found in the literature [5]. Mode of touch was confounded with the sensitivity of the receptor surface in Gibson's [2] study, since the fingertips are far more sensitive than the surface of the palm. However, Heller [6] showed that active touch maintains a superiority, even when the sensitivity of the receptor surface is comparable. Furthermore, Heller, Rogers and Perry [7] reported better numeral recognition using the Optacon, when the subjects were allowed to move their preferred index finger actively over the vibrotactile display.

There is a constraint on the advantages of active touch. It is hardly universal. There are a number of clear cases where passive touch can yield exceptionally high levels of recognition accuracy, especially when the stimuli are limited in scope or are highly familiar. Thus, a passive form of touch can be found when numbers or letters are drawn on the skin of the palm or fingers. It is relatively easy for subjects to name these patterns when they are drawn on the skin, and blindfolded sighted individuals can read words, if the letters are drawn large enough, and there is enough time between patterns. Passive touch is subject to after sensations when stimuli are drawn on the skin. Note that the digit span is relatively normal for drawing on the skin, given slow rates of presentation [8]. Deaf-blind people learn to communicate with sighted persons using this method, and one sees failures in identifying numerals printed on the fingers in parietal damage [9].

High performance with passive touch is possible, but is likely dependent upon familiarity and experience with the stimuli. This aids us to resolve the apparently conflicting results. Certainly, the advantages of active touch evaporate when we print letters or numbers on the skin. The answer involves familiarity, practice and skill. If one does not have categorical information about the nature of the stimuli that are passively presented, then recognition fails. Performance suffers when letters are printed with an irregular and unfamiliar writing style. These data support the notion that practice and skills are valuable to help overcome some of the limitations of passive touch.

We should also point out that the notion of active *versus* passive touch is often theoretically vague. Kinaesthesis can be active or passive, and intention is very difficult to operationalise. Nonetheless, there has been considerable interest in this theoretical issue, with a large number of studies in the literature [4].

Hand movements and haptic exploration

Lederman and Klatzky [10] showed that according to the type of information that people want to extract from an object, they executed different hand movements. They called these stereotyped hand movements exploratory procedures (EPs). Lederman and Klatzky [10] analysed the videotaped hand movement patterns executed by their participants while they intended to match objects in terms of different object attributes. Lederman and Klatzky proposed six different exploratory procedures. These authors reported that observers moved their hand systematically as a function of the attribute they were asked to look for. For example, when observers wanted to judge the weight of a certain object, they lifted the object from a supporting surface (*unsupported holding*

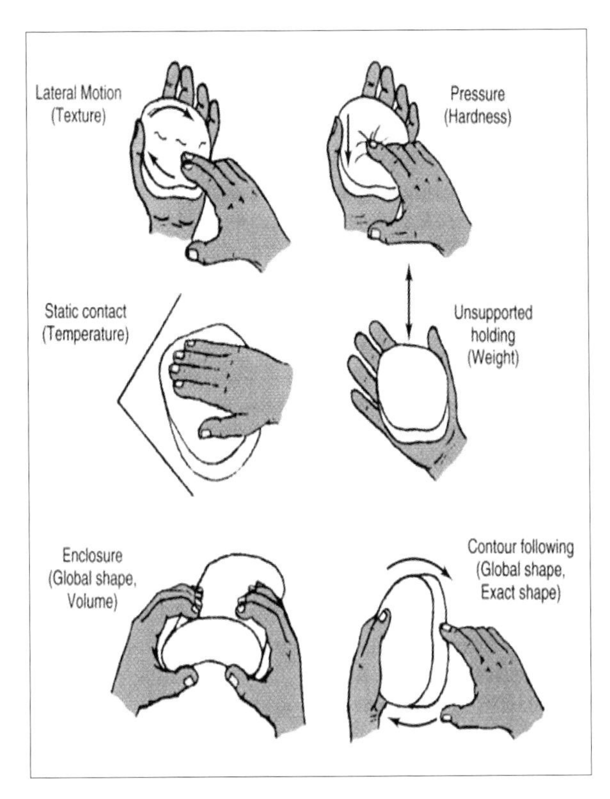

FIGURE 1.
Exploratory stereotyped hand movements (EPs) described by Lederman and Klatzky (1987). In parenthesis is shown the associated attribute for each EP. Copyright Elsevier.

movement). In contrast, when the attribute of interest was texture, they moved their hand back-and-forth across the surface performing a *lateral motion*. When the property of interest was the hardness of a surface or object, the movement was *pressure* but when observers intended to perceive the temperature of a stimulus, they perform a *static contact movement*. The last two main hand movements described were *enclosure* and *contour following*. The former was related to global shape and volume while the latter was performed to perceive the exact form of an object. Figure 1 shows the six main exploratory movements (EPs) just described. Hand movements are considered by these investigators as the 'windows' to explore the underlying representations of objects and the mental processes that deal with these representations. In a further study, Lederman and Klatzky [11] differentiated Enclosure (*Global enclosure*) from a special type of enclosure called *Enclosure (part)*. This movement was associated with the extraction of shape information and consisted of moulding the fingers to adapt them to a secondary part of a certain object.

Based on the better performance in terms of accuracy and speed, Klatzky and Lederman [12] proposed that material properties are processed by touch more effectively than are geometric dimensions.

Haptic recognition of raised-line drawings and three-dimensional objects

It is easy to recognise familiar objects by touching them [13]. Blindfolded participants were presented with 100 common objects and were very accurate (96% and 99% correct for stringent and lax criteria, respectively). Observers were also very fast: 38 of the 100 objects were correctly identified in 1 sec, 25 in 2 sec, and 10 in 3 sec. Such good performance may be due to the fact that common 3-D objects vary in several dimensions easily perceived in a converging way by executing the appropriate exploratory procedures [10]. Indeed, the inability to recognise objects by touch would be an indication of rather severe brain damage [9]. However, the operative factor here may be whether or not the objects are familiar. Perception of pictures has been far more controversial.

In both picture and object recognition, restricting subjects to the use of a single finger degrades performance. Thus Jansson and Monaci [14] reported that object recognition is improved when subjects are allowed to feel common objects with more than one finger. In feeling tangible arrays, the use of multiple fingers of two hands can aid touch, and the haptic Muller-Lyer illusion is greatly diminished when subjects are allowed to feel the stimuli with the index fingers of two hands. Also, visual picture recognition is greatly impaired when the field of view is restricted in a manner that is analogous to artificially limiting touch to the use of a single index finger.

Lederman, Klatzky, Chataway, and Summers [15] have argued that touch is not well suited for the perception of tactile pictures, since they are two-dimensional. They claim that touch is far better at the apprehension of substance related characteristics of objects, many of which may be derived from three-dimensional information (also see [16]). Substance related aspects of objects include hardness, softness, viscosity, thermal properties, and weight. According to Lederman and her colleagues, touch suffers when attempting to interpret 2-D form, since the sense of touch is limited in spatial resolution and must process information slowly and sequentially. According to Jones and Lederman [16], tactile picture perception is only adequate when the number of alternative pictures is limited, or categorical precuing is used [17].

There are some difficulties with the notion that touch cannot do well with flat 2-D arrays, such as pictures. First, most of the subjects that have been asked to name pictures have been college students or they have been completely unfamiliar with the use of touch for this perceptual process. Blindfolded sighted undergraduates do very poorly with Braille matching, yet blind persons who are skilled readers can rapidly identify Braille words with extremely short exposures. Also, many studies of tactile picture recognition have asked for a naming response. One may see an object perfectly well, yet not know what to name it, especially if the object is unfamiliar. The same principle holds for tangible pictures. Thus, blindfolded sighted subjects lack skill and familiarity in using touch for pattern perception. It is not surprising, that late-blind subjects may show better performance in picture perception tasks than blindfolded sighted subjects [18]. Moreover, people with very low vision excelled in discriminating figure from ground in tangible pictures, and greatly exceeded the performance of blindfolded sighted participants [19].

Heller, Brackett, and Scroggs [20] tested tangible picture matching in blindfolded sighted and visually impaired subjects. The participants felt a picture and picked a match from four choices. Performance ranged from 90% correct to nearly 100% correct for visually impaired subjects. Another study that demonstrated high performance by touch involved the examination of the effect of providing categorical information about tangible pictures [17]. If subjects know that a picture is, say, a type of fruit, then picture recognition is excellent.

Researchers have wondered about the depiction of depth relations in pictures, since one would not expect that touch would be especially sensitive to perspective cues. Support for this notion was found in studies that fail to find a Ponzo illusion with haptic drawings [21]. Congenitally blind people may not anticipate the visual distortion that occurs as squares are tilted, that is, they become foreshortened. Nonetheless, blind individuals can certainly learn to interpret linear perspective, and can readily comprehend depth relations in tangible pictures. The utility of pictures is dependent upon viewpoint, with some viewpoints easier to understand than others. Whether or not a particular viewpoint is useful is likely to depend upon stimulus characteristics. For example, the top view of a head is not very helpful in many instances, while the rear views of some forms can be informative.

When we read of low performance in the perception of haptic pictures, blanket generalisations are not really appropriate. Performance often is subject to large individual differences, stimulus characteristics, viewpoint effects [22], and especially to practice and familiarity.

Object perception and dynamic touch

An exciting new approach to haptics has focused on the functional use of touch for gaining information about the layout of the environment and objects within it [23]. The "new ecological" perspective is clearly not cognitive in orientation, and aspects of it have been controversial [24]. However, it is an important addition to our knowledge of objects, how we perceive them and how we make use of them.

Humans are tool-using creatures, and this is certainly a distinguishing characteristic of our nature. We make use of canes for support, and people who are visually impaired make excellent use of them to explore their environment. However, we are also able to discover important information about tools and tool-like objects by wielding them. There is evidence that we can learn a great deal about objects by exploring them with especially useful objects (e.g., a long cane or dental probe). People are able to make judgments about the lengths of objects that they wield, and they can also perceive their height and width in the same manner. Perceivers are able to judge the properties of hand-held objects, especially those that relate to where it is best to use them to strike objects, i.e., the 'sweet spot'. According to Michaels, Weier, and Harrison [25], people are able to rate how useful a tool-like object would be for a variety of tasks, even without vision, and when limited to simply wielding the object. They argue that perceivers are sensitive to the inertial properties of objects.

Identification of surface properties by touch

Material properties such as texture and hardness are more salient for touch than object properties such as form and size, according to Klatzky and her colleagues, [26]. Of course, this depends upon the particular object property involved. Thus, sharpness of a blade is certainly salient for touch, and the impact of object geometry is probably dependent upon scale.

Material properties: texture, weight, and compliance

Texture

David Katz (1984–1953 [27]) was an early important source in the study of texture (microstructure, the fine structure of the surface). He combined phenomenological observations with interesting experimentation.

Hollins and his collaborators have conducted interesting work showing that the perception of texture is multidimensional [28, 29]. Hollins, Faldowsky, Rao and Young [29] presented a series of ecological surface textures, one by one across the index finger of the participants and asked them to sort a set of 17 stimulus surfaces into 3–7 categories based on their perceived dissimilarity. The data fit reasonably well into two robust and orthogonal dimensions: roughness–smoothness and hardness–softness with a third dimension more elusive that might reflect the compressional elasticity of the surface. In another study, Hollins, Bensmaïa, Karlof and Young [28] obtained further support for the rough–smooth and hard–soft dimensions of surface textures. The support for the third dimension (sticky–slippery), however, was not so clear and was observed only in some perceivers. In the Hollins et al. studies, the fingertip of the perceiver remained static while the stimulus was moved perpendicular to its surface at a constant speed.

More recently, Picard and colleagues [30] allowed free exploration of 40 different ecological surfaces instead of the passive stimulation condition used by Hollins et al. The results showed that the *soft–hard* dimension constituted a salient perceptual dimension in the texture space and has a hedonic character. *Thin–thick* was a likely second perceptual dimension orthogonal with the soft–hard dimension. These two dimensions were stable salient dimensions in all the stimulus sets. A third sticky–slippery dimension played a minor role in the texture space. This dimension depended on the stimuli that were included into a given set.

The richness and variety of sensations obtained from the finger pad is based on neural activity of four types of mechanoreceptors in the skin. Each one of these receptors has different sensory and perceptual functions [31]. The slowly adapting SA1 system (the Merkel-neurite complex) provides the information on which texture perception is based. Its is sensitive to edges, corners, points, and curvature resolving spatial features as small as 0.5 mm or even less [32]. This means that the haptic system is well adapted to the perception of texture. Touch and vision, however, perform similarly over a range of textural stimuli (e.g., 40–1,000 grit abrasives) but vision appears more suited to interpret larger configurations, and has advantages over touch for perception of large-scale spaces and large objects. This visual advantage may disappear when surface irregularities are very small. For finer textures, touch holds advantages [33]. Hollins and colleagues [34–36] differentiated between the perception of fine and coarse textures. They suggest that fine texture perception is mediated by vibration and depends on the PC channel.

Ballesteros, Reales, Ponce de León and García [37] studied the haptic texture space and whether the dimensionality would change when both modalities, vision and touch, are used to explore the stimuli. Two groups of young adults explored 20 ecological textured surfaces haptically or by touch and vision simultaneously. They performed a *free classification task*, a *spatial arrangement task*, and a *hedonic rating task*. The first two tasks were counterbalanced across participants while the hedonic rating task was always performed at the end of the experiment. In the *free classification task*, participants were asked to sort stimuli into groups assuming that the frequency with which two stimuli are placed in the same group is proportional to their similarity. In the *spatial arrangement task*, participants explored the surfaces with the fingertips of both hands and rearranged the stimuli so that the distances among the stimuli were proportional to their similarity [38]. The experimenter asked participants to explore the stimulus and to move it in the working space so that those that feel similar will be located close and those that feel different will be located apart. The spatial arrangement produced by a participant is shown in Figure 2.

Figure 2. Example of a participant's performance in the spatial arrangement task

The arrows indicate the distance in cm from the centre of the stimulus to each coordinate. Adapted from Ballesteros, Reales, Ponce de León, García (2005) WorldHaptics Proceedings *(WHC'05) Pisa,* IEEE Computer Society, *635–638, Los Alamitos, CA, USA.*

The results from the haptic exploration group showed that the first very salient dimension that organised the perceptual space was *smoothness/roughness*. The second dimension was *hardness/softness* and a third dimension identified as *slippery/sticky*. The bimodal exploration (haptics plus vision) produced a similar dimensional map. A nearly perfect correlation was found between both conditions (touch and touch + vision). That is, stimuli that were rated in a five-point Likert-like scale as very pleasant under haptic exploration, were also rated as very pleasant under bimodal haptic and visual exploration. Those that were rated as very unpleasant under one mode of exploration were also rated in the same way under the other mode of exploration. Hedonic ratings were positively correlated to the first dimension. Smoothest surfaces were rated as the more pleasant while rough ones were rated as more unpleasant (see Fig. 3) Note that there is evidence that touch is superior to vision at making texture judgments for very smooth surfaces, but probably not for coarse surfaces [39].

A further study with older adult participants [40] showed that the perceptual haptic texture space is very similar for young and older adults.

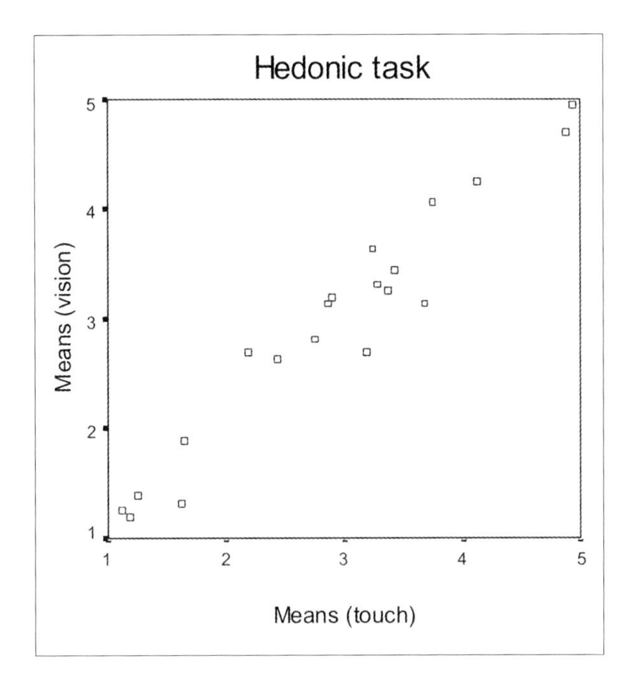

FIGURE 3. LINEAR REGRESSION BETWEEN TOUCH EXPLORATION AND VISUAL-TACTUAL SIMULTANEOUS EXPLORATION OF SURFACE TEXTURES

Adapted from Ballesteros, Reales, Ponce de León, García (2005) WorldHaptics Proceedings *(WHC'05) Pisa,* IEEE Computer Society, *635–638, Los Alamitos, CA, USA.*

Weight

The use of active touch is central for weight discrimination. The best exploratory procedure to judge the weight of an object is to lift it away from a supporting surface. This manner of exploration allows the haptic perceiver to obtain simultaneous information from cutaneous and kinaesthetic receptors. Research has shown that haptic weight perception is greatly influenced by other objects dimensions such as an object's size and an object's material.

The influence of an object's size produces the size–weight illusion. A large and a small object of the same mass do not seem to be equally heavy. That is, a large object of the same mass of a small object perceived by touch is judged lighter [41]. Ellis and Lederman [41] asked haptic observers to perform a magnitude estimation task. The stimuli were cubes of equal mass but different in volume. The results showed that unimodal haptic cues are both sufficient and necessary to obtain a bimodal vision/haptics size–weight illusion but unimodal visual cues produce only a moderate size–weight illusion.

More recently, Kawai [42, 43] has shown that the haptic size–weight illusion reflects an instance of heaviness discrimination; Kawai [43] investigated the contributions of object weight, haptic size, and density to the perception of discriminating differences in weight between pairs of cubes with cue conflicts such as that resulting from the size–weight illusion. Participants did not see the cubes and relied on the information provided by grasping the cubes with their fingertips, while attempting to discriminate possible differences in weight. Three set of seven cubes each of various weight were used (copper, aluminium, and plastic). The cubes were covered with smooth vinyl to eliminate input concerning hardness or material qualities. The results showed that the identification performance was quite accurate when density and weight both increased for the second cube or both decreased. These results were greater than those in constant-density conditions obtained previously [42], suggesting that changes in density may aid in the perception of heaviness as does weight. However, when the two cues conflicted directionally with each other, accuracy decreased greatly. According to Kawai, a person may perceive heaviness on the basis of the well-regulated relations between changes of density, size, and weight.

Ellis and Lederman [44] required that their participants use a magnitude-estimation task to assess the heaviness of equal-mass objects varying in material coverings. The results showed that weight perception is influenced by material. For example, objects with the same mass but differing in material coverings (e.g., aluminium, wood, styrofoam) are perceived to have different weights. The aluminium covered objects were judged to be the least heavy, followed by the wood, then the styrofoam objects. The authors attributed this illusion to the influence of sensory inputs.

Compliance

The research conducted to study compliance has been scarce compared to the number of studies dedicated to investigate other properties of objects and surfaces explored by touch, such as texture or shape. Compliance is the inverse of stiffness. Perceptual discrimination of compliance requires one to take into consideration not only force but also displacement. The softness of an object can be perceived in an active way through the fingers by pressing the surface or through a tool such as a pencil or any other stiff object held in the hand.

Perceived softness is the subjective assessment of an object's compliance, the amount an object deforms in response to an applied force. Tactile and kinaesthetic cues are necessary as shown by Srinivasan and LaMotte [45] using psychophysical methods. These researchers restricted sensory information to cutaneous input. The results showed that when the pressure on the finger was kept constant, cutaneous information per se was sufficient to discriminate among different rubber surfaces. In another experiment, Srinivasan and LaMotte [45] showed that observers with an anesthetised fingertip (eliminating cutaneous inputs) could not discriminate among objects of different compliance.

In another study, LaMotte [46] investigated the ability of human perceivers to discriminate softness, using a stylus under experimental conditions that differed in the motor performance required and the type of sensory information allowed. The compliant stimuli were made from transparent silicone rubber solutions mixed with varying amounts of diluents. Compliance was defined as the average slope of the approximately linear function relating displacement to force. Softness discrimination was measured under two testing procedures. Under the first, subjects were allowed to rank the softness of the 10 stimuli, under different experimental conditions. They actively palpated each specimen, in a natural way, either directly with one finger or indirectly by means of a stylus. Figure 4 illustrates different modes of contact used when ranking the softness of rubber objects of differing compliance.

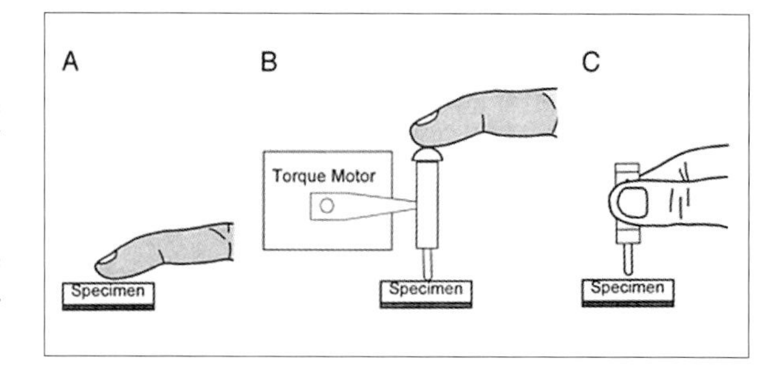

Figure 4. Different modes of contact used when ranking the softness of rubber objects of differing compliance.
*A: subjects ranked softness by actively tapping (*task 1*) or pressing (*task 2*) each specimen with the distal pad of the middle finger. B: a stylus with a tip diameter of 3 mm was mounted to a torque motor. A sphere, 10 mm diam, was mounted to the upper end of the stylus. During* task 3*, the stylus exerted an upward force of 15 g against the fingerpad and the subject tapped the stylus against the specimen. During* task 4*, the stylus exerted a downward force of 1 g after being placed by the experimenter on the specimen chosen by the subject. The subject then pressed the stylus against the specimen. C: the unconstrained stylus was held in a precision grip and was either tapped or pressed against the specimen (*task 5 *and 6, respectively).* Task 7 *was identical to 5 except that the stylus diameter was 9 mm. LaMotte, Robert H. Softness discrimination with a tool.* J Neurophysiol *83: 1777–1786, 2000.*

The accuracy of ranking softness under different methods of contact is shown in Figure 5. During active movements, kinaesthetic information provides useful cues that are not present during passive touch. These cues allow the observer to discriminate differences in object compliance not confounded by differences in applied velocity. When the tool is held in the hand, forces are applied *via* the tool to the object [46].

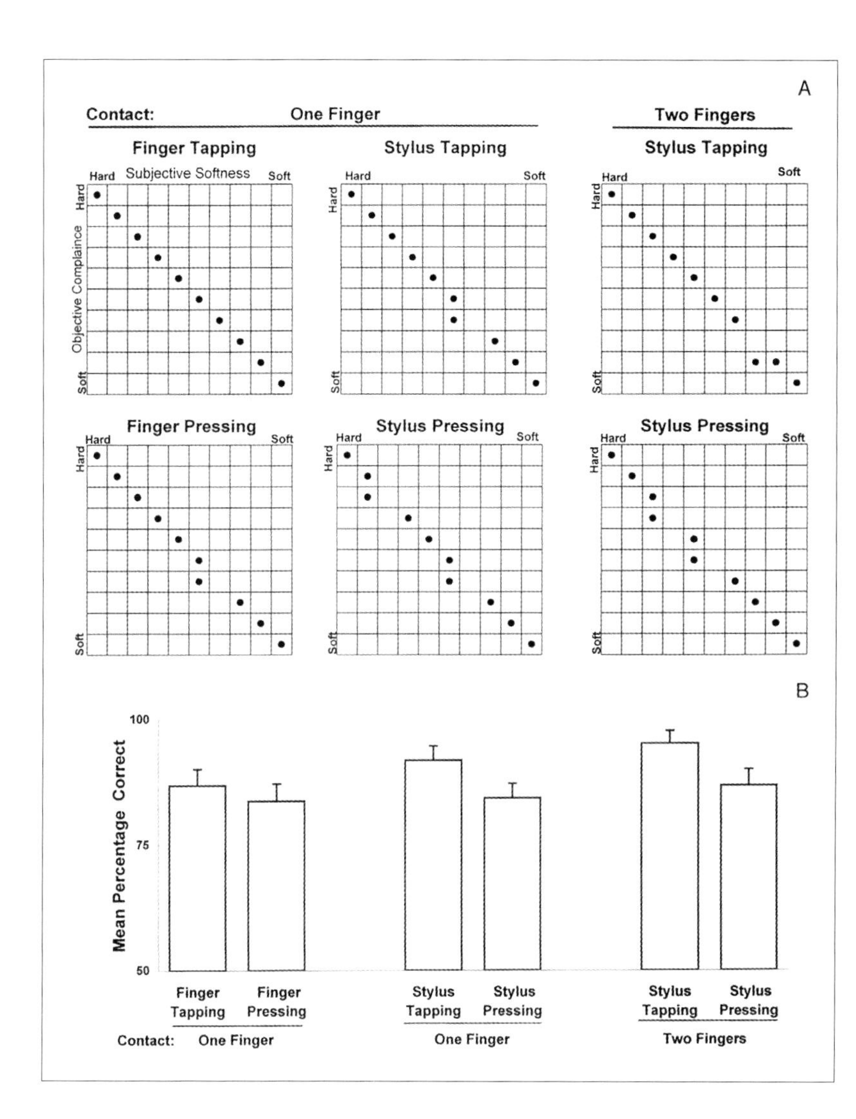

FIGURE 5. ACCURACY OF RANKING SOFTNESS UNDER DIFFERENT METHODS OF CONTACT

A: data from one subject indicate the correspondence (c) between the subjective and objective ordering of compliance. A perfect correspondence follows the diagonal from top left to bottom right. Methods of contacting the specimens were tapping or pressing the specimen either with a single finger (directly or via a stylus) or with two fingers holding a stylus in a precision grip. B: mean percentage correct for all subjects in the ranking task under each contact condition. LaMotte, Robert H. Softness discrimination with a tool. J Neurophysiol *83: 1777–1786, 2000*

Geometric properties: curvature, orientation and size

Curvature

The topic of the detection of curvature by touch has been widely investigated across different spatial sizes [47–50].

Curves are ever present in the world, yet are subject to misperception. Gibson [51] pointed out that if a person feels a curved surface repeatedly, and then traces a flat plane, the flat surface feels curved. This curvature aftereffect can occur after just a few seconds of haptic exposure [52], and can lead to misperception of the contours of forms. Moreover, people may tend to underestimate the distance between the endpoints of a curved path. This path-completion task has been studied extensively, and underestimation of path length seems to vary with extent and with irregularities in the shape of the path itself

[53]. Recently, Heller et al. [54] found that curves are subject to a horizontal–vertical illusion, with overestimation of their height compared to their width, even when both are equal.

Orientation

The orientation of a pattern may alter recognisability in vision, and it certainly can do this in touch. The study of orientation may involve interest in perception of orientation, *per se*, or the influence of orientation on haptic pattern and object perception. Note that many artificial patterns are defined by orientation, for example, numbers and letters.

It is noteworthy that stimuli at oblique orientations are a special problem for vision [55]. Just as in vision, touch is susceptible to the effects of orientation changes from the upright. Heller et al. [56] presented Braille characters to sighted, late-blind and congenitally blind subjects to study the impact of orientation on recognition. The blind subjects excelled at inverted Braille, but the congenitally blind subjects performed poorly with Braille rotated to an oblique orientation. Congenitally blind subjects have also showed lower performance with rotated patterns drawn on the skin and with rotations involving tangible pictures. One possibility is that visual images are easier to manipulate when a mental rotation is required. Another possibility is that these rotations are more difficult when context is eliminated, as in the rotation, for example of individual Braille patterns. Blind people may not show this sort of impairment when reading words. A third potential explanation of these results may derive from reduced experience coping with mental rotations in a specific concrete context. Thus, sighted individuals obviously have more experience with imaginal rotations of pictures and graphics. This helps them in solving mental rotation problems. The congenitally blind have more experience coping with left-right rotations in Braille, and performed better than sighted subjects with these problems [56].

Size

There is little doubt that people can make accurate judgments of extent using the sense of touch. However, the manner in which individuals examine patterns matters. A 'measuring' strategy was more accurate than other methods studied [57]. The measuring strategy involved subjects using their hands or fingers to mark off extent. When not prohibited from doing this, many participants will spontaneously adopt 'measuring'.

If limited to tracing with the index finger, the direction of scanning has an important impact on judged extent. Thus, numerous studies have shown that radial motions, involving tracing towards the body, are overestimated when compared with tangential scanning, where tracing does not converge upon the body. This difference in judged extent has been used to explain the horizontal–vertical illusion in touch, where people overestimate vertical extents and underestimate horizontals in inverted T and L configurations. This occurs even when the horizontal and vertical extents are equal. The illusion also occurs with curves (Heller et al., ms. submitted), and can be observed in natural, 3-D shapes, for example, with the Saint Louis Arch. However, the haptic illusion is not found when vertical Ts are gravitationally vertical. Here we may find a negative illusion in touch, with overestimation of horizontals [58].

The discrimination of symmetry by touch in two-dimensional shapes and three-dimensional objects

Symmetry is a very salient spatial property of shape. In the real world, objects and shapes very often exhibit bilateral symmetry and their right sides and left sides are similar. Bilateral symmetry in visual displays has been intensively studied as it has been considered the most salient organisational principle of a stimulus [59–62]. However, the role of bilateral symmetry in touch has been scarcely studied. Compared to vision, just a few studies have investigated the ability of active touch to discriminate this important property of shapes and objects.

Millar [63] conducted an early study using symmetric and asymmetric raised-dot displays of five and nine dots. Participants in this study performed a matching task, not a symmetry detection task. The main results were that differences in dot numerosity (texture) were detected more accurately than differences in symmetry. Millar's interpretation was that participants relied on texture differences. Haptic explorers relied on dot numerosity for spatially coding shape because the displays provided insufficient reference information for spatial coding. Millar noticed that when this happens, haptic explorers without spatial referents coded differences in texture instead of differences in shape.

Locher and Simmons [64] conducted a pioneer study on the detection of symmetry by touch. They used eight different planar polygons varying in complexity with 12–20 sides. The results showed that the detection of asymmetry was faster and more accurate than the detection of symmetry, irrespective of polygon complexity (measured by the number of sides). Moreover, accuracy was higher for asymmetric than for symmetric shapes. In another study, Simmons and Locher [65] investigated the influence of training in the detection of symmetry. They found that scanning time and number of errors decreased with training. However, symmetric shapes required double the decision time of asymmetric shapes.

Ballesteros and her colleagues [66, 67] conducted a series of experiments with small unfamiliar raised-line shapes and 3-D objects. Examples of the stimuli are displayed in Figure 6. The purpose of these studies was two-fold. First, we wanted to investigate the capacity of active touch to detect bilateral symmetry in these two types of haptic displays. Second, we wanted to explore the influence of exploration time and the use of the body-axis as a body-centred reference frame.

Haptic perception was moderately sensitive to small raised-line patters but perceivers were more accurate in detecting asymmetric shapes (81% correct under unlimited exploration time) than symmetric (61% correct) ones [66]. An important finding was that the use of two index fingers to detect the symmetry of small, raised displays improved performance relative to one finger exploration (76% vs. 63% correct symmetric detections under a forefinger and two-forefinger exploration). Moreover, the left hand was not more accurate than the right hand. Performance with the novel 3-D objects produced a different pattern of results. Three-dimensional objects convey more information as they allow a richer set of hand-movements. Furthermore, 3-D objects can be processed simultaneously by using the hand(s) to enclose the object. Bimanual exploration of 3-D objects was far more accurate and significantly facilitated processing of symmetric 3-D objects at different exploration time conditions. Moreover, in contrast to performance with the small raised-line shapes, touch was much

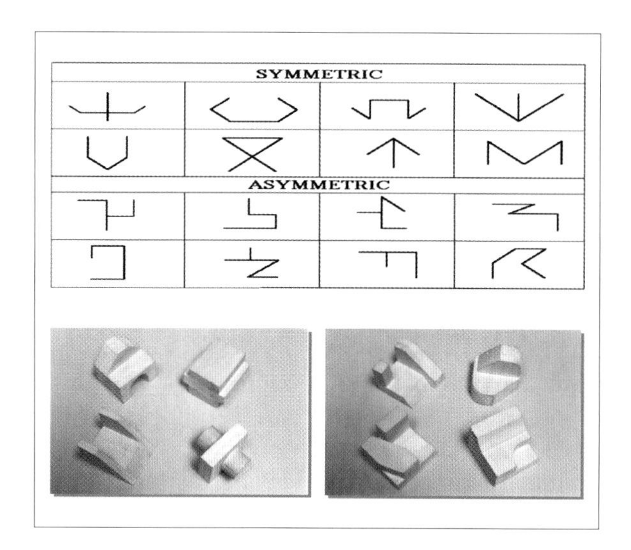

FIGURE 6.

Top: Examples of unfamiliar symmetric and asymmetric thermoformed raised-line shapes (approximately 2×2 cm size). Bottom: examples of unfamiliar symmetric and asymmetric wooden objects. Reprinted from Ballesteros, Manga, Reales (1997) Perc & Psych *59: 37–50. With permission of the Psychonomic Society.*

more accurate with symmetric objects than with asymmetric ones (93% correct vs. 86% correct for symmetric and asymmetric objects, respectively under unlimited time exploration). It is important to point out that the advantage for asymmetric compared with that for symmetric raise-line shapes was reversed with 3-D objects which allows parallel extraction of information. Note that even the bimanual mode of exploration with small raised-line shapes requires a successive mode of exploration compared to the more global simultaneous approach with objects.

Symmetry is a salient property in vision but it can also be elicited in touch when sufficient spatial reference information is provided to allow haptic inputs to be organised spatially [67]. In a new set of experiments, Ballesteros and her colleagues used small open and closed symmetric and asymmetric raised-line shapes. The task did not consist of the detection of symmetry. Instead they used an incidental or indirect task. The results showed that symmetry facilitated processing in vision even though the participants had to judge whether the shape was *open* or *closed*. The same task and stimuli did not show the symmetry effect in touch under single finger exploration and unlimited exploration time. The results cannot be attributed to poor tactual processing as accuracy in the open–closed detection task was nearly at ceiling. In contrast, a significant advantage for symmetric shapes was found when the small raised-line shapes were explored using two-forefingers, a condition that provides spatial-reference information. These findings support the idea that parallel hand movements offer haptic spatial information in relation to the body midline. This information allows the haptic explorer to judge symmetry within a body-centred frame of reference. In short, shape symmetry can facilitate haptic as well as visual processing even of small raised-line patters, without specific demands or training in the task in order to detect this spatial property of shape [67]. However, this only happened when small shapes are explored with the two-forefingers that provide body-centred spatial-reference information.

The influence of complexity and elongation in symmetry detection

The studies designed to explore the ability of touch to detect bilateral symmetry have reported a different pattern of results for raised-line shapes and 3-D objects. However, the studies used 2-D and 3-D stimuli that differed in many ways and were not directly comparable. Ballesteros and Reales [68] studied performance in symmetry discrimination using a new set of stimuli consisting of flat shapes extending in the z-axis to construct 3-D objects comparable in shape and complexity (see Box 1).

Multimodal effects and frames of reference in touch

We live in a multimodal world, and it is rare that we are artificially constrained to a single modality. Indeed, it is very difficult to completely eliminate information from a sensory system, e.g., audition. When a person is blindfolded and then immediately tested on his/her touch perception, the blindfolding does more than simply control for extraneous visual information. It deprives a sighted individual of important spatial reference information. In acting in our world, or in making spatial judgments, we function within a spatial coordinate system that serves to guide our actions and our perceptual activity. Our ability to walk in space typically depends upon visual information about the horizontal and vertical axes. We see the horizon and our spatial targets, and this helps us avoid a tendency to veer to one side while walking. Similarly, our haptic experience is normally dependent upon visual information about these spatial coordinates. When visual guidance of touch is eliminated, people may show reduced haptic performance in a variety of spatial tasks, and they certainly may be distressed at their loss of visual input. There is a large literature on sensory deprivation attesting to the impact of reduced visual experience

Box 1. Haptic discrimination of bilateral symmetry: The effect of stimulus elongation and stimulus complexity

Motivation and hypotheses

The 2-D and 3-D stimulus sets used in previous experiments differed in many ways (e.g., size, shape, number of sides, complexity). As Millar [72] pointed out, the range and type of information required for shape perception is related to the size and the type of stimuli. Small-raised shapes are difficult to code by reference to body-centred references cues while the perception of 3-D objects depends on touch and movement information.

Ballesteros and Reales [68] asked whether moving from 2-D shapes to 3-D objects with the same contour, shape, size, and complexity but extending the stimuli in the z-axis would influence the haptic perception of symmetry. For this purpose, they designed a new set of materials consisting of 2-D raised-line shapes and surfaces and 3-D short and tall objects. The hypothesis under test was that the elongation of the displays should permit a more informative exploration of the haptic displays and should facilitate the detection of symmetry. To further explore the reference-frame hypothesis, different groups of haptic explorers performed the symmetry detection task in unimanual or bimanual conditions. The reference hypothesis predicts that exploration with two hands aligned to the body axis will increase accuracy in detecting symmetry in 2-D shapes. In short, the study investigated symmetry–asymmetry discrimination performance by touch using four sets of materials specially designed to study how the third dimension influences discrimination of symmetry, a very outstanding spatial property of shapes and objects.

Methods

Participants

Thirty-two participants took part in an experimental session that lasted approximately 45 min. All have normal or corrected to normal vision and normal touch.

Materials and Equipment

The stimuli were four sets of materials. Each stimulus set was formed by nine symmetric and nine asymmetric displays matched in shape, size and complexity (number of sides in the perimeter of the shape: 3, 4, 6, 8, 12, 14, 24 and 30 sides each). Figure 7 displays examples of the four sets of materials used in the investigation. Each stimulus was glued at the centre of a card to fix the object and to prevent haptic explorers from changing its orientation while exploring the stimulus.

Figure 7.
Examples of the four sets of symmetric and asymmetric unfamiliar stimuli varying in complexity used by Ballesteros and Reales (2004). The stimuli ranged from 2-D raised polygons and raised surfaces to 3-D short and tall objects. The four sets were obtained by elongating their z-axis. From Ballesteros & Reales (2004). Visual and haptic discrimination of symmetry in unfamiliar displays extended in the z-axis. Perception *33: 315–327. Reprinted with permission from Pion Ltd.*

The stimuli were presented in a visual-haptic object tachistoscope provided with a piezoelectric board. The board used as a stimulus presentation platform was equipped with a piezosensor located at the centre of the platform used for stimulus presentation. An IBM computer interfaced with the apparatus and special software controlled the stimulus random presentation order. A voice key stopped the computer's internal clock to record latency.

Experimental design

The design was mixed-factorial design with four stimulus sets (raised-line shapes, raised surfaces, short objects, and tall objects) x two symmetry conditions (symmetric, asymmetric stimulus) x two modes of explorations (unimanual versus bimanual). The last variable was a between-subjects variable, whereas stimulus set and symmetry condition were within-subjects.

Procedure

Participants were tested in the laboratory individually and never saw the stimuli. They were to say whether the stimulus was symmetric or asymmetric. The stimulus set was counterbalanced across participants. The computer program indicated the stimulus presentation order to the experimenter. A tone alerted the participant that a stimulus was presented at the platform for exploration. Latencies were recorded automatically from the hand's first contact with the display to the vocal response. The experimenter recorded each response by pressing the keys on the computer keyboard.

Results and discussion

Figure 8 displays the accuracy results. The three-factor mixed ANOVA with mode of exploration as a between-subject factor, and symmetry and stimulus set as within-subjects factors showed that touch was more accurate under bimanual exploration. Asymmetric stimuli were judged more accurately than were symmetric stimuli ($p < 0.01$). Stimulus set was also significant ($p < 0.03$). Accuracy was higher for tall 3-D objects (91% correct) than for raised-line shapes (83% correct). The interaction between symmetry and stimulus set was also significant ($p < 0.001$). The elongation of the stimuli along the *z*-axis improved performance with symmetric stimuli only (all $ps < 0.01$).

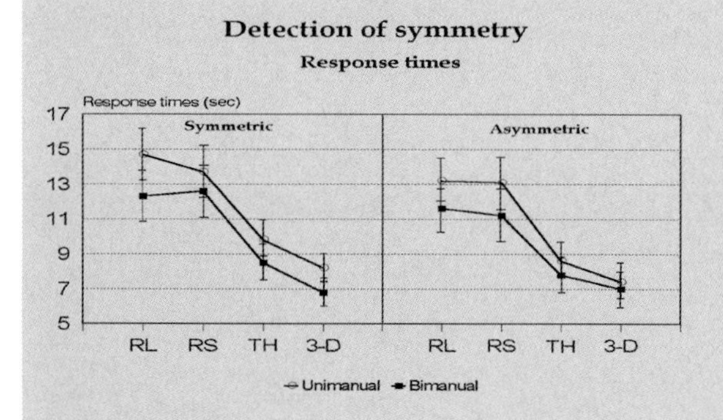

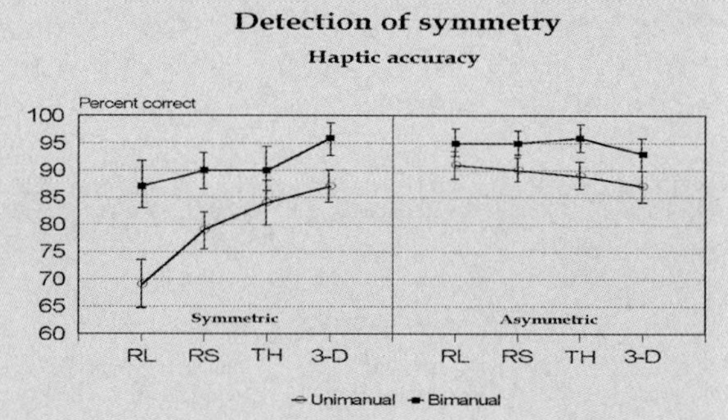

FIGURE 8.

Left: Mean accuracy for symmetric and asymmetric displays as a function of stimulus set under unimanual and bimanual exploration. Right: Mean response times for correct judgments (and standard errors) corresponding to symmetric and asymmetric stimuli and stimulus set and exploration condition. Error bars show an estimate of the standard error of the mean. From Ballesteros and Reales (2004). Visual and haptic discrimination of symmetry in unfamiliar displays extended in the z-axis. Perception *33: 315–327. Reprinted with permission from Pion Ltd*

The response-time analysis showed that the main effect of mode of exploration was significant. It was faster to discriminate symmetry/asymmetry under bimanual than unimanual exploration ($p<0.0001$). Asymmetric stimuli were detected faster than symmetric ones ($p<0.02$). Finally, stimulus set was significant ($p<0.001$). Haptic explorers were faster as the height of the stimuli increased (13 s, 12.7 s, 8.7 s and 7.3 s for raised-line shapes, raised surfaces, short objects, and tall objects, respectively). Explorers were slower for raised-line shapes, raised-surfaces and short objects than for tall objects (all *ps* < 0.001).

Manipulating stimulus height and keeping other stimulus dimensions constant, such as shape, size, and complexity, the performance with symmetric stimuli improved significantly. Moreover, active bimanual exploration produced an advantage compared to unimanual exploration for both types of stimuli, symmetric and asymmetric. This finding supports the reference-frame hypothesis. Based on the results it can be concluded that touch is a fast and accurate perceptual system in detecting bilateral symmetry, an important spatial property of flat shapes and objects. The study showed for the first time that the extension along the third dimension improves performance (accuracy and response time). Results were in agreement with findings from Ballesteros et al. [67]. Exploration with two hands aligned to the body axis facilitated symmetry discrimination.

on touch. Of course, there are cases where this sensory restriction can aid touch [69].

Vision of the hand as it touches a surface can aid in the interpretation of tangible pictures [70]. This assistance could occur for a variety of reasons. Perhaps 'non-informative' vision serves to increase the acuity of the sense of touch [71]. Merely looking at the location of one's hand in space may help improve tactile sensitivity. It is also possible that in many instances, vision provides a frame of reference that helps us judge patterns that we have examined *via* the sense of touch (also see [72]). According to Millar, we may code egocentrically, with reference to our bodies, or in an exocentric fashion, with reference to external space. Egocentric coding may sometimes assist subjects who are denied vision, as in judgments of verticality. Sight of the hand may aid recognition of Braille and improve judgments of the relative smoothness of textured surfaces [18].

Vision and touch may often work to assist each other. The visual guidance function of the sense of sight is certainly an aid to touch. Peripheral vision serves to help guide haptic exploration of objects and forms [39]. However, clear foveal vision may distract us as we feel stimuli.

Just as vision can aid touch, the two senses may not always cooperate. They can exist in conflict, and provide discrepant information to the perceiver. Thus, when confronted by a conflict between the senses of vision and touch, the outcome may depend upon the speed of processing, or the confidence we have in a sense, or in the efficiency of a sense. For judgments of the size of a stimulus, we may see reliance on either vision or touch, but this depends upon the nature and speed of the response. For a rapid, grasping pincers posture, we may get haptic dominance [73]. Texture may yield reliance on touch, since people are better at making judgments of the relative smoothness of finer surfaces with touch. Thus, we do not always find visual dominance.

Summary

Touch is a remarkable sense, and we are just beginning to discover its rather amazing abilities. Increased research interest has been directed toward the study of haptics, and the most recent trends involve a renewed interest in the relationship between touch and vision. This new study area has yielded an unravelling of prior misconceptions about haptics and neural organisation. Thus, touch is not a 'minor' sense, with serious deficiencies. Rather, it is most useful for augmenting visual information during teleoperation, and visual information can be most helpful for augmenting haptic perception. We have also witnessed recent studies that debunk a number of myths about the sense of touch. It is not invariably subservient to vision. Indeed, it is very difficult to ignore haptic input, for example, when faced with the impact of a large and heavy object upon the body surface. Older ideas in the area have tended to make overgeneralisations about visual dominance, and these sweeping generalisations should be suspect.

Note that there has been increasing interest in haptic object recognition from a neuroscience perspective. This research has tended to emphasise the notion that there are multimodal representations of space [74], and that touch is judged within a frame of reference. The relevant frame of reference may be face centred, or body centred.

Touch can operate very effectively when needed to grasp objects and discover their important features. The sense of touch often appears to operate more slowly than vision, but this is very dependent upon the skill of the perceiver and whether or not the individual can use two hands to feel objects. Scale is also important, since delay and other temporal variables may be less important for smaller objects. Blindfolded sighted individuals may not be able to identify pictures as rapidly as blind persons can, for example, and familiarity and practice clearly play a role in object identification [75]. If objects are familiar, one would expect very high accuracy and rapid object recognition using touch.

Acknowledgements

This chapter was written while Soledad Ballesteros was supported by the DGICYT grant SEJ2004-00752/ PSIC and by the European Commission, FP6-2005-NEST-Path, grant 043432. Morton A. Heller was supported by NSF grant BCS-0317293 from the program in Perception, Action and Cognition.

Further readings

Jones LA, Lederman SJ (2006) *Human hand function*. Oxford University Press

This comprehensive and well-written book is devoted to the functions of the hand as a perceptual system. All the chapters are very interesting, especially Chapter 5 on active haptic sensing.

Heller MA, Ballesteros S (2006) *Touch and Blindness: Psychology and Neuroscience*. Lawrence Erlbaum Associates

Haptic perceptual illusions

Edouard Gentaz and Yvette Hatwell

Perceptual illusions refer to systematically oriented errors in the perception of figures or scenes, and these errors are observed in almost all people. For centuries, they have been called 'opto-geometric illusions' because it was thought that they concerned only visual perception. Although visual errors are relatively rare, comments and questions about visual illusions are found as early as in the Greek and Roman literature. In scientific psychology, the theoretical and practical problems raised by these deformations have been intensively studied since the end of the 19th Century, and the elements of the figure inducing each error are now identified. However, there is no general theory explaining all the visual illusions. Instead, each figure must be analysed in order to determine the specific processes leading to the error.

The question as to whether analogous illusions occur in the haptic modality was asked only since the 1930s by the gestalt psychologists. According to these researchers, systematic errors result from the general functioning of the nervous system and particularly from the interactions (called 'field effects') between different parts of the figure. Therefore, because the same processing rules are at work in all perceptual modalities, illusions analogous to visual ones must be present in touch. Indeed, this was observed by Revesz [1] and Bean [2] who found haptically almost all visual illusions. However, further studies with better methodological controls produced some contradictory results, which will be discussed here.

The theoretical interest of the studies of haptic illusions is two-fold [3]. First, they confront purely visual explanations of visual illusions with non modality-specific theories. They can also answer the question as to whether the perceptual processes implemented in tactual perception are similar or not to those implemented in visual perception (for reviews, see [4–7]). Thus, not finding a visual illusion in touch is an argument in favour of specific haptic perceptual processes. However, observing the same illusion in vision and touch does not indicate whether the error is a result of similar and/or specific perceptual processes. To do that, it is necessary to know if the factors responsible for the presence of the tactual illusion are identical to those affecting the same phenomenon in vision. An affirmative answer brings arguments in favour of similar visual and haptic processes, whereas a negative answer favours specific visual and haptic functioning. In this latter case, the problem would be to identify these modality-specific processes.

To answer these questions, blindfolded sighted, late blind and early blind people are compared. Blindfolded sighted persons may use spatial visual representations [8–11]. As visual perception and visual mental images are generally more efficient than haptic perception and nonvisual images in representing space, blindfolded sighted subjects have an advantage over early blind subjects (for reviews, see [12–15]). But congenitally blind people benefit from greater tactile practice, and this gives them an advantage over the blindfolded sighted. Finally, late blind subjects benefit from both spatial visual representations and tactual practice.

If an illusion is present in the late blind but is absent in the early blind, it is assumed that visual experience is responsible for it. If the illusion is also present in the early blind, explanations based only on visual experience are invalidated and non-visual explanations must be found to account for the existence of the same illusion in vision and touch. These explanations may be common to both modalities, but they may also be specific to each of them, as different causes can have the same effects. Finally, one should determine whether the tactual processes implemented by blindfolded sighted subjects are different from those implemented by the early and late blind. If the same results are observed in the three groups, the existence of general haptic processes independent of the visual status will be favoured. If the early blind have consistently different results as compared with the two other populations, it will suggest that they implement original haptic processes.

The analysis of literature about haptic illusions reveals that most of the above scenarios have been observed according to the type of illusion studied. The present study will show that certain special features of touch, mainly linked to manual exploratory procedures, allow at least a partial understanding of why such contradictory results have been observed. Different features of the way an object is explored and consequently the way haptic input is coded have been invoked to account for the presence or absence of geometrical haptic illusions: a) During exploration, the size of the tactual perceptual field varies depending on whether the subject uses the inside face of an index finger only, or the whole hand, or both hands. In the first case, the lines which induce the visual error might not be perceived and the haptic illusion may not appear [3]; b) The intervention of the kinaesthetic information resulting from large exploratory movements (when the arm-hand system is involved in the perception of large stimulus) may produce spatial distortions [16–19]; c) The gravitational cues generated by the anti-gravity forces allowing the arm to be kept in the air may provide reference cues not present in vision [20–23]; d) the spatial reference frame in which the figure is coded may modify the relations between the elements of the figure and the subject [24].

This chapter will be focused on intramodal haptic illusions, i.e., on systematic perceptual errors occurring when the haptic modality is activated while the subject is deprived of vision temporarily (blindfolded sighted) or permanently (early and late totally blind). However, because the visual-haptic coordination generates too systematic errors, the second part of the study will briefly evoke some of the perceptual errors due to cross-modal interactions.

Intramodal haptic illusions

Classical geometrical illusions

Most of the literature on intramodal haptic illusions has been concentrated on three well-known visual illusions: Müller-Lyer, Vertical-Horizontal and Delboeuf. We will focus on them because of the convergent set of studies available and, as we will see further, each illusion illustrates some aspects of the theories outlined above. Other haptic illusions (Oppel-Kundt, Ponzo, Poggendorf,) will be discussed more briefly because the few studies available are often contradictory and they concern only blindfolded sighted people.

Müller-Lyer illusion

In the Müller-Lyer visual illusion (Fig. 1A and B), the perceived length of the horizontal segment is modified according to the orientation of the arrowheads situated at both ends. The segment with outward pointing wings is systematically overestimated as compared to the one (identical) with inward pointing wings, or as compared to an identical segment with no wings.

A number of studies showed that the illusion is present haptically in blindfolded sighted subjects as well as in early and late blind subjects [24–35]. In several of these studies, the subject was told to explore the segment and the arrowheads with the index finger of the dominant hand only. In this case, the intensity of the illusion is the same in the two modalities, the segment with the outward pointing arrowheads being judged 1.2 to 1.3 times longer than the one with the inward pointing arrowheads [33].

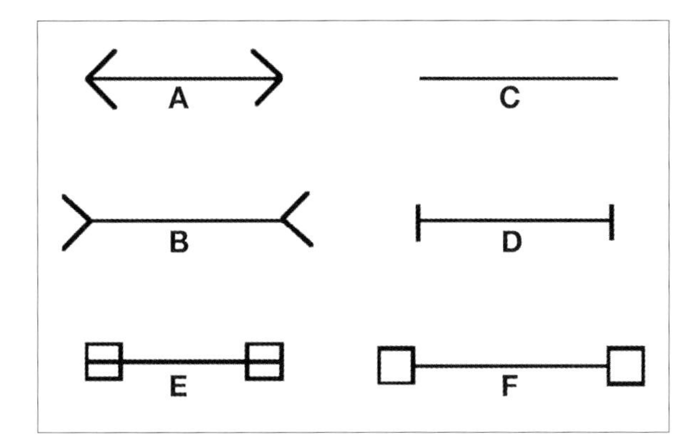

FIGURE 1. MÜLLER-LYER ILLUSION

A and B: Classical stimuli; C and D: Control stimuli proposed by Heller et al. [28]; E and F: stimuli proposed by Hatwell [26].

Heller et al. [27] compared the intensity of this illusion in blindfolded sighted, late-blind, congenitally blind and low-vision adults. Subjects were asked to use their right index fingers to feel the raised-line figure and to use their left hand to make size estimates with a ruler. The same robust haptic Müller-Lyer illusion appeared in all groups. Therefore, this haptic illusion is not dependent upon visual experience. No difference according to the visual status was also found by Calsa et al. [25] in blind adults and by Hatwell in blind children. However, when the child was asked to freely explore bimanually, Hatwell [26] obtained an illusion weaker in the haptic modality than in the visual modality in a mixed group of early and late totally blind.

In another study on blindfolded adults, Heller et al. [28] varied the mode of exploration. The authors compared the performances when tracing the figures with the right index finger, or when exploration was free but limited to the right hand, or when the subject used any two or more fingers of the right hand, or when a grasping procedure was imposed by the use of the thumb and the index fingers of the right hand. The illusion was diminished but not eliminated by tracing with the fingertip whereas the other exploratory conditions had no effect. In Experiment 2, the subjects explored with their two index fingers (avoiding tracing), and this method led to a suppression of the illusion for the wings-in figure while for the wings-out one, the error was reduced. Other experimental variations (changing the orientation of the stimuli, changing the thickness of the lines, bimodal visual-haptic presentation) had no effects, although larger size of the wings improved errors for the smallest stimulus.

On the other hand, the three factors acting on the intensity of the visual illusion have the same effect on the haptic illusion. First, without feedback, the errors in vision and haptics [36] and in haptics [37] decreased as the number of trials increased. Therefore, perceptual learning is at work both in the two modalities. Second, errors get stronger as the acute angles formed by the arrowheads and the segment to be evaluated gets smaller in vision [38] and haptics [27, 30]. Hatwell [26] reported an analogous effect by showing that if, instead of arrowheads, the segments to be evaluated were ended by small squares (in one case, these squares lengthened the segment by forming a 90° angle with it, in the other they intersected the segment end part (Fig. 1E and F), the visual error was maintained whereas the haptic error disappeared. She explained this result by the fact that the 90° angle formed by the segment and the vertical side of the end square is big enough for the finger to manage to partially exclude the square from the tactile perceptual field. As a result, the effect of the error-inducing lines is decreased or completely cancelled. This interpretation is supported by an observation of Heller et al. [28] who found no haptic errors both on the control segment with no endings (Fig. 1C) and on another control segment ended by small vertical bars (Fig. 1D).

Third, Millar and Al-Attar [24] showed that the same experimental manipulations reduce the Müller-Lyer illusion in vision and haptics to the same error level. More precisely, explicit instructions to ignore the confusing arrowheads and to use body-centred cues for spatial reference reduce the Müller-Lyer illusion in both modalities to near zero. This illusion is not reduced in the absence of instructions to use body-centred cues, even when external reference cues are present.

Finally, Gentaz, Camos, Hatwell and Jacquet [39] obtained a positive correlation between the magnitude of illusion in the visual and the haptic modalities when the same blindfolded sighted subjects performed exactly the same task in both modalities.

These findings strongly suggest that the Müller-Lyer illusion involves similar factors in both modalities. The fact that this illusion is present in the haptic modality questions purely visual explanations such as Gregory "*inappropriate adjustment of constancy*" [40]. For this author, the visual error is due to misleading depth cues such as perspective or textured background. The presence of the illusion in the early blind invalidates all explanations founded solely on visual experience. It rather suggests the presence of analogous haptic processes, independent of the visual status of the subjects. Furthermore, as different factors (learning, angle, reference instructions, correlation between visual and haptic errors) affect similarly the intensity of the visual and haptic illusions, the presence of the Müller-Lyer illusion in vision and haptics seems to result from similar visual and haptic processes.

Vertical-Horizontal illusion

In the visual Vertical-Horizontal illusion (V-H), the length of the vertical segment is overestimated as compared to the same segment in a horizontal orientation. Actually, this illusion must be decomposed in two illusions: the bisection illusion observed when one segment is bisected whereas the other is not, as in the inverted T figure (Fig. 2A) and the pure vertical illusion observed in L figure (Fig. 2B).

The V-H illusion has been observed in the haptic modality in blindfolded sighted subjects and in the early and late blind [23, 25–26, 33, 41–48]. Suzuki and Arashida [33] found that in the visual and haptic modalities, the vertical segment of an inverted T figure is perceived 1.2 times longer than the horizontal segment.

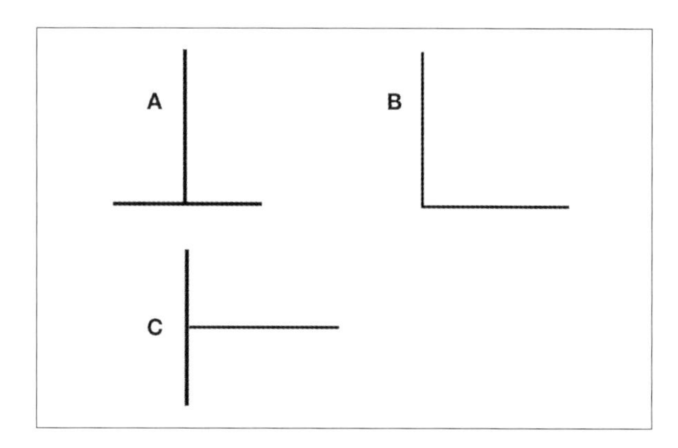

Figure 2. Vertical-horizontal illusion
A: Inverted T stimulus; B: L stimulus; C: 90°-rotated inverted T stimulus.

In the bisection illusion, research showed that the factors responsible for its presence are similar in the visual and haptic modalities. In both, the overestimation of the vertical segment is higher in the inverted T figure than in the L figure where the two segments are undivided (Fig. 2B). In this latter case, the illusion is not present either in young blind subjects freely exploring the figure [26], or in blindfolded sighted adults ([46], Exp. 2) exploring with the right index finger of one hand a L figure in which the vertical line is aligned with the body midline. In addition, it is the horizontal line that is overestimated when the inverted T figure is rotated by 90° (Fig. 2C), which divides the vertical segment in the middle whereas the horizontal segment is undivided ([26, 46], Exp. 3). According to Millar and Al-Attar ([46], Exp. 5), the reduction of illusion in the L figure and its suppression when there is no junction point can be explained by the fact that this junction point acts as an anchor added to the anchors constituted by the ends of the segments.

Millar and Al-Attar ([46], Exp. 2 and 6) compared the magnitude of the haptic inverted T illusion in blindfolded sighted adults according to whether exploration was restricted to the index finger of the right hand or was two-handed with reference cues available. In the two-handed condition, subjects used both hands to feel a rigid frame surrounding the stimulus figure in relation to the feeling of the vertical line of the figure, using similar downward movements. In addition, subjects were instructed to relate the downward scanning movements to their body midline and to the vertical sides of the frame. In the one-index finger condition, there was no external frame of reference during scanning and the subjects explored the whole figure with their index. A clear reduction in error in two-handed exploration was observed. The authors attributed it to the instruction to refer to the reference frame available in the two-handed exploration, whereas this frame was absent when only the index finger was used. Taken together, these results highlight both the role of bisection, which acts in the same way in vision as in haptics, and the role of reference frames.

By contrast, the pure vertical illusion depends on factors specific to each modality. In vision, several studies suggest that the error is due to the anisotropy of the visual field. In this theory first proposed by Künnapas [49–52], the illusion derives from a kind of framing effect. The closer a line extends toward a surrounding frame, the longer it appears. Because the retina is a horizontally oriented ellipse, vertical lines will generally be closer to the boundary of the visual field than will be the horizontal lines, and hence vertical lines will appear longer. Different studies support this purely visual explanation of the visual illusion (for a review, see [53]). On the other hand, in the haptic modality, studies have shown the role of the direction of exploratory movements [42, 48, 54]. In the early studies revealing the presence of this illusion in haptics, the figure was placed flat on a table. In this position the exploratory movement is radial for the 'vertical' segment (which is in fact in the sagittal plane) and tangential for the horizontal segment. A movement is radial when it develops in a radius having the subject at its centre and tangential when it adopts one of the tangents to these radii. Yet Cheng [55] and Wong [56] showed a tendency to overestimate the length of a radial movement compared with a tangential one, because radial movements are executed more slowly than tangential ones. Estimates concerning the length of movements are therefore affected by time cues. However, McFarland and Soechting [45] showed that the magnitude of the radial-tangential illusion in the horizontal plane was not altered when the ratio of movement times in the tangential and radial directions was changed. In order to evaluate the role played by the direction of exploratory movements in the vertical-horizontal haptic illusion, Day and Avery [54] and Deregowski and Ellis [42] presented inverted T and L figures in the fronto-parallel plane. In this case, all exploratory movements are tangential. The vertical segments were no longer overestimated in the L figure, as the two factors ('radial/tangential movement' and 'bisection') could not act. However, the overestimation of the vertical segment (although attenuated) remained present in the T figure, as it is due to the sole factor of 'bisection'.

Regarding the comparison between the magnitude of the haptic illusion observed in blindfolded sighted, early and late blind subjects, the only reported difference between these groups was in Heller and Joyner's study [43]. The size of the stimuli was estimated by the pincer posture of the index and the thumb fingers of the participant. Early blind subjects showed similar errors in the L and inverted T figures, whereas the late blind manifested a much greater overestimation of the inverted T figure. However, a further study by Heller, Brackett, Wilson, Yoneyama and Boyer ([44], Exp. 1) revealed that the above-mentioned results were due to the mode of response used. In the latter research, the participants explored with the index finger of one hand and used a sliding ruler with the other hand for size estimates. In this condition, no effect of visual status was observed.

Finally, Heller, Calcaterra, Burson and Green [23] suggested that the V-H haptic illusion is the result of a non-optimum exploration of the stimulus. They studied the effects of the implication of whole arm movements on the illusion in the inverted T figure. For example, small shapes can be explored with index finger 'tracing' movements, whereas large stimuli necessitate more ample movements including whole arm radial movements, which can lead to perceptual errors. Much of the early research in this field did not take into account the implication of the whole arm. Furthermore, the stimulus is often presented in such a way that the subjects must keep their elbow on the table during exploration or, in other studies, they must keep the elbow suspended in the air. Heller et al. [23] examined the effect of maintaining the elbow above the surface of the table. In this case, the illusion was reduced or eliminated, whereas it was present and of great amplitude when the subjects kept their elbows in the air. In a second experiment, arm movements were controlled. One group proceeded uniquely by fingertip tracing, as the arm and elbow were immobilised in a drainpipe. Another group used whole arm movements, as the arm and hand were placed in a system isolating the index finger and the arm and prevented fingertip tracing of the stimulus. Results supported the idea that the presence of the illusion is more likely when whole arm movements are possible. This shows the importance of scale in haptic perception. When the stimuli are quite small, they can be explored with the fingers. When they are bigger, they necessitate whole arm movements in order to be explored. Haptic illusions are more likely to occur in this case. Whole arm movements probably increase the impact of the gravitational cues present during exploration, and Gentaz and Hatwell [20, 21] showed the importance of these cues in haptic perception of spatial orientation in blindfolded sighted and early and late blind people.

What is the theoretical implication of these results? A common factor, bisection, affects the visual and the haptic modalities in the same way. But, contrary to the Müller-Lyer illusion, there are also factors specific to each system: the exploratory movements affect only the haptic illusion, and the anisotropy of the visual field affects only vision. The presence of the vertical illusion in the haptic modality, including in the early blind, invalidates the exclusively visual explanations of this error. Actually, this illusion is due to one perceptive factor common to the two modalities (bisection) whereas modality-specific factors, resulting in analogous errors, are also responsible for the illusion in each system.

Delboeuf illusion

In the Delboeuf visual illusion (Fig. 3), the evaluation of the size of a circle is modified if it is inserted into an exterior concentric circle. In vision [57] showed that the inner circle B is overestimated as compared with a reference isolated circle C when the B:A ratio is close to _ (A is the exterior concentric circle). This error is generally attributed to an assimilation process in which the perceived size of the small inner circle is improved by its proximity to the large external circle. Hatwell [26] showed that haptically and in free two-handed exploration, this illusion is not observed in blind children (mixed early and late blind) even in the conditions under which it is maximum in vision. Suzuki and Arashida [33] confirmed that in the same blindfolded sighted adults, the illusion concerning the inner B circle is present in vision but absent in the haptic modality. However, curiously, an illusion concerning the exterior circle A (underestimation of the exterior circle) appears in both vision and touch.

The absence of the classical Delboeuf illusion in the haptic modality may be explained by the exploratory hand movements which permit context effects to be reduced. Thus, in a free exploration situation [26] or when only exploration with the index finger of one hand is allowed [33], the subjects can use only the internal face of the index. Circle B is perceptively isolated when exploration starts at the centre of the concentric circles, and circle A disappears from the haptic perceptive field. The task is thus reduced to a comparison of B and C in which A plays no role. This interpretation is plausible but it does not explain why the illusion occurs haptically for the exterior circle. To test the role played by the exploratory procedure, exploratory movements should be recorded and the distance between A and B varied in order to modify the role of context during exploration. Whatever it may

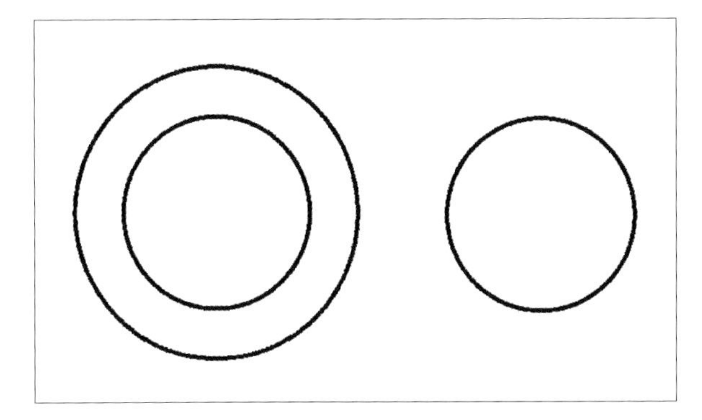

FIGURE 3. DELBOEUF ILLUSION

be, this (partial) absence of the illusion in the haptic modality confirms the analytical nature of haptic apprehension which is less susceptible to deformations because it can totally isolate one element to compare it to another.

Studies of other illusions

Studies of other illusions concern only blindfolded adults and their results are sometimes inconsistent. In the visual Oppel-Kundt illusion (Fig. 4A), the filled space is overestimated as compared to the unfilled one, although the two spaces are identical. This error is observed by Suzuki and Arashida [33] in a haptic condition in which the blindfolded subjects were not allowed to use thumb-index evaluation of the lengths. Since the lines inducing the visual error were necessarily perceived tactually, the haptic illusion was observed as in vision. In the visual Ponzo illusion (Fig. 4B), the upper horizontal bar is perceived longer than the identical lower bar. This error is due to misleading perspective and depth cues. In haptics, the results are contradictory: Whereas Suzuki and Arashida [33] observed the same illusion, this illusion was absent in Casla et al.'s [25] study. Further research is therefore needed, especially with congenitally blind people. However, following the interpretation proposed for the Delboeuf and Müller-Lyer haptic illusions, this length distortion may not be present, or weakly present, in haptics because the finger can isolate the segment to be evaluated without perceiving its troublemaking context.

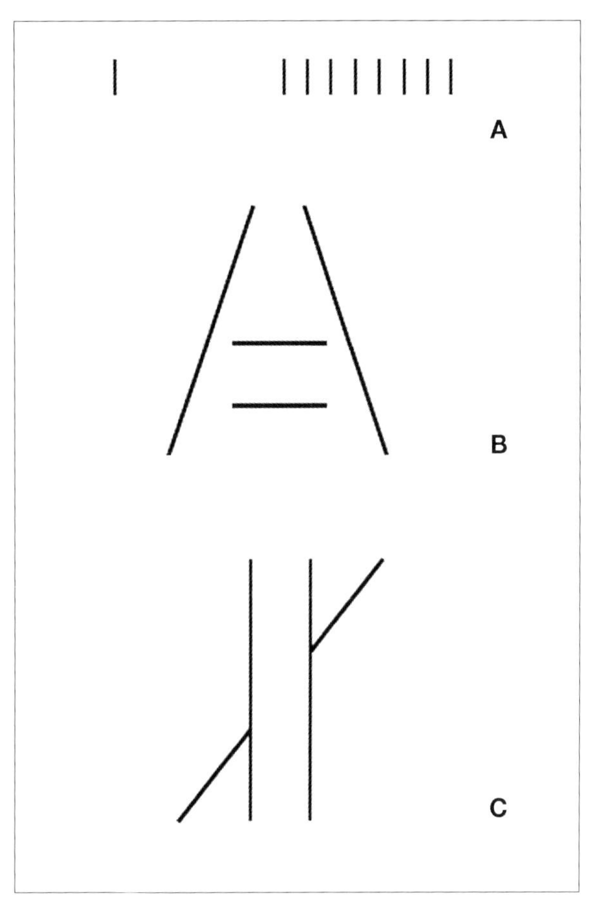

FIGURE 4.
A: Oppel-Kundt illusion; B: Ponzo illusion; C: Poggendorf illusion

In the visual Poggendorf illusion (Fig. 4C), the oblique line seems to be masked by the two vertical lines and it appears as broken. This illusion was not found in haptics by Suzuki and Arashida [33] and Wenderoth and Alais [58] despite variations in exploratory movements. By contrast, a reversed illusion emerged in Lucca et al.'s [59] study (1996). Actually, the error concerns here the perception of the continuity of an oblique intersected line. As the Gestalt laws of organisation are not always found with the same intensity in haptics as in vision [60], the interaction between the parallel lines and the oblique one may not be present in touch.

Non Euclidian spatial illusions

Most people believe that the haptically (or visually) perceived space is Euclidean and shares the same properties as physical space. This belief is illusory because several studies showed that the haptic perception of different geometrical properties is systematically distorted with respect to the physical reality. For example, Kappers [61, 62] and Kappers and Koenderink [63] investigated the haptic perception of line parallelism in both horizontal and mid-sagittal planes. In a simple task where blindfolded participants were asked to rotate a test bar in such a way that it felt as being parallel (in physical space) to a reference bar, huge systematic subject-dependent deviations were found. Observers systematically produced deviations of up to 90°. This effect was very robust and remained even after visual inspection and various ways of feedback. Unimanual and bimanual and also pointing and collinearity measurements yielded similar results. These authors suggested that the deviations reflect the use of a reference frame that is a weighted average of egocentric and allocentric reference frames. Cuijpers, Kappers and Koenderink [64, 65] performed similar parallelity and collinearity experiments in vision. That visual space is not Euclidean was already established early in the last century [66, 67], but these new studies provided some extra evidence. For example, bars at eye height separated by a visual angle of 60° have (depending on the participant) to differ by 20° in orientation in order to be perceived as parallel.

In the same vein, Sanders and Kappers [68] examined what is haptically perceived as a straight line. Blindfolded subjects were asked to construct straight lines in arranging five small magnets in a straight line between two reference magnets markers at three different distances in the horizontal plane which were approximately parallel to the front part of the body. Subjects used both hands and manipulation was unrestricted. The results were analysed in terms of a curvature parameter k, which indicated the amount of deviation from veridical. They showed that parabolic arcs are indeed good approximations to the shapes of the haptically straight lines. Overall, the authors found that the shape of the haptically straight line deviated from what would be physically straight and were generally curved away from the observer.

Perceptual errors generated by cross-modal interactions

Cross-modal interactions are intensively studied in another chapter of this book and here we will examine briefly only whether a bimodal presentation (generally vision and touch) of one or more stimuli generates perceptual errors. Two types of situations are tested. In the first, the same stimulus is presented bimodally and the performances are compared with those in which the stimulus is presented unimodally. The second situation describes more elementary cross-modal interferences. In it, the task concerns only one modality although the other modality is stimulated by a non-pertinent stimulus which should be filtered.

The Müller-Lyer and Ponzo illusions have been studied by Walkers [69] in a bimodal presentation. The lines inducing the error were drawn in black ink and could not be perceived tactually whereas the lines to be estimated were in relief. The test was either visual or haptic and in both conditions, the illusions were observed but their magnitude was slightly lower than in the pure visual illusions. This is an example of the 'visual capture' which will be described in the next paragraph. More recently, Gallace and Spence [70] studied the cross-modal consequences of viewing bimodally the Müller-Lyer figure. In their task, only the wings of the figure appeared visually on a screen. The shafts were wooden sticks presented behind the screen exactly behind the visual wings-in or wings-out drawings. The subject was asked to evaluate haptically the length of the sticks he/she could not see. A systematic error appeared but, paradoxically, its direction was reversed as compared to the classic illusion: The stick behind the wings-in drawing was judged longer and the stick behind the wings-out drawing was judged shorter than really. Whatever it may be, these bimodal studies show that visual illusions generate haptic errors when no inducing haptic lines are present.

Bimodally presented figures have often been studied in perceptual conflict situations. In them, visual and haptic data are decorrelated, generally by the use of lens. In the great number of research which followed the pilot study of Rock and Victor [71], results are consistent. In the spatial tasks implying size judgments of squares, bars, etc., a 'visual capture' or a tendency to visual capture are observed when the test is visual (e.g., [72, 73]). This means that the discordant haptic data have not been taken into account or have been taken into account very moderately (see Chapter 18 of this book). When the haptic and visual information of rotation angle of a hand-operated crank are delocalised, the weight of the dominant sense at the expenses of the other appears to increase considerably [74]. As in the case of bimodally presented illusions, the dominant visual modality generates haptic errors. When the test is haptic, a 'compromise' tends to be observed. In it, both data are processed although those provided by one modality may still be dominant.

Finally, in non-spatial tasks such as the estimation of roughness, the haptic modality dominates and a tendency toward haptic capture is sometimes observed [75, 76]. Ernst and Banks [77] proposed a statistical model to account for these results. In it, it is assumed that the organism tends to minimise the variance of the final estimate. Therefore, the more precise modality will receive a higher weighting and the less precise one will receive a lower weighting. This assumption is consistent with the results obtained, since vision is more efficient than haptics in the spatial domain and haptics is more efficient than vision in texture perception [76, 78].

In the second type of bimodal situations, the task concerns only one modality but another non-pertinent stimulus is presented simultaneously to the other modality. The question is whether cross-modal interferences will be observed. Most of these studies concern cutaneous stimulation and not haptic active touch. However, it is noteworthy that they reveal strong cross-modal interactions. For example, Pavani, Spence, and Driver [79] observed a visual capture of touch when participants had to discriminate the location of vibrotactile stimuli presented together with distractor lights having conflicting locations. Symmetrically, Spence and Walton [80] found that responses were slower and less precise when participants had to ignore task-irrelevant vibrotactile distractors in their judgment of the location of visual targets. In the same type of tasks, Caclin et al. [81] observed a tactile capture of audition when vibrotactile distractors were presented simultaneously with auditive targets in a localisation task. Finally, another cross-modal effect is observed in the Necker cube illusion. When looking monocularly at a cube, reversals occur and the cube appears sometimes as a truncated pyramid. Bruno, Jacomuzi, Bertamini and Meyer [82] showed that illusory reversals occurred even when the cube was held and grasped by the hand during visual fixation, but the number and the duration of these illusory percepts changed.

Taken together, these studies show that the strong existing cross-modal links between modalities generate perceptual errors when two modalities are stimulated simultaneously by non-congruent stimuli which are difficult to filter.

General conclusions and perspectives

All the studies presented in this chapter confirmed that a direct generalisation of processes from vision to touch should be cautious. Actually, haptic illusions may differ from visual illusions in different ways according to the type of illusion. In brief, the presence of a same Müller-Lyer illusion in haptics and vision seems to result from similar and common (at least partially) visual and haptic processes, independent of the visual status of the subjects. The Vertical-Horizontal illusion is due to one perceptive factor common to the two modalities (bisection factor) whereas modality-specific factors, resulting in analogous errors, are also responsible for the error in each system. The absence of the classical Delboeuf illusion in the haptic modality seems to result from the analytical nature of haptic apprehension. The same approach can be used to understand the non Euclidian spatial illusions: The perception of parallel lines and straight line was systematically distorted with respect to the physical reality in vision and haptics. These distortions seem due to both common and modality-specific factors. Other studies are needed to better understand all these haptic illusions.

On the other hand, the studies reviewed in this chapter showed the interest of comparing the performances of blindfolded sighted subjects with those of early and late blind. This comparison led us to reject some pure visual explanations of the visual errors. Finally, the results obtained in bimodal presentations of the tasks evidenced the strong links associating vision and haptics in sighted persons.

An original way for further studies on haptic illusions would be to examine the potential role of force cues. Indeed, the haptic perception of spatial properties relies on correlated geometrical and force cues [83]. However, most studies are focused only on stimulus geometry [5]. The question of the role of force cues was directly investigated in the haptic perception of curvature, shape and length. First, Drewing and Ernst [84] showed that the introduction of orthogonal forces in a surface creates an illusion of curvature when the force field is divergent (or convergent). Moreover, a convex surface can be perceived as a plane when conflict forces (convergent field) are generated at the surface. Second, Robles-De-La-Torre and Hayward [83] revealed that force cues overcome object geometry in the haptic perception of shape. Regardless of surface geometry, participants identified and located shape features on the basis of force cues or their correlates. Using paradoxical stimuli combining the force cues of a bump with the geometry of a hole, the authors found that participants perceived a bump. Conversely, when combining the force cues of a hole with the geometry of a bump, participant perceived a hole. Third, Wydoodt and Gentaz [85] showed that the haptic perception of a 10 cm virtual length segment was systematically underestimated when an opposition disruption was introduced during haptic exploration. Conversely, this length was overestimated when a traction disruption was introduced during haptic exploration.

Taken together, these results suggest that force cues may also play a role in the haptic illusions. However, McFarland and Soechting [45] showed that the magnitude of the radial-tangential illusion in the horizontal plane was not altered when a resistive force was added in the tangential direction. Further studies are needed to examine the respective contributions of force and geometrical cues in their occurrence.

In conclusion, there are both theoretical and practical reasons to study haptic perceptual illusions [86, 87]. The blind depend largely on touch to access spatial and graphic information. Because haptic perception is sensitive to certain illusions in a way sometimes different from visual perception, relief drawings, raised maps and general graphic illustrations intended for the blind should take this factor into account.

Summary

This chapter focused on the occurrence, in the haptic modality, of three classical geometrical visual illusions, and it discusses the nature of processes underlying these intramodal haptic illusions. The apparently contradictory results found in the literature concerning them may be explained, at least partially, by the characteristics of manual exploratory movements. The Müller-Lyer illusion is present in vision and in haptics and seems to be the result of similar processes in the two modalities. The Vertical-Horizontal illusion also exists in vision and haptics, but is due partly to similar processes (bisection) and partly to processes specific to each modality (anisotropy of the visual field, and overestimation of the radial *versus* tangential manual exploratory movements). The Delboeuf illusion seems to occur only in vision, probably because exploration by the index finger may exclude the misleading context from tactile perception. In the second part, some non Euclidean spatial illusions were presented. Finally, this chapter briefly evoked some of perceptual errors due to cross-modal interactions. Taken together, this chapter supported the crucial role of manual exploratory movements in the haptic illusions.

Selected readings

Gentaz E, Hatwell Y (2004) Geometrical haptic illusion: Role of exploratory movements in the Muller-Lyer, Vertical-Horizontal and Delboeuf illusions. *Psychonomic Bulletin and Review* 11: 31–40

Hayward V (2008) A brief taxonomy of tactile illusions and demonstrations that can be done in a hardware store. *Brain Research Bulletin* 75: 742–752

Heller, MA (2003) Haptic perceptual illusions. In: Hatwell Y, Streri A, Gentaz E (eds.): *Touching for knowing*, 161–171. John Benjamins Publishing Company, Amsterdam

Jones LA, Lederman SJ (2006) *Human hand function*. Oxford University Press, New York

Millar S (2008) *Space and Sense*. Psychology Press, Hove and New York

Haptic perception in interaction with other senses

Hannah B. Helbig and Marc O. Ernst

Introduction

Human perception is inherently multisensory: we perceive the world simultaneously with multiple senses. While strolling the farmers market, for example, we might become aware of the presence of a delicious fruit by its characteristic smell. We might use our senses of vision and touch to identify the fruit by its typical size and shape and touch it to select only that one with the distinctive soft texture that signals 'ripe'. When we take a bite of the fruit, we taste its characteristic flavour and hear a slight smacking sound which confirms that the fruit we perceive with our senses of vision, touch, audition, smell and taste is a ripe, delicious peach. That is, in the natural environment the information delivered by our sense of touch is combined with information gathered by each of the other senses to create a robust percept. Combining information from multiple systems is essential because no information-processing system, neither technical nor biological, is powerful enough to provide a precise and accurate sensory estimate under all conditions. In this chapter, we address the question how the human brain combines sensory information in order to form a robust, reliable and coherent percept of the world around us. Most of the theoretical considerations we discuss here are not confined to interactions of touch with the other senses but generalise to interactions among the other senses (e.g., visual-auditory). Therefore, we keep these considerations general but focus on examples demonstrating how haptics is combined with other sensory modalities.

Information that is perceived through different sensory pathways can be qualitatively different: the senses can provide either complementary or redundant information. Redundant sensory signals provide information about the same sort of object property (e.g., size) and are represented in a common frame of reference, in the same units. To stay with the example of the peach, vision and touch provide redundant information about the peach's size or shape. In contrast, vision and taste provide complementary information about the identity of the object.

In general, we benefit from integrating multiple sources of information. Combining complementary information is advantageous because it extends the range and variety of what can be perceived from any one sense in isolation [1] and can reduce perceptual ambiguity. Furthermore, integrating multiple sensory sources usually leads to improved perceptual performance, more precise judgments and enhances detection of stimuli (e.g., [2–11]).

But what are the mechanisms behind multisensory integration? And what happens if our senses deliver conflicting information? Imagine you grasp a coin underwater. It is a foreign coin that you have never seen before. The visual image of the coin underwater is optically distorted by the refraction of the light rays as they pass from water to air and thus looks larger as it actually is (e.g., [12, 13]) (which you are not aware of). You are asked to judge the size of the coin. Do you rely on your sense of vision, or do you trust more your sense of touch, or do you perceive something in-between the felt and seen size?

Early work suggested that vision is our dominant sense: It was argued that vision 'wins' when visual information conflicts with information from other sensory modalities (e.g., [14–19]).

For example, in a classic experiment by Rock and Victor [18] observers felt a square and looked at it through a lens system that transformed the visual input and made the square look like an oblong rectangle. Observers were asked to select an object from among a set of comparison stimuli that matched the perceived shape. The reported percept was almost completely dominated by the visual input which led to the notion that vision is our dominant sense in conditions where the senses deliver conflicting sensory information ('visual capture').

Later work converged on a more balanced view: According to the modality-appropriateness hypothesis [10, 20], the various sensory modalities are differentially well suited for different kinds of perceptual tasks and discrepancies are resolved in favour of the modality that is more appropriate (e.g., faster, more precise, more accurate) for the task at hand. According to this hypothesis, vision dominates audition for processing spatial information; touch is more appropriate than audition for the perception of texture and audition dominates vision for temporal judgments. For example, a repetitive sound (auditory flutter) presented simultaneously with a flickering light causes the rate of perceived visual flicker to shift towards the auditory flutter rate (e.g., [21, 22]).

More recently, it has been shown that the integrated percept is not simply dominated by either one or the other sensory modality. Rather, when there are several cues available to estimate a certain property – may it be cues within one sensory modality or across modalities – both cues contribute to the combined estimate and the weight of each sensory cue systematically shifts with its relative reliability (e.g., [23–38]).

Building upon this idea, recent research applied a more quantitative approach to study sensory cue integration (e.g., [2, 4, 39–45]). Optimal integration behaviour has been modelled within a Bayesian framework: According to an optimal integrator, the optimal combined estimate (Maximum Likelihood Estimate, MLE) is a linear combination of the individual unimodal estimates that are weighted by their relative reliabilities. More reliable cues are assigned a larger weight. Integrating redundant sensory information in this optimal fashion decreases the variance of the integrated estimate and yields the most reliable unbiased estimate. The Bayesian framework and the implementation of an optimal integrator within this framework are described in more detail in the next section ("The Bayesian framework and optimal integration").

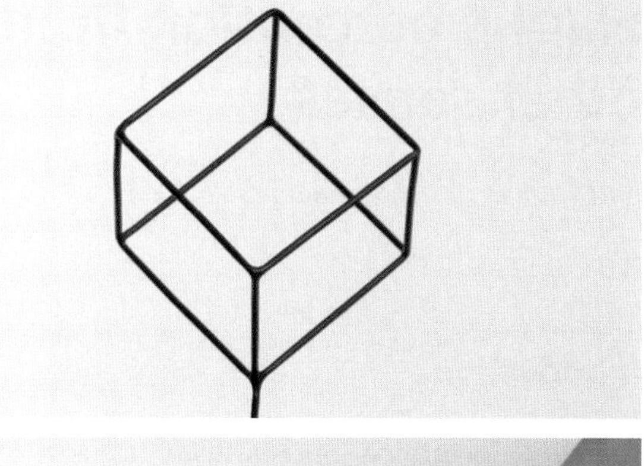

FIGURE 1.
The Necker cube is a wire-frame cube in isometric perspective that can be interpreted in two different, equally probable, ways and thus leads to a bistable percept. When a person stares at the picture, it will often seem to flip back and forth between the two valid interpretations. Haptic information can be used to disambiguate the percept.

Besides weighted averaging of sensory information, there are situations in which sensory cues are combined in a non-linear fashion (e.g., [34, 46]): 'sensory cooperation' (cues interact in a nonlinear fashion), 'disambiguation' (one cue helps resolve an ambiguity in a second cue), 'accumulation' (cues simply summate), and 'cue vetoing' (a stronger cue is selected over a weaker cue) have been observed. Depending on the nature of information, different strategies are used to combine sensory information derived from different sources. While the MLE integration model described in the section "Optimal integration model (Maximum Likelihood Estimator)" requires redundant sensory estimates [34], other forms of interaction can occur among complementary (non-redundant) cues. For example, when viewing a three-dimensional Necker cube the perceived three-dimensional structure flips forth and back between two equally likely interpretations (Fig. 1, upper panel). In this example, the visual scene can be disambiguated by complementary haptic information gathered through exploring the wire cube with the hand (Fig. 1, lower panel).

Or, for example, when a fire engine is driving past we see the typical red colour of a fire engine and hear its characteristic siren sound. Both colour and sound can be used to detect the fire engine and we are faster when we see and hear it as opposed to only seeing it. Here, information from vision and audition is accumulated (probability summation) to improve detectability. Redundant sensory cues, in turn, should only be integrated into a unified percept if they emerge from a common event or object which is generally the case when they are temporally and spatially coincident. Accordingly, it might be that when intersensory discrepancies are large one source is not taken into account for computing the final estimate, i.e., it is vetoed out [34, 46, 47].

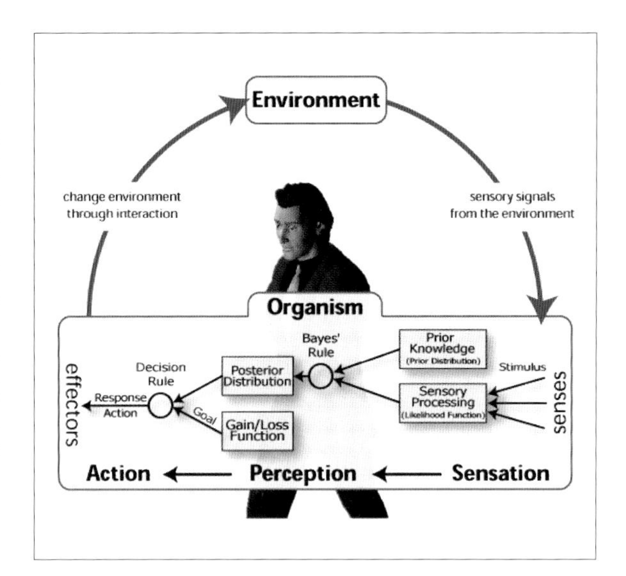

FIGURE 2. THE PERCEPTION-ACTION LOOP, INCORPORATING BAYESIAN DECISION THEORY
Reprinted with permission from [52].

The Bayesian framework and optimal integration

Multisensory integration can be modelled within an integrative theoretical framework: the Bayesian decision theory (BDT). We here introduce the Bayesian framework and show how to construct an optimal integrator (Maximum Likelihood Estimator) within this framework. Empirical data can then be tested against the optimal integration model to assess whether humans adopt optimal integration strategies. For comprehensive tutorials on Bayesian modelling see [48–51].

Bayesian Decision Theory (BDT)

The Bayesian model involves sensory information (likelihood function), prior knowledge concerning the environment (prior distribution) and perceptual decision making (gain/loss function). Sensory information and the prior knowledge are subject to uncertainty and are therefore represented by probability distributions.

To introduce BDT we will pick up the example of inferring the three-dimensional structure, St, of an object from the retinal image, I, (2-D projection onto the retina). That is, the task is to estimate the 3-D structure St given some sensory information I. Inferring a three-dimensional structure from a two-dimensional projection, however, is an ill-posed problem because an infinite num-

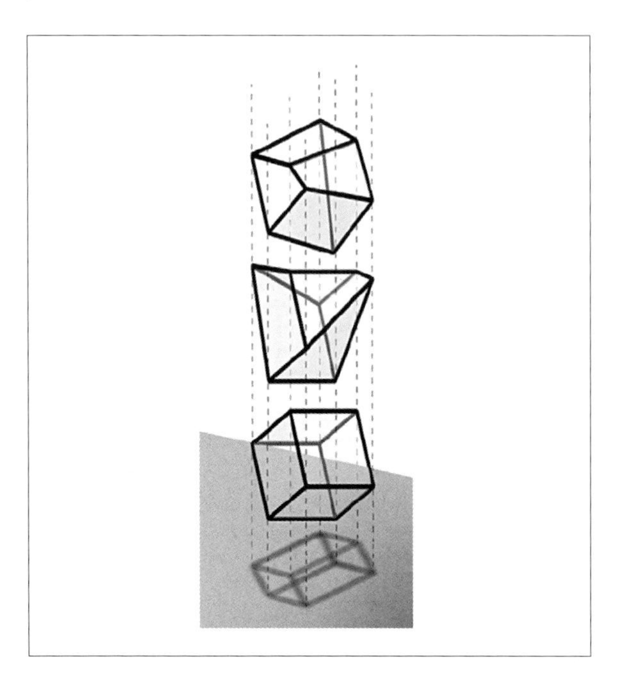

FIGURE 3.
An infinite number of 3-D configurations could produce the same projection image. This is illustrated here by the cast shadow on the tabletop. Reprinted with permission from [52].

ber of possible 3-D structures might have potentially caused the 2-D image (see Fig. 3).

As there is no unique analytical solution of the problem, the task of inferring the 3-D structure can only be solved in a probabilistic manner, that is by finding the structure St that has most likely caused the retinal projection I, which corresponds to maximising the probability distribution

$$p(St|I). \quad (1)$$

Applying Bayes' rule [53] we derive:

$$p(St|I) = \frac{p(I|St) \cdot p(St)}{p(I)} \quad (2)$$

That is, we obtain the posterior probability distribution $p(St|I)$ by probabilistically combining the likelihood function $p(I|St)$ (probability of observing the projected image I given that scene St is underlying the projected image I) and the prior distribution $p(St)$ (probability of the scene St). St does not depend on $p(I)$ and thus, $p(I)$ is just a normalisation factor that we can ignore:

$$p(St|I) \propto p(I|St) \cdot p(St) \quad (3)$$

It might be that some structures St occur more often in the real world than others and the brain makes use of such constraints. The brain has acquired knowledge about objects and shapes in the environment, for example through haptically interacting with the environment, and has built up shape priors like compactness and regularity. Therefore, considering the example illustrated in Figure 3, it is more likely that the 2-D image was caused by a cubic structure than by any of the other possible configurations (see Fig. 3).

There are many other examples showing how prior information influences our perception. Consider, for example, the circular patches shown in Figure 4A. The patches in the left, middle and right column appear to bulge out of the background (bumps), whereas those in the remaining columns appear as dimples. If you turn the book upside down, you will notice that the interpretation flips, bumps become dimples and dimples become bumps. Why is that the case? The circular patches provide ambiguous shading cues to depth, that is, there are several interpretations of the sensory data (bump, dimple, flat shaded surface). Patches that are brighter at the top part are usually interpreted as bumps, which is consistent with the assumption that the light is coming from above (e.g., [54]). The analogous assertion holds for the picture of the crater (hill) shown in Figure 4B. These examples show how prior knowledge that the light usually comes from above can strongly affect the interpretation of sensory data, even though the observer is not necessarily aware of this prior knowledge.

A further component of Bayesian Decision Theory is the decision rule that links the posterior distribution with the action taken by

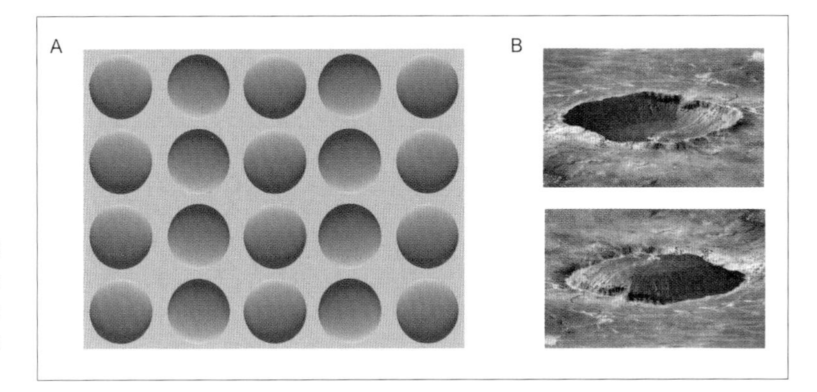

FIGURE 4.
The 'light-from-above' prior shapes the interpretation of sensory data. A: shading cues to depth. Adapted from [55]. B: picture of a crater (right side up and upside down).

the observer (see Fig. 2). For every action we might take, there are consequences. These consequences are modelled by means of a gain/loss function. According to the BDT, the observer chooses the action that maximises the expected gain. For example, it has been shown that movement planning obeys the rules of statistical decision theory (e.g., [56, 57]). In these experiments, observers were required to rapidly point to targets on a computer screen, while avoiding hitting nearby penalty zones. It was found that observers chose the aiming point so as to maximise the expected gain.

Strictly speaking, Bayesian Decision Theory as described above is a framework not a theory. In order to derive a specific theory one has to (1) compute the likelihood function, (2) define the prior distribution and combine them according to Bayes' rule to produce a posterior distribution and (3), if appropriate, specify a gain/loss function to determine the decision rule that transforms the posterior distribution into an action. These theories can then be tested experimentally. That is, the Bayesian framework is used to construct an 'ideal observer' (optimal estimator). Human performance is then compared to the ideal observer in order to assess whether humans adopt an optimal strategy to come up with a perceptual decision or action.

In the next section, we will show how the framework of BDT is applied to the problem of estimating the size of an object based on two cues to size – a visual and a haptic size cue.

Optimal integration model (Maximum Likelihood Estimator)

In this section, we will derive an optimal integration model ('ideal observer') within the Bayesian framework. This optimal integrator yields predictions that can then be compared to the performance of human observers in order to assess whether they adopt optimal integration strategies.

In multisensory perception, the observer estimates a certain property given multiple sensory cues. For example, he/she judges the size (S) based on visual and haptic size information (v, h).

The optimal estimate of an object size S given visual and haptic cues (v, h) is defined as the size S that maximises the conditional probability $p(S|v, h)$ where v and h denote the visual and haptic size cues.

Applying Bayes' rule yields:

$$p(S|v,h) \propto p(v,h|S) \cdot p(S) \qquad (4)$$

Assuming that the cues v, h are conditionally independent we arrive at

$$p(S|v,h) \propto p(v|S) \cdot p(h|S) \cdot p(S) \qquad (5)$$

Using Bayes' rule

$$p(v|S) = \frac{p(S|v) \cdot p(S)}{p(v)} \qquad (6)$$

$$p(h|S) = \frac{p(S|h) \cdot p(S)}{p(h)} \tag{7}$$

Assuming that the prior probability distributions of the size, p(S), of the visual cue, p(v), and of the haptic cue, p(h), are uniform we arrive at

$$p(S|v,h) \propto p(S|v) \cdot p(S|h). \tag{8}$$

Given that the likelihood functions p(S|v) and p(S|h) are Gaussian, the optimal estimate (maximum likelihood estimate) based on both visual and haptic cues ($\hat{S}_{vh}$) is a weighted sum of the optimal estimates based on either individual cue v and h:

$$\hat{S}_{vh} = w_v \hat{S}_v + w_h \hat{S}_h \tag{9}$$

where $\hat{S}_v$ and $\hat{S}_h$ are the maximum likelihood estimates the observer would have made from each cue in isolation (i.e., the mean of the respective Gaussian distributions) (see for example [51, 58]). The relative visual and haptic cues weights (w_v, w_h) are inversely proportional to the variance of the individual cue likelihood functions

$$w_v = \frac{\sigma_h^2}{\sigma_v^2 + \sigma_h^2} \tag{10}$$

and

$$w_h = \frac{\sigma_v^2}{\sigma_v^2 + \sigma_h^2} \tag{11}$$

and sum up to 1:

$$w_v + w_h = 1. \tag{12}$$

The variance of the joint likelihood function is given by

$$w_{vh}^2 = \frac{\sigma_v^2 \sigma_h^2}{\sigma_v^2 + \sigma_h^2} \tag{13}$$

where σ_v, σ_h and σ_{vh} are the standard deviations of the visual, haptic and combined estimate. The variance of the combined estimate is always less than the variance of either unimodal estimate:

$$\sigma_{vh}^2 \leq \min(\sigma_v^2, \sigma_h^2) \tag{14}$$

These relationships (Eqs 9–14) are used to implement an optimal integrator that comes up with the most reliable estimate (estimate with the lowest variance σ^2).

To summarise, the optimal combined estimate (MLE) is a linear combination of the individual unimodal estimates that are weighted by their relative reliabilities (inverse variances). More reliable cues are assigned a larger weight. Integrating redundant sensory information in this optimal fashion decreases the variance of the integrated estimate and yields the most reliable unbiased estimate (for an illustration see Fig. 5).

Experiments to test for optimal integration

To test whether human observers integrate multisensory information in an optimal fashion, the observers' performance is measured and compared to the predictions of the Maximum Likelihood Estimator (Eqs 9–14).

A number of empirical studies used this approach and found that the optimal cue combination rule (MLE) predicts observers' behaviour for a variety of perceptual tasks and sensory modalities (e.g., audio-visual localisation [2]; visual-haptic size/shape perception [4, 5, 40]; audio-tactile event perception [39]; visual-proprioceptive localisation [45]). That is, multisensory information is integrated in a statistically optimal fashion for a wide variety of sensory modalities and perceptual tasks.

As an example we will here present a study by Ernst and Banks [4] investigating whether humans integrate visual and haptic size information in an optimal fashion. In their study, participants performed a size discrimination task. They were asked to judge which of two objects that were presented sequentially in a two-interval forced-choice design (2-IFC), was taller. This task was performed either visually (V) or haptically

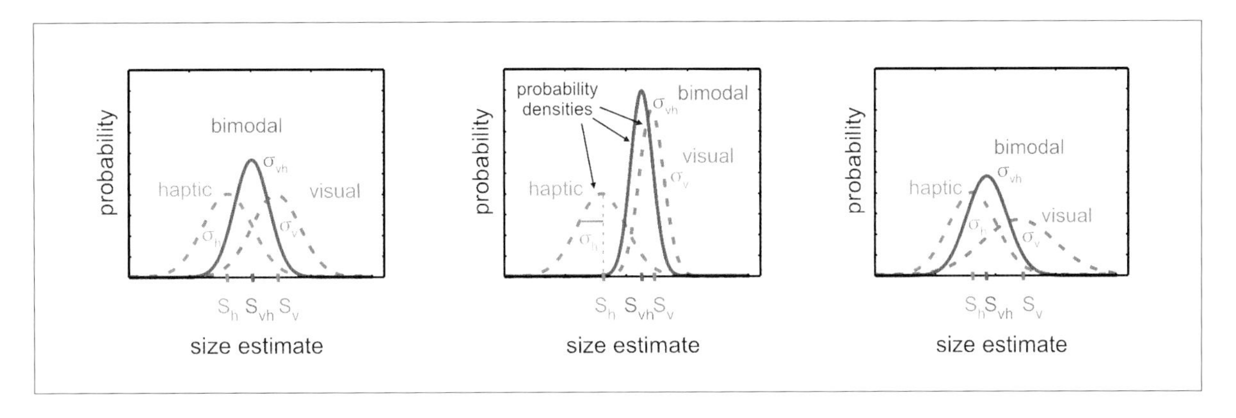

FIGURE 5. WEIGHTED INTEGRATION OF TWO UNISENSORY ESTIMATES (VISUAL AND HAPTIC) INTO A BIMODAL ESTIMATE ACCORDING TO A MAXIMUM LIKELIHOOD ESTIMATOR

Curves represent Gaussian probability densities for the visual, haptic and bimodal estimate with mean S_v, S_h and S_{vh} and variances σ_v^2, σ_h^2 and σ_{vh}^2. Three hypothetical situations are illustrated. On the left, the visual and haptic variances are equal and thus the mean of the bimodal probability density S_{vh} equals the mean of S_v and S_h (visual and haptic cue weights w_v,w_h equal 0.5). In the middle panel, the visual estimate is more reliable (lower variance of the probability density) than the haptic estimate. Thus, the mean of the bimodal probability density is shifted towards the visual estimate and vice versa *on the right where the haptic estimate is more reliable than the visual estimate. In all three situations, the variance of the bimodal estimate (red curve) is lower (smaller width of probability density) than the variances of either unisensory estimate.*

(H) alone, or by using both sensory modalities simultaneously (VH). The visual stimulus was a random-dot stereogram portraying a bar of given size. To vary the reliability of the visual stimulus, the dots that represent the surface of the bar and the background plane were randomly displaced in depth following a uniform distribution (with a range of 0%, 67%, 133%, and 200% of the depth that the bar was raised from the background plane which was 3 cm) (for an illustration see Fig. 6B). The haptic stimulus was generated using two haptic force-feedback devices (PHANTOM™ from SensAble Inc.; see Fig. 6A for details) simulating the feel of the bar.

First, the variances of the unimodal size estimates (σ_v^2 and σ_h^2) derived from visual-only and haptic-only size discrimination were determined. In the experiment, an estimate of the variance is provided by the just-noticeable difference (JND):

$$\mathrm{JND} = \sqrt{2}\cdot\sigma \qquad (15)$$ (see Fig. 7).

The unimodal variances (σ_v^2, σ_h^2) were then used to derive predictions for optimal cross-modal performance. Two predictions can be made: 1) the relative weight given to the visual and the haptic estimator can be predicted quantitatively (Eqs 10–12) and 2) predictions can be made for the variance of the combined percept (Eqs 13, 14). These predictions were derived for all four visual noise levels (0%, 67%, 133% and 200% noise) separately. If these predictions are fulfilled, it provides evidence that integration behaviour is statistically optimal.

Secondly, the actual cue weights and the cross-modal variances were measured in an experiment and compared to these predictions: Experimentally, the visual and haptic weights were derived from measurements of the point of subjective equality (PSE, 'perceived size' as determined by the 0.5 point of the psychometric functions) in a cross-modal 2-IFC size discrimination experiment. In this experiment, the visual-haptic

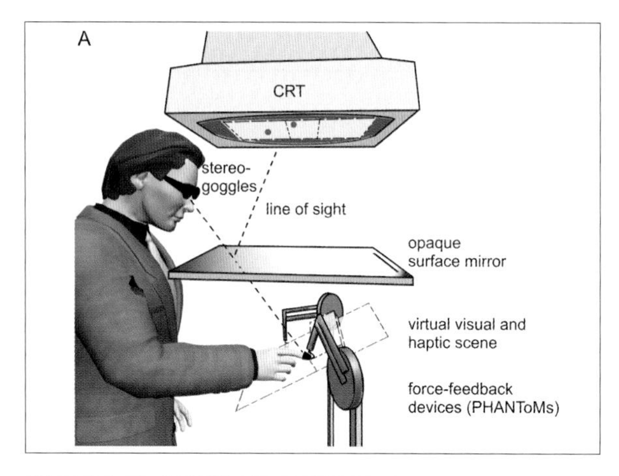

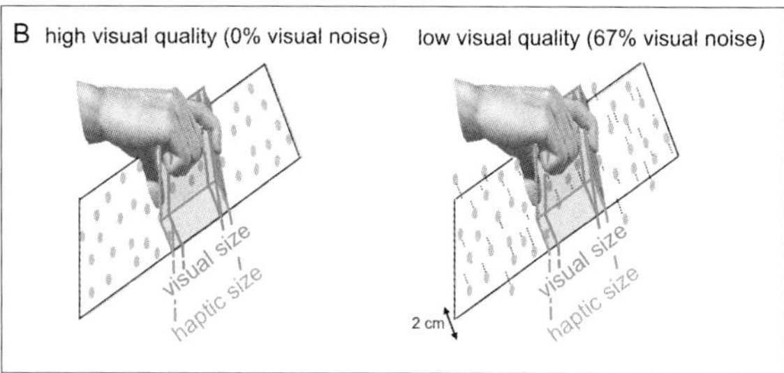

FIGURE 6.

A: Experimental setup. The visual scene is displayed on a cathode ray tube (CRT). Observers binocularly view the mirror image of the visual scene via shutter glasses that are used to present disparity. The haptic stimulus is presented using two PHANToM™ force-feeback devices, one each for the index finger and thumb. Using these devices, one can apply forces to the participants' fingers to simulate the feel of touching three-dimensional objects. B: The visual and haptic stimuli are horizontal bars. Visual and haptic scene could be controlled independently which allowed the researchers to introduce conflicts between seen and felt stimulus size. The reliability of the visual stimulus was manipulated by randomly displacing the dots that represent the bar in depth. Reprinted with permission from [52], [4].

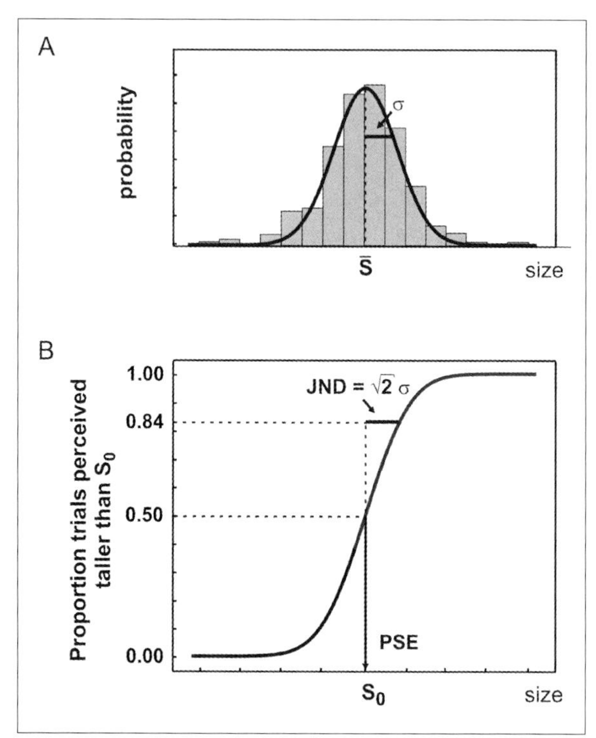

FIGURE 7.

A: Schematic illustration of the probability density function for the estimation of an object's size S. The histogram indicates the distribution of answers derived from the size-estimation process. The fitted curve is a Gaussian probability density function (with standard deviation σ and mean $\bar{S}$). B: Psychometric function derived using a two-interval forced-choice task given the probability density function for estimating the object's size from A). The just-noticeable difference (JND) derived at the 0.84 point corresponds to JND = $\sqrt{2}\sigma$. PSE is the point of subjective equality. Reprinted with permission from [59].

standard stimulus contained a small discrepancy between visually and haptically specified size ($\Delta = S_V - S_H = \pm 3$ mm and ± 6 mm). Comparison stimuli of different sizes ranging from 45 mm to 65 mm (visual and haptic sizes always consistent) were presented in order to determine the size that was perceived as being equal to the standard stimulus (the PSE). If observers rely more on their sense of vision, then the PSE is shifted towards the visual standard (S_V) and it perfectly corresponds to S_V if the visual weight equals one. If observers rely more on haptics, the PSE is shifted towards the haptic standard (S_H) and corresponds to S_H if the haptic weight

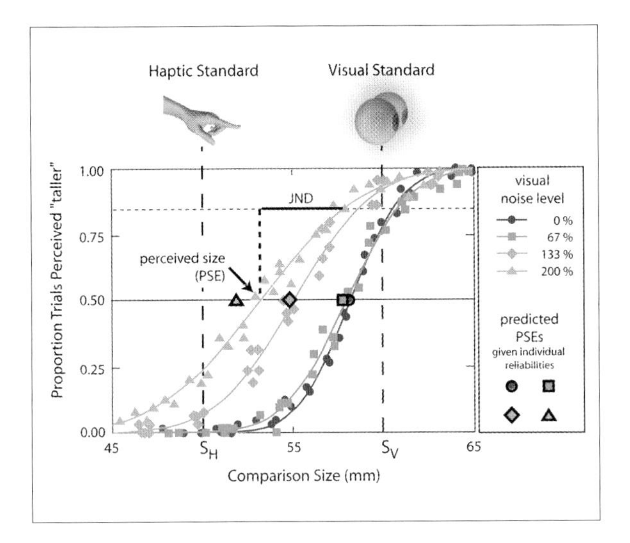

FIGURE 8.
Visual-haptic size discrimination performance determined with a two-interval forced-choice task. The visual-haptic standard stimulus contains small conflicts (here normalized to 10 mm). The relative reliabilities of the visual and haptic input were manipulated by adding noise to the visual display (0%, 67%, 133%, and 200% noise), the higher the noise level, the lower the relative reliability of the visual signal was. At all four visual noise levels, psychometric functions of bimodal size discrimination were measured. At higher visual noise levels, the perceived size, as indicated by the PSE, is more and more shifted towards the haptic standard (S_H). This demonstrates that the relative haptic weight increases as the reliability of the visual signal is reduced. The observed data (PSEs) correspond well with the predictions of an optimal integration model (MLE). Reprinted with permission from [52], [4].

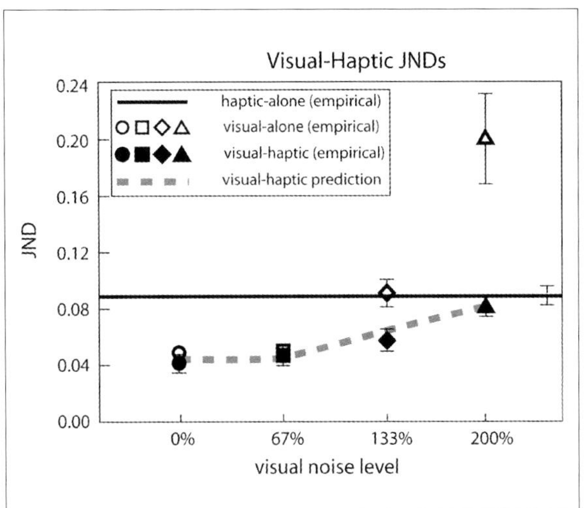

FIGURE 9.
Thresholds (JNDs) for unimodal visual, unimodal haptic, and bimodal visual-haptic size discrimination at four levels of visual noise (0%, 67%, 133%, and 200%). The reliability of the unimodal haptic estimate is independent of the visual noise level (black line). Optimal bimodal visual-haptic JNDs (grey dashed line) were predicted from the unimodal visual and haptic JNDs. The empirically determined bimodal JNDs (black symbols) are well in agreement with the prediction of the optimal integration model (MLE) and are reduced as compared to the unimodal JNDs. Adapted from [4].

equals one. Thus, the PSE is a direct measure of the relative cue weights (w_V, w_H).

As shown in Figure 8, the empirically determined PSEs were found to be in agreement with the predicted PSEs. In the 0%-noise condition (depicted in red), the PSE is close to the visual standard, demonstrating a high visual weight. In the 200% noise condition, the PSE is close to the haptic standard, demonstrating low visual but high haptic weight. With increasing noise level, there is a smooth transition from visual dominance to haptic dominance.

The correspondence of the empirical data with the predicted cue weights (and PSEs) is a first hint that information is integrated according to a Maximum Likelihood Estimator. However, different strategies may potentially have caused this result. For example, if the observer relies perfectly on either one of the two modalities at a time, but switches back and forth between relying on vision or touch in proportion to their relative reliabilities, the empirical result would mimic cue weights that also correspond to the predicted weights [43, 52]. But using this strategy, the bimodal variances (JNDs) would never become lower than the variances of the unimodal estimates. Therefore, a stronger test for optimal integration is to show that bimodal visual-haptic estimates become more reliable (lower variance) than either unimodal estimate and that the bimodal variance (JND) corresponds to the predictions derived from the MLE model (Eqs 13–15). Empirically, the bimodal visual-haptic JNDs were derived from the cross-modal psychometric functions. As can be seen in Figure 9, the empirically measured JNDs are well in agreement with the predicted JNDs. Together these results demonstrate that humans integrate visual and haptic size information in a statistically optimal fashion to come up with the most reliable visual-haptic size estimate [4].

Learning to combine sensory information

At all stages, the processes of multisensory integration can be affected by learning. For example, through sensory interactions with the environment the correlation between sensory signals or the reliability of the individual sensory cues can be learned, and the weights attributed to the different sensory systems can be changed. Sensory interactions with the environment can also shape our prior knowledge about the world. In this context, it has been observed that haptics often act as a standard by which the perceptual system learns how to combine different sources of information in a reasonable manner.

For example, Ernst et al. [60] observed that haptic feedback can change subsequent visual percepts by changing the weights given to different sources of visual information. The three-dimensional structure of objects and scenes can be inferred from a number of different visual cues (for an overview see [61]) such as texture, shading, motion, perspective and disparity (depth information provided by the difference in images of the left and right eye). In Ernst and colleagues' [60] study, participants were presented with virtual slanted planes. Stimuli were presented by means of a setup similar to the one described in the last section (see Fig. 6A). Participants viewed the visual scene through stereo goggles. Visually the slant of the plane was specified by texture and disparity information. The observer's index finger was attached to a haptic force feedback device (PHANToM) simulating the feel of the stimuli. Thus, participants could also haptically perceive the slant of the plane. In the experiment, texture and disparity cues were put in conflict so as to determine the weight initially given to either estimator (texture, disparity). In a subsequent training phase, haptic feedback was added. That is, participants saw and felt the slanted plane where the haptic slant was either consistent with the slant specified by texture or, in a second condition, with the slant specified by disparity. In a post-training experiment, it was observed that higher weight was given to texture information when, during training, the haptic feedback was consistent with the texture cue (see Fig. 10) and *vice versa* when the haptic feedback was consistent with the disparity one.

Similarly, Atkins et al. [62] studied the influence of haptic input on visual cue integration. In a virtual environment (similar to the setup described above, see Fig. 6A), participants were presented with cylinders that they could see and grasp. The depth of the elliptical cross-section of the cylinders varied while its width was held constant. Haptically, participants perceived the shape (depth) of the cylinder through haptic force feedback provided with two PHANToM devices attached to index finger and thumb of one hand. Visually, the shape of the cylinder was specified by texture and motion cues: The texture of the stimuli consisted of flat disks that were attached to the cylinder's surface. The shape of the cylinder affects the size, density and compression of the texture elements (disks) in the two-dimensional image (e.g., [63, 64]). Thus, there were texture cues to shape. In addition, the disks moved horizontally along the surface of the cylinder and were therefore providing motion information

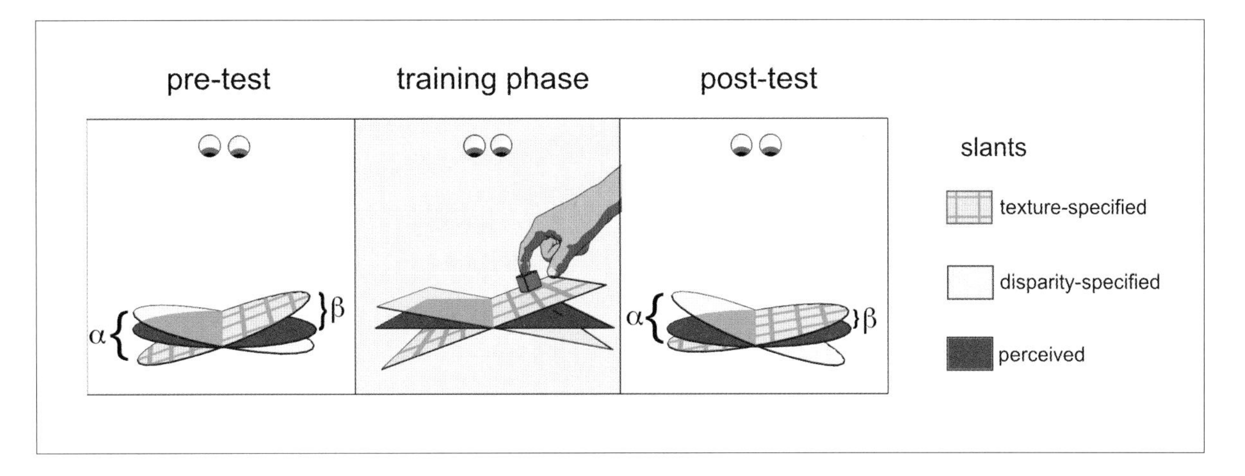

FIGURE 10. EXPERIMENTAL DESIGN

The pre- and post-tests were purely visual tasks. The haptic training phase occurred between the pre- and post-tests and consisted of visual and haptic stimulation. The visual plane had different texture- and disparity-specified slants. The angle between the two specified slants is the conflict angle α. The perceived slant is represented by red planes. The decrease in β between pre- and post-test indicates a shift of the perceived slant towards the texture-specified slant and thus an increase in texture weight. Adapted from [60].

about the cylinder's shape. Participants judged the depth of the cylinder. Texture and motion cues were put in conflict in order to determine the relative weights given to either cue. The experiment consisted of training and test trials. On training trials, observers haptically and visually explored the stimulus. The haptic input was consistent with either the texture or the motion information. Atkins and colleagues observed that on test trials participants relied more on texture information when they previously received texture-consistent haptic input and more on motion cues after motion-consistent training.

These two studies demonstrate that observers can use the haptic percepts as a standard against which the reliability of visual cues is judged. The reliability of the visual cues, in turn, determines how the cues are integrated, i.e., how much weight is given to either cue.

A further example demonstrating that exposure to haptic input can change the way humans interpret their environment was shown by Adams and colleagues [55]. In the previous section ('Bayesian Decision Theory'), we have seen that the combined unified percept does not only depend on the available sensory information but also on observers' prior experience of the environment. Adams and colleagues [55] found that haptic interactions with the scene can change these prior assumptions and thus, the interpretation of the sensory information. They presented shaded circular patches similar to those depicted in Figure 4. These patches can be interpreted as being either convex (bumps) or concave (dimples). Usually the patches are interpreted as being convex when they are brighter at the top which is consistent with the assumption that the light is coming from above. In Adams et al.'s study, it was examined whether haptic feedback can change the light-from-above prior. Stimuli differing in the orientation of the light peak were presented and observers made convex-concave judgments. First, the initial light prior was assessed (see Fig. 11, topmost panel). In a subsequent training phase, haptic feedback was added. Now stimuli with a light peak shifted by -30° (or +30°

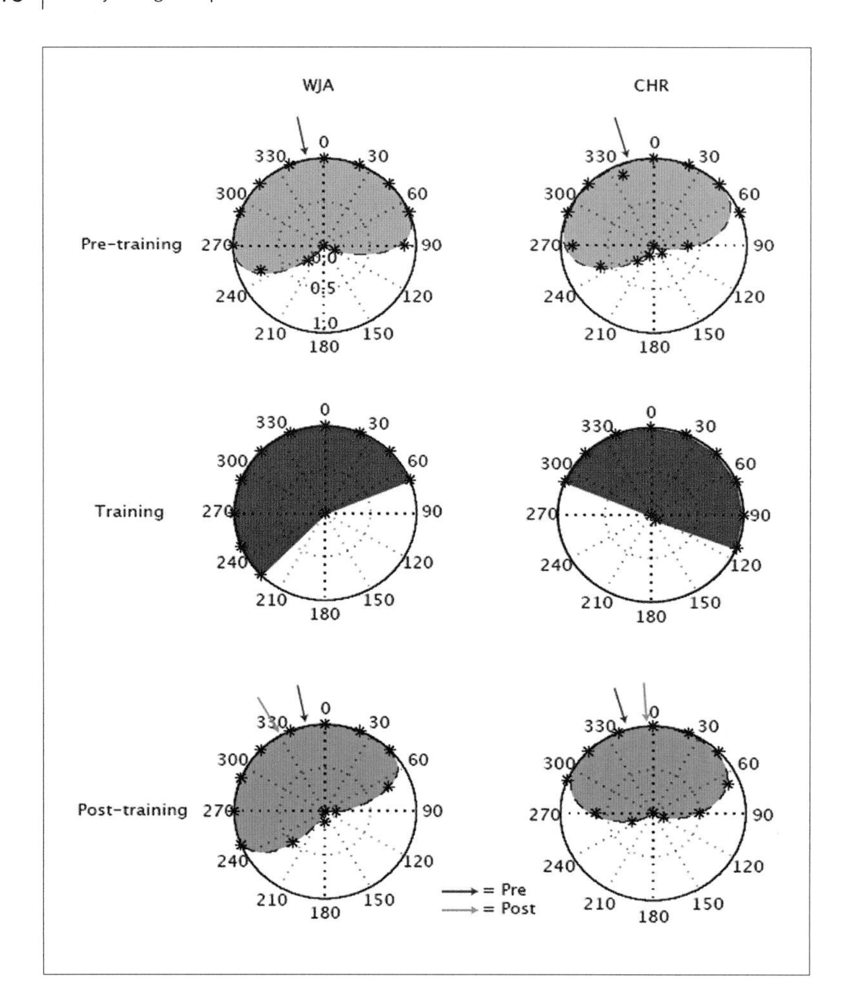

FIGURE 11.
Data for two representative observers trained with opposite shifts (–30° and +30°). 0° corresponds two stimuli brightest at the top. The proportion of stimuli perceived as convex (black stars) are fitted by a function based on two cumulative Gaussians (dashed lines) each centred at a concave-convex transition and whose average gives the light position prior (pre-training, blue arrow; post-training, green arrow). Adapted from [55].

in a second group) from the initial light prior were combined with haptic feedback signalling that the stimulus is convex. It was found that after exposure to the haptic feedback, the light prior was shifted towards the light peak orientation trained to be convex. Importantly, these results are also observed when different types of stimuli are used in the post-training session (for more details see [55]). Therefore, one can conclude that haptic feedback can change our prior assumptions.

One further issue to be raised in this context is the question whether we can also learn to combine arbitrary signals that are usually unrelated and therefore not normally combined. More specifically, do we learn to integrate visual and haptic signals from the co-occurrence of these formerly unrelated signals? This question was investigated in a recent study by Ernst [65] using stimuli that are independent under natural conditions, namely the luminance of an object (visual dimension) and its stiffness (haptic dimension). In a virtual environment, participants are exposed to conditions where bright objects always felt hard and dark objects were soft. That is, in the training session luminance and stiffness were highly correlated. Ernst [65] found that after extensive training observers learned to integrate these formerly unrelated stimuli from the natural statistical relationship.

To summarise, together these studies demonstrate that observers learn about the statistical regularities of the environment by haptically interacting with the world and adjust their perception to these statistics. Hence, haptics can teach the perceptual system how to combine different sources of information in a reasonable manner.

How does the system decide whether or not to combine sensory signals?

How does the perceptual system decide whether or not to integrate sensory signals into a combined percept?

It has been shown that bimodal integration depends on structural similarity between the bimodal sensory input, such as the degree of temporal synchrony (e.g., [66–71]) or the amount of spatial separation (e.g., [66, 72–76]). Temporal synchrony and spatial coincidence are strong cues indicating whether or not two signals refer to the same object or event and should therefore be integrated. For example, several studies have shown that perception of stimuli in one modality is influenced by the simultaneous presentation of task-irrelevant stimuli in another modality. This mutual biasing effect however is diminished when stimuli are presented in temporal asynchrony exceeding a temporal integration window of about 100–200 ms (e.g., [77–79]). Studies investigating the biasing effect of one modality on another as a function of spatial separation between the sensory sources (e.g., [74, 80–82]) found that integration subsides with large discrepancies between object properties, which renders it likely that the signals arise from different sources.

Besides such structural properties, there are also other factors indicating that two signals come from a common source. Observers may for example know that two signals belong together because they were instructed that this is the case. Or they assume that two signals belong together because they know from experience that these signals usually originate from the same cause, as for example speech sound (auditory signal) and lip movements (visual signal). The question is whether such knowledge about a common source can also help multisensory integration?

In agreement with this hypothesis, some classic cross-modal phenomena (e.g., prism adaptation (adaptation to a visual proprioceptive mismatch), ventriloquism (visual bias of auditory localisation)) have been shown to depend on the participants' 'belief' that samples arise from the same object ('unity assumption', [20]) (for a review see [20, 83]). For example, Miller [84] demonstrated the influence of instructions about object unity on the relative contribution of vision and touch to the perceived shape. He presented participants with two objects differing in shape. The two were presented in a box, above and below an occluder. Participants viewed the object on top of the occluder while simultaneously grasping an unseen object on the bottom of the box (below the occluder). The perceived shape of the felt object was virtually not influenced by vision. However, if subjects were (misleadingly) told that they see and feel "identical halves of the same object" the haptic shape percept was dominated by vision. Similarly, Welch [85] has shown that whether or not adaptation to a visual-proprioceptive conflict occurs depends on whether observers believe that differing information came from the same object. The visual-proprioceptive conflict was created by virtue of a prism that laterally displaces the visual image. The belief that seen and felt information came from the same object (here the seen and felt position of the participants' hand) led to adaptation to the visual-proprioceptive discrepancy. However, when participants actually looked at their own hand through displacing prisms but were misled to believe that they see someone else's hand (and thus there was no reason to relate the felt position of the hand to the seen position of the seemingly foreign hand), prism adaptation was diminished.

Thus, a physical discrepancy between vision and proprioception alone was insufficient for adaptation to occur; necessary was the subjects' belief that both sensory signals come from the same hand. A further study demonstrating an effect of the unity assumption on multisensory integration was conducted by Warren and colleagues [86]. They presented participants with spatially discrepant visual and auditory signals and studied the effect of task-irrelevant visual or auditory information on the perceived location of the signals in the target modality. The effect of vision on audition was larger when subjects were instructed that visual and auditory information come from a common source than when they were told that the two signals come from different sources [86]. In contrast to these results, there are other studies that did not observe any influence of task instructions about object identity. For example, Radeau and Bertelson [87, 88] did not find any difference between adaptation after-effects observed when subjects were told that the origin of the visual and auditory input was either the same or different.

Besides instructions as to whether two signals arise from a common object, the realism of the display was discussed as a factor influencing the strength of the unity assumption. A number of studies (e.g., [67, 74, 76]) examined whether integration is stronger in more 'compelling' stimulus situations that enhance the strength of the observers' assumption of unity (for a review see [20]). For instance, Jack and Thurlow [67] found larger ventriloquism effects (visual bias of perceived auditory location) for puppets with an intact face moving its mouth in sync with the speech than for static puppets with eyes and nose removed. Similarly, the number of same-origin judgments of spatially discrepant visual and auditory stimuli increased for a highly realistic stimulus scenario such as a 'whistling' tea kettle (auditory signal) emitting steam (visual signal) as opposed to a less realistic situation, e.g., when simple lights and bells were used as visual and auditory stimuli [74]. In contrast, Radeau and Bertelson [69, 89] did not observe any influence of the realism of the display on after-effects of ventriloquism.

In summary, from these results it remains controversial whether prior knowledge that two signals emanate from the same object can affect multisensory integration.

In a more recent study, Helbig and Ernst [90] studied whether knowledge about a common source can promote multisensory integration even when the signals are presented at discrepant locations. In one condition, participants had direct view of the object they touched. In a second condition, mirrors were used to create a spatial separation between the seen and the felt object. Participants saw the mirror and their hand in the mirror exploring the object and thus knew that they were seeing and touching the same object. To determine the visual-haptic interaction they created a conflict between the seen and the felt shape using an optically distorting lens that made the rectangular object look like a square. Participants judged the shape of the probe by selecting a comparison object matching in shape. A mutually biasing effect of shape information from vision and touch was found independent of whether participants directly looked at the object they touched or whether the seen and the felt object information was spatially separated with the aid of a mirror. This finding suggests that prior knowledge about object identity is a facilitatory factor of multisensory integration which can even overcome large spatial discrepancies that lead to a breakdown of integration otherwise.

In the recent past, Körding and colleagues [91] addressed the issue of how human observers decide which cues correspond to the same source and whether or not to combine them using the Bayesian approach. They constructed an ideal-observer model that infers the causal structure, i.e., assesses whether two sensory cues come from a common object or event. This model makes predictions as to when observers should process two cues independently and under which circumstances they should attribute two cues to a common event and thus fuse them into a combined percept. Human behaviour (here performance on auditory-visual localisation tasks) was found to be well predicted by this model.

Summary

Touch is combined with the other sensory modalities to form a coherent and robust percept of the world. Merging of sensory information is beneficial in that it can speed reactions, reduce perceptual ambiguity and make perceptual judgments more precise. The mechanisms underlying sensory integration can be modelled within a Bayesian framework. Empirical studies revealed that redundant sensory information is integrated in a Bayes-optimal fashion according to a Maximum Likelihood estimator that minimises the variance (increases the reliability) of the integrated percept. Moreover, touch allows for active exploration of the world. Through such haptic interactions with the environment the perceptual system can learn how to interpret the world from its statistical regularities. It seems that active touch can act as a standard against which the other senses are compared to judge their reliability and to resolve conflicts.

Selected readings

Calvert GA, Spence C, Stein BE (2004) The Handbook of Multisensory Processes. Cambridge, MA: The MIT Press

Ernst MO (2005) A Bayesian view on multimodal cue integration. In: G Knoblich, M Grosjean, I Thornton, M Shiffrar (eds.): Human body perception from the inside out, chapter 6. New York, NY: Oxford University Press, 105-131

Ernst MO, Bülthoff HH (2004) Merging the senses into a robust percept. Trends in Cognitive Sciences 8(4): 162–169

Mamassian P, Landy MS, Maloney LT (2002) Bayesian modeling of visual perception. In: RPN Rao, BA Olshausen, MS Lewicki (eds.): Probabilistic Models of the Brain: Perception and Neural Function. Cambridge, MA: MIT Press, 13-36

Spence C, Driver J (2004) Crossmodal Space and Crossmodal Attention. New York, NY: Oxford University Press

Haptically evoked activation of visual cortex

19

Simon Lacey and K. Sathian

Introduction

Thanks to the acceleration of research into multisensory processing in recent years, the idea that the brain is organized around parallel processing of separate sensory inputs is giving way to the concept of a 'metamodal' brain [1] organized around particular tasks in a modality-independent fashion. For example, it is now apparent that many cortical areas previously thought to be specialized for processing specific aspects of visual input are also activated during analogous haptic or tactile tasks. In this chapter, we review the circumstances in which haptic or tactile tasks are associated with visual cortical activity and what this means for our internal representation of the external world, i.e., how information about objects and events in the external world is stored in memory. By 'haptic', we mean tasks involving active motor exploration of stimuli, whereas 'tactile' refers here to tasks in which stimuli are applied to the passive finger or other skin surface.

Visual cortical involvement in touch

Gratings and dot-patterns

The first indication that tactile input could activate extrastriate visual cortex came from our positron emission tomographic (PET) study [2]. In this study, participants had to discriminate the orientation of gratings presented to the immobilized right index fingerpad. In a control task, participants discriminated the width of the grooves on the gratings. A contrast between these tasks revealed activity that was selective for tactile discrimination of grating orientation in the left extrastriate visual cortex, at a focus close to the parieto-occipital fissure. This same focus is known to be active during both visual grating orientation discrimination [3] and spatial imagery [4]. We therefore concluded that activation of this focus during tactile discrimination of grating orientation reflected spatial processes that are common to vision and touch [2]. This study suggested, for the first time, that task-selective visual cortex responds to a tactile analog of the relevant visual task and that visual cortical activity might be independent of the sensory modality in which a task is presented.

A subsequent functional magnetic resonance imaging (fMRI) study confirmed this left parieto-occipital activation during tactile discrimination of grating orientation when contrasted with discrimination of grating groove width, and also revealed additional activations on this contrast in the right postcentral sulcus (PCS) and the left anterior intraparietal sulcus (aIPS) [5]. Another fMRI study also found left-lateralized activity in the aIPS during discrimination of grating orientation on a contrast with discrimination of small changes in grating location [6], whereas a third study in which the control task was discrimination of grating roughness revealed a right-lateralized PCS-aIPS complex that was selectively active during grating orientation discrimination [7]. This last study, moreover, demonstrated multisensory task-related activity in the right aIPS, which was also activated by visual discrimination of grating orientation compared to discrimination of grating color.

Transcranial magnetic stimulation (TMS) studies have shown that these tactually-evoked visual cortical activations reflect functional involvement in tactile perception, rather than being an epiphenomenal byproduct. TMS involves the application of a magnetic pulse to the scalp that can temporarily disrupt the function of the immediately underlying cortex, producing a 'virtual lesion' [8]. If TMS over a particular area results in impaired task performance, we can infer that the targeted area is necessary for task performance [9]. Applying single-pulse TMS to the left parieto-occipital region found by Sathian and colleagues [2], 180 ms after tactile stimulus onset, significantly impaired tactile discrimination of grating orientation, but not grating groove width [10]. TMS at other sites did not affect performance on either task. In another study, repetitive TMS (rTMS) was used in a study of subjective estimates of tactile interdot distance and roughness [11]. The psychometric functions differ for these two kinds of estimates: perceived interdot distance increases monotonically as physical interdot distance increases up to 8 mm, while perceived roughness peaks around 3 mm. This dissociation was reflected in the rTMS effects: rTMS over medial occipital cortex disrupted estimates of tactile distance but not roughness, whereas the reverse was true for rTMS over primary somatosensory cortex. Furthermore, a congenitally blind patient with bilateral occipital lesions was impaired at estimating tactile interdot distances but not roughness [11]. However, a corresponding fMRI study failed to show a similar dissociation: activity in primary visual cortex was weak and non-selective, distance-selective activity was found only in the left IPS, and no roughness-selective activation was found [12]. The reasons underlying this disparity between the fMRI study and the corresponding TMS/lesion studies are not clear at this time.

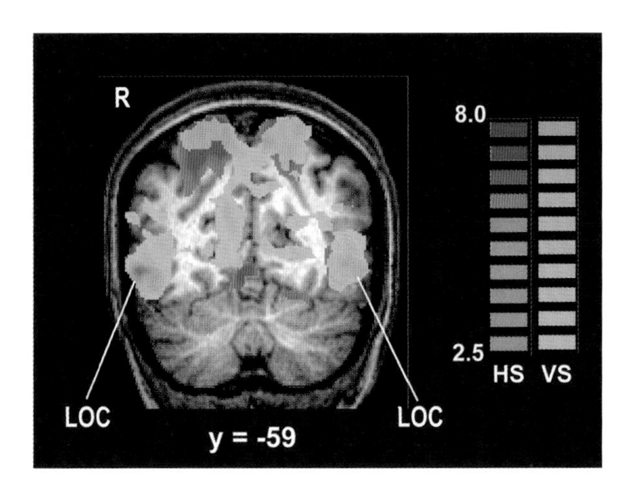

Figure 1.
Brain regions showing selective activity for perception of visual shape (VS, green) and haptic shape (HS, red) displayed on a coronal slice through the brain image (the Talairach y-coordinate for this slice is shown). Bisensory (visuo-haptic) overlap of shape-selectivity (yellow) can be seen in the lateral occipital complex (LOC, identified by pointers) and also in the intraparietal sulcus (at the top). The brain is shown in radiologic convention in which the right hemisphere (R) is on the left. The color t-scales reflect the statistical significance of shape-selectivity, in each modality, relative to the control condition (texture perception). Based on the data of Stilla and Sathian [15].

Shape

Several cortical areas have been implicated in visuo-haptic shape processing, principally the lateral occipital complex (LOC). The LOC is a visually shape-selective area in the ventral visual pathway [13] that is also shape-selective during haptic perception of 3-D objects [14–16] (Fig. 1). This bisensory shape processing is confined to a sub-region of the LOC, termed LOtv [17]. The LOC is active even during tactile processing of 2-D forms, as contrasted with both gap detection [18] and bar orientation [19]. The overlap of visual and haptic shape-selective activity in LOtv suggests the possibility of a shared representation of shape. The finding that the magnitudes of visually- and haptically-evoked shape activations are significantly correlated across subjects in the right but not the left LOC [15] suggests that the site of a shared representation across vision and touch might be the right LOC. The LOtv appears to be primarily concerned with object shape, since this sub-region did not respond to object-specific auditory cues, for example the sound of a cat or a train [17]. However, when subjects are trained in the use of sensory substitution devices that convert information about object shape into soundscapes, the LOtv becomes responsive to the encoded shape information; this does not occur when sounds are learned merely as arbitrary associations with objects [20]. Like the TMS studies reviewed above, case studies of patients with actual lesions suggest that the LOC is necessary not only for visual, but also for haptic shape perception. For example, a patient with a lesion to the left occipito-temporal cortex, likely including the LOC, had tactile agnosia (inability to recognize objects) in addition to visual agnosia, although both somatosensory cortex and basic somatosensation were spared [21]. Another patient was impaired at learning novel objects by either vision or touch following bilateral lesions of the LOC (see [22] for a review).

Other regions that exhibit multisensory shape selectivity include the PCS bilaterally, the aIPS and posterior and ventral parts of the intraparietal sulcus (pIPS and vIPS) [15]. Of these parietal cortical regions, the PCS corresponds to Brodmann's area 2 of primary somatosensory cortex (S1) [23], a region generally assumed to be exclusively somatosensory. Thus, the finding of multisensory selectivity for shape in this region was quite surprising. The remaining regions along the IPS are part of classical multisensory cortex, and indeed multisensory shape-selectivity had been reported earlier in or near some of these regions [24, 25]. It should be noted, too, that the left pIPS, similar to the right LOC, showed correlated activation magnitudes for visual and haptic shape processing [15]. The exact roles of the PCS, parts of the IPS and LOC in multisensory shape processing remain to be fully worked out. One possibility is that the pIPS foci might mediate visual imagery ([25]; Deshpande, Hu, Stilla and Sathian, unpublished observations) whereas the LOC might house modality-independent representations of shape.

Texture

Until recently, it was thought that the functional recruitment of visual cortex into haptic perception was limited to macrospatial features such as object shape, rather than microspatial features such as texture. This accords with the studies reviewed above which showed that visual cortex is functionally involved during tactile discrimination of grating orientation and interdot distance but not grating groove width or roughness [3, 11, 12] and with the minimal LOC activation found during tactile microspatial discrimination [27]. Furthermore, psychophysical studies suggest that visual processing is preferentially associated with macrospatial tactile tasks [28] and that vision is more sensitive to macrospatial properties than touch while the reverse is true for microspatial properties [29]. However, microspatial tactile discrimination of the direction of offset of the central dot in a three-dot array appears to engage at least some of the areas noted earlier in this chapter to be involved in multisensory shape processing – these areas include the aIPS and pIPS bilaterally, and the left PCS [27]. Interestingly, in a study of visual and haptic texture perception, haptic texture selectivity was found in the right medial occipital cortex (MOC): this area appeared to be mainly in the second visual area (V2), overlapping with an area in V1 (primary visual cortex) that was visually texture-selective, the visuo-haptic overlap being on the V1/V2 border [15]. However, we should recall that rTMS over MOC did not disrupt roughness estimates [11] as might be expected if it is texture-selective. In addition, the magnitudes of visual and haptic MOC texture activations were not significantly correlated, so that it may not be possible to speak of a common representation of texture [15]. This study extended findings of haptic activation of visual cortex into early visual processing and was also the first to show visual areas to be selectively active in a tactile microspatial task (texture perception).

Motion

The human middle temporal (MT) complex, an area known to be involved in processing visual motion information, is also activated by tactile motion stimuli, even when there is no explicit task [30–32]. This common activation suggests a shared representation of motion across vision and touch, an idea that is supported by psychophysical studies. For example, an ambiguous visual motion display can be disambiguated by tactile motion perception [30] even when presentation of the visual motion display is delayed, thus forcing reliance on the mental representation of motion [33]. When the direction of motion in the display is unambiguous but incongruent between vision and touch, visual motion disrupts perception of tactile motion [34], suggesting that visual and tactile input are differently weighted. It is worth noting that, unlike the auditory activation of the LOC only under special circumstances, the human MT complex is activated by auditory motion [35]. Recently, it was reported that vibrotactile stimuli activate the medial superior temporal area (MST), a region within the MT complex involved in processing visual motion, but not area MT itself [36]. Why vibrotactile stimuli should activate a motion-sensitive area remains unclear, but the reason may have to do with vibratory motion of the skin in response to such stimuli.

Visual imagery or modality-independent representation?

An intuitive explanation for the involvement of visual cortex in non-visual tasks is that such cross-modal recruitment reflects mediation by visual imagery; in other words that feeling an object also involves visualizing it in the 'mind's eye' [2]. Such visualization may be a specific instance of a more general strategy for processing complex tasks, in which the most suitable sensory modality is pressed into action [37]. Consistent with this visual imagery explanation, the LOC is active during visual imagery, in addition to being active during visual and haptic shape perception. The left LOC is preferentially activated during retrieval of geometric or material object properties from memory in the absence of both visual and haptic input [38] and in both blind and sighted individuals during mental imagery of familiar object shape derived from haptic or visual experience, respectively [39]. Secondary tasks that have been shown to interfere with visual imagery also impair cross-modal memory for unfamiliar but not familiar objects [40]. In addition, ratings of the vividness of visual imagery were strongly positively correlated with individual differences in haptic shape-selective activation magnitudes in the right LOC [16]. It has been argued that haptic shape perception is not mediated by visual imagery because LOC activity during visual imagery is substantially less than during haptic shape perception [14]. However, this study did not verify that participants were maintaining their visual images during the entire scan and this may account for the reduced imagery-related activation that was observed.

Since vision and touch encode object properties common to both modalities – for example, size, shape, and the relative positions of different object features – an alternative explanation is that vision and touch result in a single shared representation that is multisensory [41]. Although some have used the term 'amodal' [42] to refer to modality-independent representations, we believe that this term is best reserved for information that cannot be derived directly from sensory systems and that must be represented linguistically or perhaps propositionally. Instead, we prefer the use of the term 'multisensory', by which we mean a representation that can be encoded and retrieved by multiple sensory systems and which retains the modality 'tags' of the associated inputs [41]. The existence of widespread multisensory overlap in cortical shape processing, the correlation between the magnitudes of visual and haptic shape-selective activity in some of these overlap regions, and the effects of lesions (reviewed in the previous section) all point to a neural representation of shape that is common to both vision and touch. This view is also supported by studies of cross-modal priming effects (the extent to which visual recognition is facilitated by prior haptic experience). Behavioral studies have shown that the size of this cross-modal facilitatory effect is equal to that of the within-modal effect, suggesting a common representation [43–45]. Similarly, neuroimaging using fMRI has shown that LOC activity increases when viewing objects that were previously explored haptically (i.e., that had been 'primed') compared to viewing non-primed objects [46]. Furthermore, category-selectivity for faces and man-made objects in human inferotemporal cortex is relatively independent of whether the input modality is visual or haptic [47]. Early-blind individuals in this study had similar patterns of category-selective responses during haptic object perception, although blind and sighted groups differed somewhat in the cortical location of the category-selective responses.

Other work has suggested that a multisensory representation may be encoded in a modality-independent spatial format. Visual imagery can be divided into 'object imagery', referring to images that are pictorial and deal with the literal appearance of objects in terms of shape, color, brightness, etc., while 'spatial imagery' refers to more schematic images dealing with the spatial relations of objects and their component parts and with spatial transformations [48–50]. This view is supported by recent work showing that spatial, but not object, imagery scores were correlated with accuracy on cross-modal object identification but not within-modal object identification [51]. In the same vein, a concurrent spatial task was found to interfere with cross-modal object recognition whereas a concurrent non-spatial task had no measurable effect (Lacey and Campbell, unpublished observations). Both visual and haptic object recognition appear to be viewpoint-dependent within-modally, since recognition performance was reduced when objects were rotated 180° away from the learned view [51, 52]. By contrast, cross-modal recognition was unaffected by rotation around any of the three major axes, whether objects were learned visually and tested haptically or *vice versa* [51]. This finding of viewpoint-independence suggests mediation of cross-modal information transfer *via* a high-level modality-independent representation. However, where such a representation is housed remains unclear. As far as visual object representations are concerned, the IPS appears to exhibit viewpoint-independent responses [53], but results from studies of the LOC have been more equivocal, reporting both viewpoint-independent [53–55] and viewpoint-dependent [56] responses.

If vision and touch engage a common representational system, then we would expect to see this reflected in similarities between visual and haptic processing of these representations. For example, the time taken to scan a mental image increases with the spatial distance to be inspected for visual imagery [57, 58] and also when scanning haptically derived representations [59]. This implies that both visually- and haptically-derived images preserve spatial metric information and that similar, if not identical, imagery processes operate in both modalities [59]. Similarly, the time taken to judge whether two objects are the same or mirror-images increases nearly linearly with increasing angular disparity between the stimuli for mental rotation of visual [60] as well as haptic stimuli [61–64]. A similar principle was found to hold for tactile stimuli in relation to the angular disparity between the stimulus and a canonical angle [18, 65], with associated activity in the left aIPS [18]. This area is also active during mental rotation of visual stimuli [66], further supporting the idea that similar spatial imagery processes operate on both visually and haptically derived representations. These similarities between visually and haptically derived representations extend to both the early- and late-blind as well as the sighted [59, 62]. The blind may be slower than the sighted at haptic mental rotation [61], but controlling for greater variability in the use of frames of reference by the sighted suggests that this difference may be more apparent than real [63].

If tactile activation of visual cortex reflects engagement of modality-independent representa-

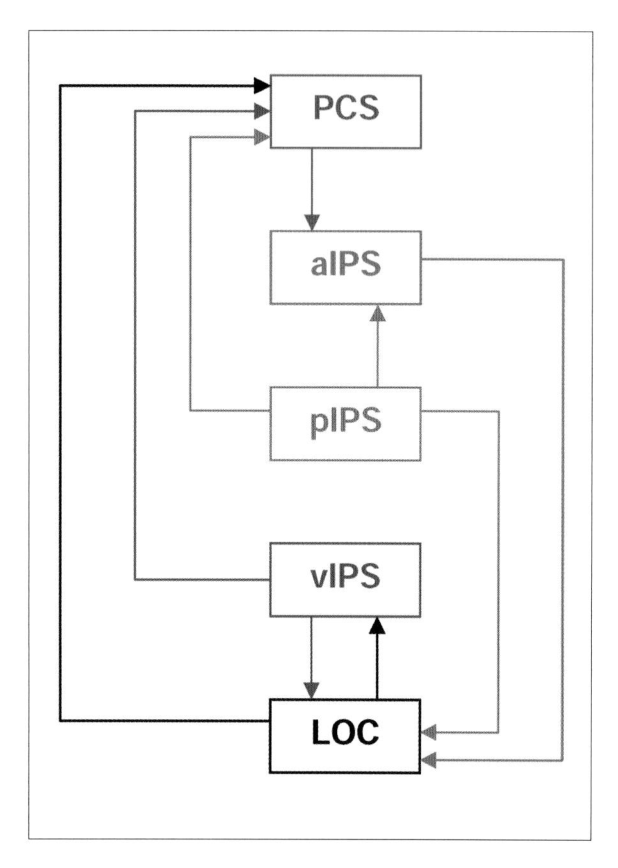

FIGURE 2. MODEL OF EFFECTIVE CONNECTIVITY BETWEEN SELECTED REGIONS DEMONSTRATING HAPTIC SHAPE-SELECTIVE ACTIVITY

The model includes bottom-up pathways originating in the postcentral sulcus (PCS) and directed into the lateral occipital complex (LOC), as well as top-down pathways in the reverse direction. The model suggests bidirectional flow of information between somatosensory and visual cortical areas, some direct and some mediated through regions along the intraparietal sulcus (aIPS: anterior intraparietal sulcus; pIPS: posterior intraparietal sulcus; vIPS: ventral intraparietal sulcus). (Reproduced from [67] by kind permission of Elsevier Limited)

tions, there should be direct, bottom-up projections from somatosensory areas into the visual cortical areas that are implicated in tactile perception. By contrast, if this tactile activation reflects visual imagery, there should be top-down projections from prefrontal and parietal cortex into these visual areas [26]. These possibilities can be investigated either by studying connectivity in experimental animals, or by analyzing human neuroimaging data to reveal effective connectivity. In fact, using structural equation modeling, based on the correlation matrix between the time courses of fMRI activity in various regions, we have found evidence for both bottom-up paths from the PCS, a part of primary somatosensory cortex (see above), and top-down paths from parts of the IPS into the LOC [67] (Fig. 2). Much other work has revealed multisensory inputs into early sensory areas traditionally considered unisensory, including V1 [68, 69] and auditory cortex [70]. Analysis of the laminar profile of these inputs also indicates that there are top-down [68, 69, 71] and bottom-up [71] pathways. There has been very limited neurophysiologic study of the cross-modal activation of visual cortex in behaving monkeys. An example is a study showing that neurons in visual area V4, but not V1, were selective for the orientation of a tactile grating, but only when this was to be matched to a subsequently presented visual grating and not when the tactile grating was task-irrelevant [72]. The fact that this cross-modal activation occurred only when it was necessary for task performance indicates that the tactile responses depended on top-down inputs. Conversely, S1 neurons in monkeys are visually responsive during a visuo-haptic matching of grating orientation [73]. Thus, the sum total of the evidence to date indicates the probable existence of multisensory representations that are flexibly accessible *via* both vision and touch, and that potential neural substrates exist for visuo-haptic interactions through bottom-up sensory inputs as well as top-down processes such as visual imagery.

Summary

Tactile and haptic perceptual tasks recruit visual cortical areas ranging from early processing areas such as V2 to high-level areas such as the

LOC, as well as classical multisensory areas such as those in the IPS. TMS studies and case studies of patient lesions show that such recruitment of visual cortex is actually necessary for performance of at least some tactile tasks. The cross-modal activity of visual cortex likely reflects modality-independent representations of objects and other stimuli such as motion, and may also involve visual imagery under some circumstances. Such findings increase support for the idea of a 'metamodal' brain organized around task processing rather than separate sensory streams. Further research is required, however, into the precise nature of visuo-haptic interactions and the underlying representational systems.

Acknowledgements

Current research support to KS from the National Eye Institute at the NIH, the National Science Foundation and the Veterans Administration is gratefully acknowledged.

Selected Readings

Calvert G, Spence C, Stein BE (eds) (2004) *The handbook of multisensory processes*. MIT Press, Massachusetts, Chapters 8, 44 and 45

Heller MA, Ballesteros S (eds) (2006) *Touch and blindness: psychology and neuroscience*. Lawrence Erlbaum Associates, Mahwah, NJ, Chapters 7–9

Rieser JJ, Ashmead DH, Ebner FF, Corn AL (eds) (2007) *Blindness and brain plasticity in navigation and object recognition*. Lawrence Erlbaum Associates, Mahwah, NJ

Haptic perception and synaesthesia

Jamie Ward, Michael J. Banissy and Clare N. Jonas

20

Synaesthesia involving haptic perception has been less well documented than other forms of synaesthesia. There are several possibilities why this might be. Firstly, it may well be less common than other types of synaesthesia. Day [1] reports that only 4.0% of synaesthetes report coloured touch and 0.8% report vision-to-touch, compared to 68.8% reporting coloured graphemes (note: these are percentages of synaesthetes, not percentages of general population). A second reason is that researchers don't always ask about touch and synaesthetes don't always volunteer it. We made a chance discovery of someone who experiences tactile sensations on her own body when watching someone else being touched as a result of an email request about other forms of synaesthesia. We have since found that other synaesthetes have it too but they didn't report it until prompted because they considered it 'normal' (i.e., they assumed everyone else had it). Preliminary findings suggest that this could be just as common as grapheme-colour synaesthesia, once considered to be the most prevalent type of synaesthesia [2]. This chapter will review research on these types of synaesthesia and consider their neural and cognitive basis. We will draw particular attention to similarities between synaesthetic perception and multi-sensory perception involving vision and touch.

Synaesthetic perceptions have three defining features. They are *conscious* percept-like experiences that are *involuntary* and are *elicited* by a stimulus that is not normally associated with this experience [3]. The synaesthetic percept co-exists with the percept of the inducing stimulus rather than over-riding it. Thus, a tactile stimulus may be perceived as a tactile sensation plus a colour sensation. As such, some have regarded synaesthesia as a special instance of the 'binding problem' in which two stimulus features (one veridical and one illusory) are combined into a unitary experience [4]. The stimulus that elicits the experience has been termed the 'inducer' and the experience itself has been termed the 'concurrent'. Here we use the terminology of referring to different types of synaesthesia in terms of inducer-concurrent pairs separated with a hyphen. Thus, touch-colour synaesthesia refers to tactile inducers eliciting a concurrent experience of colour, and vision-touch synaesthesia refers to a visual inducer eliciting a tactile experience. We have also colloquially referred to the latter as 'mirror touch' synaesthesia given its similarity to mirror systems for action.

Causes and explanations of synaesthesia

Synaesthesia exists in developmental and acquired forms. The developmental form is reported to exist throughout the lifespan and is believed to have a genetic component [2, 5]. It was once considered to be more common in women than men [2], although recent studies cast doubt on this [6]. Different types of synaesthesia co-exist within families and individual synaesthetes often have multiple types of synaesthesia (i.e., multiple inducer-concurrent pairings). Synaesthetic experiences can be temporarily acquired after ingestion of certain hallucinogenic substances [7]. Synaesthesia may also be acquired after sensory deafferentation. For example, sound-vision synaesthesia has been noted to occur in some cases of acquired blindness [8] presumably as a result of cross-modal plasticity that acts as a compensatory mechanism. Loss of haptic perception, following paralysis or amputation, can also be associated with synaesthesia as in the case of visually-induced phantom limb sensations. In this review we will consider both acquired and developmental forms of synaesthesia.

Developmental forms of synaesthesia are typically explained in terms of aberrant connectivity resulting from differences in gene expression [9, 10]. One recent study, using diffusion tensor imaging (DTI), reported greater coherence of white matter tracts in a number of brain regions in grapheme-colour synaesthetes [11]. While some researchers have emphasised that the aberrant connectivity might be relatively localised, restricted to adjacent cortical regions [10], others have postulated a role of long-range connectivity, for example between primary auditory and visual cortices [9]. While there is evidence of long-range multi-sensory connections in the primate brain [12], this does not adequately account for the most common types of synaesthesia. Synaesthesia often involves inducers that are not strictly 'sensory' (i.e., words, numbers, etc.). The adjacency assumption offers some explanation for this. For example, the presence of both grapheme recognition and colour perception mechanisms in left fusiform cortex could be explained by increased localised connectivity between these regions [10].

Ward [13] offers an account of both acquired and developmental forms of synaesthesia in terms of adaptations to normal mechanisms of multi-sensory perception. All types of synaesthesia may be associated with activation of multi-sensory processes *via* a unimodal stimulus (e.g., involved in spatial and temporal binding), but the nature of the inducer-concurrent pairings will depend on the cause of the synaesthesia. Developmental synaesthesia will tend to result in types of synaesthesia in which the inducer and concurrent are processed nearby [10] but acquired synaesthesia reflects neural adjustments (e.g., removal of inhibition, increased synaptic density) within pre-existing multi-sensory circuits. Behavioural evidence for shared mechanisms between synaesthetic perception and normal multi-sensory perception comes from the fact the mapping between inducer and concurrent is non-arbitrary. For example, synaesthetes tend to perceive higher pitched notes as lighter and smaller [14] and non-synaesthetes show evidence of the same tendency when asked to discriminate pitch and ignore surface lightness, or *vice versa* [15]. Comparable evidence in the domain of vision and touch is considered later.

Vision-touch synaesthesia

In this particular type of synaesthesia, the inducer is 'observed touch' rather than vision *per se*. More specifically, in many cases the inducer should more properly be described as 'observed bodily touch'. As such, this type of synaesthesia bears resemblance to the non-synaesthetic literature demonstrating that non-informative observation of body parts can affect haptic perception [16, 17].

Acquired synaesthesia in phantom limbs

Amputation or paralysis of a limb is frequently accompanied by tactile, painful or motoric sensations in the location of the missing limb – a so-called phantom limb. This may be the result of intra-modal plasticity such that stimulation of a region of the cortex surrounding the neural representation of the missing limb (e.g., the somatosensory face area) comes to activate the adjacent neural representation of the missing limb [18]. This dual-percept of tactile stimulation of a veridical and illusory body part resembles synaesthesia, but may not be a true case of synaesthesia given that the inducer and concurrent are of the same kind. However, it is also possible to induce phantom tactile sensations *via* the visual sense and this constitutes a true type of synaesthesia.

Ramachandran and Rogers-Ramachandran [19] used mirrors to create a duplicate image of the amputated hand/arm based on a reflection of the patient's existing hand/arm. Movement or observed touch to the intact hand produced a parallel sensation in the visualised phantom. This effect may require little or no plastic reorganisation given that viewing the normal limb through prisms can distort its felt orientation [20].

Similar effects have been reported following stroke. Case DN had loss of sensation in the left side of his body and was unable to feel any kind of tactile stimulation to the left side of his body if it was hidden from view [21]. However, as soon as the tactile stimulation became visible, DN was able to feel it. Moreover, if DN believed his left arm was being touched (by watching a previously recorded video of his arm being touched but being told that it was a real time live feedback), he reported being able to feel his arm being touched, even when the experimenter was not in fact touching his arm.

There is also one case of acquired 'mirror pain' in which observed pain was experienced as pain to himself [22]. The patient was known to have widespread cancer but the specific effects of tumours on the brain, if any, were not documented. Similar symptoms have now been well-documented in developmental cases of 'mirror touch'.

Developmental 'mirror touch' synaesthesia

Blakemore et al. [23] reported a single case, C., of vision-touch or 'mirror touch' synaesthesia. C reports that she experiences tactile sensations on her own body whenever she observes another person being touched. She does not report tactile sensations when watching inanimate objects being touched. This suggests a possible dissociation between passive and active touch given that a body-part is still implicated when an object is touched (i.e., the finger tip). An fMRI study of C and a group of 12 control participants was performed in which participants watched movie clips of a person being touched on the face or neck, or of an object being touched (e.g., the 'head' and 'neck' of a fan). A comparison of bodily *versus* object touch revealed a network of regions including primary and secondary somatosensory cortex, premotor region (including Broca's area) and the superior temporal sulcus. This may constitute a mirror system for touch analogous to that reported for action. The synaesthete showed hyper-activity in a number of regions in this network (including primary somatosensory cortex) and additional activity in the anterior insula. This suggests that this type of synaesthesia reflects hyper-activity within the mirror touch network.

Banissy and Ward [24] recently reported a behavioural study of 10 synaesthetes, including C. Additional synaesthetes (N = 8) were recruited from within our existing sample of synaesthetic volunteers, and other cases (N = 4) have subsequently been recruited from the undergraduate population (following distribution of over 300 questionnaires). Although all reported similar experiences to C, there were some important differences. C reports that observed touch to a left cheek results in a tactile percept on her right cheek (as if looking in a mirror; specular correspondence). However, other synaesthetes report an anatomical correspondence such that observed touch to a left cheek produces felt touch on the left cheek. A small number also report tactile experiences when objects are touched, although the extent to which the object must resemble (or imply) a body part is unknown. Indeed, a different fMRI study has noted somatosensory activation in response to observed touch to objects (rolls of paper) that resembled legs [25]. Banissy and Ward [24] applied actual tactile sensations to the synaesthetes while they simultaneously observed touch to another person on a computer monitor. The study is summarised in Figure 1. Their task was to report the location of the real touch (left, right, both, none) ignoring the observed touch and any synaesthetic touch triggered by it. The observed and applied touch could either be spatially congruent or incongruent with each other. Synaesthetes, but not control participants, were faster in the congruent relative to the incongruent condition. Moreover, the errors of synaesthetes implied that synaesthetic touch was often mistaken for real touch.

Research currently in progress suggests that the phenomenon is specific to observed bodily

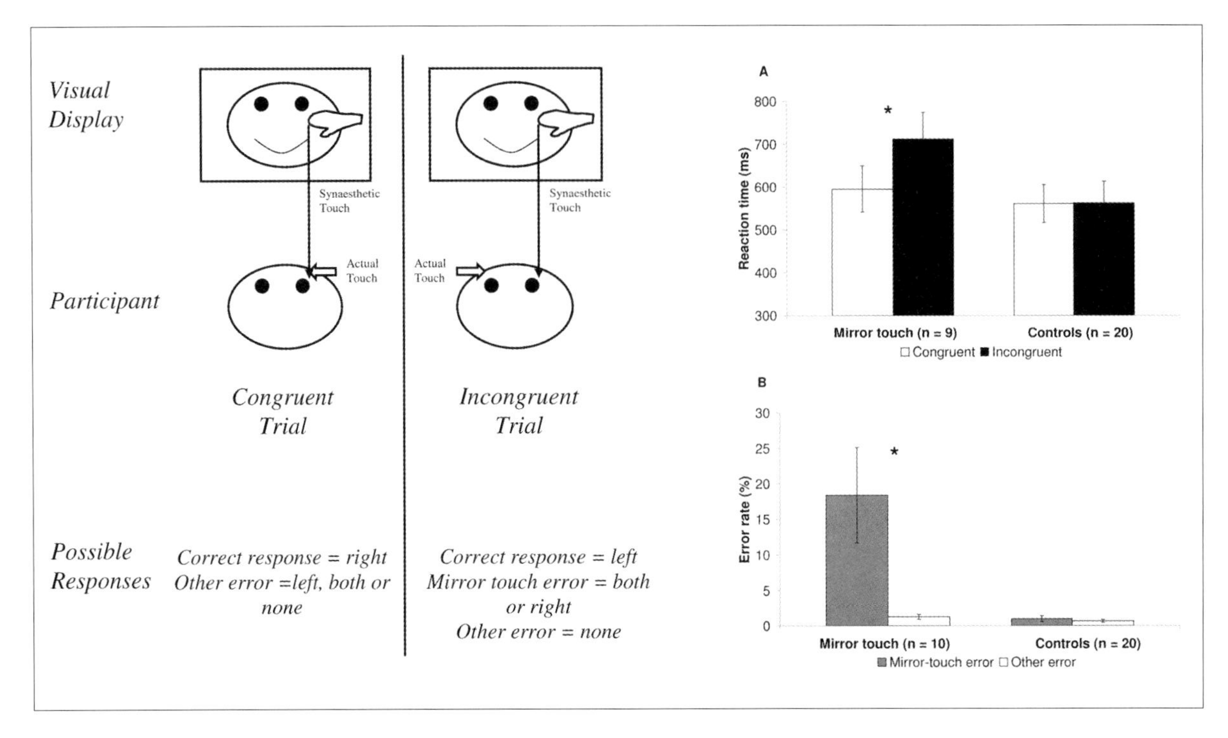

FIGURE 1.
Summary of Banissy and Ward [24]. Schematic example of congruent and incongruent trials for a vision-touch synaesthete reporting specular correspondence (note that in the actual experiment, photographs of people being touched were used). Mean reaction times (A) and percentage error rates (B) for vision-touch synaesthetes and non-synaesthetic controls when detecting a site touched on one's own face while observing touch being applied to another person's face (= p<.05).*

touch. Observed touch to objects does not produce the same pattern of results [24]. Moreover, spatial cueing to an observed body part (a flash of light) does not produce these results.

Future research will determine whether bodily touch really needs to be observed or can be merely implied (e.g., if the touch is heard rather than seen, or if the point of contact is hidden).

Given that mirror-touch synaesthesia is regarded as an exaggerated form of normal observed bodily touch, it can be used as a tool to investigate vision-touch interactions more generally. It can also be used to explore the role of this perceptual network in other aspects of cognition. Banissy and Ward [24] found that mirror touch synaesthetes scored higher on an 'emotional reactivity' measure of empathy but not a cognitive measure of empathy (thinking about others' feelings). This result was maintained when they were compared to a group of grapheme-colour synaesthetes who don't report mirror touch, suggesting that it is specific to this particular profile rather than a general feature of synaesthesia.

Touch-vision synaesthesia

In this type of synaesthesia, haptic perception results in visualised photisms which are sometimes coloured.

Acquired cases

The only known case of acquired touch-vision synaesthesia was documented by Armel and Ramachandran [26]. Tapping and other haptic stimuli induced synaesthetic visual sensations that were projected on to the spatial location of the body part (e.g., hands) irrespective of where the body part is located in space (e.g., if the left hand was in the right side of space then the photism would appear on the right). They measured the intensity of tactile stimulation required to induce visual photisms and noted that they were easier to induce (i.e., lower threshold) when the hand was in front of the person relative to behind the head. The threshold for eliciting photisms tended to be higher than the threshold for eliciting a tactile percept, and the thresholds were reliable over time.

Although the participant was blind, synaesthetic vision may be preferentially elicited when the inducer is also 'in view'. This suggests a body-based spatial reference frame that incorporates information about gaze and/or head orientation. Indeed, normal cross-modal influences of touch upon vision appear to operate on a spatial reference frame that encodes the position of the hand in external space having taken into account current gaze direction [27].

Developmental cases

In the historical literature, there are a number of brief descriptions of touch-vision synaesthesia. Smith [28] notes a synaesthete for whom: "*Hard objects when touched are dark in color; soft objects are light. If she feels something hot a dark gray is seen, if a piece of ice is grasped, white is promptly visible. A dull pain is a dark lead color, a sharp pain is of a light steel color*" (page 260). Dudycha and Dudycha [29] note a case of pain-colour, although tactile percepts are not noted to induce colour and Whipple [30] notes a case of touch-sound synaesthesia.

There are a couple of cases of blind synaesthetes for whom Braille reading triggers colour. Thomas Cutsforth notes how coloured photisms appear under his finger tips as he reads [31], and Steven and Blakemore [32] briefly document a more recent case, JF. JF experiences an array of coloured dots, "*like an LED display*", both when he touches Braille and when he thinks about Braille characters. The spatial configuration of dots, rather than specific meaning, dictates the colour. The letter I, number 9, and musical note A-quaver all have the same geometric arrangement and have the same colour. Touching other textures or objects does not evoke colour. Both cases are likely to have possessed synaesthesia prior to becoming blind. Their blindness results in the fact that graphemes are processed haptically rather than visually. If we were to all read with our fingers then touch-vision synaesthesia would probably be far more common, although it would effectively be a different outward manifestation of grapheme-colour synaesthesia.

As part of ongoing research, we have recently studied two cases (TV and EB) who report coloured visual sensations to touch. Both are normally sighted, and report a wide variety of other types of synaesthesia. Interestingly, the cases report different spatial phenomenology with regards to the location of touch. TV reports that colours are projected on to the location of the touched body part (as in Thomas Cutsforth and the case reported by Armel and Ramachandran), whereas EB reports coloured photisms internally in her mind's eye. A comparable distinction exists for other types of synaesthesia such as grapheme-colour [33]. In order to determine the factors that influence their synaesthesia a battery of 40 different tactile stimuli was constructed and administered on two separate occasions. The stimuli were varied along four different dimensions: temperature, pressure, roughness-smoothness (active touch of sandpaper), and flutter (vibration frequency of a solenoid tapper). Each set of stimuli were applied to two different locations on the hand (e.g., different fingers, or palm *versus* finger tip) that have previously been shown to differ in sensitivity. The synaesthetes were asked to choose their synaesthetic colour from a Munsell palette

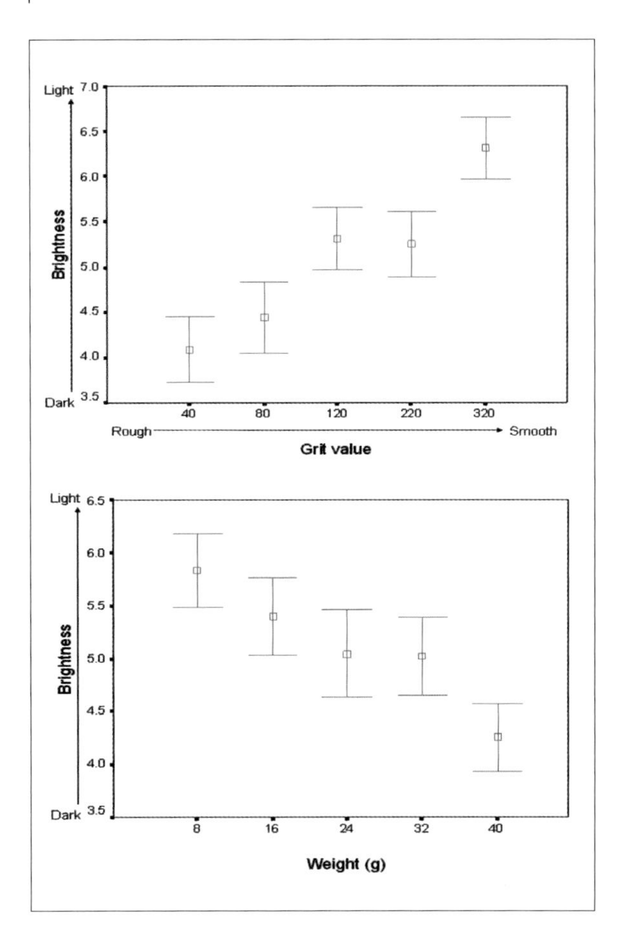

FIGURE 2.
The relationship between roughness-smoothness and brightness (top) and weight and brightness (bottom) for twelve non-synaesthetes asked to choose colours for tactile stimuli. Boxes represent mean brightness and error bars ± one standard error of the mean.

and a group of twelve controls were asked to choose the 'best' colour to match the sensation. All tactile stimuli were applied out of view.

The inclusion of the control group enabled us to consider whether there are 'universal' multi-sensory rules that dictate how visual and haptic dimensions tend to be aligned, such as those noted between vision and audition [15]. Martino and Marks [34] report a relationship between vibrotactile frequency and luminance such that lower frequency vibrations are darker than higher frequency ones. Participants are faster at making decisions about luminance or frequency when the dimensions of a multi-sensory stimulus match relative to when they don't match. In order to explore similar trends, the Munsell colour dimensions of chroma (from grey scale to saturated), luminance (dark to light) and hue (circularly varying from red to yellow, green and purple) served as dependent variables and were contrasted with the four tactile dimensions explored here. The only significant effects were between roughness and luminance (rougher textures are darker) and pressure and luminance (higher pressure is darker). This is illustrated in Figure 2. The site of stimulation (and hence tactile sensitivity) did not modulate the selected colours, although it is possible that such effects could be obtained at sites other than the hands. Similar effects were noted in the two synaesthetic participants. Further research is needed to explore the automaticity of the phenomenon using variants of the synaesthetic Stroop paradigm [35].

Summary

Although synaesthesia involving haptic perception has not been extensively documented, recent research suggests that it may be far more common than once believed. There is evidence from developmental synaesthesia that vision-touch (or mirror touch) synaesthesia depends upon over-activity of a more general mirror system for touch, present in non-synaesthetes too. Similarly, there is evidence that developmental touch-vision synaesthesia may be dependant on some general cross-modal associations (e.g., between felt roughness and surface lightness). These types of synaesthesia can be used to explore the nature of multi-sensory processes involving touch. For example, some aspects of empathy may depend on the mirror-touch system and synaesthesia may reveal different spatial processes that mediate between the senses (e.g., anatomical *versus* specular representations).

Acquired forms of synaesthesia are assumed to reflect cross-modal plasticity within pre-existing multi-sensory processes and therefore also provide important evidence concerning the multi-sensory nature of haptic perception.

Selected readings

Banissy M, Ward J (2007) Mirror-touch synaesthesia is linked to empathy. *Nature Neuroscience* 10: 815–816

Ramachandran VS, Blakeslee S (1998) *Phantoms in the brain*. William Morrow, New York

Ward J (2008) *The frog who croaked blue: Synesthesia and the mixing of the senses*. Routledge, Oxford

Websites

Author's website (www.syn.sussex.ac.uk)
UK Synaesthesia Association (www.uksynaesthesia.com)
American Synaesthesia Association (www.synesthesia.info)

Haptic perception in sexuality

Kurt Seikowski and Sabine Gollek

21

Haptic perception and the functions of the skin

The connections between haptic perception and sexuality arise initially from the actual realisation organ of this form of perception – the skin. The skin itself, however, fulfils various functions, which are related more or less to differing aspects of haptic perception. This is illustrated by two different approaches to skin function which have a number of similarities (Tab. 1).

With respect to our query, the following skin functions appear to be important for the connection between haptic perception and sexuality: The skin, as an organ of *borders* or *contact* with the environment, or as *protection against stimuli*, protects the body against environmental influences such as temperature and humidity fluctuations. In sexual contacts, a partner's cold or sweaty hand on one's own skin is usually experienced as reducing lust and evoking rejection. Physical contact which is warm and not unpleasantly damp opens the body for sexual stimulation.

The terms describing the skin as a *sensory organ* and the skin-self as *intersensoriality* or as the *basis for sexual excitement* cover the perception of tactile sensations like cold and warmth, burning, itching, tickling, prickling – that is, all the qualities directly connected with sexual sensations and which are mediated by *touch*. Sexual contacts can in a sense 'get under one's skin', whereby the erogenous zones play a special role.

Anzieu [1] discusses at this point two further functions which are relevant with respect to impaired sexual development. Under the skin-self as a *system of tactile sensory traces*, he sees the totality of pleasant and unpleasant skin experiences, which represent a sort of information system for the outside world. Touching which was experienced as inadequate was stored and actualised in later forms of physical contacts.

Table 1. Functions of the skin

On the analogy Borelli [35] (Seikowski and Haustein [36])	According to Anzieu [1]
1. Skin as a limit organ between environment and one's own	1. Skin-self as a support (cohesions) of the psyche
2. Skin as a contact organ to the environment	2. Skin-self as a protection against stimuli
3. Skin as a sense organ	3. Skin-self as a container of the external sense organs
4. Skin as an impression organ for the viewer	4. Skin-self as individuation
5. Skin as an expression organ for the presentation with regard to the environment	5. Skin-self as intersensoriality (sense of touch)
6. Skin as sexual jewellery	6. Skin-self as a basis of sexual excitement
7. Skin as social protection	7. Skin-self as a libidinous charging
	8. Skin-self as a system of haptic sensory traces
	9. Skin-self as a self-destruction mechanism

Thus, for example, it can even come to 'skin rejection' of a person whom one believes one loves. Due to unpleasant prior experience, the skin remains 'distrustful'. According to Anzieu [1] this prior experience, which was experienced as traumatising, may even lead to expression of physical contact conflicts on the skin: the skin-self as a *self-destructive mechanism* in the sense of self-mutilations of the skin. More will be said of this phenomenon later.

Touching as a connector between haptic perception and sexuality

At the latest since the experiments of Harlow et al. [2, 3] on Rhesus monkeys which showed how much more important warmth and touching by a fake mother was than a wire construction with food for the development of the young, there is no longer any doubt that touching with all of its pleasant (warmth, gentleness) and unpleasant (violence, pain, cold) properties influences the haptic perception of a mammal [4]. This recognition has also become established in everyday speech [5]: one has the 'right touch' or the 'wrong touch'; one get 'in touch' or 'in contact' with others; many people have to be 'handled with velvet gloves', others are 'thin-skinned' or 'thick-skinned'. In certain situations, we want to know how something 'feels', deep emotional experiences 'touch' us, etc.

Sexual contacts without touching are hard to imagine. Touching appears to be an essential mediator factor for functioning sexuality. But this does not mean only the purely physical contact. Three different qualities are connected with touching, if sexuality is considered. Touches can indicate tenderness. Here, gentle caresses can be experienced either as soothing by the other or progress to sensual caressing which are (generally) experienced as lustful or elicit a 'prickling' between two people. Likewise, caresses can have an erotic character – bound with the desire to act out sexual lust [6]. For this reason, the phenomenon of 'touching' appears to us to be most suitable to illustrate the relationship between haptic perception and sexuality.

Touching and sexual development

It was observed in an experiment by Harlow et al. [2] that motherless dams (monkey mothers that had grown up without their own mothers) never adopted the normal sexual-enticing positions or reacted to them. This gives rise to the assumption of a touching deficit which may lead to later sexual impairment in these female monkeys. Montagu [5] is of the opinion that appropriate 'mothering' is necessary to promote the development of a normal sex life. He transfers these relationships to people as well, and postulates that sufficient physical contact in childhood is essential for the development of normal sexuality.

In this respect, it appears helpful to rate touching differentially under the aspect of development.

Touching as a form of affection and acceptance

This form of touching already plays a role during pregnancy, in that the expectant mother and other reference persons stroke the abdomen, to which the child itself can already react. Recent studies have even shown that this form of acceptance can lead to erections in male foetuses as early as the 26th week of pregnancy [7].

During normal childhood development, ones own child is caressed and embraced. There is also more or less lustful physical contact in caring for the child. Montagu [5] assumes that later more pronounced sexual desire among men is partly because of an advantage over girls that the sex organ is outside the body. This means that in hygienic measures for a male infant would result in considerably more cutaneous stimulations than would be the case for girls.

Touching as affection and acceptance has not only a care-giving function (e.g., warmth and tenderness as a necessary prerequisite to general physical development), but is also an expression that the child is *loved*. It experiences that it is *lovable* [8].

There are a large number of case reports which address an attention deficit and rejection of the child by the parental reference person. Schmidt-Seibeth [9] found how experience of deficits, cleanliness fanaticism, antisexualism, conflict solutions using one's own children and physical avoidance behaviour in childhood can lead later to sexual impairments and rejection of one's own body, as well as to the inability of allowing touching by others. She draws an inverse conclusion in postulating that happy parents raise children capable of loving. Worm [10] also reports on the problem adults have in being embraced when they experienced the embrace of their own mother as unpleasant.

Touching as a threat

Beatings, pain and experienced violence are a negative variant of touching. Such a contact form can lead to considerable impairment in development, out of which psychosomatic illnesses may arise (for more details, see Egle et al. [11]). Touching is often experienced as a threat. Punishment by touching is a skin experience through which later touching is perceived with distrust and not with openness (Skin-Ich as a system of tactile sensory traces). This does not, however, rule out the longing for tender caressing. Here, too, there are many case reports in which especially women with childhood touching deficits enter into sexual contacts with men, although they actually are only seeking the warming, secure body contact. Sexual intercourse is, in a sense, the price of making up for lost physical affection [5, 11]. Experience has shown that these relationships do not function for any length of time, since the two sides have differing motives for physical contact and conflicts are preprogrammed.

It must also be mentioned that touching can also be experienced as a threat when religious influences – e.g., during puberty – are important. Incongruence arises between rearing with often limiting proclamations and possible lustful touching experience, but the result may be rejection of one's own body, which is not without consequences for the later acceptance of one's own sexuality.

Touching as an inadequate form of physical contact

Experiences of physical violence make touching appear as a threat. From the child's point of view, they are unpleasant, but not infrequently understandable. More difficult to classify are contacts which are generally termed sexual abuse in prepubescent children. Especially when such touching is undertaken by an accepted reference person, the consequences are often disastrous, as Ferenczi could already impressively demonstrate in 1933 [12]. The beloved reference person is idealised. The child feels that there is 'something wrong' with this form of touching, but he usually blames himself. Ferenczi discussed in this connection the defence mechanism of repetition compulsion: In order to understand what really happened, the child provokes a similar situation again to re-establish his inner equilibrium. But this is unsuccessful due to his own prepubescent development. And more – prepubescent amnesia not infrequently develops [13]. The abuse is forgotten (suppressed). As an adult, however, sexual impairments arise with origins no longer accessible to the consciousness. The entire range of these consequences could also recently be empirically supported [11, 14–16]. However, it must be mentioned that sexually inadequate contacts in childhood do not lead exclusively to negative consequences. It was found in non-clinical groups that the proportion of negative developmental consequences is considerably lower and that children apparently have a variety of coping mechanisms which protect against later sexual disorders [14, 17, 18]. In a more recent retrospective study, it was shown that unwanted sexual contacts among 16-year old heterosexual boys were experienced as negative when the contacts were made by men [19].

Impeded touching

Early skin diseases may represent an impediment for physical contact. Neurodermitis often arises in early childhood. Touching this diseased skin is rather unpleasant for the reference person. Later sexual disorders are often preprogrammed. For example, Niemeier et al. [20] found that sexuality among adult patients with neurodermitis and psoriasis is impaired. They are caressed less often, women have fewer orgasms. However, there was no difference in coitus frequency compared to a control group with healthy skin. Similar results were found for the relationship between burns and sexuality [21].

The problem of *self-mutilating acts* is a vicious circle for the patient's own sexuality, because two aspects are usually mixed in these disorders (Skin-Ich as self-destructive mechanism). Thus, self-injuries to the skin are not infrequently the consequence of inadequate touching forms (sexual abuse) and an expression of inner tension, which can be interpreted as guilt feelings [4, 22].

Self-injury is a form of autoaggression. Skin which is thus disfigured (often on the arms, legs or face) reduces the physical attractiveness for sexual contacts. These persons are in a tense area between unresolved sexual touching and the desire for touching in the form of warmth and security. Uncertainties with respect to one's own need for physical contact are the result.

Touching in the sexual behaviour of adults

Even though sexual scientists consider very critically the influence of the modern consumption-oriented society on sexuality and are of the opinion that there is an increasing lack of time available for touching, the modern digital computer technique limits the communicative character of sexuality, sexual disorders continue to increase [23, 24] and a development from sexuality in partnership toward more sexuality without partnership is possible [25], the essential basic elements remain in most sexual contacts. Touching in sexual contact is still desired and experienced as pleasant by both women and men. If touching is initially a form of physical acceptance (tenderness and sensuality), it increases sexual tension on stimulation of erogenous zones (eroticism), and touching the other (or oneself) is the mediating factor even in reaching an orgasm.

It could be demonstrated in various studies that the role of touching in sexual contacts does not decrease with increasing age, but rather increases [26, 27]. Touching thus has an essential place in the sexual contacts of adults and is important for health and wellbeing. However, this is not yet reflected in the WHO definition of sexual health [28], but it should possibly be taken into account in future definitions. An opposite trend associated with the performance ideals of the consumer society appears to be spreading and affects especially the men. On the one hand, there is an increasing medicalisation of male sexuality (for example Viagra) [29]. On the other hand, tumescent body autoinjection therapy (SKAT), vacuum pumps, penis prostheses and recently Viagra are supposed to restore the sexuality of 'imperfect masculinity'. Male sexuality is reduced to the erect penis – sexuality without touching. Or can touching only be experienced as lustful when the man has an erection? Urologists and psychologists have come into vociferous conflict [6].

Meanwhile, the new media have created a new problem for us: Men are increasingly consuming sex *via* the Internet. They masturbate so often thereby that there is a considerable loss of interest in their own partners. Real sexual contacts with the partner don't happen, so impeded touching is the consequence [30].

Touching in sexual therapy

This discussion has, however, led to a differentiated view of sexual problems and disorders. One-sided somatisation or psychologisation is definitely inappropriate. But there is no doubt that touching is important in sexuality. Even in

Table 2. Example of a couple-related sexual therapy at loss of libido of the woman

Therapy step	Aim
1. Focusing I – touching of the complete body without erogenous zones at sexual intercourse ban	- Experiencing of touches as pleasant without pressure to do well, e.g., of warmth and security
2. Focusing II – touching the complete body with erogenous zones at sexual intercourse ban	- Experiencing of touches as sexual tension without pressure to do well
3. Touching the complete body including sexual stimulation	- Experiencing of touching as a direct sexual stimulation
4. Finding out of sexual fantasies	- Experiencing of additional aspects for the increase of lust
5. Slow, but no imperious introducing	- Only now the attempt of a vaginal stimulation is carried out
6. Observation of the emotional interaction of the couple before sexual contacts (e.g., no forcing sexual contacts at tiredness etc.)	- Understanding of the sexuality as a holistic phenomenon

erection disorders with a more organic basis, non-vaginal sexuality would be possible, since the capacity for orgasm often remains intact and not every man wants to take medications for years to achieve an erection. Such non-vaginal sexuality, however, lives from touching. The highest degree of touching, by the way, is found in Asian forms of love-making – such as Tantra sex [31].

Touching has played a large role in the treatment of sexual functional disorders of men and women since the 1960s. Masters and Johnson [32] developed various behaviour-therapeutic programs, which have in common the promoting of mutual touching in the first two therapy steps [33]. Table 2 shows an example for such a therapy form in the treatment of libido deficiency in a woman with a partner and who suffers herself from the problem. The exercises are used as homework. In these, touching plays a particular role in the first two therapy steps (Focusing I and Focusing II). The pressure to perform is removed by initially forbidding sexual intercourse, which leads in the second step to possibly becoming open for sexual sensations.

Of course, such methods do not work in every case. Often, it is necessary in the diagnostic phase to differentiate why the sexual disorder has arisen, why for example, one's own body is taboo [34], what touching traumatisations played a role. These methods, which are given to the patient as a kind of homework and are based on various contents of touching, enable the patient to experience body contact as pleasant in a sexuality-related, performance-free atmosphere (assuming that the partner performs the 'exercises', too) and to learn to accept sexuality as positive for his/her own wellbeing.

Summary

Touching is the central connector between haptic perception and sexuality. It establishes contact to another or to one's one person *via* the skin. In this, the skin fulfils various functions which are decisive to the way in which touching is experienced. How various forms of touching in childhood were experienced, processed and remembered is of decisive importance for sexual experience as an adult. Touching as a form of affection and acceptance give the child the feeling of being loved and being worthy of love. Touching as a danger, for example in physical abuse, causes considerable detriment to the instinctive trust in physical contact with other people. Touching can also be experienced as an inadequate form of physical contact (sexual abuse). A lack of touching in childhood leads later to misunderstandings with respect to feelings of tenderness and sexual excitement as an adult. Touching can also lead to self-destructive acts against one's own person, from which sexual limitations may later arise.

An example taken from behaviour-therapeutic oriented sexual therapy is used to demonstrate how sexual disorders can be overcome with the aid of positive body sensations mediated by touching.

Selected readings

Gieler U (1988) Skin and body experience. In: Brähler E (Ed.): *Body experience. The subjective dimension of psyche and soma*. Berlin, Heidelberg, New York, London, Paris, Tokyo, Hong Kong, Barcelona, Budapest: Springer, 62–73

Masters WH, Johnson VE (1966) *Human sexual response*. Livingstone: Churchill

Montagu A (1986) *Touching. The human significance of the skin*. 3rd Edition, London: Harper Perennial

Rathus SA, Nevid JS, Fichner-Rathus L (2007) *Human sexuality in a world of diversity*, 7th edition, Allyn and Bacon

Website

www.worldsexology.org

22

Haptic perception in space travel

Helen E. Ross

Introduction

Humans evolved to live on Earth, and to move around under the influence of Earth's gravitational force. The acceleration of gravity is defined as 1 g and equals 9.81 m/sec^2. It is technically known as 1 g_z, because the force acts through the z axis – through the head and feet for an upright human (Fig. 1). Humans (like other animals) can easily cope with the changed patterns of the various accelerative forces involved in running, jumping and swimming. Mechanised travel causes more difficulty, because it involves large variations in accelerative forces in one or more axes. Humans can adapt to these changes to some extent, but they usually show an initial impairment in motor skills, and may suffer from motion sickness.

Astronauts in orbital spaceflight live in a *microgravity* (low g) environment, which is close to zero gravity (0 g) because the acceleration of the spacecraft cancels out that of Earth's gravity. There is thus no constant accelerative force in any axis, and no gravitational 'up' or 'down'. The same effect can be produced for about 20–30 s during parabolic flight, but this is preceded and followed by about 20 s of *hypergravity* (high g) of up to 2 g (Fig. 2). Repeated parabolas offer an opportunity to examine perceptual and motor changes during both high g and low g, and compare them with performance under 1 g during straight and level flying. However, it is difficult to adapt to rapidly changing g levels, and only prolonged spaceflight allows for the study of long-term adaptation to microgravity.

Changes in g affect many aspects of human physiology, only some of which are relevant to haptic perception. Perceptual-motor performance is usually slower in orbital or parabolic flight than on the ground, and there may be several reasons for this: a floating or poorly restrained astronaut has difficulty executing manual tasks; microgravity may directly affect the control of limb movement; and the general stresses of spaceflight may affect cognitive and other functions. Current research suggests that perceptual-motor performance is impaired rather than cognitive performance [1]. This chapter concentrates on the direct effects of microgravity on hand control and haptic perception.

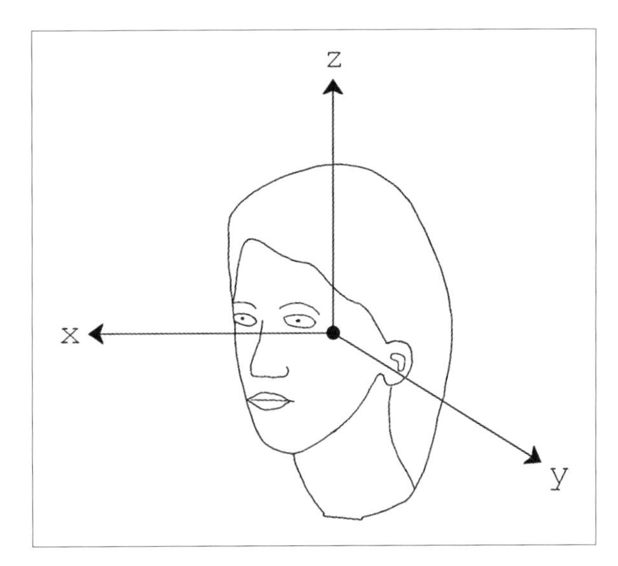

FIGURE 1.
The principal axes in relation to an upright head. Earth's gravitational force acts through the z axis.

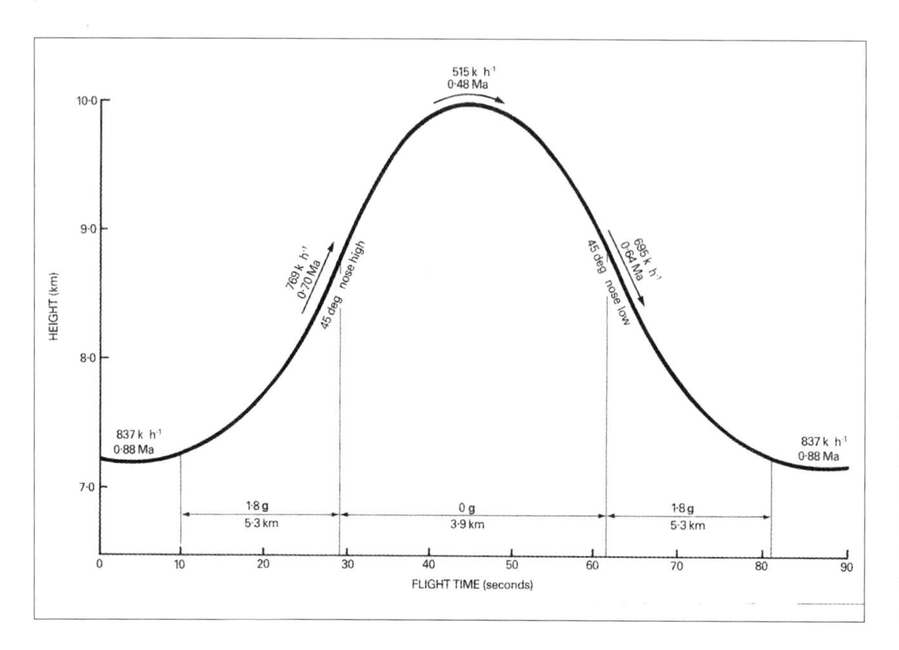

Figure 2.
Flight profile for a typical parabola in NASA's KC-135 aircraft. 20–30 s of near 0 g is preceded and followed by about 20 s of 1.8 g. (Figure courtesy of NASA)

Bodily orientation and the haptic system

There is a two-way relationship between the haptic system and bodily orientation: tactile information can assist knowledge of bodily orientation, and knowledge of limb position affects motor skills. The tactile contribution to orientation is discussed in this section.

The ability to perceive one's orientation with respect to gravity and to visual surrounds depends on several sources of sensory information. The vestibular system (otoliths and semicircular canals), the haptic system and the visual system all contribute. The changed situation in space travel is summarised by the Space Studies Board [2]. On Earth, the otoliths give information about static head orientation with respect to gravity; but in spaceflight they cannot do so. The semicircular canals respond normally to rotary acceleration under microgravity, but the relation between the canals and the otoliths is disturbed, leading to various visual and bodily illusions of tilt and motion. In the absence of reliable vestibular information, astronauts may take their sense of orientation from the visual orientation of the cabin, or from the orientation of their own body [3]. The latter strategy is less likely to cause illusions and discomfort [4].

The haptic system undoubtedly contributes to the sense of orientation on Earth [5], but microgravity reduces cutaneous pressure and also reduces the loading on the joints and muscles. Thus the haptic system contributes less to orientation under microgravity than it does under 1 g [6]. A similar haptic reduction applies to divers under water, because buoyancy reduces tactile pressure [7]. Nevertheless, tactile information can make some contribution to bodily orientation under microgravity. Visually induced illusions of bodily rotation are normally enhanced in spaceflight through lack of tactile contact, but the illusions can be reduced if pressure is put on the soles of the feet by means of bungee straps [8]. Pressure on the chest or back also assists knowledge of bodily orientation during the 0 g phase of parabolic flight [9]. These findings do not demonstrate a role for the hands, but it is likely that tactile hand pressure can assist veridical perception in spaceflight. Certainly, divers under water can reduce alternobaric vertigo (a rotary illusion) by clinging to a rock or other stable structure.

Knowledge of arm position

Knowledge of arm position and arm movements is degraded under microgravity [10, 11], perhaps because there is a decrease in muscle spindle sensitivity. Degradation can show both as a greater variable error, and as a specific directional error. It could be argued that 0 g should make the arm feel lower than it is, so that an astronaut would raise his arm too high if asked to produce a certain angle. An error of this sort was noted by Veringa [12], who found that arm position was raised too high in space. Ross and Farkin [13] found a similar effect in the 0 g phase of parabolic flight, when subjects attempted to extend their arm at an angle of 45 or 90° deg in relation to their trunk. The opposite effect might be predicted for high g, but the authors found no such effect in parabolic flight. This may be because the level of 1.8 g (or sometimes considerably less) is not large physiologically; whereas the change from 1 g to 0 g is dramatic. Similarly Bock [14] found no effect on arm positioning ability for a simulated 2.24 g load, while simulated microgravity (i.e., the buoyancy of water) produced an upward error and greater variability. He argued that high g had no effect because the motor system compensates efficiently for external weight loads.

Several authors [15–17] have used a different task – pointing at the memorised location of targets in the absence of vision. This task involves both knowledge of limb position and knowledge of target location. Large errors were shown by astronauts in space, compared to ground performance. Watt [17] found that the errors in space were largest when the astronauts continuously kept their eyes shut; he argued that the main reason for pointing errors was lack of knowledge of target position rather than that of limb position.

Motor skills

Knowledge of arm position is particularly important as it affects the ability to reach for objects and perform skilled tasks. Eye-hand coordination can go wrong if there is a mistake concerning either the visual location of the target or the sensorimotor system of the arm and hand. Visual location errors have been shown to occur during changes in g, as in the oculogravic (changing g_x) and oculoagravic (changing g_z) illusions [18], or during head tilt under constant high or low g. However, visual errors are normally absent under prolonged microgravity when the head is kept upright, so reaching errors must be due to errors within some part of the sensorimotor system. The arm is normally lifted up against the 1 g force of Earth's gravity, and lowered down in line with it. If a raised arm feels to be lower than it actually is under 0 g, ballistic (rapid and unchecked) movements should overshoot their targets in an upwards direction, and undershoot (remain too high) in a downwards direction. The reverse could be predicted under high g.

Aiming movements

The speed and accuracy of aimed arm movements is usually found to deteriorate under altered g [15, 18–20], but the direction of the errors is not always as predicted above. For example, Ross [21] found some inconsistent effects in different ballistic aiming experiments in parabolic flight. Subjects were strapped in their seats and made pencil marks on a sheet attached to the back of the seat in front of them. Five subjects showed the predicted effects when aiming blind at previously viewed targets: movement in the upwards direction produced overshooting under 0 g and undershooting under 1.8 g, and movement downwards produced the opposite errors. However, eight subjects showed less clear effects when performing a paced reciprocal tapping task, in which they tapped up and down at high speed while aiming with vision at horizontal lines. The results were generally similar to the former blind ballistic experiment, but there was no constant error when aiming downwards under high g. Variability was highest under 0 g then 1.8 g then 1 g.

The clearest effects were obtained with novice subjects. Learning was very rapid, even during the varied g of parabolic flight. Feedback from the sight of the errors may have encouraged learning.

These results seem fairly straightforward, but not all experimenters have found the same effects. For example, Whiteside [18] found undershooting for two subjects who aimed ballistically straight ahead under 0 g in parabolic flight; and overshooting for one subject under 2 g in the human centrifuge. He explained these unexpected results by reflex eye movements to altered g levels causing displaced visual locations. However, Bock et al. [19] tested eight subjects in parabolic flight, and required them to point straight ahead with closed eyes at the remembered locations of targets viewed only under 1 g, so that any eye movements induced by changes in g were irrelevant. They found that subjects pointed higher under high g than 1 g, and higher still under low g. The explanation of the overshooting under high g remains controversial. Some possible reasons for the conflicting findings of different experimenters and test procedures are discussed by Bock [22].

Tracking, grasping and complex movements

Skilled tasks are more complex than simple aimed movements. They often involve tracking a moving target, or moving an object over a curved path. They may entail prehension (grasping), in which the subject moves the hand towards the object and arranges the thumb and fingers to fit the object's shape and size. Subjects can learn the appropriate skills for a novel force environment, though different components of the skills adapt at different rates. Sangals et al. [23] tested a cosmonaut on a step-tracking task over a three week space mission, and found only a slight reduction in accuracy during and after the flight; there were, however, changes in the movement patterns in-flight and a general slowing down of movements post-flight. Bock et al. [20] tested six astronauts on a tracking task over a 14 day space flight: they found that the speed of tracking was unaffected under microgravity, but the accuracy deteriorated. Flanagan et al. [24] found that subjects could learn to move objects with novel dynamic properties along a prescribed path, but they were quicker to adjust their grip force than they were to adjust their hand trajectory. The authors argued that subjects could learn to predict the consequences of their actions (as measured by grip force) before they could learn to control their movements. Hermsdörfer et al. [25] found a very close coupling between grip force and load force under varying g levels in parabolic flight. They had subjects move objects vertically and horizontally during the normal, hypergravity and microgravity phases. The subjects gave similar vertical movements under normal gravity and hypergravity, with both load force and grip force at maximum at the lower turning point and at minimum at the upper turning point (Fig. 3). That is to be expected because the weight (gravitational load) of the object is added to that of the mass (inertial load) under both normal gravity and hypergravity, and the sum of the loads varies with the direction of movement. Under microgravity, however, there is no weight (no gravitational load), but the inertial load is the same as under normal gravity; there is thus no difference in load for different directions of vertical movement. Under microgravity the subjects gave a maximum load force and grip force at both the lower and upper turning points (Fig. 3). During horizontal movements in the y axis, the weight vector is orthogonal to the inertial load, so there should be no change in the sum of loads with left or right movement direction. As expected, the subjects' horizontal movements under microgravity were similar to their vertical movements. The high correlation between grip force and load force in all conditions suggests an automatic adjustment to changes in g level. However, the adjustment was imperfect because the relation between grip force and load force varied with the g level; the grip force was higher under 2 g than 1 g, whereas under 0 g the grip force increased more slowly with the load force.

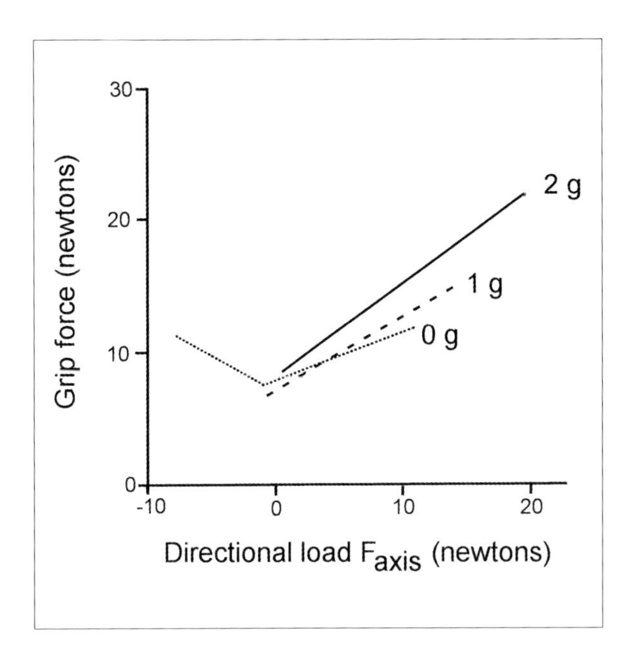

FIGURE 3.

Relation between grip force and load force for vertical hand movements while gripping an object under 2 g, 1 g and 0 g in parabolic flight. Under 2 g and 1 g both grip and load forces are at minimum at the upper turning point and at maximum at the lower turning point. Under 0 g they are at maximum at both the upper and lower turning points, giving two branches to the function. The figure shows the mean results of two subjects. (Redrawn from two figures in Hermsdörfer et al. [25])

Heaviness

An interesting question is the apparent heaviness of one's own body, and other objects, under high or low g levels. Weight is a force, and it equals mass times acceleration. There is normally a constant background acceleration of 1 g_z, in addition to any other acceleration applied in lifting an object; if the background acceleration is increased to 2 g_z the weight is doubled, and if it is reduced to zero the weight becomes zero. However, experiments on apparent heaviness show that the perceived changes are only about half what would be predicted by the true changes in weight or load force [26]. This may be because perceptual adaptation occurs to a constant change in body weight or to that of familiar objects. Alternatively, we may try to judge the constant mass of objects rather than their changing weight. Mass is defined as the ratio of weight to acceleration. Weight can be perceived passively as a pressure force acting on the skin, but knowledge of acceleration is more complicated; for that, it is necessary to know the constant background level and the additional acceleration applied in lifting or moving the object. The latter is particularly difficult, as it involves monitoring the command signals given to the arm and the resultant movement. Humans are fairly good at this in a 1 g environment, but become confused in altered force environments. In the latter situation they rely more on the load force.

It is interesting to note that judgements about weight or mass are seriously affected by changes in g, whereas other parameters such as grip force and movement patterns adjust automatically to a large extent. The cognitive brain does not seem to have accurate access to these automatic processes.

Discrimination

A related question is the ability to discriminate differences in the mass of objects under altered g levels. In a 1 g environment this ability is usually called 'weight discrimination', though technically it is a combination of weight and mass discrimination. It is measured by the Weber fraction (the difference threshold divided by the average stimulus intensity). The Weber fraction is approximately constant in the middle intensity ranges, being about 0.05 to 0.08 for unpractised subjects, but rises for light weights below about 100 grams [27]. The act of lifting an object applies an additional acceleration to that of Earth's gravity, and thus provides the information necessary for judging mass in addition to weight. The same is true in a high g environment. Under microgravity, however, there is no background weight information and mass must be judged through knowledge of the applied acceleration and the reactive force. An experiment in parabolic flight [26] showed that discrimination was impaired in both the high g and low g phases compared to 1 g, the discrimination thresholds rising by factors of 1.8 and 2.1, respectively.

Further discrimination experiments were conducted by Ross and colleagues in the NASA/ESA Spacelab 1 Mission in 1983 [28] and the NASA/German D1 Mission in 1985 [29], when astronauts compared the mass of isoinertial balls of 3 cm diameter which varied in mass from 50 to 64 grams (Fig. 4). These experiments showed discrimination to be rather better than in the 0 g phase of parabolic flight, thresholds rising by a factor of 1.2 to 1.9 depending on the conditions. The relative improvement was partly due to improved techniques for shaking the test objects, high acceleration shakes giving the best performance under 0 g and low acceleration shakes under 1 g. Adaptation during the days in flight was probably also a factor, since there was an after effect on return to Earth, when weight discrimination was impaired for two or three days. Analysis of videos of hand movements showed that astronauts imparted higher accelerations to the balls in spaceflight than pre-flight and lower accelerations immediately post-flight than pre-flight [29]. They were attempting to produce the same hand movements under all conditions, so this result suggests that they only partially adapted to microgravity, and were left with a brief after effect on return to Earth. Analysis of post-flight videos for one astronaut showed that errors of judgement were usually accompanied by abnormal hand movements. For correct judgements, heavier masses were accompanied by a slightly lower peak acceleration; but for incorrect judgements they were accompanied by a markedly higher peak acceleration (Fig. 5). If this result is general, it implies that subjects are unaware of the true acceleration they are imparting, and assume that the reactive force is due to the mass of the ball rather than to their changed hand movements. This explanation is similar to the 'reinterpretation hypothesis' of Bock et al. [30]: changes in weight caused by changes in g are interpreted as changes in mass rather than as changes in the force environment.

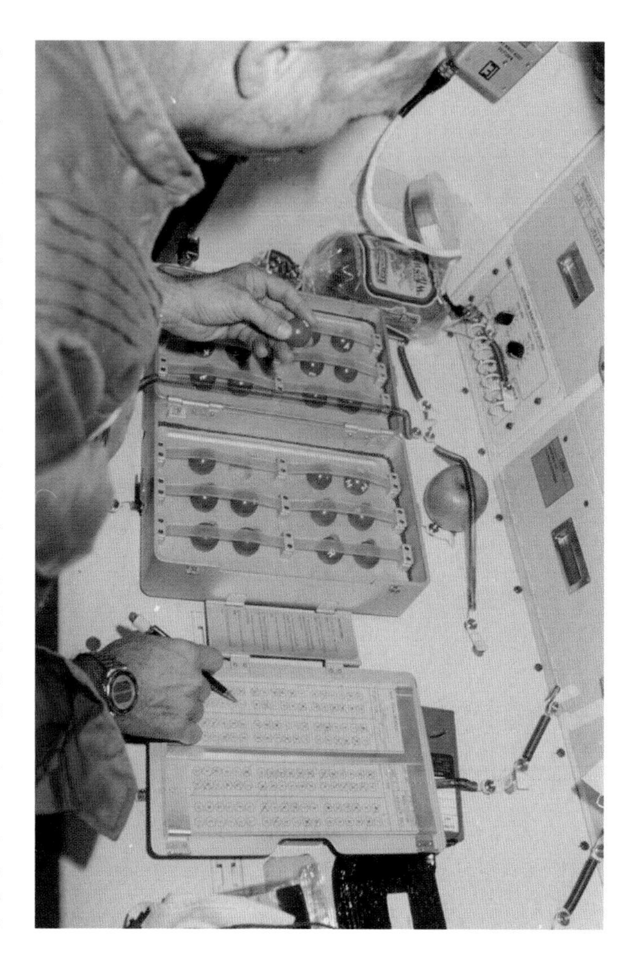

FIGURE 4.
An astronaut performs the mass discrimination experiment on Spacelab 1. (Photo courtesy of NASA)

There is also evidence that the passive pressure sense adapts under microgravity, since astronauts report that their clothes feel heavy on return to Earth [29]. Post-flight errors in mass discrimination may therefore be caused by a combination of changes in the pressure sense and unmonitored changes in hand movements.

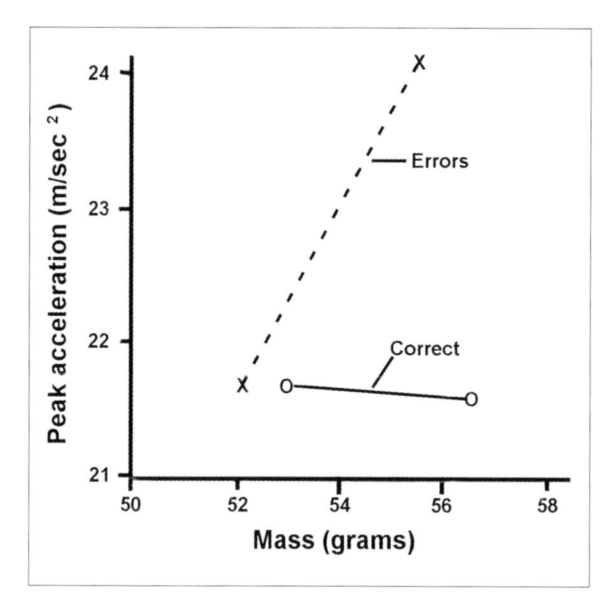

FIGURE 5.

Relation between the mass of the object and the peak acceleration imparted by an astronaut post-flight, for correct and incorrect discriminations of mass. The astronaut wrongly interprets a higher peak acceleration as being caused by a lighter mass instead of by a more vigorous hand movement. The data were collected one day after return from a space mission. The figure shows the means of all nine incorrect judgements and nine correct judgements between similar masses. The peak accelerations were calculated from video analysis of the distance and timing of hand movements while shaking balls of varied mass. (Unreported data from Ross et al. [29])

Haptic interfaces

Increasing use is made of teleoperated devices both in industrial operations and for training purposes [31]. Remotely operated vehicles (ROVs), sensory devices (e.g., video cameras) and manipulators are frequently used under water [7]. Robotic arms are particularly useful for positioning objects in space, while virtual environments are useful for training astronauts for various space activities.

In the past, many robotic devices were autonomous systems that worked without human interaction. More recently, interactive (telerobotic) systems have been developed. A haptic interface is one which allows manual interaction with a teleoperated remote system. Such interfaces are usually designed to receive motor commands from a human hand, and to send back appropriate tactile information (such as force and position). The human operator thus feels as though he is manipulating a real object. The success of these systems depends on matching the performance of the device with human haptic abilities [32]. If the operator is working in an unusual force environment, both the object and the operator's arm will be subject to unusual forces, and the feel of the interaction will be different from normal ground operations. For example, Bock [33] found that subjects reduced their grip aperture when grasping for virtual objects under both high and low g. An operator in water or low g may also have difficulty in getting himself securely tethered while working at the interface. However, astronauts are well trained at manipulations both in 'real' environments (such as parabolic flights and neutral buoyancy water tanks) and in 'virtual' environments with haptic interfaces. Astronauts quickly adapt to the conditions of space travel for many motor skills. They have successfully used robotic arms to conduct repairs outside the spacecraft. There are thus no fundamental problems for the use of haptic interfaces in space travel.

The study of haptic performance in space travel will continue to be important in the 21st Century for two reasons. One reason concerns pure science: knowledge about performance under microgravity assists the understanding of normal haptic function and neurophysiology. The other reason is practical: so long as mankind indulges in spaceflight, it is advisable to know as much as possible about human performance under prolonged microgravity, and the after effects on return to normal gravity. Astronauts are now spending considerable time in the International Space Station, and in the future there may be a manned mission to Mars.

Haptic performance remains critical to the success of many space activities.

Summary

Humans rapidly learn to adjust their movements to altered force environments, including the near zero gravity of orbital spaceflight, and the alternating high and low g of parabolic flight. Adaptation is not complete, and skilled manual tasks are usually performed more slowly or with more errors under 0 g than 1 g. Knowledge of arm position is reduced under 0 g, and subjects feel their arm to be lower than it is. Aiming movements become more variable, but the specific direction of errors is controversial. Detailed analysis of arm movements shows slight differences in the pattern of movement under altered force environments. When moving an object during parabolic flight, grip force adjusts automatically to the load force; but the relation between grip and load force differs at different g levels. Despite small differences in movement patterns, astronauts can carry out skilled motor tasks in space without serious impairment. Some perceptual-motor tasks are, however, seriously impaired. The ability to estimate the mass of objects is reduced, with judgements being influenced by the load force. Objects feel too light under 0 g and too heavy under 2 g, though not to the full extent of the change in load force. The ability to discriminate between differences in mass also deteriorates under 0 g, though the deterioration is reduced if astronauts impart high acceleration shakes to the objects. Post-flight errors of mass discrimination are correlated with abnormal hand movements, which imply that subjects fail to monitor their command signals and consequently impute a change in hand movements to a change in mass. The study of haptic performance in space travel will continue to be important both for practical reasons and for the scientific understanding of the effects of varied force environments on perceptual-motor skills.

Acknowledgements

I am grateful to the Medical Research Council, the Royal Society, the European Space Agency, DFVLR, the Leverhulme Trust, the Carnegie Trust and the Wellcome Trust for financial support for space experiments; and to Otmar Bock and Lynette Jones for helpful suggestions for this chapter.

Selected Readings

Bock O (1998) Problems of sensorimotor coordination in weightlessness. *Brain Research Reviews* 28: 155–160

Jones LA (1986) Perception of force and weight: Theory and research. *Psychological Bulletin* 100: 29–42

Hermsdörfer J, Marquardt C, Philipp J, Zierdt A, Nowak D, Glasauer S, Mai N (2000) Moving weightless objects: Grip force control during microgravity. *Exp Brain Res* 132: 52–64

Sangals J, Heuer H, Manzey D, Lorenz B (1999) Changed visuomotor transformations during and after prolonged microgravity. *Exp Brain Res* 129: 378–390

IV. Clinical and neuropsychological aspects

Phantom sensations

Thomas Weiss

Phantom perception

Phantom perceptions are an intriguing mystery that captured and still captures the attention of many people: those who are amputated, their relatives, healthcare providers, and even the public. Indeed, it is a mystery: How is it possible to feel sensations and/or motions in a limb or a body part that has obviously been surgically removed? Unfortunately, there is no clear answer to this question. Phantom sensations and especially phantom limb pain remain medical nightmares. Moreover, we can currently not explain why phantom sensations occur in some amputees but do not in others. In the present chapter, we will first report on locations, characteristics, and descriptions of phantom sensations. We then will report on patho-physiological mechanisms that possibly might lead to phantom sensations. Finally, therapeutic options and possible future directions of research and treatment will be given.

Locations, characteristics, descriptions of phantom sensations

As a result of amputation of an extremity in adults, nearly all individuals experience phantom sensations in those parts of the body that are now absent. Many amputees report on a progression of their feelings over time. Immediately after amputation, most patients claim on feelings of a phantom limb that mimics the removed body part in size and detail. Sometimes, this phantom is reported to remain in an unusual position as it occurred to be as the result of the previous injury. This phantom body part usually moves gradually into the distal end of the residual limb, a process that is called telescoping. During this telescoping, the residual limb usually progressively loses details in somatosensory detail from proximal to distal. Figure 1 illustrates this process.

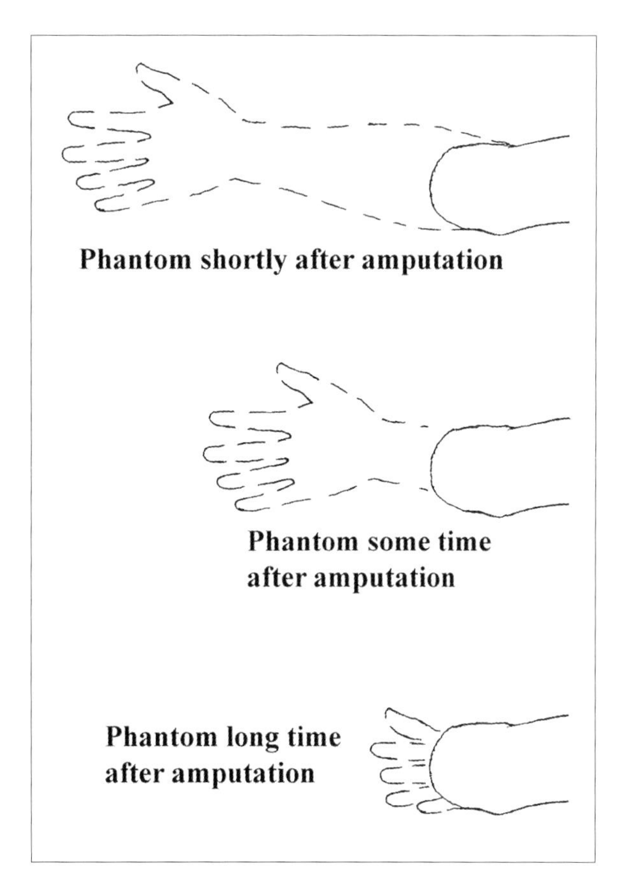

FIGURE 1. PROGRESSIVE TELESCOPING OF THE PHANTOM OF A HAND OVER TIME

Interviews revealed that telescoping occurred at some point in time after amputation in nearly two-thirds of the amputees interviewed [1, 2]. Furthermore, according to these studies, amputees reported on normal sensations of the limb including the whole spectrum of somatosensory sensations, i.e., temperature, proprioception, mechanical sensations like pressure or vibration. Moreover, most patients reported also on the perceived ability to control motion of the phantom limb. Motions could be voluntary, but spontaneous, automatic reflexive as in motions relating to losing balance [3]. However, telescoping usually does not occur in patients that report on a special phantom sensation, i.e., phantom limb pain.

50–85% of amputees [4–6] experience moderate to strong phantom limb pain within weeks after amputation that often outlasts many years and turns into chronic phantom pain. Phantom limb pain occurs in around 80% of amputees of a limb. Phantom pains have also been observed as a consequence of the loss of other body parts such as the breast or tooth [5, 7]; it also can occur as a consequence of a spinal cord injury [8].

Phantom pain has a whole spectrum of features. Some patients describe that the phantom pain is limited to simple, short-lasting and rarely occurring painful shocks in the missing body part. Other patients perceive a constant, excruciatingly painful experience during which the individual has a vivid and intense perception of the missing body part. In a survey of 5,000 amputees, Sherman et al. [4] reported that 18% of amputees had continuous phantom pain, 33% had daily episodes. The intensity of the perceived pain shows a different pattern over time as compared to telescoping: Half of the patients reported on slightly decreasing pain over time, the other half described no change or even an increasing pattern over time [5]. Phantom limb pain seems to be more severe in the distal portions of the phantom. It is described by a number of characteristics such as stabbing, throbbing, burning, cramping [4–6].

Phantom limb pain seems to be more frequent when the amputation occurs in adulthood, less frequent in child amputees and virtually nonexistent in congenital absence of a limb [9, 10]. However, different studies have presented conflicting data. Thus, Krane and Heller [11] investigated 24 individuals who underwent amputation in childhood or adolescence in a retrospective survey. These authors found a prevalence rate of 92% for phantom pain that persisted for months to years. 36% reported currently on phantom pain that does not change. The phantom pain was described as sharp, tingling, stabbing and uncomfortable, i.e., with the same adjectives that use persons amputated as adults. Additionally, while early studies [12] found no evidence for phantoms and phantom limb pain in children born without a limb, more recent studies reported on phantom limbs in individuals born with congenital absence of limbs that was existent at the time of investigation [10, 13, 14] or developed after a surgical manipulation [9]. These data provide support for the suggestion that phantom limbs represent the perceptual correlate of an innate neural substrate of the body experience [15, 16]. However, Dettmers and co-workers [17] showed that anatomical reorganisation in the primary somatosensory and primary motor cortex might occur which is coincident with the absence of phantom limb pain when amputation occurred early in life. Data from larger surveys suggest that the incidence of phantom sensations and phantom pain is lower in individuals born with congenitally absent limb as compared to adolescent amputees [16]. In summary, there are open questions to what extent the innate neural substrate will be reorganised functionally and anatomically after amputation and which factors influence on this reorganisation.

Somatosensory phenomena that can be evoked in the phantom are called referred sensations. Referred sensations are painful and non-painful sensations within the phantom that can be elicited by stimulating body areas with different stimuli (e.g., touch, vibration, warm, heat, pain). Referred sensations might even occur in patients with only relatively small amputations like the amputation of a finger [18]. Referred sensations might be evoked from body areas that can be adjacent to but also far removed from the amputated limb. Sometimes, these sensations showed a point-to-point correspondence between stimulation sites and areas of sensation, i.e., stimulations of skin areas on the face with projections to the phantom in arm amputees [19, 20]. With this pioneer work, Ramachandran and co-workers [19, 20] described referred sensations as perceptual correlates of central functional reorganisation and inspired intensive research into processes of central functional reorganisation. Later on, it has been shown that topographical referred phantom sensations occur in only a part of amputees [21], whereas non-topographical referred sensations are common [22]. Referred phantom sensations can be elicited most often by stimulating adjacent to amputation body regions, e.g., from the stump, shoulder, and/or face in arm amputees. However, such sensations can be elicited from body sides both ipsi- and contralaterally to amputation [22]. In rare cases, referred sensations in arm amputees were reported to be elicited from stimulations of the toe [23] or by movements of a leg [22], i.e., from body part very distant from the amputation site. Points from which referred sensations could be elicited as well as their somatosensory characteristics remain stable for shorter periods of time (e.g., 60 min [22]). Most often, referred sensations are reported to be of the same quality as the stimulus applied, but there are reports that the sensation might occur in another submodality of the somatosensory system. Thus, both vibration and heat has been reported to evoke tingling sensations [22] or light mechanical von-Frey hair stimulation to evoke phantom limb pain [18, 22, 24]. In contrast to the short-term constancy of referred sensation, the topography of points from which sensations could be elicited and the somatosensory characteristics change dramatically over longer periods of time, e.g., 4 weeks [24].

In an important microneurographic study carried out by Nyström and Hagbarth [25] tapping of end of the stump was performed in two amputees. Tapping was associated with an increased pain sensation. Interestingly, Nyström and Hagbarth [25] also reported on an increased activity in afferent C-fibres. This and other findings motivated a search for the potential sources of phantom sensations and phantom limb pain; possible mechanisms leading to these phenomena will be considered in the next section.

Summing up, phantom sensations occur in nearly all patients after amputation. These sensations include the whole spectrum of somatosensory sensations, including temperature, proprioception, mechanical sensations, and pain. Phantom limb pain remains a medical problem. There are different characteristics, time courses, and factors influencing on the perception of phantoms and phantom limb pain. This suggests that different mechanisms contribute to the experience of phantom sensations and phantom limb pain.

Pathophysiological mechanisms contributing to phantom sensations

Peripheral changes

Amputees often report pain and sensitivity to vibration, changes in temperature or air pressure, or touch in the stump, scar, and/or the phantom. These phenomena were associated with a classic feature following complete or partial nerve injury: the Tinel sign. Tinel sign refers to the local or referred pain that occurs in response to brief mechanical stimulation of an injured nerve. Referred pain is defined by a pain occurring in the innervation territory of a damaged nerve. For example, we all felt uncomfortable feelings and pain in the little finger after stimulation of the ulnar nerve at the epicondylus medialis of the humerus at elbow. Similarly to this example, it has been postulated that the stimulation of the injured nerve might evoke phantom sensation. Axotomised afferent neurons show retrograde degeneration and shrinking following experimental or traumatic injury of nerves, which primarily involve unmyelinated C-fibres [26]. Then, terminal swelling and regenerative sprouting of the injured axon end occurs. In case of amputation, this regeneration of the axotomised nerves is ineffective so that the regenerating nerve fibres form traumatic neuromas in the residual limb. These neuromas may be superficial or deep. They often display spontaneous activity but might also show abnormal evoked activity to mechanical and chemical stimuli [27]. Thus, ectopic discharges from stump neuromas represent a source of abnormal afferent input to the spinal cord and a potential mechanism for spontaneous and evoked phantom sensations including phantom pain [28]. There are interesting differences between the different fibre types in peripheral nerves concerning the occurrence of ectopic discharges. Ectopic discharge from myelinated axons seems to begin earlier and tends to be rhythmic whereas C-fibres tend to show slow, irregular patterns [29]. Ectopic discharge seems to be due to alterations in the electrical properties of cellular membranes. These alterations involve the upregulation or novel expression, and altered trafficking of voltage-sensitive sodium channels and decreased potassium channel expression, as well as altered transduction molecules for mechano-, heat, and cold sensitivity in the neuromas [30, 31]. These alterations might result in an increased excitability of injured nerves. In addition, experimental injury causes the expression of novel receptors in the neuroma that are sensitive to cytokines, amines, and other biologically active substances, which might enhance nociceptive processing [29]. Furthermore, non-functional connections between axons, so-called ephapses, might occur and also contribute to a spontaneous activity [32, 33]. As mentioned earlier, Nyström and Hagbarth [25] found increased spontaneous activity as well as an increased activity in afferent C-fibres during tapping on neuromas using microneurographic methods. Both spontaneous and evoked activities were associated with increased pain sensations. Further, anaesthetic blockade of neuromas was found to eliminate spontaneous as well as evoked nerve activity (but not ongoing phantom limb pain).

However, phantom limb pain is often present soon after amputation, clearly before a neuroma could have formed [34]. Therefore, additional sources should contribute to the genesis and maintenance of phantom sensations. Searching for such sources, another site where ectopic discharges might occur has been found in the dorsal root ganglion (DRG). Ectopic discharges in the DRG might occur due to cross-excitation between neighbouring neurons in the DRG, significantly amplifying the overall ectopic barrage [29]. It has been demonstrated that spontaneous as well as evoked sympathetic discharge can elicit and exacerbate ectopic neuronal activity at the level of the DRG [35]. Clearly, this phenomenon could account for the frequent exacerbation of phantom pain at times of emotional distress [36]. Furthermore, there might be an injury-induced sensory–sympathetic coupling with sympathetic sprouting into the DRGs, a phenomenon that has been described in animal studies [29]. This finding of a sympathetic maintenance of phantom limb pain is supported by the two notions:

1. Some patients suffering from phantom limb profit from agents blocking systemic adrenergic transmission. These agents reduce their phantom limb pain [37].
2. Injection of adrenaline into a neuroma has been shown to increase phantom limb pain in some amputees [37].

In summary, the regenerative activity of the organism after an amputation leads to maladaptive processes including the development of neuromas, alterations in ion channel expressions, ectopic activities, ephapses, etc., that are able to explain the occurrence of phantom sensations and, especially phantom limb pain. Furthermore, damaged and reorganising nerve endings as well as altered activity in the dorsal root ganglia represent potential sources for phantom sensations and phantom limb pain.

Changes in the spinal cord

It has been repeatedly demonstrated that local anaesthesia of the stump, the nerve plexus, or epidural anaesthesia does not eliminate ongoing phantom limb pain in all amputees [38, 39]. For example, Birbaumer et al. [38] using anaesthesia of the brachial plexus in upper arm amputees found that 50% of the amputees became pain-free due to the anaesthesia suggesting that their pain is generated by causes from below the level of anaesthesia, i.e., by peripheral reasons. However, there was no pain relief reported for the other 50% of the patients suggesting an involvement of more central factors. Naturally, the next level in the processing of somatosensory information, the spinal cord, was hypothesised to contribute to the genesis and maintenance of phantom limb pain. Indeed, anecdotal evidence from human amputees suggests an involvement of the spinal cord. It was reported that phantom pain suddenly occurred during spinal anaesthesia in patients who were pain-free before the procedure [40]. Indeed, central hyperexcitability might be triggered by nerve injury, as occurs during amputation. Increased firing of the dorsal horn neurons, structural changes at the central endings of the primary sensory neurons, and reduced spinal cord inhibitory processes have all been reported as consequences of nerve injury [41, 42]. For example, inhibitory GABA (γ-aminobutyric acid)-containing and glycinergic interneurons in lamina I of the spinal cord might change from having an inhibitory to an excitatory effect under the influence of brain-derived neurotrophic factor (BDNF) released from activated microglia [43], thereby contributing to a hyperexcitable spinal cord. Furthermore, inhibitory interneurons especially in the superficial layers of the spinal cord (lamina II) could be destroyed by rapid ectopic discharge or other effects of axotomy [42].

Hyperexcitability of spinal cord circuitry following major nerve damage might not only be a consequence of GABA- and glycinergic disinhibition, but also a response of the downregulation of opioid receptors on both primary afferent endings and intrinsic spinal neurons [44]. In addition, cholecystokinin, an endogenous inhibitor of the opiate receptor, is upregulated in injured tissue, thereby exacerbating this effect of disinhibition [45].

An additional spinal mechanism constitutes changes that might trigger abnormal firing of the spinal cord neurons to supraspinal centres. There is a whole cascade of biological events occurring in the spinal cord after peripheral nerve damage. One part of this sensitisation depends on facilitation to the primary afferent neurotransmitter glutamate primarily evoked *via* the effects of activation of *N*-methyl-D-aspartate (NMDA) receptors. This receptor is known to be crucially involved in a basic mechanism able to change synaptic efficacy, i.e., long-term potentiation [46]. Another part of sensitisation results from changes in the spinal circuitry so that low-threshold mechano-receptive afferents might become functionally connected to ascending spinal projection neurons carrying nociceptive information to supraspinal centres. Thus, substance P that is normally expressed only by C-afferents and Aδ afferents might be expressed by mechano-receptive Aβ fibres. These changes might permit ectopic or normal activity in Aβ fibres to trigger and/or maintain central sensitisation. Obviously, normally innocuous Aβ-fibre input from the periphery or ectopic input might contribute to phantom pain sensations [47].

A mechanism that seems to be of special relevance to phantom phenomena at different levels of central processing is the invasion of regions of the spinal cord that are functionally vacated by injured afferents. Figure 2 illustrates this phenomenon for the primary somatosensory cortex S1. In the spinal cord, S1, and other central structures, neurons with receptive fields in deafferented (or absent) skin regions expand their receptive fields to the skin adjacent to the denervated part of the limb. Stimulating the skin adjacent to the denervated part of the limb, activation occurs not only in neurons with receptive fields in this area but also in the neurons with the expanded receptive field. This leads to a shift of activity from these adjacent areas into regions of the spinal cord that previously served the part of the limb that was functionally deafferented by the nerve lesion [48] (see Fig. 2). This reorganisation of the spinal map of the limb is possibly due to the unmasking of previously

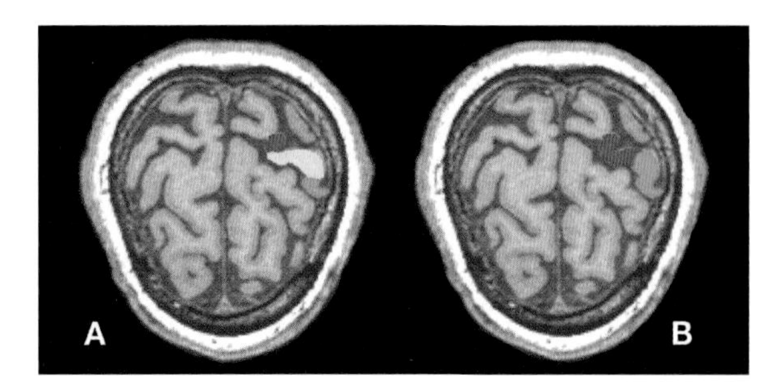

FIGURE 2.
Changes of the functional organization within the primary somatosensory cortex S1 after amputation of an arm. Red – functional representation of the stump; yellow – functional representation of the arm; blue – functional representation of the head; A – before amputation; B – after amputation. Note that the functional representation zone of the arm in S1 has been occupied by the neighbouring representations.

silent connections, long-term potentiation, and sprouting.

Summing up spinal changes, experimental data demonstrated clear evidence for changes within the spinal cord circuitry after amputation able to explain phantom sensations, although direct evidence for spinal changes in human amputees is rather indirect.

Central changes: brainstem, thalamus, cortex

A number of observations in amputees indicate that supraspinal changes could be important for the genesis and maintenance of phantom sensations. Such observations belong not only to amputees but also to other patients with deafferented parts of the body. Thus, paraplegic individuals with high spinal cord injuries may experience pain in the lower part of the body [49], i.e., a body region to which no afferent or efferent innervations exist. Another observation that point to a contribution of supraspinal levels to phantom sensations has already been mentioned. Spinal anaesthesia does not always eliminate existing phantom limb pain [38]. While it is possible that spinal changes contribute to or even drive supraspinal changes, the extent of this contribution is currently elusive.

Supraspinal changes related to phantom limb pain involve the brainstem, thalamus, and different areas of the cerebral cortex. There are different lines of evidence supporting supraspinal changes. Thus, thalamic recordings and thalamic stimulation in human amputees have revealed that reorganisational changes might also occur at the thalamic level [50]. These changes are closely related to phantom sensations including phantom limb pain [50]. Further, axonal sprouting in the somatosensory cortex is involved in the reorganisational changes observed in amputated monkeys [51]. Some studies in monkeys have shown that changes in the cortex might be relayed from the brainstem and thalamus [52–54]. Other studies, however, demonstrated that alterations at the subcortical level could be relayed from the cortex due to its strong efferent connections to the thalamus, brainstem and spinal cord [55–56] or do not involve corticothalamic changes [51].

Cortical reorganisation

The contribution of cortical reorganisation has first been investigated in studies in adult monkeys. These studies found changes in the functional and structural architecture of the primary somatosensory cortex (S1) subsequently to amputation and deafferentation. For example, microelectrode recordings revealed that an amputation of digits in an adult owl monkey led to an invasion of adjacent areas into the representation zone of the deafferented fingers in S1 [57]. The extent of cortical reorganisation was even greater after dorsal rhizotomy, i.e., the destruction of dorsal roots near the spinal cord [58]. Such a procedure leads to a deafferentation of the skin of the associated spinal segments. After rhizotomy of segments C6-Th1, the representation of the cheek in S1 occupies the cortical representation of the arm and the hand in S1, i.e., a range of several centimetres within S1. As already stated, this phenomenon has been suggested to represent the neurophysiological basis of referred sensations in the phantom [19, 20].

Meanwhile, similar changes in S1 have also been found in human amputees. Imaging studies in upper extremity amputees revealed a shift of the representation of the mouth in S1 into the hand representation [59, 60]. With respect to referred phantom sensations, evidence suggests that the functional reorganisation in S1 is less related to non-painful perceptions, but rather have a close association to phantom limb pain. Flor et al. [61] were the first that found an impressively close relationship between reorganisation in S1 and phantom limb pain (r=.93) explaining more than 86% of variance. These authors show that the larger the shift of the functional representation of the mouth in the S1 that formerly represented the arm, the larger phantom limb pain. This relationship between amount of functional reorganisation in S1 and the amount of phantom limb pain has been replicated meanwhile in several studies even when these later studies found slightly smaller linear correlations [21, 62]. It should be mentioned that functional reorganisation of S1was associated with increased phantom limb pain, telescoping, non-painful stump sensations, and painful referred sensation [21]. However, it was unrelated to non-painful phantom sensations, non-painful referred sensation elicited by painful or non-painful stimulation, painful referred sensation elicited by non-painful stimulation, perception thresholds, and stump pain [21]. Therefore, it might be hypothesised that painful and non-painful phantom phenomena are mediated by different neural substrates.

Furthermore, functional reorganisation has also been found within the primary motor cortex M1 [63, 64]. Interestingly, there is a similar relationship between functional reorganisation in M1 and the magnitude of phantom limb pain [65–67] as described for the association between reorganisation in S1 and phantom limb pain.

A further interesting observation relates to a selective loss of unmyelinated nerve fibres with low conduction velocity, i.e., C-fibres. C-fibres seem to have a special role in the maintenance of cortical maps [68]. Consequently, reorganisation within S1 and probably M1 might be enhanced by the selective loss of C-fibres that occurs after amputation [29, 68]. For human amputees, we currently do not know the concrete impact of selective C-fibre deafferentation to phantom limb pain. Possibly, newly developed methods allowing investigation of C-fibres selectively (microneurography, stimulation of tiny skin area [69]) might give answers to these questions.

The extent of functional and even anatomical cortical reorganisation after amputation seems to depend on the developmental stage of the CNS. Dettmers et al. [17] have described that the characteristic shape of the central sulcus, especially in the hand region, is changed when amputation occurred early in life. Similarly, subjects with limb amputation exhibited a decrease in grey matter of the posterolateral thalamus contralateral to the side of the amputation [70]. The thalamic grey matter differences were positively correlated with the time span after the amputation but not with the frequency or magnitude of coexisting phantom pain. Phantom limb pain was unrelated to thalamic structural variations, but was positively correlated to a decrease in brain areas related to the processing of pain [70]. These observations are in good agreement with the observation of rare phantom sensations in these patients. The relationship of age of amputation and its consequences on reorganisation in different parts of the central nervous system has been systematically investigated in rats. These investigations revealed that amputation in adulthood leads mainly to cortical changes whereas amputation at a younger age causes more comprehensive reorganisation along the neuraxis [71]. These and human studies also revealed that several stages of cortical reorganization can be differentiated [72–75]. The first stage relates to the unmasking of normally inhibited connections. This process occurs immediately after an amputation or deafferentation. For example, we [75] found that functional reorganisation occurred as a consequence of deafferentation of median and radial nerves within a few minutes after deafferentation (Fig. 3). Functional

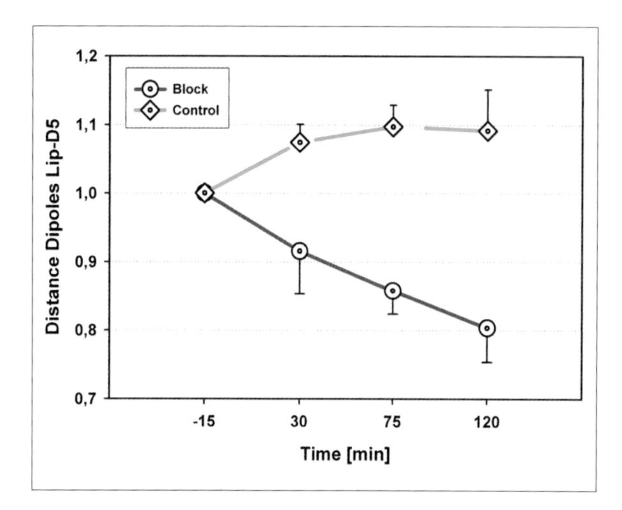

Figure 3.

Rapid reorganization of the hand's functional representation within minutes to hours after blockade of the medial and radial nerves. The decrease of the distance between representations of the lip and the little finger (D5) might be interpreted as the occupation of representation of the deafferented fingers.

reorganisation occurred in S1 occupying up to 50% of the undeafferented representation within hours. Additionally, it was shown that reorganisation was associated with a disinhibition (using transcranial magnetic stimulation TMS to the primary motor cortex).

The loss of GABA-mediated inhibition possibly prepares the next stage, an excitatory activation that also occurred rapidly and might result secondarily from the loss of inhibition [76, 77]. Several mechanisms might contribute to the increase in excitability [78]:

- decreased inhibitory inputs
- removal of inhibition from excitatory inputs
- increased release of excitatory neurotransmitters
- increased density of post synaptic receptors
- changes in conductance of the neuronal membrane
- unmasking of latent excitatory synapses [79].

Further stages involve changes such as long-term potentiation, alterations in synaptic strength, and axonal sprouting. Finally, use-dependent or non-use dependent plasticity might lead to additional changes based on Hebbian learning [80, 81].

Alterations in sensory and motor feedback

As has been described in the previous paragraphs, parallel changes occur in the primary somatosensory as well as in the primary motor cortex [21, 58–67]. It has been suggested that phantom limb pain might be related to the incongruence of motor intention and sensory feedback in the primary cortices as well as in frontal and parietal brain regions [82]. In line with this suggestion, several studies have used a mirror therapy to organise an illusionary sensory input from the absent limb [83–87]. One of these studies has found a reduction in phantom pain with increasing activation of M1 [85]. Another study supporting the hypothesis of incongruence between motor intension and sensory input as a possible source of phantom sensations used a mirror to create a discrepancy between actual and seen movements, i.e., the movements seen in the mirror were discrepant from the movements behind the mirror when both arms performed asynchronous movements. As a consequence, painful and non-painful paraesthesias in the incongruent movement condition were reported [87]. It was suggested that sensorimotor incongruence might cause abnormal sensations as seen in many neuropathic pain syndromes including phantom limb pain [86, 87].

Sensory and pain memory

It is known for a long time that the primary somatosensory cortex S1 is involved in the processing of somatosensation and pain. Concerning pain, S1 seems to be involved in the lateral thalamo-cortical pain processing system that is important for sensory–discriminative aspects of pain [88–90]. It has been suggested that S1 is, at least, important for the memory of pain. An extended hypothesis claims that S1 might be a place where parts of the pain memory are placed. In line with this hypothesis, several studies found evoked phantom limb pain in response to stimulations within S1 [91]. Combining the pain memory hypothesis and the central changes at different levels of the CNS described above, the following hypothesis for the genesis and maintenance of phantom limb pain can be forwarded [6]. A somatosensory and a pain memory might be established before and/or during amputation with a neural substrate at spinal and supraspinal levels, especially in S1. An amputation and the subsequent deafferentation and occupation of the deafferented zone (e.g., in S1) by neighbouring input might activate neurons that are part of these pain memory networks. The activation of the cortical zone previously representing the amputated limb by neighbouring peripheral input might then be referred to phantom limb and might be interpreted as phantom sensation and phantom limb pain [6, 18, 73, 92]. Several lines of evidence support this hypothesis. First, phantom limb pain has repeatedly been reported to be similar to the pain that existed in the limb prior to amputation. Depending on type and time of assessment [93–95], 10% to 79% of amputees report on phantom limb pain that remembers the pain directly before amputation. Similarities were reported concerning characteristics, locations of pain, as well as position of the limb [2–5]. Second, pre-existing sensitivities in the processing of nociceptive stimulation (e.g., pain pressure thresholds before the amputation) predict phantom limb pain after the amputation [96]. Third, some studies report on an abolishment of phantom limb pain after a surgical removal of parts of S1. Forth, phantom limb pain might be evoked in response to selective stimulations of S1 [91]. All these data suggest that special pain memories established prior to the amputation might be important for the genesis and maintenance of phantom limb pain [6]. Furthermore, it has been proposed that increasing chronicity is correlated with increasing changes in the representation zone in S1. Data from patients with chronic low back pain and chronic regional pain syndrome (type I, i.e., without obvious nerve lesions) [97–99] support this view. In these patients, long-lasting noxious input has demonstrated to lead to long-term changes at the central, and especially at the cortical, levels that affect the later processing of somatosensory input. This type of memory seems not to require conscious processes to occur so that these pain memories might be implicit. Moreover, data on patients with chronic regional pain syndrome type I also showed that a reduction of symptoms was correlated with a re-reorganisation at the central level, i.e., a normalisation of the functional organisation in S1 [99].

Type of analgesia

Several prospective studies [94, 95] have shown that chronic pain before the amputation predicts later phantom limb pain. These studies included mainly amputees with long-standing pain problems prior to amputation, in whose pain memories can have developed over a long period of time. However, there are only a few studies on traumatic amputees. In these amputees, additional factors relating to type of surgery, anaesthesia, or pre- and postoperative pain could possibly influence on the genesis of phantom sensations and phantom limb pain. Thus, from the pain memory hypothesis one might suppose that, beside central anaesthesia, a global afferent barrage of nociceptive input into the CNS might prevent the patient from the occurrence of later phantom limb pain. Therefore, peripheral anaesthesia was added for some time before and during surgery to prevent central sensitisation (pre-emptive analgesia [100]). While there were positive effects in uncontrolled studies, a controlled study in amputees showed that pre-emptive analgesia did not significantly reduce the incidence of phantom limb pain in a longer range of time [101]. A recent study with the addition of an NMDA receptor antagonist in the peri-operative phase extended the beneficial effects into several postoperative weeks [102]. This effect could be due to the erasure of pre-existing somatosensory pain memories [6], and/or the prevention or prolongation of post-amputationally occurring central changes explained in previous paragraphs. On the one hand, the role of insufficient pre- and postoperative pain management has probably been underestimated [103]. On the other hand, we are sceptical whether or not pre-emptive analgesia is able to reduce phantom sensations and phantom limb pain on a long-term prospect. In our opinion, the deafferented representations at different levels of the CNS will be reorganised on the basis of maladaptive Hebbian learning mechanisms in a long-term prospect.

Affective and motivational aspects of pain

Processing of nociceptive input not only includes a somatosensory-discriminative component of pain but also an affective-motivational component. The processing of this component is probably mediated by the insula, anterior cingulate, frontal cortices [86, 87, 104, 105]. It is likely that central reorganisation following amputation occurs not only for areas involved in sensory–discriminative aspects of pain, but also in those brain regions that mediate affective–motivational aspects of pain [106]. Thus, the amount of hypnotically-induced phantom limb pain correlated closely to the activation of the anterior cingulate cortex [107]. In animal studies, a potentiation of sensory responses in the anterior cingulate cortex was observed after limb amputation [108]. The role of affective and motivational factors in phantom limb pain is also underscored by the fact that several studies could predict the severity of phantom limb pain from coping-related variables and depression [6, 8].

Summing up, there exist several central mechanisms that probably contribute to the genesis and maintenance of phantom sensations and phantom limb pain (Fig. 4). These mechanisms include functional reorganisation at different levels of the somatosensory system and of the pain matrix. Functional reorganisation includes different stages supported by different physiological mechanisms. The functional reorganisation might also involve structures that are part of the memory network. Activation of the special pain memories established prior to or during the amputation might, therefore, become important for the genesis and maintenance of phantom limb pain.

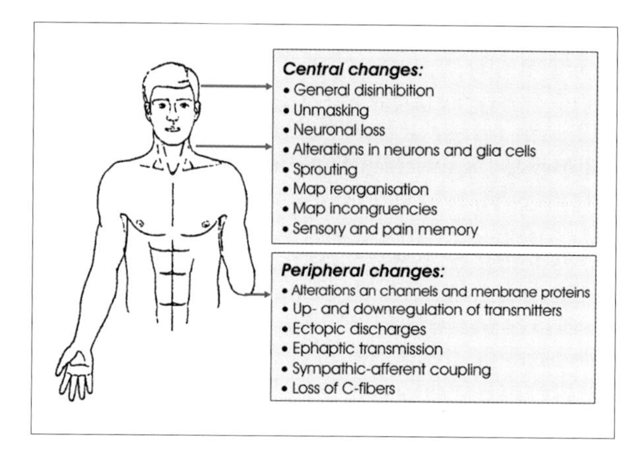

FIGURE 4.
Summary of changes possibly involved into the generation and maintenance of phantom sensations.

Implications from the pathophysiological mechanisms for the treatment of phantom pain

More than 60 different types of treatments have been described to treat phantom limb pain. However, large surveys of amputees have impressively demonstrated that most currently available treatments for phantom limb pain are ineffective in producing a permanent or a long-lasting substantial reduction of phantom limb pain [4, 5, 109]. For example, in a large survey of more than 4,000 amputees, less than 1% reported on permanent or a long-lasting substantial reduction of phantom pain [4, 5]. Several factors might contribute to this phenomenon. One important factor might be that we have only limited possibilities to counteract the central mechanisms. It might be postulate that an effective treatment of phantom limb pain that is based on central reorganisation should include measures able to reduce this maladaptive plasticity.

One idea to account for a central mechanism was to reduce the incongruence of motor intention and sensory feedback in the primary cortices and other cortical regions by means of a mirror therapy. This therapy was used to provide illusionary sensory input from the absent limb [84–86]. The mirror is used to trick the brain into perceiving movement of the phantom when the intact limb is moved. Mirror treatment has shown to be effective in reducing phantom limb pain [84–86]. The efficiency for the reduction of phantom limb pain remains to be demonstrated in a well-controlled trial.

Another idea to reduce phantom limb pain was to re-reorganise the amputation-induced plasticity. This idea was conducted on the basis of work in healthy animals and humans on stimulation-induced plasticity. Extensive, behaviourally relevant stimulation of a body part has been shown to lead to alterations in its functional representation [110, 111]. Thus the cortical representation of the digits of the left hand in S1 was larger in string players than in controls [111]. Moreover, blind human Braille readers were found to have larger functional representations of those fingers that were used for Braille reading [112, 113]. Extensive training of the impaired extremity in stroke patients was found to enhance the functional representation of the trained extremity in the primary motor cortex M1 [114, 115]. All these data suggest that intensive training of the remaining stump of an amputee could be one way to treat patients with centrally-generated phantom pain.

Several studies found indirect evidence for this suggestion. For example, we found in a retrospective study that patients wearing a somatosensory effective mechanical Sauerbruch prosthesis exhibited a significant and large decrease in amount of phantom limb pain as compared to amputees wearing a cosmetic prosthesis [116]. In patients with phantom limb pain, intensive use of a myoelectric prosthesis was found to be positively correlated with both reduced phantom limb pain and reduced cortical reorganisation [117]. An alternative approach was used by Flor et al. [118] who applied a behaviourally relevant electrical stimulation to the stump. The basic idea of this treatment was to provide input into the cortical representation of the amputated limb from closely neighboured skin areas to re-reorganise changes that occurred subsequent to the amputation. In this study, a 2 week training consisting of daily sessions to discriminate several electrical stimuli led both to a significant reduction of phantom limb pain and to a significant reversal of cortical reorganisation. A control group of patients with standard medical treatment and general psychological counselling for the same time period did not show similar changes.

Similar effects on phantom pain and cortical activation were reported for imagined movement of the phantom [85]. Imagined movements seems to use both above-mentioned strategies, i.e., to provide congruency of intended movements and its imagined sensory feedback and to re-reorganise the cortical representation areas using the representations for imagined movements. Furthermore, pharmacological interventions which relay on the neuroplastic changes for pain memory formation might also be effective. Possible candidates include NMDA receptor antagonists, GABA agonists, anticonvulsive agents, calcitonine, and others [6, 102]. In fact, all these approaches are based on the reduction of central reorganisation, reduce discrepancies between somatosensory and motor systems, and/or hyperexcitability contributing to centrally generated phantom sensations and pain.

In summary, phantom limb pain is still a medical problem. The therapy is oriented on the supposed pathophysiological mechanisms of phantom limb pain. However, it remains difficult to account for the central mechanisms of phantom pain. Recent research provides first evidence of clinically important effects for treatments aimed to re-reorganise the maladaptive functional reorganisation at different levels of the CNS.

Summary

Phantom sensations are still a scientifically unsolved mystery that captures attention of many people. Phantom sensations occur sometimes as a result of illusions, but in nearly all patients after amputation. These sensations include the whole spectrum of somatosensory sensations including pain. Different mechanisms in the periphery, the spinal cord, and higher regions of the central nervous system seem to contribute to the genesis and maintenance of phantom sensations. Due to these different mechanisms, more than 50 different therapeutic approaches have been used to reduce phantom pain; however, it still remains a medical problem. Noteworthy, functional reorganization at different levels of the somatosensory system and the pain matrix with different stages of reorganization has been described in the last years. It still remains difficult to account for the functional reorganization at different levels of the central nervous system. However, there are first approaches that try to counteract this maladaptative functional reorganization with clinically important effects in reducing phantom pain.

Selected readings

Sherman RA (1997) *Phantom pain*. New York: Plenum

Flor H, Nikolajsen L, Staehelin Jensen T (2006) Phantom limb pain: a case of maladaptive CNS plasticity? *Nat Rev Neurosci* 7: 873–881

Kuiken TA, Marasco PD, Lock BA, Harden RN, Dewald JPA (2007) Redirection of cutaneous sensation from the hand to the chest skin of human amputees with targeted reinnervation. *Proc Natl Acad Sci USA* 104: 20061–20066

The neuroscience and phenomenology of sensory loss

Jonathan Cole

Introduction

Historically deafferentation has been seen in the context of a late effect of syphilis, (before the spirochaete's effects were largely killed off by antibiotics, at least in Western societies). One of the consequences of the chronic late stage of the disease is atrophy of the dorsal columns of the spinal cord, leading to loss of the tracts carrying touch and proprioception. This in turn led to the characteristic broad based staggering gait and uncontrolled ataxic movements. Dispassionate accounts of the consequences of this may have been made more difficult by the fact that with ataxia comes intractable lightning pains.[1] That such deafferentation could arise from other peripheral and central causes of loss of proprioception was less clear.[2]

This chapter will consider neurological causes of haptic sensory loss, from peripheral diseases leading to losses of large myelinated sensory nerve fibres alone to more central conditions including dorsal column loss and cortical damage. Perhaps the purest form of loss of movement and position sensation and cutaneous touch is seen in large fibre peripheral neuronopathy. This syndrome was first described by Schaumburg and colleagues in 1980 [1].[3] Since then a number of different causes has been found, though their original post infective one remains the paradigm and, being non-progressive, has allowed the most long-term study and an understanding of the importance of afferents in haptic function. The chapter will therefore end with an analysis of function in subjects with this syndrome.

Before this, however, will be given a short précis of how sensation is now tested, a century after Head's techniques (see Chapter 3).

The testing of somatic sensation

When sensation was tested by Henry Head at the turn of the 19th and 20th century it took so long and was so detailed that such a schedule would have been impossible clinically. Faster techniques became accepted. Light touch was tested with cotton wool gently stroked across the skin, pain using a pin or hypodermic syringe needle, temperature with a cold metal object and movement and position sense by small movements of the digits, either fingers or toes. Two point discrimination did require special retractors and higher cortical sensation was sometime tested *via* the drawing of numbers on the outstretched palm and rarely by placing an object in the hand for manipulation in a manner Head would have recognised. These methods were adequate for many clinical situations, when related to clinical presentation and further investigations, whether of nerve conduction, blood tests for diabetes, B12 deficiency, etc., and radiology. But they could not be considered quantitative and there was a clear need for better testing. Fortunately methods of quantitative sensory testing (QST) have been developed and are becoming increasingly accepted and developed to be rapid and reproducible for widespread clinical use (ideally when part of assessment of patients by clinical history and examination, neurophysiology (nerve conduction studies and EMG), laser evoked potentials, neuroimaging and even skin biopsy).

QST has suffered because there has been little agreement about their standard use and because of the length of time needed to perform them as well as because of problems with reproducibility. Rolke et al. [2] have tried to address this by using a battery of tests of all aspects of sensation within approximately 30 min.

They tested thermal sensation using a small probe (7.84 cm^2) placed on the skin of the hand and foot which can be cooled or warmed rapidly. Cold and warm detection thresholds are measured using stimuli which changed in a ramped manner (1° per second). In addition the thresholds for heat pain and cold pain were also determined. All tests were done three times.

Mechanical detection thresholds were determined with von Frey filaments of 0.25–512 mN made without a sharp edge which might activate nociceptors. The threshold was determined as the geometric mean of five series of ascending and descending intensities. Mechanical pain thresholds were measured with a set of seven pinprick stimulators exerting 8 to 512 mN again using the mean of a series of five ascending and descending intensities.

They also tested psychophysical ratings of some stimuli, with mechanical pain sensitivity being determined using the same weighted pinprick stimuli as for the mechanical pain threshold but this time applied in an order five times each and the subject asked to give a pain rating for each. Three light tactile stimulators (cotton wisp, cotton wool and a brush) were used to assess dynamic mechanical allodynia by stroking them across the skin in a single stroke over 1–2 cm and subjects asked to rate the pain as for pinprick.

Temporal summation or wind up of a pinprick was also tested using 10 stimuli at 1 per second and the subjects asked to rate the pain at the end numerically and this compared with single pricks at the same site. Vibration detection thresholds were tested using a tuning fork (64 Hz) placed on a bony prominence. Pressure pain thresholds were determined with a pressure gauge device with an area of 1 cm2 and a pressure of up to 20 kg/cm2 with an increasing stimulus repeated three times. Rolke et al. then employed statistical analyses of their data, before using Z scores to compare each subject's results with control data. This group accepts that QST remains a psychophysical test with some of the problems inherent in this but hopes to bring it in to more standard and standardised clinical practice by reducing the time for comprehensive testing.

In an earlier analysis of QST by the American Academy of Neurology [3], its clinical use, efficacy and safety was assessed. Because of differences in technique, in normal ranges and in reproducibility, they were concerned that it should not be the sole criteria for diagnosing pathology and also recommended that it was not used in medico-legal cases. On the more positive side they did suggest that QST is probably or possibly useful in identifying small and large fibre sensory neuropathies associated with diabetic neuropathy, small fibre neuropathies, uraemic neuropathies and demyelinating neuropathy. As the techniques become more refined and widely agreed upon this list may increase.

Causes of large fibre neuropathy

Diseases of the peripheral nerves usually involve all fibre types with both large and small fibres being affected. It is rare therefore to have neuropathies of the large fibres alone, affecting touch and proprioception selectively. In large series it occurs in less than 5% or so. It is seen, however, in several conditions which fall into two broad areas. The first group involves a toxic agent and/or selective vulnerability to that agent (or its lack) in the axon, in cisplatin and vitamin B6 neuropathies, in Freidreich's ataxia, and in vitamin E deficiency (for review see [4]). In the second group an immune reaction appears to target the large fibre cell bodies in the dorsal root ganglia. This appears to be the case in carcinomatous sensory neuropathy, IgM neuropathy, Sjogren's Syndrome and acute sensory neuronopathy. The onset can be insidious and progressive, as in some carcinomatous and Sjogren's Syndrome related neuronopathy, or can be acute and severe, with all large sensory nerve cells being destroyed in days, as in some acute sensory neuronopathy syndrome. In most large fibre axonopathies sensory loss predominates, but in some neuronopathy syndromes ataxia is the main and presenting symptom, suggesting a selective loss of proprioceptive afferents.

Selective toxicity of large sensory nerve fibres

In large doses vitamin B6 is neurotoxic to large sensory fibres or their cell bodies. It is reversible only if the doses are stopped in time. If severe then there are also symptoms of central and peripheral nervous system affects, with autonomic disturbances, lethargy, respiratory depression [5].

Neuropathy associated with cisplatin therapy, mainly used in gynaecological cancers, was reported soon after the drug was employed. Some early patients presented with parasthesiae and sensory loss in a classical glove and sticking distribution later, but as the syndrome became more recognised it has been shown to be primarily a neuronopathy. Post mortem studies have shown necrosis of the large fibre cell bodies in the dorsal root ganglia as well as loss of peripheral axons. In a sample of 21 patients post mortem, Gregg et al. [6] found that tissue platinum levels were highest in the dorsal root ganglia and lowest in tissue protected by the blood–brain barrier. They suggested a linear relationship between platinum levels and cumulative dose and that cisplatin was retained indefinitely in a neurotoxic form. If the development of neuronopathy with cisplatin therapy is dependent on dose and exposure duration, then this might limit its usefulness, at least in those for whom survival times are long. More recently, though, it has suggested that limited prolonged use may be possible, with careful monitoring [7].

In inherited spinocerebellar atrophies, of which Friedreich's Ataxia (FA) is the most common, the ataxia is associated with peripheral loss of large sensory fibres over and above any cerebellar deficit. It is a progressive autosomal recessive disease which, in addition to degeneration in the central and peripheral nervous system, is also associated with cardiomyopathy, skeletal abnormalities and increased risk of diabetes.[4]

Immune mediated neuronopathy

Autoimmune neuropathies often present with ataxia rather than cutaneous sensory loss. They may be caused by sensory variants of the Guillain-Barre syndrome, sensory neuronopathy syndromes, subsets of immunoglobulin M paraproteinaemic neuropathy, paraneoplastic neuropathy and the neuropathy associated with Sjogren's syndrome. The targets for autoantibodies are varied, but include gangliosides, myelin associated glycoprotein, Hu antigen and extractable nuclear antigens [8]. The major site of pathology in autoimmune ataxic neuropathies is the dorsal root ganglion, but dorsal roots and peripheral nerve myelin and axons may also be affected. The discovery of the auto-immune origin for some of these neuropathies has led to trials of immune suppressant therapy, though with mixed effects as yet.

IgM associated sensory neuropathy is predominantly an axonopathy which presents with ataxia but also with distal large and small fibre symptoms. It is associated with serum IgM binding GalNAc-GD1a & GM2 gangliosides and causes, unusually for these diseases, a demyelinating neuropathy. It is slowly progressive over years and may be treated with IVIg [9].

Paraneoplastic sensory neuropathy occurs almost always in relation to smokers and small cell carcinoma of the lung, though at time of presentation of the neuropathy no tumour is evident in approximately 50%. The presentation can involve pain, suggesting small fibre involvement and include motor nerve involvement too in up to 50% in some samples [10]. Round 80% are positive for anti-Hu antibodies [11]. It can also be associated with other nervous system dysfunction including LEMS, cerebellar ataxia and autonomic problems. Tumour resection sometimes improves the neuropathy, though median survival is poor, with around 20% surviving 3 years [12].

Large fibre sensory neuropathy is a rare but well known neurological problem seen with Sjogren's syndrome. In one series 11 out of 15 presented with numbness and parasthesia.

Most followed a rather indolent but progressive course, despite treatment with steroids, cyclophosphamide or intravenous immunoglobulins [13].

The acute sensory neuronopathy syndrome was first described in 1980 by Sterman et al. in three adults who, after an infection (and antibiotics), rapidly and irreversibly developed sensory loss and ataxia. They were followed for 5 years without evidence of coexistent neoplasia or immunological disease. Subsequently more patients with this condition have been described in clinical series. It is to these few subjects that the rest of the chapter will be concerned.

Consequences of large fibre sensory loss

One of the earliest studies of complete deafferentation was by Rothwell et al. [14]. This subject had suffered a severe peripheral pan-sensory neuropathy, with loss of pain and temperature, as well as large fibre function in the arm and leg some months previously. His motor power was almost unaffected. He was able to produce a very wide range of finger movements in the lab very well, and could move individual fingers and make outline figures in the air without vision. More detailed tests showed that he could also move his thumb accurately through three different distances at three different speeds, and could produce three different levels of force at his thumb pad. He was also able to match forces applied to the thumb. But his hands were relatively useless to him in daily life. He was unable to grasp a pen and write, to fasten his shirt buttons or to hold a cup in one hand. They concluded that his difficulties lay in the absence of any reflex correction, an inability to sustain constant levels of muscle contraction without visual feedback over more than 1–2 s and an inability to maintain longer sequences of motor programmes without vision.

This subject was studied relatively soon after the illness and had a pan-sensory deafferentation affecting distal parts of the arm and leg. In contrast two subjects, GL [15] and IW [16, 17], had the acute sensory neuronopathy syndrome with complete loss of movement and position sense and of cutaneous light touch with clinically preserved pain, temperature and peripheral motor nerve function. Each was in their early adult life and each has now been studied for over 30 years during which time there has been no recovery of peripheral nerve fibre function. These subjects show some important differences from the subject studied by Rothwell et al. and indeed from each other.

Taking the latter first GL was deafferented from the lower face downwards, so she has no sensation of touch in her lower mouth and none from the neck, where up to 50% of the muscle spindles lie, whereas IW's level is C3, so that he has normal sensation from the neck. GL was married and with a son at the time of the illness, so she adapted to life in a wheelchair and carried on as best she could. IW was single and spent 17 months as an in-patient in a rehabilitation hospital, learning to move and to stop ataxic movement before returning to his life, walking, driving and living and working independently. Studying these subjects soon after their illness would have shown that without feedback there was a complete loss of controlled movement and that even the experimental motor tasks shown by Rothwell et al. might have been absent.

Initially, for instance, without movement and position sense and touch, IW was immediately completely unable to move in any controlled fashion. Once, when lying on a bed without seeing his body, he remembers feeling that he was floating, since he could no longer feel contact on the bed; this was terrifying since he could not control it. He required full nursing care for weeks. But over the next months and years both GL and IW recovered motor abilities using mental concentration and visual feedback or supervision.[5]

For instance, after a few weeks IW decided to try to control movement. Lying in bed he asked his stomach to contract, reasoning that this would life him up. Nothing happened; his arms were across his chest. So he flopped the arms out and tried again. This worked and he sat up on the bed. He was so ecstatic he yelled out, and then immediately collapsed; without constant mental concentration movement dissolved. That simple act informed the whole of his subsequent rehabilitation. He knew that he could move, if he thought about the movement and gave the moving part visual supervision. The downside was the sheer amount of mental concentration he would have to expend for the rest of his life. He learnt to feed himself, preferring hot food gone cold to being fed, he learnt to dress and then, after a year, he stood and then over the next months learnt to walk. When he left hospital nearly 2 years after the acute illness, and with no neurological recovery, he was ready to work and live independently, walking and driving, telling his new work colleagues that he had a problem with his back – for how do you explain loss of movement and position sense and cutaneous touch?

By this time he had learnt all the necessary movements for daily life, though these are never referred to as skills. All require mental effort and he reflects that he never knows whether he can do one day to the next. Without automaticity of movement it depends on his mental concentration. If he has a head cold, for instance, he retires to bed unable to think enough to coordinate movement; he cannot drink much alcohol or the same thing happens. Recently after several days ill and sleepless with earache he could not even coordinate movement of his arm towards a cup; even after 30 years of successful movement without peripheral feedback this realisation of how labile and fragile his movements remain was frightening. Having said that, he has had some return of – or linking to – previous motor programmes. He relates that walking now, on a level surface in the light with no slippery surface or wind and no people around who might jostle him (and no head cold) requires round 50% of his mental concentration whereas at first it took 100%. He could not, and cannot, daydream when walking or standing.[6]

Both GL and IW have been very generous with their time and effort towards science and there are many studies of their motor functions (see many of these papers at jacquespaillard.apinc.org/deafferented). Both GL and IW can make movements of a set duration and amplitude [15, 17]. But, in addition, they can distinguish weights placed in their hands, with eyes open, though whether this is due to remaining peripheral afferents (of muscle fatigue, tension, etc.) or due to a central effect *via* vestibular afferents is unclear [18]. IW can also maintain a force without feedback, unlike Rothwell et al.'s subject. He also writes and feeds himself and has found a way of doing up buttons, driving (using hand controls) and doing all required for daily living [16].[7] These differences reflect the later time after the deafferentation of the studies (from over 12 years onwards for IW and slightly less for GL).

Movement can be divided into locomotor, instrumental and gestural in type. GL is happy to live from a chair. IW in contrast wanted to return to the world of the walking and so expended huge effort in learning to stand and then walk. Interestingly he finds standing still far more difficult than walking steadily. More recently however, he has developed chronic back pain and finds that this is made worse by standing. He therefore lives far more from a chair to reduce the pain rather than because of the deafferentation. Both subjects, once they had learnt to feed themselves, etc. (instrumental action), decide to recover gesture, since they (independently) wanted to show emotional embodied expression. So they learnt to do this consciously though with the years gesture may now be the most automatic of the movements they make [19, 20]. Automaticity may reflect in part the way in which gesture is part of a thought/language/gesture system but also that gesture has to be accurate in time and shape but not in place. Both IW and GL find it far easier to make movements accurate in time than in relation to the external world; to gesture is far easier than coordinating movement to pick up a cup. Since haptic touch requires constant interplay between hands and object this is a particularly difficult challenge for them and one which has meant huge adjustments in their behaviour.

Very early on IW realised that he could no longer coordinate all his fingers in gripping and grasping movements, there were too many of them and too many joints and combinations for him to coordinate and concentrate on. In any case he did not need them all; by using the thumb and fingers 1 and 2 he found he was able to do all he needed to, while keeping his fourth and fifth finger out the way. His grasps are not as normal; he tends to approach objects with a wide open hand and then close on them with a stereotyped thumb/finger grip. Firm objects are relatively simple to hold, once a good force is reached and held. He does have to be careful, however, that his fingers are not trapped by the handle on the far side of a mug which he would not feel and might not see. Every such movement requires calculation and preparation. Often he sits before making such a movement and will run through his movements in his mind before making them, visualising his solution to a motor problem. Most difficult are plastic cups, where he has to look to make sure he has gripped it without crushing.[8] He writes by clamping fingers round the pen and using forearm muscles to move the hand, rather than have the fingers shape the letters.

Empirical work has shown that GL and IW have not unexpected deficits in aspects of touch during dynamic grip tasks. Grip force during grasping tasks requires knowledge of the object in terms of weight and friction on its surface, and of internal models of force production. To hold an object and move it up and down GL used high levels of force, to avoid dropping it but not scaled to the force required and did not alter this during movements up and down [21]. In further studies GL made no modulation of grip force during movements of an object; IW did make some of these, though the modulations were reduced (he had no knowledge of making any such modulation). Without sensory feedback the normal internal motor programme changing force with movement as an object is moved up and down was either degraded or abolished (Hermsdorfer et al.: *Neurorehabilitation & Neural Repair*, in press).

There are some tasks IW has not found a way of doing. He cannot feel in his pocket, so does not keep change or his keys there. He does not pick up change, instead just sweeping off a surface into his palm and then placing it in a bag. His choice of food is determined to an extent by how difficult it is to eat; fish is out if there are fiddly bones to dissect for instance. Bereft of active touch and or exploration with his hand without vision, he finds ways round such tasks. It is probably fair to suggest that though he moves his fingers with astonishing refinement, given their deafferentation, their movements are relatively slow and stereotyped. But given the fact that many people would not notice anything unusual and that few if any would guess the deficit under which he moves his recovery in movements of the hand, like his recovery in walking, is astonishing.

Conclusions

There are a number of different pathologies which lead to loss of touch and proprioception, so depriving the haptic system of peripherally originating feedback. It is by studying such pathologies that the importance of feedback is understood. But, equally important is to understand the importance of the time since the loss and the possibilities for rehabilitation.

If born without such feedback it is doubtful if good motor function could be developed, since the absence of the experience of moving and the lack of any such internal motor programmes might outweigh the beneficial effects of youth for such motor learning. This, however, remains a thought experiment, fortunately. What is clear is that subjects with loss of sensory feedback are initially overwhelmed by loss of all coordinated movement of their limbs and trunk, but that with time, mental concentration and visual supervision, useful and indeed near normal function for most motor tasks can be recovered. This shows the power of individuals to reconstruct their lives in presence of huge neurological loss. Since it seems likely that degrees of proprioceptive loss after stroke or in neuropathy might be underestimated, the ways used by such subjects with extreme deafferentation are of potentially wider interest too.

Without feedback there are several movements which have remained impossible. Neither GL nor IW can walk very fast, though their reaction times and speeds of simpler movements are normal. Cognitive control of coordinated and complex movements has a time penalty. Though they can use their hands and fingers for most actions like feeding and writing, shaving and putting on make up, they have had to find new simpler ways of finger coordination. And without peripheral feedback active touch, as first described by Charles Bell is impossible; their deficits show the intimate and essential relation between movement and sensation required for this haptic function.

Notes

1 One famous account of living with syphilis is *In the Land of Pain* by Alphone Daudet, translated by Julian Barnes (2002) Random House/Jonathan Cape, London.

2 Proprioception is a rather imprecise term, being used by some to relate to the perception of movement and position sense and by others as afferent information underpinning these percepts but not all perceived [22].

3 Large fibre loss can be due to a peripheral axonal loss, a neuropathy, or in the acute neuronopathy syndrome, damage to the cell body of the neuron in the dorsal root ganglia, hence 'neuronopathy'. Schaumburg actually developed an animal model of this by giving large does of vitamin B6 [5].

4 Friedreich's ataxia is caused by expansion of a GAA triplet located within the first intron of the frataxin gene on chromosome 9q13. Frataxin is a mitochondrial protein that plays a role in iron homeostasis, with deficiency resulting in mitochondrial iron accumulation, defects in specific mitochondrial enzymes and free-radical mediated cell death [23]. Though the precise pathological mechanism is unknown its dependence on a defect in oxidation allows potential treatment with antioxidants.

5 GL with her additional higher deafferentation does appear to use visual feedback for many movements, making them quite slow. IW in contrast seems to have accessed or developed (forward) motor programmes which allow for more complex movements which are under a looser visual supervision [24].

6 I once saw IW stagger when walking; he suddenly saw a pretty girl and his concentration drifted.

7 There have been several TV documentaries about these subjects of which BBC2 Horizon's '*The Man Who Lost His Body*', directed by Chris Rawlence in 1998 is the easiest to track down.

8 Once at work, Ian was given such a cup by a girl he fancied. He had always avoided them and so did not know how to grasp it. When he picked it up he squashed it and spilt the drink. So he took lots of such cups home that night and spent hours learning how to pick them up. His motor skills and abilities are the result of similar prolonged mental concentration and trial and error.

Selected readings

Cole J (1991) *Pride and a Daily Marathon*. London: Duckworth (reprinted by The MIT Press, London and Cambridge, MA, 1995)

Forget R, Lamarre Y (1987) Rapid elbow flexion in the absence of proprioceptive and cutaneous feedback. *Human Neurobiol* 6: 27–37

Rolke R, Magerl W, Andrews Campbell K, Schalber C, Caspari S, Birklein F, Treede RD (2006) Quantitative sensory testing: a comprehensive protocol for clinical trials. *Eur J Pain* 10: 77–86

Rothwell JC, Traub MM, Day BL, Obeso JA, Marsden CD, Thomas PK (1982) Manual motor performance in a deafferented man. *Brain* 105: 515–542

Sterman AB, Schaumburg HH, Asbury AK (1980) The acute sensory neuronopathy syndrome; a distinct clinical entity. *Ann Neurol* 7 (4): 354–358

Focal dystonia: diagnostic, therapy, rehabilitation

25

Eckart O. Altenmüller and Hans-Christian Jabusch

Definition, classification, symptoms and diagnostic

Definition

The general term dystonia is used to describe a syndrome characterised by involuntary sustained muscle contractions, frequently causing twisting and repetitive movements, or abnormal postures [1, 2]. If these symptoms are restricted to one body part, the syndrome is termed 'focal dystonia'. In task specific focal dystonia the most prominent characteristic is the degradation and loss of voluntary control of highly overlearned complex and skilled movement patterns in a specific sensory-motor task.

Classification

Dystonia can be classified by age of onset, cause, or by distribution of the body parts affected. Dystonia localised to a single body part such as the hand or neck is referred to as focal. *Focal dystonia* is by far the most frequent form of dystonia, accounting to about 90% of all dystonia syndromes. Dystonia localised to two contiguous body parts is referred to as *segmental dystonia*. Dystonia affecting body parts that are not next to each other is referred to as *multifocal dystonia*. Dystonia affecting one segment and another body part is classified as *generalised dystonia*. If it affects only one half of the body it is called *hemidystonia*.

Focal dystonia may be classified according to four criteria: age of onset, aetiology, affected body region and severity of symptoms [1]. According to Fahn, classification by age of onset represents the best prognostic indicator as to whether there will be a spread of dystonic symptoms to other body parts. While it is acknowledged that an age-criterion is rather arbitrary, onset before 28 (median of 9 years) is classified as early-onset primary dystonia and thereafter as late-onset dystonia (median of 45 years).

Second, current classification for aetiology divides focal dystonia into just two major categories, idiopathic or primary (including familial and sporadic forms), and symptomatic or secondary [1]. Secondary focal dystonia can be caused by structural abnormalities of the brain, or by metabolic disorders. For example, focal dystonia may occur as an early sign of Wilson's disease, a defect of copper metabolism that causes abnormal liver function and central nervous system symptoms such as tremor, and dystonia. Patients taking medications for psychiatric diseases such as schizophrenia or psychosis may develop dystonia as a drug reaction, e.g., after medication with Haloperidol. Rarely, focal dystonia may occur as a psychogenic dystonia in psychiatric disorders. Dystonia may also be associated with other neurological disorders. These are classified as dystonia-plus syndromes. Dystonia may be associated with Parkinson's disease or with myoclonus, another movement disorder which is characterised by involuntary muscle jerking. Focal dystonia may be part of other neurodegenerative disorders, for example Huntington's disease. For an in-depth presentation of the subdivisions of focal dystonia by etiologic classification, see [2].

Third, focal dystonia may be classified according to the affected region or the task involved in the dystonic movement: spasmodic torticollis (cervical dystonia), blepharospasm (eyelids), oromandibular dystonia, spastic dysphonia (vocal folds), or writer's or musician's cramp. For writer's or musician's cramp, the dystonic symptoms may affect a single finger up to an entire arm. With respect to the task specifically involved, embouchure dystonia may affect coordination of lips, tongues and breathing in brass and wind players, whereas pianist's cramp and violinist's cramp affect the control of hand movements in musicians when playing these instruments. Other activities producing dystonic movements involve playing golf (the 'yips') or dart ('dartism'), and – less frequently – playing tennis. Rarely, task specific dystonia occurs in other highly skilled hand movements, for example in watchmakers, dentists or surgeons.

Fourth, classification according to severity of symptoms distinguishes four stages. These stages are reflected in the most common clinical scaling instrument, the dystonia-severity scale [3].

a) *Simple focal dystonia*: Dystonic movements are only present during a specific task, e.g., exclusively during writing
b) *Complex focal dystonia*: Symptoms occur in multiple tasks, e.g., also when using utensils for eating, or when buttoning shirts
c) *Dystonic cramps*: Sustained muscle contractions, abnormal movements and postures of the body part occur spontaneously and are more or less continuously observable.
d) *Progressive dystonic cramps*: Sustained muscle contractions, abnormal movements and postures tend to spread to adjacent body parts

Some authors consider *focal tremor* as a special form of focal dystonia. It may occur as a predominant symptom as cervical tremor. Task specific focal tremor is occasionally seen as embouchure tremor or as task specific tremor of the bowing arm in string-players. Rosenbaum and Jankovic [4] found task-specific tremors in 10 (36%) out of 28 patients, eight of whom had task-specific focal dystonia. Commonly, focal tremor is a sustained postural and kinetic tremor at around 7 Hz frequencies with mostly irregular tremor amplitude. There is no tremor during rest [5].

Symptoms

The most prominent symptom of focal dystonia is loss of voluntary control and muscular coordination of movements, be it control of the head position in cervical dystonia, of eye-blinks in blepharospasm, or of skilled hand movements in focal hand dystonia [1, 2]. Typically, focal dystonia occurs without pain, although muscle aching can present after prolonged spasms. Lack of pain distinguishes it from repetitive strain injury or occupational fatigue syndrome. It is important to make this distinction bearing in mind that on the other hand prolonged pain syndromes may lead to symptomatic dystonia, possibly due to the degradation of sensory receptive fields in the somatosensory cortex (see below). The loss of muscular coordination is frequently accompanied by a co-contraction of antagonist muscle groups. For example in writer's cramp, the co-activation of wrist flexor and wrist extensor muscles is regularly observed. Dystonia can be worsened by stress and anxiety, whereas it may be relieved with relaxation and sleep. In Figure 1, typical postures of dystonic movements in musicians are shown.

In musicians with hand dystonia, an association exists between the instrument group and the localisation of focal dystonia. In instruments with different workload, different complexity of movements or different temporospatial precision for both hands, focal dystonia appears more often in the more heavily used hand. Keyboard musicians (piano, organ, harpsichord) and those with plucked instruments (guitar, e-bass) are primarily affected in the right hand. All these instruments are characterised by a higher workload in the right hand. Additionally, guitar playing requires higher temporospatial precision in the right hand compared to the left hand. Bowed string players who have a higher workload and complexity of movements in the left hand are predominantly affected in the left hand [6].

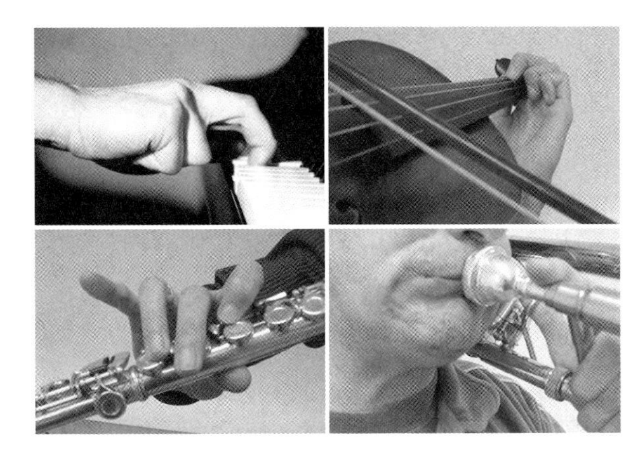

FIGURE 1. TYPICAL PATTERNS OF DYSTONIC POSTURE IN A PIANIST, A VIOLINIST, A FLUTIST AND A TROMBONE PLAYER.

Besides the dystonic movements, clinical neurological examination reveals no other signs of pathology. However, when testing more specifically, subtle sensory symptoms can be detected. Patients with focal hand dystonia have poor spatial resolution thresholds over affected hand areas with reduced accuracy of localisation [7]. They have less salient sensations evoked by touch stimulation, degraded form perception, reduced kinaesthetic and haptic abilities and problems with temporal-spatial processing [8, 9]. It should be noted that all these sensory deficits are subtle, mainly detected in group statistics, and that there is a large overlap with healthy controls. Up to now, a sensory testing battery with sufficient sensitivity and specificity to distinguish between patients with dystonia and healthy controls at an individual level is still lacking. However, though subtle, all of these deficits are consistent with a de-differentiated representation of somatosensory feedback information.

Another clinical sign, which hints toward an important role of sensory-motor integration in the pathophysiology of focal dystonia, is the 'sensory trick' phenomenon. In patients with cervical dystonia touching the face contralateral, but also ipsilateral to the direction of head rotation can reduce or abolish involuntary muscle activity [10]. In a similar way, patients with writer's cramp may benefit from modified surfaces of their writing tools. Musicians with musician's cramp frequently experience marked improvement of fine motor control when playing with a latex glove, thus changing the somatosensory input information [11].

Demographics

Focal dystonia follows Parkinson's disease and essential tremor as the most frequent movement disorder. Prevalence is estimated as 29.5 per 100,000 in the USA and 6.1 per 100,000 in Japan [12, 13]. In professional musicians, prevalence of musician's cramp is estimated up to 1% [11]. The higher prevalence in musicians may be due to a threshold phenomenon: Professional musicians with disturbances of fine motor control feel more disabled and seek medical advice at an earlier stage as compared to other professions.

There is a clear gender bias in prevalence of different focal dystonias. Blepharospasm and cervical dystonia occur more frequently in females than males (2, 3:1, respectively 1, 4:1). In contrast, writer's cramp occurs with 1, 3:1 more frequently in males. The most prominent gender imbalance is observed in musician's cramp: In Germany 89% of the patients are males [14].

Diagnostic

The diagnosis of dystonia is clinical and is usually made by an experienced neurologist. It is important that patients with task specific focal dystonias are seen when executing the specific task. Therefore, patients with musician's cramp should always be examined when playing the specific instrument. Video and audio documentation is recommended in order to analyse the actual symptoms properly and to document changes in symptoms during follow up.

Investigation of patients with dystonia will usually involve a physical examination and medical history to reveal secondary causes such as drug exposure or other family members affected, suggesting a genetic cause. Brain imaging may be performed in order to search for a structural abnormality causing the symptoms. Symptomatic forms of focal dystonia have been observed following head trauma or trauma of the cervical roots. In extremely rare cases, cortical as well as subcortical brain tumours may become symptomatic by dystonia-like symptoms. Laboratory testing may reveal abnormalities of copper metabolism associated with Wilson's disease. Genetic testing for the DYT1 gene is usually not performed unless there is a family history of similar symptoms.

Aetiology of focal dystonia and the role of the somatosensory system

Until today, the aetiology of focal hand dystonia is not completely understood, but is probably multifactorial. It may develop in individuals with a genetic history of dystonia [15] or as a consequence of alterations in the basal ganglia circuitry [16, 17]. Without going into the details, most studies of focal dystonia reveal abnormalities in three main areas: a) reduced inhibition in the motor system at cortical, subcortical and spinal levels, b) reduced sensory perception and integration, and 3) impaired sensorimotor integration. The latter changes are mainly believed to originate from dysfunctional brain plasticity. Such a dysfunctional plasticity has been described in the sensory thalamus [18]. Disorganised motor somatotopy could be found in the putamen of patients suffering from writers' cramp [19]. Finally, there is growing evidence for an abnormal cortical processing of sensory information as well as degraded representation of motor functions in patients with focal dystonia. In monkeys, repetitive movements induced symptoms of focal hand dystonia and a distortion of the cortical somatosensory representation [20] suggesting that practice-induced alterations in cortical processing may play a role in focal hand dystonia.

Altered sensory perception and mapping in focal dystonia

Several studies have demonstrated that the ability to perceive two stimuli as temporally or spatially separate is impaired in patients with focal hand dystonia [7, 8]. A recent study compared spatial sensory discrimination abilities in patients with primary generalised DYT1 dystonia and other forms of dystonia, especially focal hand dystonia. Interestingly, only the focal hand dystonia group showed an increased spatial discrimination threshold, pronounced in the dominant hand, whereas no abnormalities of spatial discrimination were found in DYT1 cases [21]. This finding suggests that altered sensory processing might play a different pathophysiological role in these conditions.

Using somatosensory evoked potential technology, it was demonstrated that in the somatosensory cortex the topographical location of sensory inputs from individual fingers overlap more in patients with writer's cramp than in healthy controls [22]. Similar observations have been made using magnetoencephalography [23]. Elbert et al. [24] showed that there is an overlap of the representational zones of the digits in primary somatosensory cortex for the affected hand of musicians with dystonia compared with the representations of the digits in non-musician control subjects (see Fig. 2).

Since in healthy musicians, an increase of sensory finger representations has been described and interpreted as adaptive plastic changes to conform to the current needs and experiences of the individual [25], it could be speculated that these changes develop too far in musicians suffering from dystonia, shifting brain plasticity from a beneficial to a mal-adaptation [1]. On the behavioural level this notion is supported by the fact that healthy musicians have lowered two-point discrimination thresholds as compared to non-musician controls [26], whereas two point-discrimination thresholds are increased in musicians suffering from dystonia.

It should be kept in mind that these mapping studies done at rest have to be interpreted cau-

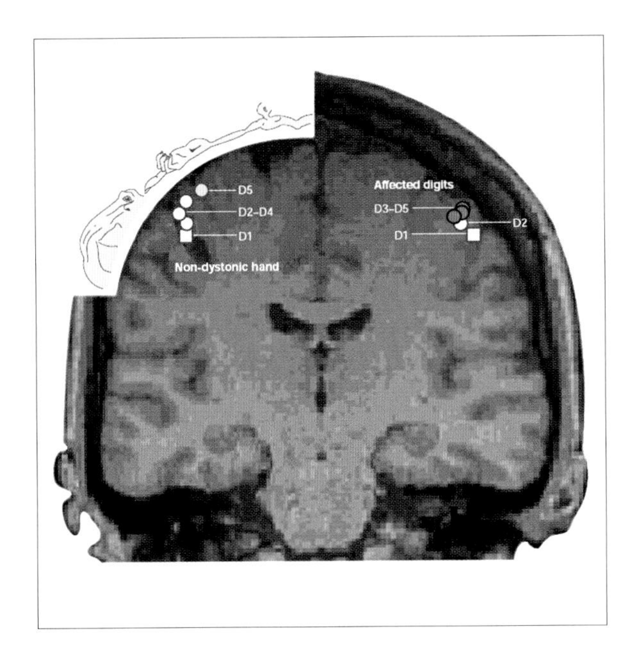

Figure 2. Neuronal correlates of dysfunctional plasticity

Fusion of the somatosensory representation of single digits of the hand in a musician suffering from focal dystonia. The best fitting dipoles explaining the evoked magnetic fields following sensory stimulation of single fingers are shown projected on the individual's MRI. Whereas for the non-affected hand the typical homuncular organisation (see inset) reveals a distance of about 2.5 cm between the sources for the thumb and the little finger (white square and brown circle on the left) the somatosensory representations of the fingers on the dystonic side are blurred resulting from a fusion of the neural networks which process incoming sensory stimuli from different fingers (red circles) (Modified from Elbert et al. [24] and Münte et al. [55] with permission).

tiously. Braun et al. [28] investigated dynamic alterations in the functional organisation of the primary somatosensory cortex (SI) during writing in writer's cramp patients and healthy controls with neuromagnetic source imaging. Similar to the studies cited above, they found a reduced distance between cortical representations of thumb and little finger of the affected hand in patients when tested at rest. However, when patients were tested during writing, they showed normal modulation of the functional organisation of SI and the cortical distances between representations of thumb and little finger were increased to normal range. Thus, despite alterations in the organisation of primary somatosensory cortex in writer's cramp, the capability of SI to adapt dynamically to different tasks is astonishingly not impaired [28].

Local pain and intensified sensory input due to various origins like nerve entrapment or trauma or overuse have been described as potential triggers of dystonia [29, 30, 11]. In a large group of musicians suffering from focal dystonia, local pain preceded focal dystonia in 9% of the patients [6]. Interestingly, there are obvious parallels of abnormal cortical processing of sensory information and cortical reorganisation in patients with chronic pain and those with focal dystonia [31, 32]. This suggests parallels in the pathophysiological mechanisms in these selected groups of patients.

Sensory-motor integration in focal dystonia

Rosenkranz et al. used focal vibratory input to individual hand muscles to produce sensory input while the excitability of corticospinal outputs to the vibrated and other hand muscles was evaluated with transcranial magnetic stimulation [33]. In musicians with dystonia, focal vibratory input to hand muscles resulted in abnormally suppressed short-latency intracortical inhibition of all other hand muscles examined, independent of their functional connectivity with the vibrated muscles. In healthy non-musicians, such vibratory input increased short-latency intracortical inhibition of neighbour muscles. In healthy musicians, however, vibratory input resulted in a suppressed short-latency intracortical inhibition only of those neighbour muscles functionally connected with the vibrated muscles. In patients with writer's cramp, short-latency intracortical inhibition of neighbour muscles was unchanged. It was concluded that the pathophysiology of musician's dystonia differs from that of writer's cramp [34]. Interestingly, short proprioceptive training with directed attention to vibratory stimuli applied to muscles of the thumb could partially reverse the pathological pattern in patients with musician's dystonia, but not in patients suffering from writer's cramp [35].

The abovementioned findings of an increased workload and complexity of movements as well as increased spatial and temporal sensorimotor requirements being related to musician's dystonia harmonise with the hypothesis of a pathomechanism which includes practice-induced alterations. A higher workload in affected musicians might additionally be reflected by the observation of a more perfectionist attitude in dystonic musicians – even before onset of dystonia – compared to other musicians [36] and by the finding of a preponderance of soloists (51%) in our sample of dystonic musicians [6]. A perfectionist attitude as well as an outstanding professional position might point at a different and more intense working behaviour resulting in a higher workload in those musicians who develop focal dystonia

Therapy and rehabilitation

General treatment strategies

Treatment for focal dystonia is mostly symptomatic and depends on the type of dystonia. Optimally, a multidisciplinary team involving neurologists, psychologists, and physical and occupational therapists should interact. Dystonia that is associated or caused by known aetiologies such as antipsychotic drugs or Wilson's disease may be improved by treating the underlying disease. The medical treatments available may be grouped into oral medications, injections with botulinum toxin and surgical modalities. Furthermore several retraining techniques are established for rehabilitation of focal hand dystonia.

Various oral medications have been used for the symptomatic treatment of focal dystonia. These medications affect different neurochemical systems thought to be important in causing dystonia. In the special case of Dopa-responsive dystonia which is generally an autosomal dominant disorder, the synthesis of dopamine is compromised. The majority of patients suffering from Dopa-responsive dystonia experience a long-term benefit from treatment with levodopa with continued clinical stability for many years or even decades [37]. The anticholinergic drug trihexiphenidyl has proven to be the most effective agent in musician's hand dystonia [38]. Occasionally, patients may benefit from baclofen, or antiepileptic drugs such as phenytoin, or primidon. Some patients show marked improvement of hand dystonia after medication with tetrahydrocannabinol. The impact on motor control is in keeping with the physiological findings of cannabinoid receptor binding in the globus pallidus and the suggestion that cannabinoid receptor stimulation might reduce overactivity of the lateral globus pallidus with the consequence of reduced dystonic symptoms [39].

Chemical denervation using botulinum toxin has been used for many forms of focal dystonia with considerable success. Botulinum toxin blocks the transmission of nerve impulses to the muscle and weakens the overactive muscles involved. Botulinum toxin injections are presently considered as the therapy of choice in cervical dystonia and blepharospasm (for a review see [40]). Results in focal hand dystonia and musician's cramp depend on the injection technique and on the precise localisation of the dystonic muscles. Especially in musician's cramp, focal dystonia usually involves dystonic movements – for example the involuntary flexion of the middle finger – and compensating movements, for example the extension of the adjacent index and ring fingers. Erroneous injection into the compensating extensor muscles consequently will lead to a deterioration of symptoms. With correct identification of dystonic movements and application of an electromyographical guided injection technique for proper localisation of the respective muscular fascicles, results of botulinum toxin injections are frequently satisfying. However, it should be emphasised that this treatment is symptomatic, requiring the patients to return at regular intervals of about 3–5 months to receive the injections.

In our series, injections of botulinum toxin were applied in 71 musicians suffering from hand dystonia. 57% of patients reported long-term improvement [6]. A meta-analysis of the patients' data revealed that botulinum-toxin treatment was successful in those patients in which primary dystonic movements could be clearly distinguished from secondary compensatory movements. This was difficult when compensatory movements were more pronounced than primary dystonic movements. The best outcome was reported after injections in forearm muscles. Musicians with embouchure dystonia and with dystonia affecting the upper arm and shoulder muscles did not benefit from botulinum-toxin treatment [6, 41].

As a promising novel approach deep brain stimulation of the globus pallidus or of the thalamus has been proven to be useful in selected patients with disabling symptoms. The long-term benefit of these invasive therapies on symptoms of dystonia has yet to be validated [42].

Rehabilitation of focal hand dystonia

In the last decade, based on the hypothesis of aberrant learning and maladaptive plasticity, several treatment protocols were designed in order to address a reversal of the central nervous system degradation that accounts for one origin and expression of hand dystonia. These therapies aim at re-differentiation of the disturbed somatotopic hand representations.

The principles of neuroplasticity are most closely aligned to constraint induced training, immobilisation, learning based sensorimotor training, and retraining at the instrument. In a study of 11 musicians with focal hand dystonia, the unaffected fingers of the dystonic hands were constrained with splints [43]. For 2 weeks, 2–3 h a day, the patients were trained to use the dystonic fingers on fine motor tasks emphasising instrumental play. When control of the most involved digits improved, then gradually the splints on the less involved digits were removed and practice at the instrument was increased. Eight out of 11 musicians showed an improvement of their motor skills. These results were supported by a brain imaging study, demonstrating that patients with focal hand dystonia showed some reorganisation of receptive fields in the somatosensory cortex after treatment [44].

Based on the evidence that topographical representation of the hand is abnormal in patients with focal hand dystonia, immobilisation of the hand has been proposed as an appropriate treatment at the first signs of dystonia. In individuals without dystonia, limb immobilisation leads to a shrinkage of the cortical representation of the immobilised limb [45]. Thus, the objective of immobilisation in patients with limb dystonia is to place the affected finger, hand, wrist, and forearm into a static splint to stop all movement in order to decrease the size of the representational area of the involved hand [46]: 19 patients wore the splint 24 h a day for 4–6 weeks. Subsequently, a careful retraining and physiotherapy was applied. The outcome of this study was heterogeneous: in six out of 19 patients, there was marked improvement and in the other group, no or only minor improvements could be observed. Immobilisation seemed more effective in young patients with a recent onset (< 1 year) of dystonic movements related to an overuse syndrome. It therefore was hypothesised that patients with a long history of dystonia had a more extensive degradation in the sensory motor representation. Furthermore, symptomatic dystonia triggered by chronic pain may be based on a different pathophysiological mechanism as compared to idiopathic hand dystonia [47]. Learning-based sensorimotor training (LBST) is another approach to rehabilitation for patients with focal hand dystonia [48, 49]. The objective of LBST is to re-define spatial and temporal processing capacities in the sensory and motor cortices in order to restore task specific skills, e.g., writing. Different exercise strategies address sub-tasks of the target tasks emphasising different aspects of sensory feedback (i.e., somatosensation, proprioception, kinaesthesia, haptics). The patients initially make relatively simple distinctions about relatively large stimuli, with task difficulty increasing as each patient masters progressively more difficult distinctions. One set of tasks includes reading Braille or embossed letters, making roughness estimations, and performing grating orientation discriminations. A series of haptic tasks requires the subjects to manipulate and identify surface features and shapes of three-dimensional objects held in their hands.

This training protocol has been tested in a series of patients with focal hand dystonia (for a review see [47]). Subjects met with a physical therapist for 1–2 h per week, and were trained on multiple tasks designed to engage the broad types of sensory inputs that provide feedback for fine hand motor control. After appropriate instruction and demonstration in the clinic, subjects were instructed to practice the tasks for 30–60 min/day at home. As a rule, to achieve effective reversals of the focal hand dystonia condition to a level at which the trainee could return to work, training was continued for 6–12 weeks. In the first study that focused on this training protocol, all 12 trained patients showed improvements in performance ability on a variety of tests of sensory discrimination, fine motor accuracy and speed, strength, flexibility and functional independence [50]. Motor control improved to an average of 70–94% of normal. All but three patients returned to their usual work.

Zeuner et al. [51] also reported improvement in motor control in patients with focal hand dystonia following sensory training: 60% of patients suffering from writer's cramp shortened the time they needed to write a standard paragraph after they had learned Braille reading. Improved sensory perception correlated positively with improvement on the Fahn dystonia scale [51]. Three patients who continued training for 1 year showed further improvement in writing, in sensory perception and in self-rating scales [52].

Pedagogical retraining has been applied in patients with all forms of musician's dystonia. They comprised a variety of behavioural approaches which were taking place under the supervision of instructors and included elements based on the following principles (e.g., [53]): 1) movements of affected body parts were limited to a level of tempo and force at which the dystonic movement would not occur; 2) compensatory movements (e.g., of adjacent fingers) were avoided, partially under the application of splints; 3) instant visual feedback with mirrors or monitors helped patients to recognise dystonic and non-dystonic movements; 4) body awareness techniques (e.g., Feldenkrais®) were applied to increase the patient's perception of non-dystonic movements. In common to all these therapies is the 'long-term' approach. For example, in a large series of 145 musicians an average rehabilitation time of 24 months, after which 35 musicians returned to playing in public, was reported [54].

In our series of musicians suffering from focal dystonia, pedagogical retraining was applied in 24 patients according to the aforementioned guidelines. 12 patients (50%) experienced improvement. Patients reporting an improvement had undergone this treatment for an average of 28 months (range 3–72). Two patients with embouchure dystonia took part in pedagogical retraining and reported no improvement [38].

Summary

Focal dystonia is still difficult to treat or to rehabilitate. With the available mostly symptomatic medical therapies, the majority of patients manage to stay in their profession, many of them, however, with substantial compromises. There is a pressing need for novel therapies. The reported rehabilitation strategies aiming at reversal of maladaptive plastic changes are promising; however, they still require considerable compliance and stamina for several years from the side of the patients without guarantee of success. Long-term outcome of these rehabilitative strategies remains to be investigated in order to confirm the beneficial effects. For example, from the eight musician patients having improved under constraint induced therapy, four returned to injections with botulinum-toxin. Since phenomenological and epidemiological data and results from electrophysiological studies and from follow-up studies imply a behavioural component in the development as well as in the treatment especially of musician's dystonia, future research is required to identify 'beneficial behaviour' on the instrument. This might, possibly, also be of help for finding strategies with the particular aim of prevention of focal dystonia.

Selected readings

Altenmüller E (2006) The end of the song: Robert Schumann's focal dystonia. In: E Altenmüller, J Kesselring, M Wiesendanger (eds): *Music, motor control and the brain.* Oxford University Press, Oxford

Byl NN, Priori A (2006) The development of focal dystonia in musicians as a consequence of maladaptive plasticity: implications for intervention. In: E Altenmüller, J Kesselring, M Wiesendanger (eds): *Music, motor control and the brain.* Oxford University Press, Oxford

Jabusch HC, Altenmüller E (2006) Epidemiology, phenomenology and therapy of musician's cramp. In: E Altenmüller, J Kesselring, M Wiesendanger (eds): *Music, motor control and the brain.* Oxford University Press, Oxford

Jankovic J (ed) (2005) *Dystonia.* Seminars in Clinical Neurology, Vol 3. Demos Medical Publishing LLC, New York

Topka H, Jankovic J, Dichgans J (2002) Dyskinesias. In: T Brandt, L Caplan, J Dichgans, HC Diener, C Kennard (eds): *Neurological disorders: course and treatment.* Academic Press, New York

Important websites

Dystonia Medical Research Foundation, USA:
http://www.dystonia-foundation.org/

European Dystonia Federation:
http://www.dystonia-europe.org/europe/

Self-injurious behavior

Armando R. Favazza

26

Self-injurious behavior (SIB) has been around for a long time. Many caves in Southern France contain hand imprints on their walls and in one cave, at Gargas, the 20,000 year old imprints display the absence of all tips except for the thumb. The 5th century B.C.E. historian Herodotus described the actions of a probably psychotic Spartan leader, Cleomenes, who mutilated him by slicing his flesh into strips with a knife; starting with his shins he worked upwards to his thighs, hips, and sides until he reached his abdomen which he chopped into mincemeat. The Gospel of Mark 5:5 describes a repetitive self-injurer, a man who "*night and day would cry aloud among the tombs and on the hillsides and cut himself with stones.*"

Despite these early references and a number of cases and small scale studies in the 20th century, SIB has only recently become the object of focused psychiatric scrutiny. Traditionally, it has been trivialized (wrist-cutting), misidentified (suicide attempt), and regarded solely as a criterion of borderline personality disorder.

It was generally regarded as grotesque and senseless. In the words of a highly respected psychiatrist, "*The typical clinician (including myself) treating a patient who self-mutilates is often left feeling a combination of helpless, horrified, guilty, furious, betrayed, disgusted, and sad*" [1].

With the publication of the book *Bodies Under Siege* in 1987, and especially in the 1996 second edition which was subtitled '*Self-mutilation and Body Modification in Culture and Psychiatry*', Favazza stripped away the mysterious aura that had surrounded SIB and demonstrated its purposefulness in culturally sanctioned rituals as well as in deviant, pathological disorders [2]. He showed that SIB is not alien to the human condition but rather is culturally and psychologically embedded in the profound, elemental experiences of healing, spirituality, and social amity. Culturally sanctioned and deviant SIB serve an identical purpose, namely an attempt to correct or prevent a pathological or destabilizing condition that threatens the community, the individual, or both.

Mental homeostasis and stress

Homeostasis is the property of organisms to self-regulate themselves in order to provide an appropriate milieu for cells and body tissue to function properly. Homeostasis may be threatened by stressful stimuli that can be anatomical, physical, chemical, physiological, or mental. Psychiatrists are especially concerned about disruptions in the homeostasis of the central nervous system which is the most complex organ system of the human body and is the site of the abstract 'organ' known as the mind. Mind implies human consciousness, but also has an unconscious component, and is manifested especially in thought, perception, emotion, will, memory and imagination.

The list of possible stressful stimuli is almost endless; some examples include childhood abuse (or the memory of it); divorce; betrayal by a significant person; death of a loved one, friend, or pet; loss of a job, financial security, or property; unwanted or forced movement to a new locale; victimization; witnessing or experiencing a traumatic event (or the memory of it); incarceration; public embarrassment; debilitating pain; and declining health due to such causes as a cerebrovascular accident (stroke), cancer (especially in the brain), Parkinson's disease, endocrine disorders, HIV-AIDS, sleep apnea, alcoholism, arthritis, and multiple sclerosis.

When stressful stimuli threaten the brain's homeostasis the result is the development of mental symptoms which may range from mild discomfort to a disruptive loss of contact with reality (psychosis) and even suicide. Just as the body has automatic defense mechanisms that attempt to re-establish homeostasis, e.g. certain cells in the blood that fight infection may be released, so too does the mind have automatic defense mechanisms that work to minimize the effects of stress. One defensive process, for example, may cause a person to rationalize the downward slide into a primitive, unstable psychosis. Such persons may reconstruct their experiences by developing a delusional, explanatory system in which they believe that they are being persecuted by aliens or the police. Delusional persons lack insight and lead miserable lives but feel some consolation because they feel that they truly understand what is happening to them. Repression is another automatic process that serves to push conflictual mental stimuli into the unconscious, away from mental awareness.

Adaptation

No one is immune from stress throughout a lifetime. Minor stress can usually be controlled by such activities as drinking a little alcohol; smoking a cigarette (nicotine is a marvelous drug with an ability to calm down a person or to elevate a mild depressed mood); taking a vacation; exercise; exposure to sunlight; listening to calming music; or meditation.

Touch can play a role in stress reduction. Children can be calmed by putting a 'pacifier' in their mouths or by cradling them in one's arms. First the mother's breast comforts the stressed infant, then the infant's own hand which is placed in its mouth, then other objects such as a baby blanket or a teddy bear. These objects become very special to the infant who assumes control over them, endows them with a unique vitality, gets upset if anyone removes them, cuddles them lovingly, and sometimes hates them. These special 'transitional objects' are the very stuff of illusion that exist mentally somewhere between inner subjective reality and outer objective reality. As the child matures, the special objects are neither mourned nor forgotten, but simply lose their meaning.

Massage is one of the most used methods of dealing with stress which commonly causes muscular tension, bodily aches, and headaches (when scalp and neck muscles contract). The effects of massage are enhanced by a soothing environment with calming music and aromas. Rocking with or without a chair may reduce stress. Sexual stimulation may decrease anxiety or may raise lowered levels of excitation.

All the examples presented are innocuous and useful for minor stress. However, when stressful stimuli reach a pathological level then different, dramatic, often devastating methods must be used to achieve adaptive homeostasis. Perhaps the most startling and counter-intuitive method of adaptation involves self-mutilation or, as it is more commonly called nowadays, self-injurious behavior.

Self-injurious behavior

SIB is defined as the direct, deliberate alteration or destruction of body tissue without conscious suicidal intent. In fact, it is the opposite of suicidal behavior whose goal is death. People who engage in SIB want to stay alive but they want to be free from the pathological symptoms produced by severe stress. SIB does not include the ingestion of poisons or overdosing on pills where the dynamics are different. In the case of ingesting pills or poison, the harm that is caused is unpredictable, uncertain, ambiguous, and basically invisible. With SIB the degree of self-harm is clear, unambiguous, predictable as to course, highly visible, and often results in sustained or permanent visible bodily disfigurement [3]. SIB is a surface phenomenon and depends upon touch.

Major SIB

This involves the destruction of vitally important tissue and organs such as eyeballs, genitals, fingers, and limbs. It is not a common behavior but neither is it rare. It is most associated with psychosis (often in conjunction with schizophrenia, major depression, and bipolar disorder), drug intoxication, and transsexualism.

A prototypical eye enucleator is a young man who is imprisoned and depressed. If the depression deepens, psychosis develops, and the man reads the Bible (almost every jail cell contains a copy); he may stumble across an almost identical passage in the Gospels of Mark 6: 22–23 and Matthew 5: 28–29. These verses state that an offending eye or hand must be cut out because it is better to enter heaven with only part of a body than to descend to hell with an intact body. In one such case a 24-year-old psychologically depressed man was in jail for a parole violation. He claimed that he saw differently from each eye and that his left eye offended him. After reading Matthew's words in the Bible and seeing a vision of Christ and a gold ball in front of his left eye, he pulled out his left eyeball in order to purify himself. He stated that he belonged to the "*youthful cosmologically enlightened movement*". Although he had planned to cut off his hand "*to quiet his mother*", he decided to take out his eye because hate rays were emanating from it. After the act he developed a Christ-like affect and demeanor and claimed that he was cleansed, at peace, and happy.

Self-enucleation in order to rid the body of an evil spirit is a psychotic elaboration of the ancient, widespread, worldwide folk belief in the 'evil eye'. The 'spirit infested' eye of a psychotic person becomes a terrifying, aggressive organ that can change the course of events, maim and kill others, and force one to look at tempting or forbidden persons or things. By isolating the evil spirit in an eye, the psychotic person gains some control over terror and aggression. Should all other attempts fail to control the spirit, the desperate act of enucleation can remove it from the person's body. In normal religious life the struggle with devils may be won through prayer but the demon of psychosis demands a more radical tactic.

Self-castration by men usually involves religious and/or sexual themes. In one religious case a 23-year-old paranoid schizophrenic man suffered from nightmares and terrors, became withdrawn, and read the Bible incessantly after he was refused admission to a Catholic seminary. He focused on a passage in the Gospel of Matthew 19: 12 "*For there are eunuchs who were born thus from their mother's womb, and there are eunuchs who were made eunuchs by men, and there are eunuchs who have made themselves eunuchs for the kingdom of heaven's sake. He who is able to accept it, let him accept it.*" When both his dog and an injured bird that he had rescued died, he became acutely psychotic. He stated that his penis and his aunt's clitoris "*were standing out at each other*". He then cut off his penis with a razor blade. The act was sudden and painless and was followed by a sense of instant relief.

Another case involved a 29-year-old man with a history of recurrent major depression with psychotic features. He insisted that he was heterosexual yet worked as a homosexual prostitute. He waded into the ocean and cut off his scrotum and testicles with a butcher knife. He brought his testicles home and handed them to his mother whom he felt had 'half died' at his birth. By presenting her with his testicles he intended to give back to her the life she had given him at birth. After the castration his anxiety, depression, and many of his delusions diminished. He expressed no regrets and was especially happy at the loss of some facial hair.

An example of SIB following drug intoxication is an 18-year-old youth who was found wandering in the street nude with his right eyeball in his hand. He had taken LSD for four consecutive days, during which he was forced into a homosexual episode. He then felt that he was going to die, and that the devil controlled his mind, and that he should obey the Bible and pluck out his eye because he had offended God. He said, "*My mind was weak because of the LSD so that the devil possessed me. Now I've got the devil out of my mind since I plucked my right eye out.*"

The theme of SIB and transsexualism is seen in the case of a 25-year-old man who had been convinced since childhood that he really was a female. He went to school and became a licensed practical nurse. He then went to work for an urologist and assisted in numerous operations. After taking some surgical instruments home he neatly dissected his testicles, wrapped them in a towel, and carried them to a university hospital emergency room where he politely asked the surgical resident to complete the job.

Stereotypic SIB

This involves behaviors that are monotonously repetitive and even may have a rhythmic pattern. It is usually impossible to ascertain any symbolic meaning, thought content, or associated emotional affect with the behaviors which are often carried out in the presence of others. Stereotypic self-injuries seem to be driven by a primarily biological imperative to reach some sort of homeostasis by harming themselves shamelessly and without guile.

The most common behavior is head banging. It may cause seizures, vomiting, and even detachment of the retina. The highest risk was reported in an autistic, mentally retarded boy who banged his head 5,400 times an hour. The most likely stresses that head-banging tries to alleviate are understimulation of the brain, a perceived lack of attention by caretakers, and the frustrations that some mentally retarded persons face due to their inability to make sense of and to deal with their environment. A fanciful speculation is that head banging is an attempt to re-experience the comfort of hearing a mother's heartbeat.

Other types of stereotypic SIB include self-slapping and punching of the head and face, throat and eye gouging, self-biting and severe skin scratching. A number of neuropsychiatric syndromes are associated with this type of SIB. The rare Lesch-Nyhan syndrome, for example, found only in males, is caused by an inborn error of purine metabolism which leads to an excess of uric acid in the blood. Affected children may develop strange movements of their limbs, mental retardation, and compulsively self-aggressive behavior. They bite off their lip tissue as well as parts of their tongue and fingers. The mutilations are remarkable for their awesome rapidity; upon release from restraints, a child's hand may go instantly to his mouth where it is savagely bitten.

Many people are familiar with Tourette's syndrome which is characterized by multiple simple and complex tics (sudden, recurrent, stereotyped movements or vocal sounds such as uncontrolled vulgar words). In about a third of cases SIB occurs such as head banging, self-hitting, tooth extraction, eyeball pressing, sticking objects in the eyeballs, joint dislocation, and biting of the lips, cheek, and tongue.

Compulsive SIB

This involves behaviors that occur many times daily, are repetitive, ritualistic, and almost automatic because the self-injurers may not even be totally aware of what they are doing. The most common disorder is trichotillomania which involves pulling out one's hair from anywhere on the body, but especially from the scalp, eyebrows, and eyelashes. Early onset cases often are benign and have a short course. Late onset (early adolescence) cases may follow a severe, chronic course with co-morbid nail biting, skin scratching, and anxiety and mood disorders. Plucked hairs may be eaten or piled up and then discarded. Hair pulling episodes may vary from 4–5 h sessions in which hundreds of hairs may be plucked to brief episodes in which only a few hairs at a time are pulled out but which re-occur dozens of times a day. Swallowed material may form hairballs in the stomach, leading to anemia, abdominal pain, gastric bleeding, nausea, vomiting, bowel obstructions, and perforation of the stomach or bowel.

Hair pulling usually occurs in private but sometimes it occurs in the presence of others.

It increases both during periods of stress and of relaxation. Most hair pullers report mounting tension or gratification with the behavior which may occur automatically without elaborate thought content or emotion. Other examples of compulsive SIB are biting finger and toe nails to the quick, and severe skin scratching. Some people have tactile hallucinations and believe that their skin is infested with parasites or other organisms. They may then devote their lives to trying to eliminate the imaginary objects by gouging their skin or putting lye and other caustic chemicals on their skin.

Impulsive SIB

This involves behaviors such as skin cutting, carving, and burning; breaking bones in the hands or feet; and interfering with wound healing. This type of SIB is somewhat more common in females, typically starts in early adolescence, and has its highest incidence in persons from age 15–25 years. It may occur sporadically or may become an 'addictive' repetitive behavior. When repetitive it may develop into a morbid Deliberate Self-Harm Syndrome in which self-injurers brood about harming themselves and self-identify as 'cutters' or 'burners'. Other impulsive behaviors such as bulimia and kleptomania may emerge. The syndrome often lasts for 10–15 years of misery with a lot of social morbidity because afflicted persons are embarrassed by extensive scars all over their body. They may attempt suicide, typically by an overdose of pills, secondary to the demoralization that results from their inability to control their SIB.

The mental illnesses associated with impulsive SIB are generalized anxiety, panic disorder, post-traumatic stress disorder, moderate (dysthymic) and major depression, bipolar disorder, dissociative disorders, and antisocial, borderline, and histrionic personality disorders. The highest prevalence is found in persons living in a repressive environment such as prisons or some locked-down state hospitals and juvenile detention facilities.

The reason why people engage in impulsive SIB is that it provides rapid relief from a host of distressing thoughts and emotions. Unfortunately the relief lasts several days or, less frequently, several weeks. In a 1989 study most subjects found relief after accidentally injuring themselves [4], but nowadays most cutters are familiar with the behaviors due to widespread media coverage. There are many situations in which the SIB occurs; among the themes most commonly articulated by patients are the following:

Tension release

People experiencing corrosive mounting tension and anxiety liken the relief they get from cutting to popping a balloon or lancing a boil. A 39-year-old school teacher said, "*Before every act of self-mutilation I felt emotionally overwhelmed. The sight of my blood seems to release unbearable tension. At first a bruise or scratch were effective, but later it took more blood to ease the explosive tension. Now I cut my veins to get results.*"

Relieving emotional deadness

Persons, many of whom learned to dissociate their sense of self from their bodies as a coping mechanism during periods of childhood abuse, many automatically experience dissociation when stressed as adults. They report emotional deadness, a diminished ability to experience normal sensations, estrangement from the environment, an altered perception of time, detachment from their bodies, and a sense of unreality. Only by cutting and seeing the blood on their skin can they feel real again.

Establishing control

Persons troubled by a sense of loss of control, e.g. rapidly fluctuating, 'roller coaster' emotions and racing thoughts, may regain stability through SIB. A 32-year-old salesman said, "*When I feel hyper my mind races and I can't sleep, then I almost always hit my legs and feet with a hammer.*

I feel the jolt of pain but it's okay because I am in control of it. Then I calm down."

Feeling secure and unique

An 18-year-old waitress said, "*Cutting, burning, and poking needles in my arms is a security for me because I know that if all else fails and leaves me feeling emotionless and empty, the pain and blood will always be there for me.*" A 24-year-old college student admitted, "*I cut myself because I need to be special. I was always taken for granted, invisible. Now, although I rarely expose my scars, I feel a smug pride. I'm not eager to give it up. Take it away form me, and I'm like everyone else.*" Some cutters liken the flow of blood over their body to a pleasant warm bath or a comforting baby blanket.

Overcoming depression

A 33-year-old homemaker explained, "*My father tried to have sex with me when I was 22 years old, and I felt like a frightened child. I became very depressed and full of hate for myself. After I cut myself, I felt relief.*" Persons who are guilt ridden often report feeling better after SIB because they feel they are paying for their 'crimes'.

Dealing with pressure from multiple personalities

Persons with the rare condition, multiple personality disorder, may attribute their SIB to the activities of one or more of their personalities. A 39-year-old nurse said, "*The personalities punish one another. They sometimes use intravenous needles to get the blood running. Occasionally you wonder why for a moment, but the why isn't important. You first have to do it to keep them satisfied. Hit your head with a hair brush. Scrape your vagina with a fish scaler. Sara (a personality) couldn't hurt my stepfather, so she hurt me. She cut my wrist with a big long knife to punish me because I'm weak.*"

Becoming euphoric

A 24-year-old male prisoner confessed, "*When I cut myself, I get such a high feeling. It feels so good, and I feel deprived when I can't do it. I have never found a substitute for this high feeling.*" A 30-year-old college instructor said, "*Self-mutilation allows me to live on the edge. It is titillation to see just how far I can go and how much real pain I can endure. Too often I am afraid to make choices. When I cut I can be swept away on a tidal wave of feelings.*"

Enhancing sexuality

While most cutters act to diminish sexual feelings, a few say the opposite. A college student said, "*When I stick open scissors up my vagina, the cold, hard steel gives me an explosive orgasm. And the blood makes it even more exciting.*"

Relief from alienation

A 28-year-old librarian said, "*When 'it' strikes, cutting is the only thing that provides relief; 'it' being a frantic, desperate, profound sense of alienation from the rest of the world, primarily loved ones. Sometimes I think that a dose of the good things – loving, hugging – would do it, but it's simpler to reach for a razor blade.*"

Influencing others

Sometimes impulsive SIB is used not for primary relief of anxiety, depression, etc., but rather for the secondary gain of influencing others who feel uneasy or threatened by the behavior. Prisoners, for example, who want to be switched to a different cell may cut themselves until the jailers give in to their demands. In a different, not uncommon scenario, a 23-year-old secretary said, "*I cut myself – it wasn't a suicide attempt – and told my boyfriend who was talking about breaking up our relationship, 'This is how deeply you have hurt me.'*"

The themes that I have presented do not exhaust all the possible reasons that people impulsively engage in SIB but they are fairly representative. Each person has his or her own story to tell.

Summary

A wide range of external and internal, physical and mental stimuli can stress people and cause mild to severe disease. When automatic physical and mental processes fail to restore equilibrium and homeostasis, people do all sorts of things to make themselves feel better. Short of suicide, the most dramatic and drastic attempts involve SIB which in all cases depends upon touching the body in a morbid way. The stabilizing effects of aggressively touching, even attacking one's body appear to be counter-intuitive but from a broad historical and cultural perspective SIB is central to many thickly layered and meaningful rituals. Think of the many saints who have mortified their bodies in order to achieve a deeper, more authentic spiritual relationship with God. Or consider the aboriginal boys and girls who to this day in coming-of-age ceremonies allow their teeth to be knocked out, their skin to be scarified, and their genitals cut as a demonstration of their willingness to give up their childish ways and to preserve the social stability of their tribe by accepting new roles as adults. Look at shamans who, in order to develop the capacity to heal others, undergo a ghastly process of evisceration and dismemberment in dream sequences that are perceived as a vivid reality. Mentally ill self-injurers seek what we all seek: an ordered life, spiritual peace, and a healthy mind in a healthy body. Their heavy-handed methods are upsetting to those of us who try to achieve these goals in a more tranquil manner, but the methods rest firmly on the dimly perceived bedrock of the human experience. The healing touch may sometimes be bloody.

Selected readings

Walsh B (2006) *Treating self-injury*. New York: Guilford

Conterio K, Lader W (1998) *Bodily harm*. New York: Hyperion

Favazza A (1998) The coming of age of self-mutilation. *J Nervous Mental Disorders* 186: 259–286

Haptic perception in infancy and first acquisition of object words: Developmental and clinical approach

Christiane Kiese-Himmel

Introduction

To recognise the properties of physical objects and build a mental representation of them is a prerequisite for the acquisition of words referring to objects, words that hold a key role in early lexical development. Object words account for about 60% of nominals. Words that name solid objects encode the first set of meaning, they can be understood by a healthy infant beginning between the 10th and 12th month and, soon after, be reproduced in child-specific language [1–6]. At first object words are generalised to new instances on the basis of perceivable similarities.

A healthy infant possesses an innate cognitive predisposition, which allows the encoding and retention of sensory input about concrete objects. They can be experienced through several sensory organs in accord, since they are tangible and perceptually stable. According to J.J. Gibson [7, 8], objects demand touch and haptic exploration due to their 'affordance' (a salient property of an object), independent of the individual's ability to recognise those possibilities."*Affordance links perception to action, as it links a creature to its environment. It links both to cognition, because it relates to meaning*" (E.J. Gibson [9] p. 4) Thus infants form object representations from real-world experience viewing and touching objects.

To categorise an object, certain perceptible information of elementary properties, such as shape or texture, has to be present. "*It is just the properties of the objects themselves that are relevant when the child begins to discover regularities in the physical world*" (Sinclair [10] p. 123). These perceptual features will become semantic features in time. The information differs depending on whether a child experiences the physical world passively tactile (passive touch, without any muscular effort) or through action (active or dynamic touch, also called haptics, obtaining not only cutaneous but also proprioceptive information from receptors in muscles, tendons and joints). Categorisation is the first evidence for conceptual organisation; such concept development plays a large role in (object) word learning. But children must also learn categories of actions. Because information about salient properties of an object is stored in sensory and motor systems active when that information was acquired, objects concepts belonging to different categories are represented in partially distinct neural networks [11].

At the age of 4–6 months infants begin to develop the categories 'similar' *versus* 'dissimilar'. Fundamental determinants of concept foundation are between-category contrast (low similarity between specimens of different categories) and within-category similarity (high perceptual similarity between specimens of the same category). Experiments with infants show that even at a very young age infants posses a concept of objects ('preverbal concept') – long before they know the corresponding words [12, 13]. The ability to differentiate between object categories is positively correlated with the development of the mental lexicon for nouns [14, 15].

Although word learning is a very active area in the research of natural first language acquisi-

Box 1. Categories

allow for the classification of sensations and form the basis of human cognition. Infants use perception to gain knowledge of object properties. The understanding of the semantics of object words requires sensorimotor experiences with objects, which are categorised as differing from others, similar or contextually related, and which are defined for a particular representational format. The infant's examination of its environment is promoted by movement, considering the sensory gain through active touch and exploration. The infant is capable of experiencing movement as an effect of an action, but may not yet anticipate it.

tion, the specific role that sense of touch plays in shared cognitive processes has so far scarcely been investigated. This specifically refers to how and in how much this modality shapes the relation of lexical-semantic development and cognition in early childhood.

This chapter will show that haptics plays an eminent role in the relation of sensorimotor activities and achievements in the object-noun domain in the young infant. Many characteristics of objects can be perceived without vision by haptic exploration of those objects.

Selected readings

Li SC, Jordanova M, Lindenberger U (1998) From good senses to good sense: A link between tactile information processing and intelligence. *Intelligence* 26: 99–122

Pruden SM, Hirsh-Pasek, K, Golinkoff RM, Hennon EA (2006) The birth of words: Ten-month-olds learn words through perceptual salience. *Child Dev* 77: 266–280

Roth WM, Lawless DV (2002) How does the body get into mind? *Hum Stud* 25: 333–358

Sheets-Johnstone M (2000) Kinetic tactile-kinesthetic bodies: Ontogenetical foundations of apprenticeship learning. *Hum Stud* 23: 343–370

Stankov L, Seizova-Cajic T, Roberts RD (2001) Tactile and kinesthetic perceptual processes within the taxonomy of human cognitive abilities. *Intelligence* 29: 1–29

Relation between sensory and motor functions during explorative touch: Developmental approach

In the sensorimotor developmental period in infants, they gain unimodal and sensorimotor experiences that form the basis of conceptual precursors: 'sensorimotor schemas'. Sensorimotor schemas are procedural forms of representation. The two essential sensory links between cognition and language are the skin as the greatest and most complex sense organ that pick up tactile information and the hands as the tool for exploration, to recognise, to grasp and to manipulate objects in the environment. The haptic perception and memory processing is present from birth [16]. According to evolutionary theory, the hand is the most remarkable organic tool that distinguishes primates, especially humans. The density of receptors in the finger tips is (next to that of the lips) the highest compared to other regions of the body. The thumb serves a significant function (its opposability allows for secure fixation of objects, its flexibility allows the rotation of objects in the palm in many directions). Object size and finger span (especially the distance between thumb and pointer finger) are related, since grasping objects of different sizes implies difference in finger span.

Soon babies are grasping, comes exploration. For Révész [17] the grasping of an infant is a precursor of action, before basic manual activities, i.e., passing objects from hand to hand, occur.

Box 2. Hands

The hand (as a natural exploratory tool) has two sub-systems:

- the *sensory* to perceive and experience the physical world, as well as
- the *motor* to seize and explore objects.

The cognitive system processes information from sensory and motor systems to identify an object.

Exploratory skills increase with maturation and with practice. The earliest form of haptic exploration is mouthing, manual exploration appears significantly later in the development.

Oral exploration

The mouth is a fundamental part of the haptic exploration system in infants. Oral exploration ('mouthing') encompasses the analysis of objects with the help of lips, the tongue and its receptors, as well as the mucous membrane of the frontal palate. Oral exploration of objects occurs by having the object in the mouth or actively by turning the object into different positions, chewing, licking, biting, rubbing, and gnawing. The tongue is sensitive to shape, texture, consistency and temperature of objects. Oral exploration is particularly pronounced within the first 3 months of life and decreases gradually, until it ceases around 18 months in a healthy infant. Blind children exhibit a significantly longer mouthing phase. Oral stereognostic ability – an indicator for oral sensory functions – is related to age, as older subjects score significantly higher than younger.

Selected readings

Juberg DR, Alfano K, Coughlin RJ, Thompson KM (2001) An observational study of object mouthing behaviour by young children. *Pediatrics* 107: 135–142

Manual exploration

The hands are sensory organs that process information about objects. "*Hand movements can serve as 'windows', through which it is possible to learn about the underlying representation of objects in memory and the processes by which such representations are derived and utilised*" [18, p. 342] .

Infants utilise specific manual exploratory procedures that newborns are not capable of, as they do not possess the motor maturity necessary to select them. "*An exploratory procedure is a stereotyped pattern of hand movement that maximises the sensory input corresponding to a certain object property*" [19, p. 140]. Lederman and Klatzky amended: "... *a stereotyped movement pattern having certain characteristics that are invariant and others that are highly typical*" [18, p. 344]. Through the use of exploratory procedures the infant can perceive a variety of object characteristics, such as substance object properties (temperature, texture, weight determined by density), structural properties (volume/global shape, exact shape, weight determined by volume) as well as functional properties (motion of a part of the object or specialised function).

The thermal receptors of a passively resting hand can receive temperature stimuli, volume and global shape of an object can be experienced through enclosure of the object, lateral motion (rubbing) facilitates detection of texture information, hardness of an object can be experienced through pressure, weight through unsupported holding, and following contours allows comprehension of the exact shape [18].

The haptic system is superior primarily concerning the perception of three-dimensional objects [20, 21]. The ability to grasp as well as the use of several or all fingers provide additional details about the object by integrating proprioceptive information such as finger posture, pressure, direction, velocity of movement and variations on the pressure force of the exploring fingertips. These exploratory procedures form the base for haptic category development. Children learn haptically-based categories in an analytic mode (and not in terms of overall similarity). The exploratory tool "hand" becomes a manipulatory tool, especially in conjunction with the second hand. Haptic perception seems to be more precise and efficient when two hands are used. Somatosensory representations that guide manual exploration do not develop until after 6 months of age.

Selected readings

Conolly KJ (ed) (1999) *The psychobiology of the hand.* Cambridge University Press, Cambridge

Exner C (1989) Development of hand functions. In: P Pratt, A Allen (eds): *Occupational therapy for children.* Mosby, St. Louis, MO, 235–259

Exner C (1992) In-hand manipulation skills. In: JC Case-Smith, C Pehoski (eds): *Development of hand skills in the child.* American Occupational Therapy Association, Rockville, MD, 35–46

Iverson JM, Thelen E (1999) Hand, mouth, and brain. *JCS* 6: 19-40

Klatzky RL, Lederman SJ (1988) The intelligent hand. In: G Bower (ed): *The psychology of learning and motivation.* Vol. 21. Academic Press, New York, 121–151

The developmental course of haptic sensitivity to object properties

The infant's sensorimotor activities become increasingly organised during the first year of life. Bushnell and Boudreau ([19], updated in 1998 [22]) have outlined an approximate developmental timetable (Tab. 1) with their 'double-filter model' of haptic perception of object properties during infancy, relying on findings of developmental studies from the literature and on the framework of Klatzky and co-workers. The two filters of temporal determination of haptic sensory development are constraints resulting from the development in the manual-motor area as well as overall attention of the child. In so far, elder children show more specified exploratory behaviour than younger ones, and in this respect they are able to perceive texture, hardness and weight more precisely or to integrate information across time as occurs while following contours to perceive configurational shape.

A motor prerequisite is the increasing development of motor abilities and the control of arms, hands and fingers during the first year, after the palmar grasp reflex of the neonate (an automatic closing of the fingers when the palm is stimulated) ceases around the third month of life. Consequently, neonates are expected to be haptically blind to most object qualities and coordination. Very young infants can experience temperature and size of small objects because they possess the raw exploratory functions (static contact; enclosure), but may not be capable of gliding motions or differentiated bi-manual activities which the haptic perception of exact shape requires. However, neonates already use their hands as active exploratory tools. Jouen and Molina [23] demonstrated that the neonate cyclical manual activity of alternating opening and closing of the hand may be considered a "*general exploratory movement pattern*". This is sufficiently variable to offer a diversity in the object properties explored, and suggests the activity of neonates' hands are not merely under the control of an archaic, rigid palmar reflex. Molina and Jouen [24, 25] found different patterns of manual pressure to a smooth *versus* a granular object in 3-day-old newborns (Hand Pressure Frequency HPF, supposed to be non-accessible to a palmar grasp function). Their results suggest that the cyclical manual activity may also be sufficient to obtain some information about texture, as neo-

Table 1. Development of haptic perception of object properties (from [19]; updated [22])

First appearance	Object property
First 3 months	Temperature, size (volume)
From 4–10 months	Hardness, texture, weight
Around 10 months	Configurational shape

nates are able to modulate their HPF according to the texture property, but the authors fail to document representation memory. Striano and Bushnell [26] confirmed in three studies that 3-month-old infants are able to perceive material properties including weight differences with their hands, but do not use specialised hand movements and do so only under appropriate conditions which hold their attention.

By 4 to 6 months of age, infants can reach and grasp objects effectively and they begin to show fingering to haptically perceive hardness, texture and weight according Bushnell and Boudreau [19]. However, Adamson-Macedo and Barnes [27] found in a small developmental study that fingering or 'proto fingering' emerges before 3 months of age (as early as the first week of postnatal life), and that healthy, fullterm newborns were able to modulate their manual activity when they were provided with objects that had surface texture. Bushnell and Boudreau [19] characterise the hand movements of 9–10 months old babies as complementary bi-manual activities. Their manual behaviour is no longer repetitive and stereotypical; one hand supports or operates while the other hand is manipulating. The study of Stack and Tsonis [28] gives empirical evidence that texture discrimination occurs beginning with the 7th month, since it is required for sensitive exploratory procedures.

Bushnell and Boudreau [19], providing a cognitive point of view, also emphasise the influence of attention development on haptic perception of specific object properties as mentioned before. Changes in exploration are related to changes in attentional mechanisms. Thus very young infants direct their attention toward such properties that give direct sensory feedback. Only once objects have been analysed according to their differing meaning/purpose connections a functional feedback will begin to direct activity. Thus temperature, hardness and texture feedback are attractive within a merely sensory context ('aesthetic relevance'), weight and configurational shape, however, achieve functional relevance. Blind children acquire adequate exploratory procedures for the varying object properties, as Schellingerhout et al. [29] show for texture perception in blind children between the ages of 8 and 24 months.

Selected readings

Bourgeois K, Kwahar A, Neal SA, Lockman J (2005) Infant manual exploration of objects, surfaces and their interrelations. *Infancy* 8: 233–252

Corbetta D, Thelen E (1996) The developmental origins of bimanual coordination: A dynamic perspective. *J Exp Psychol: Hum Percept Performance* 22: 502–522

Streri A, Féron J (2005) The development of haptic abilities in very young infants: From perception to cognition. *Infant Behav Dev* 28: 290–304

Prelinguistic representation and words

The foundation of the object noun vocabulary is abilities that refer to concept development or the

Box 3. Representation format

sensorimotor schema:	A procedural form of representation, neither a concept nor a symbol.
image-schema:	Represents any object moving on any trajectory through space

linking of preverbal concepts with nouns in early childhood [15, 30].

Perceptual categories

Prelexical categories are formed through perceptual learning in the first year of life. At the earliest stage of representation, processing of object characteristics is primarily tactile, visual and haptic, a combined representation of properties of how a thing looks and feels. Initially these are always characteristics of physical appearance and/or kinetic properties, which allow children to identify an object as part of a perceptual category.

Studies on visual object categorisation identified shape of an object and its surfaces and edges as crucial properties [31-33]. Younger children reach an understanding of physical object categorisation using the characteristic 'shape', before they understand how substance-related attribute categories are organised. So far no experimental design has clarified which object property is used to identify an object haptically; though it could be 'surface texture'. Children are highly sensitive to surface textural differences of objects and use them during categorisation.

After a child knows an object's physical properties, the child wants to discover its function. Perception of actions begins in the second half of the first year of life. While physical objects are seen as a reference for object words, action verbs correspond with motor actions/events. Thus the conceptual foundations for verb learning require to discriminate actions and to form categories. An object can be, for example, pulled, pushed, shaken, hit, rotated, waved, banged, scratched, or squeezed [34] ('examining'). This uncovers and triggers cause-effect relationships. The semantic memory builds on perception of physical object properties and actions performed on or through them. Sensorimotor and perceptual skills provide procedural knowledge. So the first verbs describe actions or movement events.

Conceptual categories: Image-schemas

Categorisation in the second half of the first year of life is not only based on perceptual but also conceptual similarities, for example objects that can move by themselves. Perceptual categorisation, which is based on processing of incoming sensory information, perhaps of the similarity of objects, has to be separated from conceptual categorisation, which classifies objects based on the perception of what an object does respectively on functional, structural and causal relationships. This quote of American psychologists Leslie Rescorla and Jennifer Mirak [35] succinctly describes that, "*Vocabulary learning builds on the child's knowledge about objects, actions, locations, properties, and stages gained as a result of sensorimotor development*" (p. 75).

Jean Mandler [36, 37] explains how infants go from stimulus representation ('sensorimotor schema') to first language symbols: by the innate, active mechanism of perceptual meaning analysis. "*Perceptual analysis involves a redescription of spatial structure and of the structure of motion that is abstracted primarily from vision, touch, and one's own movements*" [36, p. 591]. Beginning with the second half of the first year of life, information gained from bodily experience with the physical world and actions with an object

Box 4.

Two kinds of object categorisation in infancy:

- perceptual categorisation	allows for object identification
- conceptual categorisation	allows for inductive conclusions

('sensorimotor schema') are converted to a schematic format with a semi-pictographic nature: 'image-schema'. "*Image-schemas can be defined as dynamic analog representations of spatial relations and movement in space*" (Mandler [36], p. 591). Image-schemas are not visual representations, they are primary conceptual dimensions ('conceptual primitives'). From a developmental psychology point of view, they are the representation of perception content until the symbolic system 'language' is implemented. Mandler's 'image-schemas' constitute a necessary base for acquiring verbs, because they require the analysis of events (as well as objects and actions) in a specific subset of meaning components. Mandler makes no statement as to the nature of the shift from image-schemas to language.

Naming by the parents

The innate ability for perceptual meaning analysis can be facilitated by parental linguistic input. Initially, parents name an object as a whole. In later stages of development the infant recognises words within the acoustic range of the environment and connects them with present non-linguistic representations of objects and actions. Representations of objects in terms of properties as size, shape, texture, among others, provide a foundation of object names. Thus, a child's early understanding of word meanings is not completely non-linguistic, as they are shaped by language.

For example, when a child has grasped the global meaning of the category 'animal' and discovers the difference between a 'dog' and a 'cat', parents will take a picture book to explain that an animal with a particular appearance (here dog) has a different name than an animal with another appearance (here cat), that a dog barks and a cat purrs. Now the child can subdivide the global class 'animal' in the basic-level categories 'cat' and 'dog' [38]. Accordingly, properties of objects, their parts, relations between them and other properties are highlighted. To do this, parents offer a new word for each known object. A presentation by Pauen [30] showed that objects presented using naming were examined significantly longer – independent from whether the name was a global or basic-level category. To fully comprehend a word's meaning both perceptual and conceptual characteristics resulting from experiences with the reference object of that word are needed. The shift from perceptual to conceptual may explain why early word learning is so slow.

Linguistic symbols for concepts

A combination of perceptual categories (implicit knowledge) with image-schemas (explicit knowledge) forms the foundation for generating concepts which culminates in the highest level of representation: an arbitrary symbol (word). Language constitutes a system of arbitrary symbols.

By 18 months of age, a toddler's vocabulary has grown to approximately 50 words. The following fast word learning [4] is not just due to achievements in categorisation, but also to such acquisition strategies as 'knowledge-based constraints' [39–41]. That means children are predisposed to entertain certain hypotheses about word meaning over others. This specifically describes how nouns forming a dominant category are connected to concrete reference objects ('noun category-bias'; [2]).

Summary

Each thing speaks for itself: movement, shape, size and other characteristics symbolise its concrete appearance. Touch and temperature information from contact with an object are sensory references. The cortex stores this information, represents and categorises it. Exploratory procedures with mouth and hands provide first haptic representations and basic action knowledge. The references of an object word are connected through perceptual properties and functional characteristics of an object. Parallel to sensorimotor learning, the development of conceptual domains through perceptual meaning analysis begins and computational or representational resources are constructed by the child. Altogether, the development starts in a sensory modality, i.e., haptic schemas (references are percepts), followed by image-schemas in the second half of the first year of life (references are mental images), culminating in arbitrary symbols (words). The object representation memory is primarily visual, tactile and haptic, speech representation is acoustic and haptic.

Physical appearance does not suffice to explain object categorisation on a global level, causal and functional properties play a crucial role. 6-month-old infants are able to make meaningful distinctions between the global category 'animate' (defined by biological motion) and 'inanimate objects'. They seem to know that 'inanimate objects' may only move once an action is performed upon them. This conceptual divide may be the fundamental factor in discriminating such global classes like 'animal' or 'vehicle' at about 7 months of age, which is followed by basic-level categories, i.e., dog, fish or car. When children are about 15 to 18 months old, they begin to categorise objects in new ways. Especially words applied to an object by the mother (father) invite the infant to form categories.

Relation between haptic perception and developmental language disorders: Clinical approach

Children who experience difficulties with natural language acquisition might be expected to manifest a deficit in non-linguistic forms of representation. Language processing is affected by a deficit in conceptual development. Besides tactile perception, haptic perception involving the integration of cutaneous and proprioceptive information is a link between linguistic symbolic representations (words) and non-linguistic symbolic representations.

An impaired haptic perception has been observed in different clinical groups of children, particularly those with Down's syndrome [42], autism or fragile X-syndrome [43]. Autistic children exhibit stereotypical (i.e. repetitive) hand movements and unusual object manipulations that pose an obstacle for specific touch experiences, i.e., object properties. They have poor proprioception and are clumsy. Additionally, there are children who are under-reactive or over-reactive to social touch and demonstrate inefficient sensory processing [44-46]. Children exhibiting such dysfunctions do not receive sufficient uni-sensory information, so they exhibit excessive mouthing of objects. Clinical evidence suggests that these problems are associated with developmental disorders, i.e. speech and language impairments. Oral-motor and manual-motor skills differentiated autistic children from typically developing children [47].

Johnston et al. [48] found that language-impaired children aged between 5 and 8 1/2 years (26 boys, 7 girls) without gross neurological findings made significantly more errors in tactile information processing bilaterally (simultactgnosia, finger identification, and graphaesthesia of standardised figures drawn on the dorsum of each hand) while they were blindfolded than normal children controlled for age, performance IQ and socioeconomic status (19 boys, 18 girls). They demonstrated tactile perceptual deficits in children with developmental language disorders, especially problems in processing dichhaptic stimuli. A discrimination function analysis identified six variables, of which two were sensory tactile (dichhaptic finger identification and double simultaneous stimuli face hand) that correctly classified 87% of the population into their respective disordered and normal groups (see also [49]).

Kamhi [50] demonstrated that 5-year-old language-impaired children with normal nonverbal intelligence (mental age between 4;6 and 6;0 years) performed significantly lower on a non-linguistic symbolic task (haptic recognition) than normally developing peers, one matched for mental age and the other for MLU (MLU = Mean Length of Utterances). The subjects had to blindly feel geometric shapes with their hand and then to point to a visual drawing of the corresponding shape that matched the form in their hand. The language-impaired subjects consistently performed better on this task than MLU-matched, normally developing control subjects but more poorly than mental age-matched controls. Kamhi et al. [51] repeated this finding in young schoolchildren: 10 language-impaired children and 10 language-normals matched for performance. For the language-impaired subjects, a high, significant correlation was obtained between receptive vocabulary (on the Peabody Picture Vocabulary Test) and the haptic task. For the controls, a moderate, significant coefficient was seen. As Kamhi et al. [51, p.174/175] suggested, "*There does appear to be some relationship between linguistic and nonlinguistic symbolic abilities*".

Vocabulary development is a commonly used measure of language development in empirical studies. In contrast to control subjects (57 language-normal children of the same age) specific language-impaired children (n = 25) between 3 and 5 years old were found to score, on average, significantly poorer in a standardised expressive vocabulary test of object names and verbs (linguistic symbolic representations) and in a tactile recognition (hand graphaesthesia) as well as in a stereognosis task of familiar objects (non-linguistic symbolic representations) without the influence of vision [52]. Graphaesthesia tasks were consistently more difficult to perform than manual stereognosis tasks. No correlation was found between tactile respectively haptic perception and expressive vocabulary size in language-normal children, but a significant one in language-impaired children.

A formerly clinical sample of 25 children with developmental language disorders (mean age: 4.11 y, SD 10.6 months) showed in elementary school age (mean age: 8;7 y, SD 7.1 months) a below-normal haptic performance in 32% of all cases compared to 25 normal students in regular elementary school without a former or current speech/language disorder and with nearly identical non-verbal intelligence. The control subjects were matched with the formerly clinical group for gender and age [53]. The tactile performance without visual information combined the following test dimensions: touch pressure sensitivity, two-point discrimination, motion perception 'up/down', tactile stimulus localisation, tactile discrimination of sharp/blunt, graphaesthesia of geometric forms and digits and object stereognosis. Those children with former speech/language development disorders had no problem with such elementary performances as touch pressure sensitivity, two-point discrimination, and motion perception. Complex abilities that require different cognitive capabilities, such as tactile short-term memory, showed a very different picture: for discrimination between the touch qualities sharp/blunt, tactile localisation, and graphaesthesia. 40% of the formerly clinical group still exhibited noticeable language difficulties in standardised tests (deviation > 1 SD of mean, T = 50). Of these, 80% showed a reduced haptic global performance. The two children with the lowest language performance also had the lowest haptic sum score. Mean language and mean haptic global performance correlated significantly in the formerly clinical group (0.45, $p < 0.05$), but not in the control group (0.08).

This formerly clinical group performed poorly as well in a manual haptic shape discrimination task with one hand (right and left) without visual control compared to age- and gender-matched normal individuals of the same nonverbal intelligence [54]. Haptic discrimination was measured with the 'Seguin Formboard' on which the children were required to place ten different wooden geometric forms varying in shape in their appropriate holes (cross, star, ellipse, rhombus, circle, semicircle, triangle, rectangle, square and hexagon). Both groups differed significantly in their mean quantitative performances in favour of the controls. Results were not age-dependent. Qualitative analysis revealed a significant difference in haptic discrimination of the pointed shapes, probably caused by inadequate exploration procedures.

De la Peña et al. [55] found in a discrimination task with one hand (Seguin Formboard) of linguistically retarded children (deficient language production and comprehension with nonverbal intelligence in normal range) that the haptic within-modal task proved to be the most difficult condition compared to the visual-within-modal and the visual-haptic condition. The cross-modal condition was not significantly more difficult than the visual-within-modal task. The authors explained the results with the differential memory factor for the haptic and visual within-modality task. The eye can simultaneously compare the two forms faster than one hand. The input of the hand stays in short-term memory for synthesising an image to make a judgment.

Accardi [56] investigated the relationship between linguistic and proprioceptive ability of preschool children with and without speech/language impairments. Proprioceptive ability was measured with the 'Torrance Test of Creativity in Action and Movement', which includes three areas of creative movement: fluency, originality, and imagination. The 'Manual Expression Subtest' (of the 'Illinois Test of Psycholinguistic Abilities') was used to measure conceptual learning. Children with and without speech/language impairments did not differ significantly for proprioceptive ability. A significant relationship existed between imagination as one aspect of proprioceptive ability and conceptual learning.

Gestures represent actions based on proprioceptive schemas ('muscle memory'). 19 specific language-impaired children (mean age 9.9, SD 1.9 years) exhibited a dyspraxic performance in a test of meaningful gesture production (representational gestures [57]). They performed similarly to their younger, normally developing motor-ability-matched peers aged between five and six years. This could be observed in miming of six transitive gestures which required the use of an object (brush teeth with a toothbrush, comb hair with a comb, eat ice cream with a spoon, hit a nail with a hammer, cut paper with scissors, write with a pencil), but not on tasks requiring copying of unfamiliar single and multiple hand postures. The term 'dyspraxic' gives occasion to suggest deficits in neurological maturity of development rather than neurological impairments.

Results of the study of Fabbro et al. [58] have indicated that the inter-hemispheric (callosal) transfer of tactile information was significantly impaired in children with receptive and expressive developmental language disorders (aged 7–12 years) in comparison with controls.

Broesterhuizen [59] reported that, besides other variables, the interrelated abilities of fine motor coordination for hand and mouth were important in predicting oral receptive vocabulary in 70 pre-lingually deaf children 3.5 to 4.25 years old and in 135 pre-lingually deaf children 4–6 years old.

It should be mentioned that oral haptic skills seem to be substantial in speech production [60-64] and that somatosensory feedback plays a considerably role in controlling speech production. Children whose oral touch system is dysfunctional are likely to have difficulty learning oral-motor skills like articulation, because they do not receive adequate information from their receptors. Deviations in articulation may be associated with poorer oral stereognosis [65], but the opposite may not necessarily hold true. The temporal processing of simultaneity/non-simultaneity in the haptic modality differentiated children with oral clefts of various types [66].

The research so far suggests that many language-impaired children fail to organise their haptic experiences of the world and have basic

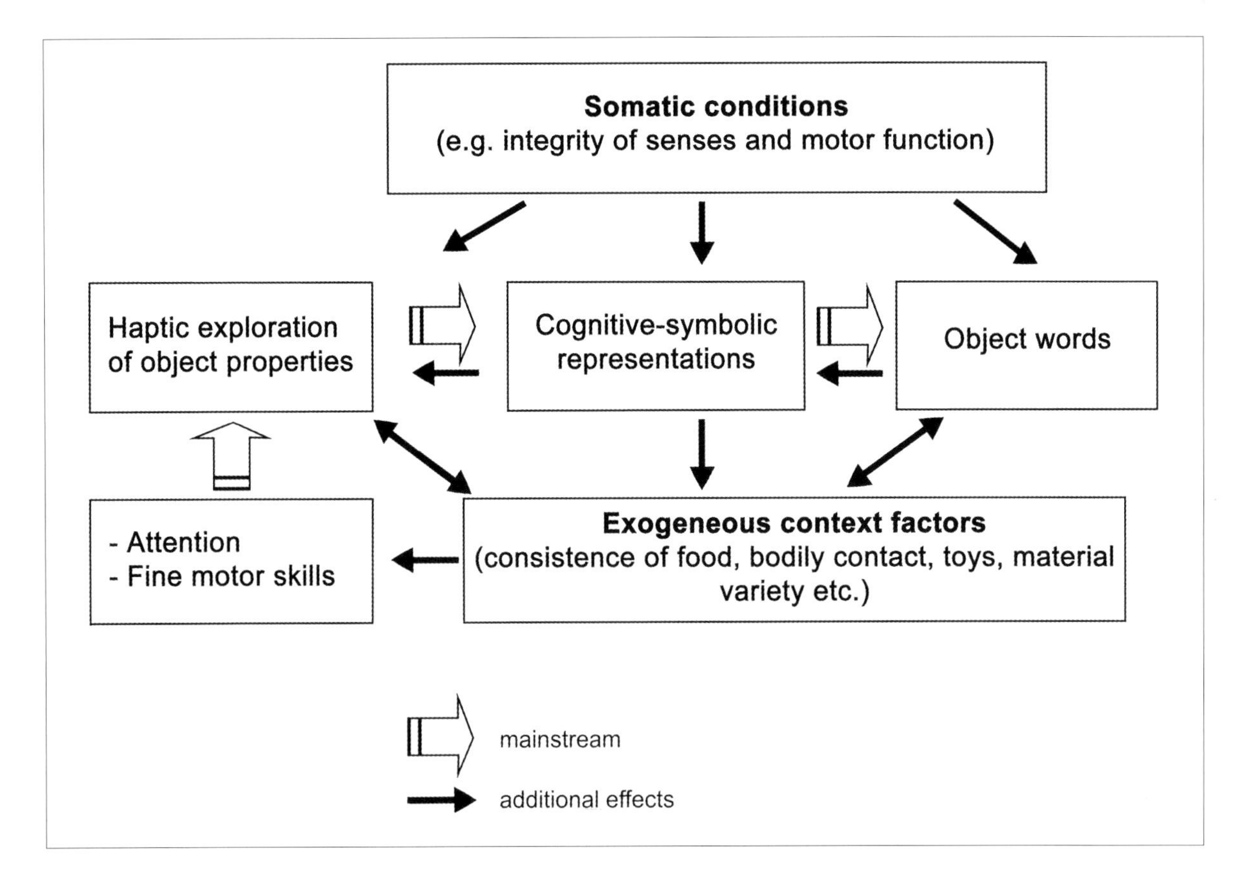

FIGURE 1. WORKING MODEL: DEVELOPMENTAL RELATION OF HAPTIC OBJECT EXPLORATION AND OBJECT WORDS

perceptual deficits, which are related to higher level language processing. Several studies were performed in this domain to investigate the influence of tactile perceptual functioning on language, mostly vocabulary. The cited studies confirmed relationships between developmental language disorders, defined by vocabulary deficits among other symptoms, and measures of haptic perception or components of fine motor-skills (correlational analyses and comparison of means), but they do not allow for the establishment of a causal relation. The obtained correlations may be a result of some third variable that operates on both haptic and lexical levels, such as a representational code. Furthermore, variables such as an intact status of the senses or a stimulating physical and social environment that an infant experiences may operate on tactile, haptic and lexical development (see Fig. 1). Finally, aside from a representational code, attentional and fine motor skills are also involved (cp. 'double filter model' [19]).

Selected readings

Baranek GT, Foster LG, Berkson G (1997) Tactile defensivness and stereotyped behaviors. *Am J Occup Ther* 51: 91–95

Royeen C, Mu K (2003) Stability of tactile defensiveness across cultures: European and American children's responses to the Touch Inventory for Elementary School Aged Children (TIE). *Occup Ther Int* 10: 165–174

Summary

Haptic processing subserve perception and action. During the first year of life, an infant's developing perceptual, motor and cognitive capabilities are coupled. Touch experiences are largely involved in cognitive development. Infants can perceive and understand the physical world with their hands alone.

Object names map to physical objects; they appear to be the most prevalent words in infants' early lexicons in the period of single-word utterances. The nature of their representation is complex. A concept is formed by transforming a sensorimotor schema into an image-schema. Conceptual knowledge is the most important aspect in how to generalise an object's name. Concerning the haptic system for word-learning, it is absolutely essential in the cognitive canon of modal representation for non-verbal categorisation, since infants interactively integrate object information from all sensory modes, especially the haptic mode – the first functioning system *in utero* is touch. Haptics offer a sensorimotor organising structure for non-verbal categorisation in early learning of object words. Because objects are discrete, they are more easily conceptualized.

There are at this time no plausible correlational hypotheses for any other types of word and object categorisation. Verbs enter the individual vocabulary much later than nouns and at a slower rate [4], since it is harder to form concepts for actions as opposed to objects. Semantic representation of objects and actions are different, because noun-meaning is more restrictive than verb-meaning. Compared to objects, actions are intangible and transient. The relationship between linguistic input and non-linguistic categorisation has not been sufficiently specified.

Conclusions and clinical implications for diagnostics and remediation

Until now diagnostics of developmental language disorders has largely focused on the beginning of spontaneous speech in children ('late talkers'), which often caused a late onset of indicated language therapy. The goal of preventive diagnostics in clinical developmental psychology should be to recognise valid and reliable markers for a risk of disordered language development. Since object words are not an isolated achievement, but a developmentally dependent, functional end-product, they can serve as a criterion of language development (main outcome measure).

The presented model in Figure 1 shows the suggested pathway to acquire object words. It postulates the infant's ability to perceive by haptic exploration as a cognitive measure for early object naming. Both somatic conditions and environmental factors influence categorisation underlying the learning of object words. Infants' physical and social environments are structured in ways that facilitate or limit these achievements in daily experience (e.g., playing with toys, experience exploring and manipulating objects). One can assume that insufficient processing of tactile and haptic input from the hand (early indication: haptic object exploration) results in an impaired development of conceptual abilities, followed by deficits in the lexical-semantic development of object words. Subsequently, specific early intervention, for example by massage therapy for skin conditions, tactile stimulation, vibrotactile stimulation, touch therapy, training of fine motor skills, such as in-hand manipulation, hand use in bilateral manipulation, stereognosis or other dimensions could be an efficient way to reduce tactile and haptic processing disorders and thus to prevent developmental language impairments.

However, children with innate dysmelia do not necessarily develop impairments in language or cognition. They have a peripheral physical handicap. Their perception deficits are inherently physical and do not amount to central perception or processing deficits. Other touch-sensitive organs, like feet, lips or tongue, can compensate for missing hands and provide active sensory input for the development of sensorimotor schemas. Especially the mouth with palate and tongue can provide discrimination with the same (or higher than) accuracy as hands. Studies with blind and seeing children suggest that haptic exploration strategies do not require visual experiences, thus the visual mode is not essentially required for object recognition [67]. Although identification of an object may occur faster and with less effort using the visual mode, it is potentially possible through haptic perception alone.

It needs to be taken into account that the findings reported represent evidence for speech/language deficits derived from group data. Therefore in clinical practice it is necessary to screen or investigate each language-impaired child's tactile and haptic profile of strengths and weaknesses. The evaluation should consist of a developmental, medical, and sensory history with assessments of sensory defensiveness, observations like stereotypical behaviours or tactile defensiveness, a structured interview and age-based testing. Moreover, the individual's visual and auditory perceptual abilities have to be tested to check whether an individual has a general or a specific processing deficit (differential diagnosis).

In Germany, there is a 32-question behaviour-rating-scale available for standardised screening of infants' tactile-proprioceptive responsivity: the DEF-TK [68], and a multidimensional, standardised test for tactile-proprioceptive perception that can be administered in the age range from 3.6 to 6.0 years: the TAKIWA. The TAKIWA [69] consists of 47 items over seven sub-tests in left- and right-handed testing: Object stereognosis, touch localisation in hands and forearms, two-point discrimination, stereognosis of object properties, uni- and dichhaptic finger identification, touch pressure sensitivity, graphaesthesia. In the USA, sub-tests of the Sensory Integration and Praxis Tests [70] are commonly used to this end.

Future work should focus on sensorimotor components of manual exploration to learn more about the child's cognitive processing system. The relationship between haptics and its preconditions and object words must be investigated more precisely in a prospective longitudinal study (developmental follow-up). It still remains a challenge to discover under which circumstances the representation of haptically explored object properties will be critical for object naming. In the light of recent work, the role of haptics in causing symptoms of object word deficits deserves further consideration. Finally, a main question continues to be posed by the specificity of these symptoms. Because the precise nature of pre-lexical impairments is not completely understood, much clinical and research work remains to be done.

Summary

Perception is considered to be one precursor to first language acquisition. Increasingly, the tactile and the haptic modality become the focus of attention and researchers aim to specify the nature of possible associations. The present chapter focuses on the natural acquisition of the meaning of open class content words referring to perceivable objects, actions, attributes, especially 'words for things' as the building blocks of language through which we represent our world and communicate with others [71, p. 266] and on the possible tactile and haptic link between object words and categorisation (concept organisation) in infants. Haptic discrimination can be viewed as a prerequisite for higher cognitive processes that underlie symbolic development in infants. After the introduction, I addressed how haptic experiences allow the child to achieve sensory facets for conceptual organisation as well as lexical-semantic representations (developmental approach).

"*It is proposed that language impairment should no longer be viewed as a strictly linguistic deficit, but rather as a deficit associated with a delay in the development of a number of representational abilities*" [64, 71, p. 229]. Building on this statement, I then described how poor tactile and haptic performances are associated with a disordered language development in children and presented some behavioural evidence from clinical studies (clinical approach). After all the suggested role of the tactile-kinesthetic systems guiding the early object name learning (in developing a conceptual base onto which object naming can be mapped) will be provided. The last part illustrates conclusions and implications for the clinical field in diagnostics and intervention.

Selected readings

Mauer DM (1999) Issues and applications of sensory integration theory and treatment with children with language disorders. *Lang Speech Hear Service in Schools* 30: 383–392

Haptic perception in anorexia nervosa

Martin Grunwald

Introduction

In the mid 1990s we, at the Institute for Neurophysiology at the Friedrich Schiller University in Jena, Germany, conducted psycho-physiological trials into brain electrical changes during haptic examination [1, 2]. The experimental task consisted of having subjects haptically explore sunken reliefs with closed eyes and then reproduce the image presented upon them afterwards with their eyes open. The aim was to analyse the stimuli-dependant changes in the subjects EEG. The more than 30 probands in this trial had, as expected, no major problems meeting this demand. However, much to our surprise, one test subject was completely unable to perform the task. Her exploratory times were much longer than the average. Despite this, the drawn reproductions that she produced showed that her perception of the stimuli's entire structure was completely false and distorted. The subject, an intelligent, female university student in her third semester, had a good school record and was, in no way, neurologically conspicuous. She was, however, extremely thin and her skin's texture was abnormal. We wanted to explain these unexpected individual findings and followed a series of theoretical considerations and experimental studies which led us to the field of clinical-experimental psychology and an extreme mental disorder – anorexia nervosa.

Anorexia nervosa (AN) is one of the most severe mental disorders. For the most part, this illness affects young girls and 10–15% dies as a result of the physical effects over the course of time. The illness is characterised by extremely low body weight and with an obsessive fear of gaining weight. Anorexia nervosa patients often control body weight by voluntary starvation, vomiting, purging, excessive exercise, sports or other weight control measures [3–5]. The illness often begins in puberty and the affected do not, as a rule, believe that they are sick despite their being extremely underweight. In various experimental studies it has been repeatedly observed that patients with anorexia tend to over-estimate the size or fatness of their own bodies [6–8]. Disturbances of body image are among the most important predictors for clinical severity of AN. However, current therapeutic approaches are relatively inefficient to this parameter [9, 10].

The reasons for the development of anorexia nervosa are still not fully explained. It is unanimously agreed that multifactorial conditions are responsible. Alongside psychological and social factors, genetic and patho-physiological processes are being discussed. Because body image and body schema distortion represent a critical factor in anorexia nervosa (and because the reasons for the disorder are still not completely understood as well as the fact that there is still no truly effective therapy for the disease) these aspects of the illness should be granted special attention. Especially body schema represents an absolutely central and basal aspect of one's mental bodily representation.

In the following article, we attempt to address the question of whether or not there are connections between body schema distortion and haptic perception in anorexia nervosa. Experimental and neuro-physiological data will be discussed and, in conclusion, a treatment attempt with the goal of reorganisation of the body schema in an anorexic patient is presented.

Body schema and body image

The scientific attempt to describe mental representation, the conscious and unconscious depiction and perception of one's own body, has a long history. Over the course of its development, in which various scientific disciplines have participated, a variety of concepts, and accordingly, terminologies have been developed which are not used uniformly. This is especially true of 'body image' and 'body schema' which are used differently not only in the German speaking world but also in English speaking circles. A comprehensive analysis of the history of terms is impossible here but more information on the subject can be found in [11, 12]. For the purposes of this article, only the basic principles and features of the usage of these terms and their current situation will be presented. This digression is necessary in order to correctly place our findings from anorexia nervosa patients into the current concepts of body schema and body image.

The German physiologist Hermann Munk (1839–1912) was the first to suspect that our body is depicted, as a mental representation, in the brain – particularly in the parietal cortex [13]. In 1908, the psychiatrist and neurologist, Arnold Pick (1851–1924) reported neurodegenerative illnesses in which the patients were no longer able to point out certain parts of their own body on demand. Pick suspected that the body representation in these patients was disturbed [14, 15]. Head and Holmes [16] established the term 'body schema'. They supposed that any sensory input generated a constantly changing postural model of one's own body which actively monitors body position and movements. Paul Ferdinand Schilder (1886–1940) adopted the term from Head and Holmes and introduced the term 'Körperschema' (body schema) [17] to the German language. However, Schilder also used the term 'Körperbild' (body image) as a synonym for 'Körperschema' as a way of describing the mental representation of one's own body. It is unclear, in both cases, if the terms body schema and body image are meant to describe processes that are available to the conscious mind.

Following scientists from psychology, philosophy, medicine and brain research added more terms to the discussion of the mental depiction of the body: body concept, body experience, body perception, body image and schema. etc. The great confusion of the terms has, only in the last few years, led to an effort to create a uniform terminology to be used in all of these fields. With this in mind, and following H.B. Coslett [18], Buxbaum et al. [19] and Kammers et al. [20] one can use the term *body schema* to refer to an on line, real-time abstract, internal representation of one's own body in space which is derived from sensory input (including muscle, haptic, cutaneous, vestibular, tactile, visual, and auditory). This representation provides a three-dimensional, temporal, dynamic representation of the body in space and biomechanical properties of one's body, which articulates with motor systems in the genesis of action. The term *body image*, in contrast, refers to a conscious representation of the body; the private body-related concepts of oneself (attitudes, thoughts about and feelings towards one's own body) as well as conceptual knowledge about the body. Aspects of body image can be represented verbally or cognitively.

The two terms, therefore, essentially differentiate themselves in purpose and content of that which they are meant to describe. As a result, the methodological possibilities of scientific examination of related disorders are also correspondingly different. In the case of the analysis of body image distortions it is possible to use perceptions and verbal judgements. Within the framework of trials using questionnaires or visual projection processes patients are able to judge which attitudes they have towards their body and whether they associate positive or negative feelings with it. In this way, it is also possible to generate estimations and judgements regarding their own body's spatial dimensions. These methods are used primarily for the examination of body image distortions in patients with eating disorders but also in patients with depression [21, 22], body dysmorphic disorder [23], adiposity [24], anxiety [25], HIV [26], etc. In the field of non-clinical body image research, cultural comparative studies are being used more and more [27–29].

In contrast, the analysis of body schema distortions is much more difficult as this concept describes the internal, mental aspect of bodily representation and this is not consciously accessible by the examined person. Accordingly, we do not have the possibility of directly coding body schema processes linguistically but must, rather, conclude that – based upon behavioural and perceptional changes – underlying disturbances of the body schema might possibly exist. Particularly comprehensive studies regarding this come to us from the field of neurology. Body schema disturbances of different sorts have been observed especially in cases of damage to the inferior parietal lobe [30]. Perceptional and behavioural disturbances differ according to which side of the parietal lobe is damaged.

Neglect patients show, for example, a complete disregard for the left half of their body (contralateral neglect) in cases where there is a lesion in the right parietal cortex. They do not perceive their left arm and only shave the right side of their faces. Not only are the extremities on the left side of the body not perceived but, also, any external objects which appear in the left half of the patients' visual fields are ignored. Such a neglect patient draws only the right side of a clock's face and not the left half and tactile stimulation on the body's left side cannot be detected. What's more the patients do not understand why they are in need of medical treatment as they do not believe that they are ill (Anosagnosia) [18, 31].

Patients with left parietal lesions show a typical form of body schema disorder: Autotopagnosia. In this case, patients make localisation errors when asked to point to specific body parts. Finger Agnosia appears when the left parietal lobe – especially the angular gyrus – is affected. In such cases, patients are unable to identify the finger touched by an experimenter if they have their eyes closed. Body schema disturbances are also found in patients who have undergone the entire or partial removal of a limb. Many patients still feel the presence of the extremity (or the removed portion thereof) after the amputation (see chapter by Weiss).

The different qualities of the body schema disturbances in cases of left- or right side damage to the parietal lobe can be explained with the following simple, functional model: Kolb and Whishaw [31] and Haggard [32] summarised, that the parietal lobe of the left hemisphere may contain an abstract body representation used for purposes of localisation whereas the right hemisphere, specially the right parietal lobe may correlate multisensory information to maintain a 'sense of bodily self'.

Body schema and body image in anorexia nervosa

Hilde Bruch [33] was one of the first to point out the importance of body image in cases of anorexia nervosa. Subsequent studies have – using various measurement methods (projective processes, self-assessment using questionnaires, video distortion techniques = Somatomorphic Matrix [34]) – consistently reported that body image is predominantly coupled with negative emotions in patients with anorexia nervosa. They view their bodies as unpleasant, unattractive, ugly, fat and bloated. What's more it is widely accepted that patients with anorexia nervosa tend to overestimate their own body dimensions [35–39]. This body image distortion can be observed particularly strikingly when one confronts an anorexia nervosa patient with their own reflection in a full length mirror. The patients report seeing a fat and ugly body. They are not willing (or able to) recognise their objectively emaciated body as such but, rather, because they feel that their body is fat, they *see* their reflection (body) as being fat as well. Unfortunately this body image disturbance continues to exist even after several years of therapy and also after the patient has gained weight. The causes of this body image distortion remain unclear and the question of which relationships exist between body image disturbance in anorexia nervosa and body schema is also unsolved. That these relationships have not been researched enough is surely accountable to the fact that, so far, scientists researching clinical anorexia nervosa have predominantly concentrated on the research of body image disturbance as it was possible to fall back on accordant methods of examination. Examinations into the neuro-biological and neuro-physiological basis of the development of anorexia nervosa have only recently been started. For this reason it is understandable that so little attention has been paid to the analysis of body schema disturbances in cases of anorexia nervosa.

Right parietal – right hemisphere dysfunction in anorexia nervosa

A prerequisite for the adequate representation of body schema is the neuronal integration of complex sensorimotor information from the whole body [40]. This processing, in turn, is dependent on peripheral sensorical structures and their inputs as well as on subcortical and cortical processing [41–44]. From a psycho-physiological point of view, body schema is thus formed through cortical processing of proprioceptive, interoceptive, vestibular, tactual and haptical information. This leads to the conclusion that the processing of tactil-haptic information is vital for body schema.

Currently, we know very little about the significance and function of tactil-haptic information processing in shaping body schema. There are indications, however, that the cortical representation of body schema hinges on elementary tactil-haptic information and, furthermore, on how this information is processed [32, 45, 46]. Haptic perceptions are distinguished from tactile tasks by their active explorative movements of the exploring limb. The resulting changes in the receptors of the skin, muscles, tendons and joints lead to successive information about the explicitly/implicitly explored object. This information should be integrated to explain the precise spatial characteristics and the texture of the explored object. It can be postulated that an inadequate cortical integration of this information should result in deficits in haptic perception. Thus, in patients with astereognosis, deficits in body schema and haptic perception are associated with a lesion of the right parietal cortex [31]. These patients are unable to explore and explain the structure of a common object haptically with their eyes closed.

Florin [47] observed that healthy controls and patients suffering from anorexia nervosa (AN) and bulimia nervosa react differently to two-point discriminations on various parts of the body. Lautenbacher et al. [48] were able to prove that AN patients are significantly less sensitive to pain than healthy controls. It can thus be supposed that a disturbance in body schema goes along with a distorted tactil-haptic perception in patients suffering from anorexia nervosa. In 1984, Kinsborne and Bemporad [49] hypothesised that a dysfunction of the right hemisphere of the brain, especially of the right parietal cortex is evident in patients with AN. They assumed the dysfunction to be involved in the perception of a distorted body schema ('anorexic's neglect'). On the basis of this assumption, Rovet et al. [50], Pendleton-Jones et al. [51] and Bradley et al. [52] conducted neuropsychological studies exploring perceptual-cognitive functions especially of the right hemisphere in patients with AN. Bradley et al. [52] found changes in event-related brain electrical potentials (ERPs) during perceptual-cognitive tasks supporting the hypothesis of a right parietal dysfunction in patients with AN. Finally, they found significant differences in ERP amplitudes between an AN group and a control group (CO) in verbal as well as in nonverbal tasks. Patients with AN showed no left-right asymmetry for the P3-amplitude in a nonverbal task. However, neither the studies of Bradley et al. [52] nor Pendleton-Jones et al. [51] found significant differences during neuropsychological examination, that is, no cognitive deficits were detected in patients with AN.

In contrast to these observations, other studies [53–55] showed deficits in perceptual-cognitive tasks in patients with AN that could not only be explained by deficits of the right hemisphere. The question remains whether patients with AN show deficits in perceptual-cognitive tasks based on deficits of the right hemisphere. Taking into account the electrophysiological results obtained by Bradley et al. [52] it might be supposed that the neuropsychological inventories used in studies were not sensitive enough to explore the right hemispheric deficits in patients with AN. It is well known that individuals with AN usually have higher IQs than age-matched healthy controls [56–57]. Thus, the higher IQ might interfere with the perceptual-cognitive dimension and thereby hide the right hemispheric deficits. In order to test the hypothesis of perceptive-cognitive deficits in patients with AN it seems necessary to develop a test which does not allow the use of previous experience and usual strategies to solve the task. Furthermore, such a test should be based on sensory integration and explicit spatial orientation because these processes are known to involve the right hemisphere to a greater extent [31].

Based on these assumptions, our study set out to examine possible differences between anorectic patients and healthy controls in fulfilling haptic tasks of varying degrees of complexity. We began with the premise that AN patients would recognise simple and familiar geometric designs in sunken reliefs just as well as healthy controls using haptic exploration. We predicted, however, that, due to diminished somatosensoric integrative ability of the right parietal lobe, the anorectic patients would have problems reproducing complex haptic stimuli. In addition we sought to determine whether EEGs from anorectic patients and healthy controls would show any discrepancies between the two groups during haptic explorations and state of rest intervals.

It has been shown that activity of frontal and parietal cortex increases during haptic exploration tasks [1, 58]. Task-dependent changes during haptic exploration tasks were observed over frontal and parietal regions particularly for the theta-band. Changes in spectral theta-power activity covariate with specific kinds of information processing and memory, as well as with processes of memory loading within the scope of different cognitive paradigms [59–61]. For this reason, we suspected that spectral EEG-power (μV) of the theta frequency band would be significantly lower during haptic exploration than during the state of rest in both groups. As a result of increased perceptive-cognitive demand we expected theta-power over the right parietal cortex in patients with AN to be clearly lower during testing than in the healthy control group. Furthermore, we started out with the premise that the haptic reproductions and spectral EEG-power would be independent from weight gain in anorectics.

Sunken relief paradigm: haptic deficits and brain electrical differences in right parietal lobe

We presented anorectic patients and healthy controls six different sunken reliefs and instructed them to close their eyes and recognise the patterns by haptically exploring the stimuli with both hands. Afterwards the patterns were to be recreated (drawn) with open eyes. The test subjects' 19 channel EEGs were recorded during the procedure (methodological details can be seen in Grey Box 1).

The reproductions submitted by the AN group were of considerably poorer quality than those of the control group in T_0 and in T_1. There were no significant differences in the reproductions between T_0 and T_1 in the AN group. Figure 2 shows the reproductions of the control group and the reproductions of the AN group at both testings (T_0 and T_1). The mean rating scores applied in evaluating the reproductions differ substantially between the two groups. Within the healthy control group the quality of reproductions differed only marginally. The linear regression between Body Mass Index (BMI) and quality of reproductions showed no significant relationship.

Box 1.

Haptic tasks

Sunken relief paradigm

The haptic task consisted of exploring six individual sunken reliefs (13 cm x 13 cm), which were presented to the participants in random order (Fig. 1a and b). All participants were asked to palpate the haptic stimuli with both hands while keeping their eyes closed. Following the haptic explorations all participants were asked to reproduce the structure of the stimuli as closely as possible on a piece of paper with their eyes open. Optimal positioning of the stimuli in relation to the fingers was allowed due to an adjustable holder. During haptic exploration the forearms rested on a wide base in order to allow free movement to the fingers only. No arm and shoulder movements were made during haptic exploration. The exploration time per stimulus was not limited. With the help of a strategically placed screen the participants were prevented from gathering visual information about the stimuli. The participants were not given any feedback on the quality of their reproductions or the stimulus structure. The exploration time per stimulus was registered by means of pressure sensors (in seconds). The participants were allowed to familiarise themselves with the haptic material by looking at one sample stimulus and practicing the haptic exploration task for a duration of 1 min prior to the experiment. Patients were tested after hospital admission (T_0) and again 1 month after being released from the hospital (T_1). The time interval between initial testing (T_0) and follow-up (T_1) ranged from 8 to 23 months, with a mean of 14.5 months (SD = 5.7). The mean Body Mass Index (BMI) for the AN group was 15.24 (SD = 1.27) at T_0 and 16.60 (SD = 1.71) at T_1. Mean BMI of the healthy control group (CO) was 22.16 (SD = 3.01). The IQ was measured by HAWIK (Tewes [62]) at the beginning of testing (Anorexia group: mean = 115.20, SD = 7.98, Control group: mean = 114.69, SD = 13.52).More details in [65–67].

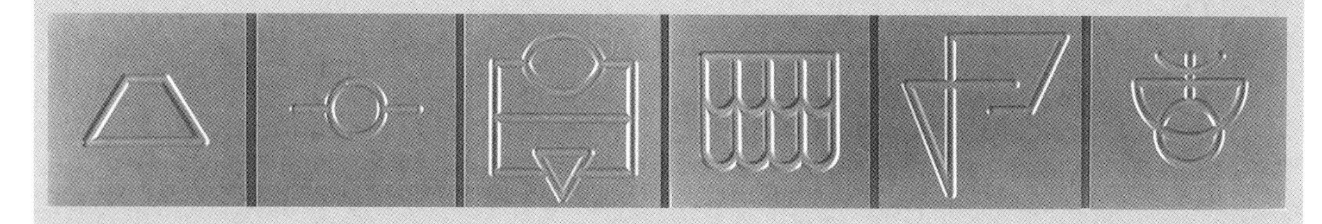

Figure 1a.
Six haptic sunken reliefs of varying degrees of complexity made of hard plastic. The engraving design is 7 mm wide and 3 mm deep.

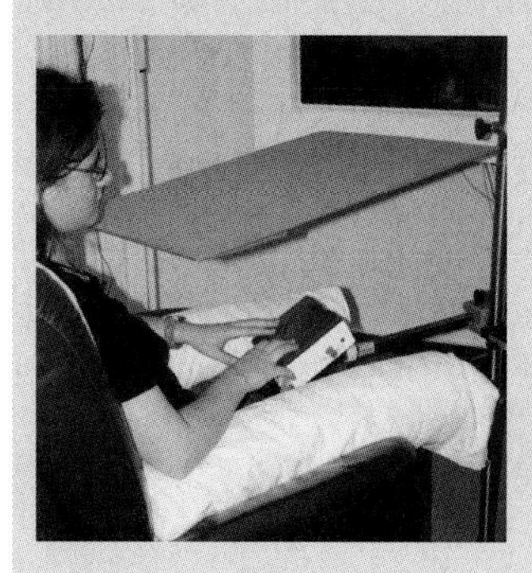

Figure 1b.
Position of the test subject during haptic exploration of one sunken relief.

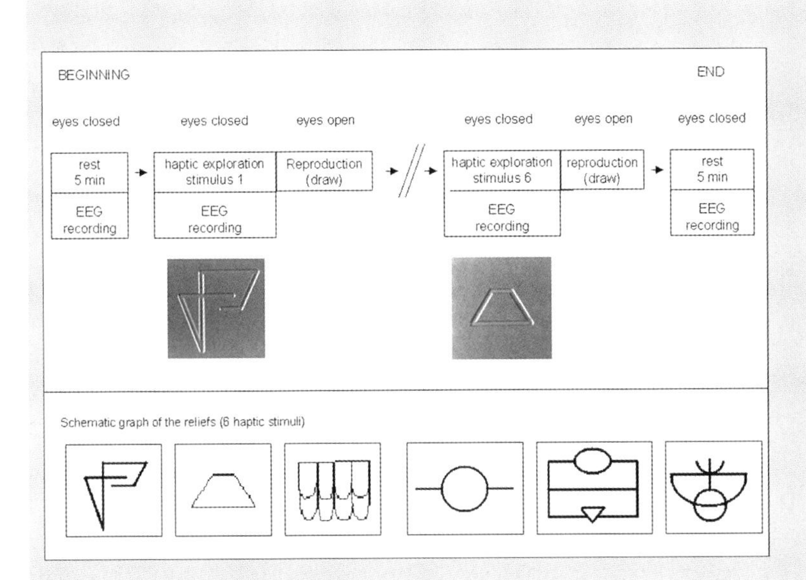

FIGURE 1C.
Test schematic. EEG data was recorded from anorectic patients and controls during rest and haptic explorations. The six sunken reliefs were presented to the test subjects in random sequence. The stimuli were explored with both hands, eyes closed.

EEG Recording and theta-power evaluation

A 19-channel digital EEG (linked ears) was continuously recorded during rest and haptic tasks on 10 anorectic patients and 10 healthy controls. In accordance with the international 10–20 system, Ag-AgCl electrodes were attached to the scalp in the standard electrode positions. Movements of the right eye were monitored by horizontal and vertical electrodes. The signals of these electrodes were recorded on separate channels. The EEG was conducted in a Faraday Cage with a digital EEG-system by Walter Graphtek (Bad Oldesloe, Germany). The sampling rate was 333 Hz with a time constant of 0.3 s. Artefact-free segments of 1.53 s (512 samples/channel) of the state of rest as well as of haptic exploration intervals were chosen and substantiated by cross correlation analysis between relevant frontal EEG and EOG electrodes ($r_{crit.} < 0.5$). Mean spectral power density was calculated as the mean amplitude of the spectral lines of the theta band (4–8 Hz).

The significant differences between rest and haptic exploration intervals, between the different groups and between T_0–T_1 in the AN group are documented in a probability map as applied by Rappelsberger et al. [63, 64]. A blank square indicates a significantly lower resp. decrease of power (significance level $p < 0.05$) whereas a black square indicates a significantly higher resp. increase of power (significance level $p < 0.05$) *via* the corresponding electrode in comparison to the statistical hypothesis. The size of the squares corresponds with the reached significance level. The significance level $p < 0.05$ is shown by a small square and significance level $p < 0.01$ is shown by a large square. For the comparison between the state of rest and haptic exploration task, a blank square means a significant decrease of theta-power during haptic exploration in contrast to the state of rest for the respective electrode. For the comparison between AN and CO, a blank square means a significant lower theta-power in AN patients than in CO for the respective electrode.

Evaluation of reproductions

For evaluation of the reproductions a scale from 1 to 4 was used (1 = correct reproduction of stimulus; 2 = correct reproduction of stimulus with one to three mistakes; 3 = failure to reproduce stimuli adequately, correct reproduction of single elements only; 4 = failure to reproduce stimulus or single elements correctly).

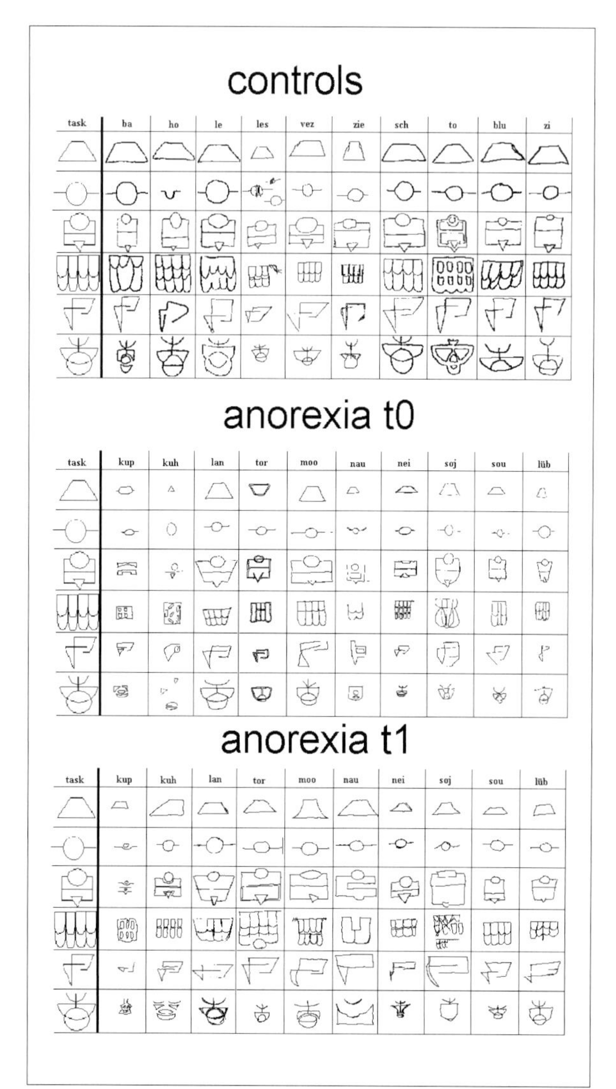

Figure 2.
Reproductions submitted by healthy controls and by patients with anorexia nervosa a T_0 and T_1 (reduced by 170%).

Anorexic patients required the same exploration time for haptic exploration tasks as healthy control subjects. After weight gain (T1) the exploration time of patients with AN were even shorter than during the first test although they do not differ significantly.

The analysis of spectral EEG power of both groups showed significant decrease in power data in the theta frequency band during haptic exploration compared to the rest intervals (Fig. 3a). In both groups the significant reductions in the theta band (indicated by blank squares in the probability map) are spread globally across the entire cortex. The most significant difference between the two groups is evident in the area of the frontal electrodes where the healthy controls showed no major variations in theta-power. In AN patients, however, an increase could be observed over frontal electrodes. In addition, decreases of theta-power in AN patients were concentrated in the right hemisphere.

The comparison of spectral EEG power between controls and anorectic patients during haptic exploration showed major differences between the groups in both T_0 and T_1. Theta-power was lower in anorectic patients (blank squares) than in the healthy controls (Fig. 3b). Especially interesting is the lower spectral power over the right hemisphere and over right parietal regions. The comparison of spectral EEG power between first and second measurements in the AN group showed only minor variations during haptic exploration in comparison to the healthy controls (Fig. 3b). AN patients showed significantly lower theta-power during haptic exploration than CO in T_0 and in T_1 over parietal and central electrodes.

In another part of our study [68] we analysed the interhemispheric brain electrical asymmetries between healthy controls and patients with anorexia nervosa during a resting period and during haptic tasks in theta frequency band (4–8 Hz). We hypothesise that brain electrical asymmetry within the theta band in AN patients performing haptic perception tasks changes significantly over central and parietal somatosensory regions at the time of starvation (T_0) as well as after weight gain (T_1).

Our EEG data clearly demonstrates significant asymmetric interhemispheric theta-power for patients with anorexia nervosa (see Fig. 4). Firstly, during rest, theta asymmetry in AN patients was observed in an acute stage of

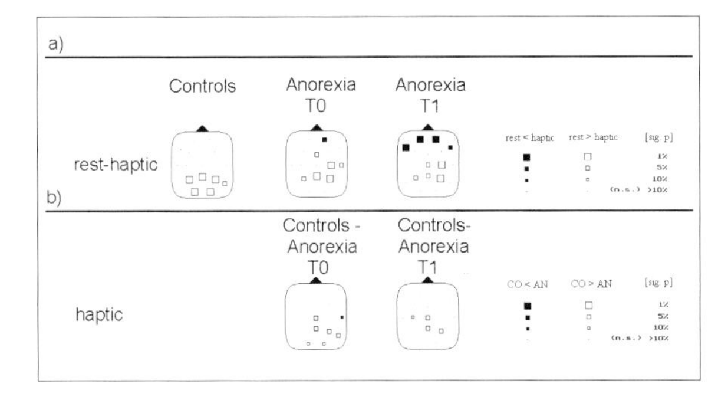

FIGURE 3.
a) Probability map power differences between state of rest and during haptic exploration for CO and AN patients before (T_0) and after weight gain (T_1). Black squares indicate a significant increase of theta-power during haptic exploration compared to the state of rest. Blank squares indicate a significant decrease of theta-power during haptic exploration compared to the state of rest.
b) Probability map power differences between CO and AN patients during haptic exploration. Black squares indicate significantly higher power in AN patients compared to CO during the haptic exploration. Blank squares indicate significantly lower power in the AN patients compared to CO during the haptic exploration.

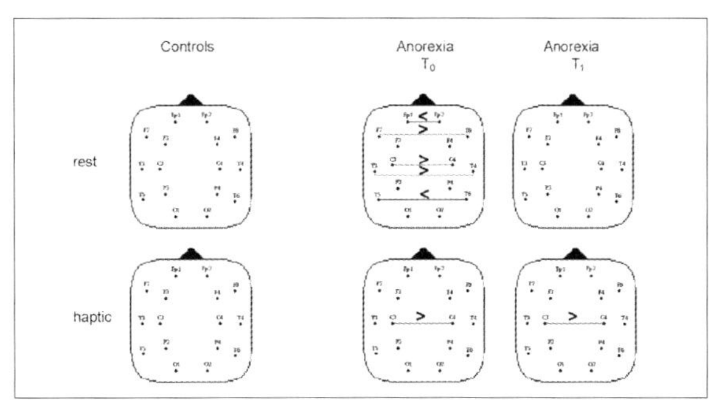

FIGURE 4.
Significant interhemispheric asymmetries at the critical alpha-level ($\alpha'=0.001$). A "<" or ">" indicates a significantly lower or higher theta-power in the pair of electrodes, respectively.

starvation (T_0) but not after weight gain (T_1). Secondly, theta asymmetry was observed in AN patients over central regions in an acute stage of starvation (T_0) as well as after weight gain (T_1) while subjects performed haptic exploration tasks. Thirdly, in the control group, no significant brain asymmetry has been found during rest nor while performing haptic tasks.

This EEG data provided evidence that the brain electrical activity in the theta-band over the right hemisphere, especially over the right parietal cortex is reduced in patients with AN during haptic perception. The differences in AN patients and controls between T_0 and T_1 are relatively stable. The theta-power over the right parietal cortex during haptic perception is markedly lower in AN patients compared with healthy controls. This indicates that this brain region processes the incoming information inadequately. The lower theta-power of AN patients can be interpreted as an irregular activation in the process of multisensoric integration. The small changes of topographic distribution of theta-power differences over the two times of measurement indicate that these findings are independent of weight. The relatively clear topographic arrangement of activation differences affirms the hypothesis of right parietal dysfunction in AN patients.

The lower quality of reproduction of haptic stimuli submitted by patients with AN points to an altered ability in processing perceptions and somatosensorical integrations. Our study, in fact, showed that different qualities of reproduction exist within the AN group. The findings demonstrated that patients with AN have greater problems with complex haptic information than healthy controls. Based on studies in the field of Gestaltpsychology, simple geometric figures are identified on the basis of only a few basic characteristics without comprehensive perceptual-cognitive operations. With increasing complexity, however, greater demand is placed on somatosensoric integration abilities, short-term memory processing, and selective attention. It can thus be deduced that patients with AN are unable to forge the complex relations of individual stimuli elements into an overall concept. The haptic requirements of complex stimuli call for the simultaneous sensory integration of a multitude of pieces of information about space and dimensions. We know from neuropsychological studies that these types of tasks are organised in the parietal cortex. Lesions of the parietal cortex can result in disturbances of tactual-haptic perception (i.e., tactile agnosia, tactile aphasia). It is possible that the poorer reproduction abilities displayed by patients with AN originate from a functional disorder of the right parietal lobe.

The observation concerning the – weight gain-independent – poorer performance in a haptic-perceptual task may be an indication of a general deficit of integrative somatosensory processing in, at least, some patients with AN. Therefore, the heterogeneity of reproduction quality in the AN group may indicate a different perceptual-cognitive and cortical development in patients with AN.

Neuropsychological test of right parietal dysfunction in anorexia nervosa – Angle Paradigm (*Haptimeter*)

The experimental design of the sunken relief sheets provided only data of a nominal scale level due to the visual rating, which must be considered as a disadvantage. For this low data level concerning the assessment of haptic perception in AN patients, we have been looking for a new experimental design which allowed the measurement of the disturbed multisensory integration process on an interval scale. Thus, we developed an experimental paradigm which provided behavioural data of an interval scale level. With regard to the task types of our previous studies, the new experimental paradigm should also consist of bimanual haptic tasks which demand complex multisensory integration capacity of the subjects without visual information. In light of the direct access model (Springer and Deutsch [69]) the test should be organised in a way that the capacity of the right and left hemisphere varies during the same type of tasks. In this way, a varying capacity of the right and left hemisphere should be obtained while solving the tasks.

With respect to our previous studies we expect for bimanual haptic tasks that AN patients will perform worse if the capacities of the right parietal cortex are exceeded. This effect should become obvious in a task type in which target information (respectively information about a nominal value) have to be organised at the same time as operations of comparison ('target-actual-value-comparison') in the right hemisphere with use of the right parietal cortex. If the bimanual haptic task allows the use of the resources of both hemispheres, then we expect that there will be no distinctive differences between the performance of AN patients and the control group.

We developed an apparatus (the so called *Haptimeter*)[1], to which two adjustable, digital angle legs are attached. The investigator set one of these legs to a defined position and the test subject was to reproduce this angle on the second leg (with closed eyes). Two types of tasks were defined depending on whether the subject was to use their left or right hands (right hand tasks and left hand tasks). See details of the experimental procedure with the Haptimeter in Grey Box 2 and [70].

The differences between nominal value and actual value without considering the direction of the deviation were used for the data evaluation. Our investigation revealed a clear difference between the AN group and healthy controls, because there was, generally, a higher difference between nominal values and actual values in the AN group.

But this result was seen only in the right side tasks. The *post hoc* analysis revealed a significantly higher angle difference in the AN group compared to controls for the right side tasks. This effect did not occur in the left side tasks.

Concerning the right side tasks, the nominal values deviate from the actual values by 5.38° (SD = 4.49°) in the AN group and by 4.17°

1 We later discovered that a similar experimental structure had already been developed by Emil von Skramlik (1937).

Box 2.

Angle paradigm with the Haptimeter

Haptimeter

In our experiment we used the following experimental setting outlined in Figure 5 by example of a right-parallel task. It consisted of an on/off switch to measure the time needed to fulfil the assignment. The results of the time measurements were shown by a digital data display. The angle position was assessed by a digital measuring instrument with an exactness of one hundredth of a degree provided by the company NESTLE (Dornstetten, Germany) as well as a separate display on which the deviations of the angles were shown. Two metal bars (5 mm×10 mm×240 mm) served as angle legs. The distance from the table to the end of the angle legs was 28.7 cm in the position of 90°. The distance between the angle pivots was 28 cm. After the angles were adjusted the data were recorded manually by the experimenter. During the exploration the left hand was only allowed to touch the left angle leg, and the right hand was only allowed to touch the right angle leg. No crossovers or both-handed exploration or touch of the opposite angle leg and the tabletop at the same time were allowed. Both hands were to leave the table and explore the angle position simultaneously. The target angle leg was not moveable by the test person. The exploration was done by an up-and-down movement of one or more touching fingers on the angle leg. The hands were to be returned to the body as soon as they finished the readjustment of the moveable angle.

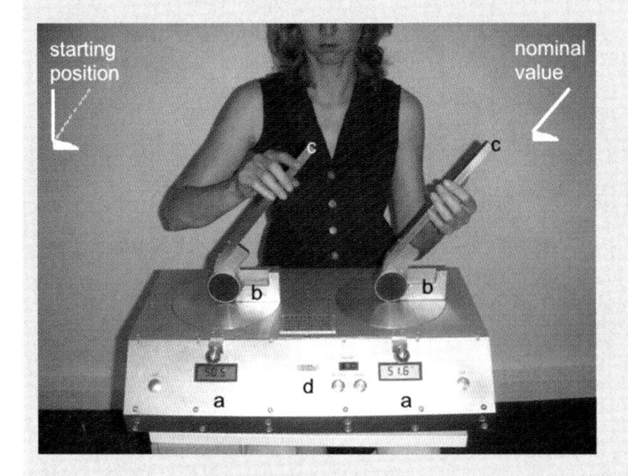

Figure 5.
Experimental arrangement with the Haptimeter *(shown by a right-parallel task): a) Digital data display, b) Digital measuring instrument for the angle position, c) Angle leg, d) Digital data display for time measurement.*

Task procedure

The experimental design consisted of two angle legs, one of which had to be readjusted to a given angle adjustment. We distinguished between two task types – a mirror task type and a parallel task type. The assignment of the mirror task type was to readjust a given angle adjustment (target angle leg) with the respective other angle leg (task angle leg) in a mirrored way, whereas the assignment of the parallel task type was to readjust the adjustment of the target angle with the task angle leg in a parallel way. Furthermore, we distinguished within the task types between right side and left side tasks. In right side tasks the left angle was locked and subjects were asked to readjust the right angle. In left side tasks the left angle had to be readjusted to the locked right angle. All in all, they were to fulfil four different tasks: a right-parallel task and a left-parallel task, as well as a right-mirror task and a left-mirror task. Each task again consisted of five different degree adjustments. No time limit was given and no visual feedback was provided. The task types are schematically represented in Figure 6.

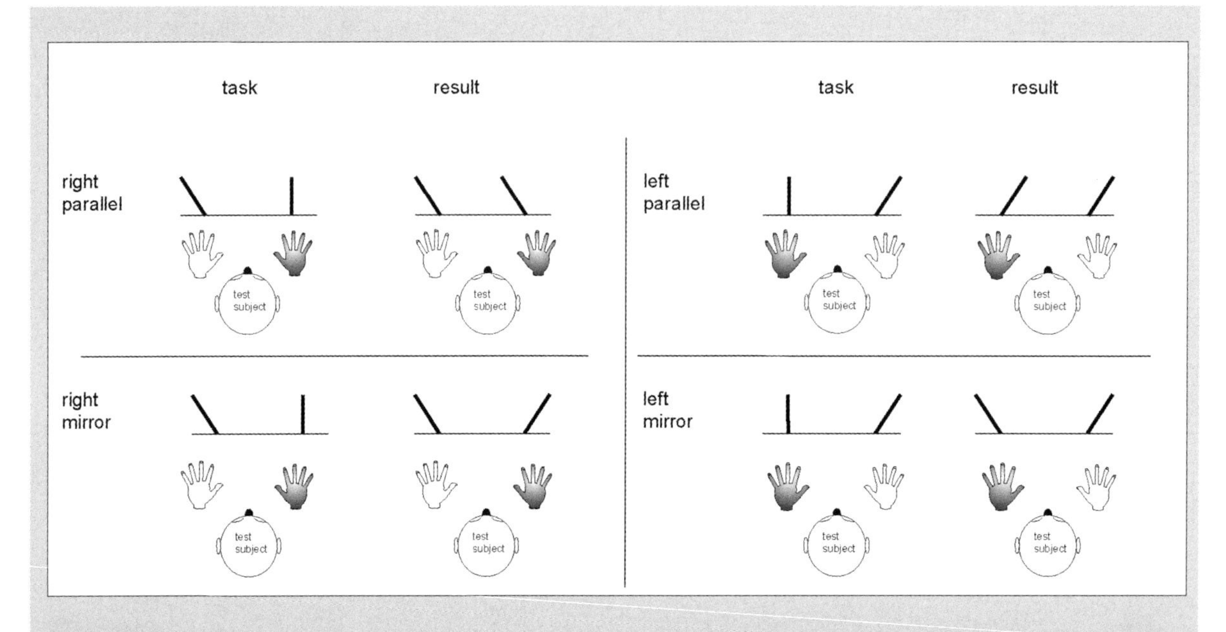

FIGURE 6.
Schematic description of the angle paradigm tasks. Upper line: The task angle has to be readjusted in a parallel way. Lower line: The task angle has to be readjusted in a mirror-like way.

The starting position of all angle legs to be adjusted by the subjects was 90°. All participants had the opportunity to familiarise themselves with the assignments during four training tasks – one of each task type. The subjects received visual feedback as well as information about their results in degrees of deviation. Afterwards, each subject was blindfolded and her hands rested on the table. Then, the experimenter adjusted the first task. Figure 5 shows the left angle leg (as seen by the test person) which was adjusted to a defined amount (nominal value). The right angle leg had a starting position of 90°. Next, the subject was asked to adjust the right angle leg with the right hand parallel to the (left) target angle leg. Then, the experimenter read the readjusted angle (actual value) and adjusted the next task. The nominal values for the right side tasks (mirror and parallel) were: 135°, 158°, 125°, 165°, 145°, and the nominal values for the left side tasks (mirror and parallel) were: 45°, 22°, 65°, 15°, 35°. Thus, five different angle adjustments had to be performed in each task type. All subjects had to solve the tasks of one task type in the same order, but the order of the task types varied. The exploration times (the time needed for the readjustment of the angle leg) and the difference between nominal value and actual value without consideration of the direction of the deviation were used for the data evaluation.

(SD = 3.92°) in the control group, while there were no group differences in the need of time.

Our results from the angle paradigm can be explained on the grounds of the direct access model [69]. In this model it is assumed that sensory information is processed dominantly in that hemisphere which firstly receives the information. It is known that sensory and motor information of the hands including information of its joints and muscles reach the ipsilateral but mainly the contralateral fields of the motoric and sensomotoric cortex [71–74]. Thus, infor-

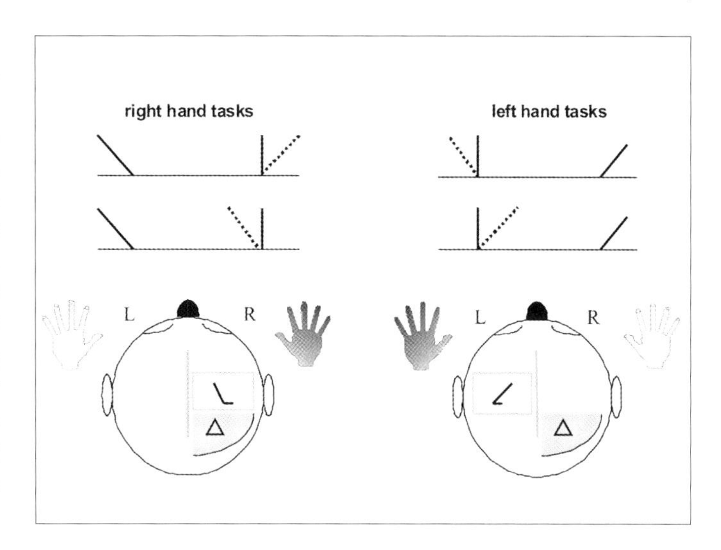

FIGURE 7.
Schematic description of the representation of the nominal value information in the right, respectively left, hemisphere relating to the task side. In right side tasks, the nominal value is encoded in the right hemisphere, whereas in left side tasks, the nominal value is encoded in the left hemisphere. The right parietal cortex serves as a processor for the integration of haptic information and compares the nominal to the actual value. This illustration points out the higher capacitive strain of the right parietal lobe during the right side tasks compared to the left side tasks. AN patients had poorer test performances in right side tasks compared to healthy controls.

mation about movement and position reach the contralateral hemisphere by involving the thalamic nucleus. Moreover, it is known that the information is processed by the motoric as well as the somatosensory areas of the anterior and posterior parietal lobe. Interhemispheric and intrahemispheric connections exist between these regions providing an afferent and efferent exchange of information [75]. The posterior parietal lobe provides sensory information to the frontal cortex [31, 76]. Furthermore, connections exist between the posterior parietal lobe and the hippocampus, the thalamus, the spinal cord, and the basal ganglia [77]. Particularly the right posterior parietal lobe organises multisensory integration and is responsible for the production of spatial percepts [78–80]. This warrants the assumption that the right posterior parietal lobe plays a crucial role in the comparison of nominal and actual values.

For right side tasks, this means that information about the nominal values of the left hand is mainly represented in the right hemisphere. Simultaneously, the dynamic actual value is compared to the constant nominal value by co-working of the sensomotoric and the parietal cortex of the right hemisphere. Thus, in the right hemisphere two processes are done simultaneously when it comes to solving right side tasks (Fig. 7). By way of contrast, during left side tasks, nominal values are encoded in the left hemisphere since hereby the nominal value is assessed by the right hand and the actual value is readjusted by the left hand. The operations of comparison as well as the control of the readjustment movements of the left hand occur in the right hemisphere which receives the nominal value information by the commissures. Thus, task sides are substantially different concerning the capacitive strain of the right parietal cortex in proceeding and integrating the multisensory information. Right side tasks require the organised analysis of the nominal values and of the deviation values from the right parietal lobe as well as of the decision procedures in cooperation with the frontal cortex. All these processes will be affected if a functionally disturbed right parietal lobe does not provide enough processing resources. Thus, the distinctively increased angle differences as shown in the AN group can be interpreted as a consequence of this disturbance.

On the other hand, the demand structure of the left side tasks leaves more resources to the right parietal lobe because the right parietal lobe does not have to encode the nominal value. The nominal value is encoded in the left hemisphere and reaches the right posterior parietal lobe by the commissures [81, 82].

Conclusions and a new sensory treatment for body schema reorganisation in anorexia nervosa with a neoprene diving suit

Our results support the presumption that not only a distortion of body image exists in cases of anorexia nervosa but, rather, of body schema as well. The notable EEG changes in the parietal cortex during haptic demands, the poor reproduction performances with the sunken reliefs and also the large deviations in the angle paradigm suggest that multi-sensory integration is disturbed in patients with anorexia nervosa. This impairment of multi-sensory integration is probably a direct result of a functional disturbance in the right parietal cortex. However, frontocortical regions could also be involved in the integration disruption. If one follows this hypothesis then, in anorexia nervosa, a body schema distortion exists based on a functional deficit of the right parietal region. It remains unclear, though, whether the suggested functional deficit of the right hemisphere in AN patients is a result of the disease or whether it can be seen as an origin for AN. It is also unclear whether hormonal dysfunctions or genetic factors might lead to the functional deficit of the right hemisphere. Due to the lack of data, we still do not know whether similar functional deficits exist in all anorectic patients, i.e., we do not know whether the observed effects are gender specific or not.

It is possible that the functional deficit is developed during growth in childhood. There may be a connection between social and physical stimulation in early childhood and the functional development of multi-sensory integration in the right parietal cortex. If so, violent experiences or a lack of physical contact could support the brain's misdevelopment. Various studies have proved that the processing of tactile and haptic stimuli leads to relevant changes in bodily representation and body schema [32, 83]. One may also suppose that prenatal or genetic influences lead to this misdevelopment.

Regardless of the reasons for the described, multi-sensory integration disturbances in AN, one must ask the question of how body schema distortion may be positively adjusted. As this basal structure of our bodily self is not linguistically or cognitively accessible, an intervention strategy which uses another form of stimulation must be used in order to reorganise the body schema distortion. The aim of this should be to create an entire-body, sensory stimulation which is above threshold. This stimulation would not be generated passively (e.g., with massage) but rather should be generated by the patients themselves through movement. The decisive and functional point is the direct dependence of the bodily stimulation on the physical movements of the patient. This above-threshold bodily stimulation should be temporally and spatially coherent (agree with) the movements made by the patient. This form of stimulation is comparable to the close physical contact such as that which normally exists between a mother and her newborn child.

To achieve a similarly intense form of bodily stimulation we conducted a pilot study in which we used a conventional, custom-made neoprene diving suit [84].

A female long-term anorectic patient (19 years old) wore the diving suit over her underclothes for 1 h, three times per day. Body temperature was measured auxillarily before and after wearing the diving suit. The patient kept a diary to record events of the day and specific bodily feelings while she was wearing the suit. The pilot treatment lasted 15 weeks with the whole project lasting 14 months. During this time, body weight was recorded 39 times (Fig. 8). Background EEG (eyes closed, everyday clothes on without diving suit) was recorded five times to evaluate EEG theta power. Quality of body representation was assessed repeatedly using the 'angle paradigm'. Standardised, clinical questionnaires focusing on eating disorders were carried out at the beginning and the end of the project.

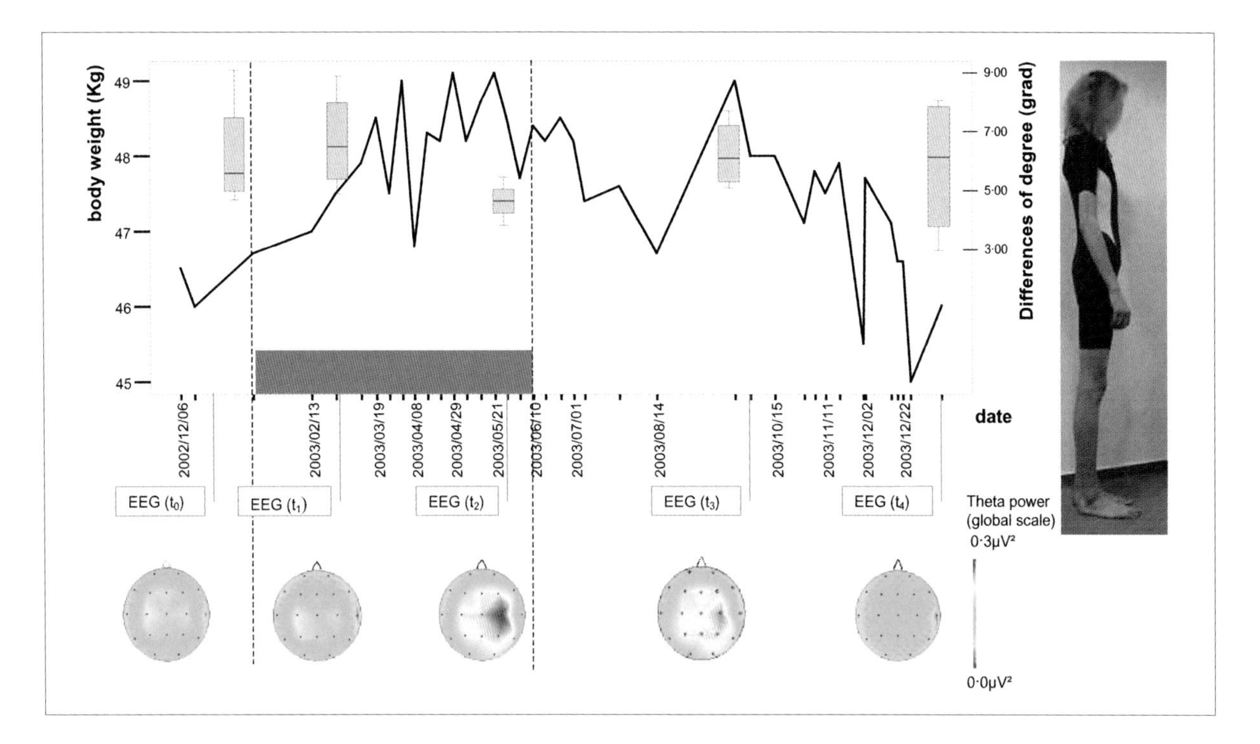

FIGURE 8.
Body weight, deviations in the angle paradigm, and EEG theta (3.5–7.5 Hz) activity of patient A.B. over 14 months. Upper part: Pattern of body weight (solid line, ordinate on the left side) and deviations of angle between nominal and reproduced angles in the 'angle paradigm' at five points in time (T_0–T_4) (grey bars, ordinate on the right side). The red bar indicates the time during which the patient used the diving suit (for 1 h three times per day). Small yellow bars indicate that the patient had a common cold. Note that highest weight and smallest differences between nominal and reproduced values of angles were observed at the second measurement (T_2), i.e., during the treatment.
Lower part: Power-maps of background EEG theta activity with eyes closed at five points in time (T_0–T_4). A strong activation of the right hemisphere could be observed at (T_2), i.e., during the treatment. The patient wearing the diving suit (at T_0) is depicted on the right.

Since the age of 14, she has been suffering from anorexia nervosa. Before our treatment, she underwent two in-patient and three ambulatory psychotherapies. No medications (except contraceptives) were being taken by the patient during the study. No neurological or any other psychiatric diseases were diagnosed. During the time of investigation the patient was actively taking part in her arts and media classes at university.

The patient described no mental or physical side effects from wearing the diving suit.

Body weight increased significantly during the period in which she wore the diving suit. EEG theta activity showed low activity, with left-hemispheric dominance at the beginning of investigation (T_0 and T_1). At the end of pilot treatment (T_2), theta activity showed a remarkable enhancement, including a shift to the right hemisphere. This change reversed at the end of the project (T_4). Analysis of angle deviations revealed strong deviations at T_0, smallest deviations at T_2, and reversing deviations at T_3–T_4 (Fig. 8).

How is this information to be interpreted? We assume that significant weight gain and smaller angle deviations at T_2 are consequences of an improved body schema representation due to an activation of the right hemisphere by the increased somatosensory stimulation while wearing the diving suit. Theta activation, body weight, and body schema approximate initial levels after the end of treatment. Therefore, we assume a close causal relationship between these factors. Alternatively, thermal insulation of the neoprene suit might have influenced these factors.

Although this case involved only one patient, the data collected supports the belief that it is possible, with this treatment, to positively change the body schema disturbance without medication or side effects. In order to achieve stable effects it is surely necessary to generate the bodily stimulation in younger patients within the framework of a complex clinical, therapeutic setting. Fortunately, our colleagues at the child and youth psychiatry department of the Charite (Berlin) have integrated the diving suit into their bodily-therapeutic program and report very positive effects. It must be left to future research to determine which role the bodily stimulation and which the psychotherapy plays in these effects. It is, however, hoped that future clinical-experimental research will look more closely at and take into account the various social and biological factors that influence the development of the body schema. A key to the complex origins of eating disorders and other physical disturbances undoubtedly lies within.

Summary

Eating disorders, especially anorexia nervosa, are extreme psychological disturbances which are very difficult to treat. Around 10% of the affected die as a result of the disease despite intense psycho-therapy and treatment. It is known that patients with anorexia nervosa have a strong distortion of their own body image. They overestimate their body's dimensions and view it as ugly or fat despite extreme emaciation or bodily cachexia. Regardless of various indications from the field of neurology, very little research has been conducted into whether a body schema disturbance is present in anorexia nervosa. In contrast to body image, body schema is a dynamic process of depiction of one's own body and its space time structure which is not consciously accessible. Disturbances in the body schema cannot be directly analysed in words or with cognitive processes. Perceptive and behavioural disturbances can indicate the presence of a body schema disturbance. From various neurological disorders, we are aware that body schema distortion is connected to a functional defect of the right parietal lobe.

Based upon the hypothesis that a body schema disturbance is present in anorexia nervosa, we examined haptical perception performances using two different paradigms (sunken relief paradigm, angle paradigm) and corresponding brain electrical activities. Patients with anorexia nervosa showed strong impairments in the haptical perception as well as characteristic changes in the brain electrical activities in the right hemisphere – especially in the right parietal cortex. The effects remained visible in patients with anorexia nervosa during long-term observation also after the patient had gained weight. These and other findings speak for the assumption that a body schema disturbance in anorexia nervosa is based in a right parietal functional defect. Multisensory integration capabilities of the right parietal lobe are especially impaired.

With the goal of a reorganisation of the body schema, a conventional, made-to-measure neoprene suit was used within the framework of a pilot study. A long-term patient wore this suit three times per day for 1 h over 14 weeks. Body weight, EEG and other clinical-neuropsychological data were collected before, during and after the application phase. When the suit was being worn the patient's body weight improved and a clear activation of the EEG in the right hemisphere of the brain was observed. The neuropsychological data showed a clear improvement of the multisensory integration abilities when the suit was worn. Despite the low temporal stability of these positive changes, this trial makes it clear that, with relatively little effort, stimulation of the functions in the right hemisphere, which have positive effects on the basal structure of the body schema, can be achieved. Future application trials, especially in younger patients, will have to show how sustainable this stimulation concept is for the reorganisation of the body schema in the case of anorexia nervosa and other mental disorders.

Acknowledgements

This paper was supported by the Deutsche Forschungsinitiative Eßstörungen e.V., and the Faculty of Medicine, Paul-Flechsig-Institute for Brain Research, University of Leipzig.

Selected Readings

Blakeslee S, Blakeslee M (2007) *The body has a mind of its own: how body maps in your brain help you do (almost) everything better*. Random House, New York

Bruch H (2001) *Golden cage: the enigma of anorexia nervosa*. Harvard University Press, Cambridge, US

Greenfield L, Herzog DB, Strober M (2006) *Thin*. Chronicle Books LLC, San Francisco

Gregory RL (ed) (2004) *The Oxford companion to the mind*. Oxford University Press, Oxford

V.
Haptic interfaces and devices

History of haptic interface

Hiroo Iwata

Introduction

It is well known that sense of touch is inevitable for understanding the real world. The use of force feedback to enhance computer-human interaction has often been discussed. A haptic interface is a feedback device that generates sensation to the skin and muscles, including a sense of touch, weight and rigidity. Compared to ordinary visual and auditory sensations, haptics is difficult to synthesise. Visual and auditory sensations are gathered by specialised organs, the eyes and ears. On the other hand, a sensation of force can occur at any part of the human body, and is therefore inseparable from actual physical contact. These characteristics lead to many difficulties when developing a haptic interface. Visual and auditory media are widely used in everyday life, although little application of haptic interface is used for information media.

Haptic interface presents synthetic stimulation to proprioception and skin sensation. Proprioception is complemented by mechanoreceptors of skeletal articulations and muscles. There are three types of joint position receptors: free nerve ending as well as Ruffini and Pacinian corpuscles. Ruffini corpuscle detects static force. On the other hand, Pacinian corpuscle has a function to measure acceleration of the joint angle. Position and motion of the human body is perceived by these receptors. Force sensation is derived from mechanoreceptors of muscles; muscle spindles and Goldi tendons. These receptors detect contact forces applied by an obstacle in the environment.

Skin sensation is derived from mechanoreceptors and thermorecepters of skin. Sense of touch is evoked by those receptors. Mechanoreceptors of skin are classified into four types: Merkel disks, Ruffini capsules, Meissner corpuscles, and Pacinian corpuscles. These receptors detect edge of object, skin stretch, velocity, and vibration respectively.

The tactile display that stimulates skin sensation is a well-known technology. It has been applied to communication aids for blind person as well as master system of teleoperators. A sense of vibration is relatively easy to produce, and a good deal of work has been done using vibration displays [1, 2]. The micro-pin array is also used for tactile displays. Such a device has enabled the provision of a teltaction and communication aid for blind persons [3, 4]. It has the ability to convey texture or 2-D geometry [5].

The major role of tactile display is to convey sense of fine texture of object's surface. The latest research activities of tactile display focus on selective stimulation of mechanoreceptors of skin. As mentioned at the beginning of this section, there are four types of mechanoreceptors in the skin: Merkel disks, Ruffini Capsules, Meissner Corpuscles, and Pacinian Corpuscles. Stimulating these receptors selectively, various tactile sensations such as roughness or slip can be presented. Micro air jet [6] and micro electrode array [7] are used for selective stimulation.

These tactile display technologies cannot stimulate proprioception, although it is inevitable to understand the real world as well as virtual environment. External force should be applied to stimulate mechanoreceptors of skeletal articulations and muscles. Device for generation of such force has difficulty in its implementation. This chapter focuses on history of development of haptic interface that stimulate proprioception.

FIGURE 1. GROPE-I

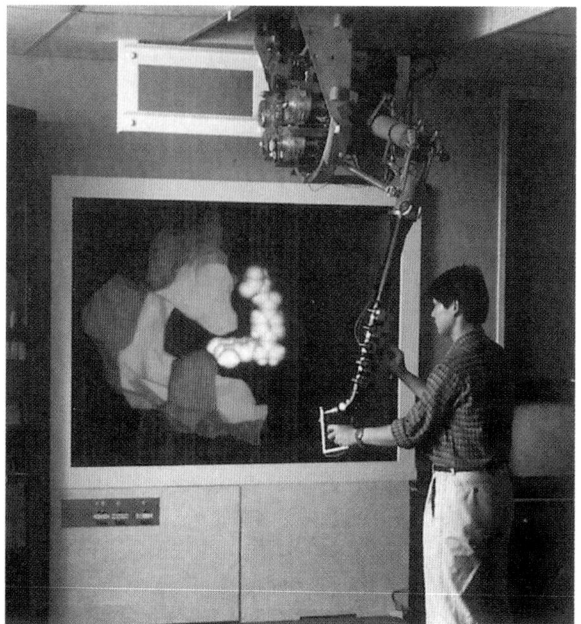

FIGURE 2. GROPE-III

1990: Epoch of haptic interface

There were several roots of haptic interface and they were coincidently published in 1990. Brooks began in 1967 a project to develop visual+haptic display for 6-D force fields of interacting protein molecules [8, 9]. The first haptic system, GROPE-1, employed 2-D movable platform (Fig. 1). A small knob was attached to the platform that can be positioned within a horizontal plane two inches square. Potentiometers sense its x and y position; servomotors exert x and y force. Both are connected to the computer driving an associated visual display. This was the first haptic device that present virtual environment. Later they introduced the Argonne ARM, a 6DOF (degree of freedom) force reflective teleoperator. It was integrated with stereoscopic large screen (Fig. 2). The system was developed for the molecular docking task. The final system, GROPE-III, proved that force significantly contributes to the task giving the lowest potential energy of the docked combination.

Minsky developed a haptic system, called Sandpaper, designed for experimenting with feeling texture [10]. The force display technology used in the system was a motor-driven 2DOF joystick (Fig. 3). The software created very small virtual springs which pull the user's hand toward low regions and away from high regions of a texture's depth map. It also created feel-able physics such as variable viscosity soups, springs, and yo-yos. The research was the beginning of theory of haptic rendering.

In 1988, the author started research into design of haptic interface for natural interaction in virtual environment. We proposed a concept of the Desktop Force Display and the first prototype was published in 1990 [11, 12]. The device applies force to the fingertips as well as the palm. Figure 4 shows overall view of the system. It was the first exoskeleton designed for haptic interaction in virtual environment.

FIGURE 3. SANDPAPER SYSTEM

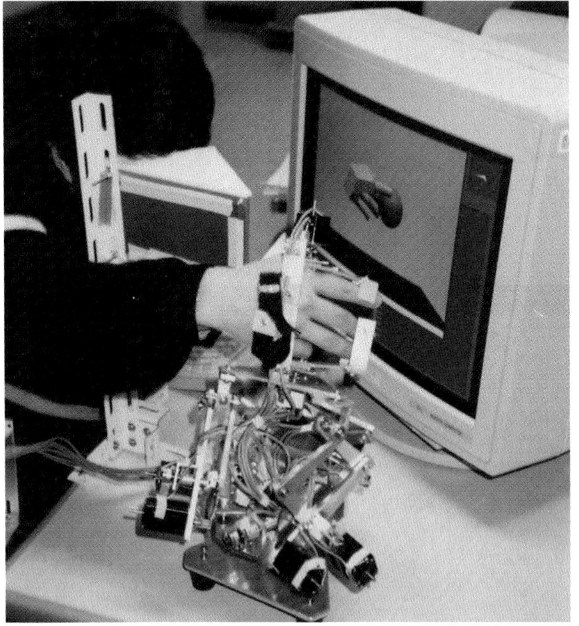

FIGURE 4. DESKTOP FORCE DISPLAY

Exoskeleton (1990~)

An exoskeleton is a set of actuators attached to a hand or a body. In the field of robotics research, exoskeletons have often been used as master-manipulators for tele-operations. However, most master-manipulators entail a large amount of hardware and therefore have a high cost, which restricts their application areas. Compact hardware is needed in order to use them in human-computer interactions.

The Desktop Force Display was a compact exoskeleton suitable for desktop use. Force sensation contains six dimensional information: three dimensional force and three dimensional torque. The core element of the force display is 6 DOF parallel manipulator. The typical design feature of parallel manipulators is an octahedron called 'Stewart platform'. In this mechanism, a top triangular platform and a base triangular platform are connected by six length-controllable cylinders. This compact hardware has the ability to carry a large payload. The structure, however, has some practical disadvantages in its small working volume and its lack of back-driveability (reduction of friction) of the mechanism. In our system, three sets of parallelogram linkages (pantograph) are employed instead of linear actuators. Each pantograph is driven by two DC motors. Each motor is powered by a PWM (Pulse Width Modulation) amplifier. The top end of the pantograph is connected with a vertex of the top platform by a spherical joint. This mechanical configuration has the same advantages as an octahedron mechanism. The pantograph mechanism improves the working volume and backdriveability of the parallel manipulator. The inertia of motion parts of the manipulator is so small that compensation is not needed.

The working space of the centre of the top platform is a spherical volume whose diameter is approximately 30 cm. Each joint angle of the manipulator is measured by potentiometers.

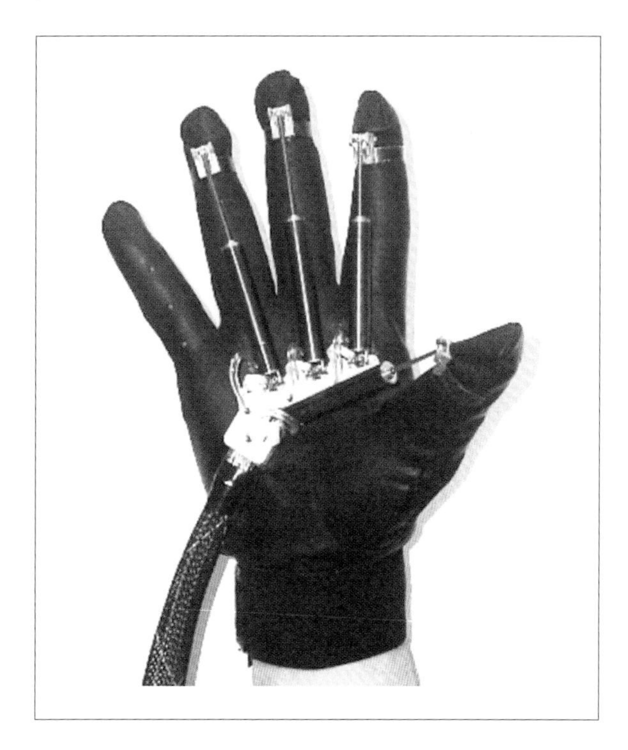

FIGURE 5. RUTGERS MASTER

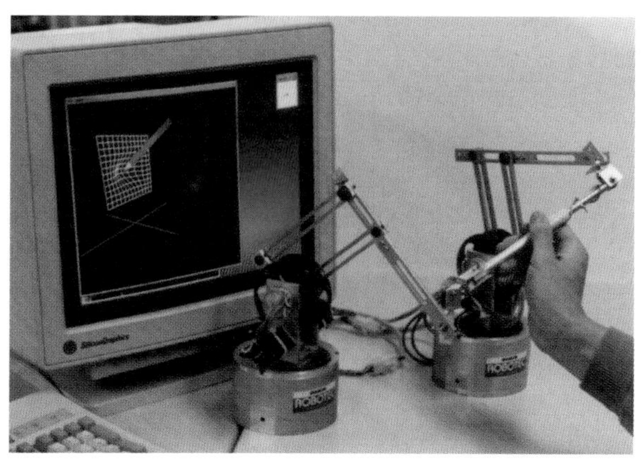

FIGURE 6. PEN-BASED FORCE DISPLAY

Linearity of the potentiometers is 1%. The maximum payload of the manipulator is 2.3 Kg, which is more than a typical hand.

The top platform of the parallel manipulator is fixed to the palm of the operator by a U-shaped attachment, which enables the operator to move the hand and fingers independently. Three actuators are set coaxially with the first joint of the thumb, forefinger and middle finger of the operator. The last three fingers work together. DC servo motors are employed for each actuator.

Another exoskeleton for virtual environment was developed by Burdea [13]. The Rutgers Master is a lightweight and portable exoskeleton using small pneumatic cylinders was developed to apply the force to the fingertips (Fig. 5). The device cannot generate grounded force such as wall or gravity, but effectively simulates grasping virtual objects.

Tool-handling-type haptic interface (1993~)

Users of exoskeletons feel troublesome when they put or off these devices. This disadvantage obstructs practical use of force displays. Tool-handling-type is a method of implementation of force display without glove-like device. A pen-based force display is proposed as an alternative device [14]. A 6 DOF force reflective master manipulator which has pen-shaped grip was developed. Users are familiar to a pen in their everyday life. Most of the human intellectual works are done with a pen. People use spatulas or rakes for modelling solid objects. These devices have stick-shaped grips similar to a pen. In this aspect, the pen-based force display is easily applied to design of 3-D shapes.

Human hand has an ability of 6 DOF motion in 3-D space. In case a 6 DOF master manipulator is built using serial joints, each joint must support the weight of upper joints. This characteristic leads large hardware of the manipulator. We use parallel mechanism in order to reduce size and weight of the manipulator. The pen-based force display employs two 3 DOF manipulators. Both end of the pen are connected to these manipulators. Total degree-of-freedom of the force display is six. 3 DOF force and 3 DOF torque are applied at

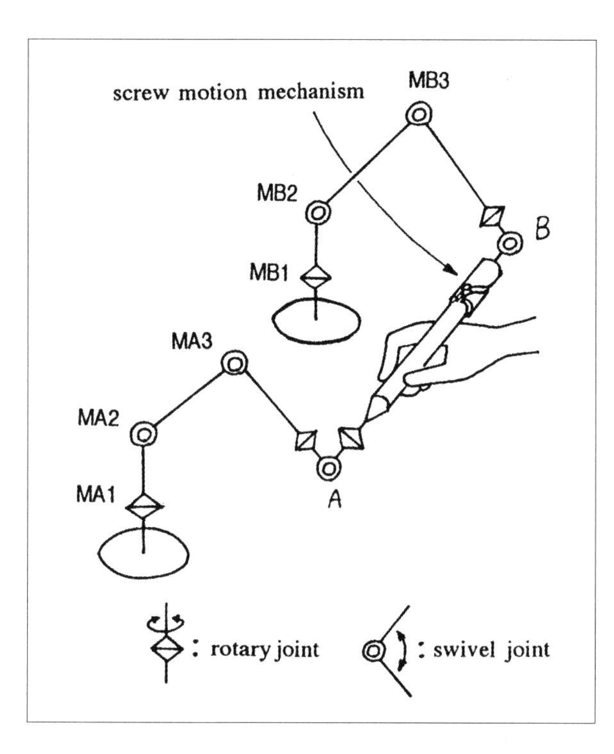

FIGURE 7. MECHANICAL CONFIGURATION OF PEN-BASED FORCE DISPLAY

the pen. Overall view of the force display is shown in Figure 6. Each 3 DOF manipulator is composed of pantograph link. By this mechanism, the pen is free from the weight of the actuators.

Figure 7 shows a diagram of mechanical configuration of the force display. Joints MA1, MA2, MA3, MB1, MB2 and MB3 are equipped with DC motors and potentiometers. Other joints move passively. The position of joint A and B are measured by potentiometers. Three dimensional force vector is applied at the joint A and B. The joint A determines the position of the pen point, and the joint B determines the orientation of the pen. Working space of the pen point is a part of a spherical volume whose diameter is 44 cm. The rotational angle around the axis of the pen is determined by the distance between the joint A and B. A screw motion mechanism converts rotational motion of the pen into transition of the distance between the joint A and B.

Applied force and torque at the pen is generated by combination of forces at the point A and B. In case these forces have the same direction, translational force is applied to the user's hand. If direction of the forces are reverse, torque around the yaw axis or the pitch axis is generated. If two forces are opposite, torque around the roll axis is generated by the screw motion mechanism.

The tool-handling-type haptic interface cannot present force between fingers but it has advantages in practical use as mentioned at the beginning of this section. Therefore, this type of haptic interface is suitable for commercial products.

Massie and Salisbury developed the PHANToM, which has a 3 DOF pantograph [15]. A thimble with a gimbal is connected to the end of the pantograph, which can then apply a 3 DOF force to the fingertips. The PHANToM became one of the most popular commercially available haptic interfaces.

Object-oriented-type haptic interface (1994~)

The object-oriented-type is a radical idea for the design of a haptic interface. The device moves or deforms to simulate the shapes of virtual objects. A user of the device can physically contact with the virtual object by its surface.

An example of this type can be found in Tachi's work named 'HapticSpace' [16]. Their device consists of a shape approximation prop mounted on a manipulator. Figure 8 shows overall view of the system. The position of the fingertip is measured and the prop moves to provide a contact point for the virtual object. McNeely proposed an idea named 'Robotic Graphics' [17], which is similar to Tachi's method. Hirose developed a surface display that creates a contact surface using a 4×4 linear actuator array [18]. The device simulates an edge or a vertex of a virtual object.

The author demonstrated the haptic interfaces to a number of people, and found that some of them were unable to fully experience virtual

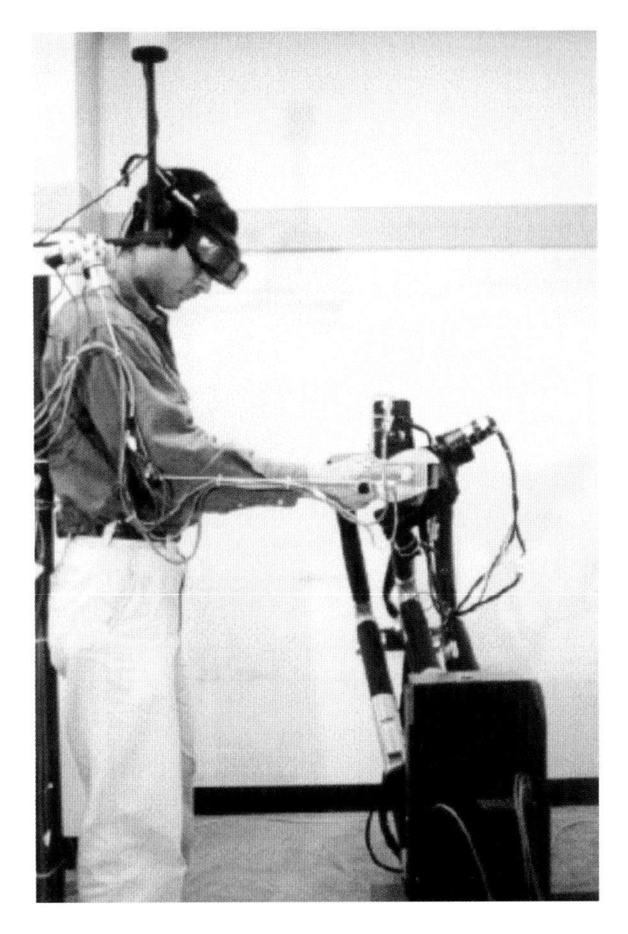

FIGURE 8. HAPTICSPACE

objects through the medium of synthesized haptic sensation. There seem to be two reasons for this phenomenon. Firstly, these haptic interfaces only allow the users to touch the virtual object at a single point or at a group of points. These contact points are not spatially continuous, due to the hardware configuration of the haptic interfaces. The user feels a reaction force thorough a grip or thimble. Exoskeletons provide more contact points, but these are achieved by using Velcro bands attached to a specific part of the user's fingers, which are not continuous. Therefore, these devices cannot recreate a natural interaction sensation when compared to manual manipulation in the real world.

The second reason why they fail to perceive the sensation is related to a combination of the visual and haptic displays. A visual image is usually combined with a haptic interface by using a conventional CRT or projection screen. Thus, the user receives visual and haptic sensation through different displays, and therefore has to integrate the visual and haptic images in his/her brain. Some users, especially elderly people, have difficulty in this integration process.

Considering these problems, new interface devices have been developed. The project is named 'FEELEX'; the word FEELEX is derived from a conjunction of 'feel' and 'flex'. The major goals of this project are:

1) to provide a spatially continuous surface that enables users to feel virtual objects using any part of the fingers or even the whole palm.
2) to provide visual and haptic sensations simultaneously using a single device that doesn't oblige the user to wear any extra apparatus.

A new configuration of visual/haptic display was designed to achieve these goals. The FEELEX is composed of a flexible screen, an array of actuators, and a projector. The flexible screen is deformed by the actuators in order to simulate the shape of virtual objects. An image of the virtual objects is projected onto the surface of the flexible screen. Deformation of the screen converts the 2-D image from the projector into a solid image. This configuration enables the user to touch the image directly using any part of their hand. The actuators are equipped with force sensors to measure the force applied by the user. The hardness of the virtual object is determined by the relationship between the measured force and its position of the screen. If the virtual object is soft, a large deformation is caused by a small applied force.

The first FEELEX was developed in 1997. It was designed to enable double-handed interaction using the whole of the palms. Therefore, the optimum size of the screen was determined to be 24 cm×24 cm. The screen is connected to a linear actuator array that deforms its shape. Each lin-

FIGURE 9. FEELEX

ear actuator is composed of a screw mechanism driven by a DC motor. The screw mechanism converts the rotation of an axis of the motor to the linear motion of a rod. The motor must generate both motion and a reaction force on the screen. The diameter of the smallest motor that can drive the screen is 4 cm. Therefore, a 6×6 linear actuator array can be set under the screen. The deformable screen is made of a rubber plate and a white nylon cloth. The thickness of the rubber is 3 mm. Figure 9 shows an overall view of the device.

The screw mechanism of the linear actuator has a self-lock function that maintains its position while the motor power is off. Hard virtual wall is difficult to simulate tool-handling-type force displays. Considerable motor power is required to generate the reaction force from the virtual wall, which often leads to uncomfortable vibrations. The screw mechanism is free from this problem. A soft wall can be represented by the computer-controlled motion of the linear actuators based on the data from the force sensors. A force sensor is set at the top of each linear actuator. Two strain gauges are used as a force sensor. The strain gauge detects small displacements of the top end of the linear actuator caused by the force applied by the user. The position of the top end of the linear actuator is measured by an optical encoder connected to the axis of the DC motor. The maximum stroke of the linear actuator is 80 mm, and the maximum speed is 100 mm/s.

Summary

This chapter is an overview of the history of haptic interfaces for virtual environment. Although research into haptic interface is growing rapidly, its methods are still in a preliminary state.

Haptics is indispensable for human interaction in the real world. However, haptics is not commonly used in the field of human-computer-interaction. Although there are several commercially available haptic interface, they are expensive and have limitation in its function. Image display has a history of over 100 years. Today, image displays, such as TV or movie, are used in everyday life. On the other hand, haptic interface has only a 10-year history. There are hazards to overcome for popular use of haptic interface. However, haptic interface is a new frontier of media technology and it will definitely contribute to human life.

Selected readings

Burdea G (1996) *Force and touch feedback for virtual reality*. John Wiley & Sons, INC

Sears A, Jacko J (2008) *The human-computer interaction handbook*. Lawrence Erlbaum Associates, 229–245

Principles of haptic perception in virtual environments

Gabriel Robles-De-La-Torre

Introduction

During haptic interaction with *everyday environments, haptic perception relies on sensory signals arising from mechanical signals such as contact forces, torques, movement of objects and limbs, mass or weight of objects, stiffness of materials, geometry of objects, etc.* (Fig. 1a). In contrast, *haptic perception in Virtual Environments (VEs) relies on sensory signals arising from computer-controlled mechanical signals produced by haptic interfaces* (see Fig. 1b, the online animation [1] under Selected Readings and Websites, and [1, 2]). Haptic interfaces are programmable systems, which can reproduce mechanical signals that are normally experienced when haptically exploring real, everyday environments. Perhaps more importantly, haptic interfaces can create combinations of mechanical signals that *do not have* counterparts in real environments. This allows creating haptic VEs in which entirely new haptic sensory experiences are possible. As a result, it becomes feasible to investigate haptic perception and related phenomena, such as motor control, in entirely new ways. In this regard, interfaces do for haptic perception research what computer graphics does for human vision research. The importance of haptic technology extends beyond scientific research. This technology opens the door to new applications in a variety of fields.

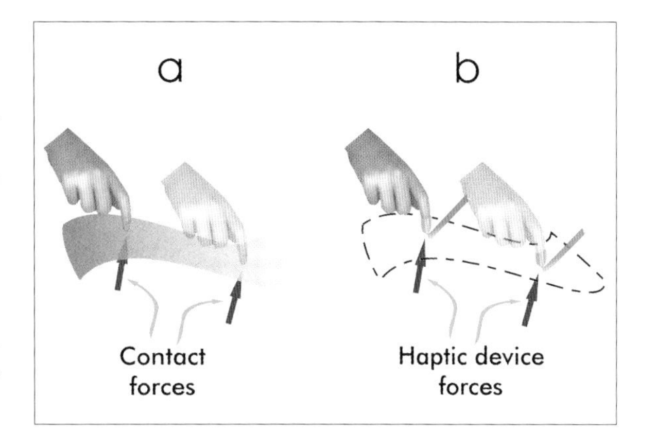

Figure 1.
(a) Haptic perception in everyday environments, *as when a perceiver explores a surface (the dark object here), involves mechanical variables such as contact forces, which arise from the physics of the haptic interaction scenario. (b) In contrast, haptic perception in* virtual environments *involves mechanical variables generated through haptic interfaces (here in blue, only part of the interface is pictured). Perceivers use haptic manipulandums (the golden, thimble-like part attached to the interface) to interact with interfaces. Under computer control, haptic interfaces produce mechanical variables (device forces) defining haptic virtual objects (the dashed surface). Note that during interaction with the virtual object, the perceiver's hand moves in an empty area of space.*

The main objective of this chapter is to discuss the essentials for effective use of haptic VEs in perception research and applications involving user testing. To illustrate this, the chapter also discusses some recent haptic perception discoveries in which haptic VEs played a key role. This chapter does not review the growing perception literature that uses haptic VEs. A full treatment of this literature would require an entire book of its own. Also, an important part of this literature can be consulted in several other chapters of this book. In this chapter, practitioners from fields such as neuroscience will find information to understand the similarities and differences between real and virtual haptic environments. Such practitioners will also find important information about designing and conducting perception experiments involving haptic VEs. This information would also benefit practitioners from areas such as engineering who are interested in haptic perception assessment in specific applications. The close interplay of basic research and engineering in haptic VEs has important implications for perception research and for haptic interface design. This is illustrated with examples of how haptic technology research contributes to our basic understanding of perception, and also of how perception research contributes to practical applications of haptic technology.

Haptic interaction, mechanical variables and haptic signals

Haptic perception relies on sensory signals arising from haptic interaction with a real or virtual environment. Haptic interaction involves mechanical (or haptic) variables such as forces, torques, masses, motions, stiffness of materials, etc. Note that mechanical variables arise from the environment, but also from the body of the person (i.e., the *perceiver*) that haptically interacts with the environment. For example, haptic interaction involves motion of limbs, as when actively moving a hand/arm when exploring a surface (Fig. 1a). When touching objects, contact forces between limbs and objects coexist with limb movement, as when lightly pressing down on a rigid surface while exploring it (Fig. 1a). In this example, mechanical variables involving the perceiver's body (hand/limb motion) are related to mechanical variables (contact forces) arising in part from mechanical characteristics of the environment (the surface's rigid material). This interplay between mechanical variables arising from environment and perceiver is an essential characteristic of haptic interaction, and has important roles in haptic perception. More generally, this interplay relates a perceiver's *actions* (such as precisely-controlled hand movements) to the *environment's reactions* (such as the motion of an object when pushed by a perceiver). Note that the interplay is bi-directional, as the environment's reactions may have an effect on subsequent actions by the perceiver. For example, when haptically exploring an object, the object may deform when squeezed or move when pushed. As a result, subsequent haptic exploration actions would need to be adjusted accordingly.

During haptic interaction, a given mechanical variable may or may not supply information that contributes to haptic perception. Throughout this chapter, a mechanical or haptic variable that supplies important information for haptic perception will be interchangeably called a *mechanical* or *haptic signal*. One or more mechanical variables or haptic signals define a *haptic stimulus*. A haptic environment consists of one or more haptic stimuli, which define the haptic properties of entities such as objects. A haptic object or environment is *virtual* if it is created through haptic technology. A haptic environment is *real* if it is not created through haptic technology. To further clarify important terminology, 'mechanical variable' and 'haptic variable' will be used interchangeably throughout.

The physics of haptic interaction and its importance in haptic perception

Haptic interaction is a process subject to the applicable laws of physics, such as those of dynamics [3]. This is because the laws of physics quantitatively describe the behaviour and characteristics of mechanical signals and variables present during haptic interaction. Physics describes haptic variables and signals in terms of mathematics. As haptic perception relies on haptic interaction, a thorough understanding of the relevant physics is essential to investigate how different mechanical signals contribute to haptic perception.

A full, quantitative characterisation of haptic interaction in terms of physics may become difficult in some cases, especially when many haptic signals/variables are simultaneously present. This may happen, for example, when haptic interaction involves multiple fingers, or when objects react to perceiver actions in complex ways. However, characterising haptic interaction in terms of physics allows identifying such complexities in detail and, more importantly, allows simplifying the haptic interaction scenario so it becomes tractable and remains meaningful for haptic perception experiments. As we will see, this is especially important for experiments using haptic technology, because *the physics of haptic interaction is a major basis for creating haptic VEs.*

Haptic virtual environments and perception research: the essentials

Haptic perception in real and virtual environments relies on the physical aspects of haptic interaction, but also on how the perceiver's nervous system processes information arising from interaction. However, in this chapter we will concentrate on factors directly related to mechanical variables and signals present during haptic interaction.

A full understanding of haptic perception requires controlling the haptic stimuli occurring during interaction with an environment. Typically, this involves systematically controlling and/or varying the haptic signals/variables that define the stimuli. For example, when investigating haptic perception of shape, important factors such as forces experienced when touching objects should be controlled and/or systematically varied.

This is a challenging task in general. Important mechanical signals/variables have been traditionally difficult to measure and control during haptic perception experiments. For example, contact force control during haptic exploration is possible to achieve in some cases (e.g. [4]), but has been difficult to obtain in general. As a result, many important aspects of haptic perception have remained barely explored until very recently.

Today, advances in haptic technology [1, 2, 5] allow exerting considerable control over important mechanical variables in haptic perception experiments. In this chapter, only the essentials of haptic technology will be discussed. This discussion will concentrate on the aspects of the technology that are of greater importance for perception research. For more information on haptic technology, consult the chapters by Bergamasco, Hayward and Hirzinger in this volume.

Haptic technology allows producing computer-controlled haptic signals/variables that a perceiver experiences through a variety of tools called manipulandums, which resemble thimbles (Figs 1b and 3b), pens (Fig. 2b), plates (Fig. 4), as well as joysticks, driving wheels, etc. [1]. To experience haptic variables, a perceiver wears or handles the manipulandums. Manipulandums are physically attached to computer-controlled mechanisms, which generate the haptic variables. Mechanisms and manipulandums are essential parts of a *haptic device* [1, 2]. The region of space in which a manipulandum can move is the haptic device's *workspace*. Note that the workspace is physically constrained by the device's mechanism. That is, the manipulandum can only move as far as the mechanism allows. Typical workspaces are three-dimensional, with volumes in the order of 100 cm^3 or larger.

Generally speaking, a haptic device can be classified as passive or active. Passive devices include those in which a perceiver applies energy to the device (for example, through applied forces and motions), and computer-controlled dissipation of this energy is provided by the device [1]. In contrast, *active devices* supply computer-controlled energy to perceivers, typically in the form of forces [1, 2]. The engineering of an active device includes the following basic components:

a) Actuators, such as electric motors, provide the energy needed to generate haptic signals such as forces. Actuators apply this energy to the device's mechanism and manipulandum.

b) Sensors, measuring the current state of the device, which typically includes the workspace position of the manipulandum and its orientation.
c) Power sources and electronics to drive the actuators and operate the sensors.
d) Electronics and control software enabling communication between the device and external control computer(s).
e) Software toolkits to control the device.

Software written with these toolkits runs on the external control computer, allowing real-time, programmable control of the device's haptic signals. This software implements *haptic rendering algorithms*. These algorithms precisely define the characteristics of haptic VEs and haptic *virtual objects* (VOs). Haptic rendering software is analogous to graphics rendering software. Together, haptic device hardware and software define a *haptic interface*. Virtual objects and environments created through haptic interfaces are 'virtual' because they are purely computational entities which are, so to speak, sculpted through the haptic device. Intuitively, this works as follows for isotonic, active haptic devices [1], in which the position of a manipulandum is sensed and used to compute programmable forces. Consider the case of a perceiver that uses a pen to poke into a flexible rubber surface (Fig. 2a). The physics of the contact between pen and surface involves several mechanical variables, among them contact forces and hand/pen displacements. When poking into the surface, the perceiver experiences the contact forces through the pen. The surface deforms when poked, and the perceiver's hand and pen are simultaneously displaced.

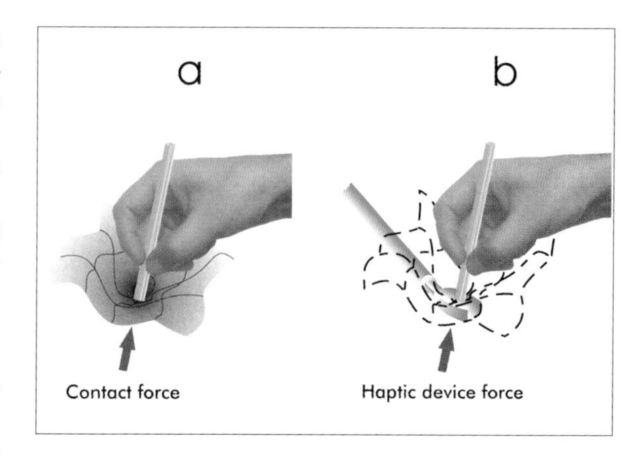

FIGURE 2.
A perceiver interacts with (a) real and (b) virtual deformable objects through a pen-like manipulandum. Both interaction scenarios involve similar mechanical variables, such as hand displacements and forces.

This real haptic interaction scenario can be reproduced through a haptic interface. For this, the perceiver holds a pen-like haptic manipulandum (Fig. 2b). The position of the manipulandum's tip is sensed in real-time (a thousand times *per second* or more) by the haptic device's sensor electronics. This positional information is sent to the external computer that controls the device. The positional information is monitored by the computer's haptic rendering software. When the perceiver moves the manipulandum's tip into the workspace location in which the haptic rendering software defines a virtual surface, the perceiver contacts the virtual surface and must experience adequate, simulated contact forces. The haptic rendering software computes these forces by using a *contact model* [5] (for example, a set of equations) of the mechanical interaction between, in this case, a pen and a *virtual rubber surface* (Fig. 2b, dashed surface). The rendering software then sends force control commands to the haptic device, and the device's actuators physically produce the computed contact forces. The perceiver experiences these forces through the manipulandum (Fig. 2b). When the manipulandum is withdrawn from the workspace zone in which the virtual surface is defined, the rendering software stops device forces. The perceiver no longer touches the virtual surface. Note that, excluding the manipulandum (and, indirectly, the device), the perceiver is not touching any physical object at all, but is only experiencing device forces. For an illustration of haptic rendering, watch the online animation [1] under Selected Readings and Websites at the end of the chapter.

Isotonic haptic devices are currently the most common type of device. Other haptic devices, known as *isometric devices*, measure forces applied by a perceiver to produce manipulandum displacements [1]. Unless otherwise noted, in the rest of this chapter the term 'haptic device' will refer to isotonic devices, and 'haptic interface' will refer to interfaces that include isotonic devices.

The importance of haptic devices in perception research

From the point of view of perception research, perhaps the most important characteristic of haptic devices is their capability to generate programmable haptic variables and signals that relate to perceiver's actions. Four key reasons for the importance of this capability are:

i) It allows reproducing real-world haptic signals to create, for example, virtual versions of real objects.
ii) It allows to dynamically change the mechanical properties of VOs. This allows creating, for example, VOs that, on command, change their shape and/or physical size (e.g., growing or shrinking), or become stiffer or softer to the touch. Other possibilities include VOs that move around the workspace in experimenter-controlled ways, or multiple VOs that interact with one another and also with the perceiver.
iii) It allows creating VOs that *do not exist in nature*. For example, it is possible to create haptic interaction scenarios in which mechanical variables relate in normally impossible ways to perceiver actions.
iv) In principle, it allows relating perceiver actions to haptic signals in a quantitative manner. For example, computer-controlled haptic signals experienced by perceivers (as well as the movement of the haptic manipulandum under perceiver control), can be recorded in the external control computer for detailed analysis.

Such capabilities are probably impossible to achieve without haptic technology. These capabilities offer many opportunities for perception research, especially the capability to create normally impossible haptic objects/environments. Using these in perception research has allowed discovering new, important characteristics of haptic perception and, more generally, of brain function. An example of this will be discussed later in this chapter.

Haptic interface characteristics and their importance for perception experiments

Haptic interaction with real objects may or may not involve tools, while interaction with VOs nearly always involves tools (manipulandums) at the present time. Understanding the characteristics of these tools and related device engineering allows using them effectively for designing haptic perception experiments and other applications.

Haptic interaction in the real world is extremely rich in terms of the mechanical variables typically present. As a consequence, a haptic device that can reproduce all aspects of real haptic interaction is not feasible at the present time. Haptic devices can currently generate only certain haptic variables or signals, particularly forces. Haptic interface engineering largely defines the characteristics of the virtual haptic interaction scenarios that can be simulated. Some of these characteristics relate to device mechanics, actuators and/or sensors, some to rendering software, and some to both. Additional hardware and software may compensate for an interface's limitations, or expand interface capabilities, as we will see later in the chapter.

In what follows, a discussion of key aspects of haptic interface engineering will be presented. The discussion will highlight the importance of interface engineering in perception research and perceiver assessment. This discussion applies fully to isotonic haptic devices. However, excepting those issues related to contact force production, this discussion also applies to active devices in general.

Tools are used to interact with haptic virtual objects

Many human activities involve tool use. Handwriting is an example. However, many other activities are tool-free, and involve direct contact between skin and objects. Object palpation is an example of this. However, when using haptic interfaces to simulate activities that are normally tool-free, a manipulandum still needs to be used. This is because haptic variables created with interfaces need to be mechanically delivered to the perceiver through manipulandums. However, manipulandum use may have an effect on haptic perception. Much of the rich cutaneous information that is present in real-world haptic interaction is not available when exploring VOs through manipulandums. For example, when sliding a fingertip along the surface of a real object, there is substantial cutaneous information available to perceivers. This includes skin deformation and/or indentation related to surface features such as texture, shape and the surface's material properties (for example, surface stiffness). More generally, skin deformation is related to local stress fields arising from fingertip-surface mechanical interaction. Although pioneering research work to provide such cutaneous information is underway (e.g., [6]), analogous cutaneous information is generally not available when interacting with virtual surfaces through manipulandums. However, perceivers using manipulandums have access to cutaneous information arising from the mechanics of skin-manipulandum interaction. Depending on the manipulandum and the task at hand, this cutaneous information may approximate the cutaneous information available when using analogous tools to perform real-world tasks. For example, similar cutaneous information may be available when touching real objects with a pen (Fig. 2a), and when touching VOs with a pen-like manipulandum (Fig. 2b).

Normally present thermal information (such as that available when touching a cold steel surface) is also typically absent during interaction with VOs. Also, manipulandums have always some inertia and, typically, weight, which are not present in a real, tool-free haptic scenario. These manipulandum-related differences during interaction with real and virtual objects should be taken into account when using haptic technology in perception research. Experiments may systematically explore how these differences may affect perception in specific situations. However, note that manipulandum-related differences during interaction with real and virtual objects are not disadvantages *per se*. They also offer experimental opportunities. For example, simple manipulandums can be used to learn how perception is affected by loss of cutaneous cues. Then, more sophisticated interaction with VOs can be introduced, for example, by systematically using different manipulandums and/or devices providing computer-controlled cutaneous information, such as the one described in [6].

A fixed number of independently-actuated forces or torques are available for virtual object interaction

Many haptic devices generate only one computer-controlled force vector, typically with three Cartesian components. This force is applied at a single manipulandum location. This location typically defines also the manipulandum-VO contact point. In the example above, computer-controlled force is applied at the tip of the pen-like manipulandum (Fig. 2b). In addition to computer-controlled force, some devices also generate torques along different axes.

Real-world haptic interaction typically involves many independent, or nearly independent, forces acting together. For example, consider the case in which a perceiver grasps a small, rigid cylinder with the thumb and index finger (Fig. 3a). During interaction, different forces are applied to each finger. When simulating an analogous two-finger interaction with a VO, a common solution involves two haptic devices (Fig. 3b). Each device generates one independently-controlled force. For more complex virtual interaction scenarios, some multifinger haptic devices have been developed [1, 7].

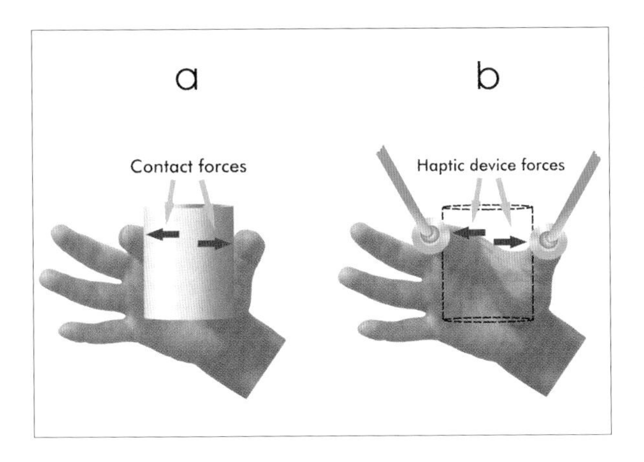

Figure 3.
A perceiver grasps a real cylinder (a) and a virtual one (b). In the real case, the perceiver experiences contact forces. The virtual cylinder is created with forces generated with two haptic devices (the blue machines in b); only part of the interfaces is pictured. Each device has its own manipulandum (the golden, thimble-like parts).

Haptic devices generate mechanical variables and signals with finite magnitudes

Isotonic haptic devices generate mechanical signals through actuators such as electric motors. The actuators transform one form of energy (e.g., electricity) into mechanical energy. This energy manifests itself as the device's mechanical signals (e.g., forces or torques). Actuators have a limited mechanical-energy output capability. As a result, only a finite range of, for example, force can be generated. Maximum forces vary from device to device. Typically, maximum forces are in the order of ten Newton, which is close to the weight of 1 litre (or quart) of water. As a consequence, VOs created with isotonic devices tend to feel somewhat soft, spongy. In contrast, isometric devices can produce rigid (highly stiff) VOs. However, it is difficult to use an isometric device to simulate touching an empty space. This is much simpler to do with isotonic devices, because they tend to feel very light when handled. Typically, perceivers can produce considerably more force than isotonic haptic devices. As a result, a perceiver can literally penetrate a haptic VO when applying enough force.

However, note that finite device forces are not a disadvantage *per se*. Perfectly rigid objects are idealisations. Many real objects may noticeably deform depending on the forces you apply during manipulation. Also, if rigid objects are needed for an experiment, haptic augmented reality setups may be used (see later in this chapter).

Haptic device forces vary in discrete steps

Typically, device forces are represented with a limited number of bits. As a result, forces vary in discrete steps. The magnitude of the smallest force step varies depending on the interface, but values in the order of 10 mN (or better) can be achieved [8].

Only selected aspects of haptic interaction are typically sensed through a haptic interface

Haptic interfaces have a limited number of sensors. During operation, interfaces typically sense the three-dimensional workspace position of one point on the manipulandum. Typically, the device applies forces to the manipulandum at this same location. Over time, changes in this point's position reflect actions performed by the perceiver. Depending on the interface, other aspects of interaction may be sensed. This may include the spatial orientation of the manipulandum, and/or the state of control buttons placed on the manipulandum. The resolution of sensed variables is finite, as they are represented with a limited number of bits. When sensing manipulandum position, resolutions in the order of 10 μm can be achieved [8], although common commercial devices have resolutions in the order of 40 μm [8]. This can be particularly important in applications requiring high spatial resolution, such as haptic texture rendering [8]. In addition to variables that are actually sensed, interfaces may also estimate other variables such as manipulandum velocity. Generally, such estimates should be used with caution due to

potential variability. Haptic interface manipulandums can be specially made to add extra sensing capabilities, such as torque sensing. Possible perceptual consequences of such enhancements should be thoroughly evaluated. For example, the enhanced manipulandum may have substantially increased weight.

Haptic interaction with virtual objects involves a discrete number of events executed a high number of times *per second*

A digital movie displayed on a computer screen is not the same as a natural visual scene. However, both can be remarkably similar to the eye. Unlike the natural scene, the movie consists of a series of still images that are typically displayed tens of times *per second*.

Haptic interaction with VOs has analogous aspects. During typical operation, a haptic device reads the state of its sensors, and sends the readings to a computer. The computer uses the readings as inputs to a haptic rendering program, which in turn sends control commands back to the device to generate haptic signals such as forces. To create VOs, these sensing, communications and force-production events are executed repeatedly. These events constitute the *haptic rendering control cycle* (HRCC). The execution rate of this HRCC depends mainly on i) haptic rendering computational load, ii) the processing speed and resources of the computer that executes the rendering, iii) the communication latency between control computer and haptic device and, iv) the computational load from any other active tasks running on the control computer, including network and operating system overhead. Significant delays introduced at any of these elements will affect the HRCC execution rate.

The HRCC execution rate is clearly an important variable for perception experiments. In general, the HRCC is executed with a frequency of 1 kHz or faster. Slower execution rates result in substantial, easily noticeable delays between perceiver actions and device force production. This negatively impacts the perceived realism of many VOs. However, the perceived realism of VOs is a complex, multidimensional phenomenon that requires further research. An adequate HRCC execution rate depends on factors such as hand velocity during haptic interaction, the temporal and spatial response of the human haptic system to a variety of stimuli such as displacements and forces, the characteristics of specific VOs, as well as the tasks in which the objects are used (e.g. [8]). A good rule of thumb is to execute the HRCC as fast as possible. Depending on the haptic device and the computer hardware available, it is possible to execute the HRCC at 4 kHz or faster.

For haptic perception experiments, a high, constant, rock-solid HRCC execution rate is extremely desirable in general. Otherwise, VO quality could vary during experiments, with perceptual consequences that can be difficult to assess. This is a demanding, but achievable objective, particularly when using control computers with highly-customisable operating systems (OSs) such as Linux, or real-time OSs.

Note that some cycle variability may be unavoidable, and generally tolerable, during deployment of haptic technology in applications other than perception experiments. However, this depends on the application.

The events happening when touching real and virtual objects: similarities and differences

This discussion applies to VOs created with isotonic haptic devices. Although similar events occur when touching real or VOs, there are important differences which will be discussed here in more detail. For simplicity, this discussion will consider a perceiver that uses a single fingertip to interact with a real or VO. It will also be assumed that real or VOs do not move away when touched, although their surface may deform when touched. Real and VOs are assumed to be of a large size compared to the perceiver's fingertip. It is also assumed that objects' surfaces have texture features that are small when compared to the size of the fingertip's surface. First, the events occurring when touching a real object will be discussed. Then, the analogous events occurring when touching a VO will be presented, and the differences between the real and virtual cases will be highlighted. Note that the important events described here extend to more general situations involving interaction with multiple fingers, for example. Note also that the following discussion illustrates only the main factors involved in the relevant physics of interaction. Plain language is used instead of mathematics. This was done to make this material accessible to most readers. However, this discussion is not a substitute for a detailed analysis of the physics of a specific interaction scenario.

Events happening during interaction with real objects

i) Typically, prior to contact with the object, the perceiver's fingertip does not experience major external forces. Usually, external forces are restricted to the minimal air resistance that occurs when limbs are displaced. External forces may become significant during strong wind, when operating in a denser medium such as water, or in the presence of acceleration, as when travelling aboard aircraft. These cases will not be discussed here.
ii) In general, when a perceiver's fingertip touches an object, the perceiver actively applies forces ('perceiver-applied forces') to the object. Following the physics of the interaction, the object also applies contact forces ('object-related forces') back to the perceiver's fingertip.

Perceiver-applied forces originate mainly from perceiver's muscle contractions, and/or the mass and acceleration of perceiver's limbs. Also, due to elastic mechanical characteristics of fingertip skin and subcutaneous tissue, as well as those of tendons and muscles involved in finger, hand and arm control, some energy originating from object contact may be stored in these tissues. When released, this energy may contribute to perceiver-applied forces.

On the other hand, object-related forces depend on factors including the following ones:

a) The forces applied by the perceiver. Let us illustrate this with a special case in which a rigid object is slowly touched under nearly frictionless conditions. This is approximated, for example, when touching a large, slippery piece of ice. In this case, by Newton's Third Law, object-related contact forces are equal in magnitude to the forces applied by the perceiver when pushing down onto the object, but have the opposite direction. Note that perceiver-applied forces may vary over time: the perceiver may apply more or less force when pushing down onto the object. When this happens, as a consequence of Newton's Third Law, object-related contact forces are automatically adjusted. Object-related contact forces remains equal at all times to the forces with which the perceiver pushes down onto the object.
In the special case discussed and in general, perceiver-applied and object-related contact forces may vary not only over time, but also spatially. That is, forces are applied at all the spatial locations in which a perceiver contacts objects during haptic interaction. Such spatial locations vary during interaction. This is important, because perceivers use stereotyped exploratory procedures involving different spatial patterns of hand movement [9]. As a result, different exploratory procedures may be associated with different spatial patterns of contact force.
b) The physical properties of the object's surface. For example, when touching a rigid object that has a rough surface, significant frictional forces will be typically present as a result.
c) The mechanical state of the object, particularly at the contact area. For example, if a rubber ball was previously touched, and its shape deformed as a consequence, mechanical energy was stored in the ball's material. This energy is released in part when the ball regains its former shape. If a perceiver

touches the ball at this point, the ball's stored energy may contribute to the object-related contact forces experienced by the perceiver.

d) The mechanical properties of the perceiver's fingertip and body. For example, the fingertip may be lubricated, or the skin of the fingertip may be more or less rough. Both factors would influence frictional contact forces.

e) The mechanical interaction between the fingertip and the object. This includes all preceding factors, and also the relative movement between the object and the perceiver's fingertip. Frictional forces depend on this relative movement. Also, during movement, the total contact area between object and fingertip may vary. This would also change frictional forces.

Events happening during interaction with virtual objects

In a broad sense, touching VOs involves events that are digital (i.e., discrete) versions of events happening when touching real objects. Let us discuss them in detail.

i) Before contacting a VO, the perceiver may experience several mechanical effects resulting from haptic device characteristics. These effects may include damping, backlash and the inertia of the manipulandum [10]. Typically, there are also gravity effects related to the masses of different parts of the haptic device. Gravity effects are approximately compensated for through counterweights or software. Because of the potential impact on perceiver performance, the design of devices typically attempts to minimise these undesirable mechanical effects. However, as these effects depend on the specific interface, adequate technical information should be requested from the manufacturer.

ii) The mechanical effects in i) are still experienced during the following events. When a perceiver's fingertip touches a VO, the interface generates computer-controlled forces, which are experienced by the perceiver through the haptic manipulandum. Device forces define *virtual-object-related contact forces* (VOCFs), which in many cases are analogous to object-related forces experienced when touching real objects. The features of VOCFs are jointly determined by the engineering of the haptic interface, and by the haptic rendering software that the experimenter writes to control the interface. This means that, to a considerable extent, VOCFs can behave in whatever manner the experimenter finds useful for his/her purposes. This allows for a wide range of possibilities. Note that the haptic rendering software must precisely define *all* aspects of haptic interaction with the VO. This includes sensing manipulandum position to computationally detect when and where the perceiver contacts a VO (a process called *collision detection* [5]), and how the object will react to this contact. The software must also detect when the perceiver has ceased touching the object.

Let us examine further how haptic interface engineering contributes to similarities and differences in the events happening when touching virtual or real objects. Compare the following to the equivalent cases discussed above for real objects.

a) Typically, perceiver-applied force is not sensed during interaction with VOs. As a result, real-world relationships between perceiver-applied and object-related forces may be simplified when touching VOs. A typical strategy for this computes VOCFs from manipulandum position and a linear spring model [5]. This strategy may be reasonably applied in some situations. However, simulating rigid objects in this, or analogous ways, is problematic. For example, it is possible to increase the stiffness of the spring model to approximate a rigid object, but this would result in unstable device behaviour, which produces unwanted device vibration and audible noise [5]. This approach is also limited by the maximum force that the device can generate, which may be easily overcome by perceiver-applied forces.

Spring-model contact forces and related approaches are very useful in many applications, such as user interface design, for example. However, in perception experiments it may be necessary to use rigid objects. Note again that isometric haptic devices can simulate contact with rigid VOs [1].

b) When touching VOs, the mechanical properties of the perceiver's fingertip would not have, in general, the same interaction effects that they have when touching real objects. For example, when touching real objects, lubricating the fingertip results in reduced frictional forces. When operating a haptic interface, a lubricated fingertip will not result in reduced virtual friction forces. Note, however, that different states of fingertip lubrication could be systematically simulated through haptic rendering computations, but this would be a different experimental situation. In this case, the perceiver's fingertip would not be physically lubricated. Instead, the haptic rendering software would use a model of fingertip lubrication when computing VOCFs.

Using haptic technology in the design of perception experiments

As we have seen, haptic technology offers exciting capabilities for investigating haptic perception and interaction in general. This section will discuss specific issues related to designing and performing experiments using haptic technology. This discussion will not be exhaustive, but will attempt to present issues of wide applicability involving current and, hopefully, future developments in haptic technology.

Perceiver issues

Physical demands of experiments

As perceivers interact with a computer-controlled mechanical system during experiments, tests may tend to be physically demanding for perceivers. The experimenter should have this in mind when designing the experiments, and ensure that adequate rest breaks are periodically provided to perceivers to help minimise possible effects of fatigue. It is also desirable to limit the total duration of a testing session. This duration depends on the actual experiment but, from experience, a testing session lasting about an hour and a half is reasonable for experiments involving forces of about 1 Newton, with rest breaks after each 10–15 min of actual testing.

Perceiver safety during experiments

Haptic devices may accidentally hit perceivers if not handled properly. This could happen if a perceiver suddenly loosens his/her grip on the manipulandum while the haptic device is generating force. To avoid this, and before the experiment starts, perceivers must be carefully instructed not to loosen their grip on the manipuladum, except when told to do so by the experimenter, which will only happen in safe, tested conditions. Also, perceivers should be told not to hold the manipulandum too tightly. This would help avoid accidents and also minimise perceiver fatigue. Some interfaces will automatically turn off forces when the manipulandum moves above a certain velocity. It is necessary to ensure that such a safety mechanism is actually in place. Otherwise, it is necessary to program it into the haptic rendering software used in experiments. It is also advisable to physically locate the haptic interface at a safe distance from the perceiver's face. To further ensure perceiver safety, all experiments must be reviewed and approved by a supervisory ethics committee, and perceivers must provide informed consent in writing prior to testing.

Individual variations

There are wide individual variations in hand and finger sizes. As a result, haptic manipulandums may be more or less effectively used by some perceivers. For example, some perceivers may find it difficult to use some thimble-like manipulandums, as their fingertips may not fit or perhaps fit too loosely for effective manipulation. It may be necessary to find ways to adjust the manipulandum, and also to assess how this may affect experiments. The manufacturer of the device may offer special manipulandums to deal with this. It may be necessary also to adjust haptic interface force levels, so that perceivers with different physical strengths can perform the tests comfortably. Due to individual variations such as these, it is frequently useful to design within-subject tests for experiments.

The instructions given to perceivers and related, unwanted expectations

For successful experiments, it is critically important that perceivers receive substantially the same instructions about how to perform experimental tasks. For this, instructions should be carefully prepared prior to testing and, preferably, in writing. A full discussion of instructions is beyond the scope of the chapter, but there is some information that, in general, should not be provided to perceivers in the instructions, and which should also be avoided during perceiver recruitment. In particular, *perceivers should not be told that in the experiments they will manipulate or interact with virtual objects*. Perceivers provided with this information will have potentially undesirable expectations about the experiments, which may greatly affect their performance. For example, instead of concentrating on performing experimental tasks, some perceivers will think that some or all of the stimuli are *fake*, and try to guess which stimuli are the fake ones, or why they are fake.

It is also necessary, in general, to minimise other undesirable perceiver expectations. Such expectations can arise from allowing perceivers to visually inspect the experimental setup, including the haptic interface. As a result of this, most perceivers will understand that they will not interact with real objects, but with a machine that simulates objects. Ideally, perceivers should not be allowed to see the setup or haptic interface. This can be accomplished by using a screen or similar arrangement to block visual information. It is also important to avoid touch cues about the setup and haptic device. For example, perceivers may notice from touching the haptic manipulandum that they will not be exploring real objects. This can be avoided by using *haptic augmented reality* setups (see later in this chapter).

Other perceiver expectations or assumptions

Perceivers tend to assume that device forces reflect characteristics of VOs, instead of assuming other possibilities, such as force effects related to the mechanical properties of the manipulandum. For example, when experiencing spring-like device forces, perceivers tend to relate the forces to VO deformation. Perceivers generally do not assume that forces could result from touching a rigid VO with a springy manipulandum, for example. It is clear that many factors shape these assumptions, such as previous experience with real objects. Assumptions like these should be carefully considered by the experimenter when designing tests or analysing results, as assumptions may influence perceiver behaviour. Informally querying perceivers after testing may also help identify such assumptions. On the other hand, perceiver assumptions may be useful in areas such as perception-based haptic rendering, for example, as a potential source of ideas for new rendering methods. Perceiver assumptions may also suggest new phenomena in haptic perception and/or cognition, and in related areas.

Perceiver practice

Perceiver performance may be affected by unfamiliarity with the experimental setup. Providing enough practice trials prior to formal testing helps correct this.

Haptic interface issues

Many important interface issues have been discussed above. Here, some additional issues will be discussed. Strategies to deal with these and with previously discussed issues will be outlined.

Device noises

The level of mechanical noise present during normal device operation varies from device to device. It is necessary to ensure that perceivers do not receive unwanted experimental cues from these noises. Sometimes, screens used to block visual cues also help reduce noises. If this does not help, perceivers may wear earplugs or headsets delivering wide-band noise.

Device overheating

Device actuators such as motors may overheat, typically when using relatively high force levels during extended periods of time. Some devices include safety measures to help prevent permanent actuator damage from overheating. These measures, however, may interfere with experiments. For example, devices may include low-level software that shut them down automatically. To do this, instead of actually sensing actuator temperature, such software may estimate the energy that actuators dissipate as heat during operation, as well as estimating the time needed to cool actuators through device inactivity. As a result, physically cooling the actuators may not allow restarting device operation immediately. Simple solutions to avoid this include using lower levels of force, and avoiding situations in which large forces are continuously exerted during long periods of time. Rest breaks usually help achieve this. Generally, such problems may be easily found and solved during pilot tests. More drastic solutions may sometimes be needed, such as replacing device driving electronics [11].

Device limitations and strategies to deal with them

When evaluating the purchase of haptic devices/interfaces for experiments, up-to-date, detailed technical information should be requested from manufacturers, especially about those features that matter the most for the experiments at hand. More information may be gathered from colleagues, the literature or from online resources such as the Haptics-L mailing list and the International Society for Haptics (see [2, 3] in Selected Readings and Websites). Three important examples of device limitations will be mentioned here, with strategies to deal with them.

a) Variations in nominal resolution of manipulandum position sensing. Nominal resolution may vary in a significant, systematic way throughout the workspace [8]. Depending on the experiment, this may or may not be an issue. If this is an issue, experiments can be designed so that the manipulandum operates only in the workspace region in which resolution varies the least (typically, the centre of the workspace). If this is not feasible, then VOs should be presented in different parts of the workspace, so that overall, possible effects of nominal resolution variability are averaged out, or isolated during data analysis.
b) Nominal forces may systematically vary across the workspace. This can be due to actuator drive electronics [11], which can be corrected by replacing the electronics. Up-to-date information about this should be requested from device manufacturers.
c) Limitations when simulating the physics of real world haptic interaction. It is common to use simplified contact models when perceiver-applied forces are not sensed. An open issue is how such simplifications may affect haptic perception, as real and simplified virtual scenarios may be very different, physically and perceptually. Because of this, relating perception of VOs to perception of real objects may be difficult.

A possibility to deal with c) consists in using interfaces to create haptic VOs that coexist with real physical objects. By analogy with the equivalent case for visual displays [12], such setups might be called *Haptic Augmented Reality* (HAR) setups. HAR setups might be also called Haptic Mixed Reality setups. An example of a HAR setup is shown in Figure 4. This setup uses an isotonic device for investigating haptic perception of shape. Here, a perceiver interacts with a real, rigid object (Fig. 4). The rigid object is carefully designed and machined, so its geometrical features are known. The manipulandum has wheels, rolls on top of the rigid object, and is mechanically constrained so it always remains vertical to the object's surface, as shown in Figure 4. The manipulandum is always in contact with the object, and includes a sensor that measures the force that perceivers apply when lightly pushing down on the manipulandum's plate (this force is called the 'perceiver-applied normal force', Fig. 4). Following Newton's Third Law, this perceiver-applied force is balanced by a corresponding object-related force.

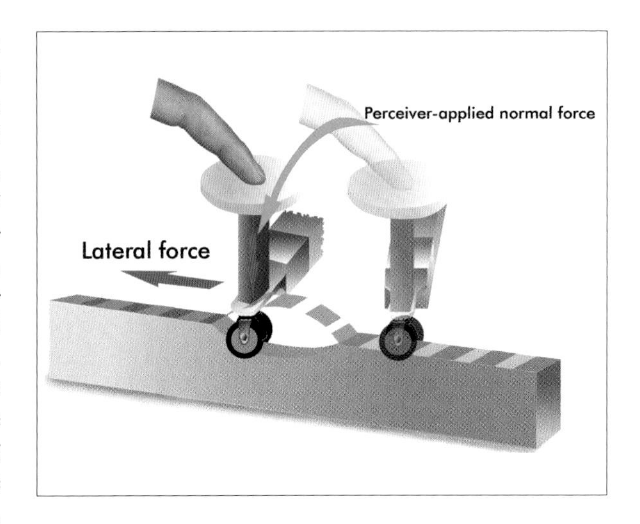

FIGURE 4.
A haptic augmented reality (HAR) setup, in which real and virtual objects coexist. A perceiver applies a normal force through his/her fingertip to hold the haptic manipulandum (golden plate) down. The perceiver rolls the attached haptic device (in blue, only part of the device is shown) on top of a real, rigid object (grey bar). Simultaneously, the haptic device generates net, computer-controlled lateral forces corresponding to a virtual bump (the red, dotted figure). This paradoxical, haptically augmented object combines the geometry of a hole (grey bar) with the lateral forces of a bump. This paradoxical object is typically perceived as a surface with a bump.

This HAR setup has several advantages: i) the physics of the interaction with the rigid object is very simple, which allows identifying the most important physical aspects of the interaction, and expressing them precisely and quantitatively; ii) this allows knowing which haptic variables to sense; iii) the setup allows selectively modifying real-world mechanical variables through the haptic interface; iv) this allows creating normally impossible, paradoxical combinations of real and virtual objects, and to use them to probe perception in new ways; v) the setup allows relating perception of real objects to that of VOs; vi) for experimental purposes, finite haptic interface forces can be used more effectively. Let us see how these advantages are achieved in the current HAR setup. From the physics of the interaction [13], it is found that, when exploring the real, rigid object in this setup, object-related lateral forces (along the horizontal direction of movement, Fig. 4), depend mainly on i) perceiver-applied normal force (Fig. 4) and ii) the local geometry of the real object (under the very low friction conditions used here). By design, this local geometry is always known. It is given by the precisely-machined surface of the real object. As the position of the manipulandum is sensed, it is always possible to recover the current local geometry in real-time. Perceiver-applied normal force (Fig. 4) is sensed in real-time also, so object-related lateral forces can always be known from the physics of the interaction and the sensed variables. As a result, haptic interface forces can be added in real-time to object-related lateral forces. The resulting net lateral force is experienced by perceivers through the manipulandum. When the haptic interface does not generate lateral force, perceivers experience only the object-related lateral forces arising from the natural interaction with the object. However, haptic interface forces can be designed so that a variety of different, paradoxical combinations of mechanical variables are achieved. For example, under haptic interface control, net lateral forces can correspond to those experienced when touching an object with a bump, even when the perceiver's hand moves along the surface of a real object with a physical hole (Fig. 4). More generally, in this way important lateral force cues can be decoupled from the geometry of real objects, and the perceptual contributions of each of these cues can be investigated separately, with surprising results. For example, the stimulus shown in Figure 4 is typically perceived as a surface with a bump. In this and other, paradoxical situations, shape perception depends on lateral force cues, and not, as previously thought, on geometrical cues such as fingertip trajectories during exploration. Such paradoxical objects are not experienced in the real world, and are probably impossible to generate without haptic interfaces or equivalent apparatus. These findings suggest that lateral force, under some conditions, is the effective stimulus for haptic shape perception. Alternatively, these findings suggest that lateral force elicits illusory haptic shapes [13, 14]. Research also suggests that these findings may apply to more general situations, for example, to elicit the perception of ascending a slope during locomotion on flat surfaces [15]. Overall, these findings indicate that lateral forces alone can elicit perception of shape. As a consequence, from the perceptual point of view, lateral forces as used here define haptic virtual objects. Therefore, this HAR setup uses real objects combined (or augmented) with lateral-force-based VOs [13]. Clearly, in addition to lateral-force VOs, other types of haptic VOs could be used in HAR setups.

Besides achieving paradoxical combinations of haptic signals, HAR setups allow overcoming other potential limitations of haptic interfaces. For example, finite haptic device forces result in VOs whose nominal boundaries can be penetrated when perceiver-applied force overcomes device forces. As a result, manipulandum trajectory under perceiver control may not accurately reflect the nominal geometry of VOs. When using HAR setups, these inaccuracies can be eliminated through the use of precisely-machined rigid objects. Also, perceivers' exposure to unwanted cues about the workings of the experimental setup can be eliminated when using HAR setups. In the example just discussed, perceivers can be thoroughly instructed about using the setup without their being aware of the haptic interface at all. This is generally difficult to do when using a haptic interface alone. Perceiver safety may also benefit from HAR setups. In the example discussed here, it is mechanically difficult that the haptic manipulandum accidentally contacts perceivers. *Finally, as HAR setups involve real and virtual objects, it is possible to investigate perception of a continuum of stimuli ranging from real objects to purely virtual ones* [13]. This allows relating perception of VOs to that of real objects, which is generally difficult to achieve otherwise.

In general, instead of using haptic interfaces to generate all the relevant mechanical variables in an experiment, interfaces can be used to selectively modify the variables present during interaction with real objects. This allows for better control of important experimental cues. In a sense, using real objects in HAR setups constrains the ways in which haptic interfaces are used. Therefore, it could be thought that HAR setups limit the applicability of haptic interfaces, for example, by severely constraining device workspace. This is not so. For example, a realistic, precise HAR setup can be used first to understand the perceptual contributions of important haptic signals. Then, this basic understanding can be used to design a purely virtual haptic environment (with no real objects involved, as in [17], for example), that approximates the features of the related HAR setup. This purely virtual environment can exploit to the full the capabilities of haptic interfaces, and can be used to test a variety of more complex situations. Perceiver performance in the HAR and purely virtual cases can be compared and further understood. For example, the HAR setup discussed above has been used as the basis to investigate the role of contact force in active and passive touch perception, in experiments involving only purely VOs that move across a

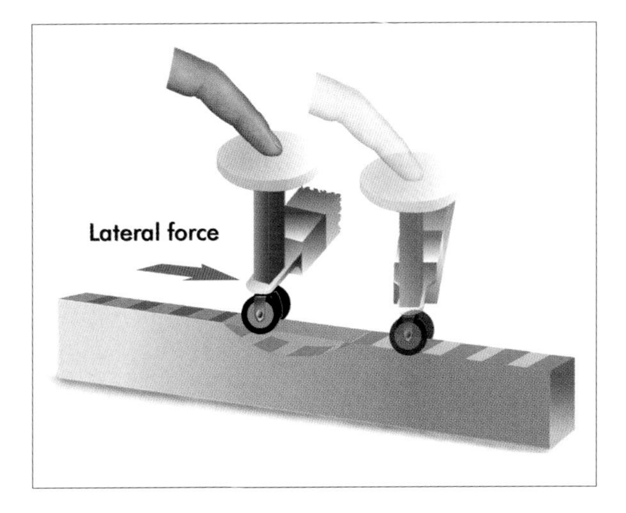

FIGURE 5.
A haptic augmented reality (HAR) setup in which a real, flat, rigid object (grey bar) is combined with net lateral forces generated through a haptic device (the blue machine, shown here only in part). The net lateral forces correspond to those of a virtual hole (the red, dotted figure). This geometrically flat, paradoxical haptic object is typically perceived as a surface with a hole.

workspace [16]. Such VOs allow controlling lateral forces during object exploration in passive and active touch situations, which is very difficult to achieve with real objects. These experiments with purely virtual objects are justified by, and rely on, results obtained with the HAR setup described above.

Extending interface capabilities through haptic perception: perception-based haptic rendering

As we have just seen, haptic perception of shape can be modified considerably through haptic device forces. In another example of this, when exploring a flat, rigid surface while simultaneously experiencing the lateral forces of a hole (Fig. 5), perceivers do not experience touching a flat surface, but one with a hole [13]. We can see that, as happens in other sensory modalities, greatly simplified stimuli can elicit compelling perception of complex objects. This has been found, for example, in haptic perception of shape [13, 14, 17, 18], and texture [19]. These findings contribute to understanding how haptic perception works, but also allow simplifying the technology needed to render haptic objects for different applications. For example, as lateral forces can elicit perception of shape when exploring flat surfaces, this means that lateral forces can *haptically render* virtual shapes, without the need to simultaneously provide perceivers with the geometrical information present in real objects. This can be achieved through devices with planar workspaces [13, 14], instead of using more complex devices with three-dimensional workspaces. This promising field, in which the properties of human perception are applied to haptic rendering, is frequently called *perception-based haptic rendering*.

Summary

Haptic VEs do for haptic perception research what computer graphics does for vision research. Haptic interaction with VEs allows investigating perception and related phenomena in totally new ways. This includes creating haptic objects that do not exist in the real world. The level of stimulus control provided by haptic VEs allows relating perception to mechanical signals in a quantitative way. This is difficult to achieve otherwise. The close interplay between haptic technology and perception research is a constant source of advances in both fields. Clearly, human perception research benefits greatly from haptic technology and, conversely, haptic technology benefits greatly from human perception research. Current and future advances in both fields offer potentially important opportunities for understanding haptic perception and related phenomena, as well as their underlying neural implementations (e.g., see [4–6] in Selected Readings and Websites). This can contribute much to highlight and understand the profound importance of these commonly underrated phenomena [20]. Haptic technology also offers considerable scientific and technological potential when used in combination with other technologies such as visual or auditory digital displays. This potential is largely untapped.

Acknowledgements

I thank Lorena Robles-De-La-Torre for preparing the figures, and Vincent Hayward for insightful comments.

Selected readings and Websites

Animated explanation of haptic technology essentials. 3 April 2007. http://www.roblesdelatorre.com/gabriel/ch.html

Haptics-L: the Electronic Mailing List for the International Haptics Community. 3 April 2007. http://www.roblesdelatorre.com/gabriel/hapticsl/

International Society for Haptics. 3 April 2007. http://www.isfh.org

Flanagan JR, Lederman SL (2001) Neurobiology: Feeling bumps and holes. *Nature* 412: 389–391

Henriques DYP, Soechting JF (2005) Approaches to the study of haptic sensing. *J Neurophysiol* 93: 3036–3043

Wexler M, van Boxtel JJA (2005) Depth perception by the active observer. *Trends Cogn Sci* 9: 431–438

31

Haptic shape cues, invariants, priors and interface design

Vincent Hayward

Introduction

Perception is often discussed by reference to cues as separate sources of information for the perceiver [1]. With vision and audition, the list of such known cues is quite extensive [2, 3]. For example, visual depth perception in humans is thought to rely on monocular, oculomotor and binocular cues. Monocular depth cues include motion parallax, color contrast, perspective, relative size, relative height, focus, occlusion, shading, texture gradient, shadows, interreflections, and others. Oculomotor cues include accommodation and convergence. Binocular cues include disparity-based stereopsis. Such collections have been also identified for other object qualities such as size or color. With audition, say for object localization, there are analogous notions, such as interaural time difference, interaural intensity differences, or spectral cues related to head-related transfer functions, in addition to monaural cues [4].

These cues are tied with the manner in which the sensory apparatus – physically and computationally – has evolved to account for the ambient physics. For example, sound localization obeys fundamental constraints related to the propagation of sound such as wavelength and speed of propagation. Nature has developed marvelous mechanisms to cope with these constraints and at the same time take advantage of them.

It is thus natural to propose that for touch, like for vision and audition, such physically and computationally specific cues must exist and can be identified. This chapter is about discussing some putative tactile cues that refer to shape as one of the object attributes that a perceiver could be interested in.

To this end, the notion of invariant will be used to identify a collection of possible tactile shape cues, and priors necessary to the processing of haptic shape are suggested from the analysis of experimental evidence. Examples of how these notions can be applied are described by looking at two specific haptic detection tasks and how stereotypical movements can be interpreted.

Displays may be thought to operate like 'mirrors' of the perceptual system. The colors channels of a LCD display 'mirror' the color channels of the visual system. The fast repetition of frames – a sampling process – 'mirrors' the computational spatiotemporal interpolation performed by the visual system – a reconstruction process. Examples such as those abound. For haptic interfaces one may adopt a similar view point and examples of how this approach can be applied are discussed later.

Before exploring these topics, general observations are made to illustrate the fundamental differences between direct touch and tool-mediated touch.

Observations on the mechanics of touch

In this chapter we discuss the case of touching rigid and stationary shapes. By this, it is meant that the touched objects do not deform nor move significantly compared to the deformation and displacements of the touching object. It is also needed to assume that when a finger slides on an object, the tangential deformation caused by slipping can be neglected. More general cases will be mentioned when needed.

Tools and fingers

It is commonly observed that haptic interaction can happen in one of two possible ways [5, 6]. Perceivers can interact with objects using tools or with direct finger contact. Forks and chopsticks, surgical instruments, or switches are examples of what is meant by tools. In these cases, the question arises of what are the haptic cues that used to extract information about particular object qualities. As far as shape is concerned, this question turns out to be more difficult to discuss when tools are used rather than bare fingers, as discussed next.

Transformations induced by tools

During haptic exploration with a tool, the information that can be extracted from an interaction is entirely contained in the displacements of the tool, whether they are large movements or small oscillations. The exclusive medium of information transmission are the movements of the tool [7]. Resting a pen on a table tells nothing about the table [8], but when there is movement, the tool first transforms the tool-object mechanical interaction into radically different mechanical events at the periphery of the perceiver [9]. There, what is potentially available is the motor activity that gives rise to the interaction and the resulting deformation of tissues in the fingers and limbs. From the perspective of the perceiver, this corresponds to a second transformation. In order to recover a given attribute, say shape, we may follow this path in the reverse order. Mathematically we could say that if f associates a shape to the movements of a tool and g the movements of a tool to the deformations of tissues that are sensed, then the brain has to invert $g \circ f$ to have access to shape, which is compute $f^{-1} \circ g^{-1}$.

Evidence that the brain is able to invert the second transformation – the tool-hand interaction – can be obtained from the observation that, by and large, similar sensations are experienced when the same tool is used against the same object but with different grips, each creating a different version of the second transformation. We may call this effect a grip-related perceptual constancy effect. Then, the first transformation caused by the tool can be inverted to recover relevant aspects of the tool-object interaction, those related to shape for the case in point. Only then can the sought-after object attribute be recovered from the properties of the tool, since the interaction depends on the tool as much as it does on the object. Here, unlike grip-related perceptual constancy, tool-related perceptual constancy is less likely to succeed, especially if the tool is inadequate such as having a curvature that is commensurate with that of the touched object.

Using intermediaries

When using a tool, the perceiver is faced with two hard, cascaded problems to solve since the variations introduced by the intermediary have impact on both these transformations. Factors that enter into the complete equation include the relative curvatures of the tool and the object at the place of contact, the relative compliance of the materials in contact, their internal structure, the structural dynamics of the tool and the nature of the interface between the tool and the hand as well as the grip used.

This analysis is general and applies also to interaction with complex mechanical devices such as switches, knobs or piano keys. In the later example, the impact of the hammer on the string is actually 'felt' although there is no direct mechanical path between the finger and the string since the hammer is in free-flight at the time of impact! [10] The impact can only be felt by inverting the dynamics of the escapement in order to anticipate the velocity of the hammer as it hits the string. A simpler example that we have studied are surgical scissors [11]. It was seen that the design of the scissors and the tissue properties both have profound effect of the information available to the surgeon. Similarly, experienced surgeons 'invert' the scissors to appreciate tissues properties, the result of which guides the next incision, since surgical cutting is the result of many small fast cyclical cuts.

Implications for the design of interfaces

Given the limitations of current technology, haptic interfaces are able to replicate real mechanical interaction with only a large degree of imprecision. One can only wonder why force feedback interfaces work so effectively [12]. This must be attributed to the brain's ability to deal with an extraordinary range of possible intervening transformations. From the analogy that displays act as mirrors of the perceptual system, one may conclude that the practical realization of force feedback displays appears to be much easier to achieve than the realization of direct contact cutaneous displays. With direct finger contact, the tool is the finger itself, so the brain benefits from a lifetime of its use and of incorporation of its properties. There must be much stricter rules that the displays must obey.

Force feedback

It is useful to reflect briefly on the notion of *force* since it is central to the concept of force feedback. First, recall that a force does not have a physical existence. It is a mere abstraction used to describe the action of one particle on another in terms of a vector. We can also use the notion of force for systems of particles that are assembled in a solid. In the later case, we can also describe forces of contact in addition to the forces acting at distance. If we consider a perceiving body, say a finger, then when we speak of force, we simply describe the action of another system of particles that acts on it to change its state, lumped into three numbers. This generalizes to torques and tractions. With the method of Lagrangian dynamics, we can in fact do away with the idea of force and think only of the trajectories of generalized coordinates [13].

When reproducing a virtual object, we may therefore regard the function of a force feedback interface to be that of causing the displacements and deformations of limb(s) and finger(s) that would be equivalent to the displacements and deformations experienced when interacting with an original object. Experimental evidence of the value of this view is provided by the success of acceleration matching techniques, equivalently movement-matching techniques [7, 14].

For the purpose of this paper, although we leave aside the problem of the analysis of the cues available with tools, we retain from this discussion the possibility of describing the cues arising from direct finger contact *entirely in terms of deformations and displacements.* Thus, in the rest of the chapter the word 'force' (or 'traction', or 'pressure') will no longer be needed.

Mechanical invariants

With direct finger contact, since there is no tool transformation involved, we can devise a more systematic approach to the analysis of cue generation as they relate to shape.

Notion of perceptual invariants

Invariants have for a long time played a central role in the study of perception [15]. There is a strong connection between the idea of perceptual cue and the notion of invariant [16, 17]. In fact, behind each depth cue listed in the introduction hides one or several invariants. Invariants can arise from three sources [18]. They can arise from mathematical properties, they can arise from physics, and they can arise from the structure of the sensory apparatus, in that they have a biological origin. The properties of straight lines, circles, or symmetry groups are examples of sources of mathematical invariants. That gravity accelerates free-falling objects at a constant rate in a uniform direction, or the speed of sound, exemplify the sort of invariants we might expect from physics [19]. The near-orthogonal geometry of the semicircular canals in the inner ear or Listing's law that governs the eyes to rotate around a common fixed axis are examples of the sensory structures that create invariants which are presumed to enhance the computational efficiency of perception [20].

Specific shape invariants

There is a number of physical phenomena related to the mechanics of contact which, collectively, provide a rich source of invariants that are highly relevant to tactile shape perception. Within limits, these invariants are generic to any touched shape. Some are briefly discussed next.

Static invariants

Contact mechanics dictates that when two objects are in contact, no matter how hard or how soft they are, they share at least one surface of contact that grows from an initial point [21]. It is a fact of geometry that it is only when the contacting objects are convex (one could be flat but not both) that the contact surface is guaranteed to be a connected component. As long as one can ignore the effects of the fingertip viscosity and hysteresis, the invariants that result may be said to be static since they do not depend on history nor speed.

A first aspect of the surface-forming phenomenon which is special to the mechanics of a finger is related to the spread pattern of this surface from an initial point [22, 23]. The finger mechanics cause the size of a contact surface to define an anatomically-related tactile yardstick of about one centimeter (this can be seen by grabbing a glass and looking directly at the contact), which changes by a factor no larger than, say, 1.5 for a flat surface, see Figure 1a. Let's call that invariant S1.

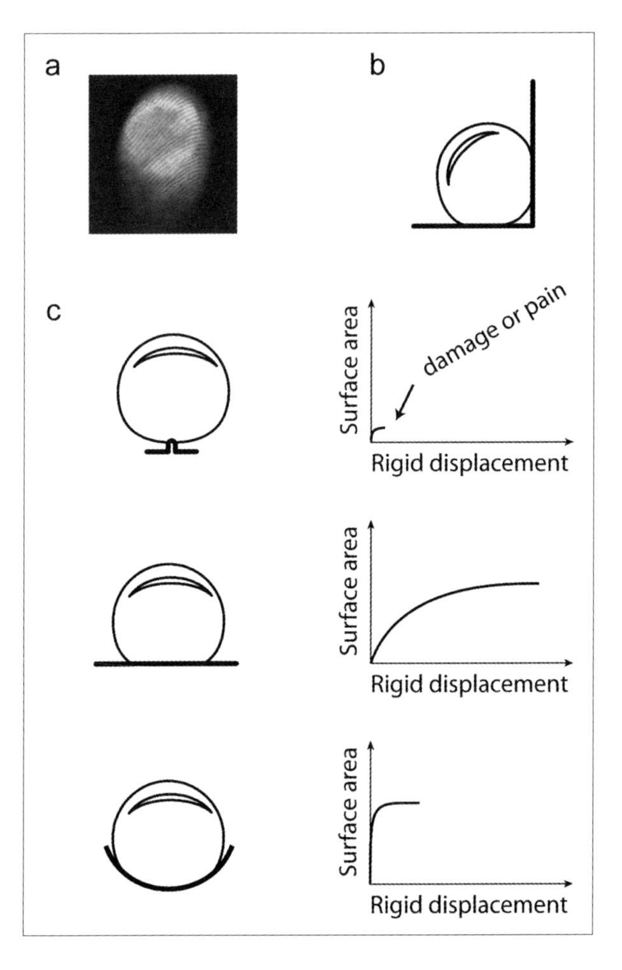

Figure 1. Static tactile invariants
View of a finger contact surface imaged through a desktop scanner (a). A sharp inside corner causes multiple disconnected contacts (b). Dependency of the contact surface relatively to the rigid displacement of a finger (c).

Since the fingertip's sensitive area is a convex surface, if the contact areas are disconnected, then the touched object must have at least one sharp concave region. It is something that can be verified by touching the inside corner of a box, see Figure 1b. More generally, not only the size but the shape of the surface contact is characteristic of the local curvature of touched surface. In particular, the eccentricity or deviation from circularity is directly related to the ratio of the principal curvatures of the touched object. Thus, a second invariant and its many variants, call them S2 collectively, is identified.

The surface-forming phenomenon creates yet another powerful sensorimotor invariant, S3, that correlates the curvatures of the object with the growth curve of the contact surface [24–26], see Figure 1c. When the object curvature is high, say a rod or a Braille dot, the surface grows fast but plateaus at a low value. If the touched surface is flat, the growth rate is at its slowest and plateaus at a higher value. If the touched surface is concave, then it grows fast but plateaus at a high value. At the limit, if the curvatures almost match, then the contact surface is instantly created and the finger rigid body displacement is almost impossible. Of course there is an infinite family of surface-forming curves, each characteristic of the curvature of the surface relative to that of the finger.

Kinematic invariants

In mechanics, one distinguishes local deformation from global deformation [21]. This is expressed by St Venant's principle. This principle states that the effects of different but statically equivalent loads are not distinguishable at distances greater than the dimension of the contact area. For example, if one grips a brass rod in the jaws of a vise with the aim of bending it, the shape of the contact areas (seen by the marks left by the jaws) has no effect on the overall shape of the bent rod. This is the source of powerful invariants and has consequences in almost all aspects of mechanical sensing.

For shape perception, this principle has the effect of mechanically separating the sources of information about a touched object into two neatly segregated categories. There is what is available inside the contact in terms of the strains developing at the surface of the skin and in the subcutaneous tissues, the details of which have absolutely no effect elsewhere. Conversely, the net displacement of a contact area, regardless of the details of its shape, provides a second category of source of information. To identify more specifically shape-related invariants and hence possible shape cues, let us consider the information that is available from the velocity of a contact surface on the finger. In considering velocities, we may call these invariants kinematic invariants.

The mechanics of the relative motion of two bodies in contact requires that each infinitesimal portion of the two surfaces is in one of two

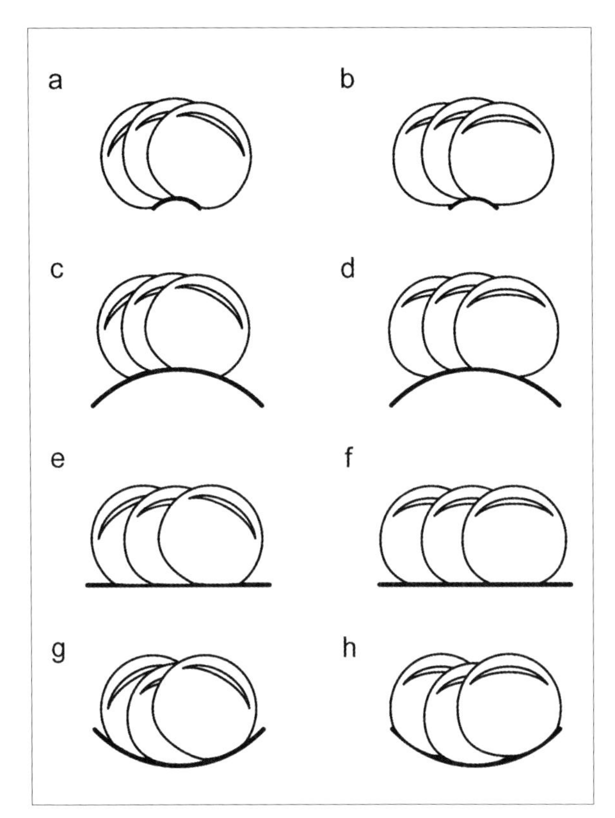

Figure 2. Kinematic tactile invariants

When a finger rolls on a surface, invariant K1 holds that the relative radii of curvature determine the velocity of the contact surface relatively to the angular velocity of the finger (a), (c), (e), (g). When a finger slips, K2 expresses that for a convex object, the lower the curvature of the object, the lower the velocity of the contact on the skin (b), (d), (f). For a concave object the relationship is inverted (h).

states [27]. Because of friction, each pair either sticks or slips. Upon initial contact the whole of the contact surface sticks. Under sufficient tangential load, the whole of the contact surface slips. With a highly deformable body such as a finger there is a transient regime during which the slipping region grows from the periphery to eventually invade the whole region [28, 29]. This and many other possible patterns are likely to be another rich source of invariants which, unfortunately, cannot be discussed here.

Leaving the transient regime aside, new invariants can be identified. Mechanics requires that when there is no slip, and of course no pivoting, there is pure rolling motion between the finger and the object. In other words, the rigid-body instantaneous velocity of the finger (a global deformation of the perceiver's body) relatively to the object is constrained by a relationship between the velocity of the contact region on the fingertip (a local deformation) determined by the relative curvatures of the finger and of the object.

Specifically, for a given angular velocity of the finger, the smaller is the curvature of a convex touched object, the slower is the velocity of the contact region, see Figure 2a and b. At the limit, for a flat surface, see Figure 2e, this velocity is essentially the effective radius of the finger times its angular velocity. If the surface is concave, Figure 2g, there is amplification up to the point where when rolling the finger on a concave surface with a radius that tends to that of the finger, the contact surface velocity tends to infinity. In a nutshell, for curvatures ranging from an infinitely curved convex surface to a concave surface of the same curvature of the finger, then the ratio of the finger angular velocity to contact surface velocity varies from zero to infinity. When the surface is flat, this ratio is the finger radius. We call the relationship linking angular velocity with contact surface velocity as a function of surface curvature invariant K1.

Now, let's look at the same cases but when there is slip. Slip must occur, for instance, when exploring an object while the orientation of the finger is kept constant. The scanning velocity must be considered and a very different type of relationship is created. For a convex object, see Figure 2b and d, for the same scanning velocity, the lower is the object curvature the lower is the velocity of the contact surface on the finger. At the limit, if the object is flat, then this velocity is zero, Figure 2f. If the object is concave this velocity approaches infinity as the object curvature approaches that of the finger. We denote this invariant K2.

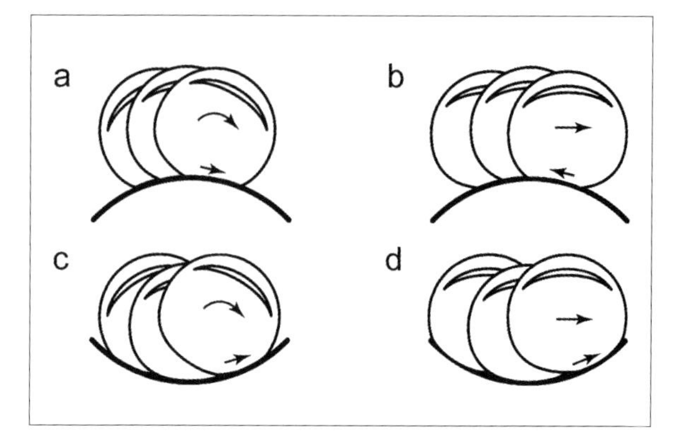

FIGURE 3. GENERIC TACTILE INVARIANTS
Comparing a rolling strategy (a), (c), and a sliding strategy (b), (d). When rolling, the direction of the velocity of the contact surface on the finger is unrelated to the shape of the object, only the magnitude is. When sliding, the relative directions of the velocities depend on the sign of the curvature of the object.

Generic invariants

The reader will also notice that the relative signs of the finger velocities and of the contact velocities changes from a convex to a concave surface in the case of sliding motion, but not in the case of rolling motion. Said another way, with suitable coordinates, the sign of the product of the velocities indicates the sign of the curvature of the touched surface, a very powerful invariant indeed denoted G1. It is generic because it holds for any magnitude of the velocities and any magnitude of the curvatures. It is known that humans can detect slip velocity accurately for many types of surfaces [30].

Experimental evidence of haptic shape priors

It is clear that when a perceiver has the experience of an object, complete information cannot be available instantly or may never be available at all. So even if information can be integrated over time to build knowledge about a scene or an object, the perceiver must rely on prior knowledge.

Notion of perceptual prior

In general, a *perceptual prior* is knowledge used by the brain to make a judgement when sensory evidence is lacking or if the likelihood of a property to hold is high enough to override contrary indication. In vision, 'light-from-above' and 'object stationarity', are well studied examples of priors [3].

In this section, we discuss indirect evidence of haptic priors that coincide with the assumptions already made. For this, we use the results of two experiments where observers experience illusory shapes based on a highly simplified set of perceptual cues [31, 32].

Shape from contact movement

The objective of a device, called the 'Morpheotron' (see Fig. 4a), is to show that the brain can effortlessly take advantage of a single, segregated shape cue [33]. It has a plate constrained to rotate around a point located inside the perceiver's finger, see Figure 4b, and allows for free exploration in a horizontal plane. The machine eliminates proprioceptive cues since the rigid movements of the finger are independent from the plate orientation. By design, under servo position-control, the flat plate rolls on the fingertip and its movements do not affect the finger rigid-body displacement. Under these conditions, invariants S2, S3, K1, and K2 are not available to the perceivers. Yet, they are able

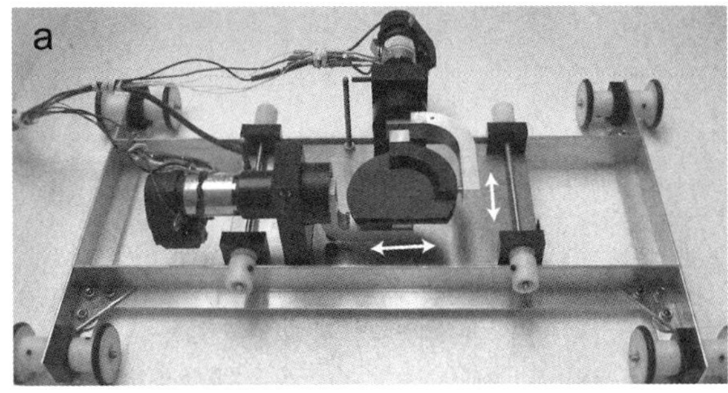

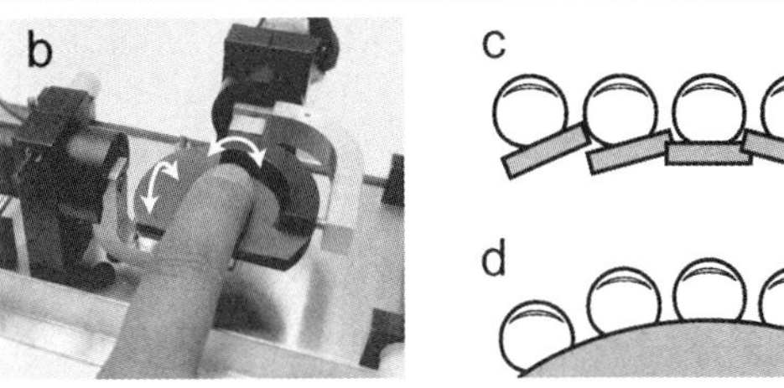

FIGURE 4. THE MORPHEOTRON

A plate free to move in a horizontal plane (a) rotates under servo-control control around a point located inside the finger (b). The contact moves on the surface of the fingertip (c). The form represented in (d) is typically experienced. Please see [6] for the various conditions in which this can happen. Please consult reference [34] for practical means to experience similar sensations.

to perform in a concavity/convexity detection task at a level equivalent to when exploring real objects [31].

From what source can subjects derive the experience of shape if the static invariants report a flat surface, if the kinematic invariants are destroyed, and there are no proprioceptive cues? Perceivers are likely to use a prior assumption of stationarity. Recall that earlier it was assumed throughout that the touched object is stationary, that the object does not deform, and that the finger deformation due to sliding can be neglected. But when observers touch the plate of the Morpheotron, either they are not aware that the plate actually moves (because of the testing conditions), or if they look, the prior that a touched object is stationary is sufficiently strong to override visual report of movement. In these conditions, the stimulation depicted in Figure 4b and c is sufficient to give the sensation of the shape represented in Figure 4d [33].

Neither K1 nor K2 are available in their native form, nevertheless, information similar to that given by K2 is made available by the device if the touched virtual object is assumed to be perfectly slippery and immobile, thus exposing the existence of three priors, that of 'object stationarity', call it P1, of 'perfect slip', call it P2, in addition to 'object rigidity', call it P3.

Shape from tangential fields

This experiment is best described by reference to Figure 5a [32]. There, a finger is represented in the act of exploring a protrusion on a surface. In terms of the cues that we have discussed so-far, a protrusion may be characterized as four consecutive changes of curvature, from zero curvature, to negative, to positive, to negative, back to zero. We could therefore invoke the entirety of the static, kinematic, and generic invariants discussed earlier to express the information potentially available to the perceiver. In addition, if the protrusion is large enough to require detectable limb movements, proprioceptive cues are also available.

Similarly to what was described in the previous section, the principle of the experiment is to eliminate most of these cues by suppressing the corresponding invariants. The apparatus, depicted in Figure 5b, eliminates all static and kinematic cues by using a flat plate constrained to follow a cam. For the observer to push the plate, say up the hill, the finger must deform laterally. In effect, the protrusion is felt, presumably as a result of a combination of proprioceptive and tactile information, originating in the skin, tendons, muscles and other places of the perceiver's anatomy that are sensitive to strain.

At this point, we see that if the perceiver feels the protrusion, the only information available are the priors P1, P2, P3, the later being violated by the apparatus, plus deformations corresponding to the direction of movement. Perfect slip, P1, is especially important since without it deformation could be attributed to, say, varying surface friction and not to shape. We can suppose the efficacy of a fourth perceptual prior, P4, which would require that the friction of a surface does not vary.

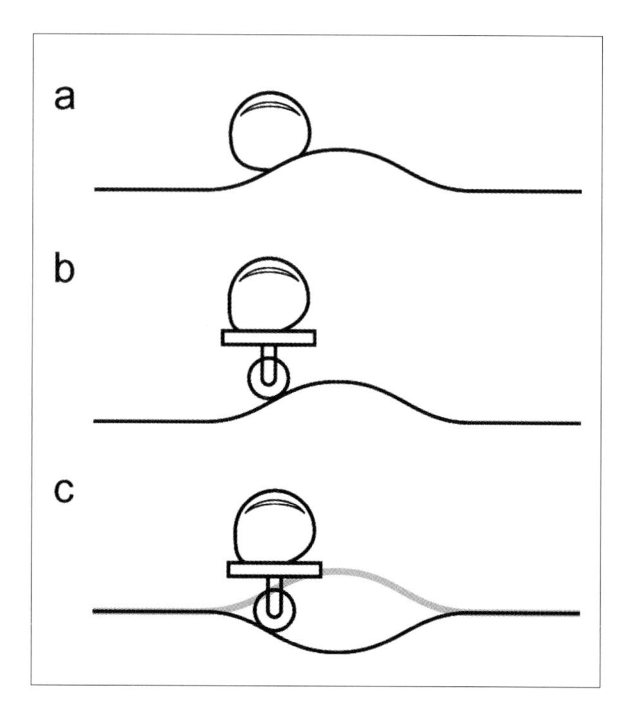

FIGURE 5. SHAPE BY TANGENTIAL FIELDS

A finger scanning a protrusion produces many cues (a). A plate and a constraining mechanism eliminates all but two cues (b). Experimentally, proprioceptive cues arising from rigid-body motion can be separated from tactile cues arising from local and global deformation of tissues (c). Please see reference [34] for practical means to experience similar sensations.

The experiment shows that, under appropriate conditions, the information available from proprioception can be overridden by that of other sources, including that of the four priors that we have identified. To demonstrate this, the apparatus, using a combination of sensors and actuators, causes deformations that corresponded to exploring a shape as illustrated by the grey line in Figure 5c but with rigid-body displacements corresponding to the black line [32].

What the perceiver typically feels is the shape represented in gray rather than that in black. It can be further shown that these sources of information are cues that are integrated according to a Maximum-Likelihood-Estimation model for cue integration [35].

While the figure represents deformation in the finger only for illustration purposes, there is anecdotal evidence that redundant information is available in other ways since the illusion occurs equally well when using other parts of the hand such as the back of the hand, the wrist, or the hard knuckles [32].

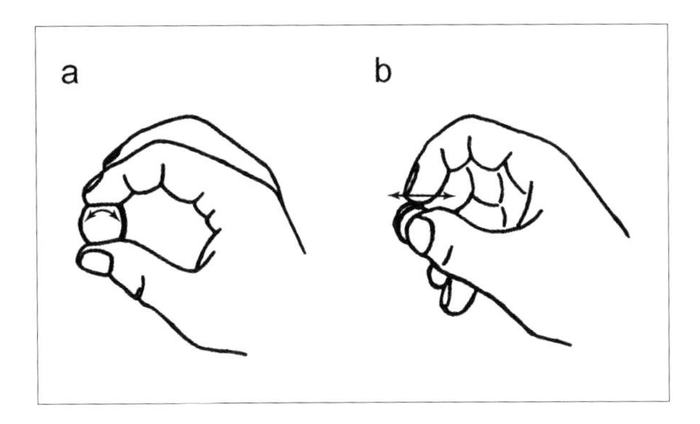

FIGURE 6.
Unsuccessful attempt to use kinematic invariant K1 (a). Invariant K2 can be invoked although the index finger does not have to have a fixed orientation (b). Its kinematics are presumed to be available in order to relate the contact surface velocity to that of the index in a version of invariant K2.

Applications

In actual everyday manipulation and exploration, the conditions that we have employed to characterize some invariants, the priors, and their attending shape cues, rarely occur in the simplified forms described so-far, and many more indeed could exist. For example, tactile information can be combined with proprioceptive inputs. This can be shown by eliminating cutaneous sensitivity by anesthesia [36]. In another example, the static relative location of the surface of contact in a pinch grasp is shown to determine motor behavior when lifting an object [37]. This later example shows that the presently considered single-finger tactile shape cues can be generalized to multiple fingers.

In the next subsections, the notion of shape invariant will be exemplified, first in the context of stereotypical movements. Such movements have been described for the detection of a specific of objects qualities [38]. Then, a special but interesting task, that of the detection of flatness, will be discussed in the light of the shape invariants and of their associated cues.

Examples of stereotypical movements

Consider the task of sizing a coin. A static grasp is certainly not an optimal strategy. Yet, there is more information available in a static pinch grasp, see Figure 6a, than that coming solely from the separation between the two fingers. Invariant S3 can be invoked: a higher curvature corresponds to a smaller contact surface. In addition, there are typical movements that may be employed to detect the curvature and hence the size of the coin in question. A version of invariant K1 may be called upon by rolling the coin between two fingers as indicated in Figure 6a. However, it can be seen in this case that K1 is ineffective since the velocity of the contact region is independent from the coin's radius. On the other hand, a version of invariant K2 can be triggered by holding the coin between the thumb and the third finger, and using the index to explore its curvature. This is shown in Figure 6b. It is likely that the later movement is more efficient and more often used than the former.

Detection of flatness

The notion of tactile flatness is natural to us. Yet, for an observer, on what ground can flatness be decided? Humans are known to be able to detect very shallow protrusions and hence very small curvatures that cause surfaces to deviate from flatness [39]. The invariants we have discussed can all be used to detect flatness. Let us take them one by one.

For a flat surface, the final size, invariant S1, the shape of the surface, S2, and the growth pattern of the surface of contact, S3, are all characteristic. We cannot expect the corresponding cues to be highly reliable but there is evidence that static touch can detect flatness, even with one single finger [25]. Kinetic invariant K1 is a possibility, yet an unlikely one, since its sensitivity to flatness is not great in the case of rolling a finger on a flat surface. On the other hand, invariant K2, available when sliding, is a very appropriate one since, mathematically, the only surfaces that can give rise to zero velocity of contact surface on the finger are the Reuleaux surfaces, that is, the flat, spherical, cylindrical, revolute, helicoidal, and prismatic surfaces [40]. In addition, its precision can be increased by using multiple contact areas within the hand. In particular three fingers are ideal since a three-finger touch will constraint the hand posture appropriately. It is easy to find other cues that eliminate all possibilities but a flat surface. The generic invariant G1 is not applicable in the case of a flat surface. It is the combination of these, and probably other cues that signify flatness to the observer. By these observations we can justify why the two, and most likely, the three-finger scanning posture is typically preferred to appreciate the flatness of a surface, see Figure 7.

The conceptual analogy with visual straightness is interesting. Visually, an invariant for straightness comes from the mathematical property that a straight line is invariant under very general transformations that resist those introduced by the visual system and that is independent from any coding [18]. This also coincides with the fact that light propagates in straight lines as well (provided that the propagation milieu is homogenous). A special case is when looking in alignment with a flat surface or a straight edge to make them vanish. This is analogous to the sliding velocity being exactly equal to the scanning finger velocity in the special case where the finger(s) contact surface velocities vanish to zero. Here too, the haptic invariants are independent from any transformation and from any coding.

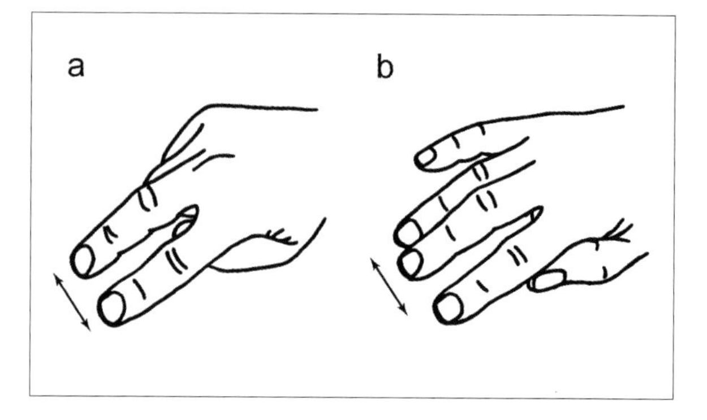

Figure 7. Appreciating flatness
Scanning with two (a) or three (b) fingers are typical strategies that can leverage invariant K2. The three-finger strategy is most appropriate since it determines the correct number of freedoms and constraints.

Devices to generate missing cues

Since the cues that we have discussed so far are eliminated when using tools or intermediaries, designers of haptic interfaces have attempted to devise systems that could produce them. In this section we look at four of these devices and discuss the invariants that they can potentially preserve.

Slip displays

What if a device was devised so that 'perfect slip' prior was replaced by actual physical experience? Then it could be possible to have a haptic interface that could produce invariant K2 efficiently. Generating slip is not a new problem. In human studies, commanding a drum to rotate and cause slip under the finger pad is the traditional method to deliver slip [30, 41, 42]. But of course, there is a desire to command omnidirectional slip for general-purpose applications in mechanical virtual environments. To date, the preferred method to achieve this is a ball constrained between motorized traction wheels [43, 44], but other approaches may exist.

Fingertip orientation displays

Similarly, designers have been working on techniques to generate the static and kinematic invariants which are eliminated by force feedback devices in much the same way a tool eliminates them through two cascaded transformations. Of course, designers will always be facing a fundamental tradeoff. The larger the number of shape cues that can be delivered by a device, the larger is its complexity. In mechanical engineering terms, complexity can be measured in terms of number of actuated degrees of freedom. For the Morpheotron, only two are needed.

The first practical realization of a device that could introduce the missing tactile cues is probably a system described in 1993 by Hirota and Hirose. The authors articulated their goal clearly: "*In this concept [of force-feedback], force is considered to be an output from the virtual object to the user. However, it is possible to adopt a different approach where, instead of force from the virtual object, the existence (or surface) of the object is simulated*" [45].

Recently, several new devices have been described which explored various design niches. The system that comprises the largest number of such actuated degrees-of-freedom, termed 'encounter-type', has nine degrees of freedom, three per finger for three-digit grasps [46]. Its design is driven by anatomical considerations regarding the mobility of the human hand. Another approach is to add a mobile spherical tactile element sliding inside a thimble attached to a force-feedback device. The result is a system with four actuated degrees of freedom [47]. The contact location can be changed according to the movement of the finger which is otherwise stimulated using a force-feedback strategy. The authors show that the addition of this extra element combined with the force feedback give the users discrimination performance comparable to that achieved in real conditions, as long as the finger is constrained to fore-aft exploratory movements. Finally, a device that enables the exploration of arbitrary surfaces in three dimensions is described in [48]. It has five actuated degrees-of-freedom and combines force-feedback with kinematic cues delivered by a flat plate rolling on the fingertip. With this system, users can achieve the detection of very low curvatures, down to 2.3 m [48].

Summary

Here is in list form the steps that were followed:

- Individual perceptual cues have been identified for vision and audition. For touch, the same should be possible. We look at shape only.
- Display technologies may be thought to mirror the perceptual system of an observer. Therefore the knowledge of cues is useful to design systems which can economically provide a desired percept.
- Priors are necessary for perception. For tactile shape, object rigidity, stationarity, as well as perfect slip are such possible priors that, for analysis purposes, are first assumed to hold.
- There is a fundamental difference between direct and tool mediated touch. Tools introduce cascaded transformations which must be inverted by the brain.
- Force feedback devices with which the brain must assume the existence of a tool are easier to realize than devices where this assumption is lifted.
- The notion of force is not needed for analysis. Only notions of displacement and deformation are needed.
- Like in vision and audition, perceptual cues are linked to invariants that have mathematical, physical, and biological origins. The same should hold for touch.
- Static invariants were identified that are related to the contact surface: its final size (S1), its

shape especially connectivity and eccentrity (S2), and the growth law as a function of rigid displacement (S3).
- Kinematic invariants are associated to rolling. The angular velocity of the finger is related to the velocity of the contact through object curvature (K1). They are also associated to sliding. Then, the velocity of the finger is related in a different way to the contact velocity also through curvature (K2).
- Generic invariant (G1) relates the relative direction (or sign) of finger and contact velocities when sliding.
- These invariants owe their existence to the physics of contact such as Hertzian surfaces, St Venant's principle or the properties of friction. They also arise from the near spherical shape of biological fingertips and from their visco-elastic properties in addition to the mathematics of contacts such as convexity, connectivity or limit cases.
- The need and the existence of at least three priors: object rigidity, stationarity, and perfect slip can be evidenced experimentally.
- Stereotypical movements can be related to invariants and to the cues they provide.
- Special cases such as flatness detection are related to invariants that have conceptual analogies in vision.
- Haptic devices are being developed which can uphold these invariants and hence provide relevant shape cues.

In conclusion, it is noted that visual or audio displays can take many forms, from store-front LED banners to IMAX theaters, from telephones to wave-field synthesis systems. As long the information delivered is relevant to the task at hand and the signal types and noise ratios match the perceptual mechanisms at play, the display will operate successfully. Perceivers excel at taking advantage of any available cues. The same applies to haptic displays.

In this chapter we have looked at some of the cues that are relevant to haptic shape perception of low convexity objects. There certainly exist many others, but these provide a solid foundation from which more complex ones can be created, particularly by involving several fingers rather than one.

Acknowledgements

This research was supported by a discovery grant from the Natural Sciences and Engineering Council of Canada. The author would like to thank McGill University for a sabbatical leave.

Selected readings

Gibson JJ (1962) Observations on active touch. *Psychological Review* 69(6): 477–491

Jones LA, Lederman SJ (2006) *Human hand function*. Oxford University Press, Oxford

Pont SC, Kappers AML, Koenderink JJ (1999) Similar mechanisms underlie curvature comparison by static and dynamic touch. *Perception & Psychophysics* 61(5): 874–894

Goodwin AW, John KT, Marceglia AH (1991) Tactile discrimination of curvature by humans using only cutaneous information from the fingerpads. *Experimental Brain Research* 86(3): 663–672

Fearing RS, Binford TO (1988) Using a cylindrical tactile sensor for determining curvature. In: *Proc. IEEE International Conference on Robotics and Automation*, 765–771

Design guidelines for generating force feedback on fingertips using haptic interfaces

Carlo Alberto Avizzano, Antonio Frisoli and Massimo Bergamasco

Introduction

Manipulation and grasping have key importance in most types of interactions between humans and the world surrounding them [1, 2]. Even if almost all existing haptic interfaces provide a user interaction based on a single contact point, an increased number of contact points, not only allows to display a more natural haptic interaction [3, 4], but also improves the quality of interaction that users can perform in the environment. Haptic exploration is highly dependent on the number of fingers used for exploration of common objects [5], the largest difference appearing between the 'one finger' and the 'two fingers' conditions [6], and as proven by Jansson et al. [7] by the ability to discriminate a precise tactile pattern during the exploration. In [8], we found an experimental confirmation of this hypothesis: the haptic exploration do not improve with the increase of contact points, from one to two fingers. This suggests that the restriction imposed on the fingerpad contact region can blunt the haptic perception of shape and so indicates that local haptic cues play an important role in haptic perception of shape. Factors that can account for the observed performance in these experiments are lack of physical location of the contact on the fingerpad and lack of geometrical information on the orientation of the contact area, that constitute interesting insights and suggestions for the design of haptic displays.

Multipoint haptics [9, 10] are devices that can simultaneously interact with the user through more than one contact point. These systems allow both force and torque feedback during the simulation of dexterous manipulation and complex manoeuvring of virtual objects and can improve the interaction in several applications, e.g., assembly and disassembly in virtual prototyping [11, 12], medical palpation during simulated physical examination of patients [13] and many other ones.

In this chapter we present different approaches to improve the quality of haptic feedback during virtual manipulation of objects. Four different aspects of modelling perception and manipulation are proposed and investigated through conducted experimental studies.

Initially we investigate the capabilities of using a haptic system for grasping and manipulating virtual objects, by means of a two contact points haptic device. The work also investigates the relationship between human prehension and features of the physical model of the grasped object, finding out how grasping in virtual conditions present higher forces and safety margins than in real conditions. A possible motivation of this observed difference is due to the limitations of kinesthetic devices in stimulating local mechanoreceptors.

A possible improvement in this sense to kinesthetic haptic devices is presented, where we wonder whether local haptic cues provided at the fingertip can improve haptic perception of shape. We adopt a prototype of a new encountered haptics, allowing haptic exploration of three dimensional shapes, and show how discrimination threshold for curvature perception can be significantly improved by providing both kinesthetic and local haptic cues at the contact point.

We also introduce an alternative approach to enhance haptic perception, by using haptic illusions to elicit haptic sensations, and discuss potential applications and ways to simplify the design of future haptic devices.

Finally we evaluate the usage of high frequency vibrotactile tactors to render the feeling of contact during the manipulation. Such a device consists in an active digital glove integrated with an array of vibrotactile actuators placed at level of fingers phalanxes. The device is equipped with embedded electronics that allows to control the motors directly from the virtual environment.

An investigation of manipulation capabilities in virtual environments

Grasping an object allows us to identify some of its properties (geometry, material, surface textures, [14]), change its physical state (position in space, internal structure) and use it for mediated interaction with other objects. The possibility of interacting with more than one point of contact is fundamental for the manipulation of objects in virtual environments. Humans unconsciously use suboptimal [15] algorithms for the prehension of objects when performing tasks with their hands. For instance, during a peg-in-hole task, they precisely adjust the relative position and interaction force between the peg and hole. Johansson and Westling conducted a series of experiments, relating tactile information to grip force when performing a lifting task [16]. The ability to adjust grip force appears to be independent of the surface friction characteristics, but further studies from the same authors confirmed that this is not true for the case of objects with different curvature, and propose an active role of rotational friction for the stabilisation of the grip [17]. The nature of contact during slip provides important tactile cues regarding features on the surface as well as the nature of movement of the object, and can explain how humans take advantage of slip sensitivity when perceiving objects.

In this study, we used a GRAB haptic system [18], composed of two identical robotic arms, to provide the force-feedback for the two fingers during simulated grasping operations. The user can operate the device by inserting his fingers in two thimbles placed on the end-effectors of both the arms, as show in Figure 1, so that both single hand (thumb and index) and two hands (right and left indexes of two hands) interaction are possible.

A set of rubber thimbles of different sizes allow any finger size to properly fit in the device. Each arm has six degrees of freedom (dofs), of which the first three, required to track the position of the fingertip in the space, are actuated, while the last three, required to track its orientation, are passive. While the user is grasping and manipulating objects in the virtual environment, the device is controlled by multiple concurrent threads, from the fastest internal 1 KHz haptic loop to slower external collision detection loops, physical modelling, up to the slower 20 Hz graphical loop (see Fig. 1). The collision detection is actually implemented using an external module based on bounding volume hierarchy and running at about 100 Hz. The simulation of the dynamics of the objects is achieved through a dynamic simulator that is carried out through a separate thread running at about 200 Hz.

Whenever an object is grasped in virtual environment by the user's fingers, it is virtually tied to the contact points through a couple of springs representing the virtual stiffness of the environment. The application allows the experiment to change the working parameters in terms of size, weight and surface stiffness. While the size was kept fixed, in our experiments we interchanged weights from 0.1 to 0.5 Kg, and surface stiffness from 500 to 2,000 N/m.

The control loop also implemented a linear friction model who generates the force information for determining the object motion. The position of each contact point is measured directly by the haptic interface (x_h) and the relative feedback force (F) is computed through a constraint-based proxy method with friction, based on [19, 20]. The haptic rendering algorithm computes the position of an additional proxy point x_p, lying

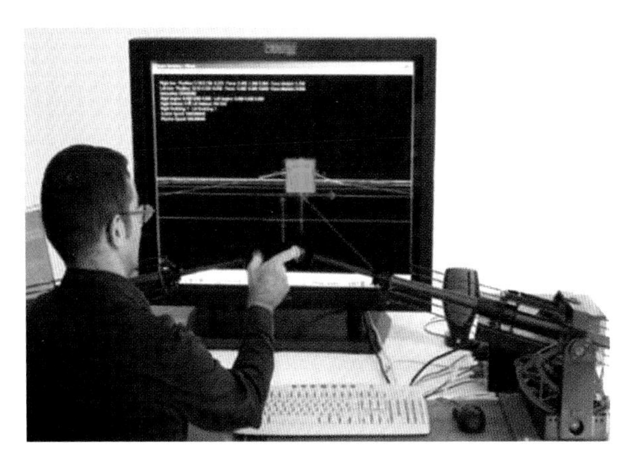

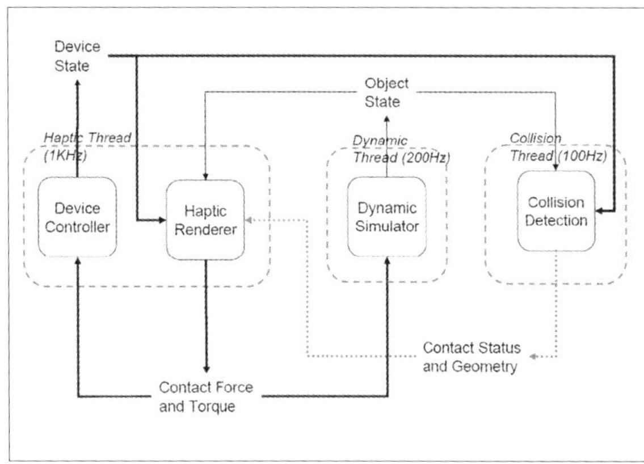

FIGURE 1.
The device during a typical grasping procedure (left) and the associated force rendering integration scheme

on the object geometry, and on the basis of current xh position of the haptic interface and contact geometry (C) the force is generated through a direct rendering method $F = G(x_h, x_p, C)$, using the elastic coupling between the proxy x_p and the interface position x_h.

The quality of object grasping was investigated by means of several numerical experiments. Firstly the force sequences during grasping, such as the ones shown in Figure 2, were investigated. Three phases can be identified: the first is the lifting phase when the user starts to grasp the object, the second is the holding phase and finally there is the releasing phase when the object starts to slide. In the diagrams the green and red lines represent the position of the contact point and the object along the Y axis (aligned along the gravity vector). During the lifting and holding phases the two positions have a constant difference that depends on the grasping point on the object, then in the releasing phase they diverge because the object is falling. The black line shows the status of the friction that is zero when there is no contact, one during the non-slip state and two in the slip state: it is clear that during the lifting and holding phases the proxy is in non-slip state because it is firm between the fingers and, when the object is released, initially it changes to slip mode and then the contact is lost. Finally the blue line represents the grasping force that has an increasing and varying behaviour during the lifting phase, but it is almost constant during the holding phase.

In order to assess the pick and place operations, grasping information were compared to available data on human grasping of real objects [15, 21]. The influence of weight on static grip has been experimentally studied by [15], where safety margins for grasping for prevention of slipping are analysed. The safety margin is defined as the difference between the grip force and the slip force, that is the minimum grip force required for preventing slipping. Three healthy right-handed men, aged between 27 and 35 years, served as subjects for the study. The subjects sat on a height-adjustable chair. In this position the subject might hold with his right hand the two thimbles connected to the haptic interfaces, and respectively wear them on the thumb and right index of his hand. A wide visualisation screen was placed in front of the screen and a desktop, where during the experiment the subject was invited to place its elbow. A sequence of 27 objects was presented twice to each subject, for

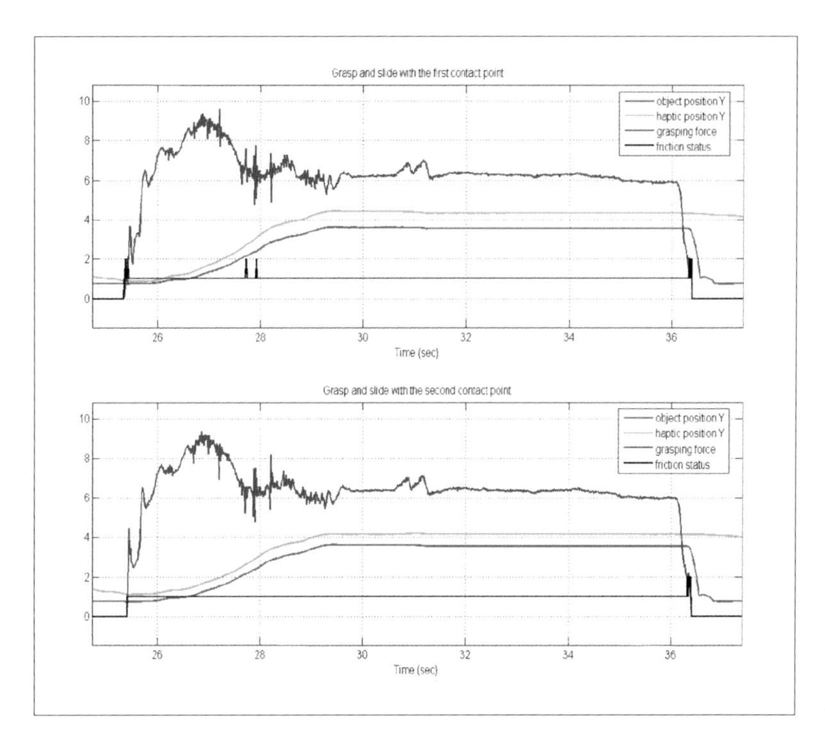

FIGURE 2. FORCE AND POSITION DURING GRASPING, LIFTING, HOLDING AND RELEASE OF A VIRTUAL OBJECT

a total of 54 runs performed by each subject. All the objects in the randomised sequence were cubes with the same geometry, with pseudorandom changes in the weight m (0.1, 0.2, 0.4 Kg), in the friction coefficient, both static μ_s (0.4, 0.8, 1.2) and dynamic μ_d (0.3, 0.6, 1.1), and in the stiffness k (0.5, 1, 2 N/mm). In each randomised sequence all the possible combinations of weight, friction and stiffness, without repetition, were presented to the subject. The values of μ_d were associated to μ_s. The experiments were conducted with only one grasping condition, with the object hold between index and thumb tip of the same hand. Values for friction coefficients were assumed by [22] where experimental values of linear friction are reported between index tip and different materials, equal respectively to 0.42, 0.61 and 1.67 for rayon, suede and sandpaper. Each subject was asked to grasp and lift the object using index and thumb and then to slowly release the object letting it falling down.

A significant correlation was found between the values of the gripping force F_n and stiffness, weight and friction values. The value of grip force F_n was found to be significantly positively correlated with mass and stiffness, while negatively with friction value ($p < 0.001$). Greater gripper forces are required for holding heavier weights and stiffer objects, while lower gripper forces are required for higher friction values. In [15] it was found that the relative safety margin, defined as the safety margin in percent of the grip force, was about constant during lifting with increase of weights, was almost constant with change of weight. The calculation of the safety margin in the case of virtual manipulation allows to make an interesting comparison. As it is shown in the logarithmic plot in Figure 3 below in the case of virtual manipulation, the safety margin tends to be reduced with increasing weight of the lifted mass.

This can be explained by the larger dispersion of grip forces observed for lower mass values. In fact, due to the absence of local sensation of slip, it was more difficult to discriminate the weight of lighter objects. Moreover lighter

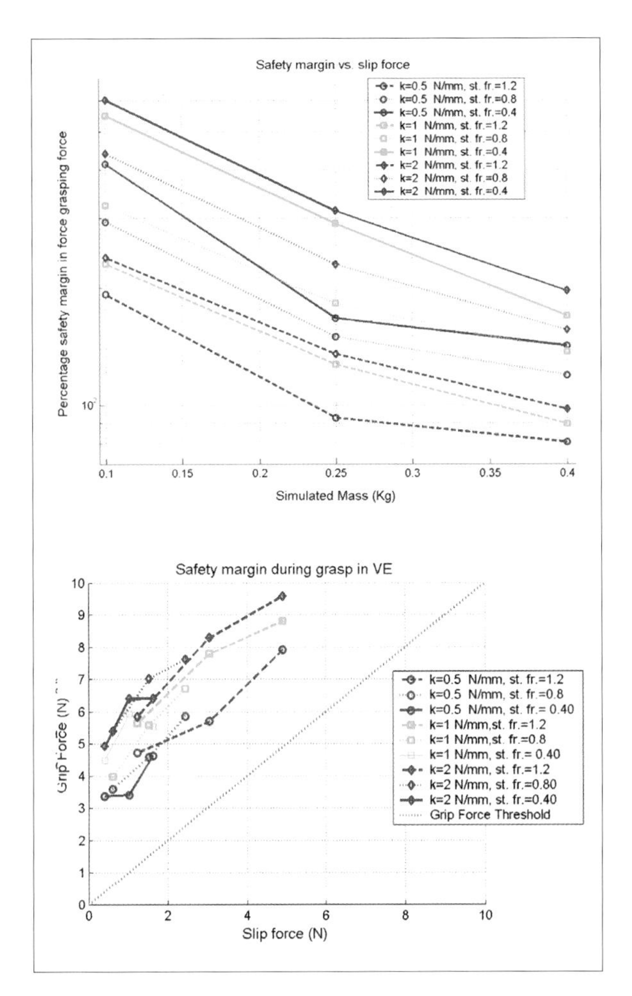

FIGURE 3.
SAFETY MARGINS IN MANIPULATION OF A VIRTUAL OBJECT

objects required a smaller resolution in the control of force (ΔF), that is limited by the position resolution of the device ΔX, according to the law $\Delta F = k\Delta X$, where k is the simulated contact stiffness. This is confirmed by the finding that better safety margins are obtained for lower values of the stiffness, as it is evident from the plot above, where grip forces are plotted *versus* slip forces. While the minimum required grip force is represented by the diagonal line, experimental data can be clustered in three main groups according to the value of contact stiffness during the simulation. The evaluation of the performance during grasping and weight lifting, has shown that the simulation produces outcomes that are similar to experimental findings on real objects.

Overall from this study it results that kinesthetic haptic devices when used to simulate operations of grasping blunt the haptic perception in such a way that safety margins adopted by human subjects are greater than in reality.

Enhancing haptic perception by directional and geometrical local cues

A way to improve the local perception of shape and grip forces at the contact points is to elicit a direct stimulation of the mechanoreceptors at the fingertip, enriching the kinesthetic force feedback usually provided with traditional haptic devices. Different solutions have been proposed to validate the effects of an encountered haptics. In [23] the shape recognition is due either to the perception of slipping of the fingerpad over the object surface or to the displacement of the contact area over the fingerpad. In [24] preliminary tests reveal that relative motion can be used to render haptic sensation. In [23], a new haptic device is presented which integrates grounded point-force display with the presentation of contact location over the fingerpad area. The second approach considers that recognition of shape is linked to the perception of the orientation of the object surface at the contact points. Hayward et al. [25] demonstrated how curvature discrimination can be carried out through a device providing only directional cues at the level of the fingerpad, without any kinesthetic information and moreover with a planar motion of the finger. This concept is also exploited to build robotic systems that can orient mobile surfaces on the tangent planes to the virtual object that is simulated, only at the contact points with the finger [26]. In the solution presented therein two haptic devices were differently used to support and track user finger and to present force feedback to the finger tip.

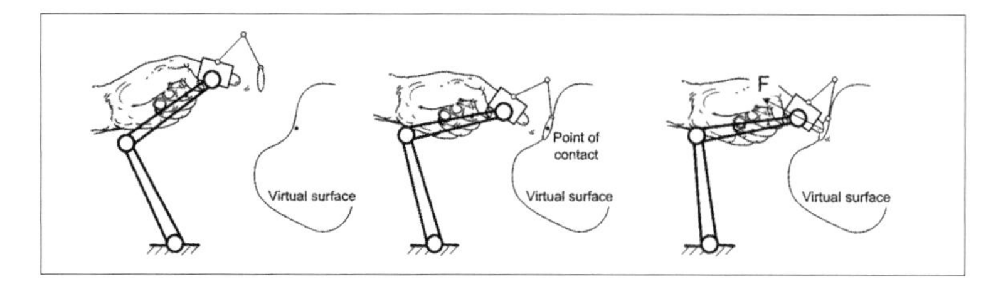

FIGURE 4. CONCEPTUAL SCHEME OF THE DEVICE

In this section we describe a new type of encountered haptic system, composed of a fingertip haptic interface in the shape of a plate [27] integrated with a tracking device. The interaction among the user and the device goes by means of an active control of the haptic plate which is moved in correspondence of the virtual objects to be touched, while the tracking system plays the role of accurately monitoring the finger position. Suppose the user is interacting with a virtual object: when the finger is out of the surface of the object, the plate is kept far apart from the fingertip. When the finger touches the virtual surface, the plate comes into contact with the fingertip with an orientation determined by the geometric normal of the explored surface, as shown in Figure 4.

Fingertip haptic interface

The supporting haptic interface is a pure translational parallel manipulator with three degrees of freedom (DoF). A classic impedance haptic control scheme was adopted, generating a force proportional to the penetration in the virtual surface. As shown in Figure 5, the fingertip haptic interface was devised to bring the final plate into contact with the fingertip with different orientations, according to the direction of the perpendicular to the virtual surface in the point of contact. Moreover, the contact can occur at different points of the fingertip surface, depending on its orientation in respect of the virtual surface. These requirements can be satisfied by a kinematics with five DoF, three translational and two rotational ones. A hybrid kinematics, consisting of a first parallel translational stage and a second parallel rotational stage, resulted the most suitable solution. The translational stage has the same kinematics of the supporting haptic interface, with 3-UPU legs. In each leg the cable connected to the motor and a compression spring are mounted aligned to the centers of the universal joints. The spring works in opposition with the motor, in order to generate the required actuation force and to guarantee a pre-load on the cable.

The control was implemented with local position controllers at the joint level. An inverse kinematic module was used to convert the desired position expressed in cartesian coordinates to the corresponding joint coordinates. The nonlinear term due to the spring pre-load and to the weight of the device was compensated, by a feedforward term in the control loop.

Can local haptic cues improve haptic perception?

Performances were estimated by simulating the contact with a virtual sphere having a radius of 70 mm. Fingertip positions and the interaction forces were monitored during the interaction: when the finger is out of the sphere, the platform is moved far apart of a given offset from the finger. When the finger comes in contact with the sphere, the two positions coincide, meaning that the plate is in contact with the finger. In Figure 6, the black continuous line represents a scaled representation of force (with an offset only for the purpose of superimposing it to the plot) generated by the supporting haptic interface: the force is null when the finger is out of the sphere.

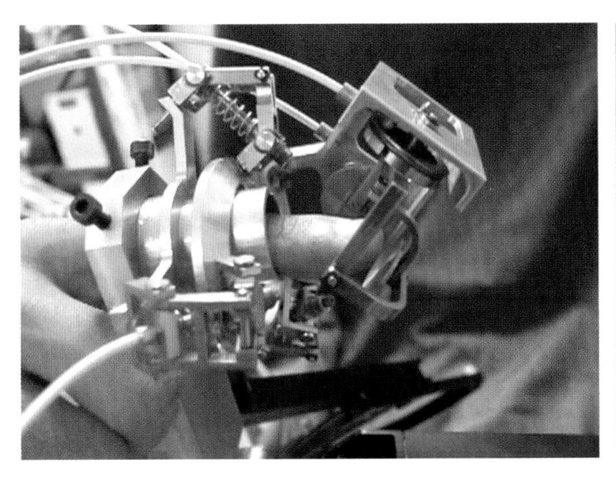

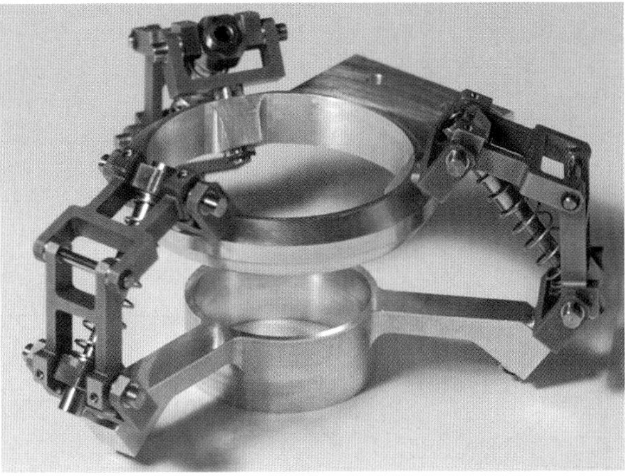

FIGURE 5. PERCRO ENCOUNTERED HAPTIC DEVICE, DETAILS OF THE FINGERTIP PLATE AND TRACKER

A psychophysics assessment was carried out in order to measure the discrimination threshold in the perception of curvature, with only kinesthetic devices and with the additional of informative geometric local cues, through the new device. The procedure was based on the Theory of Signal Detection (TSD). Four participants were recruited for the experiment, three males and one female. They were completely novices to haptic interfaces and did not present any dysfunction of the fingers. The same-different procedure of TSD was adopted for the determination of the difference threshold. The test consisted in exploring in the virtual environment a pair of curved surfaces. The exploration was carried out in a restricted workspace, consisting in a vertical cylinder with a diameter of 25 mm. Figure 7 shows a planar scheme of the displayed haptic cues (virtual spheres with given curvature) and their position in the space, relatively to the device. The curvature of the two presented surfaces could be either the same or different. The two stimuli were randomly presented to the observers; each series was composed by 100 trials with the same probability to have different or equal stimuli. The observer's task was to judge if the curvature of the two surfaces was different or the same.

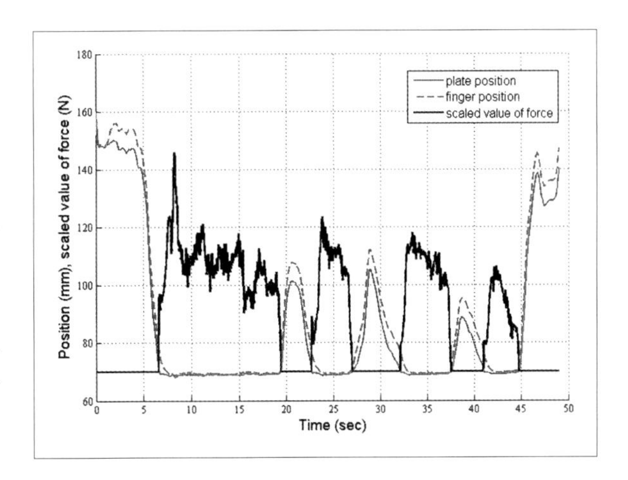

FIGURE 6. EXPERIMENTAL PLOT SHOWING THE RESPONSE OF THE TWO DEVICES

The test was carried out in two different modalities, A and B: in condition A both the kinesthetic and the local geometry haptic cues were provided to the observers, while in condition B only the kinesthetic feedback was provided. In modality A, the mobile platform of the fingertip device was kept in contact with the fingerpad when the user was in contact with the surface,

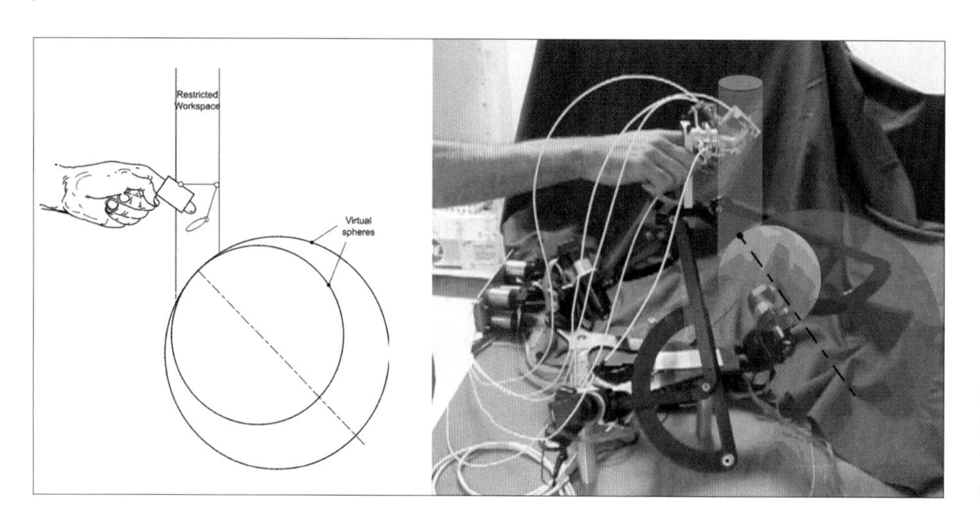

FIGURE 7. REPRESENTATION OF THE DISPLAYED HAPTIC CUES

with an orientation tangent to the displayed virtual surface. As a result the observer was perceiving in the contact point the indentation of the platform, oriented along the direction of the normal force applied by kinesthetic device. In the second modality the mobile platform was substituted by a fixed thimble, into which the user was required to insert its finger. In this case the only haptic cue applied to the fingerpad was the force perpendicular to the virtual surface generated by the supporting haptic interface; no local geometry information was provided. The sequence of the series in the two modality changed for each subject, in order to minimise the influence of the participant's learning for the procedure.

For each series the hit rate p_h and the false alarm rate p_f were calculated. The hit rate corresponds to the 'different' responses percentage when the two surfaces of the pair had a different curvature, while the false alarm rate is the percentage of 'different' responses for equal surfaces. The rates are converted to z-score of the normal distribution and the sensitivity measure d' is calculated as the difference between the two values:

$$d' = Z_h - Z_f \quad (1)$$

The Just Noticeable Difference (JND) was identified as the difference between curvatures for which d' was equal to 1, according to the criterion most commonly adopted in literature. Three series were presented to each participant, changing the value of the difference between the two possible curvatures, according to the values reported in Table 1.

For each Δ of curvature, the sensitivity measure d' was obtained and the three points were linearly interpolated, as shown in the plot of Figure 8 for one subject in both the two conditions. Finally the JND was calculated from the interpolating function, defining the overall JND is defined as the mean of the participants' JNDs.

The JND values resulted 1.51 m^{-1} in modality A and 2.62 m^{-1} in modality B, with a statistically significant difference ($p < 0.05$). For each subject the improvement in curvature discrimination using the new device was evident. This allows us to conclude that an enhancement of haptic perception of shape cues can be reached by complementing pure kinesthetic feedback with haptic cues, applied locally at the fingertip, informative of the contact geometry, and that this mechanism appears to be a fundamental and physiological component of haptic perception of real shapes.

Table 1. Presented curvature values

	Curvatures	Δ curvature
Series 1	5 m^{-1} / 6 m^{-1}	1 m^{-1}
Series 2	4.5 m^{-1} / 6 m^{-1}	1.5 m^{-1}
Series 3	4 m^{-1} / 6 m^{-1}	2 m^{-1}

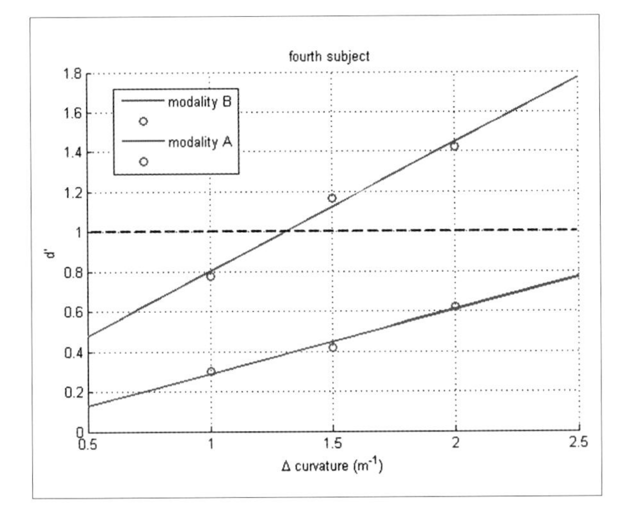

Figure 8. Experimental results for one subject: modality A (red upper line) and B (blue lower line).

Virtually altered force feedback to generate illusions in fingertips exploration

Another alternative approach to enhance haptic perception is to adopt haptic rendering algorithms that can generate suitable haptic illusions during active exploration. In fact, the level of sensation provided with kinesthetic haptic devices below fingers is realistic, but still far to be exact: an important open issue that occurs then [28, 29] is to understand to which extent such system can be employed for real tasks where high sensitivity is needed, such as the ability to render the force feedback of thin objects such as needles and, in general, of objects having spatially sharp features. As matter of fact, during haptic interaction some information is lost due to bandwidth and stiffness limitation of the mechanical devices, moreover, as smaller the device and its mechanical parts are, the softer the result. It has been found that force feedback information alone can elicit complex perception relating to haptic texture and shape [30–32]. Such perceptions can be considered as haptic perceptual illusion of texture and shape. In what follows we report available results on the ability of using haptic effects to render the sensation of sharp smaller objects.

Experiment design

The reduced reliability of sensations produced by haptics was already proven in the past. For instance, the use of lateral forces [33, 34] already showed the existence of illusory effects that may be generated on fingertips. In order to understand the level of human perception sensitivity in fine exploration tasks, a haptic to vision matching design was employed.

The experiment was designed in the following way. A kinesthetic haptic interface was employed. The users placed the right index finger in the device thimble in order to touch the 3-D virtual surface. Hence the user may move his/her finger all over a virtual surface modelled in software. During such a motion he/she may perceive geometric features (such as corners, edges, curvatures). During the experiment the user was asked to distinguish among sizes, distances and spatial relationships among elements. Additionally during the experiment subjects were instructed to maintain their finger inserted into the thimble all over the time and to keep the same standing orientation, subject had to explore the virtual surface sideways (left to right). This instruction allowed them to maintain the properties of the exploration independent from the feature position in the space, while preventing accidental contact with other parts

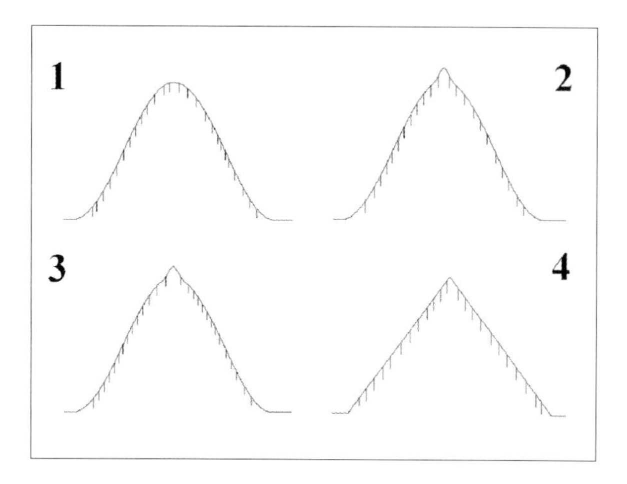

FIGURE 9. SELECTED SHAPE PROFILES

of the mechanical device. Subjects explored the surface at their own pace without any time constraint. Each exploration had to be performed with closed eyes.

During the experiment a series of haptic surfaces was presented. The surfaces, having different shapes, were presented in a randomized order. Each presentation defined a trial and the shape of the surface was changed between different trials. When subjects finished haptically exploring each shape, they matched the haptic shape to a visually displayed profile. A set of shape profiles was visually shown to subjects on a computer screen placed in front of the subject, as shown in Figure 9. Each profile had a number. Subjects used a computer keyboard to enter the number of the profile that they believed to be closest to the profile of the shape that was haptically explored. If they were not completely certain about the shape, subjects were instructed to give their best guess. The following variables were saved during the experiment: trajectory of device thimble during exploration, trial exploration time, haptic shape presented and visual shape matched. A test consisted of 90 trials, and typically lasted for 25 mins. Ten right-handed subjects ages 21–37 participated in the experiment. All of them had previous experience with haptic interface but were navice as to the purpose of the analysis carried out.

Five different haptic surfaces were used in this experiment: 1) a sinusoidal segment, 2) a sinusoidal segment with a lateral force-based [33, 34] illusory Gaussian shape, 3) a sawtooth segment, 4) a sinusoidal segment with a small sawtooth bump, and 5) a sinusoidal segment with a small Gaussian bump. The force-feedback and geometrical features of each surface are described below. A surface was rendered with lateral (Fx, along the x-axis) and vertical forces (Fy, along y-axis). The vertical forces were served to maintain the vertical position of the haptic manipulandum as close as possible to the target geometry. The virtual surfaces were displayed within a 310 mm workspace defined along the x-axis. The centre of the feature (c) was randomly placed and changed among trials. All features were 10 mm large and 3 mm high. The haptic control loop was performed at high frequency (2.5 KHz). The mathematical definition of the shapes was described in [35].

Table 2 shows subjects' performance in the haptics-to-vision matching task. The table relates the frequency with which a given haptic surface was matched to one of the visual profiles described before. This frequency is expressed as a percent of the overall matching performance for all subjects.

Subjects consistently matched the haptic surface SineSeg to the sinusoidal shape (Row 1, Column 1). However, subjects' matching performance was completely different when exploring SineLFGauss (Row 2, Column 2). Note how SineSeg and SineLFGauss had the same geometry.

It is striking how the Saw haptic surface was very frequently matched to the sinusoidal shape segment (Row 3, Column 1). But the converse was not true: the haptic sinusoidal segment (SineSeg) was rarely matched to the sawtooth segment (Row 1, Column 4). This helps highlight the difficulties of consistently rendering a good sawtooth shape by using a literal approach. Even though the stimulus had an approximation to a real, sharp edge, the results suggest that

TABLE 2. AVERAGE HAPTIC TO VISUAL MATCHING FOR ALL SUBJECTS

The highlighted cells indicate the visual shapes to which the haptic shapes should be ideally matched. For each haptic shape, the difference in matching performance is statistically significant (ANOVA, p<0.01)

Haptic shapes	Matched visual shape (% of all subjects)			
	Sine	Sine and small gauss-ian	Sine and small Saw	Saw
SineSeg (Profile 1)	**96.7**	2.2	0.0	1.1
Sine LFGauss (Profile 2)	6.1	**28.9**	48.9	16.1
Saw (Profile 4)	45.6	14.4	3.3	**36.7**
SineSaw (Profile 3)	4.4	16.7	**68.9**	10.0
SineGauss (Profile 2)	6.1	**60.0**	9.4	24.5

subjects did not perceive an object with a sharp feature.

This contrasts with subjects' matching performance for SineLFGauss (Row 2). This haptic surface was rarely confused with the sinusoid. It was sometimes classified into different categories by some subjects, but overall it was mostly matched to the sinusoidal surface with a small Gaussian bump (Fig. 9, Profile 2) and to the sinusoidal surface with a small sawtooth bump (Fig. 9, Profile 3). This suggests that SineLFGauss was perceived by subjects as a sinusoidal segment with a sharp feature (Fig. 9, Profiles 2 and 3), rather than as a large sawtooth shape (such as the one in Fig. 9, Profile 4).

In contrast to the matching performance for the haptic Saw (Row 3), haptic surfaces SineSaw and SineGauss were rarely matched (4.4% and 6.1%, respectively) to the visual sinusoidal segment (Column 1, Rows 4 and 5). This suggests that the perception of sharp features depends on the context in which they are presented. For example, when exploring haptic shapes SineLFGauss, SineSaw and SineGauss, there was a decrease in force as the top of the sinusoidal position of the stimuli was reached, and then the sharp feature or the Gaussian lateral force provided a large increase in force. Compare this to the constant forces experienced when ascending/descending the slopes of Saw (see 'Haptic Shapes' in Methods). This may not be surprising from the perceptual point of view, but it is potentially useful for haptic rendering purposes. Finally, subjects consistently matched SineSaw and SineGauss to the equivalent visual shapes: 68.9% and 60% of the time, respectively.

Subjects' finger trajectories during object exploration were compared to the ideal geometries of each haptic surface through Mean Squared Errors (MSEs). This allowed examining how close subjects' finger trajectories were to the geometry of each stimulus. Figure 10 shows a typical trajectory when exploring a SineSaw stimulus. Subjects' finger trajectories had a slight vertical offset relative to the ideal geometry of the objects. This was due to the finite stiffness used (which in this experiment was set to 2 N/mm). The offset was eliminated before computing MSEs. MSEs were calculated within the range

MSE analysis is summarised in Table 3. Results are expressed as a percent of the trials in which a given trajectory was closest to the geometry of a haptic shape. For example, consider all the trials in which SineSeg was explored by subjects (Row 1). The table shows that in 98.33% of those trials, subjects' finger trajectories were closest to the ideal SineSeg geometry. The table also indicates that, very rarely, subjects' finger trajectories were closest to the ideal geometries of Saw, SineSaw or SineGauss.

Table 3 helps to assess the difficulty to render the geometry of some sharp features. While rendering the geometry of SineSeg and Saw was simple, this was not the case for SineSaw and

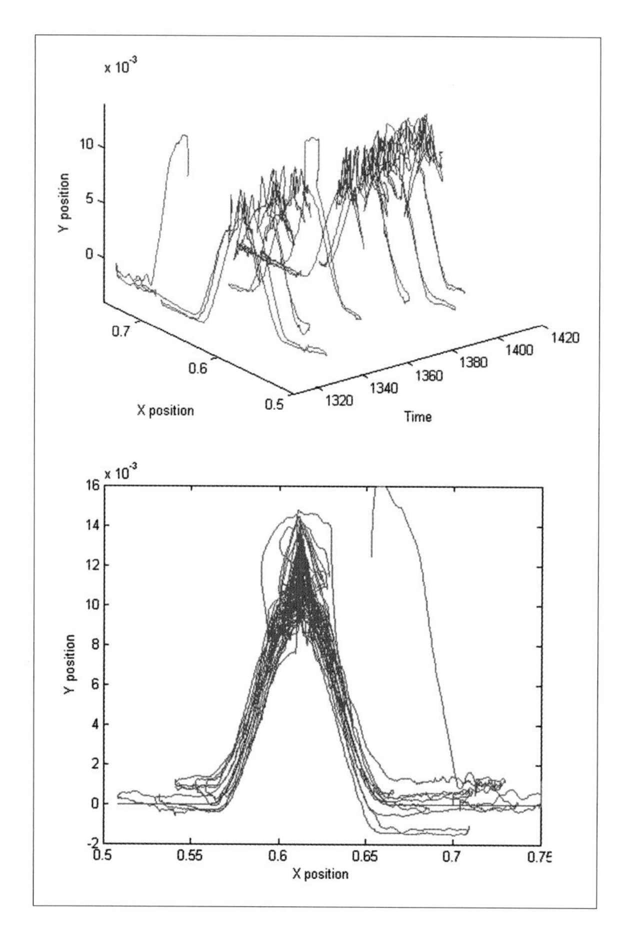

FIGURE 10.
A user's fingertip trajectory while exploring a SineSaw shape (Subject 1, Trial 34) and the trajectory data corrected for vertical offset and ready for MSE computation.

TABLE 3. COMPARISON OF THE EXPLORATION TRAJECTORIES USED BY SUBJECTS TO THE GEOMETRY OF THE HAPTIC SHAPES.
Note that SineLFGauss stimuli had the same geometry as the Sinusoidal segment (SineSeg). For each haptic shape, the difference in the trajectories followed is statistically significant (ANOVA, $p<0.01$).

Haptic shapes Explored	Closest geometry (% of total trials for each haptic shape)			
	SineSeg	Saw	SineSaw	Sine Gauss
SineSeg	98.33	0.56	0.56	0.56
Sine LFGauss	91.67	1.67	4.44	2.22
Saw	11.10	92.78	1.11	5.00
SineSaw	15.00	26.67	56.11	2.22
SineGauss	2.22	36.67	21.11	40.00

SineGauss (Rows 4 and 5). Subjects' finger trajectory when exploring SineSaw and SineGauss did not, in general, closely follow the target geometry of these surfaces. In contrast, rendering the geometry of SineLFGauss was possible (Row 2). As mentioned above, SineLFGauss had the same geometry as SineSeg (a smooth sinusoid), but was consistently matched to visual shapes with sharp features. The difficulties to render the geometry of SineSaw and SineGauss suggest that maintaining surface contact with these objects during exploration was made difficult by the surfaces' sharp features, which is something commonly observed when rendering these in general. In contrast, Table 3 suggests that a good surface contact with SineLFGauss (the surface with an illusory Gaussian bump) was simply achieved.

Two major points are suggested by the second table also. The first is that the geometry of Saw was accurately rendered (Tab. 3, Row 3, Column 2), but this did not result in subjects consistently matching this haptic surface to the visual Sawtooth profile (Tab. 2, Row 3, Column 4). In contrast, the haptic Saw was consistently matched to the sinusoidal visual profile (Tab. 2, Row 3, Column 1). This suggests that accurate rendering of the geometry of sharp objects does not always result in subjects perceiving the sharp features of objects.

The second major point is that, even when the geometry is not very accurately rendered (perhaps due to unstable contact with the surface of the object due to spatially sharp features), subjects still could be able to perform an accurate haptic to visual match. This is suggested by the matching performance for SineSaw (Tab. 2, Row 4, Column 3).

The results suggest that lateral-force-based haptic shape illusions (such as the one used in SineLFGauss) can be combined with a smooth object geometry to haptically render sharp features of objects. The results also suggest that this can be achieved while maintaining a stable contact with the object, which is what happens, in general, during haptic interaction with real objects with spatially sharp features. It is not possible to stress enough the desirability of such stable interaction with haptically rendered, spatially sharp objects, particularly in applications such as surgery simulators. However, the variability found during haptic to visual matching for SineLFGauss suggests that there are more factors determining subject perception in these cases. For example, subjects may have had a bias toward choosing one of the three visual figures, or perhaps subjects' expectations regarding the haptic features of stimuli were modulated by the visual shapes. More research is needed to clarify these possibilities.

Vibrotactile haptic feedback in manipulation and exploration in virtual environments

Vibrotactile feedback cues can significantly enhance touch perception for virtual environment applications with minimal design complexity and cost [36]. From a developmental psychological perspective, touch plays an essential role in our perceptual construction of spatial environmental layout. Combining touch and vision allows the simultaneous extraction of perceptive process invariants, crucial for establishing the reciprocal connections that allow for higher order perception and categorisation of objects and environments [37].

To date, there are several types of data-glove commercially available [38–40], but not so many researchers [41] or [42] have integrated this glove with prototypical vibrotactile pads. For the most of commercial data-gloves previously cited, quantitative assessment of rigid range of motion (RoM) is required and a measuring procedure must be done [43] and [44].

The data-glove application presented in this section represent a unique combination of absolute goniometric and tactile stimulators (see Fig. 11). The data-glove is based on 'Hall Effect' goniometric sensors [45] and a specific mechanical design that allows the device to be worn by anyone without requiring a specific pre-calibration procedure. The prototype can track any human gesture (grasping simple objects, touching surface, etc.) in order to perform psychophysics experiments and rehabilitation procedures.

The data-glove is equipped with at least two sensors of different length for each finger; the vibrotactile actuators are displaced on the back of each fingertip to stimulate the human cutaneous receptors when the virtual hand comes in contact with an object in the virtual environment; they measure the angular displacement of proximal (MCP) and medial phalanxes (PIP) with respect to the back of the hand. The difference between the two signals can be implemented *via* software in order to obtain the relative angular displacement between the two phalanxes. The adduction–abduction movement of each MCP finger is not measured, because it is not needed to perform the main gestures useful for the foreseen applications. A third sensor has been added to the thumb; it bends in a plane normal to the flexo-extension of the other two sensors.

Goniometric sensor

The working principle of the goniometric sensor relies on the fact that a flexible beam having a deformed elastic line lying on a plane has the property that the longitudinal elongation of the fibers depends linearly on their curvature and their distance from the neutral axis of the beam. The total elongation of the fiber is a function of

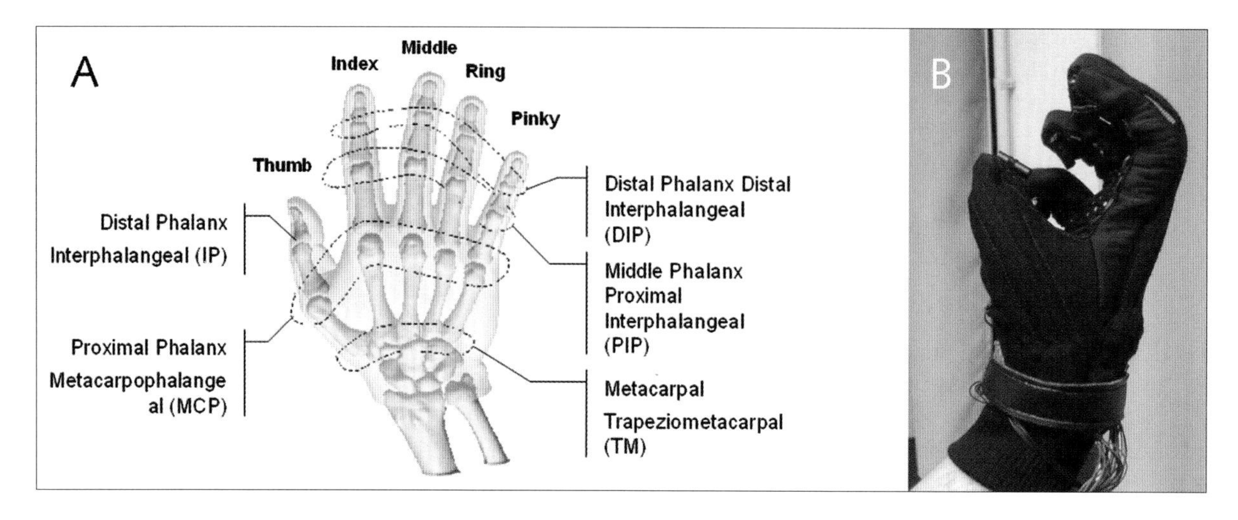

FIGURE 11. HAND JOINTS REPRESENTATION AND PERCRO DATA-GLOVE

the angle between the two endpoints of the flexible beam and it is independent from its specific elastic line. Hall effect sensor measures the intensity of a magnetic field produced by a magnet attached to the movable end of the wire.

The goniometric sensor is composed of only four parts: a commercial cylindrical permanent magnet, a commercial miniaturised Hall Effect sensor with a built-in signal amplifier, a multi-wire flexible steel cable; a flexible thin beam made of plastic with a square cross section a longitudinal hole. The beam ends with a casing for the magnet and the Hall Effect sensor [45].

The sensor (see Fig. 12) is composed of a transducing bulb and a sensing flexing bar. When a relative bending angle d to a beam element of length dL is imposed, the length of the neutral fiber of the beam remains unchanged while the fiber, positioned at a distance e from the neutral axis, changes its length by a quantity of:

$$dL = e \cdot d\theta \tag{2}$$

Integrating this variation along the entire beam, in case of constant e, we obtain:

$$L = \int_O^L dL = e \int_O^{\Delta\vartheta} d\vartheta = e \cdot \Delta\vartheta \tag{3}$$

Therefore, the total elongation of the fiber is a function of the angle between the two endpoints of the flexible beam and it is independent from its specific elastic line.

One time calibration and test have shown high and reliable stability of sensor measurements (Fig. 13) both in terms of accuracy and performances: 180° range of motion, ±1.5° of accuracy (worst condition).

Vibromechanical stimulator

Neural mechanisms that involves the sensation of touch have been studied extensively for many years. These studies demonstrate the sensory capacity of a human by functional proprieties of the sense organs in the skin, rather by mechanisms within the central nervous system [46].

The tactile units in the skin area of the human hand are of four different types: two fast adapting, RA (Meissner corpuscles) and PC (Pacinian corpuscles), and two slowly adapting types (Merkel cells). The slowly adapting are sensitive to low frequency stimulation (10 Hz) and primarily encode pressure, texture and

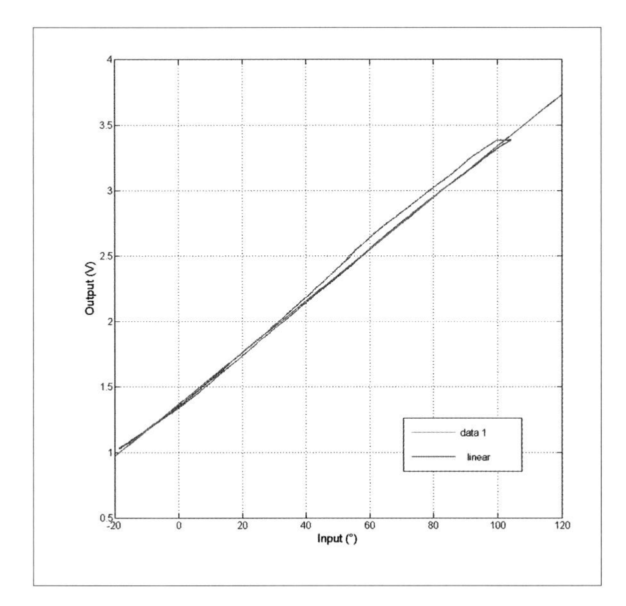

FIGURE 12. MAIN COMPONENTS OF THE GONIOMETRIC SENSOR

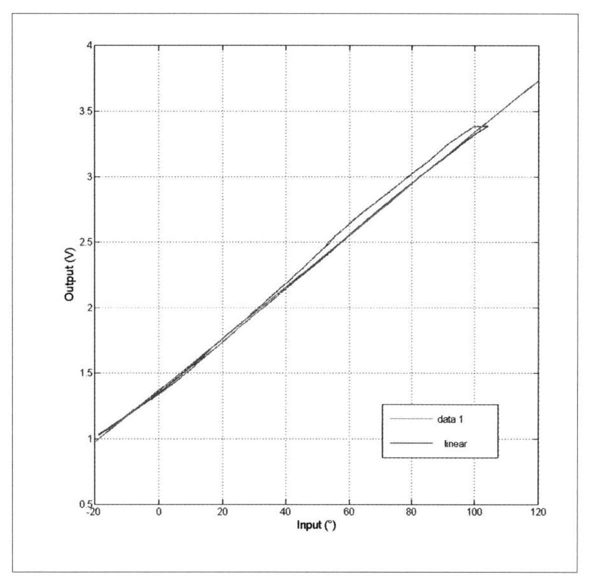

FIGURE 13. CALIBRATION PERFORMANCES (TENSION *VERSUS* ANGLE)

form of the object. The Meissner corpuscles are most sensitive to vibrotactile frequencies of 30 Hz and response to the flutter, slip, and motion of objects. The Pacinian corpuscles are most sensitive to high frequency vibration centred around 200 Hz.

In order to stimulate the Pacinian corpuscles vibrating motors have been attached on the palm-side of each fingertip. These actuators consist of small motors commonly used as vibrational alarms in pagers, mobile phones and many vibrotactile game controllers, which can be made to rotate at different speeds, and so different frequencies.

Vibration intensity is controlled varying the voltage through a PWM command between 0 and 5 V. Due to the voltage constant of such actuators, at the maximum voltage command the motors, with an eccentric mass (0.16 gr) mounted on its shaft, the frequency reaches a theoretical value of about 500 Hz.

Acquisition and control interface: The hardware control architecture acquires the analogue signals from the goniometric sensors, and convert them to digital signals, then send the digital data to a host computer *via* a serial communication protocol.

Two types of microcontrollers (C) PIC18LF4420 and PIC18LF443 have been used to develop a master-slave mini-network; each of them operates with an external oscillator of 40 Mhz. 11 analogue inputs are used to read the input signals from the goniometric sensors using 12 bits ADCs. The communication between the C and the host computer is performed through RS232 interface at 115,200 bps.

The communication among the master and slaves is based on the Serial Peripheral Interface (SPI) that guarantees 12 MBit/s. Each slave can activate four motors with different PWM signals through a Darlington array.

Rehabilitation test scenario: A test scenario which includes a whole hand avatar has been implemented. The environment includes a dining table, plates and kitchen utensils. A rehabilitation main task (to move and pick-up the objects

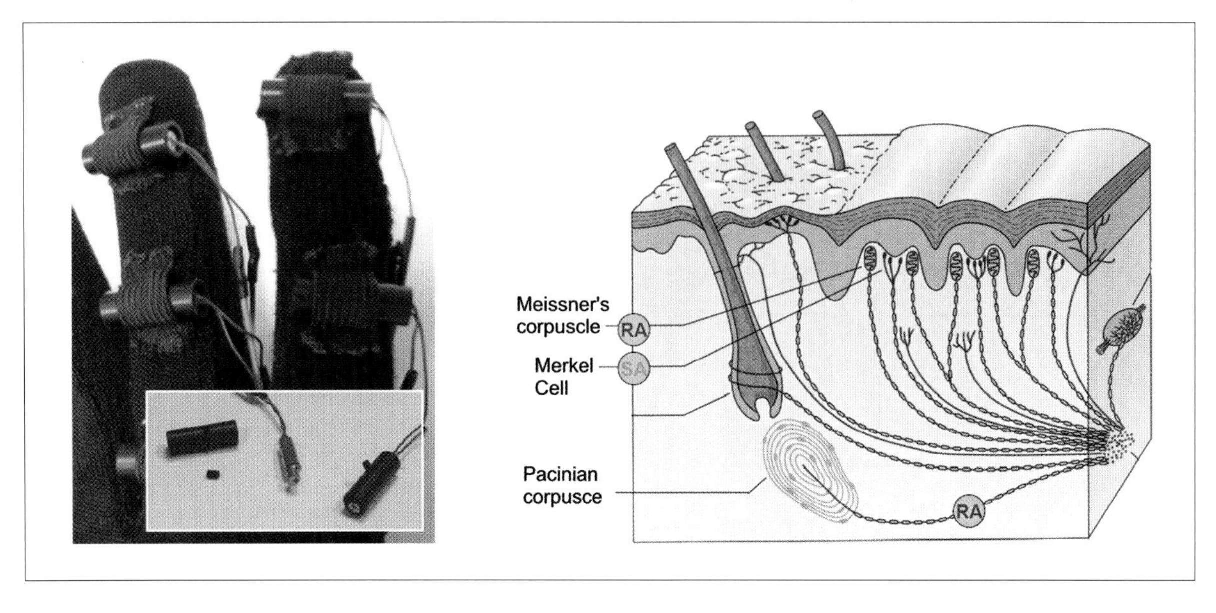

FIGURE 14. VIBROMECHANICAL ACTUATORS AND HUMAN HAND RECEPTORS

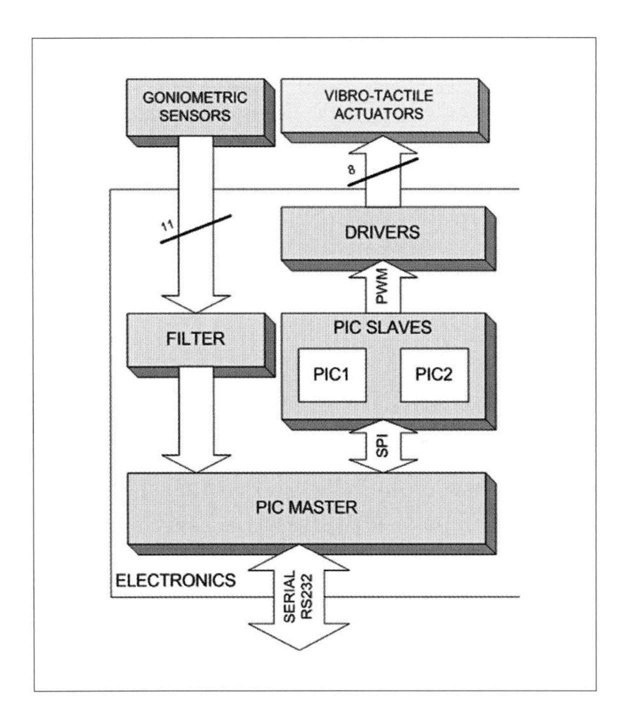

FIGURE 15. DATA-GLOVE ELECTRONIC ARCHITECTURE

around the table) was designed to estimate how a user feels the tactile sensation made by hand-object contact.

A graphic representation of the hand and the kitchen environment was developed and implemented through a virtual 3-D platform called XVR [47, 48].

To allow the user to sweep the VE, a 6 DoF magnetic tracker by Polhemus has been embedded on the data-glove. Such a sensor acquires the position and the orientation of the user's hand in respect to a physical receiver.

For this application, the angular positions of the skeletal joints are associated to the obtained angles, and then 3-D hand meshes are deformed applying geometrical transformations to the point of the mesh associated to the corresponding point of the skeleton, as shown in the application example of Figure 16.

The integration with Physics AGEIA has been performed for providing a more realistic dynamic interaction with the virtual environment and for relaying back to the user information about the

FIGURE 16. VIRTUAL HAND AND KITCHEN ENVIRONMENT
(Courtesy of Graphics and Hypermedia Lab at the University of Cyprus)

contact of the virtual hand fingers against the objects.

Summary

One of the most demanded technology for the human haptic perception is the development of highly immersive manipulative tools. To date, the development of new haptic devices for the control of human hand motion, namely hand exoskeletons, is close to the performance extreme that can technologically be achieved. However such devices have far to go to show the user with a good quality force feedback. Limits in the range of motion, friction, power, force, weight and several other factors make a real improvement in the design of a high quality system difficult to be achieved in a short period of time.

Fortunately, when considering manipulation, a lot of new discoveries have been achieved from the neurosciences by investigating not only the quality of representation but also the levels of perception. The original technical approach of the best rendering device for a realistic feeling, has been replaced with a more reasonable approach of the most immersive device for a believable feeling.

The present chapter has reviewed a set of four new methodologies to deliver force feedback at the level of human fingers. Two of them are based on software and control concepts, while the other two are based on the development of new devices that may overcome the above-mentioned limitations.

Control concepts included a new method to represent friction to achieve a better control human grasping and manipulation, as well as an investigation of the relationships among haptic and proprio perception in order to produce perceptive illusions.

New devices included the analysis of an encountered haptic device that only gets in contact when the user touches an object in the virtual environment, and a haptic glove that only produces the high frequencies of the contact feeling when the user touches objects in a virtual scene.

Each methodology has been designed, described and tested with qualification experiments in order to determine the level of believability that can be achieved when they are introduced into a virtual environment.

The results are very promising and show new potential design guidelines that may be applied when developing applications for specific contexts.

Selected readings

Kapandji IA (1982) *The physiology of the joints, volume I: upper limbs.* Elsevier

Johansson RS, Westling G (1984) Roles of glabrous skin receptors and sensorimotor memory in automatic control of precision grip when lifting rougher or more slippery objects. *Exp Brain Res* 56(3): 550–564

Zilles CB, Salisbury JK (1995) A constraint-based god-object method for haptic display. *Proc. IEE/RSJ International Conference on Intelligent Robots and Systems* 146–151

Bergamasco M, Salsedo F, Fontana M, Tarri F, Avizzano CA, Frisoli A, Ruffaldi E, Marcheschi S (2007) High performance haptic device for force rendering in textile exploration. *The Visual Computer* 23: 1–11

Robles-De-La-Torre G, Hayward V (2001) Force can overcome object geometry in the perception of shape through active touch. *Nature* 412: 445–448

Suggested websites

Haptic devices and VRSystems
www.percro.org

Modeling of interaction in training with VE
www.skills-ip.eu

Enactive interfaces
www.enactivenetwork.org

Presence and virtual reality
www.presenccia.org

VR development environment
www.vrmedia.com

Haptic rendering and control

Carsten Preusche, Thomas Hulin and Gerd Hirzinger

Introduction

In haptic simulations a human operator is coupled to a virtual environment, in such a way that the user is able to perceive the scene with his/her sense of touch. In Figure 1 the interaction paths between user and virtual world are shown. To obtain a good level of immersion, at least two interaction paths must be provided for haptic simulations, namely the visual and the haptic path. In the upper part the visual path is shown, in which a visual rendering algorithm computes images from a virtual model. These images are displayed to the user by a visual display. The user can modify the viewpoint by moving the virtual camera, e.g., by using a spacemouse or, if a tracking system is used by moving his/her head. The lower part presents the haptic path, which is in focus of the present book. Similar to the visual path, this path contains a software and hardware module, i.e., a haptic rendering algorithm and a haptic interface. As both these modules are bilateral, no additional device is needed for human interaction.

For exploration tasks, the visual path is of great importance, but not exclusively. Haptic information can increase immersion in virtual reality setups and contribute to intuitive operations. Especially manipulation tasks are much more intuitive if haptic feedback is provided besides visualisation. For several tasks – including virtual assembly verification and training of mechanics – the required time can be considerably decreased if haptic feedback is provided. Haptic interaction with virtual objects is based on a feedback loop, which is minimally composed of:

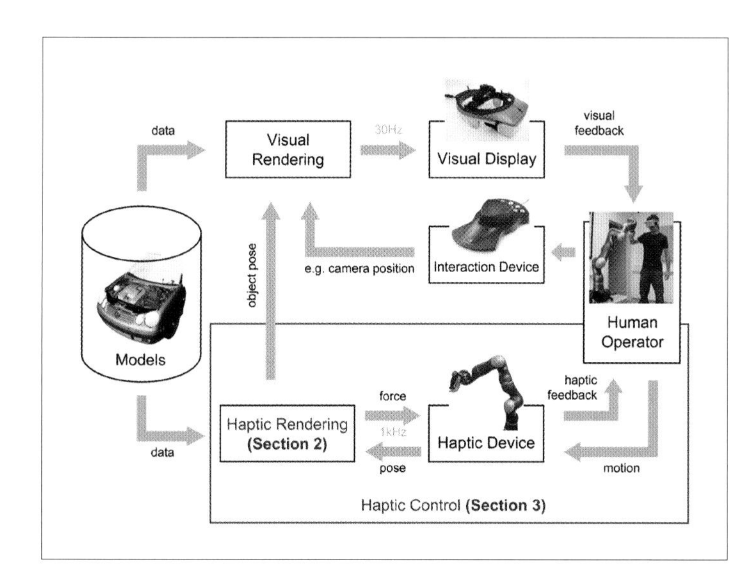

FIGURE 1. SCHEME OF USER INTERACTION WITH A VIRTUAL SCENE BY VISUAL AND HAPTIC MEANS

- digital models of the virtual scene
- a haptic rendering module
- a haptic interface
- control of the haptic interface
- a human operator

Except the digital model, which is loaded by the haptic rendering module, all interactions between the haptic subsystems are bidirectional, i.e., information flows from and to each module. This also means that the human operator closes the control loop with the haptic interface and the rendering module. The parameterisation of the elements within the haptic path depends on two main factors; the human perception of haptic information and the technical limitations of the haptic interface itself (see Chapter 29 and Chapter 32). Immersive simulations with realistic force feedback require stability of the haptic system, and moreover its transparency. As broadly accepted in the scientific community, the update rate of the loop must be at least 1 kHz for immersive haptic simulations. In a fully transparent system the user cannot distinguish whether he/she interacts with a real or a virtual environment. In the field of control for haptic rendering, the concept of passivity is very common despite the fact that it yields in conservative controllers [1, 2].

This chapter concentrates on algorithms for haptic rendering and control of haptic systems. Haptic rendering algorithms compute responses for human interaction from a virtual scene. Hereby, the data representation of the model depends on the rendering algorithm itself. The section on *Haptic rendering* gives an overview of existing methods and goes into details of the Voxmap-PointShell Algorithm as an example for a fast method for computing collision forces. In the section *Haptic control* several control approaches for the haptic loop are presented. Starting with a general overview of control methods, stability analysis based on the parameter-space approach is investigated in detail. Finally, a case study for using haptic feedback for assembly verification is given in *Case study: haptic assembly verification*. The section *Conclusions and further readings* gives some concluding remarks and starting points for further reading.

Haptic rendering

In analogy to visual rendering, which generates a dedicated view of a scene depending on the virtual camera position, haptic rendering computes reaction forces from the virtual scene, depending on the current position input of a human operator. This computed result is displayed by a haptic interface back to the user, who can feel the reaction of the virtual world. So, haptic rendering algorithms compute the resulting interaction forces depending on the current position of the user's hand or tool in the scene.

The rendering process is generally divided into two parts: First, collision detection and secondly, computation of reaction forces. The overall computation time for these calculations needs to be predictable and shorter than 1ms to meet the requirement of 1 kHz update rate of the controller.

The next section gives a detailed overview of the haptic rendering process, and section *Voxmap-PointShell Algorithm* describes the Voxmap-PointShell Algorithm as an exemplary haptic rendering approach.

Overview

As mentioned before, haptic rendering consists of two computational steps (Fig. 2). In the first step collision detection is performed, i.e., for the new (updated) position of the user's hand or tool all collisions with the virtual scene are computed. This process is generally very time-consuming, as has been evaluated in [3]. Hereby the force computation determines the type of information, which has to correspond to the result of the collision detection. Having determined the collisions, the resulting forces from these collisions are calculated in a second phase. These forces are then displayed to the human user via the haptic interface, controlled by a haptic controller (section

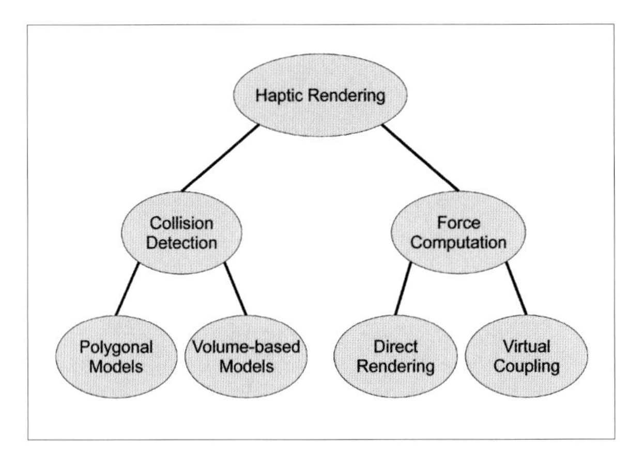

FIGURE 2. TWO PHASES OF THE HAPTIC RENDERING PROCESS: *collision detection and force computation*

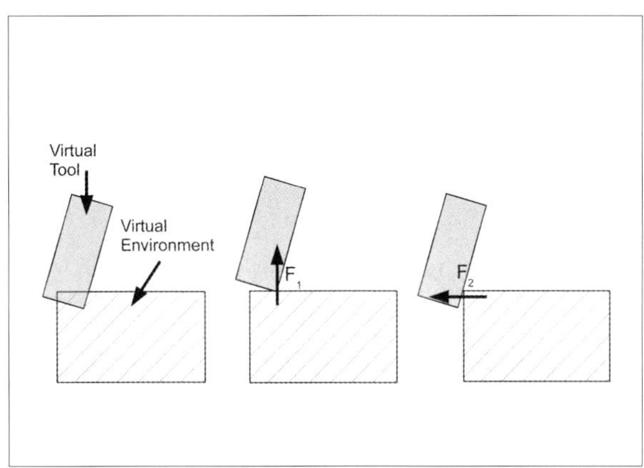

FIGURE 3.
Illustrating the problem of haptic rendering with arbitrary tools (schematic view in 2-D)

Haptic control). To obtain an optimal feedback, the haptic algorithm must fit to the properties of the used haptic interface, e.g., degree of freedom or resolution.

The first haptic rendering algorithms only used a single point of interaction, representing the tool tip of the haptic interface, (see [4, 5]). This method and its improvements allow, even with low computing power, to render forces from the point contact of the user's fingertip with the virtual environment using the so-called HIP (Haptic Interaction Point) as a proxy. Although this approach is sufficient for haptic exploration of virtual scenes, it cannot be used for more sophisticated interaction tasks, like virtual training or assembly verification.

In these tasks the tools have more complex geometrical shapes, and besides the forces also interaction torques have to be generated. Consequently, a six-dimensional force/torque reaction vector needs to be rendered. In the following we will therefore concentrate on haptic rendering of arbitrary shaped interaction tools.

Collision detection

The haptic rendering loop starts with detecting collisions in the virtual scene, once the geometrical description of the scene is updated. For solid non-deformable objects this description is the geometrical relation or relative pose between virtual objects. Usually the pose (position and orientation) of an object is given by a transformation matrix ${}_0T^{Obj}$ with respect to the origin.

After this update a new static description of the virtual scene exists, in which all intersections (collisions) between objects have to be detected. For efficiency reason the virtual world can be separated into a static part and a dynamic part. The static part will not be affected by the update of the object poses, so no collisions between objects within the static part can occur. The dynamic objects will change their pose in the virtual world, so the algorithm needs to test for collisions between all dynamic objects and the static scene, including collisions between the dynamic objects, if there is more than one.

In general the use of virtual tools to interact with the virtual scene increases the complexity of determining collisions correctly. This includes the correct configuration of the virtual tool (see Fig. 3) and the integration of several possible contact points of the same virtual object with the environment. To compute the correct response for the tool, furthermore the history of motion is required, although only few existing methods take into account this information. The methods for collision detection depend on the data representation of the objects. Starting from polygonal models, which are used for visual rendering, also non-polygonal models are used and have been developed for haptic rendering. Especially volume-based models seem to be suitable for a fast rendering with a predictable computation time.

Polygonal models

The collision detection based on polygonal models has been under research for many years by the computer graphics and robotics communities. In these applications there are weaker requirements compared to haptic rendering, regarding update rate (30 Hz for graphics) or accuracy (having a safety margin for robot collision avoidance, which allows the use of simplified models). Additionally, not all the information needed for haptic rendering is computed by these algorithms. For example, contact points are also needed as the penetration depth. A good survey on collision detection using polygonal models, including public domain software packages, is given in [6]. If applying these methods, additional computation is required to determine contact point and penetration depth. One solution for computing the penetration depth is the Gilbert-Johnson-Keerti algorithm [7]. As this algorithm is able to handle only convex objects, Gregory et al. [8] extended the Lin-Canny closest points algorithm [9] to perform collision detection for six Degree of Freedom (DoF) haptic rendering.

Collision detection based on polygonal models becomes slow if the complexity and the level of detail of the models increase. Therefore hierarchical models are used to speed up the algorithms. The octree is the most popular hierarchy for structured polygonal models [10].

Volume-based models

Large and complex geometries lead to a huge number of polygons such that the computation time for collision detection increases drastically. Representing the virtual objects by voxels (3-D pixels) makes possible another group of collision detection algorithms, whose computation time is independent of the number of polygons (see page 415). Beside voxels the algorithm uses a second type of representation, namely a cloud of points on the surface of virtual objects.

Force computation

The second step of haptic rendering is force computation. In this step a six dimensional force/torque vector is computed for each colliding object pair. Independent of the algorithm used to compute this reaction vector, two solutions can be distinguished for determining the final force response – *direct rendering* and *virtual coupling*.

With direct rendering the computed force/torque vector is directly commanded to the haptic interface, whereas with virtual coupling a spring-damper system is introduced between a virtual handle and a virtual object. The force/torque vector is determined by the elongation of the spring, in translation and orientation. The concept of virtual coupling was introduced by Colgate et al. [11] to limit the stiffness of virtual contacts and to enable stable interaction. On the other hand virtual coupling decreases transparency for haptic rendering, as it artificially softens virtual collisions.

There are several approaches to compute collision forces. The following lines describe the most relevant approaches for force computation of haptic rendering.

Penalty-based

Penalty-based algorithms use either the penetration depth of two colliding objects [1, 12] or the intersection volume [13, 14] to determine collision forces. In fact, most haptic algorithms are penalty-based algorithms. Also the Voxmap-PointShell Algorithm belongs to the category of penalty-based algorithms. As will be seen later, this algorithm uses a point cloud representation of virtual objects and determines the penetration depth for each point to obtain a collision force. Another approach uses NURBS (Non-Uniform

Rational B-Spline) surfaces and implicit surfaces that come directly from CAD (Computer Aided Design) systems or modelling software [15]. This method overcomes the inherent approximation error of polygonal models and therefore has the advantage of being more precise than algorithms based on polygonal objects. Yet, all penalty-based approaches result in a force behaviour which is directly proportional to the intersection depth.

Impulse-based

Another group of haptic algorithms generates an impulse each time a collision occurs [16, 17]. Most often impulse-based algorithms are combined with penalty-based algorithms. The impulse-based algorithm generates the impression of hard contacts, whereas the penalty-based algorithm computes the collision force that counteracts intersections of objects. McNeely et al. [18] introduced a braking force for their Voxmap-PointShell Algorithm, provoking a pre-contact force to reduce the penetration caused by the momentum of the dynamic object and the haptic device. Kuchenbecker et al. [19] displays pre-recorded force profiles during collisions, to simulate material properties. The problem of impulse-based algorithms in terms of stability is that impulsive forces do not always decelerate objects, but can also introduce energy in excess if they are badly designed. Nevertheless, they are able to increase the realism of virtual collisions.

Constraint-based

The constraint-based approach was one of the first methods for haptic rendering, because these algorithms are computational cheap, and made possible haptic rendering more than ten years ago. The method was introduced by Zilles and Salisbury in 1995 [4], enabling three-DoF haptic rendering. To overcome numerical problems of the polygonal objects, Ruspini et al. enhanced this approach [20]. Yet, only a few articles exist about constraint-based algorithms, as it is difficult to implement constraint-based six-DoF haptic rendering algorithms. With increasing computational power of computers, penalty-based algorithms replaced the constraint-based ones.

Further effects

After the first haptic rendering algorithms were developed, researchers tried to enhance these algorithms and render more physical effects than only collision forces. Therefore, effects like deformations [21–23], friction [24] or texture properties [25, 26] were investigated and included in their algorithms. Definitely, rendering those effects increases immersion of haptic rendering simulations. Yet, up to now there is no algorithm that is able to render all those effects in parallel.

Voxmap-PointShell Algorithm

The Voxmap-PointShell™ (VPS) Algorithm, which was introduced by McNeely in 1999 [18], allows for six-DoF collision force computation of two virtual objects at a guaranteed rate of 1 kHz. It can be used for both direct haptic rendering and virtual coupling. Instead of computing intersections of polygonal meshes, this algorithm is based on its own haptic data structures, which divide the virtual scene into a dynamic object and a static scene. It uses a *pointshell* for the dynamic object and a *voxelmap* for the scene (see Fig. 4).

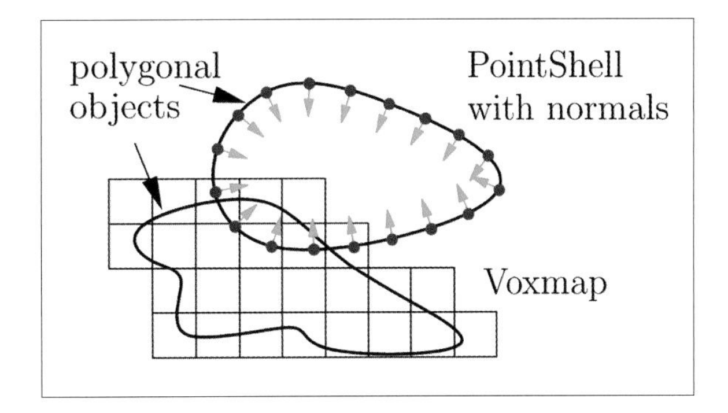

FIGURE 4. SCHEMATICAL ILLUSTRATION OF A POINTSHELL COLLIDING WITH A VOXELMAP

FIGURE 5.
Polygonal model and pointshell representation of a car battery with 11,491 pointshell points; Bottom: Polygonal model and voxelmap of a motor compartment of a Volkswagen Polo. The visualised voxels are the 1,267,652 surface voxels, of a total of 38,165,880 voxels in the whole voxelmap. For both the pointshell and the voxelmap the resolution is 5 mm.

The *pointshell* is a cloud of points located on the surface of the moving object. Each point has a normal vector perpendicular to the surface, pointing inside the object. The *voxelmap* results as space discretisation of the static scene. It represents the scene through a large number of voxels (volume pixels). Each voxel that collides with the surface of the original scene object is marked as surface-voxel. Furthermore, the voxelmap distinguishes between voxels that are inside the object (inside-voxels) and those that are in free space (outside-voxels). Figure 5 shows the pointshell of a car battery and the voxelmap of a car's motor compartment at a resolution of 5 mm.

The VPS algorithm performs two tasks for computing collision forces, namely collision detection and actual force computation. For the first task every single pointshell point is transformed into the coordinates of the voxelmap and the voxels containing pointshell points are determined. If the voxelmap is stored as one-dimensional array in computer memory, as it is in some VPS implementations, the voxel's address results directly from the location of the pointshell point. The point is definitely colliding, if it is located within an inside-voxel. If it is inside a surface voxel, the vector product between the points normal and the distance from the centre of the voxel to the location of the point must be evaluated to derive, if the point is colliding [18]. For the original implementation of the VPS algorithm the voxelmap is stored in a hierarchical tree structure, in order to save memory. On the other hand, this kind of voxelmap representation makes the collision test slower. Nevertheless, checking one pointshell point for collision is extremely fast due to its simplicity.

For the actual computation of the collision force, the normals of all colliding points are multiplied with their penetration depth. Summing up these resulting forces of each colliding pointshell point, transformed into the origin of the pointshell, yields the resulting collision force and torque.

The required time for computing collision forces depends mainly on the number of pointshell points and is quite – although not completely – independent of the collision situation. The worst case for computation time is reached if all points are colliding. Furthermore, the time is independent of the complexity of the static scene, i.e., the number of voxels. On a standard PC it is possible to perform haptic rendering at guaranteed 1 kHz rate with 8,000 pointshell points1. (Rendering was performed on a PC with an Intel Pentium 4 CPU with 3.20 GHz.)

Several studies suggest improvements of the VPS algorithm. A more accurate pointshell and smother force computation is introduced in [27]. McNeely investigated in [28] the use of distance fields, geometrical awareness and temporal coherence, to increase speed and accuracy of the algorithm. Recent studies by Barbic modified the VPS algorithm so that deformations of objects can be simulated. Related to the VPS algorithm is also the approach of representing objects by implicit spheres [29], to increase the speed of collision tests with octrees. The VPS algorithm is furthermore ideally suited for parallel computation, as its collision detection and force computation contain sequential loops. This becomes especially interesting nowadays, with the increase in the number of computers that are equipped with multi-core CPUs.

Haptic control

Haptic rendering algorithms compute haptic feedback directly or by means of virtual coupling [11]. Yet both approaches compute a force that is displayed by a haptic device to a human operator. Position-force characteristics of those calculations are often similar to linear springs. For direct rendering the force is proportional to the penetration depth of two colliding virtual objects. Similarly, for virtual coupling, the larger the distance of a god object to the haptic device, the greater is the force that is fed back to the human. Although there are other approaches that do not result in spring-like force behaviour – like the impulsive forces approach [17] – haptic setups with spring-like force characteristics are the majority, by far.

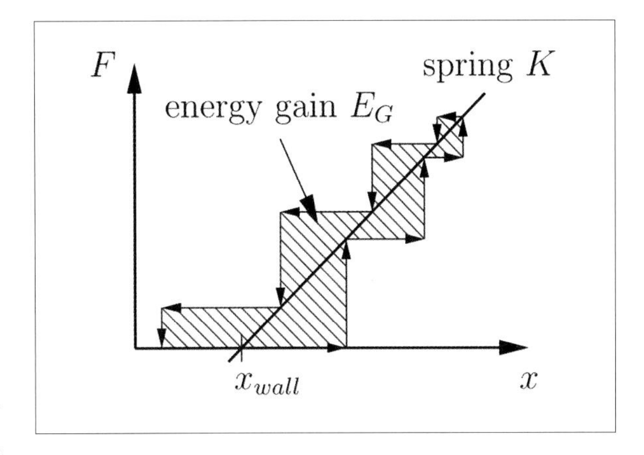

Figure 6.
A collision with a virtual wall represented by a discrete-time spring generates energy (shaded region)

The problem of these setups in terms of stability is the active behaviour of virtual springs modelled as gain of penetration depth or distance. More precisely, the activity originates in the combination of this gain with the time discretisation that comes along when using digital controllers for haptic devices.

This fact becomes obvious in Figure 6. Due to the discrete-time characteristics, the force generated by a haptic device acts during one sampling step on the device's mass, and is only updated at the next sampling step. Therefore, energy is generated, which is in contrast to a physical (continuous-time) spring which is passive. The amount of energy, also denoted as energy gain [30], is determined by the shaded area in Figure 6, defined by the exceeded force multiplied with the way this force acts.

The energy gain increases with the stiffness of the wall. For generating realistic impressions of hard contacts the stiffness of virtual surfaces must be at least 2,000 N/m [31]. Thus, the resulting energy gain can easily become huge and the haptic device starts oscillating. The haptic impression is completely disturbed. But even worse, an unstable haptic device is harmful for human operators.

In most haptic setups, a virtual damper is added to the virtual wall in order to stabilise the system again. As it is generally implemented as negative gain on the velocity of the haptic device, the control equivalent of the virtual wall is a discrete-time PD-controller. Yet, this virtual damper is not always able to make the system stable, as is shown in the following sections.

Passivity-based control

A well-known approach for stabilising a haptic system is to guarantee passivity of its elements. These are in general a human operator and a haptic device that is displaying forces computed by a haptic rendering algorithm. The human operator is assumed to be passive for the high frequencies of haptic device dynamics, because active movements of humans are below 10 Hz [32]. Hence, to obtain stability it is sufficient to make only the haptic device passive. Colgate and Schenkel [1] derived a condition for passivity of a haptic device that is displaying a virtual wall,

$$b > \frac{KT}{2} + |B|, \tag{1}$$

with b being the physical damping of the device, K and B stiffness and damping of the virtual wall, and T the sampling period. Surprisingly, the damping of the virtual wall does not contribute to passivity. The virtual damper even decreases the range of passive virtual stiffness.

Although haptic systems that fulfil the passivity condition are stable, that condition is not the optimum criterion for stable haptic rendering, as passivity is a conservative condition in terms of stability [1]. A virtual wall which stiffness is higher than admitted by the passivity condition, may still result in stable haptic interaction. Hence, numerous publications try to find less conservative criterions and controllers for haptic systems. Some of them are investigated in the following subsections.

Time domain passivity control

A much less conservative approach for stability for haptic rendering was suggested by Hannaford and Ryu [2]. They introduced a time-domain passivity controller that adjusts a variable virtual damping in dependency of the energy gain, which is calculated by a so-called passivity observer. In other words, the virtual damping is set in such a way that it dissipates exactly the amount of energy in excess. To smoothen the damping force they enhanced their time-domain passivity controller by a reference energy following method [33]. This avoids exciting high frequency modes of the haptic device. Instead of using the virtual damper implemented as negative velocity gain, the passivity controller approach can be adapted to physical damping. In [34] eddy current viscous dampers, mounted on a haptic device, dissipate the energy gain. Problematically on electrical dampers is their huge size and weight, which is contrasting design paradigms of haptic devices. Yet, one problem of the time-domain passivity controller, independent of the kind of damping, is that the commanded force never corresponds exactly to the applied force of the haptic device, which is affected by several physical effects, e.g., motor dynamics. Therefore, it is not possible to determine the exact amount energy gain. But, still this approach has proven to work well on several haptic systems.

Other approaches

Some other approaches are also based on passivity. Stramigioli et at. [35] suggested the use of the port-Hamiltonian approach for stabilising haptic interaction. This approach splits the interconnection between the virtual wall and the real world into effort-flow pairs. Therefore, they combined discrete- and continuous-time port-Hamiltonian systems. The studies in [36] investigated the design of virtual coupling, requiring that the human operator and the virtual environment are passive. A lot of related research for passivity-based analysis has also been performed in the field of bilateral telemanipulation (e.g., [37]).

Stability-based control

Although the passivity-based approaches try to be less conservative than Colgate's passivity condition, they are still not able to define an exact and non-conservative stability condition or control law. This section investigates stability for a system that consists of a human operator grabbing an impedance type haptic device. Some studies investigate nonlinear effects on the stability of haptic systems [38, 39], but this section focuses on linear stability analysis.

The exact stability region for haptic walls represented by a virtual spring-damper system was first determined by Salcudean and Vlaar [40]. They considered their haptic device as a simple mass, actuated by a one sample-step delayed force. For this simplified control loop, they found the stability boundary inside a normalised parameter plane. The human operator was ignored for the stability analysis, as he/she tends to stabilise the system [36, 41]. This approach was enhanced by the authors of [42], through considering also time delay as parameter. A more complex model of the haptic system including physical damping was investigated in [43]. It was observed that physical damping is increasing the stable region. This result and the derived stability boundaries are in accordance with previous experiments [44, 45]. A similar approach [41], without normalised parameters and time delay, resulted in the same stability boundaries. Yet, that article introduced a linear stability condition, which was recently generalised in [46] to admit time delay. Recent results [47] prove that this linear condition still holds if a human operator is included in the analysis. This section summarises the results obtained by linear analyses of haptic systems.

Linear model of haptic control loop

As shown in Figure 7, the haptic device is modelled as damped mass. The dynamics of the device's actuators are neglected, such that the force, F, of the virtual world can be assumed to be constant over one sampling period, T.

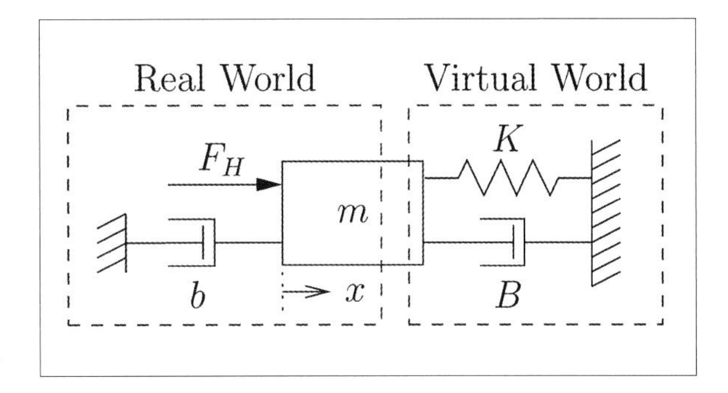

FIGURE 7.
The haptic device modelled as a damped mass m *combined with a virtual spring and damper*

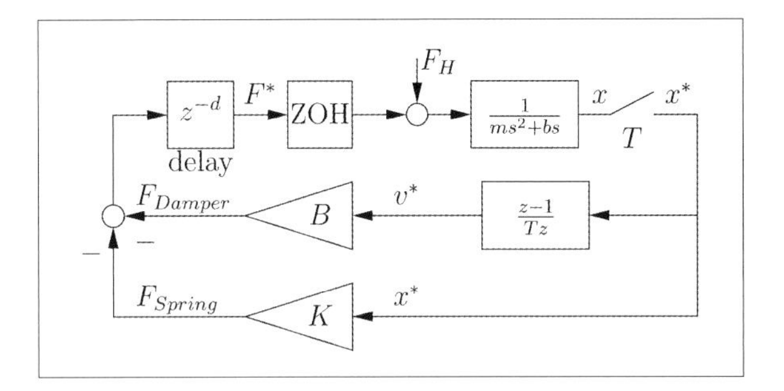

FIGURE 8. CONTROL LOOP OF THE SYSTEM

For impedance control the worst-case scenario for stability is the situation when the operator is not grabbing the haptic device [36, 41]. Thus, the human operator is not taken into account in the following analyses. A system which is then found to be stable will also be stable if the operator is interacting with the device.

The haptic device displays a virtual wall which is composed of a virtual spring, K, and damper, B (discrete-time PD-controller). The force of the virtual wall is a by t_d delayed discrete-time signal with sampling time, T, and delay factor $d = t_d/T$.

Nonlinear effects like static friction, or quantisation and saturation of sensors and actuators are not taken into account. With these assumptions the control loop shown in Figure 8 can be set up easily. It contains continuous-time (haptic device's mass and damping) and discrete-time (virtual wall) elements. Thus, before a stability analysis can be performed, all blocks of the control loop have to be transformed into the same time domain. In analogy to [43], the next subsection transforms the continuous- into discrete-time blocks.

Remark 1 *This section analyses stability for virtual walls modelled by a bilateral spring-damper system. But this analysis holds also for the case of a human pushing against a virtual wall $F_H > 0$, such that the position of the mass is not leaving the wall, as this is equivalent to a system with a bilateral wall. If bouncing occurs, i.e., the mass is leaving a unilateral wall during an oscillation cycle, the authors expect the found stability condition to be conservative, because the destabilising discrete-time elements would not affect a whole oscillation cycle. Yet, this case is not investigated inside this chapter.*

Exact discrete model

The exact discrete-time equivalent of the continuous-time block in the control loop in Figure 8 can be determined by some calculations, assuming that the input force F_H is constant during one sampling step. The way to determine the exact discrete-time equivalent is shown in detail in [43]. Here, only the result is given as:

$$1/(ms^2 + bs) \quad \triangleq \quad \frac{Tb(z-c) + m(1-z)(1-c)}{b^2(z-1)(z-c)} \tag{2}$$

with

$$c = e^{-Tb/m}. \tag{3}$$

With this result, the transfer function $G(z)$ from force F_H^* to position x^* can easily be determined to be:

$$G_x(z) = \Big(\big(-Tb + m(1-c)\big)z + \big(Tbc - m(1-c)\big)\Big)Tz^{1+d}/p(z) \tag{4}$$

with the characteristic polynomial

$$\begin{aligned} p(z) = &\Big(\big(m(1-c) - Tb\big)(KT + B)\Big)z^2 \\ &+\Big(\big(Tbc - m(1-c)\big)KT + \big(Tb + Tbc - 2m(1-c)\big) \\ &-Tb^2(z-c)(z-1)z^{1+d} + \big(-Tbc + m(1-c)\big)B \end{aligned} \tag{5}$$

Parameter	Variable	Dimensionless variable
Sampling period	T	-
Mass	m	-
Virtual stiffness	K	$\alpha = \frac{KT^2}{m}$
Virtual damping	B	$\beta = \frac{BT}{m}$
Physical damping	b	$\delta = \frac{bT}{m}$
Delay	t_d	$d = \frac{t_d}{T}$

TABLE 1. NORMALISED PARAMETERS

where K is the virtual stiffness, B the virtual damping and d the delay factor. The authors of [43] also introduced normalised parameters to simplify their characteristic polynomial according to Table 1. These normalisation rules simplify the characteristic polynomial (5) to:

$$\begin{aligned} p(z) \;=\; &(1-c-\delta)(\alpha+\beta)\,z^2 \\ +\; &\big((c\delta - 1 + c)\alpha + ((1+c)\delta - 2(1-c))\beta\big)z \\ -\; &\delta^2(z-c)(z-1)z^{1+d} + (1-c-c\delta)\beta \end{aligned} \tag{6}$$

with

$$c = e^{-\delta}. \tag{7}$$

The normalised characteristic polynomial $p(z)$ depends only on the four dimensionless parameters α, β, δ and d. The mass, m, and the sampling time, T, dropped out.

Stability boundaries

The previous subsection derived a normalised characteristic polynomial of the control loop in Figure 8, using the normalised parameters of Table 1. A stability check of the investigated system can easily be performed by computing the zeros of the discrete-time characteristic polynomial function (6). The system is stable if all roots of $p(z)$ are located inside the unit circle. The stable region is obtained if two parameters of $p(z)$ are gridded, and all the others are fixed.

For practical purposes, the boundaries are plotted in the (α, β)-plane in Figure 9, because

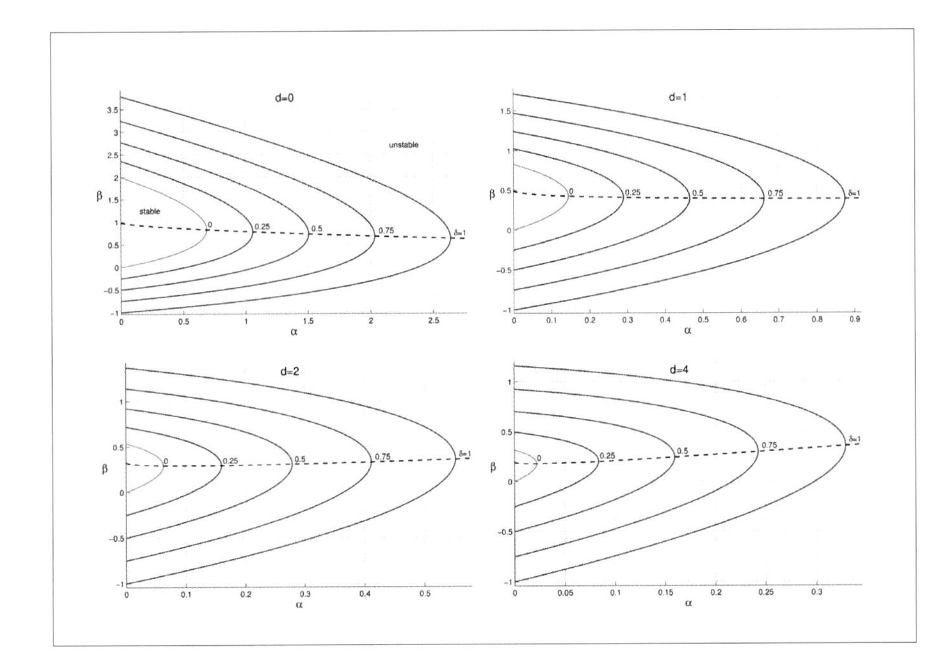

FIGURE 9.
The stability boundaries in the (α,β)-plane for d ∈ [0, 1, 2, 4] and δ ∈ [0, 0.25, 0.5, 0.75, 1]. The dashed lines represent the paths of maximum virtual stiffness.

these are the two parameters that are modified in the haptic simulation. The figure shows the stability boundaries for fixed values of the delay factor $d \in [0, 1, 2, 4]$ and a physical damping of $\delta \in [0, 0.25, 0.5, 0.75, 1]$. The poles and thus the stability boundaries in the plane of the normalised parameters are independent of the haptic device's mass and the time delay.

The stable region is increasing for higher physical damping δ. For negative α the system becomes unstable.

The lower starting point of each boundary is located at $(\alpha, \beta) = (0, -\delta)$, independent of the delay factor, d. This condition becomes obvious when looking at the stability boundaries for $\alpha = 0$ in the (α, β)-plane in Figure 10: the stable region is bounded by the line $\beta = -\delta$. Formulating this line as necessary stability condition with the non-normalised parameters yields:

$$B \geq -b. \tag{8}$$

For zero stiffness $K = 0$, the virtual damping must be greater than the negative physical damping.

Remark 2 *A special case occurs for $d = 0$ and $d \geq 2.3$. In the (α, β)-plane a line appears which is limiting the stable region besides the parabola-like stability boundary (see Fig. 11). This line appears only for $\delta = 0$.*

Relation to passivity

Colgate's passivity condition for $d = 0$ derived in [1]:

$$b > \frac{KT}{2} + |B| \tag{9}$$

can be rewritten with the normalised parameters:

$$\delta > \frac{\alpha}{2} + |\beta| \tag{10}$$

for $T > 0$ and $m > 0$. This condition is visualised as dash-dotted lines and compared to the stability boundaries in Figure 12. As was expected, the passive region is a subset of the stable region for $d = 0$. This corresponds to the fact that the passivity condition is conservative [1].

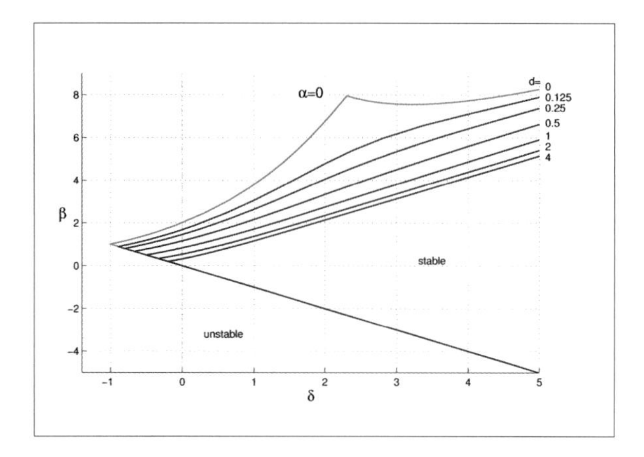

FIGURE 10.
The stability boundaries in the (δ,β)-plane for δ=0 and d ∈ [0, 0.25, 0.5, 1, 2, 4]

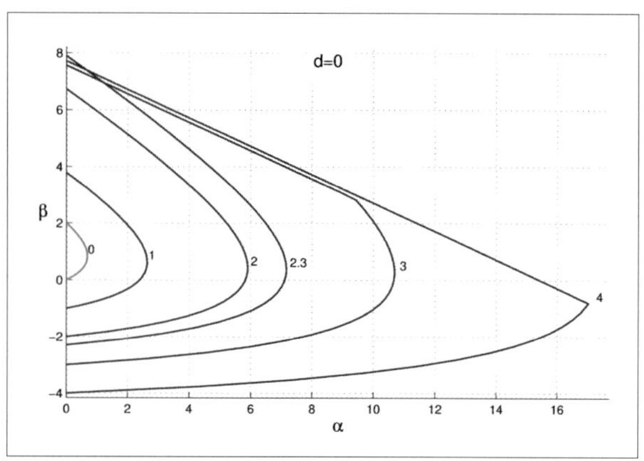

FIGURE 11.
The stability boundaries in the (α,β)-plane for the case d=0 and large δ. A line with slope –0.5 is limiting the stable region from above for δ > 2.3

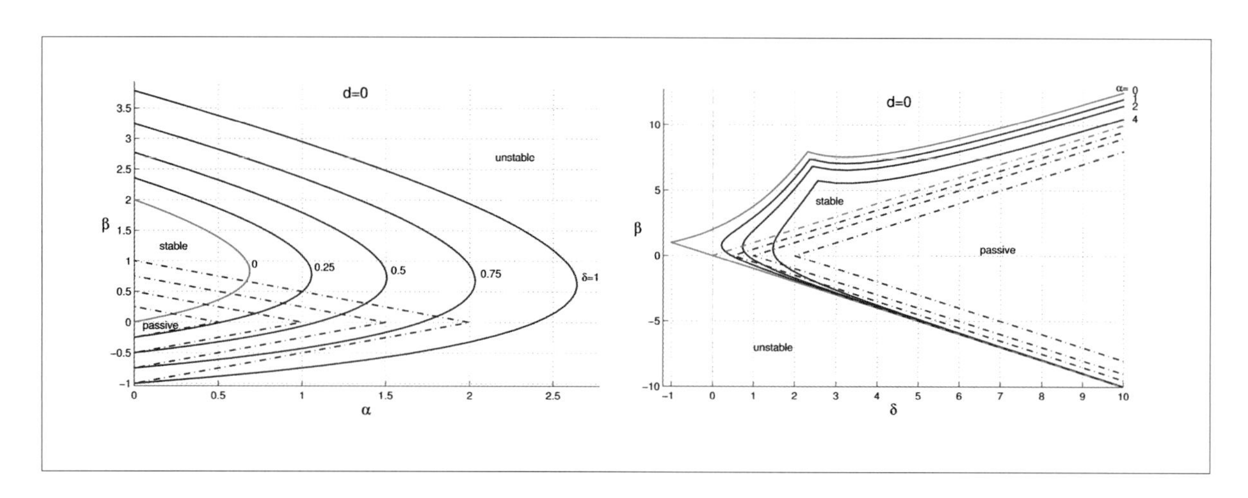

FIGURE 12.
The stability boundaries (solid lines) compared to Colgate's passivity condition (dash-dotted lines) for d=0. Left: boundaries in the (α,β)-plane for δ ∈ [0, 0.25, 0.5, 0.75, 1]; right: boundaries in the (δ,β)-plane for α ∈ [0, 1, 2, 4].

Note, the passivity condition holds only for d=0. As in most haptic systems time delay occurs due to various reasons, the Colgate's passivity condition represents only a boundary for simulations where d = 0.

Linear stability condition

For a real haptic device it is valid to simplify the exact stability boundaries presented in previous paragraphs [46]. Linearising the boundaries around small α and β yields:

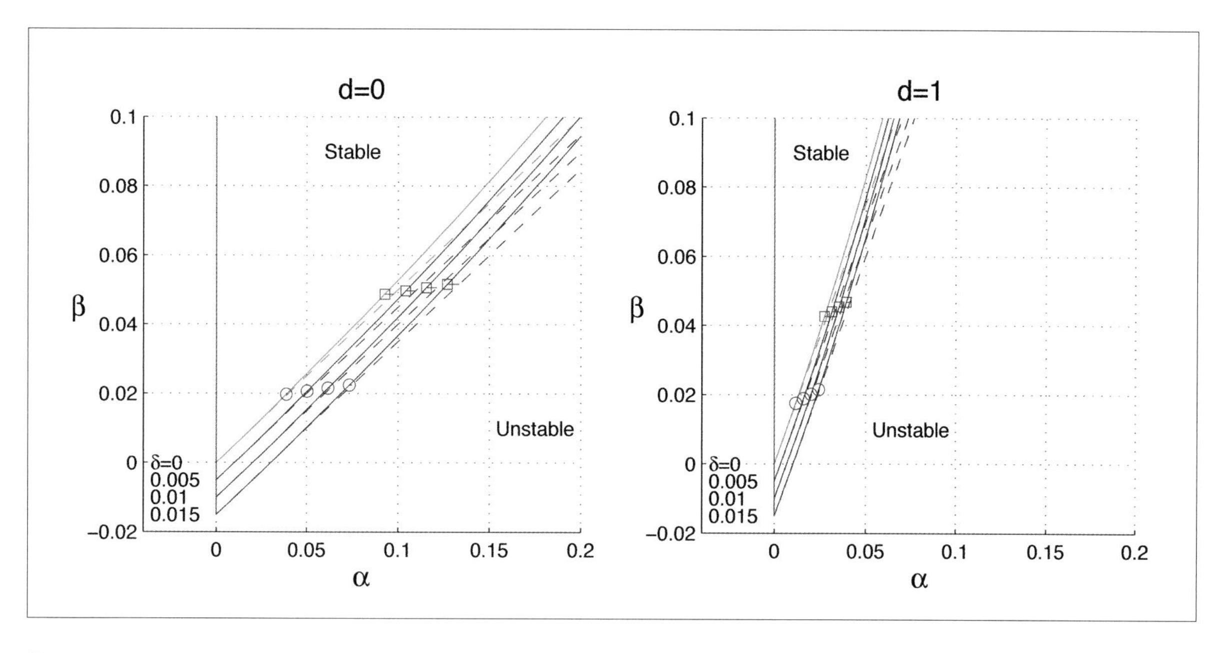

FIGURE 13.
Validity of the linear stability condition equation (11). Exact stability boundaries (solid) for places where the relative error of the linear condition (11) (dashed) is equal to 2% (circles) and equal to 5% (squares), for δ ∈ [0, 0.005, 0.01, 0.015] and d=0 (left) resp. d=1 (right)

$$\alpha < \frac{\delta + \beta}{0.5 + d}. \tag{11}$$

Rearranging the stability equation which relates the real parameters of the system,

$$K < \frac{b + B}{\frac{T}{2} + t_d}, \tag{12}$$

and taking into account that the effect of the sampling and hold in the control loop can be approximated by a delay of half the sampling period $T/2$, stability condition (11) can be interpreted with the following statement:

$$\text{Critical stiffness} = \frac{\sum \text{Damping}}{\sum \text{Delay}}. \tag{13}$$

The following lines compare linear stability condition (11) to the exact boundaries. Following [46], the relative error of the virtual stiffness α between both boundaries is considered. This error is defined as:

$$|(\alpha_{lin} - \alpha)/\alpha| \tag{14}$$

for α and α_{lin} located on the exact, respectively linear stability boundary for the same β. Figure 13 shows the exact (solid) and the linear condition (dashed) in the same plots. It visualises also the points at which the relative error reaches a level of 2% (circles) and 5% (squares).

Although the valid range of the linear condition ends at numerically small values on the α-axis, the non-normalised values for the virtual stiffness at which the linear condition holds are huge. This is due to the square of the sampling rate in the normalisation rule of α and the typically fast sampling frequency ($T \leq 1$ ms) in haptic systems. Recent results show that this linear condition even holds if a human operator is included in the analysis.

Experimental results

The DLR Light-Weight Robot III (Fig. 14) has been used to demonstrate experimentally that the theoretical results hold [47]. This robot is a seven-DoF robot arm with carbon fibre grid structure links. Though it weighs only 14 kg, it is able to handle payloads of 14 kg throughout the whole dynamic range. The electronics, including the power converters, is integrated into the robot arm.

Every joint has an internal controller which compensates gravity and Coulomb friction. Since high-resolution position sensors are used to measure link orientation (quantisation q≈20"), non-linear effects can be neglected.

A virtual wall was implemented in the third axis of the robot, indicated by the rotating angle ϕ in Figure 14. The environment was implemented using a computer connected to the robot via Ethernet. The sampling rate was 1 kHz and the overall loop contained a delay of 5 ms. Figure 15 shows the experimental results, introducing several fixed values for the virtual damping. A set of experiments was performed with only the system delay of 5 ms, while additional delays were artificially introduced into the loop to obtain an overall delay of 6 and 10 ms (left). A very large delay was also introduced in the system in order to receive a curved stability boundary for t_d = 55 ms (right). The theoretical behaviour is depicted with dotted lines. The experimental stability boundaries fit the linear

FIGURE 14. THIRD GENERATION OF THE DLR LIGHT-WEIGHT ROBOT ARM [55]

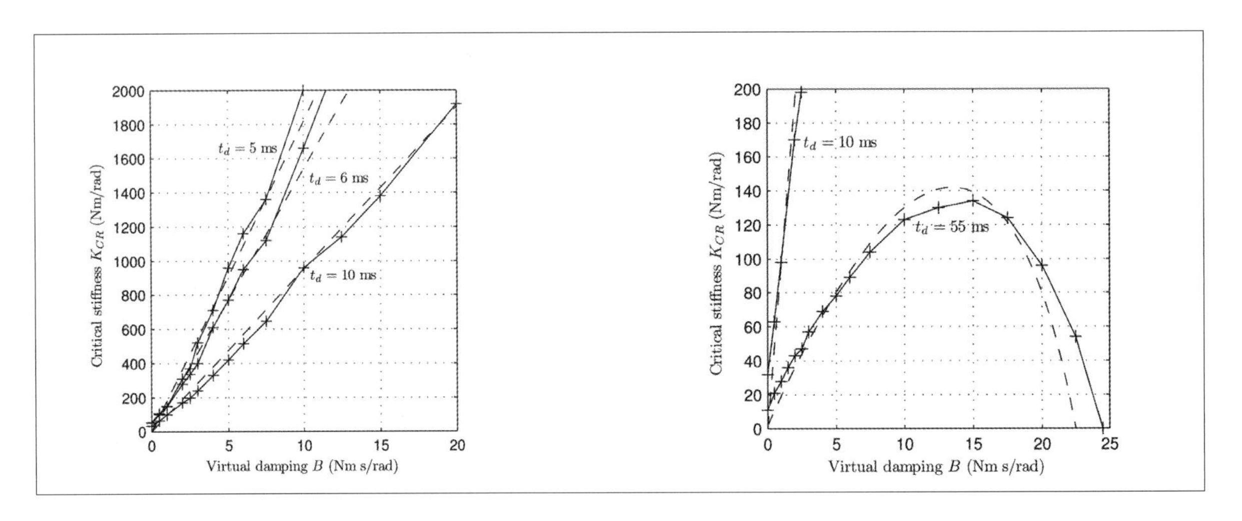

FIGURE 15.

Experimental stability boundaries (solid) compared the theoretical boundaries (dashed) for delay $t_d \in$ *[5, 6, 10] ms (left), and* $t_d \in$ *[10, 55] ms (right)*

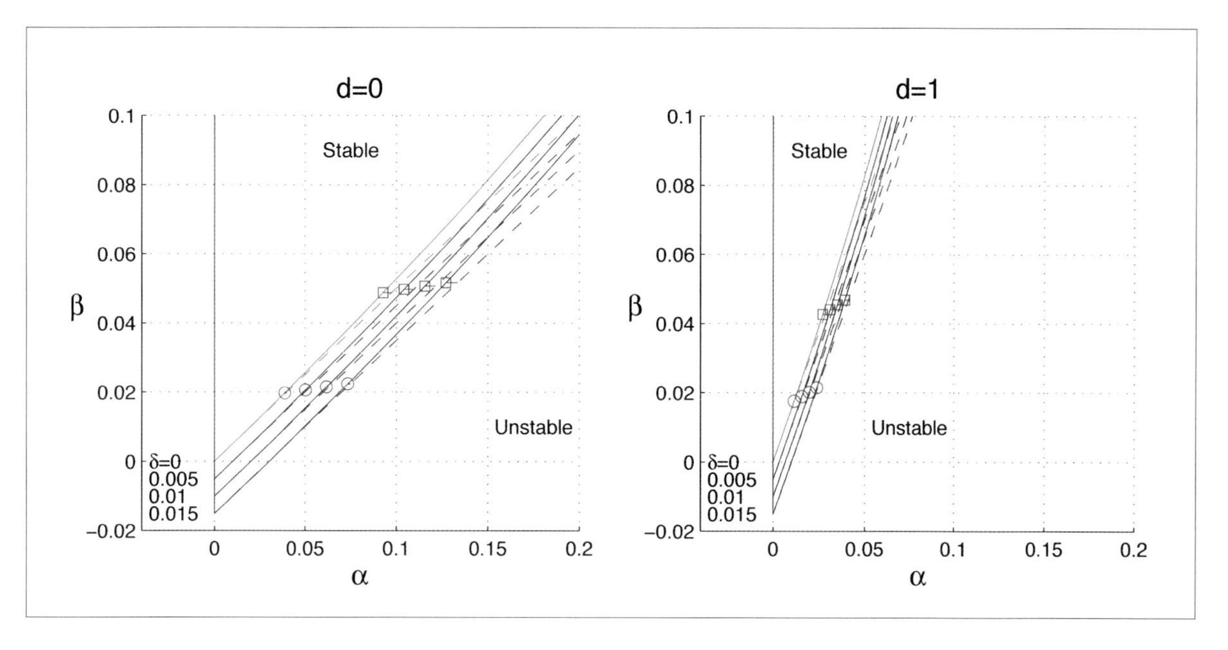

FIGURE 13.
Validity of the linear stability condition equation (11). Exact stability boundaries (solid) for places where the relative error of the linear condition (11) (dashed) is equal to 2% (circles) and equal to 5% (squares), for δ ∈ [0, 0.005, 0.01, 0.015] and d=0 (left) resp. d=1 (right)

$$\alpha < \frac{\delta + \beta}{0.5 + d}. \tag{11}$$

Rearranging the stability equation which relates the real parameters of the system,

$$K < \frac{b + B}{\frac{T}{2} + t_d}, \tag{12}$$

and taking into account that the effect of the sampling and hold in the control loop can be approximated by a delay of half the sampling period $T/2$, stability condition (11) can be interpreted with the following statement:

$$\text{Critical stiffness} = \frac{\sum \text{Damping}}{\sum \text{Delay}}. \tag{13}$$

The following lines compare linear stability condition (11) to the exact boundaries. Following [46], the relative error of the virtual stiffness α between both boundaries is considered. This error is defined as:

$$|(\alpha_{lin} - \alpha)/\alpha| \tag{14}$$

for α and α_{lin} located on the exact, respectively linear stability boundary for the same β. Figure 13 shows the exact (solid) and the linear condition (dashed) in the same plots. It visualises also the points at which the relative error reaches a level of 2% (circles) and 5% (squares).

Although the valid range of the linear condition ends at numerically small values on the α-axis, the non-normalised values for the virtual stiffness at which the linear condition holds are huge. This is due to the square of the sampling rate in the normalisation rule of α and the typically fast sampling frequency ($T \leq 1$ ms) in haptic systems. Recent results show that this linear condition even holds if a human operator is included in the analysis.

Experimental results

The DLR Light-Weight Robot III (Fig. 14) has been used to demonstrate experimentally that the theoretical results hold [47]. This robot is a seven-DoF robot arm with carbon fibre grid structure links. Though it weighs only 14 kg, it is able to handle payloads of 14 kg throughout the whole dynamic range. The electronics, including the power converters, is integrated into the robot arm.

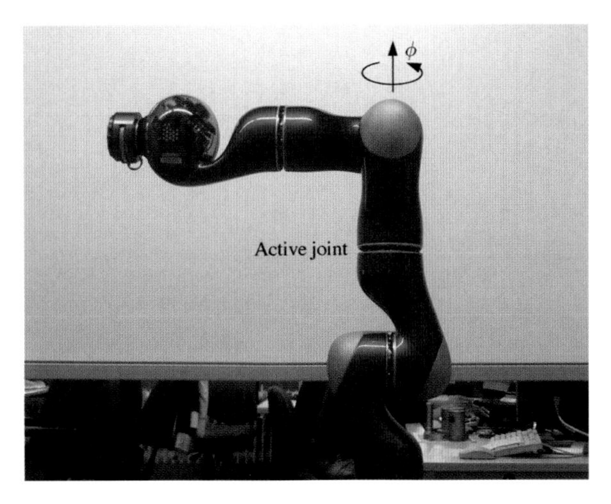

FIGURE 14. THIRD GENERATION OF THE DLR LIGHT-WEIGHT ROBOT ARM [55]

Every joint has an internal controller which compensates gravity and Coulomb friction. Since high-resolution position sensors are used to measure link orientation (quantisation q≈20"), non-linear effects can be neglected.

A virtual wall was implemented in the third axis of the robot, indicated by the rotating angle ϕ in Figure 14. The environment was implemented using a computer connected to the robot via Ethernet. The sampling rate was 1 kHz and the overall loop contained a delay of 5 ms. Figure 15 shows the experimental results, introducing several fixed values for the virtual damping. A set of experiments was performed with only the system delay of 5 ms, while additional delays were artificially introduced into the loop to obtain an overall delay of 6 and 10 ms (left). A very large delay was also introduced in the system in order to receive a curved stability boundary for t_d = 55 ms (right). The theoretical behaviour is depicted with dotted lines. The experimental stability boundaries fit the linear

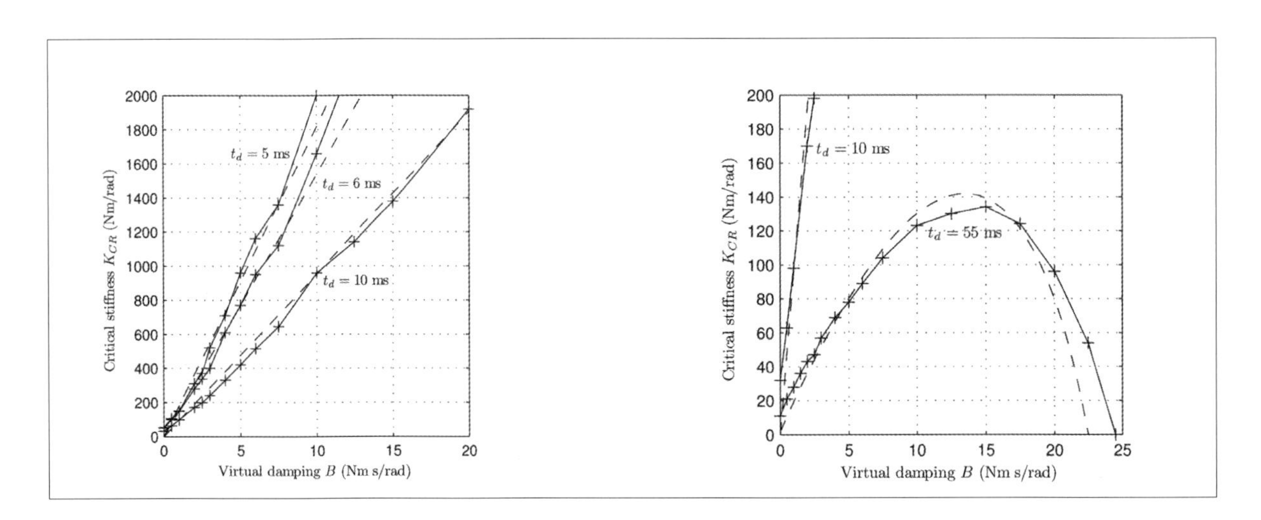

FIGURE 15.
Experimental stability boundaries (solid) compared the theoretical boundaries (dashed) for delay $t_d \in [5, 6, 10]$ *ms (left), and* $t_d \in [10, 55]$ *ms (right)*

condition remarkably well. The theoretical stability curve for $t_d = 55$ ms has been computed using the inertia of the robot configuration shown in Figure 14, which is approximately $0.8\ \mathrm{kg \cdot m^2}$.

Case study: haptic assembly verification

The use of digital mock-ups for virtual product creation is state-of-the-art. Yet, hardware mock-ups are still needed to verify maintainability of mechanical products. Haptic feedback has the potential to replace the use of such hardware mock-up by simulation studies.

This section describes a scenario for virtual assembly verification in the automotive sector [48]. The main goal is to find out on a virtual prototype, whether it will be possible for a mechanic to dis-/assemble a part on the real product. The second step is to optimise the virtual prototype and the assembly task itself in terms of maintainability and assembly order. During such a simulation, collision forces are calculated and displayed to a human operator *via* a haptic interface, see Figure 16.

The Voxmap-PointShell (VPS) Algorithm is used as a fast algorithm for calculation of collision forces. This algorithm is well suited for computing collision forces in such scenarios, which is mainly composed of non-deformable metal or plastic parts. It can also handle very complex models such as a motor compartment built of numerous parts.

The DLR Light-Weight Robot III, which is described in the previous section, is used as the haptic interface [49]. The advantage of this robot compared to other haptic devices is the combination of its large workspace, which is similar to the workspace of a human arm, and its powerful actuators that enable the robot to handle payloads of 14 kg over the whole dynamic range. This case-study on virtual assembly verification proves that such simulations have many advantages once haptic feedback is integrated in the development process of complex machines:

Figure 16. Assembly verification with visual and haptic feedback.
The DLR light-weight robot displays forces computed by the VPS Algorithm.

- Enables improved design of machines with respect to maintainability
- Makes use of hardware mock-ups superfluous
- Speeds up the development process

Conclusions and further reading

A general introduction to haptic rendering in virtual environments is given in [32]. In [50] and [51] a more recent survey about haptics – rendering, control and devices – can be found. This chapter investigated two main issues for haptic rendering, namely haptic rendering algorithms and control of haptic systems. On each issue a short literature overview was given and one suited approach investigated. A more detailed survey for haptic rendering algorithms can be found in [52]. As haptic rendering algorithm, the VPS Algorithm was introduced, and exact linear stability analysis was performed to obtain a highly transparent control law.

The VPS algorithm is able to meet the challenging requirements for haptic rendering. It can compute accurately collision forces, and even for extremely complex virtual objects guarantee the required update rate of 1 kHz. Therefore, it is suited for generating realistic impressions for haptic simulations, see also [28].

An overview of the most common approaches on the control of haptic systems was given in this chapter, which are mainly based on passivity. More precise analyses in terms of stability were thoroughly investigated. Through linear stability analysis of a haptic system, exact stability boundaries were obtained. A linear stability condition was derived by linearising these boundaries. This easy to use stability condition has proved to hold well for real haptic systems and enable highly transparent haptic interaction with virtual environments [53].

Present developments are promising and show a steady increase in the efforts of enhancing the functionality of haptic rendering algorithms. The trend goes to simulating deformable objects, textures and other physical effects. Some of these effects will be realised through a combination of haptic rendering with physical engines. Others require a more sophisticated approach for haptic algorithms. Another field that receives increasing attention is collaborative work. In the near future multimodal collaboration on virtual scenarios, including haptics, will be widely employed. This technology will allow for even faster exchange of knowledge and more intuitive communication.

These improvements do not only rely on enhanced haptic algorithms, but also on modern control strategies of haptic interfaces. Still missing is, for example, a stability analysis that includes non-linear effects like friction or encoder resolution and the dynamics of a human operator at the same time.

Summary

Haptic algorithms compute forces from a virtual scene that are displayed via haptic interfaces to humans. This chapter focuses on suited algorithms and on control approaches for stable haptic interaction and highly transparent haptic rendering.

In haptic control systems, a fixed update rate is crucial for stability of the control loop. Thus, haptic rendering has to fulfil the requirement of deterministic computation time. Furthermore, high transparency requires high spatial resolution of the device's sensors and of the haptic algorithm. Haptic rendering based on the Voxmap-PointShell approach has the advantage of guaranteed update rates for collision detection and high resolution models, and is therefore suitable for haptic interaction. Stability of haptic interaction limits the displayable stiffness of contacts in a virtual scene. Thus, a non-conservative stability condition is essential for high-transparent haptic interaction. The presented method is based on exact calculation of stability boundaries, which is in contrast to the widespread but conservative passivity-based approaches for haptic rendering.

Acknowledgements

The authors would like to thank Volkswagen for making available the digital model of the Volkswagen Polo.

Haptic perception in human robotic systems

Heinz Wörn, Catherina R. Burghart, Karsten Weiß and Dirk Göger

Introduction

Why is haptic perception essential in human robotic systems? This question is often posed in connection with humanoid robots. First of all, humanoid robots are intended to assist people in a typical human environment. A person expects a humanoid robot to think, move, act, and communicate in a human-like manner. This also includes the usage of typical human senses like vision, hearing and tactile sensing. Second, a robot manipulating items in an unstructured environment like a person's home needs to have some haptic feedback: information whether an item is firmly grasped or sliding is important for handling objects. Third, different kinds of haptic and tactile feedback are required for moving and acting in a human-built environment: collision detection just as well as haptic feedback for actions or control by a human operator.

Different types of haptic feedback require an acquisition of environmental data by different kinds of haptic sensors. Most commonly, force torque sensors in the wrists or other joints of the robot are used. In addition, tactile input is achieved by sensor matrices based on various physical working principles. Sometimes, data inputs of depth perception and surface perception have to be combined to achieve an optimal haptic sensing.

This chapter mainly focuses on tactile sensing of a humanoid robot, illustrating the working principle of resistive sensor arrays used as an artificial sensitive skin, classifying different types of contact, and giving an insight into a current project of a humanoid robot.

Tactile sensor system

A tactile sensor system normally consists of discrete sensor cells, so called 'texels'. They are arranged in homogeneous matrices, detecting an applied load profile. For data acquisition, the measurement converter is connected to a local intelligence, a sensor controller, digitising the sensor signals and pre-processing them. A host system processes the data made available by the controller and extracts characteristics. The data can be used, e.g., for reactive control of a robot. The measurement principles of the tactile sensor cells found in literature are based on three major classes: optical, capacitive and resistive effects. Optical sensors commonly utilise force dependent absorption or reflection of light beams [1]. They are very insensitive against corrosion and electromagnetic disturbances. For high area applications, like covering a whole robot system, the interconnection between the texels, commonly done by PMMA fibre cables, and the detection circuit becomes too complex. Another common approach for optical tactile sensors is to measure the scattered light from a lightened transparent polymer material by using a CCD camera [2]. These sensors can be used for tactile object recognition and orientation sensing in grippers, but are not suitable for surface covering.

Capacitive sensors can be found as single cell sensors as well as tactile sensor matrices with low drift and a high reproducibility [3]. They utilise the change of capacitance between two electrodes covering a deformable dielectric. These changes in the range of a few femto farads are very difficult to detect, therefore a complex signal conditioning electronic is needed. Due to the required high sensitivity of the electronics, capacitive sensor systems in general are very sensitive against electromagnetic interferences. Another approach integrates a capacitive texel matrix into a single chip with the signal conditioning. This greatly improves the interference robustness and enables high resolution tactile sensing [3]. Like optical tactile sensors, these sensors are suitable for gripping purposes but not for surface covering.

Of particular interest for service robotics are resistive tactile sensor systems. They utilise the effect that the resistivity of the interface between two surfaces changes according to the applied load. This was first discovered by the French electrical engineer Theodore du Moncel in the late 19th Century. He discovered that an electrical current flowing between a sooty metal plate and a nail is modulated by acoustic waves. Based on this conclusion, he invented the carbon microphone which revolutionised telephony [4].

Resistive tactile sensing

Today, instead of a sooty metal plate, conductive polymers are used for resistive tactile sensors. The load-dependent change in resistivity is acquired using an electrode matrix and – in comparison to that of capacitive sensors – simple signal conditioning electronics. It shows a hyperbolic style characteristic between the applied load and the electrical resistance. This nonlinear behaviour is of special interest, e.g., in collision detection, since for light-weighted contacts to its surface the sensor is more sensitive than at high loads – the measurement range is expanded. In addition, resistive tactile sensors in general are very robust on overpressure, shock and vibration due to its simple construction. Different resistive sensor systems can be found in literature and in industrial applications. Commonly, the electrodes for detecting the resistance are mounted on the facing sides of the sensor material co-axial to the direction the load will be applied, as shown in Figure 1.

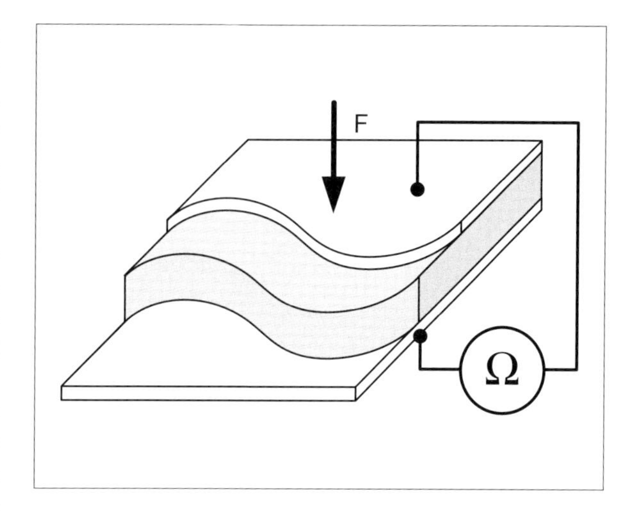

FIGURE 1. DOUBLE SIDED CONTACTING OF THE SENSOR MATERIAL

This arrangement enables an easy construction of tactile sensor matrices by adding horizontally and vertically aligned electrode stripes onto the sensor material, realising a network of force sensitive resistors. Those sensor matrices are proposed (e.g., in [5–7]).

While contacting the sensor material from both sides, the load has to be applied over the upper electrode. This is unfavourable, since the sensor material is usually flexible, whereby the upper electrode is exposed to a bending stress reducing the life time of the sensor. Therefore, we proposed another construction of the sensor cells by placing both electrodes on the back side of the sensor material, bypassing the fatigue problem.

With an arrangement as shown in Figure 2, even foams can be used as a sensor material without bending the electrodes by applying a load to the sensor cell, provided that the sensor is mounted on a rigid surface. At the Institute of Process Control and Robotics at the University of Karlsruhe a resistive tactile sensor system for service robots has been developed [8]. It con-

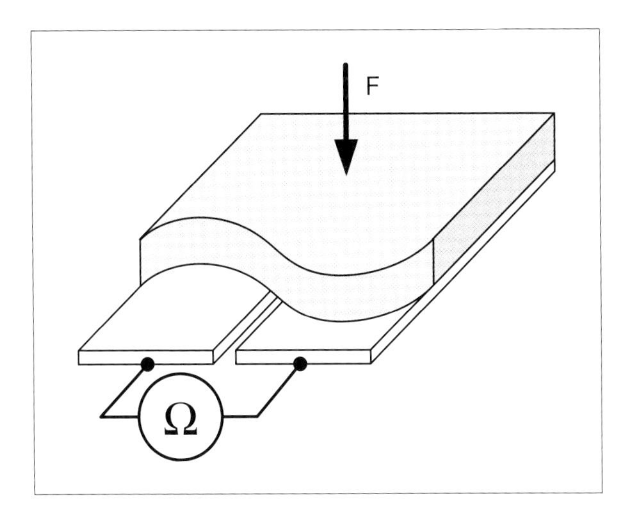

FIGURE 2. SINGLE SIDED CONTACTING OF THE SENSOR MATERIAL

sists of a flexible sensor matrix with 230 texels and a spatial resolution of 16 mm. The sensor was developed for human-machine interaction with an anthropomorphic robot arm with seven degrees of freedom and is used for manually guidance of the robot arm (reactive control), for collision detection and classification. Carbon enriched foam, which changes its electrical resistance locally according to the applied load was used as sensor material. The resistivity is measured using an electrode array on the back side of the foam. This electrode array is constructed in a way that the sensor material has not to be structured – in opposite to other approaches a full material can be used covering the whole electrode matrix. Therefore, the sensor matrix is very robust. Since the electrodes are realised on a flexible polyimide carrier, the sensor can be easily adapted to nearly any surface bended in one dimension, making it ideal for covering a robot's surface.

Even though there are many approaches to construct resistive tactile sensor systems in literature, the exact working principle is little investigated. To further improve the sensors, the effect of the pressure dependent change in electrical resistance was analysed, which can be found in resistive sensor material. Based on these results, a physical model of a resistive tactile sensor cell was generated thus helping to understand its working principle.

To realise the desired volume conductivity of the sensor material, commonly carbon blacks are used. Reference [9] describes the use of other semi-conductive particles such as molybdenum, antimony, ferrous sulfide or carborundum, too. The concentration of semi-conductive particles in the polymer binder has to be high enough to achieve a surface resistance lower than approximately 10 kΩ/sq, but on the other hand it must not be too high to interfere with the mechanical properties of the binder polymer. Unfortunately, it is not possible to calculate the volume and surface resistance in dependence of the concentration of the semiconductor particles theoretically, because it is based on many empiric factors like particle size, surface area, chemical processes on the surface and even on the binder polymer used [10]. Thus, the manufacturing of an adequate sensor material is currently done experimentally by mixing probes with different loadings of particles, curing them and measuring the electrical resistance in an iterative process.

Working principle of resistive tactile sensor cells

As already investigated and published by Weiss et al. [11], the sensors working principle depends on an interface effect between the metal electrodes and the structured sensor material (Fig. 3). The resistance between the common electrode and a sensor cell electrode is a function of the applied load and time. Every sensor cell needs one wire so the sensor array's cells have to be multiplexed. This technique leads to very accurate pictures of the applied pressure profile and minimises crosstalk between the sensor cells as well. A certain trade-off has to be paid in the high amount of connections to the cell array. Figure 4 depicts the electrical circuit of a resistive sensor cell.

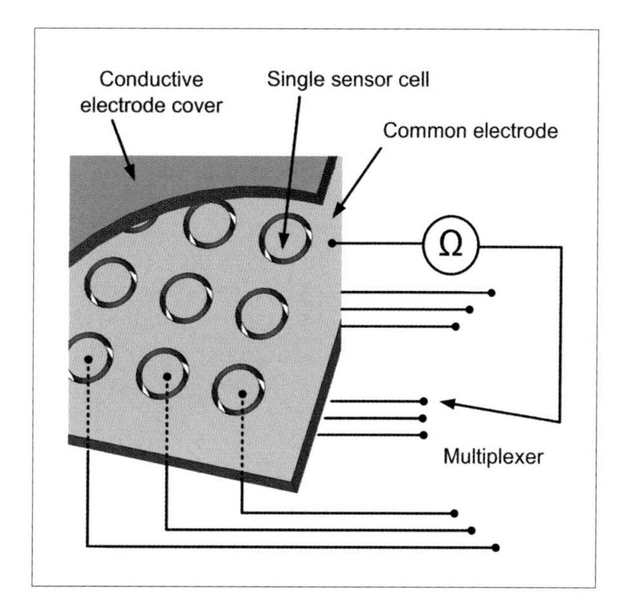

FIGURE 3. WORKING PRINCIPLE OF A RESISTIVE TACTILE SENSOR CELL

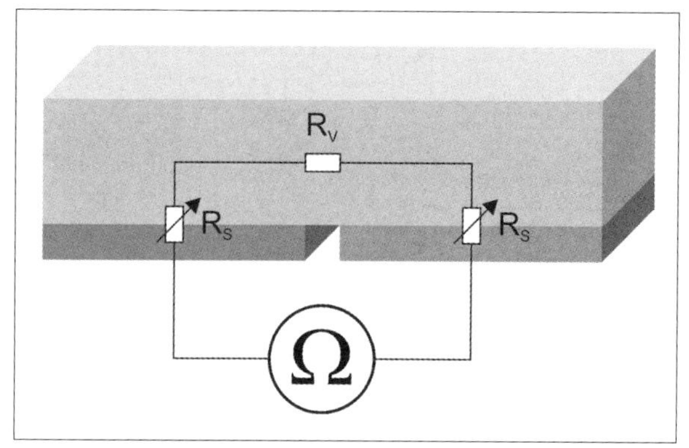

FIGURE 4. ELECTRICAL CIRCUIT OF A RESISTIVE SENSOR CELL

Sensor model

As described above, the sensor's working principle mainly depends on a surface effect between the metal electrodes and the carbon black filled polymer [12]. The measured resistance is a function of the contact area between the electrode and the foam. When no load is applied to the system, the foam has no or only few contact points with the electrodes, so the resistance of the cell is very high. When a load is applied to the sensor cell, a deformation of the foam occurs. The contact area increases and as result the resistance of the cell drops. Polymers show a very high undesired creeping effect when a constant pressure is applied. Of course this effect has a high impact on the measured signal; a constant load leads to a constantly decreasing resistance. As a consequence of this influence, the measurement system interprets this drop in resistance as an increasing load. Measurements show nearly a doubling of the initial value in 120 s at a load of 4 kg (Fig. 5, compares the reference signal with the raw sensor signal). To cope with this circumstance, we built a simple sensor model (Fig. 5) which can be computed on the data acquisition system in real time. To build a simple model, some assumptions and simplifications have to be made.

The first assumption is that the measured cell resistance R is only the surface resistance between the polymer foam and the sensor cell electrodes. In fact, this resistance includes the surface resistance between the common electrode, the sensing electrode and the volume resistance (RV) of the foam. The applied load has only a minor influence on the volume resistance. The influence of the volume resistance can be neglected, because RV is much lower than the surface resistances. This is achieved by keeping the gap between the common electrode and the sensing electrode small (0.2 mm for this sensor).

The resistance R is a function of the contact area AC between the polymer and the metal electrode. The measured surface resistance can be thought as a shunt circuit consisting of small discrete resistances. Their number increases with applied load, therefore R is a measure of the contacted area. The function between the contacted area AC and the strain ε depends on the structure of the foam surface, its mechani-

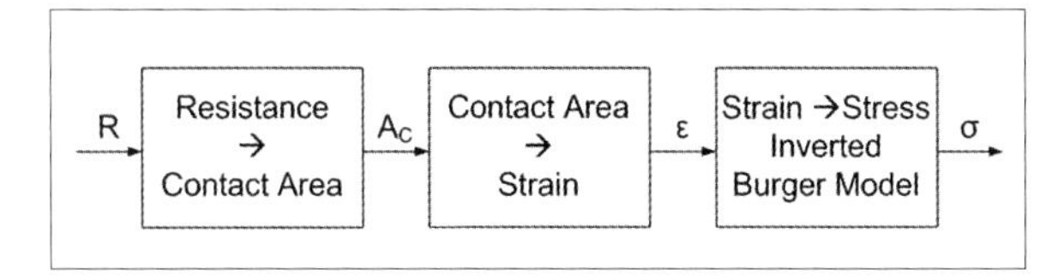

FIGURE 5. SIMPLIFIED INVERSE SENSOR MODEL

cal properties and variations in the electrode surfaces. Due to variations in behaviour between different sensors, the relationship between AC and R respectively R and ε must be acquired by measurement for every sensor.

The visco-elastic behaviour of polymers is often modelled using a sufficient number of elastic elements (springs, parameter R) and viscous elements (dashpots, parameter η). Assuming a linear visco-elastic behaviour the mechanical behaviour can be modelled as Burger model, a series connection of Maxwell and Voigt model. The Burger model is a four parameter model containing two springs and two dashpots. The Laplace transformation of the differential equation which characterises the Burger model can be expressed as

$$\sigma = \varepsilon \frac{(\lambda_1 S + 1) R_1 R_2 \lambda_2 S}{R_1 (\lambda_1 S + 1)(\lambda_2 S + 1) + R_2 \lambda_2 S}$$

with the relaxation time being $\lambda k = \eta k / Rk$ [13]. Knowing the strain and assuming an inverted Burger model G-1 to be valid the stress σ applied to the system can be computed. Additional assumptions are necessary for the model to be valid. First, it is assumed that the sensor material has not been loaded before. The second assumption is, that at the time $t = 0$ no load is applied to the system. This is true, if there is a sufficient time gap between different applications of load onto the sensor. Using simulation software to identify the parameters, it is not necessary to have the physical parameters of the foam.

The model parameters have been identified by minimising the squared error between the reference signal (Fig. 6) and the signal calculated by the model. The model works quite well in the desired working range. The difference between the raw sensor signal (Fig. 6, red) and the signal calculated by the model (green) is significant. The model's output nearly matches the reference signal (blue).

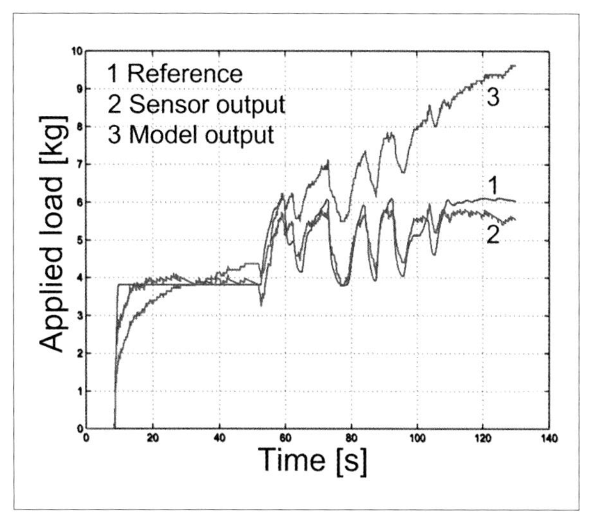

FIGURE 6. SIGNALS: REFERENCE SIGNAL (BLUE), RESULTING OUTPUT TACTILE SENSOR (RED), OUTPUT OF THE SENSOR CELL MODEL (GREEN)

Applications

The tactile sensors presented above are used on a humanoid robot system within the German Humanoid project [14, 15]. Here, the robot is equipped with tactile sensor pads on its upper arms forearms, and shoulders as can be seen in Figure 7.

The robot is used in a typical human everyday environment: home and kitchen. It can either act autonomously or interact with a person. As it is intended to enable intuitive interactions between human and robot, the robotic system uses tactile and haptic information in various forms and for different purposes, which are described in detail in the following paragraphs.

FIGURE 7. THE ROBOT ARMAR IIIA EQUIPPED WITH TACTILE SENSORS

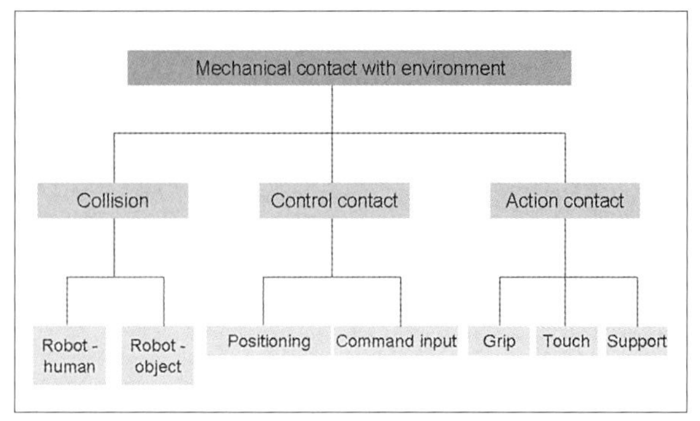

FIGURE 8. DIFFERENT CLASSES AND SUB-CLASSES OF CONTACTS

Classification of haptic sensation

One model of artificial haptic perception is the human sense of touch. It is often considered to only consist of tactile sensing of the skin, but proprioception or depth sensibility also forms an integral part of haptic perception. It includes the senses of force, position and movement of the different parts of the human body. A classification of a detected contact between the robot and its environment can be derived from the acquired tactile sensor data, the acquired proprioperceptive data, and additional context information (i.e., information about the task the robot is performing, the robot's state and the environment).

In the following classifications a sensor system comprising surface and depth sensibility of a robot is considered. Surface sensibility is achieved by using tactile sensors in foil design applied to the robot arm and hand with different spatial resolution. The palm of the robot hand and the finger tips are supplied with sensors at relevant points to support hand control. Depth sensibility is realised by a force torque sensor in the wrist.

The main purpose is the classification of contacts according to the quality of intention. There are three classes of contacts which cover all possibilities of interaction between a robot and its environment: collisions, control contacts and task contacts (Fig. 8). These main classes can be split into generic subclasses, which can be partly divided again in specific contact patterns, e.g., the sub-class *grip* can be divided in a taxonomy proposed by Cutkovsky and Wright [16] with the help of related contact patterns.

Collisions can be described as contacts not intended at all, neither by the robot nor by its environment. The distinction into collisions of the robot with humans or objects is useful if the contacts cause different reactions of the robotic system.

Control contacts are initiated by the robot's environment with the purpose to give information or commands to the robot. *Positioning* is an intuitive interaction to control a robot by applying contacts and forces. A person just grasps the forearm of the robot and guides it to a desired position. Another example for *control contacts* are tactile signals performed on a sensor pad in

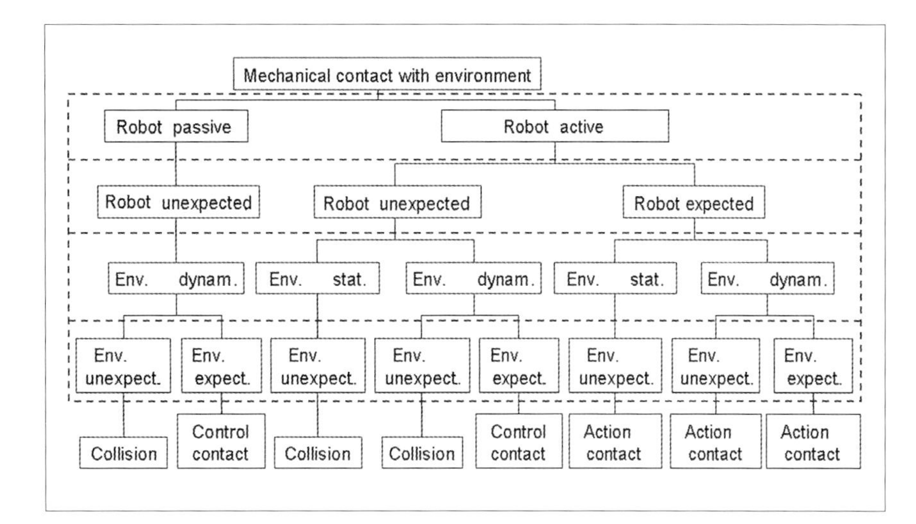

FIGURE 9. DISTINCTION OF CONTACT CLASSES ON BASIS OF PARAMETERS FOR MODE OF OPERATION, EXPECTATION, STATE AND INTENTION OF ENVIRONMENT

order to communicate commands to the robot, i.e., in noisy environments or just to arouse attention.

Task contacts are contacts that are related to the actual task of the robot and therefore they are expected and mainly initiated by the robot. In the case of *grasps* these are specific contact patterns for each particular grasp. *Exploration contacts* are initiated by the robot and expected in a certain area of its body, e.g., a fingertip. They are also contact patterns a robot detects when exploring the shape of an unknown object with the help of its tactile sensors on its hand.

Support describes contacts, which are expected as part of a support task, e.g., the robot supporting a handicapped person with its arm.

A formal description for classifying contacts between the robot and its environment involves the following parameters:

- *Robot's mode of operation:* This mode specifies whether the robot has to perform a task or not (it is active or passive). If the robot is passive it is in stand-by-mode and is waiting for an input ('idle mode'). If it is active, it has to perform a task with a related contact pattern; *no contact* (Fig. 9) also represents a contact pattern, e.g., for the task 'move from point A to point B'.
- *Expectation of robot:* The robot's expectation describes whether the perceived contact is expected or unexpected. If the robot is active a contact is expected if it is matching the related contact pattern; if not, it is unexpected. For this purpose the contact classification unit of the robot control has to retrieve the appropriate contact patterns from a knowledge base. If the robot is passive, all contacts are unexpected.
- *State of environment:* The state of the environment defines whether there are autonomously moving objects (humans or other robots) in the workspace of the robot (dynamical) or not (static). If this information is not available, the default status of the environment has to be dynamical for safety reasons.
- *Intention of environment:* A contact can be intended or unintended by the environment.

As can be seen in Figure 9, the three types of contact – *collisions*, *control contacts* and *task contacts* – can be distinguished on basis of the four parameters *operation mode*, *expectation*, *state of the environment* and *intention*. For further classification characteristics of sensor signals have to be used, i.e., to describe specific patterns of tactile sensor data. If no additional information is available to distinguish between *collision* and

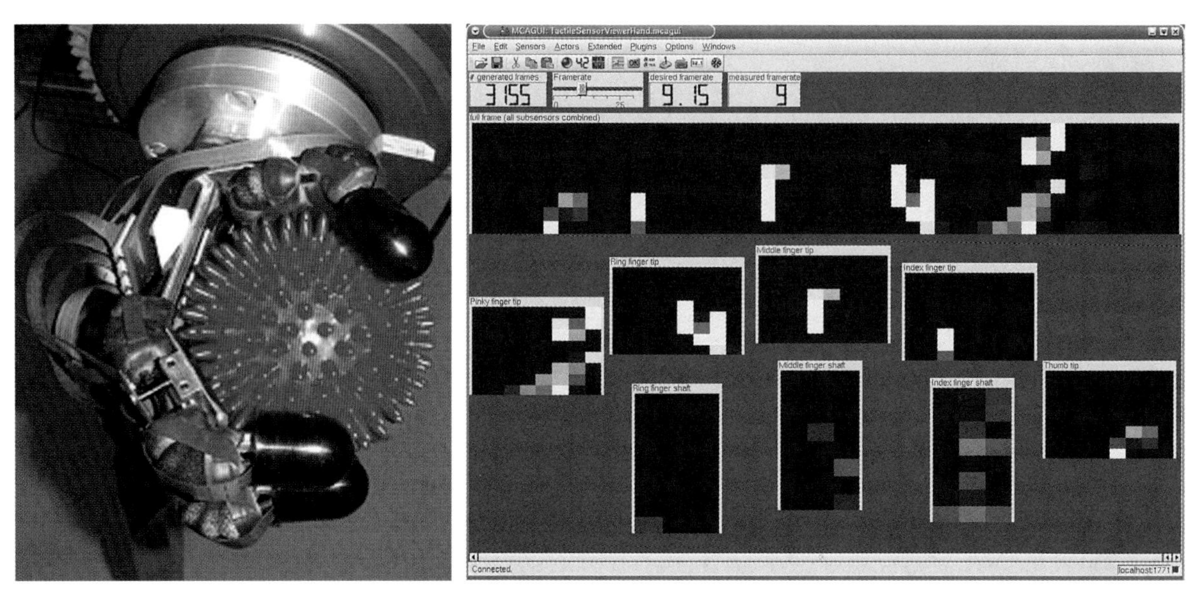

FIGURE 10. TACTILE PATTERN OF EIGHT TACTILE SENSOR MATRICES IN AN ANTHROPOMORPHIC ROBOT HAND GRASPING A PRICKLY BALL

control contact, measured acceleration, forces in joints and the probability of the sensed tactile pattern (Fig. 10) can serve these means.

Classification of haptic and tactile co-operation

Haptic co-operation between a person and a robotic system is characterised by a direct physical coupling between both partners which can either use inputs by force-toque-sensors of by tactile sensor pads. Based on measured data of the applied forces and resulting torques a reactive behaviour of the robot is computed and performed.

Haptic and tactile co-operations can be further distinguished: first of all, robot and person can touch each other directly (person grasps robot arm) or can be coupled by an object. In case of direct and intentional coupling, each of the co-operation partners can either guide or support the other, or limit the movements of the other partner. If the co-operation is based on a rigid kinematic chain one partner firmly grasps the other in order to manually guide the partner to a desired position. In case of a loose kinematic chain the coupling can easily be ended by a quickly moving or twisting partner. Examples for this specific type of haptic/tactile co-operation are a person touching (not grasping) a robot's arm at the forearm or upper arm in order to correct the performance of a movement, if the robot's arm is equipped with an artificial skin (tactile sensor matrices). *Vice versa* a robot can support a handicapped person with its arm without grasping the person.

If an object is used to physically connect human being and robot, the object can either be just passed from one co-operation partner to the other or both partners manipulate the object. In the first case the robot requires haptic feedback in order to detect, if a person has grasped the object, is tucking at it and the robot can safely release the object out of its hand. The same applies if the robot passes the object onto a person's hand without the person grasping the object. Here again, haptic feedback is required to detect a resistance, i.e., by the object reaching the person's outstretched hand. *Vice versa* the robot needs tactile information in order to detect, if it has grasped an object presented or thrown by a person. Additionally, haptic data from a force-torque-sensor in the robot's wrist serve as information, that a person has withdrawn his or her hand on which the object grasped by the robot was presented.

Robot and human being can co-operatively manipulate an object in an aligned, complementary or independent manner. They can, i.e., push or pull an object into the same direction or open a bottle with each co-operation partner turning the lid or bottle into the opposite direction of the other. An example for an independent co-operation manner using haptics and vision is a robot filling a glass placed on a tray which is carried by a person with a jug of water.

All the above mentioned haptic couplings are typically employed in household scenarios as performed within the German Humanoid Project. Some couplings can be found in industry as well, i.e., if a robot is carrying a heavy load with the human partner manually guiding the robot and fitting the object in a predefined place.

Summary

Humanoid robotic systems require a sense of touch as well as haptic depth sensibility in order to move, work and manipulate in typical human everyday environments. As haptic depth sensibility is achieved with the help of force-torque-sensors this article highlighted tactile sensor principles with an emphasis on resistive force sensors for a robot's sense of touch. An artificial skin can be composed of individual pads of matrices of tactile sensors. The sensors illustrated in this chapter are resistive force sensors designed at the Institute of Process Control and Robotics at the University of Karlsruhe. They do not just measure whether there was or was not a contact between the robot and the environment, but they also supply the actual area of the sensor, which was touched (position information), as well as the intensity of the detected pressure at each sensor cell. The described working principle of the sensor mainly depends on a surface effect between the metal electrodes and the carbon black filled polymer. The measured resistivity is a function of the contact area between the electrode and the foam. When no load is applied to the system, the foam has no or only few contact points with the electrodes, thus the resistivity of the cell is very high.

Technical requirements are just one aspect of haptic perception in human robotic systems. Thus the different kinds of contacts between a human being and a humanoid robot equipped with tactile sensor matrices were presented. If the type of contact is to be automatically analysed by a cognitive robotic system for planning an appropriate reaction various parameters like the robot's mode of operation and the robot's expectation as well as the state and the intention of the environment have to be taken into account. Additionally, contact patterns are useful to distinguish and classify contacts between robot and environment.

As haptic perception of human robot systems involves both tactile perception and haptic depth perception, different forms of co-operation between a person and a robot using either force-torque sensors or tactile sensor pads were highlighted. Actually, both surface and depth perception are essential if a person intuitively interacts and co-operates with a humanoid robot in a typical everyday environment. A naïve person is not aware of the wide range of applications requiring haptic perception in robots; this chapter should have illustrated to the reader the vast variety of haptic perception in a human robot system.

Selected readings

Burghart C, Haeussling R (2005) *EvaluProceedings of the AISB'05 Conference, Workshop on Robot Companions*, Hatfield, Hertfordshire, England, April 2005

Yigit S, Burghart C, Wörn H (2004) *Co-operative carrying using pump-like constraints. IEEE-RAS Interna-*

tional Conference on Intelligent Robots and Systems, Sendai, Japan, September 2004

Kerpa O, Yigit S, Osswald D, Burghart C, Wörn H (2003) *Arm-hand-control by tactile sensing for human robot co-operation. Humanoids 03*, October 2003, Karlsruhe

Göger D, Weiß K, Burghart C, Wörn H (2006) *Sensitive skin for a humanoid robot. HCRS'06*, München, October 2006

Burghart C (2006) Human-robot-cooperation. In: W Luther and D Soeffker (eds) In: *Guidance and control of autonomous systems*, Logos-Verlag, Berlin

Recommended websites

SFB 588, German Humanoid Project, Homepage. 16.1.2008 http://www.sfb588.uni-karlsruhe.de/

Institute of Process Control and Robotics Homepage. 16.1.2008 http://wwwipr.ira.uka.de

VI.
Applications

Haptic design of vehicle interiors at AUDI

Werner Tietz

AUDI attaches great importance to the haptic design of its vehicles. Haptic design represents an essential factor in meeting the high expectations of AUDI customers. A team has been set up to assess the haptic impressions created by vehicle interiors. The team includes representatives from technical divisions who form the interface with the development departments concerned. The remainder of the team is composed of members drawn from all areas of the company who, quite deliberately, have no direct link with the technical environment concerned. By means of their own assessments and using representative customer surveys, this team produces findings that give a broad and objective reflection of how the vehicle haptics are perceived by users.

The technical departments concerned have the task of implementing the conclusions of these assessments in design and engineering, principally in surface contours and 'haptic' material characteristics. This is done by the development of measuring and testing processes, the use of special materials, such as coating the air vent operating wheels with a rubber-like anti-slip material, the design of components and materials, and the positioning of components on the basis of own and surveyed experiences. Three examples of areas of vehicle interior development in which these methods are applied are described below.

Seat comfort

A range of parameters affect the perception of how comfortable a seat is [1] (Fig. 1). Alongside the purely subjective element, which is

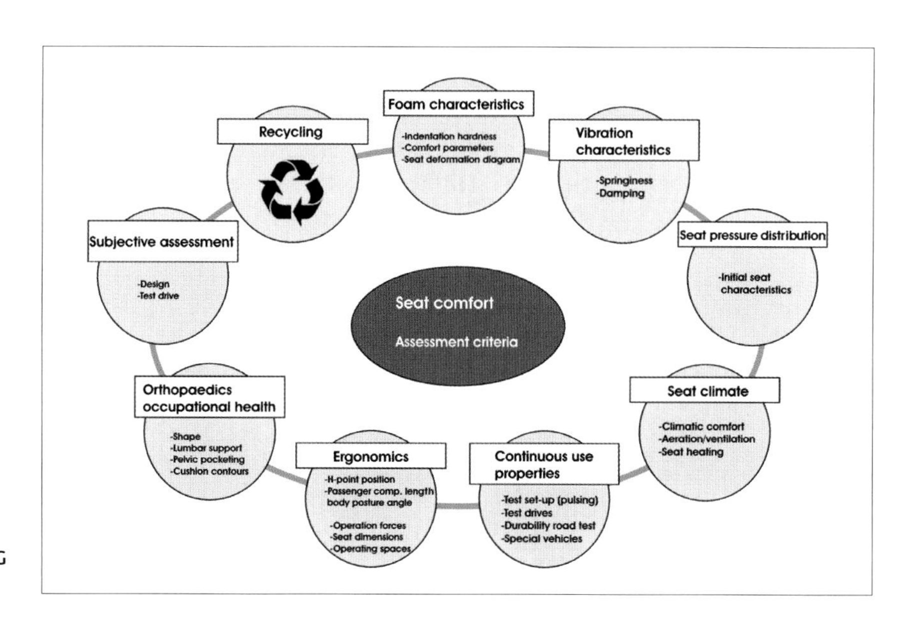

FIGURE 1. FACTORS AFFECTING SEAT COMFORT

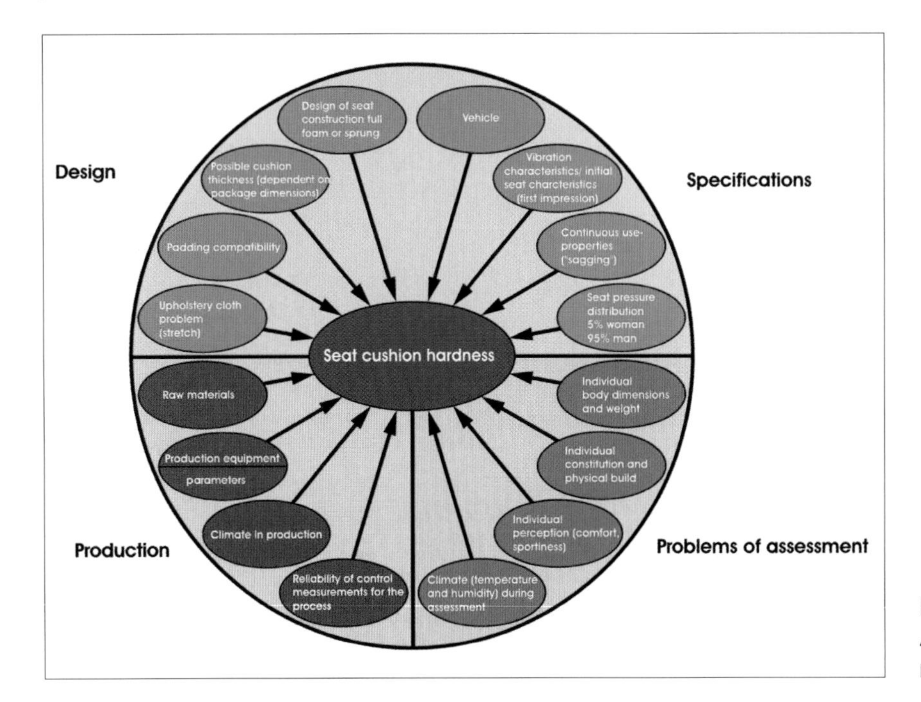

FIGURE 2. FACTORS AFFECTING SEAT CUSHION HARDNESS

often largely determined by visual impressions (design), there are measurable quantities. The features of an optimum seat design in terms of occupational health criteria have now been established to a large extent and these determine the principal parameters, such as the shape of the backrest and the lumbar support function. The ergonomics are determined by the optimum seat position in the available interior package. In this respect, the requirement to achieve a suitable level of comfort for a given vehicle class will frequently impose restrictions on the design. The continuous use characteristics of the interior are dependent on the long-term properties of the materials used. This aspect is crucial to the customer's long-term satisfaction with the vehicle and thus has an important impact on future buying decisions. The seat climate is a quantity that can be measured and compared, and is especially important for seat comfort over long journeys. Thanks to the use of new materials, major progress has been achieved in this area with regard to permeability and air exchange.

The principal seat comfort properties are, however, determined by the hardness and hardness distribution of the seat cushion. This includes the factors seat pressure distribution, vibration characteristics and the associated physical foam characteristics. The 'optimum' seat cushion hardness depends on technical conditions associated with the vehicle development and production and on the specific assessment criteria (Fig. 2).

There are criteria that are strongly dependent on the vehicle concerned and the vehicle manufacturer's specifications, and can be described directly or indirectly in the form of measured values. These parameters may vary greatly in accordance with the vehicle concept. A 'sporty' seat requires a different design to a 'comfortable' seat. Moreover, there are also conditions that are determined by the manufacturing location and the available production equipment. As seats are mainly manufactured for AUDI by system suppliers, these parameters are dependent on the selected supplier. In addition, there will be

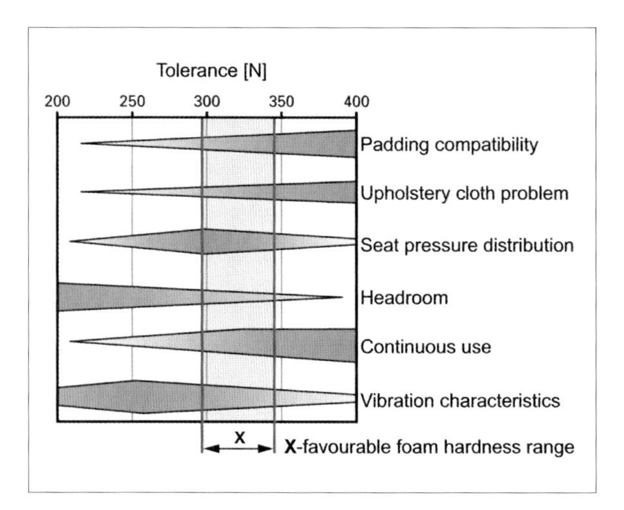

FIGURE 3. TOLERANCE FIELDS FOR SEAT CUSHION HARDNESS

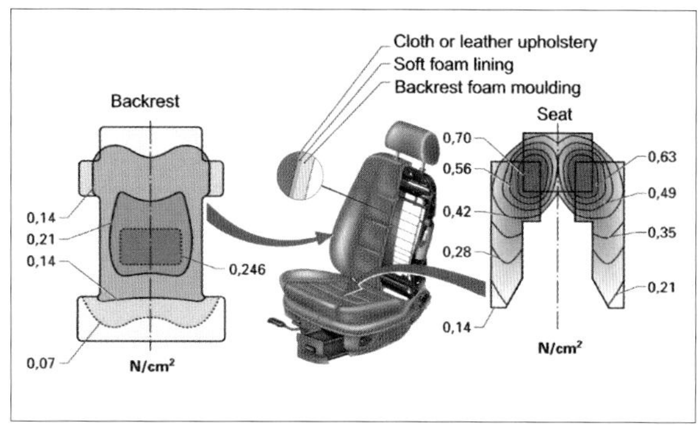

FIGURE 4. SEAT PRESSURE DISTRIBUTION (EXAMPLE)

the specific aspects of the individual case. These parameters can only be assessed by means of the surveys and comparative tests described above.

The interdependency of the parameters and their, in some cases, opposing effects on the considered property of 'seat cushion hardness' are illustrated in the example below (Fig. 3).

- A high foam hardness has the advantage that it provides a good padding compatibility. A hard foam body is easier to line with padding material.
- Critical upholstery cloths with differing stretch values in different directions also require a high foam hardness. The low padding deformation prevents anisotropic straining of the cloth.
- However, an optimum seat pressure distribution requires a specific foam hardness to achieve the ideal support for the posterior and back (Fig. 4).
- A hard seat cushion is disadvantageous with regard to headroom. As the occupant only sinks a small way into the seat, the result is less headroom.
- The continuous use properties, i.e., the preservation of the seat foam's initial properties, are also directly dependent on the foam hardness. As a rule, a comparatively hard foam has a higher fatigue strength.
- The characteristics of vibration affecting vehicle occupants also require a specific seat cushion hardness to provide the optimum damping effect on vibrations transmitted from the vehicle to the seat and to prevent occupant vibration.

The challenge facing the developer is create the optimum seat for the vehicle concerned from the large number of parameters outlined above.

Function components

At AUDI the term function components refers to moving interior units such as the glove compartment, ashtrays, drink holders, stowage compartments, etc. A large number of function components of this kind are provided for customers in modern vehicles. The vehicle forms a suitably equipped living environment. This requires these components to be manufactured to a high level of sophistication. Thus, for instance, ashtrays, drinks holders, stowage features etc., that are contained in closing compartments in AUDI vehicles are equipped with special damping in their opening mechanisms. This design ensures that the components open quietly and with an even

Component: front ashtray
(open/closing)

Subjective Assessment

Scorecard

Variant	Noise	perceived mechanical action	Force	Score
I	+	0	0	3
II	+	+	+	2
III	+	+	+	2
IV	0	-	-	4
V	0	--	0	4

FIGURE 5. AN EXAMPLE OF VARIANT ASSESSMENT – ASHTRAY DRAWER

movement. The mechanisms have to fulfil specific requirements to create opening and closing actions that feel pleasant for the customer.

Suitable function units are selected on the basis of the customer's view. To achieve this, various versions of a component function are created in the specified installation space with different opening movements, forces and kinematics. The variants are subjectively analysed through comparison by a representative group of people using quality-based assessment criteria (Fig. 5).

In the example shown, five variants of an ashtray drawer were tested. A quality assessment was made based on the criteria noise, perceived mechanical action and force:

- *Noise:* Noise refers to the operation noise, i.e., the acoustic feedback from a closing movement. Depending on the range of noise produced, this acoustic feedback may be perceived as positive or negative. In addition, acoustic feedback frequently leads to a haptic perception. A scraping noise, for instance, is automatically associated with an unsmoothness in the operating force even if no unsmoothness can be measured. Therefore, noises should only be used to produce specific effects. Thus a defined click noise at the beginning of the operation of a component in combination with haptic feedback in the form of a resistance point, for example, can produce an overall high-quality operating feel.
- *Perceived mechanical action:* Perceived mechanical action is understood to mean the running properties of a mechanism when operated. The aspect being assessed is whether it is possible to 'feel' the functioning of the individual components, such as the meshing of toothed parts or the working of a kinematic mechanism in response to the operating force.
- *Force:* This refers to the level of force applied to operate the component. An important aspect is whether all possible users can easily apply the required force when the function component is at a given position.

Measurements are conducted on the assessed sample function components. The measurements are then used to generate quantitative specifications for the production part to be developed. One example of the findings from this kind of assessment is that a fluctuation in the operating force of a drawer (e.g., for an ashtray) is perceived as unpleasant or gives an impression of inferior quality if it exceeds a certain level (Fig. 6). The figure shows the variation in force over the operation travel. The force is first increased up to the operating force level (in this case approximately 9 N) and then the drawer can be moved further with a constant force. This is achieved

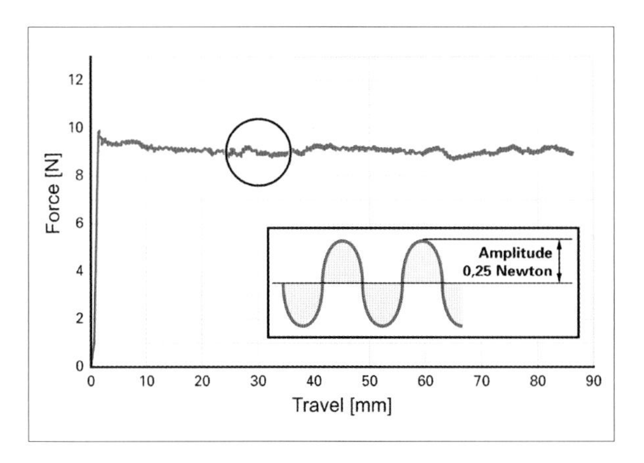

FIGURE 6. OPERATING FORCE CURVE (EXAMPLE)

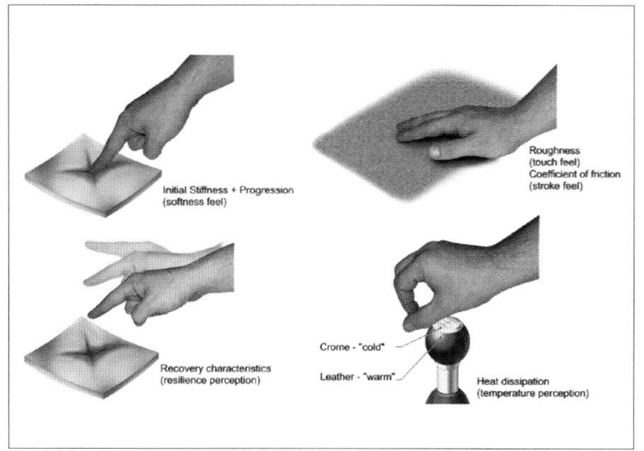

FIGURE 7. THE HAPTIC IMPRESSION

using a special design principle. The unsmoothness created by friction and the toothed parts of the synchronising guide can be clearly seen in the measurement. If the amplitudes exceed the defined limit (in this case 0.25 N), the unsmoothness becomes noticeable.

Variety is set to increase in terms of the design and manufacture of function components plus the range of components and the scope of items incorporated into vehicles. It is expected that this will be accompanied by a continuous increase in operating comfort. Customers buying vehicles in the upper segment today expect similar functionalities to those they find in their hi-fi systems.

Surfaces

In addition to the functionality and visual appeal of controls and their surfaces, customers expect interior surfaces to be pleasant to touch. Here again we face the problem of translating a subjective, normally comparative assessment of surfaces into measurable quantities that will provide the basis for implementing the required haptic qualities in technical development. This has been tackled at AUDI by examining various parameters on the basis of subjective assessments, leading to the development of a scoring scheme (Fig. 7). According to this scheme, the 'haptic impression' is characterised by four areas:

- Softness feel
- Surface resilience
- Touch/stroke feel
- Temperature perception

The 'softness feel' can be described physically as the pressure resistance of the surface as perceived when pressing down on it with one's fingertip. This can be defined in measurable quantities in terms of the *initial stiffness* and the *progression*, i.e., the force increase curve.

The 'surface resilience' is described by the surface's recovery characteristics upon unloading. If, for instance, the surface exhibits slow recovery characteristics, an impression initially remains visible on the surface after unloading. The surface does not follow the fingertip. The recovery characteristics are tested in a loading – unloading test. The curve of force against movement may yield a strong (low recovery rate) or a slight (high recovery rate) hysteresis.

The touch/stroke characteristics are mainly determined by the surface texture. Roughness is thus felt by touching a surface with your fingertip. By moving his/her hand over the surface, the tester assesses the feeling of friction between the

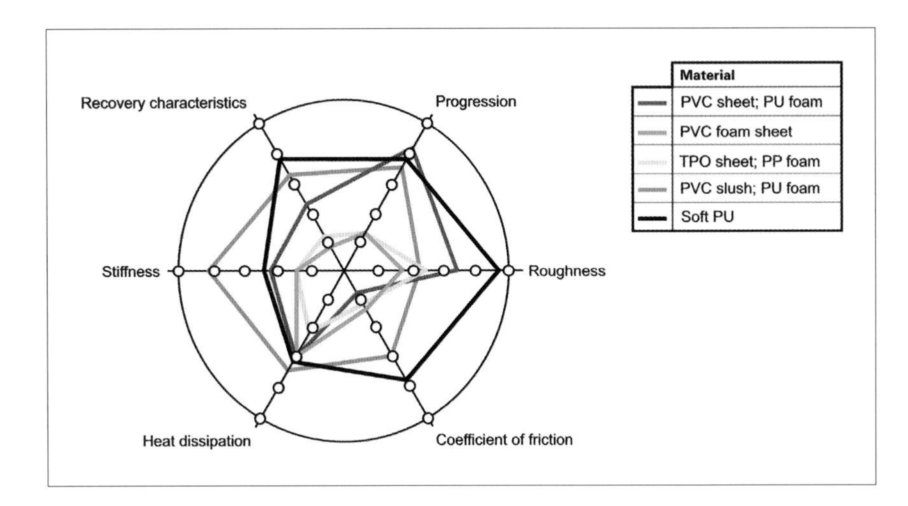

FIGURE 8. ASSESSMENT OF HAPTIC IMPRESSION BY MEASUREMENT

surface and his/her skin. Although it is possible to measure the quantities described above using suitable equipment, considerable deviations are found in the correlation between subjective perception and actual measurements.

The temperature perception reflects whether a surface feels 'warm' or 'cold'. The main factor in this is the surface's heat dissipation. This quantity is relative, i.e., it can be effectively recorded by comparative analysis. This is done by analysing the curve of the junction temperature of a medium with low heat capacity when touching the test surface. The overall analysis of the measured quantities described above is based on a modified version of existing models (ECIA [2]) and allows their representation in a six-parameter diagram (Fig. 8). This polar diagram shows the physical quantities in standardised form. A curve thus represents the 'haptic fingerprint' of a specific surface.

The method described produces an unambiguous overall characterisation of a surface. This serves as a basis for the comparison of different surfaces, which can thus provide clear objectives with regard to the surface development. This can help, for instance, in ensuring that adjacent surfaces have similar haptic qualities. In addition, a specific haptic impression may be enhanced using individual parameters. One development objective, for example, is to achieve uniformity in the softness of armrest surfaces, which can be achieved by specific alteration of the stiffness parameters.

The examples described here serve to give an impression of the complexity of the concept of 'haptics' for development of vehicle trim and equipment. With regard to the increasing model variety and ever shorter development times, the challenge of devoting suitable attention to haptics can only be successfully met using systems such as the parametrical definitions described above coupled with a forward-looking vehicle concept development.

Visual-haptic interfaces in car design at BMW

Alec Bernstein, Bernd Bader, Klaus Bengler and Hermann Künzner

General background

The increasing complexity of information in the car brings is aligned with an increase of the amount of features and a visual overload of knobs, buttons and physical devices. This complexity of features must be balanced against the driver's need for simplicity, ease of use and safety. Thus there is need of a tool that would free the driver's attention for driving, while facilitating information access in the car.

Some current automotive solutions that manage information reduce the clutter of knobs and buttons by creating a single on-screen solution. In this paper, we argue that the centralization of information is not the ultimate solution, but the right step towards an intuitive system which marries the 'feel' of information with its look.

Another factor frequently referred to in BMW's user observation studies is the existence of driving 'modes' based on conditions such as traffic, weather, passengers, etc. The actual mode of the driver highly affects the driver's window of comprehension, and with it, the ability to handle information complexity. Considering the limitations on the driver's attention span, it is necessary to create an interface which takes these modes into account. Again, the need is to reduce complexity and increase intuitiveness.

Below we give a detailed description of the development of visual-haptic interfaces in the course of time. The concept study of the Z9 is being dealt with in detail and the use of visual-haptics in various Series systems is specified by means of concrete examples.

Based on an extended notion of haptics the following article emphasizes the visual-haptic alignment as a new dimension in car-driver interface design. Advantages of aligning the two senses rather than treating them as two separate systems will be exemplified through the Turn-Push-Knob (Dreh-Drueck-Knopf) input device and a parallel information design on the screen. This system was first shown in the BMW Z9 show car (1999), has been optimized and implemented in the BMW 5 and 7 Series and developed further in research projects.

BMW Z9 show car (1999)

The Z9 user interface supports visual-haptic matching wherever possible. The physical action of the input device (for example, turning a knob left or right) is mirrored on-screen through the use of animations (rotating graphics). Pressing the 'edges' of the input device (top, bottom, left, right) activates information displayed on the corresponding edges of the screen ('direct mapping'). Finally, tactile feedback in the input device mirrors discrete steps or choices through the sense of touch. For example, when the driver needs to make a choice from a number of items, the input device has a discrete 'detent' or feeling like it is snapping into position for each item of choice.

Features like edge-driven access and tactile feedback support the driver in building a mental construct of the information space. This can result in a 'haptic memory' of the system, making it easy to use without even having to look. For example, if you know that climate controls are accessible by activating the bottom edge of the screen, you are a button-press away from your goal rather than having to actively read and decode lists of information.

The BMW approach to a visual-haptic interface

At BMW we asked ourselves how to resolve the issues of car-based information complexity in a manner consistent with the BMW brand. How does the BMW gestalt of driving dynamics, performance, handling, comfort, safety and joy translate to the use of information? How does the driver's overall haptic perception relate to the sense of well-being in the car? The answer was in the relationship of a physical input device and the design of information on a central screen.

The relationship between the visual and the haptic interface is an intricate one. It is not so much a matter of simply combining the two systems, but rather a question of appropriately aligning them. When there is a direct relationship between what you are feeling or doing and what you are seeing in response, the window of comprehension is opened. In fact, in visual-haptic matching studies, it has been shown that a perceptual inconsistency in visual and haptic stimuli results in a poorer performance [1].

Premises for design solutions

Through the careful examination of these complex issues, BMW developed the following premises as guidelines for design:

- Haptic-visual alignment reduces the energy in the action(s), improving safety, response time, and comfort for the driver.
- If the visual model does not align with the physical haptic action, two mental models must be decoded.
- If the visual model does align with the physical haptic action, only one mental model must be decoded.

Visual-haptic matching is critical for the success of the system.

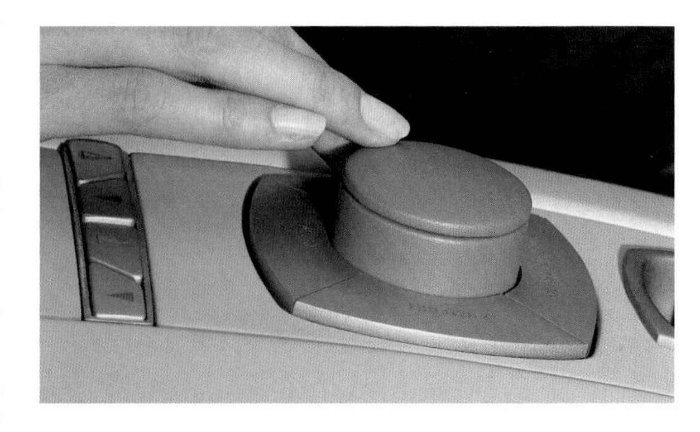

FIGURE 1. INPUT DEVICE IN BMW Z9 SHOW CAR: DDK AND 4-BUTTON ARRAY

The design solution: the input device, information design and their alignment

In the development of BMW's visual-haptic interface solution, three levels of analysis had to be considered and integrated:

- physical interaction with the haptic input device
- the graphic equivalent (on-screen information 'look' which maps to haptic 'feel')
- the visual information structure and its semantic (meaning, message)

In the following, we will describe these three levels and their interaction.

The Z9 haptic input device

The development of the Turn-Push-Knob (Dreh-Drueck-Knopf) input device (a tool which takes the importance of the haptic sense into consideration) has a long history in information ergonomics at BMW.

In the Z9 show car, this input device (consisting of a single knob which can be turned and

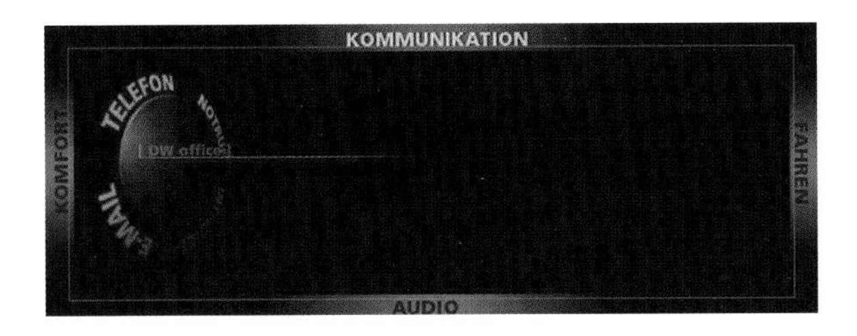

FIGURE 2. 'EDGE-DRIVEN' GRAPHIC SYSTEM IN BMW Z9 SHOW CAR CORRESPONDS TO FOUR-BUTTON ARRAY ON INPUT DEVICE

FIGURE 3. CONCENTRIC ELLIPSES REPRESENT CONTROL HIERARCHY IN THE Z9 INFORMATION SYSTEM

pushed) was extended to support an 'edge-driven' graphic interface. An array of four buttons placed around the Turn-Push-Knob gives the opportunity to have four direct access points to the system's information (Fig. 1).

The four buttons surrounding the knob are dedicated to the four main categories of information in the system: communication, driving, audio and comfort. These four buttons are not only labeled, but arranged to align with their graphic counterparts on the screen, located centrally in the car's dashboard. For example, the button *above* the Turn-Push-Knob corresponds to the top edge of the screen, the left button with the left edge of the screen, etc. (Fig. 2).

Note that these four buttons are usable at any time, no matter where the user is in the system. This direct access supports haptic memory construction, mentioned above.

The Z9 graphic equivalent: mapping visual systems to haptic actions

In the Z9 information system, the user must not only have access to the four main information categories, but also be able to browse sub-topics within these categories and select choices. We have already discussed how the four-button array on the input device is mirrored in the edges of the graphic system. Now let us discuss how the DDK relates to information access.

The input device supports browsing through choices *via turning* the knob, and selecting choices by *pushing* the knob. Therefore, much of the system navigation is controlled by simple combinations of turning and pushing. As previously mentioned, tactile feedback while turning the knob supports a sense of moving through discrete choices. A different mode of tactile feedback is provided when selecting a choice – creating a fast way of differentiating between, through the sense of touch, the actions of browsing and selecting.

Graphically, the action of browsing choices takes on the motion, through animated sequences, of the input device. Turning the DDK initiates a graphic 'turning' or animation of choices around what we call a 'control-ellipse'. Choices that affect the control or navigation of the system are arranged in an elliptical layout (thus control-ellipse) to most closely approximate a visual representation of the haptic action.

Further, because the information system handles many layers of complex information, there must be a graphic device to represent hierarchical arrangements of information. For example, if the driver wants to place a phone call to someone in their list of favorite ten callers, they must first choose 'Telephone', and then select from a secondary list of choices including a 'Top Ten' list.

Graphically, when the choice 'Telephone' is selected (by pushing on the DDK), the first level choices zoom toward the center of the control-ellipse, and are replaced by the second

FIGURE 4. INFORMATION DESIGN APPROPRIATENESS FOR INFORMATION TYPE

level choices around the outside of the ellipse (Fig. 3).

The first level choices remain visible, though blurred. This graphic method supports the notion of movement to a new level of information by '*pushing in*', while keeping a visual cue that the first level is still available.

The Z9 information design semantic

The structural design of information in the Z9 show car addresses the need for intuitiveness and comfort in a very complex information space. We discuss 'information design' as a concept which differs from its graphic and haptic manifestations, because it is the vehicle through which the two relate. It is the system which organizes how functions and features are grouped; it is the system that supports the need to handle many different information types (e.g., lists, toggles of on/off information, status displays, navigation choices, etc.). The information design must balance and integrate the haptic and the visual.

As previously discussed, the relationship of the visual and haptic systems support an intuitiveness when using the system. In order to extend this mental construct for the user, certain information design constructs were put in place.

One is the notion of a 'control/data' split. Graphically, the system separates choices that are for control (elliptical layout) and those that represent the bottom-level data (list form.) This difference in representation is made so that users know intuitively when they are choosing to control the navigation, and when they are making a choice that will make a change in their environment. For example, choosing to call Mr. Muller from your Top Ten list (see Fig. 3 above) would require selecting from a list instead of an elliptical layout.

Another information design premise is that bottom-level data varies greatly in its type. Some is list-oriented, but other types of information are much better represented using illustrations or schematics. We adhere to the principal of using the correct representation for data types, rather than requiring a single system at this level (Fig. 4).

Series concept since 2001 (BMW 7 Series)

The findings resulting from the concept study of the Z9 described above have been included in the development and design of the BMW 7 Series and can be experienced in these vehicles beginning with 2001.

A serialized Turn-Push-Knob has been implemented to operate the driver information systems. This knob can be turned, pushed and slided in eight directions. In addition there is a menu button and a programmable button available (Fig. 5). By pushing the menu button the user can easily access the main menu. The function of the programmable button can be freely chosen and allows a short-cut to entertainment sources, traffic information or brief operation instructions.

While in the Z9 concept study the sub-menus were reached *via* separate buttons with the BMW 7 Series this is accomplished by sliding the Turn-Push-Knob. The eight possible sliding directions comply with the eight menu items on the display (Fig. 6). They are made evident by means of eight arrows. So, if the knob is slid to the left, the climate menu is opened. Before that, however, an intermediate screen shows which menu is going to be accessed (see Figs 7 and 8). The color code

FIGURE 5. 7 SERIES CENTER CONSOLE

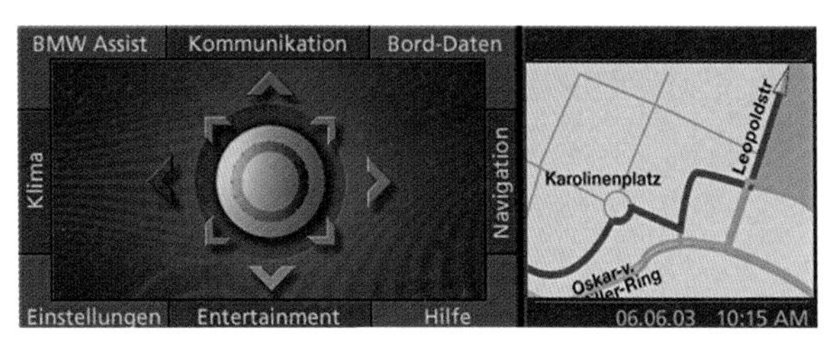

FIGURE 6. 7 SERIES MAIN MENU

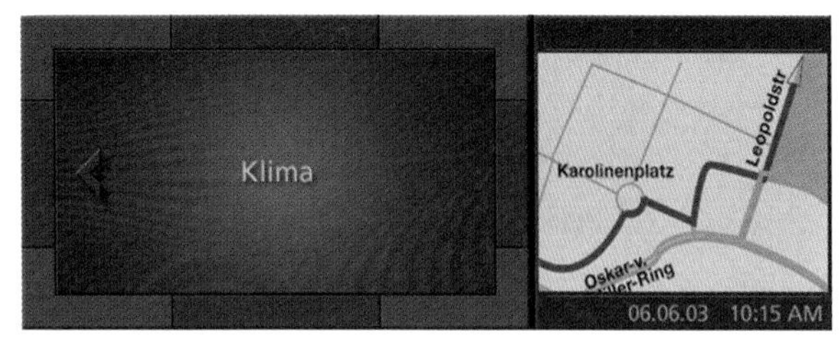

FIGURE 7. INTERMEDIATE SCREEN TO 7 SERIES CLIMATE MENU

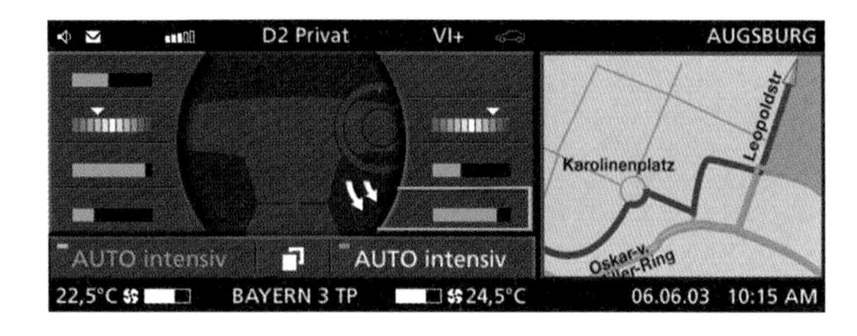

FIGURE 8. 7 SERIES CLIMATE MENU

of the arrows in the main menu (Climate red, Navigation green, a.s.o.) is maintained in the sub-menus, where it becomes the background color of the screen.

Series concept since 2003 (BMW 5 Series)

To operate the driver information systems the BMW 5 Series uses the same Turn-Push-Knob as the one in the 7 Series. Beside the Turn-Push-Knob there is a menu button for immediate access to the main menu and a button to activate the voice control system.

When compared to the BMW 7 Series the number of sliding directions with the BMW 5 Series has been reduced from eight to four. To maintain a connection between the Turn-Push-Knob knob and the main menu, the Turn-Push-Knob is represented together with the indicating arrows on the main screen. Attention has been paid to maintain the already known position and color of the sub-menus (Climate red, Navigation green, a.s.o.) (Fig. 10).

If the Turn-Push-Knob, starting from the main menu, is slid down, the user enters the Entertainment Menu. Here a miniature Turn-Push-Knob symbol in the upper right hand corner indicates the four possible slide directions as well as the menu active for the time being (Fig. 11).

Within a sub-menu two directions of movement are possible. By rotating the Turn-Push-Knob the vertically arranged items of the sub-menu are highlighted successively (DVD, CD1, CD2 a.s.o.) By sliding the Turn-Push-Knob up or down the system switches between sub-menu levels (Fig. 12).

In addition there is the possibility to directly switch from one sub-menu to another one by

FIGURE 9. 5 SERIES CENTER CONSOLE

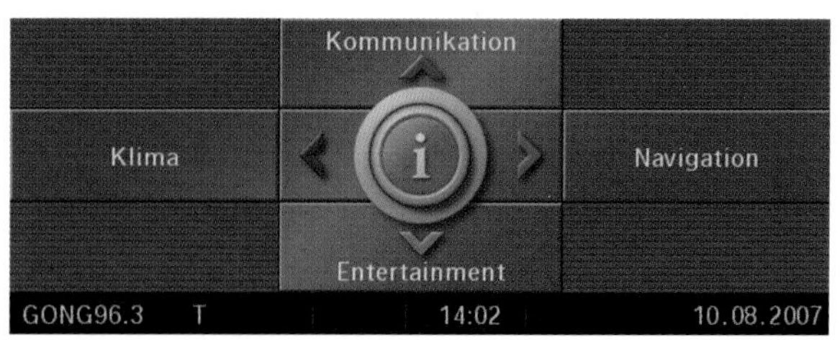

FIGURE 10. 5 SERIES MAIN MENU

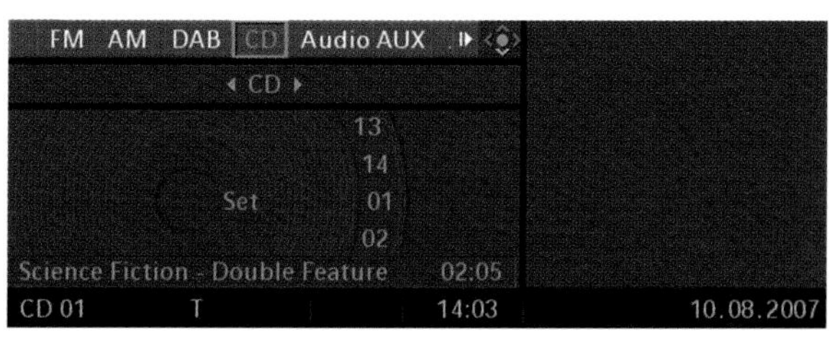

FIGURE 11. 5 SERIES ENTERTAINMENT MENU

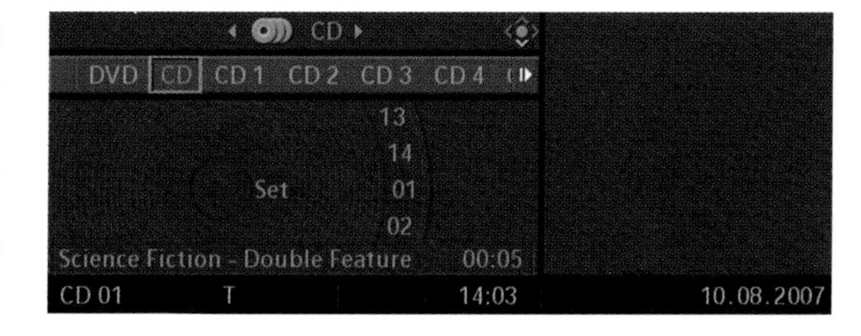

FIGURE 12. 5 SERIES ENTERTAINMENT MENU

activating the Turn-Push-Knob for a longer time (>2 s). The sliding direction to a sub-menu is always the same (climate left, navigation right, a.s.o.) and it remains available for the user at any time.

One of the main questions is whether a consistent visual-haptic design which follows the guidelines that were outlined before is adapted and learned by the user. Some experimental results shall describe the procedures that can be used to investigate such cognitive processes and the potential of adequate visual-haptic design.

In the following, learning experiments are chosen as the type of experiment that is used to investigate the interaction design. A learning experiment typically focuses on questions about acquisition of knowledge on the system and users' mental models.

If the premises hold that were described earlier in this chapter, then operational errors and time to finalize a given activity should decrease with increasing user experience.

A learning experiment was conducted with 15 users that had to solve comparable tasks in four sessions (standstill operation and two driving sessions) following one another. Whereby each session lasted for about 15 mins.

Figure 13 shows the decrease of visual control preceding the motoric action if one out of the four submenus had to be selected by sliding the knob.

This indicates that users establish a connection between the visible part of the menu structure and the related movement ('sliding the knob to a given direction') and that visual activity can be replaced by cognitive achievement while a motoric action is planned.

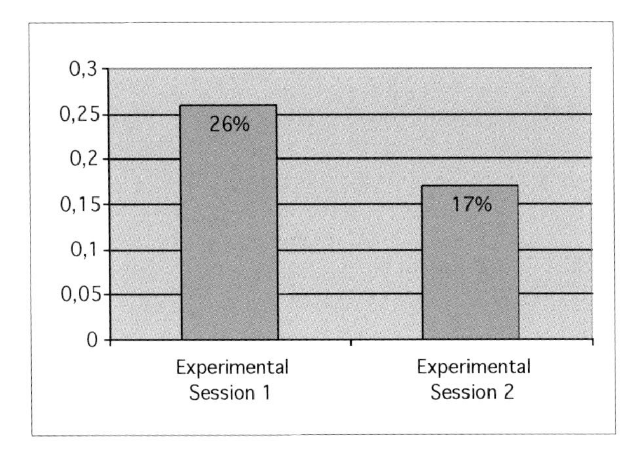

FIGURE 13. RELATIVE FREQUENCY OF SUBMENU SELECTIONS PRECEDED BY A GLANCE

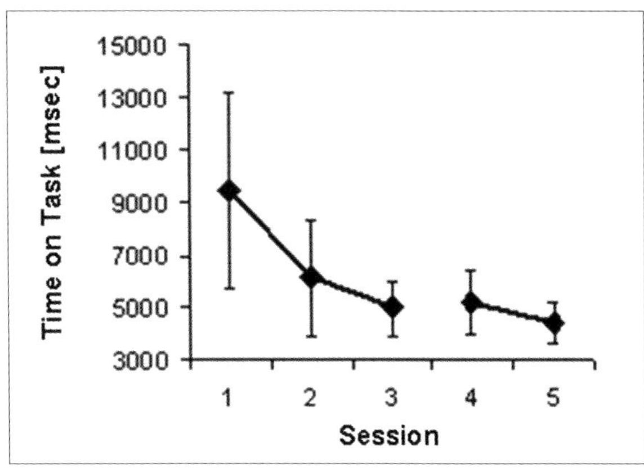

FIGURE 14. TIME TO SOLVE A MENU NAVIGATION TASK BY EXPERIMENTAL SESSION

Another experiment was conducted that was using a matrix like menu that was unfolding from left to right to get from the main menu to its submenus operated by a four-way joystick. In this case 28 subjects had to perform five sessions distributed over two days, each session lasting about 30 mins.

Figure 14 shows the increase in user performance over sessions, as the time on task decreases from ~9,000 msec to ~5,000 msec. Another interesting effect is the peak between session 3 and 4 which represents the 1 week break between day 1 and 2.

A detailed analysis shows that the performance increase and accelerated menu navigation is achieved by avoidance of wrong menu selections (40%) and increased selections/decisions (20%), which clearly shows that users are able to act faster and more efficient. This speaks for the involvement of a mental representation of the menu structure, too.

Summary

In conclusion, we have found that alignment of visual and haptic systems is critical for building intuitive information systems in the car. By creating direct mapping between 'doing' and 'seeing', there is less attention and focus required to operate the system, and therefore a greater safety, comfort and overall pleasure in the activity of driving. And this, after all, is what BMW cares about – preserving the joy of driving.

Haptics research at Daimler AG

Holger Enigk, Uli Foehl and Verena Wagner

The following article provides a brief overview of applied haptics research at Daimler AG. In order to meet the need to impart knowledge to students and do justice to the nature of this textbook greater attention will be paid to the objectives, the research object and methodology instead of reporting detailed results of specific studies. The purpose of this is to illustrate the research methods and enable their application to relevant problems.

Objectives of haptics research at Daimler AG

In many cases it is assumed that driving a vehicle is largely based on visual perception and that the majority of information processed is of a visual nature. However, one must ask whether this human distance sense is not being overestimated for certain areas of driving and what importance the sense of touch, a human close sense, has with regard to driving. In order to drive a vehicle the driver absorbs many different items of information from the environment and from the road *via* the oscillations and vibrations of his vehicle. He feels the condition of the road and the dynamic response using the seat and steering wheel when driving. At the same time he drives his vehicle using the controls: he moves the steering wheel sensitively, enters route information into the navigation system, regulates the air-conditioning, presses the accelerator and brake with his foot, opens and closes the glove compartment or hatch, etc. The sense of touch is therefore used for sensori-motor coordination [1] of driving and operating activities at the 'points of contact' between the human being and the vehicle. In order to ensure safe, accurate and precise operation, controls are necessary which are optimally adapted to the quality of human information processing, and particularly to that of the sense of touch. As a result, the *first objective* of haptics research at Daimler AG is to ensure ergonomic, reliable operation which guarantees safe and comfortable driving of the vehicle.

The *second objective* of haptics research focuses on perceived quality. Nowadays customers do not just expect a functional vehicle. The vehicle tends to be a 'work of art' which the customers perceive with all their senses. An attractive and harmonious overall picture of vehicle design and a selection of top-quality materials are absolutely essential. Consequently, questions arise like: What materials do customers regard as top-quality? What is the distinguishing feature of real leather? What do plastic surfaces have to look like if they are not to be perceived as inferior or distracting? This results in the second objective of haptics research at Daimler AG, designing controls and surface materials which give an overall impression of top quality, highlight the character of the vehicle and achieve a high level of acceptance by the customer.

The *third objective* of haptics research relates to the quality of haptics research itself. The latter has to produce valid and reliable results and specify and quantify customer requirements in detail for vehicle development. This type of haptics research is not basic research but research with a clear application. It is intended to develop top-quality products which inspire customers.

Object of haptics research within automotive research

Haptics research chiefly concentrates on the haptics of mechanical controls (handles, flaps, lids, etc.), electric controls (switches, rotary actuators, operating levers, etc.) and areas of surface materials (leather, covers, material touch lacquers, decors and trim, etc., see Fig. 1). However, these vehicle components are 'only' the target objects of optimisation and development. The actual object of haptics research within the company is rather the assessment of customers' experiences with those vehicle components depending on their method of use and application. Sensory experience itself and the features of the sense of touch in the use of vehicle components are researched systematically.

FIGURE 1. S-CLASS INTERIOR WITH DIFFERENT CONTROLS AND MATERIALS

Methodology of haptics research at Daimler AG

Haptics research at Daimler AG is located centrally in the Customer Research Center of Group Research and it addresses issues coming from all business units. In the haptics laboratory of the Customer Research Center staff consists primarily of psychologists with specialisations in experimental and perception psychology. Owing to its cross-functional orientation the haptics laboratory is closely networked with the development, design and marketing departments so intensive interdisciplinary collaboration is guaranteed.

In order to optimise the haptics of a driver control (e.g., a switch or rotary light actuator) and basically also of surface materials, one methodological approach has proved successful. It is based on five process steps (Fig. 2).

Phase 1: conceptual analysis of the problem

The objective of the first phase is to analyse the problem, compile existing information and structure the content of the project. For this purpose it is necessary to understand the functional and constructional principles of controls in terms of physical engineering. Therefore, it is helpful to analyse many different types of controls, including models of other brands, to conduct technical measurements and classify them. Analysis of the force-travel characteristic plays a key role because it is crucial to haptics perceived generally (see Figs 3 and 4). In addition, research has to be conducted into perception thresholds, i.e., absolute and differential thresholds. For example, one has to investigate what differences in force level can be perceived reliably by a human being. It is the thresholds which determine the degree of detail and the dimensions of optimisation options. If information available is inadequate, a specific investigation has to be conducted into the perception thresholds.

Phase 2: definition of customer requirements and customer-relevant criteria

With adequate technical understanding of the controls initial hypotheses can be developed determining what their haptic quality might be dependent on from a customer's point of view. In this phase the requirements must be identified which customers expect switches and materials to meet. For this purpose it is also possible to apply results of market research studies or comparative vehicle

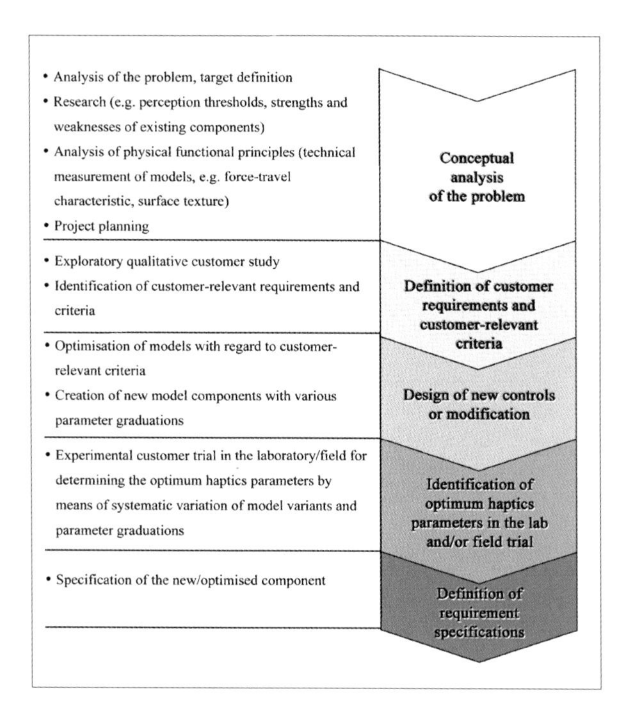

FIGURE 2. METHODOLOGY OF HAPTICS RESEARCH AT DAIMLER AG: FIVE PROCESS STEPS

studies. However, experience shows that these sources of information provide a large number of general appraisals but no specific inferences for the improvement of haptic quality. For this reason an exploratory, qualitative customer study is to be recommended, involving representatives of the target group. The subjects are requested to freely explore a selection of controls. It is crucial to find out how customers operate controls, what details they pay attention to and what criteria they apply to assess the component. When this phase has been completed with a positive outcome, initial customer-based evaluation criteria are available and one can set about designing new controls or modifying existing ones.

Phase 3: design of new controls or modification

Based on the results of the first two phases the component is optimised, particularly with regard to customer-relevant criteria and a whole range of different models is made. In particular, when considering customer-relevant criteria (e.g., force-travel characteristic) the graduation and variation must be sufficiently fine to be able to identify the best empirical blend of parameters. This takes place in the fourth phase.

Phase 4: identification of optimum haptics parameters in the lab and/or field trial

At this process stage model variants and parameter variations are investigated in an experimental customer trial in order to determine the optimum empirically. By means of systematic experimental condition variation and inferential statistical analysis of data (e.g., by means of regression, variance, cluster and factor analyses) it is possible to make accurate statements about the acceptance of the models. Evaluation is first performed in the laboratory. Depending on the type of control, a practical test in the field is necessary, i.e., on the road under real ambient conditions with visual distraction, with realistic jolts and vibrations. The field trial is useful because in the case of vibrations, for instance, involuntary operating errors and regulating inaccuracies are becoming more.

Phase 5: definition of requirement specifications

When the four phases described above have been completed with adequate success in iterative loops, a requirement specification can be compiled for the control or relevant target vehicle.

This methodology is designed in such a way that generalisable, robust results can be obtained which can be used not only for one switch but for types of switches and for a number of model series. The approach is similar for the evaluation of materials covering areas (e.g., covers, trim, etc.). Since in the case of this material type there is no operator interaction, there is often no need to conduct field trials. Physical analysis of surface materials is just as desirable in order to make generalisable statements about the influence of material properties, e.g., texture or softness, on customer acceptance and estimation of perceived quality.

The following section will discuss in greater detail the experimental procedure of Phase 4, taking a specific control as an example.

Haptics studies on push–push switches

Physical parameterisation: force-travel characteristics

One frequent subject of haptics studies is push–push switches, i.e., pressable switches which return to their original position. In the vehicle they are used, for example, to activate the air-conditioning system or to call up stored radio stations.

In the development of push–push switches a large number of different factors which have an influence on the haptic experience have to be taken into consideration. In the case of switch size and shape special care must be taken to ensure ergonomic design, allowing optimum operation at the relevant position of installation in the vehicle. The material and surface quality also determine the perception of a switch when it is touched and operated.

The force-travel characteristic plays a key role with regard to the overall haptic impression and the experienced appeal of a switch. It describes the force which has to be applied between touching the switch and triggering the pressure function along the distance to be covered. This characteristic is typically non-linear and can be divided into different sections (upper curve represented in Fig. 3). A usually convex rise in actuating force is followed by a drop in force (snap). On the follow-up journey the force rises again. The lower curve in Figure 3 illustrates the force-travel behaviour after releasing the switch. In the following, only the upper curve, the triggering process, is described in more detail.

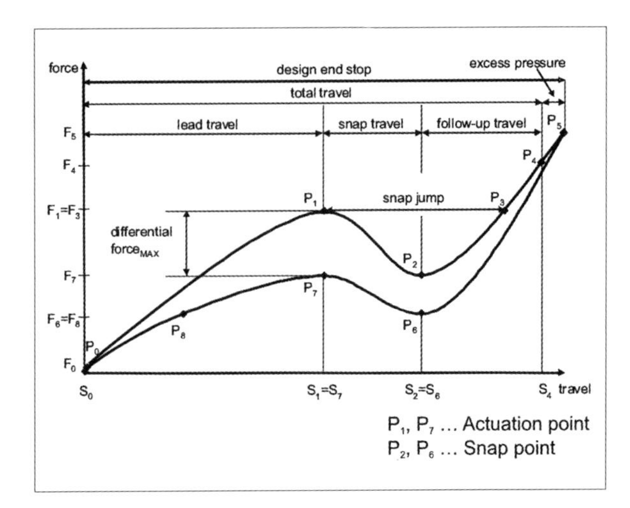

FIGURE 3. FORCE-TRAVEL CHARACTERISTIC

From the characteristic chart it is possible to derive various force and travel parameters with the aid of which the haptics of the switch can be described in physical terms. Lead travel, for example, defines the distance up to the actuation point (P_1, P_7), that is, the position at which the jump mechanism of the switch is actuated. Snap travel describes the section between the actuation point and the snap point (P_2, P_6), as of which the force characteristic rises again. Apart from the various distances of travel the forces required at the actuation point and snap point are crucial to the haptic experience.

Experimental procedure

For haptic studies in Phase 4 an experimental approach is normally envisaged which allows the comparison of a number of variants within the scope of a factorial trial design. The procedure described below is a prototypical one, which is typical of investigations into switch haptics and the evaluation of surface materials.

If for a specific target object, a new car model series for example, optimum switch actuation haptics are to be determined, several variants are typically compared with one another. In the

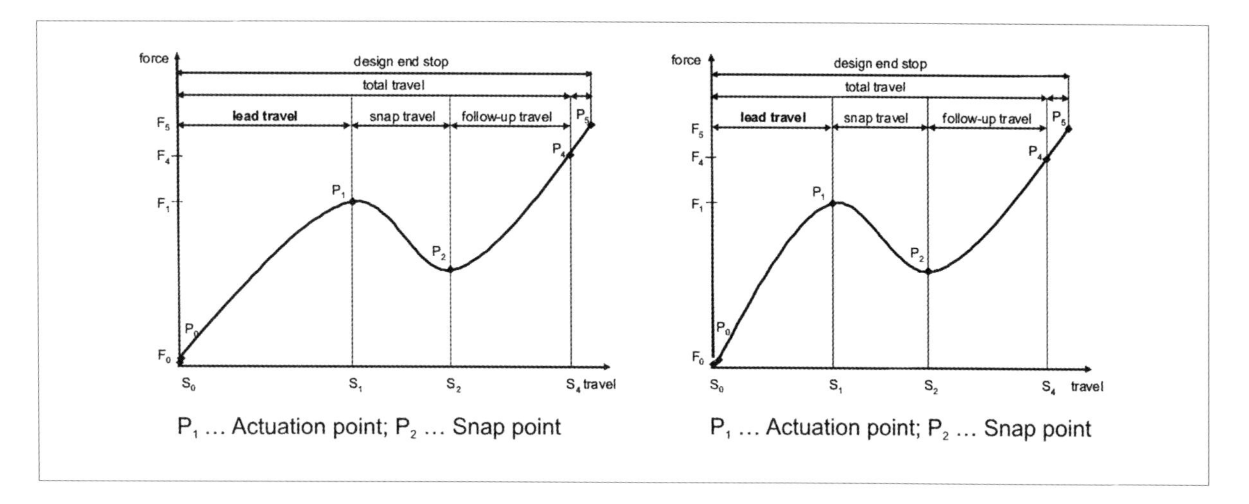

FIGURE 4. FORCE-TRAVEL CHARACTERISTIC: MODIFICATION OF LEAD TRAVEL TWICE

definition of variants the aim is to achieve maximum constancy of influencing factors which are not the direct subject of the haptic investigation, switch size or surface design, for example. The force-travel characteristic of the variants, in contrast, is varied as systematically as possible in order to clearly attribute differences in subjective assessment (dependent variable) to physical features (independent variable) and ensure internal validity. Since when operating push–push switches there is a characteristic sound, one must always examine to what extent the acoustic modality is of interest within the scope of the experiment. In as much as only decisions about haptics are to be taken, it is advisable to mask the sounds with the aid of headphones in order to avoid confusion between the haptics and acoustics of the controls.

As explained in the previous section, force-travel characteristics can be used to derive a large number of parameters which can be varied by experiment, making it necessary to focus on a small number of crucial variables. These are known from the previous process steps (Phases 1–3). A typical haptics experiment would, for example, vary two influencing factors in three stages separately from one another, which would result in a total of nine variants being investigated (e.g., three conditions for lead travel x three conditions for snap travel). Figure 4 depicts systematic variation of lead travel.

Especially with a low number of variants being evaluated the method of complete pair comparison, as used in psychophysics, is employed; that is, each variant is compared to each other variant with regard to various features. Paired comparison of always two haptic variants and repeated evaluation make it possible to reveal even minimal differences. If the number of variants being evaluated is very high, a complete pair comparison is not very practicable so they are presented one after the other and evaluated individually. Repeated presentation of the various switch types increases the quality of testing.

In order to evaluate the switches with regard to quality it is usually unipolar and bipolar rating scales or a semantic differential which is used [2]. To supplement this, qualitative information is obtained in the form of structured interviews. Table 1 contains an overview of other standard data collection tools for haptics research.

If in an investigation the expressions of switch haptics have been varied in different dimensions separately from one another, lead travel and snap travel, for example, optimum haptics parameters can be determined by means of variance analy-

Table 1: Overview of standard psychological methods for data collection in haptics investigations, classified by their level of standardisation [3]

Non-standardised methods			• Qualitative interview • Qualitative questionnaire • Thinking out loud • Expert interviews • Focus group
Standardised methods	Document analysis methods		• Company documents
	Data collection methods	Assessment methods	• Ranking • Rating • Pair comparison • Semantic differential
		Observation	• Standardised observation protocol
		Interviewing	• Standardised interview • Standardised questionnaire
		Measurement methods	• Response times
	Experimental methods		• Field experiment • Laboratory experiment

Note: for more detail: [4–8].

sis of data and these can be incorporated in the requirement specification in Phase 5.

However, it is not always possible to define the expressions of a switch variant in different dimensions. In many cases forces and distances cannot be varied separately from one another owing to construction, for example, which leads to confusion of the two variables. In such cases it is particularly important to supplement quantitative evaluation with qualitative methods in order to systematically establish causes of statistical effects in terms of content. Only when one can clarify what aspects of a force-travel variant investigated have led to positive or negative ratings is it possible to derive action recommendations for switch development.

Summary

This article presents the objectives, the object and the methodological approach involved in haptics research at Daimler AG. Haptics research pursues three objectives: 1) to ensure ergonomic, reliable operation and driving of the vehicle; 2) to provide the vehicle with a top-quality, harmonious overall design; and 3) to provide a detailed specification of customer requirements for development of the vehicle components involved. The research work focuses on the customer's requirements for controls and surface materials in the motor vehicle and his sensory perceptions in using them. Haptics research takes place at the Customer Research Center of Daimler AG on a cross-functional basis for all business units. It is distinctly characterised by perception psychology and experimentation. For the optimisation and development of new controls and surface materials a multi-stage process has proved successful which is illustrated citing examples.

Haptic design of handles

Thomas Maier

Introduction

This paper is based on a study [1] which has been carried out by the Institute for Engineering Design and Industrial Design (IKTD), Department Industrial Design Engineering and completed by further aspects regarding haptic perception.

Among other things, industrial design engineering deals with human product relations. Starting from the designed product and, respectively, the product gestalt, consisting of the sub-gestalts: assembly, shape, colour and graphics, it is at first perceived by human. An identification process is activated in the human brain by the types of perception: vision, hearing, feeling, smelling, tasting and others. This identification causes a certain human behaviour and, for this reason, a special operation and use of this product [2].

Each type of perception has different perception performances and, respectively, information gathering which is highest when seen by the sense organ 'eye'. The tactile or haptic perception performance by the skin is approximately 4×10^5 bit/s [3]. To 'feel pressure', 'feel roughness', 'feel position', 'feel movement', 'feel heat', 'feel humidity' and 'feel electrically' are tactile or haptic perceptions.

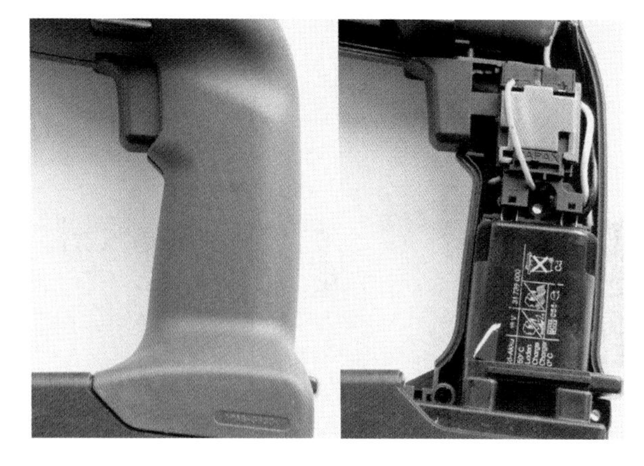

FIGURE 1.
HANDLE OF A CORDLESS SCREWDRIVER (LEFT) AND WITH FUNCTION ELEMENTS (RIGHT)

The study

The haptic design of a handle is tested by means of a cordless screwdriver (see Fig. 1). Characteristic handles of power tools must be designed according to the hand geometries of different users. In reality, this is unfortunately not always feasible, since technical restrictions, as, for example, the integrated switch and the inserted neck of the battery pack do not allow this. The aim of the study is the variable adaptation of the outer handle contour to the individual operator's hand. This adaptation to different operators with different hands can be done manually as well as automatically, for example by intelligent materials. The highest aim is a haptically optimal, individual and intelligent handle.

Sequence of study

The study is divided into the phases: task, benchmark, ergonomic analyses, compilation of requirements, conception of alternative versions and evaluation and selection. This shows the design process from the beginning to the evaluated conceptual versions.

Benchmark

The benchmark of cordless screwdrivers offered on the market shows that only one manufacturer can guarantee their adaptability by exchanging a padded element on the back of the handle, available in three different sizes [4]. Most of the other cordless screwdrivers have a rubber-coated handle with different shore-hardness and thickness in order to obtain better haptics and a higher grasping security.

The extension of the benchmark to similar areas shows that, for example, individual professional knives for butchers, sports weapons and pistols with individual handle inserts, individually adapted handles of walking sticks and gel-padded bicycle handle bars exist. Even for sports shoes and ski boots there are reversible or non reversible adaptation principles, as, for example, integrated inflatable air cushions or the use of hardening foam.

Ergonomic analysis

At first, the user of the cordless screwdriver must be demographically identified. This applies to female and male users between 18 and 65 years of age whose body height, according to DIN 33402-2 [5], ranges from the 5-percentile woman with a body height of 1,535 mm to the 95-percentile man with a body height of 1,855 mm. Due to the available measured body data, the user group was limited to Germans and North Americans. *Significant hand measurements*, depending on the body height, can be found in the anthropological atlas [6] and in another current literature [7]. A comparison of the above mentioned data collection showed that the deviations are very small. For this reason, the hand measuring data from [7] were used for further analysis (see Fig. 2).

The handle of the cordless screwdriver is the interface between the user and the machine. This is therefore the main focus of the ergonomic study and the ergonomic design, respectively the haptic design.

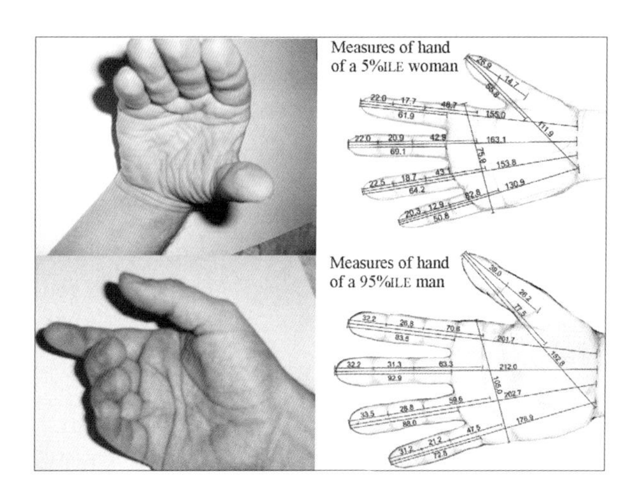

FIGURE 2. INDIVIDUAL HANDS AND MEASURES OF HAND ACCORDING TO [7]

Besides the significant hand measurements, the following parameters have a wide influence on the quality of this interface, i.e., the tactile and haptic perception:

- Work load cases 'drilling and screwing'
- Working position of the user
- Shape and dimensions of handle
- Ways of gripping
- Grasping areas and coupling degree
- Material and surface.

The cordless screwdriver is used for drilling holes and for screwing which results in the two *work load cases 'drilling and screwing'* (see Fig. 3). The forces and torques resulting from these work load cases are transmitted to the user's hand over the handle, which acts as a lever. As a reaction, the necessary operating forces and operating torques are applied by the user's hand.

The optimum *working position of the user* results from the anatomy of the hand-arm-system as well as from the work load cases (see Fig. 4). The working positions drilling and screwing into walls, ceilings, floors and with limited working space can occur.

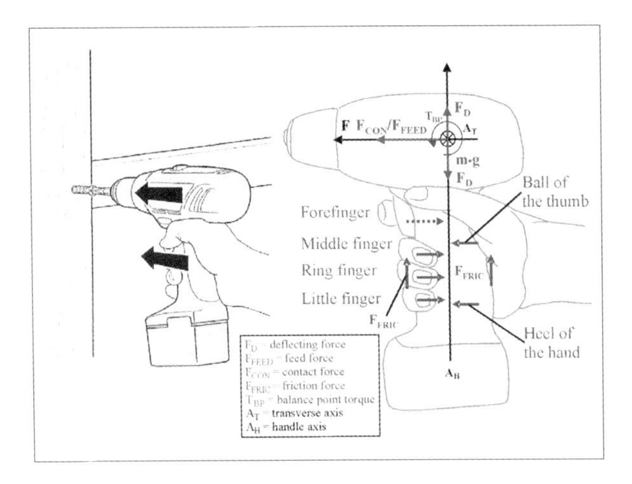

FIGURE 3. WORK LOAD CASES DRILLING AND SCREWING

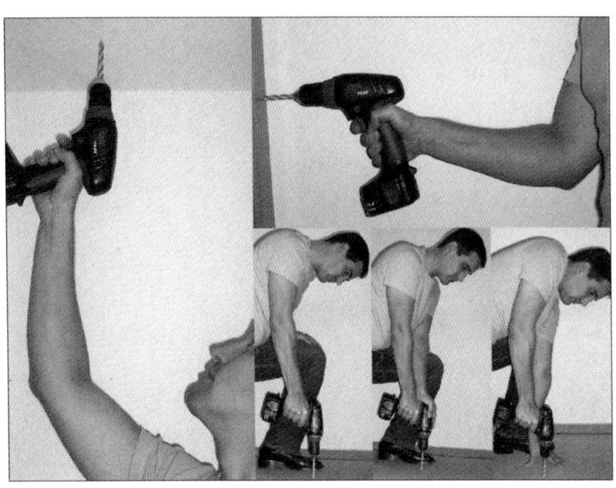

FIGURE 4. WORKING POSITIONS OF THE USER

Shape and dimensions of the handle depend on the function elements of the cordless screwdriver as, inside the handle, there must be enough space for the installation of the switch and the neck of the battery pack. Therefore, such power tools are mostly designed from the inside to the outside, which means centrifugally, with the smallest possible handle circumference.

The *ways of gripping* are very well defined and illustrated in [8]. The user's entire hand grasps the handle of the screwdriver. In doing so, the forefinger is constantly pressing the switch in operation by means of a contact handle.

The *grasping areas* of the hand result from the interface hand-handle for left- and right-handed users. Thus, different grasping areas on the handle, where the palm, the forefinger, the thumb and the rest of the fingers touch the handle (see Fig. 5) are the result of this. In this case, the *coupling degree* is basically a form closure with additional friction closure in the direction of the handle axis.

The *material* of the handle is mostly made of rubber-coated thermoplastic material. The *surface qualities* are important for the haptics. Texturised surfaces, made of structured lacquers, are only useful if a drainage effect as, for example, the drain of hand perspiration, is required. Otherwise, smooth surfaces, respectively rubber-coated hard surfaces, have a better haptic quality and are therefore more pleasing to touch.

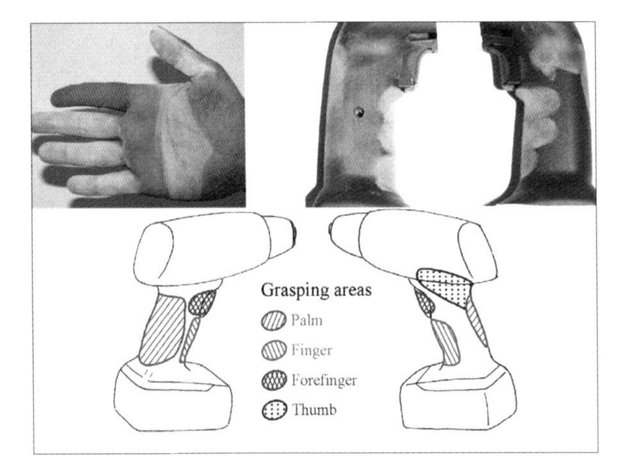

FIGURE 5. GRASPING AREAS AND COUPLING DEGREE

TABLE 1. EXCERPT OF THE REQUIREMENTS LIST

F/S	No	Requirement	Description	Quantification
		Individual rubber coating/ Customisation (variable handle gestalt)		
F	28	Reversible adaptable to several users		
S	29	Fast customisation		< 120 seconds
S	30	Adjustment of customisation avoiding unintended changing	ideal: automatic adjustment	manual adjustment by user
S	31	Customisation in steps	min. 5% ILE woman, 5% ILE man, 50% ILE man, 95% ILE man	at least 4 steps, ideal continuous
S	32	Usable with gloves	range for 5% ILE woman and 95% ILE man	oversize 2 mm to the measure of hand
S	33	Assembling joints beyond the highest-bonded grasping areas	ideal: covered joints/no joints	grasping areas according to Figure 5
		Handle (main gestalt with variable and fixed handle gestalt)		
S	34	Angle of comfort	wrist angle 75°	70°–80°
F	35	Grade of interconnection		from closure
F	36	Kind of grasping		5 finger embracing handle

Compilation of requirements

The requirements regarding the haptic design of a handle are stated in a list of requirements (see Tab. 1). The most important requirements, the human product requirements which are divided into visibility and identification as well as into operation and use, are included. As types of requirements, this list of requirements includes the fixed requirements which must be fulfilled, and the sector requirements which include a range of values [2]. In total, 44 different requirements were defined, described and quantified.

Conception of alternative versions

In total, 15 different versions have been conceived; eight of them are described more precisely in this paper (see Fig. 6). The listed requirements as well as the technical feasibility had great influence on these concepts.

Version 1 consists of a completely exchangeable individual handle of different sizes which, consequently, has the highest degree of individuality. This handle is slid from the bottom over slots in the switch housing with detachable snap-on connexion and locked. The handle is made of partly crystalline thermoplastic material, the rubber coating is a thermoplastic elastomer.

Version 2 consists of two-piece changeable handles. Primarily, the handle is the smallest basic handle for the $5\%_{ILE}$-woman. The adaptation to other user hands is effected by individually changeable handles, consisting of two half shells, which are fixed by snap-on connexions above the basic handle. Here as well, the haptics

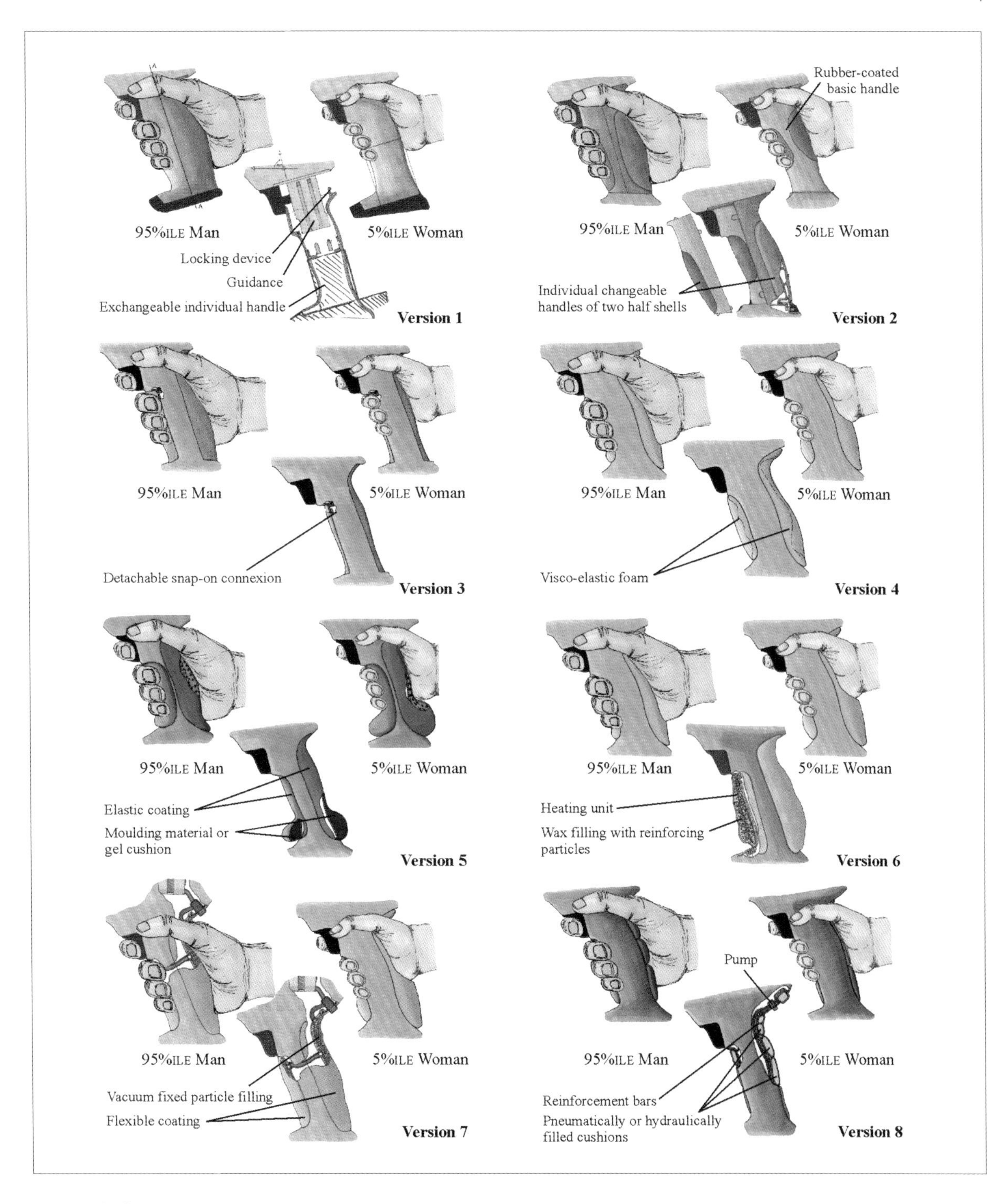

FIGURE 6. OVERVIEW OF THE ALTERNATIVE VERSIONS

is improved by a rubber coating.

Special rubber-coated elements on the front and the back of the handle are characteristic for *version 3*. Here a stepped individual adaptation to the particular user hand is feasible. The rubber-coated elements can be easily exchanged and locked by a snap-on connexion. They consist of a thermoplastic elastomer which is rubber-coated with another elastomer of different thicknesses.

Version 4 is based on the adaptation of the handle by visco-elastic foam. This foam is continuously adjustable to each user hand by compressing the foam at the position with the highest load. The foam has a temporary form memory and goes back to its original form only very slowly. For this reason, the adaptation to another user hand takes a relatively long time which is a disadvantage. Tempur, an open-pored foam, which must be additionally protected against environmental influences, is used as material.

The consequent search for a solution and a conception leads to *version 5*. In this case the adaptation to the individual user hand is effected by a moulding material or a gel cushion adjusted to the front and the back of the handle and provided with a flexible elastic coating. The gel, particle or moulding material filling of the cushions is displaced when the handle is grasping it. Thus, the circumference of the handle can be reduced.

The handle of *version 6* is assembled in a similar way. It contains a wax filling with reinforcing particles and a heating unit. The individual continuously adjustable adaptation of the handle is carried out by means of wax material reinforced by crystal balls or other particles. The wax material is softened by a heating unit and is thus adaptable to the hand. The wax material has an elastic coating to protect it from environmental influences.

Version 7 shows an interesting high-tech-solution with a vacuum fixed particle filling. The flexible coating connected to the basic handle is filled with particles. After the grasping of the handle and the individual deformation, a vacuum is built up in the coating when the switch is actuated by means of the motor of the cordless screwdriver. This vacuum remains until air is blown again into the coating by means of a valve. The vacuum sucks the coating to the basic handle, at the same time the particles of the filling are pressed together. The handle is now inherently stable! The system is completely reversible and adaptable.

Pneumatically or hydraulically filled cushions are characteristic for *version 8*. Hollow cushions are filled, and thus extended, with air or another fluid by a pump. Reinforcement bars are used to shape the hollow cushions. In order to stabilise the handle, the hollow cushions must also be flanked by the screwdriver housing. This entails a restriction of the individual adaptability.

Evaluation and selection

The evaluation of the eight alternative versions was based on the criteria of the requirements list which were already evaluated during this phase (see Tab. 2). In order to obtain a more exact differentiation of the evaluation result, all criteria related to the individual adaptability and emphasising the haptics are weighted twice. The evaluation scale ranges from ‘insufficient’, i.e., 0 points, up to the maximum ‘very good’, i.e., 4 points. In case of a criterion, which has been weighted twice, the number of points is then multiplied by the factor 2. This results in a maximum of eight points. The alternative versions were ranked as follows:

- rank 1 with 111 evaluation points – version 7
- rank 2 with 109 evaluation points each – version 2 and version 3
- rank 3 with 107 evaluation points – version 1

Summary

This study shows that the conception of a haptically optimal and individual as well as intelligent handle is possible. The potential of development concerning new adaptive materials and surfaces exists and further research in this regard is done at the Industrial Design Engineering Department.

Table 2. Evaluation of the alternative versions

No	Version number / Requirements	weighting	1	2	3	4	5	6	7	8
2	Little weight of individual rubber coating	1	4	4	2	4	2	2	3	3
4	Resistance against frost (material, principle)	1	3	3	3	4	3	3	3	3
5	Resistance against heat (material, principle)	1	3	3	3	1	4	2	4	4
6	Self-cleaning	1	4	3	3	4	4	4	4	2
12	Mountable	1	4	4	4	3	3	2	2	1
19	Selection of the mounting procedure (steps and costs of mounting)	1	3	4	4	3	2	2	2	1
21	Operation of the individual rubber coating/ Recognition of handle customisation	2	6	8	8	2	6	4	8	8
22	Impression 'stable'	1	4	4	4	0	2	4	4	2
23	Impression 'professional'	1	4	4	4	1	2	3	4	3
24	Coding in size of individual rubber coating	2	8	8	8	8	8	8	8	8
26	Angle of operation ergonomical correct for the forefinger	1	4	4	4	4	4	4	4	4
29	Fast customisation	2	8	8	8	0	8	4	8	6
30	Adjustment of customisation	2	8	8	8	0	4	8	8	8
31	Customisation in steps, ideal: infinitely variable	1	4	4	4	8	8	8	8	8
32	Usable with gloves	1	4	4	4	2	4	4	4	4
34	Angle of comfort	1	4	4	4	4	4	4	4	4
37	Operation within different grasping situations	1	4	4	4	2	2	4	4	4
39	Shock absorption	2	2	2	6	8	8	4	4	8
40	Grasping height	1	4	4	4	4	4	4	4	4
41	Ergonomical moulding of single grasping areas of the handle	1	4	4	2	3	3	3	3	3
42	Adjustable range between minimal and maximal grasping circumference	2	8	8	8	8	8	8	8	8
43	Adjustable range between minimal and maximal grasping circumference (switch)	2	8	8	8	8	8	8	8	8
44	Recyclable	1	2	2	2	1	1	1	2	2
	Summary of the evaluation points according to version		**107**	**109**	**109**	**82**	**102**	**98**	**111**	**106**
	Rank		**3**	**2**	**2**	**7**	**5**	**6**	**1**	**4**

The next step is a detailed draft including the creation of grasping patterns. From the scientific point of view, an evaluation of the results by tests with test persons and, after this, a final validation of these results must be carried out. The results of this study are further interesting research approaches within the interface design which are now described and for which we want to apply.

The author would appreciate suggestions and scientific discussions regarding this subject.

Selected readings

Dreyfuss H (2002) *The measure of man and woman. Human factors in design. Revised edition.* Alvin R. Tilley. John Wiley & Sons, New York

Salvendy G (1997) *Handbook of human factors and ergonomics.* 2. edition. John Wiley & Sons, New York

Landau K (2003) *Good practice. Ergonomie und Arbeitsgestaltung.* Ergonomia Verlag, Stuttgart

Luczak H (1998) *Arbeitswissenschaft.* 2. vollständig neubearb. Auflage. Springer-Verlag, Berlin, Heidelberg, New York

Kirchner JH, Baum E (1986) *Mensch-Maschine-Umwelt. Ergonomie für Konstrukteure, Designer, Planer und Arbeitsgestalter.* Beuth Verlag, Berlin, Köln

Vestibular sensory substitution using tongue electrotactile display

Yuri P. Danilov, Mitchell E. Tyler and Kurt A. Kaczmarek

Sensory substitution

Sensory substitution systems provide their users with environmental information through a human sensory channel different from that normally used. For example, a person who is blind may use a long cane to detect obstacles while walking and Braille or raised-line graphics to read information normally received visually. A person who is deaf may read lips to understand speech. A person without vision or hearing may use a method called Tadoma, placing his or her hands over the face and neck of a speaker to understand speech [1]. Persons with an impaired vestibular (balance) system use their hands, not primarily for mechanical support, but to sense how they move relative to their environment. Electronic sensors and tactile (touch) displays enable more sophisticated applications for sensory substitution. In this chapter we will briefly review visual and auditory sensory substitution, as well as tactile feedback in robotic systems, followed by an extended discussion of vestibular sensory substitution. A more detailed discussion of vision substitution is provided in the chapter by Ptito in this book.

For the brain to correctly interpret information from a sensory substitution device, it is not necessary for the information to be presented in the same form as the natural sensory system. One needs only to accurately code action potentials in an alternate information channel. With training, the brain learns to appropriately interpret that information and utilize it to function as it would with data from the normal natural sense.

Reported attempts to present spatial visual information *via* tactile displays date from at least the early 1900s, with serious scientific study starting in the 1960s [2–4]. These systems operate by using an electronic camera or matrix of light sensors to control the stimulation intensity on a spatially-corresponding matrix of electrical or mechanical tactile stimulators on the surface of the skin. The user perceives tactile shapes on the skin having the same shape as the visual image recorded by the camera. Blind and blindfolded users are then able to identify simple objects in a high-contrast environment, and have reported visual concepts such as distal attribution (i.e., perceptually localizing the target object out in front of the camera, rather than on the tactile display proper) [5], looming, and perspective [6], and also optic flow phenomena [7]. A vision substitution device using a forehead tactile display is being commercialized (http://www.eyeplus2.com).

Electrotactile stimulation (inducing touch sensations using carefully-controlled electric current pulses delivered *via* electrodes on the surface of the skin) has been studied extensively in many laboratories [8, 9], including ours at the University of Wisconsin-Madison. Electrotactile displays have been used to deliver tactile graphics to the fingertips [10], abdomen [7], or tongue [11]. Vision substitution research using the tongue-based tactile display developed in our lab [12] has demonstrated that the visual cortex is active in processing visual data presented to the tongue [13, 14]. This is an example of brain plasticity, the ability of the brain to reorganize its function for optimal processing of information following injury, loss of 'normal' sensory information, or other challenge [15, 16].

Auditory prostheses using electrotactile stimulation typically use the stimulation intensity on each electrode in a linear [17, 18] or two-dimensional [19] matrix to represent the sound intensity in a particular frequency range of sounds recorded by a microphone. Results have generally included increased awareness of sounds and improved lip reading ability [20, 21]. At least two devices (made by Tacticon Corporation and Sevrain-Tech, Inc.) were commercially available in the 1970s–1980s. Based on our tongue-based tactile display system, Wicab, Inc. (http://www.wicab.com) is commercializing auditory, visual, and vestibular (see below) sensory substitution systems.

The telerobotic-manipulation community has long recognized the inadequacy of strictly visual feedback, and the particular need for haptic information such as contact, grasp force, shear, and slip, which convey critical information about the state of the hand-object interaction [22]. Operators with normal motor control could incorporate hand feedback and tongue tactile stimulations from a sensate robotic gripper to create an additional haptic channel [23]. The technical term for human touch is haptic or tactual perception, which combines tactile (spatial pressure profile) and kinesthetic (joint position and torque) information [24]. Haptic human-machine interfaces are an active area of research [25].

One of the recent and successful discoveries in sensory substitution approach for clinical rehabilitation purposes is electrotactile vestibular substitution [26]. The tactile sensory channel of the skin was used to substitute missing or damaged sensation from the one of the most sophisticated natural sensory devices – vestibular sensory organs. The remainder of this chapter is an in-depth case study of electrotactile vestibular substitution.

Vestibular sensory system

To navigate correctly in three dimensional space, we should use sophisticated guidance systems that register every acceleration along three spatial axis and rotation around them. To do that technically we are using highly sophisticated systems, including 3-D accelerometers and gyroscopes under computer control to provide precise information in many areas of industry, robot technology, space and military applications. Vertebrates, including primates and humans, are using a naturally designed navigational system – vestibular.

The sensory receptors that allow us to maintain our equilibrium and balance are located in the vestibular apparatus, which consists (on each side of the head) of two otolithic organs, the utricle and the saccule and three semicircular canals. The semicircular canals, hoop-like tubular membranous structures, sense rotary acceleration and motion.

The sensory structures in these organs, called maculae, also employ hair cells, similar to those of the auditory sensory system. The 'hairs' of these cells, which consist of numerous microvilli, called stereocilia, and one cilium, called a kinocilium, are embedded in a gelatinous mass weighted by the presence of otoliths composed of protein and calcium carbonate. The gelatinous mass moves in response to gravity, bending the hair cells and initiating action potentials in the associated neurons.

Gravity pulls on the otoliths and bends the hair cells as the position of the head changes. The body responds by making subtle tone adjustments in muscles of the back and neck, which are intended to restore the head to its proper neutral, balanced position. The maculae also respond proportionally to linear acceleration and activate your awareness during acceleration or deceleration in the car or speeding lift.

The impulses generated by these hair cells are carried by the vestibular portion of the 8th cranial nerve to the cerebellum, the midbrain, and the temporal lobes of the cerebrum. The cerebellum and midbrain use this information to maintain equilibrium at a subconscious level.

The three semicircular canals are fluid-filled membranous ovals located on each side, and lie in three mutually perpendicular planes. Near its junction with the utricle, each canal has a swollen portion called the ampulla. Each ampulla contains a crista ampullaris, the sensory structure for that semicircular canal; it is composed of hair cells and supporting cells encapsulated by a cupula, a gelatinous mass. The cupula extends to the top of the ampulla and is moved back and forth by movements of the endolymph in the canal. This movement is sensed by displacement of the stereocilia of the hair cells.

The role of the semicircular canals is in sensing of rotary acceleration. Because of the inertia of the endolymph in the canals, when the position of the head is changed, fluid currents in the canals cause the deflection of the cupula and the hair cells are stimulated. The fluid currents are roughly proportional to the rate of change of velocity (i.e., to the rotary acceleration), and they result in a proportional increase or decrease (depending on the direction of head rotation) in action potential frequency.

The vestibular apparatus is an important component in several reflexes that serve to orient the body in space and maintain that orientation. Integrated responses to vestibular sensory input include balancing and steadying movements controlled by skeletal muscles, along with specific reflexes that automatically compensate for bodily motions. One such mechanism, for example, is the vestibuloocular reflex.

If the body begins to rotate and, thereby, stimulate the horizontal semicircular canals, the eyes will move slowly in a direction opposite to that of the rotation and then suddenly snap back the other way. This movement pattern, called rotatory nystagmus, aids in visual fixation and orientation and takes a place even with the eyes closed. It functions to keep the eyes fixed on a stationary point (real or imaginary) as the head rotates. As rotation continues, the relative motion of the endolymph in the semicircular canals ceases, and the nystagmus disappears. When rotation stops, the inertia of the endolymph causes it to continue in motion and again the cupulae are displaced, this time from the opposite direction.

The vestibular system is important in virtually every aspect of our daily life, because head acceleration information is essential for our adequate behavior in three-dimensional space not only through vestibular reflexes that act constantly on somatic muscles and autonomic organs, but also through various cognitive functions such as perception of self-movement [27, 28], spatial perception and memory [29–32], visual spatial constancy [33, 34], visual object motion perception [35, 36], and even locomotor navigation [37]. Vestibular input functions also include: egocentric sense of orientation, coordinate system, internal reference center, muscular tonus control, and body segment alignment [38].

Vestibular rehabilitation

Chronic balance disorders due to vestibular loss result in significant functional disability. Bilateral vestibular loss can be caused by drug toxicity, meningitis, physical damage or a number of other specific causes, but is most commonly due to unknown causes. Brainstem or cerebellar lesions due to a stroke or other trauma are the most frequent causes of central involvement.

It produces multiple problems with posture control, movement in space, including an unsteady gait and various balance-related difficulties, like oscillopsia [39–41]. An unsteady gait is especially evident at night (or in persons with low visual acuity). The loss is particularly incapacitating for elderly persons. The symptoms and signs of severe vestibular loss are well described and can include ataxia, dysmetria, rigidity, nausea, lightheadedness, episodic vertigo, veering while walking, an inability to walk in the dark, read while walking, or stand on compliant or unstable surfaces, and decreased gait velocity [42–44].

To date, the treatment of choice for individuals with moderate to profound loss is vestibular rehabilitation. Vestibular rehabilitation therapy (VRT) has been shown to be effective for numerous vestibular disorders, but has also been shown to be less effective for persons having central disorders, severe unilateral or mixed etiologies. Patients with central vestibular disorders have poorer outcomes of rehabilitation overall, are more likely to have significant balance impairment, and do not progress as quickly as those with peripheral vestibular disorders.

Thus, while therapeutic gains are highly important to the patients that *do* respond to these procedures, reports also reveal that roughly half of all vestibular disorder patients *do not* derive a meaningful benefit from VRT. Furthermore, problems of this nature are not amenable to surgical treatment, and pharmacological management of these conditions with vestibular suppressants often retards the recovery process. Consequently, these people are left to manage their lives as well as they can with inadequate therapeutic intervention.

For the current project of vestibular rehabilitation we chose subjects with bilateral vestibular loss, one of the most severe damage possible to the vestibular sensory system. All of them were identified as disabled or handicapped.

Vestibular substitution

Existing feedback devices

Several alternative devices, based on sensory substitution and biofeedback approaches, already exist or are in development. One device has recently been shown to significantly assist vestibulopathic subjects during simple postural control tasks [45–47]. Dynamic tactile displays [48, 49] have been demonstrated successfully as both auditory (e.g., TactAid [50–53]) visual prostheses (e.g., Optacon [54]) or combined audio-visual feedback [55, 56]. Moreover, much data has been collected on the properties of the skin that can be used in building a tactile display [57–59].

Usefulness of vibrotactile displays have also been demonstrated in aviation [27]. The US Navy is developing a device to evaluate the use of tactile cues to provide situational awareness information to aircraft pilots. Their tactile cue device uses externally measured angular and linear motions that are fed into a three-dimensional vibrotactile display (vest) from the aircraft's inertial navigation system. In roughly 1 h of training, experienced Navy pilots were able to perform acrobatic maneuvers like loops or barrel rolls and return to stable, level flight while completely blindfolded [60, 61].

Galvanic stimulation is another method of providing an alternate sensory input. Galvanic sensory substitution has to be distinguished from galvanic stimulation of the VIIIth nerve *via* an electrode that touches or is in very close proximity to the nerve. It has been applied transdermally by Collins and his collaborators to the temporal bone in order to stimulate the eighth nerve [62–64].

Direct stimulation of the vestibular nerve by implantable or semi-implantable electronics is also possible, and it is analogous to the cochlear implants. Like the three semicircular canals in a normal ear, such a device should at least transduce three orthogonal (or linearly separable) components of head rotation into activity on corresponding ampullary branches of the vestibular nerve [65–68].

Electrotactile vestibular substitution system (ETVSS)

It is important to notice the main difference between ETVSS and all approaches mentioned above –the source of feedback signals and point of delivery. Almost all known systems use as a main input the movement of CGP (center of gravity projection, usually taken from the force platform) or integrated signal from body located sensors. Points of delivery are usually body skin (vibrotactile systems oriented on the skin mechanoreception) or main sensory systems (signal delivered as auditory pattern or visual image).

ETVSS was developed to transmit head inertial sensory information normally provided by the vestibular system to the brain through a substitute sensory channel: the complex combination of mechanoreceptors, thermoreceptors and free nerve endings on the anterior dorsal surface of the tongue using electrotactile stimulation of the tongue.

BOX 1. ELECTROTACTILE SENSITIVITY OF THE TONGUE

The tongue is not a uniform, a homogeneous sensory interface. The range of sensation threshold voltage magnitude is highly dependent on the size and location of the area stimulated, varying by as much as 300% when 'mapping' threshold intensities for a single electrode. This variability is likely due to the differential innervations of the tongue by three cranial nerves: the lingual branch of the Trigeminal nerve (CN-V), the chorda tympani branch of the facial nerve, (CN-VII), and at the posterior region of stimulation the glossopharyngeal nerve (CN-IX). The inconsistency in sensation threshold may also be attributed to natural variations in the both the type and density of tactile receptor populations. In particular, Type-1 fast- and slow-adapting mechanotactile receptor densities are highest at the tip of the tongue, having small and well defined receptive fields. Receptor densities are slightly lower at the remaining perimeter of the tongue, and progressively decline toward the posterior and midline [73]. A sensation intensity compensation algorithm has been created to adjust the relative stimulation amplitude at each electrode to compensate for the innate variations in tactile sensitivity. The tongue electrotactile thresholds map is presented in Figure 1B.

ETVSS is the unique system that takes signal from the head (from the top of the head or the oral cavity, in close proximity of natural vestibular system) and delivers it to the head (the same location, anterior surface of the tongue).

Why the tongue?

Use of the tongue as a silent and hands-free sensory information interface could prove beneficial for many applications. The tongue, along with the lip and fingertip, differ from other body sites in their specialization for spatial acuity. The tongue is uniquely suited for electrotactile stimulation because in the protected environment of the mouth, there is no corneal or protective layer of skin typically found on external body surfaces (particularly the hands and feet), and the sensory receptors of the skin are located close to the surface. The tongue is also continuously bathed with an electrolytic solution (saliva). Because the tongue is very sensitive and highly mobile, it requires only about 3% of the voltage (10–20 V), and far less current (1–4 mA), than the fingertip to achieve equivalent sensation levels [69]. Our previous research has demonstrated that the tongue is more sensitive than the fingertip based on mechanical two-point discrimination thresholds, e.g. [70] . We have previously reported that geometric pattern perception by electrotactile stimulation on the tongue is as accurate as that for the fingertip [70, 71]. It should be noted, however, that these studies investigated the sensitivity and electrotactile stimulus dynamic range, limited their measurements to the anterior half, and none addressed ETVSS.

ETVSS

Device

A miniature two-axis accelerometer (Analog Devices ADXL202) was mounted on a low-mass plastic hard hat. Anterior-posterior and medial-lateral angular displacement data (derived by double integration of the acceleration data) were fed to a previously-developed tongue display unit (TDU) that generates a patterned stimulus on a 144-point electrotactile array (12 x 12 matrix of 1.6-mm diameter gold-plated electrodes on 2.3-mm centers) held against the superior, anterior surface of the tongue [26] (Figs 1A and 2B). Subjects readily perceived both position and motion

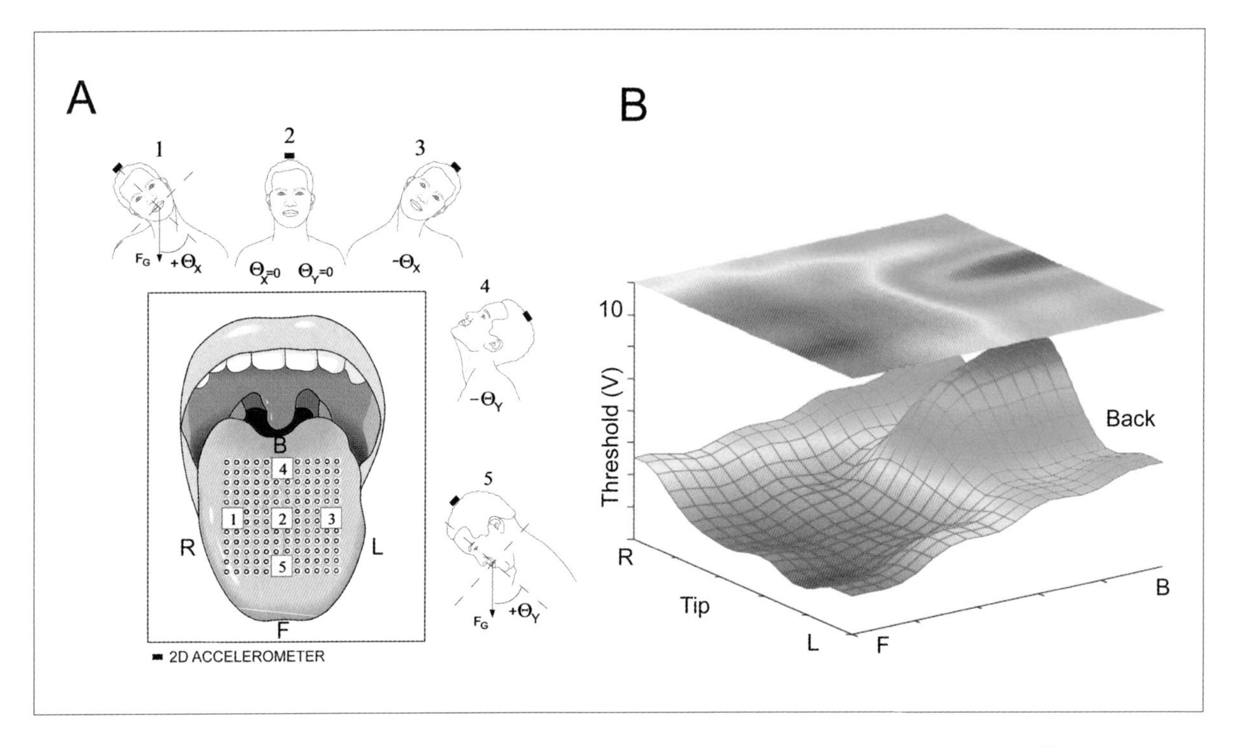

Figure 1. The scheme of tongue stimulation (A) and electrotactile threshold sensitivity map (B)
A. Conceptual diagram characterizing the relationship of head angle (measured by 2-D accelerometer, black rectangles) and resultant position of the ETVSS 'target' stimulus pattern on the tongue. The head position is represented on the tongue array by 4 (2×2 pattern) active electrodes, continuously stimulated with optimal intensity. In starting position, head is vertical and centered on the body. Correspondingly, electrotactile signal will be also centered on the electrode array (head/electrode position 2).
Angular displacement of the head in any direction will produce similar displacement of the signals on electrode array. For example, the rotation of the head to the right will produce signal movement to the right (head/electrode position 1), similarly for movements to the left (head/electrode position 3), forward (head/electrode position 1) and back (head/electrode position 1).
B. The average of nine subjects tongue electrotactile threshold sensitivity maps. The tongue of each subject was tested several times to measure minimal intensity of electrotactile stimulation capable to produce distinct sensation under each of 144 electrodes on ETVSS array. Note that sensitivity of the tongue is not uniform. Tip of the tongue and surrounding areas are the most sensitive. Sensitivity is fast decreasing to the center of the tongue and with less extent to the sides of it. The contour map on the top is projection of the 3-D surface.

of a small 'target' stimulus on the tongue display, and interpreted this information to make corrective postural adjustments, causing the target stimulus to become centred.

Signals from the accelerometer, located in the hat on top of the head, deliver position information to the brain *via* an array of gold plated electrodes in contact with the tongue. Continuous recording from the accelerometer produced the head base stabilogram (HBS). The HBS is the major component of our data recording and analysis system (Fig. 2C–D).

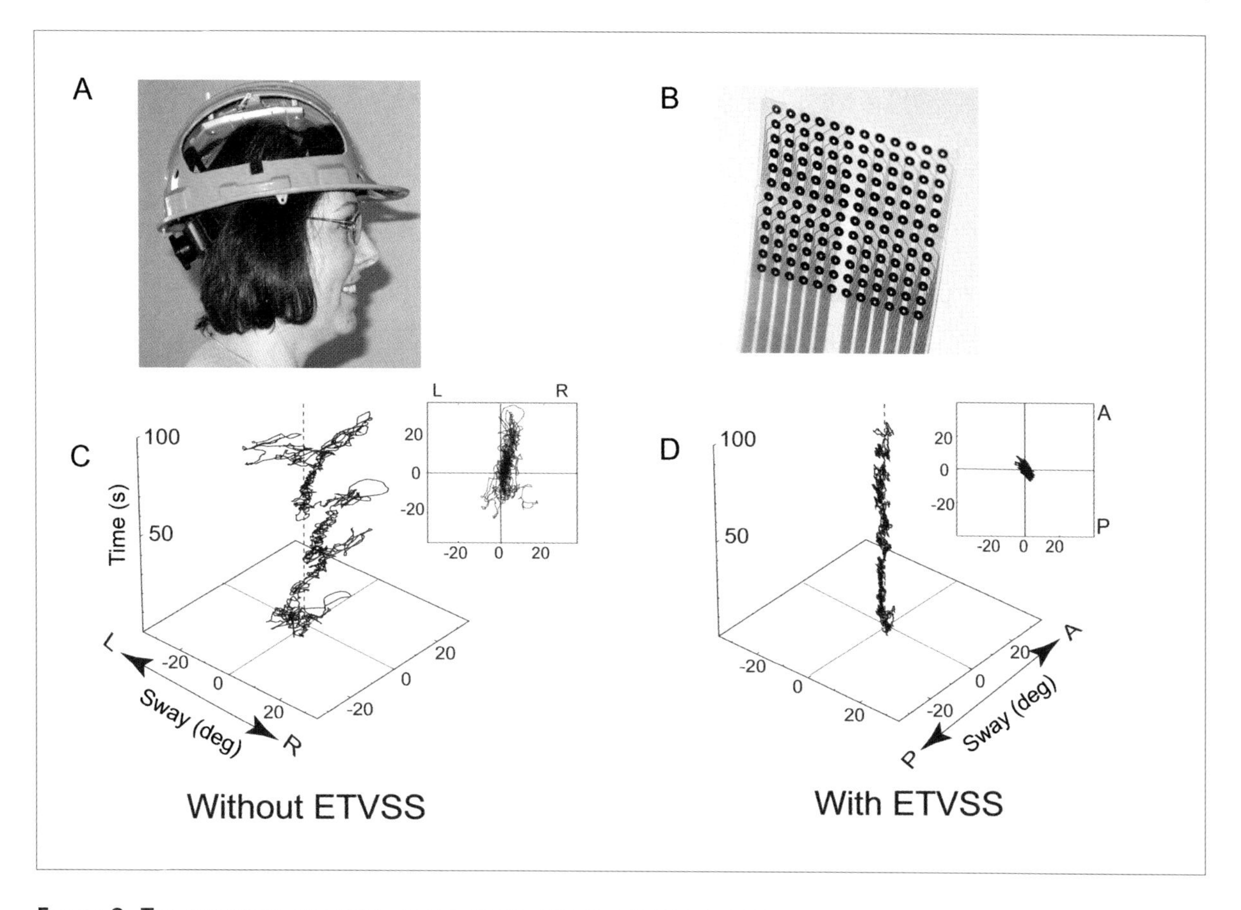

FIGURE 2. THE STABILIZATION EFFECT OF ELECTROTACTILE VESTIBULAR SUBSTITUTION SYSTEM
A. Helmet with 2-D MEMS accelerometer.
B. The 144 electrode array (3 x 3 cm).
C. 3-Dimensional graph showing subject head position without ETVSS system as a function of time (vertical axis). The graph of head displacement in both anterior (forward)/posterior (backward) (A/P), and medial-lateral (Left/Right) directions for an adult BVL (bilateral vestibular loss) subject with eyes closed and sitting upright without back support. Mean sway amplitudes: ± 3.0 deg. (L/R); ± 7.6 deg. (A/P). Note the slow drift of the mean position, and occurrence of periodic (~23 s) sway perturbations.
D. The same graph showing stabilization effect of ETVSS system. Subject is using tongue stimulation as a reference point instead of missing gravity sensation to stabilize head and upper torso in space. Insets: 'Spaghetti' plots of corresponding displacement profiles are representing 2-D projection onto a horizontal plane.

Subjects

We will use for demonstration in this chapter data recorded on subjects with bilateral vestibular loss (BVL) tested and trained using the ETVSS. The dysfunction of the participant was a result of ototoxicity from the use of the amino glycoside antibiotic gentamycin. Some people are very sensitive to this drug (about 3–5% of population). After a few hours of circulation of such drugs in the bloodstream, it starts to affect membranes of hair cells in auditory and vestibular systems. Hair cells malfunction and are finally destroyed in both vestibular apparatus, leaving central mechanisms (nerves, vestibular nuclei and vestibular cortex) intact.

Testing and training procedure

To determine abilities prior to testing, each subject completed a health questionnaire as well as a task ability questionnaire, along with the required informed consent forms. Prior to testing each individual was put through a series of baseline tests to observe their abilities with regards to balance and visual control (oscillopsia). These baseline tests were videotaped.

Prior to undergoing a 20-min trial each individual underwent a series of data captures with the ETVSS designed to obtain preliminary balance ability baselines as well as to train them in the feel and use of the system. These data captures included 100, 200 and 300-second trials both sitting and standing, eyes open and eyes closed.

Upon completion of the balance ability baselines and confirmation from them that they fully understood the EVSS system and how it operates, each individual proceeded into the 20-min trials and/or were trained to stand on soft materials or in Tandem Romberg posture. For all patients both conditions were 'unimaginable' to perform. Indeed, none of the subjects could complete more than 5–10 s stance in any conditions.

Typical testing/training included nine sessions 1.5–2 h long (depending on patient stamina and test difficulty). The shortest series a patient completed was five sessions, while the longest was 65 sessions.

Electrotactile feedback stabilization effect

As a result of training procedures with the ETVSS, all ten patients demonstrated significant improvement in balance control. However, speed and depth of balance recovery varied from subject to subject. Moreover, we found that training with the ETVSS system demonstrated not one, but rather several different effects or levels of balance recovery.

We can separate at least two groups of balance recovery effects of ETVSS training: direct balance effects (stabilization and training) and residual balance effects. In addition to balance recovery effects, we found multiple effects directly or indirectly related to the vestibular system, for example improvement of vision (oscillopsia and eye movement control) and sleep improvement.

Stabilization effect

Stabilization was observed in sitting and standing BVL subjects almost immediately (after 5–10 min of familiarization with ETVSS) and included the ability to control stable vertical posture and body alignment (sitting or standing with closed eyes) (see Fig. 2C, D and [26]).

Training effect

Some of our BVL patients, especially after long periods of compensation and extensive physical training during many years, had already developed the ability to stand straight, even with closed eyes, on hard surfaces. However, even for well-compensated BVL subjects standing on soft or uneven surfaces or stances with limited bases such as during a Tandem Romberg stance, standing was challenging, and absolutely unthinkable with closed eyes.

Using the ETVSS, BVL patients not only acquired the ability to control balance and body alignment standing on hard surfaces, but also the ability to extend the limits of their physical conditioning and balance control. As an example, standing in the Tandem Romberg stance with closed eyes became possible. After one training session of 18 training trials each 100 second

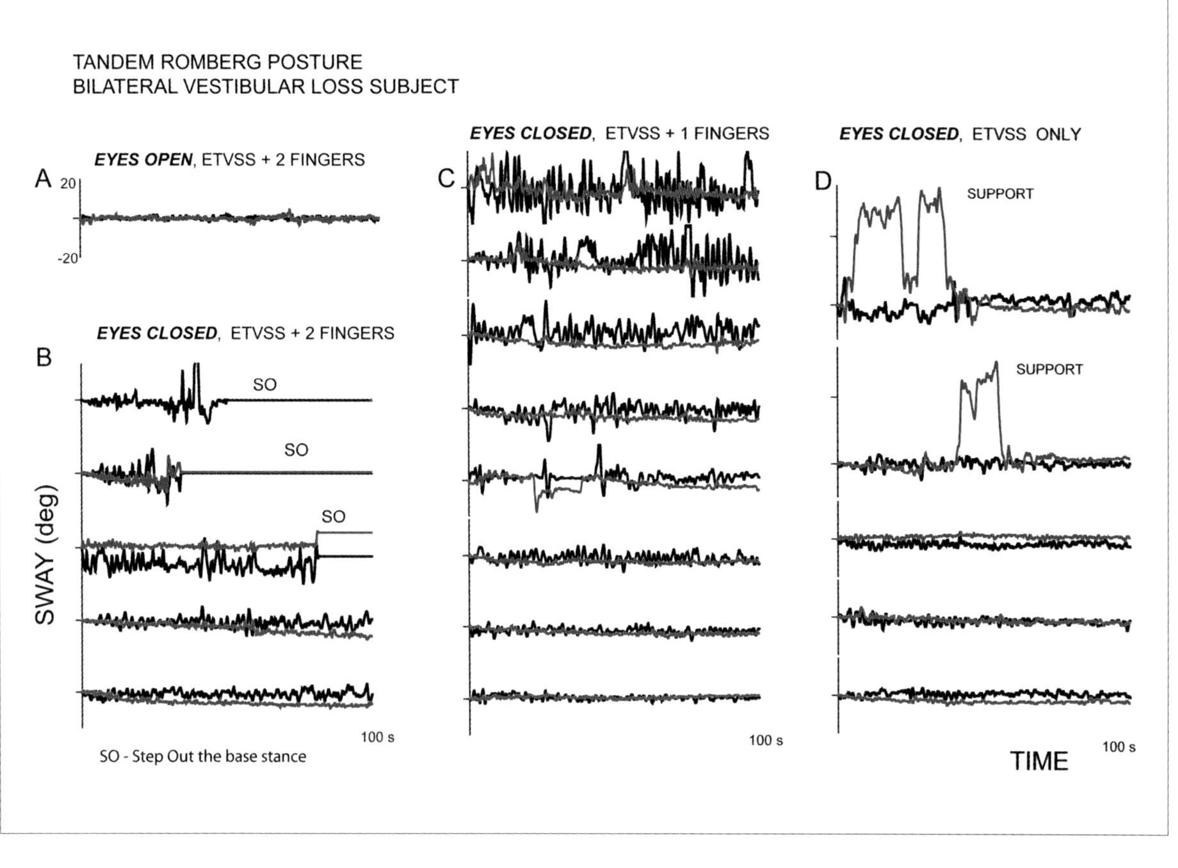

FIGURE 3. TRAINING EFFECTS OF THE ETVSS.

Initially, the BVL patient was absolutely unable to stand in Tandem Romberg position (the patient should stand with one foot in front of the other, both feet aligned strait) even with open eyes for few seconds.

Only with additional stabilization available (index fingers of both hands touching a wall, using ETVSS and keeping eyes open) it was possible to keep balance for 100 testing seconds with reasonable confidence (A). Two lines on each graph represent displacement in two directions – forward/backward and side to side. In tandem Romberg position subjects swaying with largest amplitude in a side to side direction.

B. Five sequential training trials (100 s plus 60 s resting time), performed in the same conditions as in A, but eyes closed. Note, that in the first three trials, subject lost balance and stepped out of basic position (SO). However, the last two trials were uninterrupted; the subject was capable to keep stability with eyes closed.

C. The next eight sequential trials, performed when only one index finger touching the wall, as additional external tactile support. Note the dramatic increase of amplitude of postural sway in the first two trials, sequential trial-to-trial improvement and complete lack of SO events. Posture stabilization during the last trial was almost identical to the trial in condition A.

D. The five sequential trials when subject did not use any additional external tactile stabilization. Only ETVSS system was used with eyes closed conditions. During the first two trials subject was supported by therapist (using quick light touches to the shoulder in the deltoid area) to keep posture uninterrupted when sway amplitude was critical.

In total 18 sequential trials (less than 2 h) the permanently damaged vestibular subject was capable to learn and train the body using only ETVSS to keep balance with closed eyes in one of the most difficult postures.

long (total ETVSS exposure time 30 min), a BVD patient was capable of standing in the tandem Romberg stance with closed eyes for 100 s (Figure 3).

Residual balance effects also were observed in all tested BVL patients; however strength and extent of effects significantly varied from subject to subject depending on the severity of vestibular damage, the time of subject recovery, and the length and intensity of ETVSS training.

At least three groups of residual balance effects were noted: short-term residual effects (sustained for a few minutes), long-term residual effects (sustained for 1–12 h) and a rehabilitation effect that we observed during several months of training in a single subject. All residual effects were observed after complete removal of ETVSS from the subject's mouth.

Short-term retention effect

This effect usually was observed during the initial stages of ETVSS training. Subjects were able to keep balance for some period of time, without immediately developing an abnormal sway; as it usually occurred after any other kind of external tactile stabilization, like touching a wall or table. Moreover, the length of short-term after effects was almost linearly dependent on the time of ETVSS exposure (Fig. 4A, B, C). After 100 seconds of ETVSS exposure, stabilization continued during 30–35 s, after 200 s ETVSS exposure 65–70 s and after 300 s EVSS trial the subject was able to maintain balance for more than 100 s. Short-term retention effect continued during approximately 30–70% of the EVSS exposure time [26].

Long-term retention effect

This group of effects developed after long (up to 20–40 min) sessions of ETVSS training in sitting or standing subjects and continued for a few hours. The duration of the balance improvement retention effect was much longer than after the observed short-term after effect: instead of the expected 7 min of stability (if we were to extrapolate the 30% rule on 20 min trials), we discovered from 1–6 h of improved stability (Fig. 4D, E).

During these hours BVL subjects were able to not only stand still and straight on a hard or soft surface, but were also able to accomplish completely different kinds of balance-challenging activities, like walking on a beam, standing on one leg, dancing. One subject brought a bicycle to the lab and was able to ride it. However, after a few hours all symptoms returned.

The strength of long-term retention effects was also dependent on the time of ETVSS exposure: 10 min trials were much less efficient than 20 min trials, but 40 min trials had about the same efficiency as 20 min. Usually, 20–25 min was the longest comfortable and sufficient interval for standing trials with closed eyes. Sitting trials were less effective than standing trials.

The shortest effects were observed during initial training sessions, usually 1–2 h. The longest effect after a single ETVSS exposure was 11–12 h. The average duration of long-term retention effects after single 20 min ETVSS exposure was 4–6 h.

Rehabilitation effect

We were capable of repeating two or even three 20 min ETVSS exposures to a single subject during one day. After the second exposure effects were continued on average in about 6 h. In total, after two 20-min ETVSS stabilization trials, BVL subjects were capable of feeling and behaving what they described as 'normal' for up to 10–14 h a day.

One BVL subject was trained continuously during 20 weeks, using one or two 20-min ETVSS trials a day. We found systematic improvement and gradual increase of the long-term retention effect for the same subject during consistent training. Moreover, we found that repetitive ETVSS training produced both accumulated improvement in balance control, and global recovery of the central mechanisms of the vestibular system.

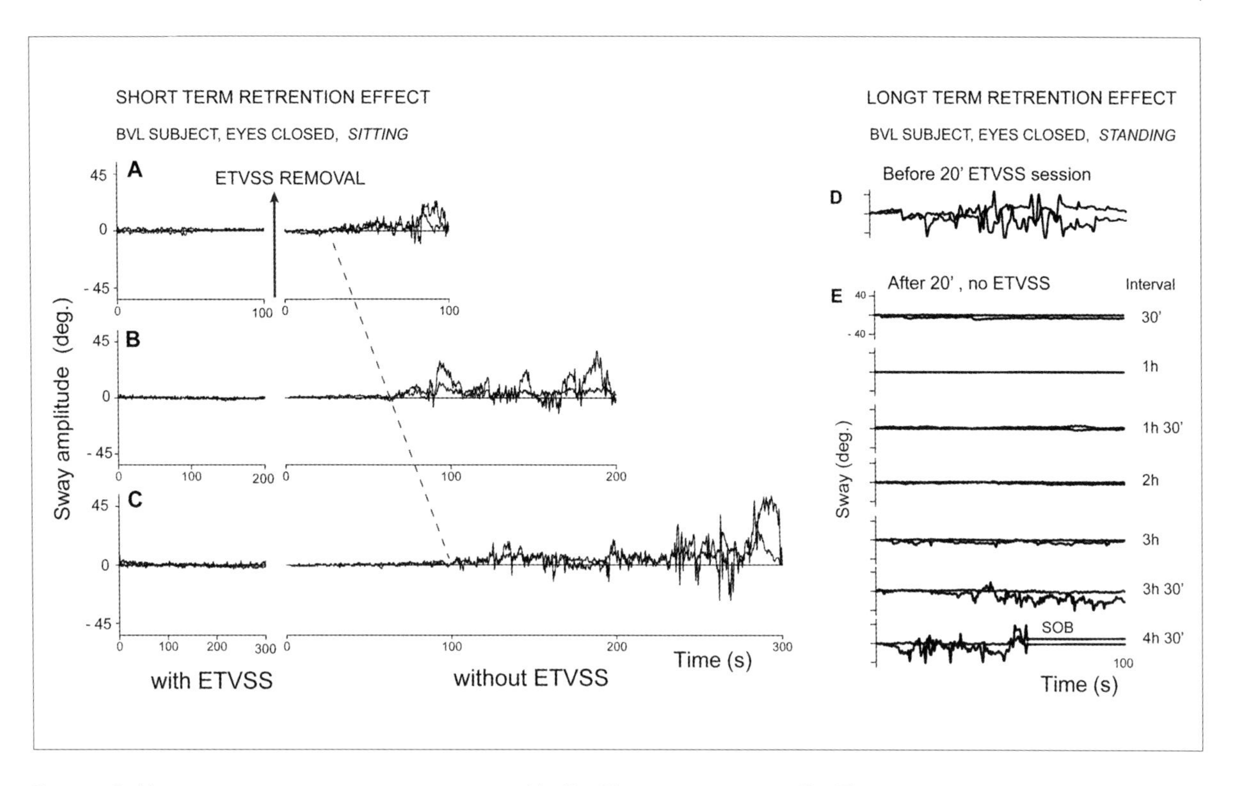

FIGURE 4. VESTIBULAR SUBSTITUTION SHORT TERM (A, B, C) AND LONG TERM (D, E) RETENTION EFFECTS

A, B, C. Results of the head stabilization using ETVSS with eyes closed in sitting conditions. Left half of each row represents trials with ETVSS, right half represent trials after tongue array removal, the post-EVSS period. A consistent and proportional period of head-postural stability without ETVSS exists across the three test durations (A – 100, B – 200, and C – 300 s for each phase). Sway oscillations consistently begin with small A/P motion at the head at approximately 30% into the post-ETVSS period, while the torso remains initially stable. Head-movement gradually increases in amplitude, and at approximately 70% of the post-ETVSS period motion begins to involve the torso, again growing in amplitude over time until the subject is unable to maintain stability.

D, E. The example of the long-term retention effects. Before the training session, the same BVL subject demonstrated strong sway standing with closed eyes in normal stance (D). Then, the subject was trained to stay still with ETVSS device during 20 min with closed eyes. After 20' training session, subject ability to keep balance was tested in normal stances with closed eyes after regular 30 and 60 min intervals, using standard 100 s testing trials (E). Note, the subject was able to keep very solid balance during first 2 h after training session, slowly worsening during another 2 h, and returned to original in-stability and sway more than 4 h after training session.

For the same BVL subject, after two months of intensive training, we completely stopped any ETVSS exposure and begin regular checking of the subject's balance and posture control. During the 14 weeks after the last ETVSS system training, the subject was able to stay perfectly still with closed eyes, while standing for 20 min on hard or soft surfaces. We considered that fact as convincing evidence of the actual rehabilitation capability of the ETVSS.

Box 2. Vestibular Disorders: Key statistics

- Vestibular disorders affect about 40% of the general population (‰122 million people in the US), usually reported as 'dizziness', at some point in their life [74].
- Second most common cause for primary care provider visits (>6 million in 2005). Prevalence increases with age: approximately 25–33% of the elderly complain of some form of dizziness, and by age 75 this proportion increases to nearly 50% [74].
- Approximately 12.5 million Americans over the age of 65 experience chronic balance or dizziness problems that do not respond to traditional treatment, and significantly interfere with their lives [75].
- The cost of medical care for patients with balance disorders has been estimated to exceed $1 billion per year in the US [76].
- Of all falls suffered by the elderly, 50% are the result of vestibular problems.
- Cost of medical treatment for falls exceeded $6.8 billion in the US (2005) [76].

Effects of vestibular loss

Chronic balance disorders due to vestibular loss result in significant functional disability. Peripheral origins of this disorder may be due to a variety of neurosensory or fluid-mechanical malfunctions. Central disorders usually result from brainstem or cerebellar lesions due to stroke or other trauma. The symptoms and signs of severe vestibular loss are well described and can include oscillopsia, ataxia, dysmetria, rigidity, nausea, lightheadedness, episodic vertigo, veering while walking, an inability to walk in the dark, read while walking, or stand on compliant or unstable surfaces, and decreased gait velocity [77].

To date, the treatment of choice for individuals with moderate to profound loss is vestibular rehabilitation. Vestibular rehabilitation therapy (VRT) has been shown to be effective for numerous vestibular disorders but has also been shown to be less effective for persons having central disorders, severe unilateral, or mixed etiologies [78, 79]. Patients with Meniere's syndrome, however, are typically not considered good candidates for VRT. Those with central vestibular disorders have poorer outcomes of rehabilitation overall, are more likely to have significant balance impairment, and do not progress as quickly as those with peripheral vestibular disorders [80–82].

These reports also reveal that roughly half of all vestibular disorder patients do not derive meaningful benefit from VRT. Consequently, therapy for this latter group consists of intensive development of compensatory strategies (forced visual fixation, use of frequent tactile waypoints, increased double-limb support, use of a cane/walker, avoidance of visually dynamic or poorly lit environments, and avoidance of physically unstable surfaces). Furthermore, problems of this nature are not amenable to surgical treatment, and pharmacological management of these conditions with vestibular suppressants often retards the recovery process. Consequently, these people are left to manage their lives as well as they can with inadequate therapeutic intervention.

Another aspect of chronic balance disorders is that they dramatically alter the sufferer's ability to engage in daily activities, which greatly diminishes their quality of life. Patients usually present with ataxia, dysmetria, tremor, which make standing or walking difficult and necessitating the use of a cane or walking stick for additional support. For the individual with severe chronic vestibular dysfunction, the situation is even more distressing in that they also frequently experience oscillopsia, nystagmus, and hearing loss. They must employ adaptive strategies such as moving from one tactile or 'touch' way-point to the next in order to regain stability and spatial orientation. Furthermore, because both vestibular and visual cues for postural control are compromised, patients with chronic balance disorders are at far greater risk of falls, and must rely on constant vigilance and restricted physical activity to avoid instability and falls.

General effects of ETVSS training

Balance improvement

Subjects experienced the return of their sense of balance, increased body control, steadiness, and a sense of being centered. The constant sense of moving was disappeared. They were able to walk unassisted, reported increased ability to walk in dark environments, to walk briskly, to walk in crowds, and to walk on patterned surfaces. They gained the ability to stand with their eyes closed with or without a soft base, to walk a straight line, to walk while looking side to side and up and down. They gained the ability to carry items, walk on uneven surfaces, walk up and down embankments, and to ride a bike. They became willing to try new things, and in general became much more physically active.

Posture, proprioception and motor control

Body movements became more fluid, confident, light, relaxed and quick. The stiffness disappeared, limbs, head and body felt lighter and less constricted. Fine motor skills returned, and gait returned to normal. Posture and body segment alignment returned to normal. Stamina and energy increased.

All these functions ultimately relate to the individual's ability to maintain employment, personal and social activities, family life and relationships leading to a great improvement in their quality of life. The results emphasized the truly rehabilitation effects of ETVSS.

Summary

It should be clear by now that sensory substitution using tactile displays, coupled with brain plasticity, provides a powerful means to provide substitute or augmentative information for rehabilitation, training, and enhanced human-machine interfaces [72]. A well-designed sensory substitution system becomes part of its experienced user, enabling direct perception of external events: balance is perceived not on the tongue, but innately; vision is perceived not on the abdomen, but in front of the camera; a prosthetic hand with tactile sensors feels and functions less like a tool, and more like a real limb. It may indeed be feasible to convert any kind of sensory information to touch, which is an underappreciated sense unless it is missing. The sense of touch is highly extensive and redundant; with approximately 2 m^2 of skin, the possibilities for tactile communication are only beginning to be explored. The challenges ahead are three-fold: we need to (1) better understand the capacity of human haptic perception to mediate new forms of information, (2) develop better haptic (force + spatial pattern) displays, and (3) develop better ways to encode multimodal sensory and supplemental information to achieve optimal human performance.

Acknowledgements

The research and development reported here was supported by an NIH SBIR grant (R44 DC004738) from the NIDCD, and from Industrial & Economic Development Research grants from the UW-Madison Graduate School.

Recommended reading

Bach-y-Rita P (1972) *Brain mechanisms in sensory substitution*. Academic, New York

Kaczmarek KA, Bach-y-Rita P (1995) Tactile displays. In: W Barfield, T Furness (eds): *Virtual environments and advanced interface design*. Oxford University Press, New York, 349–414

Kaczmarek KA, Webster JG, Bach-y-Rita P, Tompkins WJ (1991) Electrotactile and vibrotactile displays for sensory substitution systems. *IEEE Trans Biomed Eng* 38: 1–16

Loomis JM (2003) Sensory replacement and sensory substitution: Overview and prospects for the future. In: MC Roco, WS Bainbridge (eds): *Converg-*

ing technologies for improving human performance: nanotechnology, biotechnology, information technology and cognitive science. Kluwer Academic, Boston

Saunders FA (1977) Recommended procedures for electrocutaneous displays. In: FT Hambrecht, JB Reswick (eds): *Functional electrical stimulation: applications in neural prostheses*. Marcel Dekker, New York, 303–309

Szeto AYJ, Riso RR (1990) Sensory feedback using electrical stimulation of the tactile sense. In: RV Smith, JH Leslie Jr (eds): *Rehabilitation engineering*. CRC Press, Boca Raton, FL, 29–78

The blind get a taste of vision

Maurice Ptito, Daniel-Robert Chebat and Ron Kupers

Introduction

Sensory substitution and cross-modal plasticity

In sensory substitution, a given sensory modality acquires the functional properties of a missing one. This phenomenon is due to a reorganization of the sensory systems that are deprived of their normal input through a process called cross-modal plasticity [1]. 'Rewiring' studies carried out on ferrets [2] and hamsters [3] provided strong support for these phenomena. For example, lesions of central retinal targets induce the formation of new and permanent retinofugal projections into non-visual thalamic sites such as the auditory nucleus [3]. Single neurons in the auditory cortex of these rewired animals respond to visual stimuli and some of them respond equally well to auditory as to visual stimuli. Moreover, those cells that respond to visual stimuli show properties (e.g., orientation selectivity, motion and direction sensitivity) similar to those encountered in the visual cortex of normal hamsters. At the behavioral level, rewired hamsters can learn visual discrimination tasks as well as normal animals and a lesion of their auditory cortex provokes cortical blindness [4]. These data raise the question whether similar processes happen in congenitally blind humans. The absence of visual input from birth leads to the recruitment of the visual cortex by other sensory modalities such as touch or audition [5]. Most studies in man have been carried out in blind subjects who have had many years of experience with Braille reading and it is difficult to conclude on brain reorganization since the extensive reliance of these subjects on tactile or auditory stimulation may by itself result in enhanced activity in the occipital cortex [6, 7]. To avoid this bias, we took advantage of the tongue display unit, a tactile to vision sensory substitution device which does not use the fingertips or the auditory system and which is therefore equally novel for early blind, late blind and healthy controls. This device translates visual information into electrotactile stimulation which is delivered to the tongue by means of an electrode array.

The tongue display unit (TDU)

The tactile visual substitution system (TVSS) comprises the TDU, a laptop computer with custom made software and an electrode array (3×3 cm) consisting of 144 gold-plated contacts each with a 1.55 mm diameter arranged in a 12×12 matrix. The device has an update rate of 14–20 frames per second. An electrical pulse (40 µs positive pulse) is generated when a stimulus is presented. The TDU is the descendant of a series of devices developed by the late Pr. Paul Bach-y-Rita over half a century of research. Bach-y-Rita experimented with the potential of the skin as a channel for transmitting pictorial material [8]. The basic idea was to transmit real time images from a video camera to the skin *via* electro- or vibrotactile stimulation. The first systems translated the video images from the camera into vibrotactile stimuli which were delivered to the back of the subjects *via* a 20×20 matrix of solenoid vibrators, spaced 12 mm apart, mounted on the back of a dentist's chair. The 400 stimulators covered 645 cm^2 of the back and enabled subjects to recognize certain shapes and to make judgments about object orientation. These early prototypes were bulky and lacked mobility and maneuverability. Therefore, Collins and Bach-y-Rita conceived a more portable version in 1973 [9]. In this system, the lens of a light-weight camera was mounted on a pair of eyeglasses. The information captured by the camera was translated into electrotactile stimuli that were relayed in a point-to-point manner to the abdominal skin *via* a flexible electrode matrix. Case studies with such devices demonstrated that it was possible to make distance discriminations and even to catch objects in motion [10]. In the late 1990s, the device was adapted to stimulate the tongue instead of the abdomen. Thereto, the device was further miniaturized and made more aesthetically acceptable by hiding the electrode matrix inside a small dental retainer [11]. The new 49 points (7×7 array) matrix proved to be more efficient when stimulation was delivered to the tongue instead of the fingertips. Moreover, tongue stimulation only required a fraction (3%; 5–15 V) of the voltage needed for stimulation of the fingertips. A better image resolution was obtained by augmenting the number of electrodes in the matrix placed on the tongue from 49 to 144 points (see Fig. 1). Point-to-point tactile discrimination capacity of the tongue is superior to that of the skin of the back that allows more information to be presented to a smaller skin surface. The tongue representation in the primary somatosensory cortex also covers a much larger area than that of the back, making it a highly sensitive tactile organ.

The TDU was recently adapted for navigational purposes (Brainport, WICAB Inc., Wisconsin, USA). The portable system consists of a webcam connected to a laptop and a tongue stimulator array of 100 small electrodes with a diameter of 1 mm, spaced 1 mm apart and arranged in a 10×10 matrix. The entire tongue array measures 2×2 cm, and can be comfortably placed on the tongue due to its smooth and soft silicone mouthpiece, which rests on the roof of the mouth (see Figs 2, 5). This model has the advantage of being much smaller than previous versions, thereby enabling true mobility of the user for navigating and exploring his surroundings.

Figure 1. The tongue display unit (TDU) and its components.
Left: The TDU box and the tongue grid. Right: A subject being tested with the device.

Other haptic devices

Over the past two decades, haptic devices invaded many aspects of our daily lives, creating more efficient user-machine interfaces. Vibrating cell phones, video games and tactile displays have become commonplace in society today. There are also many haptic devices aimed at improving the quality of life of the blind in terms of accessibility to various types of visual information. We will present here only some of these devices that have been placed into two categories: those aimed at computer navigation, and those aimed at real world navigation.

Computer navigation

The Optacon (Optical to Tactile Converter) was one of the first set of devices aimed at helping blind people to navigate in a virtual computer world. The Optacon is an electro-mechanical system allowing blind people to view printed material that has not been transcribed into Braille or other pictorial representations. The Optacon consists of an electronic box connected to a lens and a tactile array on which the blind subject places his/her index finger. By moving the lens over a computer screen with printed text, the text on the screen is translated into tactile stimulation through a raised dots system which recreates the image in a point-to-point topographic fashion.

Just like the Optacon, the virtual haptic display (VHD) is a system aimed at transmitting haptic virtual information for blind computer users. The VHD is a computer-controlled apparatus for presenting haptic stimuli using active touch, a process by which the subject gets feedback from the device in real time, as he explores a virtual shape. The images are transferred onto a topographic, two-dimensional array of taxels (tactile-pixels) [12]. The taxels are raised or lowered according to the shape of the two-dimensional object being probed. Untrained participants were able to feel the virtual shape by placing their fingers on a carriage that is moved over the array of taxels and they were able to recognize complex 2-D shapes [12].

The Impulse engine 3000 (Immersion Corporation: see www-ais.itd.clrc.ac.uk/.../hardwick/100.html for details) is a device that permits a pictorial representation of graphics involved in computer navigation [13]. The device consists of a pen-like probe attached to three motors inside a small box. The three motors offer resistance when the pen is moved simulating the contours of an object. The motors create the illusion that the object is trapped inside the box while it is being explored with the pen probe. The representation of the object is thus magnified enhancing therefore the small details of an object.

Real world navigation

Other systems inspired by the TDU for electro-tactile stimulation use different sizes of electrode arrays, but exploit the same basic idea of translating images into electro-tactile stimulation. The Forehead Retina System (FST; EyePlusPlus, Inc. of Japan and Tachi Laboratory in the University of Tokyo) is designed to stimulate the skin of the forehead with 518 electrodes recreating the images from a camera. The system uses a high voltage of switching integrated circuit, which is normally used to drive micro-machines such as digital micro-mirror devices.

Another such device translates images from a camera in real-time into a 4×8 electrode array that is applied to the highly sensitive lip area. Each electrode only measures 700 um in diameter [14]. Just like the tongue, the lips have a dense mechanoreceptor innervation and a large somatosensory cortical representation. A psychophysical study on lip perception showed that the lips also have a high spatial discrimination capacity. Correct response rate in a spatial discrimination task was close to 85% for subjects trying the device for the first time, the upper lip being slightly more sensitive than the lower lip [15].

Like the FSD, the 'smart finger' and the 'smart tool' are part of the smart touch devices developed by the Tachi laboratory in Japan. These haptic systems are designed to inject the sense of touch into a computer interface and to provide information about the real world in real-time. Smart touch devices can also be applied to the finger used to manipulate a computer mouse. The vibrotactile pad is placed on the mouse and displays information to the finger concerning the virtual world. This type of vibrotactile technology, already commonly used in video games, can also be used in situations were high precision manipulations are needed, such as during surgery. Feedback about the texture of the object is amplified and relayed by means of electrical or vibrotactile information on the touch pad. The smart finger consists of a sensor and a vibrotactile display which is attached to a finger nail. Information about the texture of the surface is given through vibration of the device on the nail. In a different model of the device, a familiar tool (e.g., a surgical knife, or a pen) is used as a transmitter of vibrotactile information regarding the texture of the surface the sensor is in contact with and allows the user to point to objects on a computer screen in order to extract pertinent information from the stimulus.

The prototype power-assisted wheelchair is another navigational aid for blind people. In contrast with the devices described above, this machine actually makes turning and direction decisions for the user. This chair is equipped with sensors all around capable of recognizing nearby obstacles. When the user pushes the wheels of the chair, sensors predict the velocity, torque and direction of the chair. Using navigation assistance software, motors adjust the speed and direction of the chair in order to avoid obstacles. The wheelchair gives a haptic feedback of the environment [16]. The wheel hubs offer greater resistance when there is an obstacle in the vicinity and direction of the chair. This allows the user to get a feeling of the environment before coming too close to an obstacle.

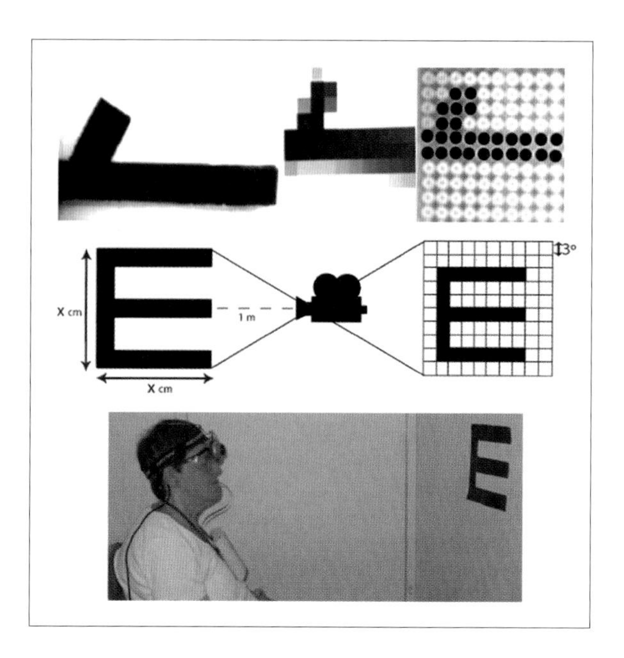

FIGURE 2. APPARATUS AND DESIGN FOR TESTING VISUAL ACUITY WITH SNELLEN TUMBLING E.
Upper portion: the image as captured by the camera, then translated topographically unto the tongue grid in the form of electro-tactile pulses. Lower portion: The subject fitted with the tongue grid for visual acuity testing.

Applications: the blind get a taste of vision

The TDU has been used to evaluate a variety of 'visual' functions in our laboratory, not only at the psychophysical level but also using various brain imaging techniques.

Tactile-'visual' acuity of the tongue

It is possible to determine the visual acuity of the tongue using Snellen letters as it is done in a classical visual clinical examination (Fig. 2). We used a modified method of limits to evaluate the tongue acuity when the subject was presented Snellen tumbling Es of various sizes. No overall significant difference was found in thresholds between early blind (1/206) and sighted controls (1/237). In the two highest visual acuity categories (1/150 and 1/90) however, we found a significantly larger proportion of early blind subjects ($p < 0.0001$) [17]. Without any training, 18% of all subjects (blind and sighted) reached the limits of visual acuity permitted by the device. WICAB is already working on creating a tongue array with more electrodes to increase the resolution of the device.

Orientation discrimination

Born blind subjects and normal blindfolded controls were trained in a task where they had to determine the orientation of the letter T applied to the tongue *via* the TDU. Before training, they underwent a familiarization period during which they were introduced to the apparatus and received a few trials. Thereafter they were scanned for the first time while performing the orientation discrimination task in a Positron Emission Tomograph (PET) examination. Next, participants were trained in the discrimination task until they reached the learning criterion of 85% or more on two consecutive days. Once criterion reached, subjects underwent a second PET scanning while performing the task. This study design is unique since it allows us to obtain functional

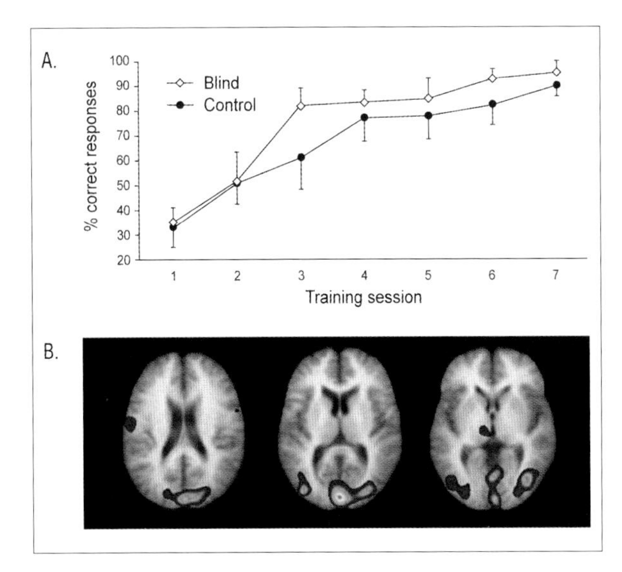

FIGURE 3. ORIENTATION DISCRIMINATION LEARNING IN CONGENITALLY BLIND AND BLINDFOLDED SEEING CONTROLS
A. Performance curves. B. Activation of the visual cortex measured with $O_{15}H_2O$ PET.

brain images before and after subjects became acquainted with the task. Our results showed that all subjects, blind and controls, learn the task at the same speed (Fig. 3). However, only the blind recruited their visual cortex post-training. No visual cortex activation was observed during the first PET scanning session, showing that the recruitment of the visual cortex is training-dependent (Fig. 3). We concluded that when deprivation of a particular sensory modality occurs at birth, the 'deprived' cortex can be recruited by other sensory modalities [18], supporting the concept of cross-modal plasticity. This is reminiscent of results obtained in numerous studies on Braille reading in which tactile input to the fingers activates the visual cortex in proficient blind Braille readers (see reviews in [1, 5, 19]).

Motion detection and motion discrimination

The visual cortex of the blind has a supra-normal perfusion at rest as shown by fluorodeoxyglucose-PET methodology [20] and confirmed in several PET water activation studies [21]. In normal seeing subjects, the visual system consists of a dorsal (where) and a ventral (what) stream [22]. The dorsal stream originates in the visual cortex and runs dorsally to extrastriate areas in the parietal and temporal lobes and is involved in motion perception. This raises the intriguing question whether the functional separation of the visual cortex in a dorsal and a ventral stream remains preserved in subjects who have never experienced any visual input. In a series of experiments, we tested whether we could activate the dorsal stream in congenitally blind subjects when presenting, *via* the TDU, 'visual' stimuli that are optimal for activating the dorsal stream. Random static or moving dots were presented *via* the TDU and subjects had to report if the stimulus was moving or static. In the following step, they had to discriminate the direction of motion along the x-axis (left to right and right to left). Blind and control subjects performed equally well in both tasks with a slight advantage for the blind. Brain imaging results showed that the dorsal stream was activated during the motion discrimination task. In the blind but not in the control subjects, extrastriate area 19, posterior parietal area 7 and the middle temporal cortex (area MT) showed statistically increased levels of cerebral blood flow [23] (Fig. 4). These results were confirmed in subsequent study using functional Magnetic Resonance Imaging (fMRI) [24].

Shape recognition

The ventral stream runs from the visual cortex to the inferior temporal lobe and is mainly involved in shape recognition. Blind and blindfolded sighted controls had to use the TDU to learn to recognize the shape of four simple geometric figures: squares, triangles, crosses and circles. After reaching learning criterion (85% correct and above), participants performed the object recognition task during an fMRI study. Results showed a significant activation of the inferotemporal cortex which is the recipient area

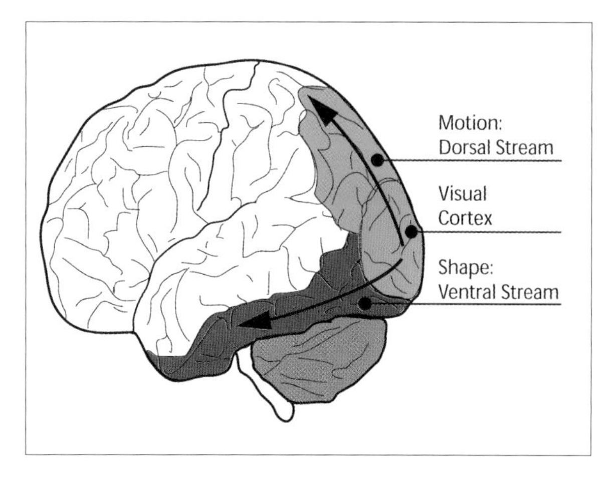

FIGURE 4.
Schematic of the two visual streams showing activations obtained with haptic motion stimuli (dorsal stream) and shapes (ventral stream) in congenital blindness.

of the visual cortex and the termination site of the ventral pathway [25] (Fig. 4).

In summary, these studies indicate that blind subjects can use haptic stimuli delivered to the tongue to grasp visual information from their environment and that the structural organization of the visual cortex into a dorsal and ventral stream seems to be preserved in the absence of any visual experience.

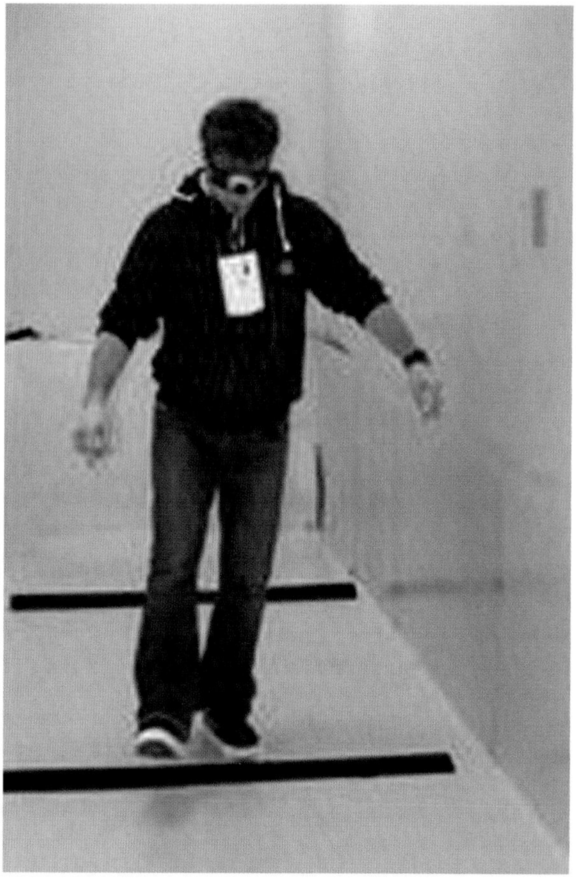

FIGURE 5. THE TDU IN THE NAVIGATION STUDIES
A subject fitted with the device in the obstacle corridor.

Navigational skills in the blind

The ability to detect, correctly identify and avoid obstacles is a crucial mobility problem faced by the visually impaired population [26]. The long cane is the most frequently used tool by blind and low vision adults to provide ground level information on the location and textural quality of obstacles [27]. The short range of the cane (1 m) and the fact that it fails to provide information concerning obstacles above waist level leaves much to be desired when navigating a novel environment or when novel obstacles appear on a well-known route. The TDU was originally designed with the objective of transmitting visual information from the environment in real-time. Even in its early days, the ultimate goal was to miniaturize the system to enable blind subjects to use it as a navigational aid. Indeed the TDU is able to deliver information about obstacles far beyond the reach of a conventional cane.

The experiments conducted in our laboratory demonstrated the efficiency of the system in delivering information about obstacles of varying sizes and difficulty. We tested the ability of blind subjects to navigate in an obstacle course with the TDU (Fig. 5). In a controlled environment, obstacles made of foam and cardboard were placed in different experimental corridors. The obstacles were made black and the corridors white in order to maximize contrast of the TDU display image. Wearing a camera mounted on a pair of eyeglasses, participants were requested to point to the obstacle (detection), move towards it and negotiate a path around it (avoidance). Several different obstacles were used to simulate an ecological environment and the kinds of obstacles that could be encountered on a sidewalk. Subjects walked five times through the same corridor while pointing and avoiding the obstacles. Performance improved after each trial and the number of errors (obstacles undetected or a collision with an obstacle) was significantly reduced during the fifth passage of the corridor. Results also demonstrated that subjects needed to recognize the obstacles (detect) in order to be able to locate them and to negotiate a path around them (avoid). Subjects were then able to navigate through a corridor with a novel configuration of obstacles with the same accuracy as they did in the practiced corridor [28]. Subjects demonstrated that the skill they had learned was transferable to a new set of obstacles. Surprisingly, there was no significant difference between sighted and blind subjects in their ability to detect and avoid obstacles. With training, subjects were able to detect obstacles up to 9 m away. The average distance of object detection in the blind group was 2 m meaning that they were able to detect obstacles that are on average double the distance away than what they could have sensed using a cane. A clip illustrating the ability of a congenitally blind subject to navigate in the obstacle course is presented at http//:www.tonguevision.blogspot.com.

We have previously demonstrated that the visual cortex of blind subjects using the TDU for various visual tasks shows a high degree of cross-modal plasticity [18]. Previous studies have also reported that the hippocampus, a structure involved in the formation of spatial cognitive maps, is capable of a certain degree of plasticity in normal seeing subjects [29]. We therefore hypothesized that the hippocampus of the blind could also be subject to a certain degree of plasticity. Our previous brain morphometry analysis had shown that the volume of the right posterior hippocampus in the blind is significantly reduced compared to that in sighted controls [30]. Despite this hippocampal alteration, congenitally blind individuals are not impaired on a spatial competence level [31] or in the formation of novel spatial maps of the environment [32]. The question therefore arises as to how blind people form spatial maps.

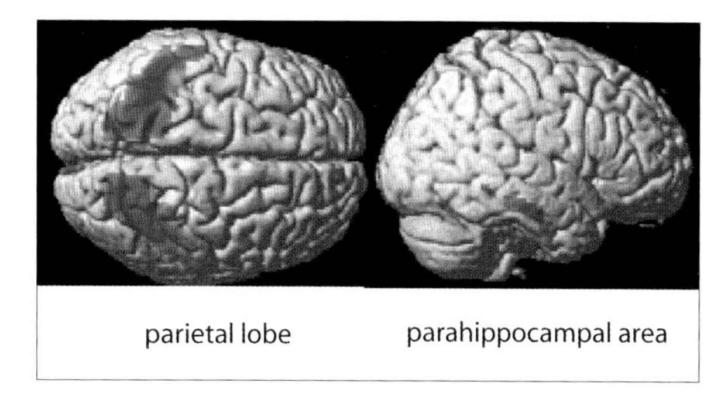

FIGURE 6.
Cortical activations obtained in a fMRI 3.0T scanner during performance in a virtual maze situation.

In order to answer this question, we used fMRI to investigate blind and sighted subjects' ability to remember and navigate through virtual hallways. We used a virtual maze task in which subjects used the TDU to move along a path. Blind participants were significantly better in path finding than blindfolded controls. At the same time, blind subjects activated several brains regions including the posterior parietal cortex and the parahippocampal place area [28] (Fig. 6). These results demonstrate that early blind people can actually form maps and in doing so, call upon structures involved in route forming.

Subjective experiences associated with visual cortex activation in the blind

We next addressed the question of the subjective character of this visual cortex activation in

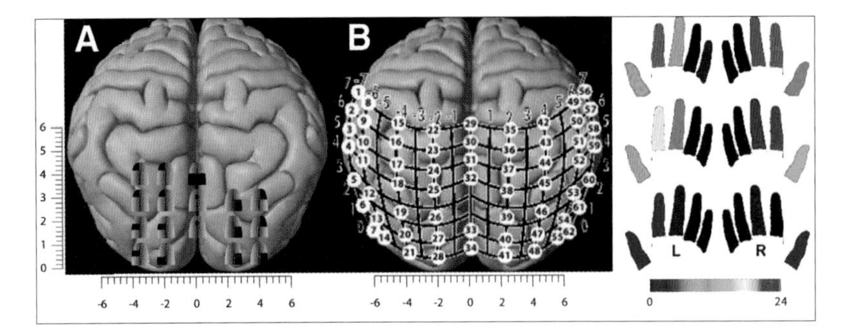

FIGURE 7.
Transcranial magnetic stimulation of the visual cortex produces tactile sensations referred to the tongue (A) and the fingers (B) of congenitally blind subjects.

the blind, by studying the subjective responses induced by transcranial magnetic stimulation (TMS) of the visually deprived and cross-modally responsive occipital cortex. More specifically, we wanted to test the possibility whether stimulation of the occipital cortex can induce subjective sensations or qualia associated with the new (tactile) input. We stimulated the occipital cortex with TMS in a systematic manner before and after training with the TDU in a group of blind and blindfolded seeing control subjects. As expected, TMS of occipital cortex in control subjects only elicited visual phosphenes. In sharp contrast, TMS in blind subjects provoked somatotopically organized 'tactile sensations' which were referred to the tongue. These were described as short-lasting experiences of distinct tingling, varying in intensity, extent and topography depending on the locus of the occipital cortex that was stimulated (Fig. 7a) [33]. The same phenomenon has been reported in the fingers of congenitally blind experts in Braille reading who had their visual cortex stimulated with TMS [34] (Fig. 7b).

Structural changes in the visual system of the blind

What happens to the visual cortex of the blind at the structural level? Our anatomical studies indicate that all afferents from the eyes to the visual cortex are largely atrophied [35]. How is this to reconcile with the findings described above that the visual cortex is still active in blind subjects? Together, this seems to suggest that the information reaching the visually deprived 'visual' cortex is coming from other sources. This hypothesis was confirmed in our brain morphometric study showing increased white matter connections in congenitally blind subjects between the frontal and the occipital lobe (fronto-occipital fasciculus) on the one hand, and between the parietal and the occipital lobe (superior longitudinal fasciculus) on the other hand. These findings give support to the cortico-cortical hypothesis according to which activation of the visual cortex of the blind comes from increased cortico-cortical pathways. An example of these enlarged connections in the blind's brain is shown in Figure 8.

Summary

The TDU is the most important haptic visual aid today. The TDU is also the only device that has been scientifically tested in a variety of situations. The TDU has proven its usefulness for detection of orientations, shapes and motion, and can be used by the blind to navigate in their environment, to scan for obstacles and even to read maps. This permits the representation of both allocentric (outer space) and egocentric space (inner space) since it allows the subject to 'perceive' stimuli at various distances from the camera mounted on the glasses or the forehead (Fig. 5). In Braille, stimuli have to be within

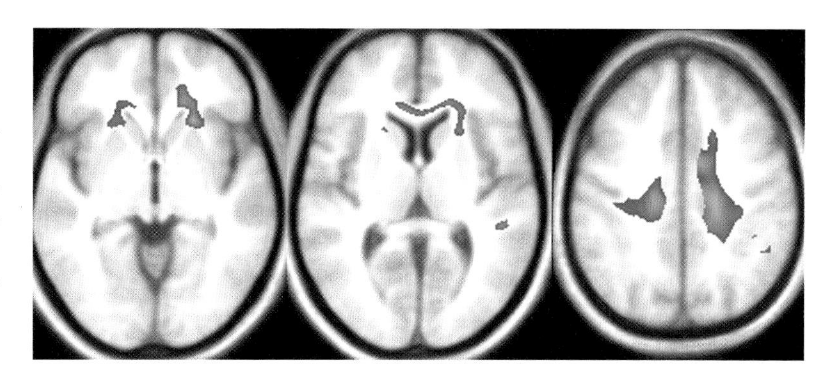

FIGURE 8.
Illustration of white matter increases in the cortico-cortical connections to the visual cortex of the blind derived from whole brain MRI voxel-based morphometry (VBM). FOF, fronto-occipital fasciculus; GCC, genu of the corpus callosum; SFL, supra longitudinal fasciculus.

reach of the subjects (arm length or cane length) whereas with the TDU distance is not an issue. This is useful for blind travelers interested in having a pictorial representation of the route they are traveling that a verbal description only could not provide in the same fashion. Having a tactile map readily available enhances the ability of blind subjects to remember a route [36]. Performance acuity on the TDU can be measured with a conventional visual acuity test. It was shown that visual acuity with the TDU in naïve subjects often reached the highest possible score permitted by the device. This suggests that the TDU has not yet reached the full potential resolution of the tongue, and that it is possible to further augment image resolution by increasing the number of electrodes on the tongue array. Wicab is currently working on developing such a device. Another advantage of the TDU is that it is hands-free allowing the subjects to use their hands for other activities. It can be used in conjunction with other navigational tools such as a cane or a guide dog. The small size of the tongue grid once wireless will make the device cosmetically desirable.

Acknowledgements

Our research is supported by the Harland Sanders Foundation and by the Danish Medical Research Council. Our thanks also to Wicab Co for generously providing the equipment used in our studies. This chapter is dedicated to Prof. Paul Bach-y-Rita for his invaluable contributions to the field of re-adaptation.

Selected readings

Merabet LB, Rizzo JF, Amedi A, Somers DC, Pascual-Leone A (2005) What blindness can tell us about seeing again: merging neuroplasticity and neuroprostheses. *Nat Rev Neurosci* 6: 71–77

Sathian K (2005) Visual cortical activity during tactile perception in the sighted and the visually deprived. *Dev Psychobiol* 46: 279–286

Sadato N (2005) How the blind "see" Braille: lessons from functional magnetic resonance imaging. *Neuroscientist* 11: 577–582

Ptito M, Schneider FC, Paulson OB, Kupers R (2008) Alterations of the visual pathways in congenital blindness. *Exp Brain Res* [Epub ahead of print].

Kupers R, Fumal A, de Noordhout AM, Gjedde A, Schoenen J, Ptito M (2006) Transcranial magnetic stimulation of the visual cortex induces somatotopically organized qualia in blind subjects. *Proc Natl Acad Sci USA* 103: 13256–13260

Tactile ground surface indicators in public places

Timm Rosburg

Introduction

Visual information is essential in traffic: traffic *lights* tell us when to cross the street. Zebra crossings signalise *visually* street sections where car drivers have to pay special attention to pedestrians. Children are taught to *look* to the left and to the right before crossing the street. Motorists are aware of the problems and hazards occurring by darkness, rain, snow or fog when range of *sight* is decreased. A lot of accident avoidance deals with the issue to *see* and to be *seen*, e.g., the failure of motorists to detect and recognise motorcycles in traffic was regarded as the predominating cause of motorcycle accidents [1].

Blind and visually impaired people have to manage their way through a world of traffic which was mainly created for people with full eyesight. According to statistics compiled by the National Center for Health Statistics of the US Centers for Disease Control and Prevention, as summarised by the American Foundation for the Blind, in 1996, there were approximately 1.3 million legally blind people in the United States. 5.5 million elderly individuals are regarded as blind or visually impaired [2]. As others, these individuals benefit from a barrier free architecture.

The idea of barrier free architecture is to provide a structure which enables everyone independently of his/her physical and mental condition to enter buildings, to move within them, and to use technical devices or other devices of daily life (as rest rooms or ticket machines) without major restrictions. Nowadays in western countries, the exclusion of handicapped people from everyday life's activity is prohibited, e.g., in the US by the 'Americans with Disabilities Act' (ADA) [3] or in Germany by the 'Equality Law of Disabled Persons' ('*Behindertengleichstellungsgesetz*') [4]. These rights were also brought up in a United Nations resolution (56/168) [5]. Two aspects of barrier free architecture should be differentiated: one is the elimination of barriers, the second is the compensation of handicaps by appropriate designing and implementation of technical devices. Tactile floor indicators are thought to compensate the loss or impairment of vision by the provision of complementary haptic information.

Tactile floor indicators

Tactile floor indicators can be used for providing directional guidance or as warning signals for potential hazards at specified locations as, e.g., at road crossings or at platform edges. When using canes, tactile information by using canes is complemented by auditory feedback. The tactile information can also be obtained more directly by contrasts in the texture underfoot. In addition, tactile floor indicators might contain strong visual contrasts in order to provide (warning and orientation) information to people with residual and full vision. Such an implementation of different modes (pictorial, verbal, tactile) for a redundant presentation of essential information is recommended by the principles of Universal Design [6].

The usage of a long cane for scanning the ground in front is described as the most common mobility aid for visually impaired pedestrians [3, 7]. By using a cane, visually impaired people are able to locate potential obstructions, provided that the obstacle is continuous to the ground or provides some form of tapping rail within 15 cm of the ground [8]. Canes also help to detect distinct changes in the surface material (e.g., from asphalt to grass) and changes in the level (e.g., steps). The principal cane technique in uncontrolled and/or unfamiliar environments is the touch technique where the cane arcs from side to side and touches points outside both shoulders [3]. An increasing number of people use a long cane with a roller tip which maintains contact to the ground. Maintaining contact to the ground facilitates the detection of distinct changes in the pavement structure.

In 2000 only 10 of 16 European countries had national guidelines for the installation of floor indicators [7]. There is a strong case to harmonise the guidelines, both within the European Community (*CEN EN 15209*) and worldwide (*ISO 23599*). However, for the current moment, inconsistencies between the guidelines of different countries have to be stated and, as a consequence, installed tactile floor indicators might differ between countries, but unfortunately there are inconsistencies within single countries and even within towns. The successful use of tactile floor indicators depends crucially on visually impaired pedestrians understanding the different meanings assigned to these indicators and being aware of the presence of such facilities [8]. A simple and intuitive structure of floor indicators, as another principle of Universal Design [6], facilitates their usability.

Floor indicators as warning system

Any sudden change in the ground level, as steps and platform edges, as well as moving vehicles represent fundamental hazards for visually impaired people, resulting in stumbling, falling, collisions, and, far too often, injuries. In order to prevent accidents of visually impaired pedestrians, the European Conference of Ministers of Transport recommended that warning surfaces should be used at pedestrians crossings, edges of rail (and other traffic) platforms, and to warn of other hazards, as steps and level crossings [9]. In order to fulfil their function as warning signals, these floor indicators should be readily distinguishable one from another [9]. In the UK and countries like Australia tactile floor indicators mostly consist of a series of truncated domes. These blister surfaces might also differ from their surroundings by colour and material. A height of 5 mm was found to be sufficient for the majority of blind people to detect the surface change [10]. Increasing the height would increase the perceptibility but would increase the risk of stumbling, and, therefore, conflicts with safety regulations: For example, in Germany the maximum height is limited to 4 mm by rules of accident prevention (*UVV VBG1*) [11].

The perceptibility and safety issues as physical characteristics of floor indicators do not represent the sole criteria for evaluating their functionality. Both the single tactile floor element and a tactile warning system, consisting of various elements, are part of a complex environment with a variety of possible hazards. The correct interpretation of warning elements is of crucial importance, as different kinds of hazard require a different behaviour of visually impaired subject. For example, crossing a cycle track usually requires a down slowing and increased attention, while crossing a street requires a stop before the curb stone and awaiting acoustic signals from the traffic or from acoustic devices at the traffic lights, signalling a continuation of the walk.

The interpretation of different kinds of hazards can be facilitated by using different kinds of tactile floor indicators, but, given that the tactile quality of the ground surface in public space is generally very varying, the options for a semiotic system of tactile warning elements are somewhat limited. As it is true for other kinds of semiotic systems, the correct interpretation of tactile warning elements depends also crucially on the abilities of the visually impaired subject. The individual need of tactile information depends on the (subjectively experienced) complexity of the place; handling of more complex places requires more information. Colour coding can be helpful for subjects with residual vision, but tactile floor indicators as warning elements have to fulfil their function unequivocally for those subjects relying on the tactile information alone.

Floor indicators as guidance and orienting system

Besides as warning elements, floor indicators can be used to provide guidance for visually impaired people through large open spaces (where no other orientation marks are available) or through complex pedestrian environments (where orientation is otherwise hard to obtain). The architectural structure of a given place alone might enable visually impaired subjects to orientate, but it can also hamper orientation. Curb stones, house walls, or lawns along footways can potentially fulfil the same function for orientation as specially designed tactile floor indicators. However, it must be possible for the visually impaired subject to follow these 'natural' elements without any problems or hazards. Curb stones suddenly reduced in height to street level might result in a loss of orientation and, as a consequence, in some potential hazard. Some architectural elements, as crescent-shaped ways with steps, frequently varying floor materials or sound-deadening wall constructions, might hamper orientation and, worse, might have some additional risks for visually impaired subjects. Tactile elements might have simultaneously a warning and orientation function. On rail platforms, tactile warning surfaces often run in parallel to the edge of the platform and provide also a guidance function.

There are no general rules for the design of floor indicators as guidance system. Manufacturers of floor indicators use different materials and designs. Böhringer listed in his report alone 18 manufacturers [12]. The guidance systems in German railway stations consist of plates of fluted material with a total width between 20 and 50 cm. The flutes run in walking direction. By using the touch technique with the cane and moving its tip transversely over the flutes, tactile (and auditory) feedback can be received (Fig. 1).

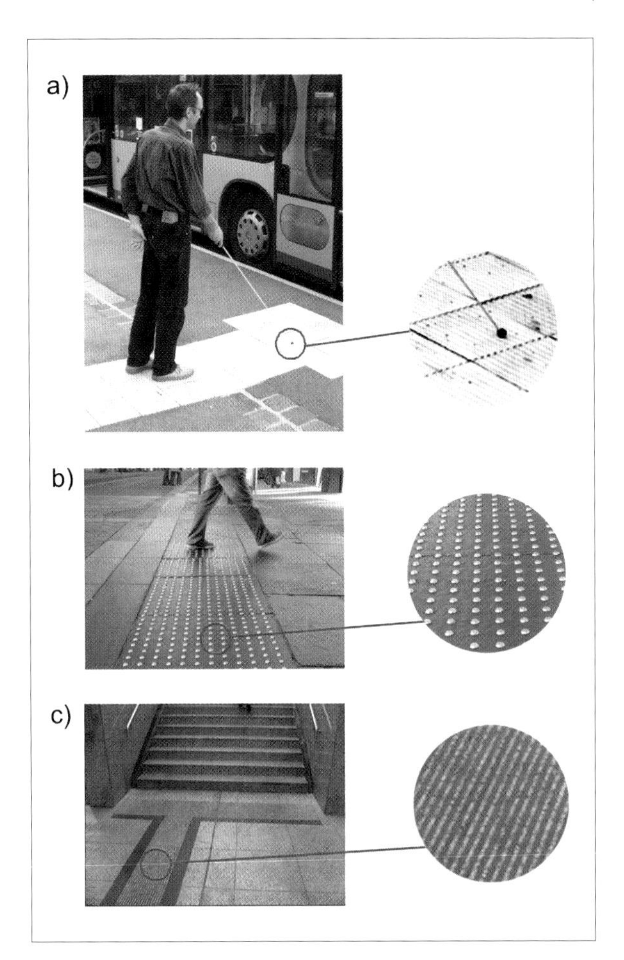

FIGURE 1. EXAMPLES OF TACTILE FLOOR INDICATORS
(a) shows floor indicators realized with fluted material at a bus stop in Leipzig (Germany); in this configuration, indicators catch the attention of a cane user when walking along the sidewalk, but indicators function little as guidance system since the tactile contrast alongside the strip is low and the attention field at the end of the strip exhibits little difference to the strip; (b) tactile floor indicators made of plates with little domes in the city centre of Newcastle upon Tyne (UK); these indicator can be perceived directly by contact with the feet and function as attention and warning elements at street crossings; (c) a tactile guidance path made of fluted material, found in the main station of Cologne (Germany); here, the fluting runs parallel to the strip, allowing orientation when walking in the same direction as the strip; the strip ends in a large attention field at the beginning of the stairs.

Whenever these tactile guidance systems come to an end or to a junction, special floor indicators should be installed, tracking the attention towards this point ('attention fields'). If the guidance system consists of fluted material, the direction of the fluting together with the width of the fluted tiles should be changed at these positions. Where necessary, the tactile guidance systems might contain tactile warning elements, as described above. So, a tactile guidance system might end in a broad warning field of tactile floor indicators oriented in parallel to the first step of stairs. Guidance systems on the stairs are unnecessary as on stairs orientation is relatively easy obtained by blind people using a cane.

Two approaches in the research on tactile floor indicators

Two aims of research on tactile floor indicators might be differentiated: one refers to the perceptibility of single elements, the second to the usability of tactile warning and guidance systems. The perceptibility of tactile floor indicators represents a basic prerequisite for their usability. The (tactile) perceptibility of tactile floor indicators is influenced not only by their physical characteristics, but also by the subjects' general ability, their past experience with tactile floor indicators, the used cane and the applied touch technique. Furthermore, subjects might differ in to what kind of degree they rely on certain modes of information (tactile by cane/tactile by foot/auditory/visual). Perceptibility might be regarded as the premiere quality criterion for tactile floor indicators, but the indicators have to fulfil other criteria, too. They have to be sufficiently safe with respect to slip hazard and risk of stumbling under all possible physical/weather conditions and they have to be sufficiently durable.

Not only the 'individual' characteristics of tactile floor indicators are of importance. If tactile floor indicators are installed, they become part of an already complex environment. As this environment already contains rich tactile information, it might happen that existing elements negatively interact with the installation of tactile floor indicators. Tactile guidance systems might, e.g., pass too close at waste baskets, bicycle racks, or ticket machines or might interfere with other kinds of paving. As the need for guidance and orientation and the amount of potential hazards vary from place to place, it is hard to come to general recommendations and regulations for the installation of tactile floor indicators. A systematic testing of each installed tactile guidance system at place (and modification of the system if required) would be desirable, but currently represents an exception. In the following, some findings of relevance, regarding the perceptibility of different floor indicators and regarding the usability of tactile warning and guidance systems, are described.

Perceptibility of different kinds of floor indicators

One of the largest studies on the perceptibility of different kinds of tactile floor indicators was conducted by Böhringer [12]. 68 blind subjects tested 21 different tactile floor indicators and 20 different kinds of attention fields, installed on a training area. The perceptibility of the indicators was differentiated between canes with a roller tip or with a tinier ceramic tip. (In daily life, most of the tested subjects preferred a roller tip.) The perceptibility was rated semi-quantitatively (ratings from 1 to 6; 1 = excellent perceptibility, 6 = completely insufficient perceptibility).

Best ratings were obtained for floor indicators providing a very clear auditory feedback. This finding underlines the importance of redundancy of the information provided by the floor indicators. Fluted tiles were rated the worse the smaller the flutes were separated. 10 mm spatial separation between the flutes turned out to be perceived only insufficiently, especially when using a cane with a roller tip. This finding is in sharp contrast to previous recommendations, claiming that such fluting can be perceived easily. Ratings differed considerably between those obtained by using a roller tip and those obtained by using a smaller ceramic tip. Perceptibility of a finely graded texture, as the 10 mm and 20 mm fluted material, was improved by using a ceramic tip, while the perceptibility of grossly graded material was better when using a roller tip. However, 20 mm fluted material was rated better than 10 mm fluted material, no matter whether a roller or ceramic tip was used. Of note, also a preference of already known tactile floor indicators was reported [12].

In another study, conducted in the Netherlands, 40 different floor indicators, mainly with ribbing profile and blisters, were tested [13]. 12 visually impaired subjects (four completely blind) rated how well different indicators could be localised, their orientation detected and how easy they could be followed. It was found that when a floor indicator was easy to locate, then it was generally also easy to determine its orientation and traverse it. The study further showed that the tactile detection of indicators by contact with the feet (sole and ankle), and tactile (and auditory) detection by cane might considerably differ for the same floor indicators. Detection by feet was found to require a coarse pattern, with an indicator profile of 3 mm minimum. A reasonable good detection by cane was possible with profile of 2 mm and even less. However, coarse pattern have always the problem of being uncomfortable for walking and causing some risks, e.g., for pedestrians with high heel shoes.

The researchers of the study propose to use multi-level tactile signalling analogue to the menu structure of software programs, with the main menu consisting of two options: floor indicators with ribbings are proposed to denote tactile pathways and blister patterns to indicate warnings. Such raised blisters are easily perceptible and identifiable with the foot and can therefore potentially also be identified by those visually impaired subjects not using a cane. Additional information might then be provided in a sub-menu. These subsidiary information can be obtained with a cane using the sweeping technique. The authors argue that those visually impaired not using canes will generally have sufficient residual sight to make sense of the environment. However, the exact sub-menu structure requires probably reconsideration, as the authors propose that also amenities should be indicated by blisters [13].

In Sweden, 15 courses of guidance surfaces, consisting of an attention field and a tactile path, were tested by 14 completely blind subjects [14]. Similar criteria for the evaluation were used as in the Dutch study. It was found that completely blind people had problems to recognise a warning surface with blister paving when it is built at the end of a guidance path. Obviously, the tactile information obtained by the cane did not allow a sufficient discrimination between the surface guidance path with ribbed or sinusoidal paving and the warning surface with blister paving. Sticking with the tip of the cane in slap edges was noted as another problem, resulting in stoppages and increased difficulties in orientation.

Problems with the detection of warning elements might be overcome with a relatively new type of attention field made of sound tiles (‘*klangtegel*’, TG Lining BV, Heiloo, NL, http://www.tglining.com). This type of tile was developed in the Netherlands and consists of a metal plate connected on a concrete tile in such a way that a gap (sound chamber) exists between the plate and the tile. As indicated by its name, contact by cane or foot produce a clearly distinguishable sound. In addition, its metal surface functions as visual cue, and domes within the plate provide some information about direction [15].

Usability of tactile warning and guidance systems

Within a project on the disability friendly design of subway stations (‘Experimental Subway Station U-Borgweg’), tactile floor warning and guidance systems were tested [16]. In contrast to the above cited studies, field experiments were conducted. These experiments took place in a number of subway stations of Hamburg (Germany) where various kinds of tactile floor indicators had already been installed. Field experiments have the advantage of higher ecological validity, as compared to experiments in artificial setups: Subjects were instructed to make their way within the subway station, and factors potentially hampering their ability to orientate (as noise and other passengers) were existent. However, such kinds of experiments are limited by the fact that only already installed warning and guidance systems can be investigated in their functional qualities. In the U-Borgweg study, the perceptibility of tactile floor indicators was investigated, but also other aspects of floor indicators, as the design of warning fields close to stairs and the required distance between warning fields and potential hazards. Blind subjects were interviewed after testing and their performance videotaped.

Tactile guidance systems installed at subway stations in Hamburg mainly consist of fluted tiles with 11 mm width, but at some stations also blister surfaces were installed as floor indicators. In the U-Borgweg study, no systematic difference between the two kinds of floor indicators was observed. However, most study subjects did not use roller tip. Generally, the perceptibility of fluted tiles varied in dependence of the surrounding material. The larger the contrast between the floor indicators and the surrounding material, the better the floor indicator was perceived. Also a strong contrast in colour between tiles with floor indicators and of their surrounding tiles improved the perceptibility, as well as the usage of different kinds of material. Surrounding tiles with a gross surface structure worsened the perceptibility of tactile floor indicators, as they reduced the perceptual contrast between fluted tiles and their surroundings.

Most participants of the U-Borgweg study had no roller tip and preferred not to drag the tip over the ground. However, for detecting floor indicators with a cane it is necessary to maintain contact to the ground with its tip. Dragging a cane with a small tip over the ground has the risk that the tip gets stuck in the floor indicators or in the splices between the tiles. In the study of Böhringer [12], subjects had ten times more problems of getting stuck with their cane when using a ceramic tip instead of a roller tip. The subjects in the Hamburg study had only minor problems with getting stuck with the tip of their cane in the 11 mm fluted tiles, but splices between the tiles were sometimes a problem and should be taken into account when installing a tactile guidance system [16].

The field experiments at subway stations revealed that the tactile attention fields at junctions were sometimes unnecessary as easy structured junctions were also detected without these fields. On the other hand, attention fields might sometimes be irritating if there are too many of them. Attention fields are not self-evident as they just 'tell' the blind user to pay attention, but without exact knowledge of the guidance system it is not clear for the user to what kind of element his/her attention has to be drawn: attention fields might indicate junctions, but they might also indicate the termination of a tactile guidance system, and the guidance system might terminate not only close to the exit of the subway station but also in front of lifts or tactile maps. As for other tactile floor indicators, it is important that attention fields can be easily perceived and are structured with a good contrast to the surrounding materials. A change in fluting on a small space (<20 cm) was often not perceptible. Junctions should be perpendicular wherever possible and/or, in halls and tunnels, tactile guidance system should run in parallel to walls [16].

If tactile floor indicators were used simultaneously as a guidance system and as a warning element at the edge of platforms, the distance between the two of them was recommended to be between one or one and a half steps (60 to 90 cm) [16]. When using truncated domes as tactile warning indicators, Japanese researchers determined the needful distance between the warning elements and the platform edge as 80 cm [17]. Some older subway and tram stations might not have enough space to provide a larger distance between the warning elements and the edge of the platform than 60 cm. The key issue is probably not the exact distance (60–100 cm), but again the possible perceptibility of tactile warning elements. The detection rate of tactile warning bands with truncated domes increased with the width of the bands from 90% for 30 cm bands to 100% of 60 cm bands [17]. The researchers of the U-Borgweg project recommended that the tactile guidance and warning elements run over the total length of the platform and that warning elements are provided also at the head of platforms [16]. Of note, there is a considerable difference between countries: While in countries like Germany and Japan tactile warnings indicate the beginning of an area of hazard, in the US the area of hazard itself, i.e., 61 cm (24 inches) from the edge of the platform, is marked by tactile floor indicators [18].

Beside the edge of platforms, stairs represent one of the main hazards in subway stations for visually impaired passengers. Generally, visually impaired subjects tend to step on the stairs wherever they hit on them, i.e., they do not look especially for the hand rail as orientation, because otherwise they would cross the stream of other passengers. Therefore, an attention field at the right side of the stairs has no benefit for the blinds. Instead, the whole width of the stairs should be indicated by tactile warning elements. Again the distance between the warning elements and the stairs should be in the range of 60–90 cm [16].

Similar experiments were conducted in a follow-up study in Hamburg [19]. The study report illustrates that reliance on tactile information can result in a loss of orientation when attention fields are not sufficiently distinguishable from the guidance path [19]. In the same experimental situation, another subject, leaving the guidance path and mostly relying on acoustic information, had no problem to orientate. Field observations like this suggest that some of the existing tactile guidance paths do not fulfil their function.

Information about tactile guidance systems

Grahmbeeck states in his lecture for the European Commission that guide paths must have a logical beginning and end [15]. Considering that tactile guidance systems might be applied in complex pedestrian environments, makes clear that the demand of logical beginnings and ends is everything but easy to fulfil. Difficulties in orientation increase with complexity of a location, both for visually impaired subjects and people with full sight. As a consequence of the complexity of an environment, also a tactile guidance path system within this environment would probably have more than one beginning and end. Therefore, it is necessary that the installation of tactile floor elements for guidance is combined with the provision of information about this guidance system, as a tactile path system is not self explaining. Information might be provided, e.g., by internet pages, tactile maps, but also by trainings, organised for example by transportation companies and performed by mobility coaches. Such assistance is especially required in environments with complex tactile path systems. What might be considered as a logical end and beginning of a tactile path is rather difficult to decide from outside, as this kind of logic does not only depend on the location but also on the demands of the subjects. An involvement of potential users in the planning process is without doubt helpful.

The ability to handle complexity differs from subject to subject, and the smaller the complexity, the more people can handle it. With regard to a tactile guidance system, this would mean keeping the complexity as low as necessary. However, a platform might have four stairs and an elevator, and, therefore, a tactile guidance system on such a platform might already exhibit a considerable complexity. This cannot be avoided and can only be handled by provision of information and training. The human ability to adopt and to learn is somewhat underrated and neglected in most of the studies. To my knowledge, no study investigated how well particular tactile floor indicators are perceived after training.

Interference with other pedestrians

The needs of all pedestrians and users of public transportation have to be taken into account for the construction of stations and sidewalks. The needs of different groups might sometimes conflict to some extent. Curb stones represent a barrier for people in wheelchairs. On the first glance it would be recommendable to countersink curb stones to a completely at-grade level. However, curb stones also help blind people using long canes to keep their orientation. In order to reduce the barrier for one group of users and not to build up a barrier for a second group, it is now recommended to countersink curb stones to a level of 3 cm which still allows blind people to use the curb stone as orientation mark, but eliminates the barrier for people in wheelchairs at the same time. Such a height of curb stones at pedestrians crossings is recommended, e.g., in the German technical norm DIN 18024 [20]. Reversely, tactile floor indicators might interfere with the needs and demands of other people with disabilities. Floor indicators might cause stumbling of elderly people in some cases. So the influence of tactile floor indicators on the walking behaviour in general should be studied [21] and controlled, too.

Outlook

The current (potential) usability of tactile floor indicators is often not high despite large financial efforts to install them in many places. The lack of guidelines has resulted in a large variability of installed floor indicators and related solutions of tactile paths. As outlined, variability *per se* lowers the usability of tactile floor indicators. To make things worse, some of the existing guidelines were heavily criticised, as standing in contrast to results of empirical studies [12], and some of the recent empirical studies were reported to be not in line with older studies [14]. The overall number of studies is relatively low, considering the tremendous need of studies in the field and considering the fact that every unusable (or unused) tactile floor indicator is a pure waste of money. It is somewhat questionable how international guidelines can be formulated without a sufficient empirical basis.

Future studies should include much larger samples as in previous times. Although most cited studies stressed the importance of human factors (as, for example, the reliance on different kinds of information or the training background), they simply did not include a sufficient number of subjects to investigate this aspect in a proper manner. As international guidelines are aspired, a multinational research program would be adequate in order to take national peculiarities into account: Installed tactile floor indicators and experiences with them might considerably differ between countries, but also the quality and quantity of mobility training. The beneficial effect of training on the usage of tactile floor elements should be considered as well. A more qualitative approach instead of mainly quantitative analyses might furthermore help to understand the needs of users as well as their difficulties in using tactile floor indicators.

New technical devices might help to improve the usability of tactile paths in the future. Navigation technique is developing rapidly. Already in the last decade, the use of computer assisted navigation aids for visually impaired and blind persons had been proposed, e.g., by the project 'Mobility of Blind and Elderly People Interacting with Computers' (MoBIC), funded by the European Community. The accuracy of navigation systems will be improved within the next five years by a new system of navigation satellites (Galileo) [22]. Its accuracy will lie in the range of a few metres. The vision for the future might be that gross navigation for visually impaired and blind persons is provided by such an electronic aid system, while tactile floor indicators help to provide the orientation. The need for a user-friendly tactile guidance and warning system would, however, be increased.

Summary

Tactile ground surface indicators are installed in order to warn people with vision impairment of potential hazards and to provide orientation help. The tactile information can be obtained by contact with a cane and more directly by the texture underfoot. The tactile information is usually complemented by acoustic information, provided by special materials or surfaces for the indicators, as well as by strong visual contrasts to the surrounding ground surfaces. Up to now, there are no internationally accepted comprehensive guidelines for tactile ground surface indicators and their installation. The usability of these indicators is influenced by a number of factors. Besides characteristics of the tactile ground surface indicator itself, the technique to use the cane and characteristics of the cane have an impact on the perceptibility and usability of the surface indicators. Furthermore, the surrounding materials and other attributes of the environment might ease or aggravate the usability. User oriented studies help to understand the needs of visually impaired people for tactile orientation and warning indicators and to improve the usability of these indicators.

Acknowledgement

The author appreciates the helpful comments of Dr. Klaus Behling and Dietmar Böhringer.

Selected readings

Bentzen BL, Barlow JM, Tabor LS (2000) *Detectable Warnings: Synthesis of U.S. and International Practice.* <http://kiewit.oregonstate.edu/pdf/acc_DW-synthesis.pdf>

42

HapticWalker – haptic foot device for gait rehabilitation

Henning Schmidt, Jörg Krüger and Stefan Hesse

Introduction

The restoration of gait for patients with impairments of the central nervous system (CNS), like, e.g., stroke, spinal cord injury (SCI) and traumatic brain injury (TBI) is an integral part of rehabilitation. The rehabilitation of CNS impairments usually takes several months at minimum and its outcome often influences whether a patient can return home or to work. Particularly stroke is the leading cause for disability in all industrialised countries; the incidence is approximately one million patients in the European Union each year [1, 2]. Modern concepts of motor learning favour a task specific training, i.e., to relearn walking, the patient should ideally train all walking movements, needed in daily life, repetitively in a physically correct manner [3]. Conventional training methods based on this approach proved to be effective, e.g., treadmill training [4], but they require great physical effort from the physiotherapists to assist the patient – so does even more training of free walking guided by at least two physiotherapists. Assisted gait movements other than walking on even floor, like for instance stair climbing, are practically almost impossible to train, due to the overstrain of the physiotherapists. Robotic haptic gait training devices may offer a solution to fill this gap and lead to an intensified patient training plus a relief for the physiotherapists from strenuous work.

Different rehabilitation robotics research groups already developed a number of robotic gait rehabilitation devices with haptic features [5, 6], all of them are based on the exoskeleton principle and need to be operated in combination with a treadmill on which the patient walks. The exoskeleton robot then substitutes the physiotherapist and moves the patients' legs. Due to their operating principle, all of these machines are restricted to training of walking on even ground and do not allow physical interaction between patient and physiotherapist during training.

In contrast, it was a major goal of the HapticWalker project to develop a robotic walking simulator, which offers training of arbitrary and freely programmable foot motions. This lead to the development of a machine based on the principle of programmable footplates. On this type of machine the patients' feet are attached to two footplates, on which he stands. The footplates are located at the end-effectors of two robot arms which carry the patients' body weight and move his feet along the foot trajectories. In addition to daily life walking trajectories like walking on a plane floor or stair climbing, the machine dynamics should allow the simulation of asynchronous events, such as stumbling or walking on rough ground, which require high acceleration capability. For application in gait rehabilitation it is essential for the machine to provide a permanent foot attachment to the footplate, i.e., it needs to comprise foot support and guidance during all phases of the gait cycle (stance and swing phase).

Even though fully guided foot motions are needed for the most part of gait rehabilitation, due to the patients' lack of ability to perform voluntary leg motions, the machine support should be gradually reduced depending on his learning success. Thereby the machine should mimic the physiotherapist's behaviour of appropriately reducing the degree of assistance in later phases of the rehabilitation training.

The HapticWalker is equipped with force/torque sensors under each footplate in order to measure the contact forces between foot and footplate continuously. These force values are used for gait analysis purposes, as a measure for the patients' learning success, and for compliance control algorithms.

For the HapticWalker control software, it was necessary to develop algorithms for the simulation of arbitrary natural foot walking trajectories regarding position and velocity profiles, because classical robot motion commands would not be suitable to perform this kind of motion task. In addition, it was essential to develop an intuitive graphical user interface that allows safe programming and operation of the machine by nontechnical personnel such as physicians or physiotherapists. Therefore, appropriate algorithms for the intuitive synthesis of gait trajectories were developed.

Locomotion interfaces

The HapticWalker was designed as a generic haptic foot device, not only to be applied for gait rehabilitation, hence it belongs to the class of so-called 'locomotion interfaces'. Haptic locomotion interfaces allow walking and other foot movements within virtual environments. In order to present kinesthetic feedback to the foot during all phases of gait, these haptic foot devices need to comprise a permanent contact between the foot and the device. Such locomotion devices belong to the group of 'programmable footplates'; they present haptic foot sensations to the user, similar to hand haptic interfaces presenting haptic sensations to the hand. The main purpose of human foot movements is to move the body into a desired direction. Hence inside a virtual world the user could use a haptic foot device in order to navigate in this environment. During daily life a human performs different kinds of walking trajectories (e.g., walking on a plane floor, up/down staircases). This takes place on different types of terrain (even or rough ground); it could also happen that one trips over an obstacle and stumbles. A haptic foot interface should be able to simulate all these and other foot movements. So far only a very small number of haptic locomotion interfaces based on the principle of programmable footplates have been developed and built [7–12] or are under development [13]. Some of them do not have permanent contact between foot and footplate, but provide foot support only during stance phase. In contrast, the presentation of haptic sensations during swing phase or the application of tractive forces to the foot requires permanent foot attachment to the footplate. Values for footplate workspace and dynamics that have been published are rather small [11, 13] or only qualitatively given and indicate slow walking speeds [8, 10].

The goal of the HapticWalker project was to design and build a haptic foot device based on the principle of programmable footplates with permanent foot machine contact and high dynamics, in order to enable walking speeds up to 5 km/h and simulate walking situations requiring very high dynamics (e.g., stumbling).

HapticWalker

Gait analysis

The analysis of different walking trajectories was the major basis for the design of the HapticWalker. Motion capturing and analysis was done particularly for walking on even floor at speeds up to 5 km/h with cadences up to 120 steps/min and stepping staircases up/down at speeds up to 2.3 km/h and cadences up to 120 steps/min [14].

Gait data was captured using a ZEBRIS ultrasonic motion capture system (see Fig. 1a, c and e). Data processing was done using the method described in [15]. Figure 1b, d and f show the measured and transformed walking trajectories for walking on plane ground, walking up and down a staircase. A detailed trajectory analysis can be found in [14]. For the three DOF base module gait data was analysed in the sagittal plane, since all major foot movements occur in

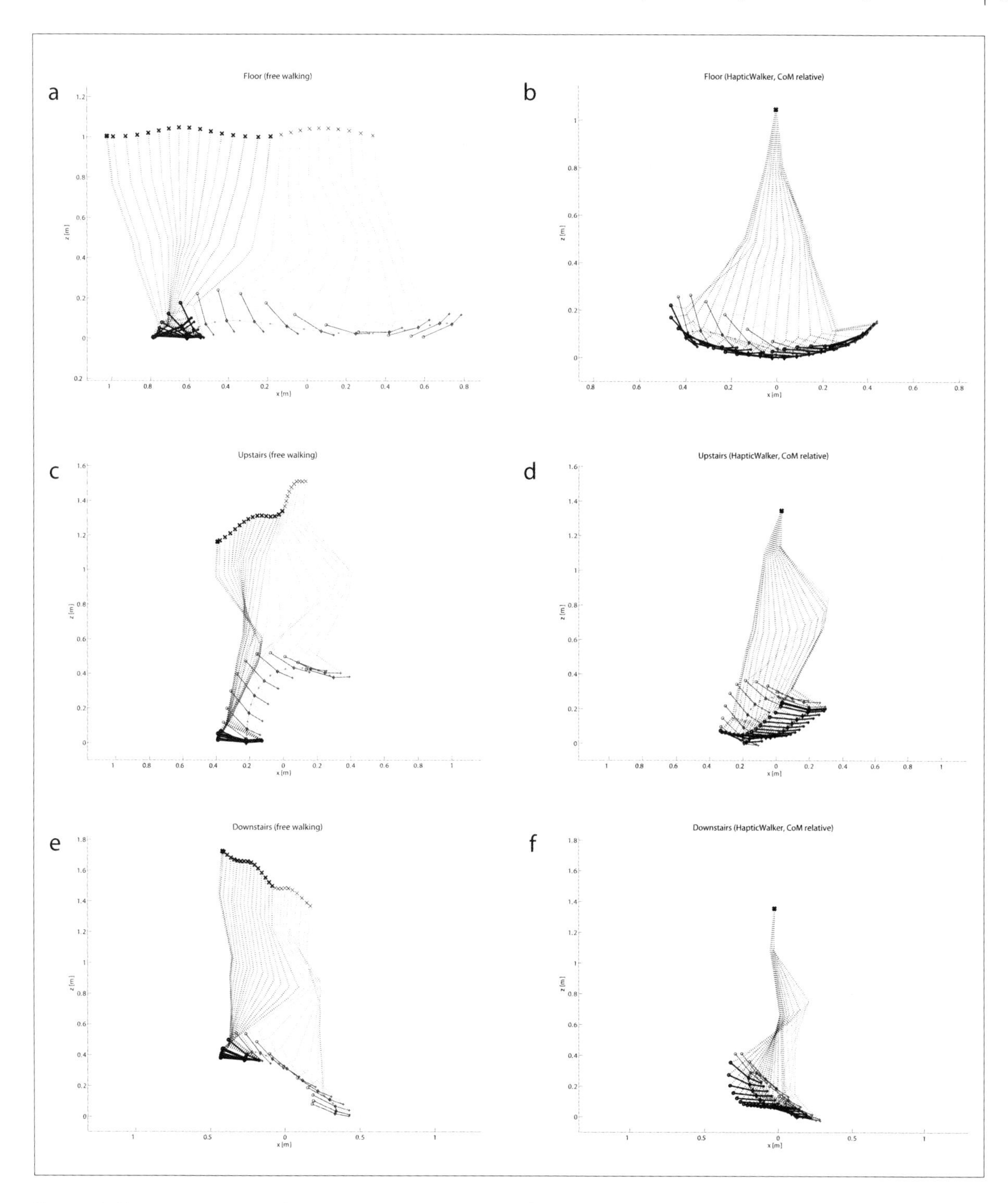

FIGURE 1. FOOT TRAJECTORIES (FLOOR, UPSTAIRS, DOWNSTAIRS)

this plane. From there the necessary workspace requirements including peak values for translatory and rotatory footplate velocities and accelerations were derived. The metatarsal joint was chosen as the reference point at the foot. The maximum metatarsal joint values covering all trajectories are:

x_{min} = 0.8 m (1)
z_{min} = 0.33 m (2)
ϕ_{min} = –75 deg …30 deg = 105 deg (hindfoot) (3)
$|v_{max}|$ = 3.28 m/s (4)
$|a_{max}|$ = 33.74 m/s^2 = 3.4 ·g (g = 9.81 m/s^2) (5)
$|\omega_{max}|$ = 462 deg/s (6)
$|\alpha_{max}|$ = 7,450 deg/s^2 (7)

These values match very closely the measurements published in the literature [16, 17]. It was decided to design the machine to be able to perform the above velocity and acceleration ranges within a foot workspace of the following size:

x_{min} = 0.9 m (8)
z_{min} = 0.4 m (9)
ϕ_{min} = –90 deg ...40 deg = 135 deg
(hindfoot: metatarsal–heel) (10)

The maximum payload for each footplate (i.e., user weight) is 150 kg.

Robot kinematics

In contrast to haptic hand devices, which can be built very lightweight, the construction of a haptic foot device based on the principle of programmable footplates does not allow this, since the full weight of the human body is to be carried. Hence relatively large and powerful robot manipulators are necessary.

In order to achieve high fidelity movements with such heavy robot arms, dynamic and powerful drives must be employed. These must be able to quickly accelerate the foot and leg, especially in the horizontal and vertical directions, as well as the mass and inertia of the robot arm and embedded drive motors. Besides the aforementioned foot workspace parameters, a number of other design aspects had to be taken into account:

- maximum passive security (i.e., all drives and mechanics must be totally enclosed)
- low user entry (i.e., easy access for wheelchair-bound patients)
- low protective hood for linear drives located on either side of the user (for access to the patient from all sides)

Several machine designs were developed and analysed. Finally the foot workspace and other design criteria were best met by hybrid serial-parallel robot kinematics. With this mechanical design, a beneficial compromise of serial (good workspace machine-size ratio) and parallel kinematics (good payload machine-mass ratio) was found. The chosen machine design was the basis for subsequent mechanical construction process and finally building of a working prototype (see Fig. 2).

The basic system comprises two three DOF robot modules, moving each foot in the sagittal plane. Foot movement along the two base axes in this plane (horizontal, vertical) is performed by linear direct drive motors, which move independently on a common rail, but are connected *via* a slider-crank system. Hence, the distal end of the small robot arm of both motors moves horizontally, if both motors move forward or backward without changing their distance. A change in distance results in a vertical movement. For the remaining degrees of freedom, the decision was made to use serial robot linkages. Inside the robot arm a rotary motor is located, which turns the footplate mounted at the distal end of the arm. The foot module contains the footplate and a six DOF force/torque sensor in order to monitor the user contact forces between foot and machine.

Each robot is based on a modular machine design where the current foot module can be substituted by a further three DOF module gaining six DOF at each footplate, thus enabling free foot movement in three-dimensional (3-D) space.

FIGURE 2. HAPTICWALKER – GAIT REHABILITATION WITH PATIENT AND PHYSIOTHERAPIST

An additional axis to support the metatarsal joint movement can be mounted as well. Drive size and dynamics of the basic three DOF sagittal plane setup were chosen such that footplate workspace and dynamics of the extended version will be no less than the values in Equations 8–10 and 4–7.

In general the machine is designed as a unidirectional motion interface. For rehabilitation purposes, this is completely sufficient, since the focus there is on the precise guidance of arbitrary walking trajectories including asynchronous walking events.

As a safety measure to prevent the user from falling and for trunk suspension (e.g., body weight support during gait rehabilitation), the walking simulator is combined with a trunk suspension module based on a patient lift as shown in Figure 2. The basic version of this suspension system is not actuated when walking on the simulator. At a later stage, the suspension system will be equipped with an active drive in order to enable lateral centre of mass (COM) movements in synchronisation with the foot movements. Vertical actuation of the suspension is not necessary with the HapticWalker, since the footplates can be programmed to move the feet along a COM-relative trajectory (see Fig. lb, d and f). This would not be possible with treadmill-bound machines.

Machine safety

Since the walking simulator employs rather powerful drives in order to achieve the aforementioned values for footplate dynamics (i.e., bandwidth) in combination with a payload capability for carrying the weight of a human, user safety is a major issue. Therefore we developed an integrated safety concept which comprises redundant levels of mechanical, electrical, and software safety components (see Fig. 3).

The basic level of safety appliances consists of purely electromechanical devices, such as manual emergency switches, plus a number of automatic emergency switches (safety release bindings (see Fig. 4) and control PC watchdog). All of them are connected to the central emergency switch circuit, which stops the machine immediately in case one or more switches are actuated. An important fact is that the emergency switch circuit is completely hard wired and runs independently from the control software. A comprehensive description of the safety concept can be found in [18].

Control software

The robot control software architecture was derived from the NASREM reference design [19] and modified for use as a robot controller. The main characteristics of this approach are the hierarchical structure of different layers which are represented by cyclic software threads. The threads are arranged according to their cycle time (low hierarchical order = small cycle time = more time critical). A data flow between different threads is accomplished *via* a system of

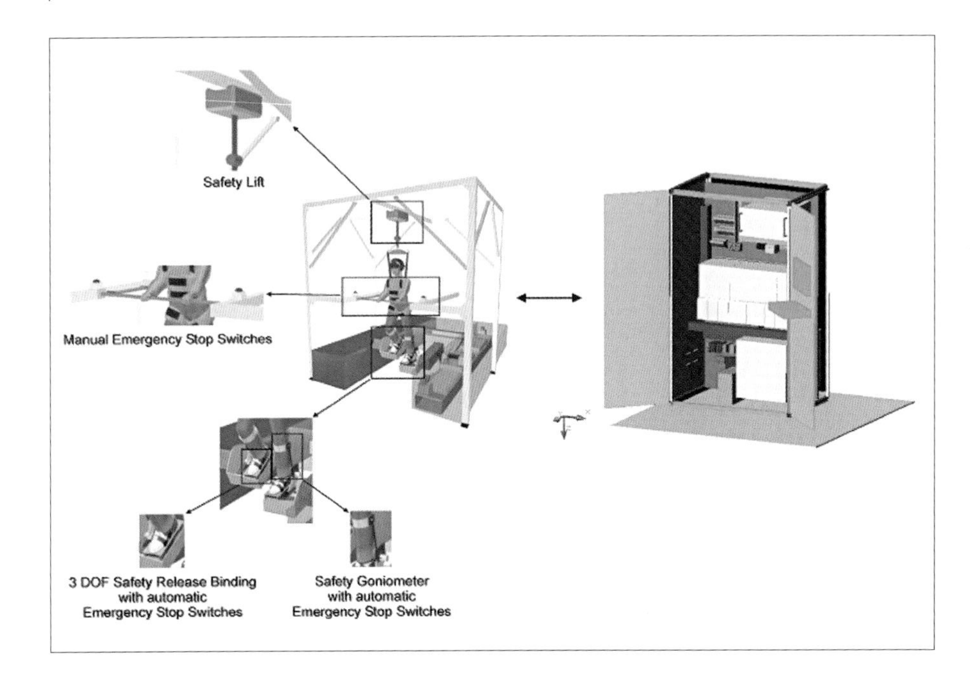

FIGURE 3. MACHINE SAFETY SYSTEM

FIGURE 4. FOOT BINDING AND SAFETY RELEASE MECHANISM

FIFO queues, which ensures data consistency (see Fig. 5).

The control software implementation was done based on Linux/RTLinux operating system. The software is a full-featured standalone multirobot control software, which was designed for use with different types of robots and robot kinematics [20]. The most time critical threads (motion generator, servo) are located in the real-time kernel module, whereas the less time critical modules (process control, user I/O, TCP/IP communication) run in soft real-time as maximum priority threads in the Linux user space. The communication between real-time kernel and user space is accomplished *via* shared memory.

In the motion generator thread, a special Fourier-based interpolation method (11) was implemented for efficient execution of cyclic foot trajectories [21], besides the well-known standard robot interpolation methods (e.g., point to point, continuous path, linear, circular).

$$x(t) = a_0 + (a_k * \cos(kt) + b_k * \sin(kt)) \qquad (11)$$

The advantages of a Fourier approximation based interpolation are:

- representation of cyclic curves with minimum number of coefficients
- continuous and differentiable higher order derivatives

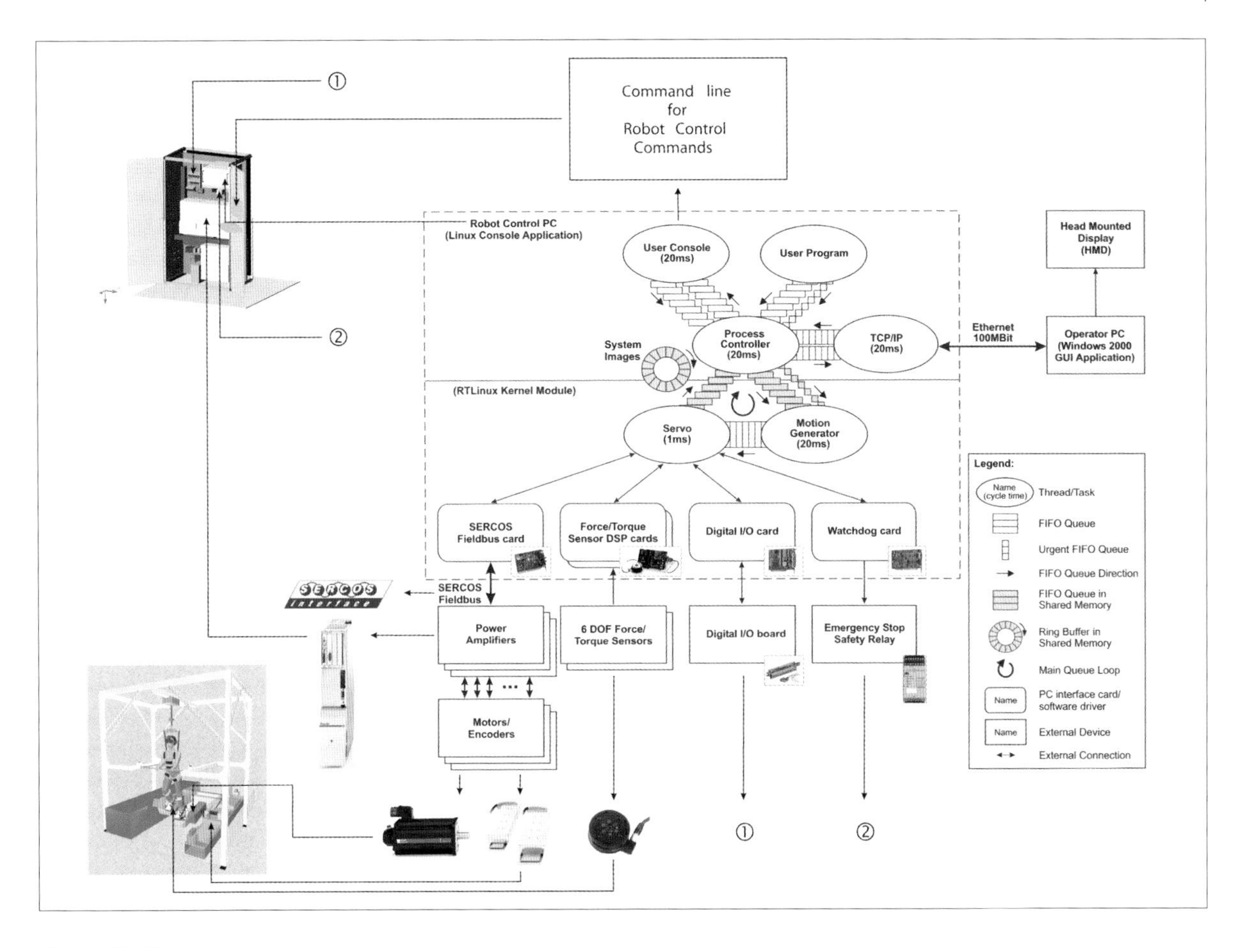

FIGURE 5. CONTROL SOFTWARE

- good path approximation already with small number of harmonics (i.e., coefficients)

This novel Cartesian interpolation method was derived from electronic cam mechanisms and allows the smooth and synchronous interpolation of arbitrary cyclic position and velocity profiles. It matches natural cyclic motions in contrast to conventional robot interpolation methods in fly-by mode. Another important improvement over conventional industrial robot controllers, which calculate only coarse pre-interpolated points inside the motion generator with subsequent linear fine interpolation, is the implementation of an interpolation of every single point of robot motion inside the motion generator thread. Hence, an improved trajectory smoothness is gained by sending continuously differentiable position, velocity, and acceleration reference values to the servo control loops.

Operator software

The operator interface consists of an especially developed MS Windows-based graphical user interface (GUI) software [20].

The chosen GUI layout (see Fig. 6) was partly derived from design layout of industrial robot programming interfaces (e.g., current Windows-

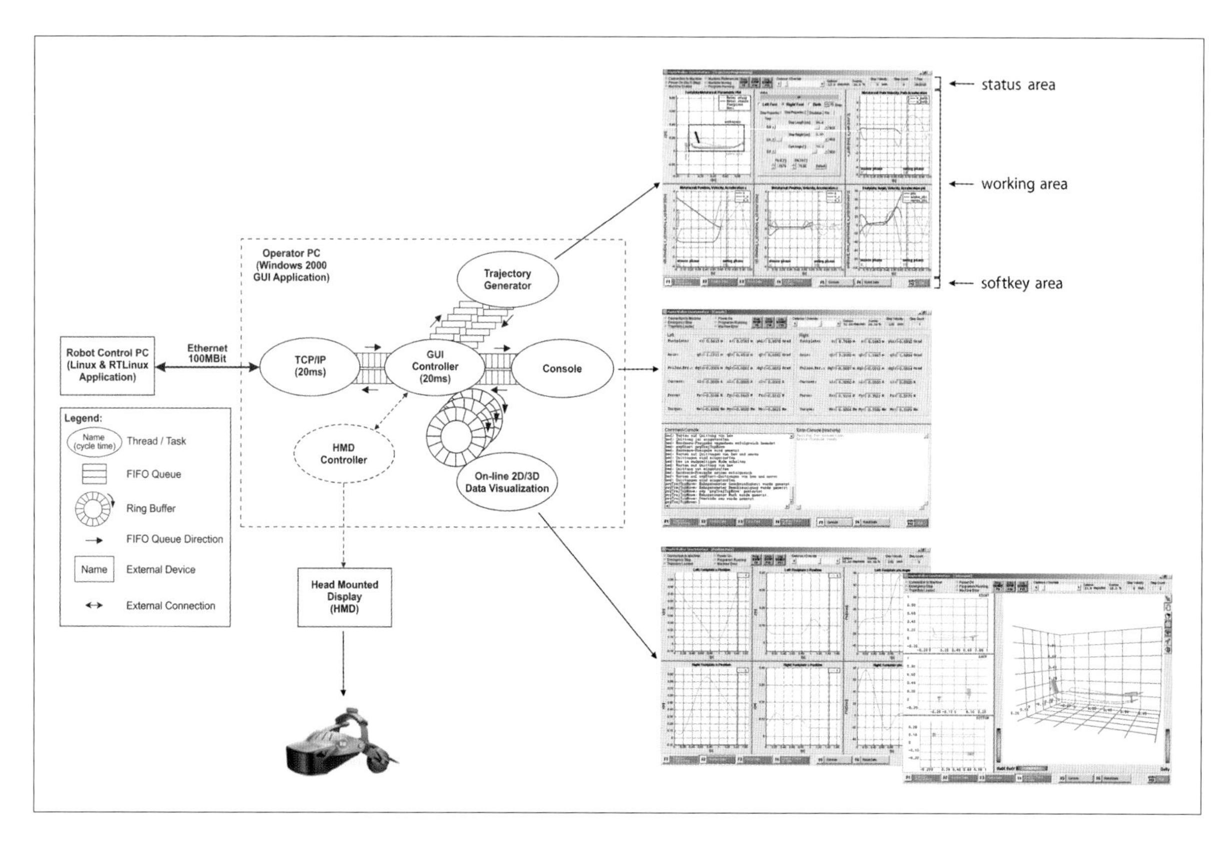

FIGURE 6.
GRAPHICAL USER INTERFACE ON OPERATOR PC
(trajectory synthesis/console/two-dimensional and three-dimensional online data visualisation)

based KUKA or SIEMENS 840D teach pendant). It is divided into three main parts: status area, working area, and soft-key area.

The trajectory generation window (Fig. 6, top) is primarily designed for use by medical operating personnel during gait rehabilitation. There the operator can synthesise foot trajectories based on the predominant gait parameters (step length, step height, cadence). The trajectory synthesis is based on the modification of an arbitrary master trajectory, which is loaded first and subsequently modified. The embedded trajectory generation algorithm then calculates a complete cyclic foot trajectory. Before executing the trajectory on the machine, the operator can do a real-time offline simulation. The algorithm for trajectory synthesis is described in [22]. Finally Fourier coefficients are calculated by fast Fourier transform. These coefficients are then transferred to the robot control PC *via* Ethernet. Besides the built-in motion generation algorithm, the GUI software will comprise a facility for data import of arbitrary foot motion capture data.

In the console window (Fig. 6, middle) the operator is given maximum flexibility to do remote machine operation and diagnosis online.

Different online two-dimensional and three-dimensional data visualisation windows (Fig. 6, bottom) visualise all captured data (position, force, torque, motor current, digital I/O) in real-time. Different two-dimensional windows comprise a scope view which automatically adjusts the data display to one gait cycle independent of the machine override currently chosen. A separate 3-D window gives an intuitive visualisation of the foot position and contact force data. There the user is presented a combined 3-D visualisation of the actual foot position and centre of pressure (COP).

Implementation and test

A prototype of the system with three DOF per foot in the sagittal plane (see Fig. 2) was built and tested successfully, including all hardware and software components.

Interpolation of walking trajectories

The robot controller was implemented on a standard industrial PC with a Pentium III/850 MHz CPU running RTLinux. The intelligent power amplifiers (BoschRexroth Indramat DIAX04), which are attached to the control PC *via* a fibre optic SERCOS fieldbus link, receive set values for position, velocity, and motor current from the servo thread at a frequency of 1 kHz. At the same rate the motion generator previously calculates all set values for all six axes. Each power amplifier comprises a classical cascaded control loop for position, velocity, and motor current. Currently we are using this control method in combination with velocity and motor current feedforward.

Figure 7 shows an example of two different interpolation methods which were implemented in the motion generator thread. Here the endeffector footplate moves from an arbitrary position in the workspace to a start position on the foot trajectory with Cartesian linear interpolation. Afterwards the foot trajectory is performed for six steps by Fourier interpolation (see Eq. 11). The sequence of motion phases is as follows:

- 0.3 … 0.8 s: Cartesian linear motion (acceleration, full speed, deceleration)
- 0.8 … 2.9 s: acceleration of Cartesian cyclic motion (automatic override adjustment by low pass filtering with critical damping characteristic)
- 2.9 … 7.0 s: Cartesian cyclic motion at full speed
- 7.0 … 9.1 s: deceleration of Cartesian cyclic motion (automatic override adjustment by low pass filtering with critical damping characteristic)

Figures 7b and d reveal the inherent step-like velocity profile of conventional Cartesian motions, no matter if the robot end-effector moves in linear, circular, or spline interpolation mode.

When running a sequence of conventional motions commands in fly-by mode, the user can prevent the robot from decelerating to zero velocity after each motion command, but the step-like shape of the velocity profile would remain and would feel like a very unnatural ‘ripple’. In contrast to this, the newly implemented cyclic Fourier-based interpolation delivers continuously differentiable velocity profiles, which feel fully natural [21].

Perturbations

Besides the training of smooth-walking trajectories, the patients’ reaction to perturbations, for example, stumbling should also be trained on the machine. Hence, we added a special switched override filter cascade in order to manipulate the velocity profile accordingly (see Fig. 8). The physiotherapist can issue a ‘stumble’-command for one of the footplates at any time, which stops the footplate instantly.

An override control algorithm ensures proper catch-up behaviour after the event, in order to get back to the previous time offset between both footplates. Footplate force sensor readings from stance and swing footplate after the event reveal information about the patients’ reaction.

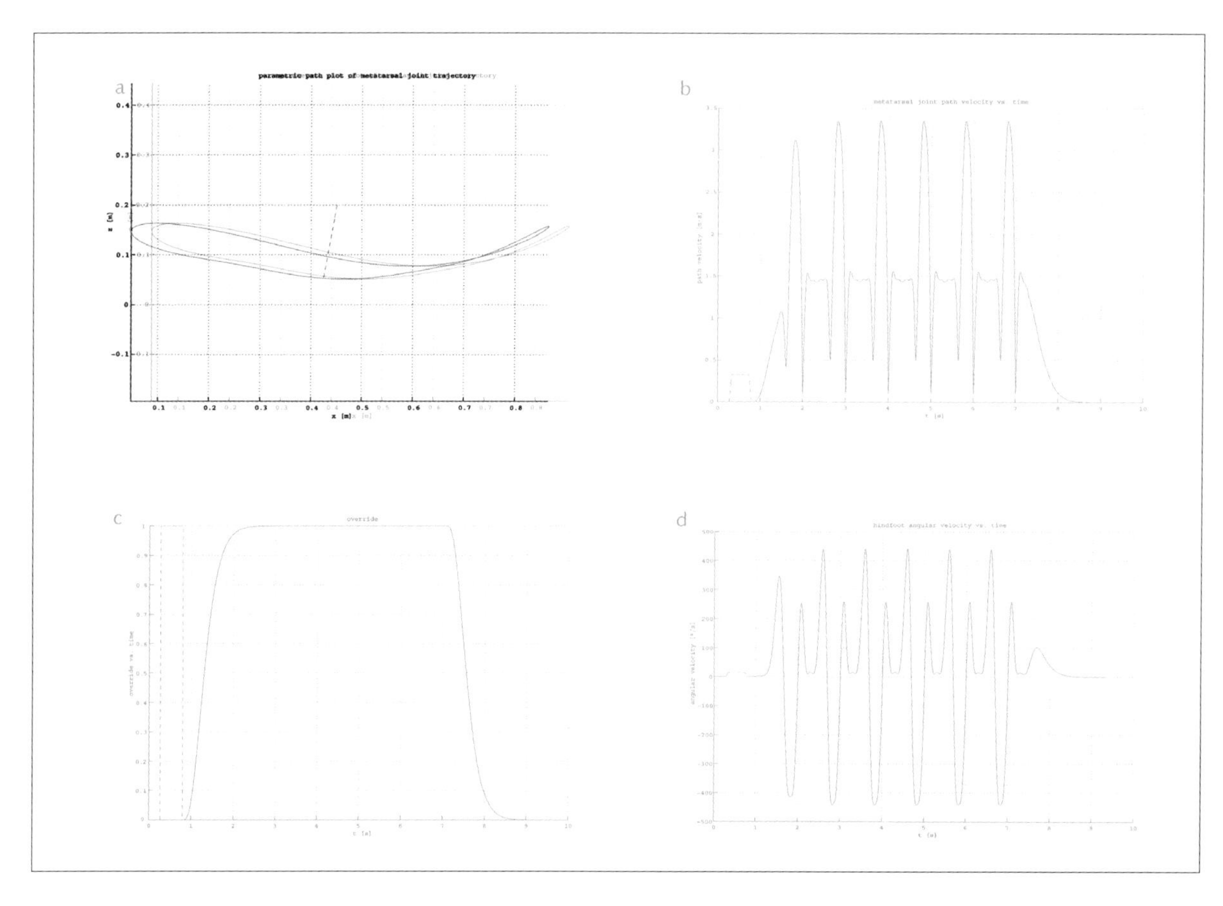

FIGURE 7. INTERPOLATION METHODS: CONVENTIONAL CARTESIAN LINEAR AND NOVEL FOURIER INTERPOLATION

Testing with healthy subjects and CNS impaired patients

Clinical trials were started after receiving full approvals by the TÜV Rheinland Medical Product Safety (German Technical Inspectorate) and the Charite University Hospital ethics board for clinical studies. So far more than 20 healthy subjects and 40 stroke patients successfully tested the machine including all its features. For patient training in the subacute phase, the clinical outcome when applying the aforementioned position-controlled motions is of primary interest, with the HapticWalker being the first device to offer an unlimited variety of possible training trajectories for fully guided natural or artificial foot motions. Compliant machine behaviour will be the next issue to investigate. Therefore a number of compliance control algorithms were implemented for testing, e.g., compliance in selected degrees of freedom, virtual spring behaviour along the guided path, full haptic device mode with free patient motion above the simulated virtual ground.

Summary

The HapticWalker is a novel generic haptic walking interface, which is based on the principle

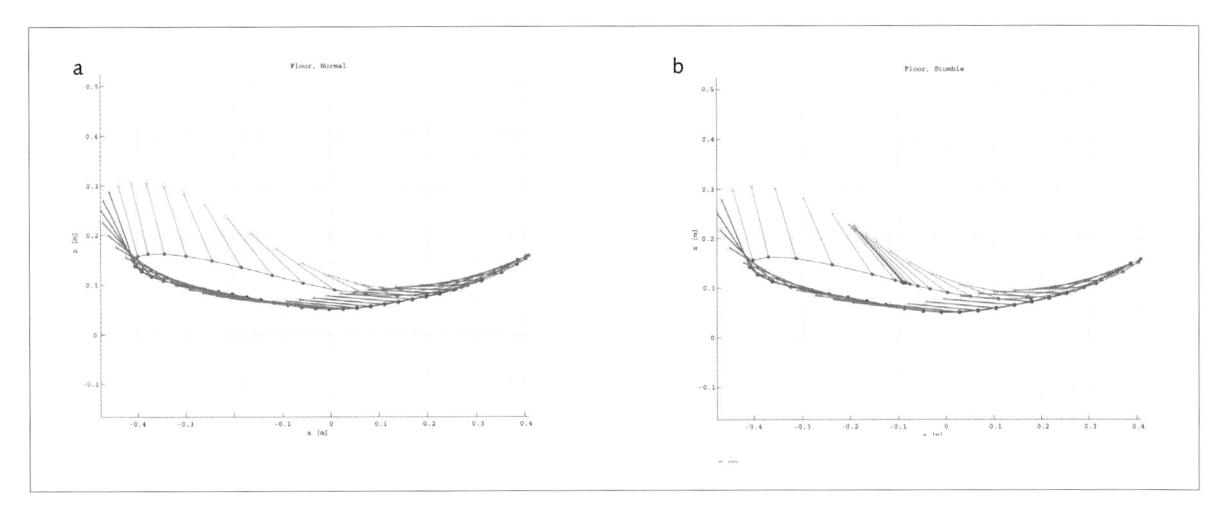

FIGURE 8. FOOT TRAJECTORY OF WALKING ON PLANE FLOOR (METATARSAL PATH TRAJECTORY AND HINDFOOT)

of programmable footplates with permanent foot machine contact. It comprises a specially developed robot kinematics, and is equipped with highly dynamic drives in order to be able to perform natural walking movements. It has been achieved to design and build a machine which covers the whole foot motion range and dynamics during all phases of gait. The machine design is based on a modular approach; the basic system comprises three DOF per foot (sagittal plane), in order to enable foot movements in the main DOF of natural foot movement. It can be extended to 6 + 1 DOF per foot without loss of the specified footplate dynamics. The analysis of different daily life gait trajectories in combination with a number of other design requirements resulting from application in gait rehabilitation served as the basis for the design of the presented machine. In addition to the mechanical machine design and construction, specially designed simulator control and operation software was developed and presented. The first development stage of a working prototype with three DOF per foot in the sagittal plane has been built and tested successfully, including all hardware and software components.

Selected readings

Hesse S (2007) *Lokomotionstherapie*. Hippocampus Verlag

Carr J, Shepherd R (2003) *Stroke rehabilitation*. Butterworth Heinemann

Winters JM, Crago PE (eds) (2000) *Biomechanics and neural control of posture and movement*. Springer Verlag

Stanney KM (ed) (2002) *Handbook of virtual environments*. Lawrence Erlbaum Associates

Recommended website

http://www.hapticwalker.de

Haptic sensing of virtual textiles

Nadia Magnenat-Thalmann and Ugo Bonanni

Introduction

Simulating the appearance and dynamic behaviour of textiles and clothes has a number of applications in the movie and entertainment business, in the textile industry, or in the artistic garment design process. For these reasons, cloth simulation has become a popular topic in the computer graphics research community, and competition is constantly increasing. During the last two decades, textile animation and rendering techniques have dramatically improved, especially in the physical behaviour and visual realism areas [1] (see Fig. 1). The interaction modalities with cloth-like deformable surfaces, however, have not followed this evolution. The most widespread methods for handling virtual textiles are traditionally based on the use of a mouse and a keyboard. But the limits imposed by old-fashioned interaction technology decrease many opportunities, as *humans, used to skin contact with clothing materials since prehistoric ages, strongly rely on their feeling of touch when handling textiles.*

Therefore, providing the sensation of touching digital textiles can significantly increase the realism and believability of the user experience. Perceiving the kinaesthetic and tactile characteristics of virtual textiles while interacting with them not only speeds up the process of handling and creating digital clothes, but also introduces a completely new way of assessing the specific surface and material properties of 3-D objects representing real products (e.g., during the online purchase of real garments). But such a visionary and sophisticated interface has not yet reached the mass consumer market. First attempts to provide the sensation of stroking virtual fabrics set important milestones in the domain of haptic rendering [2]. The proposed approach, however, did not take into consideration the physically-based, three-dimensional simulation and animation of deformable cloth surfaces. This was due to change with the advent of the HAPTEX Project [3].

FIGURE 1.
Haute Couture garments, recreated and animated from sketches [1]

The HAPTEX Project

The aim of the HAPTEX Project, a research initiative funded by the European Union, was to investigate how far we can go in providing the sense of touching three-dimensional, deformable virtual textiles. The target application scenario of the project is depicted in Figure 2.

The HAPTEX researchers' efforts resulted in a multimodal virtual reality (VR) system capable of rendering the sensation of stroking and manipulating a piece of three-dimensional digital

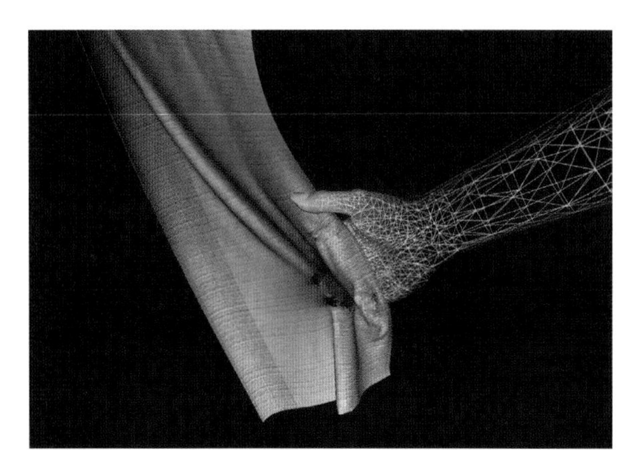

Figure 2.
The EU Project HAPTEX proposed a VR System to touch virtual textiles [3].

fabric. The sense of touching virtual textiles was provided by a new generation haptic interface integrating kinaesthetic and tactile actuators.

The HAPTEX research project was unique for several reasons:

Firstly, it produced an innovative working system which *successfully synchronised a realistic visual representation of a deformable textile with appropriate kinaesthetic and tactile stimuli* on the thumb and index finger simultaneously.

Secondly, the textile physical behaviour, the tactile rendering and the force feedback were computed by *accurate mechanical models relying on physical parameters measured from real fabrics*.

Thirdly, it generated a vast amount of expertise and know-how with a high educational impact, providing *methodologies applicable for multimodal simulation of deformable objects* [4], thus paving the ground for further research on haptic simulation and interaction.

The HAPTEX system has set an important milestone in the research of multimodal interfaces for perceiving thin deformable objects. There is every reason to believe that the necessary improvements will result from future work following the HAPTEX philosophy. This philosophy can be outlined by scientific roadmap for researching and developing multimodal simulation and interaction frameworks.

A scientific roadmap

A multimodal virtual reality experience allowing for haptic sensing of digital textiles requires the development of several separate components and finally their integration into one single system. To achieve this goal, several significant advances beyond the state of the art are necessary. Thanks to the HAPTEX experience, we can identify the main research issues as follows:

- Analysis of the real world scenario.
  - Identification of the *physical properties which are relevant for perceiving textiles* from a visual and a haptic viewpoint.
  - Development of enhanced methods for *objectively measuring the relevant physical properties* on real fabrics.
- Development of *physically based simulation techniques to model the behaviour of textile* according to the specified fabric parameters. This includes:
  - A visual display model that enables efficient and realistic rendering of the textile in real time.
  - A tactile rendering strategy which provides appropriate stimuli to the fingertip skin according to the fabric surface patterns.
  - A model computing the force feedback to the user according to the performed interaction and the specific kinaesthetic behaviour of the fabric sample.
- Design and realisation of a *haptic interface integrating a force-feedback device with tactile arrays*, being able to provide both kinaesthetic and tactile stimuli.
- *Synchronisation of the multiple sensory feedbacks* to create a consistent and physically possible experience.
- *Cross-modal validation* of the complete system to evaluate how far the interaction in virtual reality matches the real experience at different perception modalities.

Even though specifically addressing the multimodal simulation of textiles, the challenges described above represent a scientific roadmap which should be followed when tackling the visual and haptic simulation of any deformable objects.

The remainder of this chapter follows this roadmap.

Handling and measuring real textiles

How do humans evaluate the make and the quality of textiles? This question is of particular importance in the textile industry, were garments must be manufactured respecting practical production criteria while at the same time matching the consumers' expectations in terms of visual appeal and comfort. The comfort sensation of a fabric has a multi-dimensional character which is impossible to describe through a single physical property, but is commonly defined as 'fabric hand' (FH). FH refers to the sum of the sensations experienced when a fabric is touched or manipulated with the fingers [5]. It is a complex parameter and it is related to the fabric properties such as flexibility, compressibility, elasticity, resilience, density, surface contour (roughness, smoothness), surface friction and thermal character. *Fabric hand is often the fundamental aspect that determines the success or failure of a textile product.*

Subjective evaluation

The traditional method to assess FH in the apparel industry is the evaluation of the fabric sample by experts. This is done not only by considering the technical specifications of the textile, but also by sensory evaluation through handling (commonly indicated as 'subjective hand evaluation'). During this process, textiles are touched, squeezed, rubbed or otherwise handled, according to specific guidelines, such as the AATCC Evaluation Procedures [6]. *A subjective evaluation gives an insight of the textile's perceived properties from both the visual and the haptic point of view.* Such information is of utter relevance when defining how to reproduce the feeling of touching fabrics. Moreover, it is a key metric when validating the believability of the VR system by comparing subjective evaluation processes of real and virtual specimen.

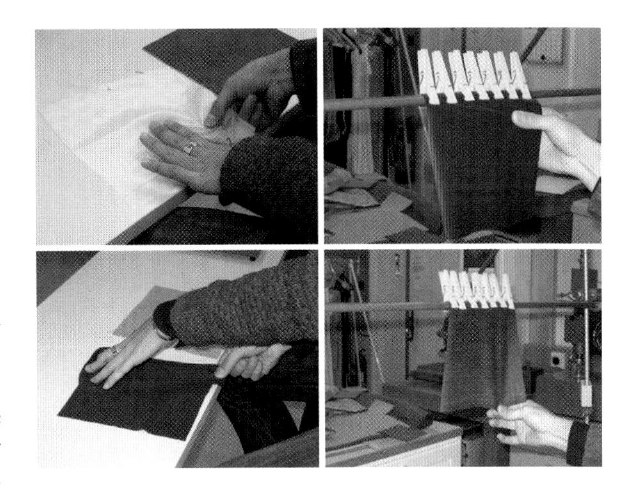

FIGURE 3.
Test arrangements of standard fabric evaluation procedures (left) adapted to match the workspace and actions allowed in the haptic simulation (right) [7]

For the comparison to be useful, real and virtual subjective evaluation of fabrics should be similar. But the complex handling procedures used by experts in textile evaluation cannot be performed on any of the haptic interfaces available today. This is mainly due to the fact that experts use both hands and all fingers over the whole textile surface while haptic VR Systems currently support only a few interaction points, such as the tips of the thumb and the index finger. Therefore, the subjective fabric evaluation needs to be simplified to allow a direct comparison of the real and the virtual process. This can be done by adapting the assessment of the real fabric to the requirements of the VR-based assessment, i.e., by evaluating the specimen using two fingers of one hand, and in the same workspace. The main mechanical properties such as tensile, shear, bending, compression, friction, surface and weight can then be assessed with the new test arrangements. Figure 3 shows the adapted evaluation procedures as described in [7].

Objective measurements

While the perceived properties of textiles are obtained through subjective evaluation, it is possible to derive objectively measured values of physical parameters from standard fabric characterisation experiments such as the 'Kawabata Evaluation System for fabrics' (KES-f), the 'Fabric Assurance by Simple Testing' (FAST) or the 'Fabric Automatic Measurement and Optimisation Universal System' (FAMOUS). These existing characterisation methods, however, do not specifically address the requirements of the simulation of textiles in virtual environments. Therefore, parameters obtained from measurements cannot be applied to a dynamic simulation in a universal way, but rather reflect the specific capturing capability of the particular measurement method, as described in [8]. Moreover, the information obtained from standard measurements might be not enough. For example, a correct representation of the dynamic mechanical behaviour of textiles requires additional information about the viscoelastic damping behaviour of the simulated fabric. To this aim, further measurement standards are being developed, such as the step-tensile method, which examines the textile elongation properties under the effect of several consecutive extension-relaxation cycles [9].

Multimodal simulation of textiles

A software framework for multimodal simulation of virtual textiles should efficiently handle realistic real-time animation of a piece of textile and provide adequate physical stimuli for visual, tactile and force rendering. The major challenge in the development of such a framework is to find the best compromise between the need to achieve a high degree of quantitative mechanical accuracy and the drastic performance requirements of interactive applications.

Many simulation systems have a strong limitation which consists in imprecise mechanical models, the main difficulty being the efficient computation of anisotropic nonlinear strain-stress behaviour of fabrics in real time. Therefore, efficient solutions for simulating the mechanical and visual behaviour of the complete textile ('large-scale'/global simulation) are needed to achieve real-time performance while preserving physical accuracy.

Another challenging task is the calculation of appropriate tactile and kinaesthetic feedback to be returned during the haptic interaction. When the fabric is virtually touched local deformations at the contact points between the virtual fabric and virtual fingers must be carefully taken into consideration in order to appropriately define the detailed contact surface and its haptic properties ('small-scale'/local simulation).

Mechanical simulation of textiles

As mentioned in the previous section, modelling the behaviour of textiles is a complex task due to the interdependence of the large number of parameters defining fabric hand such as tensile and bending elasticity, thickness and compressibility, or surface roughness and friction. For a large-scale simulation aiming at defining the global cloth behaviour at a centimetre-size resolution many of these characteristics can be ignored. The correct global nonlinear behaviour of the textile, however, must be ensured.

Global textile simulation model

Virtual fabrics can be approximated as thin 2-D surfaces and represented through a mesh of centimetre-size triangles. Hence this allows avoiding taking into consideration thickness and compressibility, surface friction and surface roughness. Further reduction of computation time can also be reached by limiting the size of the textile to be simulated, optimising the collision handling, and avoiding unnecessary calculations (of the collision detection and particle system) wherever possible.

The simulation of the remaining mechanical properties of cloth materials needs to be tackled defining a fast and stable simulation system. *Performance* is an indispensable requirement in the context of interactive haptic applications. *Stability* is needed because of possible irregular frame rates and other artefacts related to motion tracking techniques which make it necessary to implement an efficient numerical integrator.

Many real-time garment simulation systems rely on cloth simulation methods which are specifically optimised for performance. They mostly use particle systems (such as spring-mass systems) allowing simple computation of approximate mechanical models. Higher accuracy, however, can be obtained from particle systems expressing the behaviour of viscoelastic materials with the accuracy of continuum mechanics [10] while still achieving real-time performance.

The haptic sensing of virtual textiles requires to efficiently simulate the nonlinear and anisotropic behaviour of cloth materials. This behaviour is typically described as strain-stress curves measured along the weft, warp and shear deformation modes, as described in [9]. The underlying mechanical model of this approach is indeed based on continuum mechanics, but is still a particle system [11]. The simulation evaluates the strain of each triangular element according to the position and speed of the particles. It then uses the mechanical properties of the material for computing the stress of the elements, and converts the stresses back into equivalent particle forces. Figure 4 shows the draping accuracy of a piece of textile simulated using a spring-mass system (left) and an accurate particle system model (right) [9].

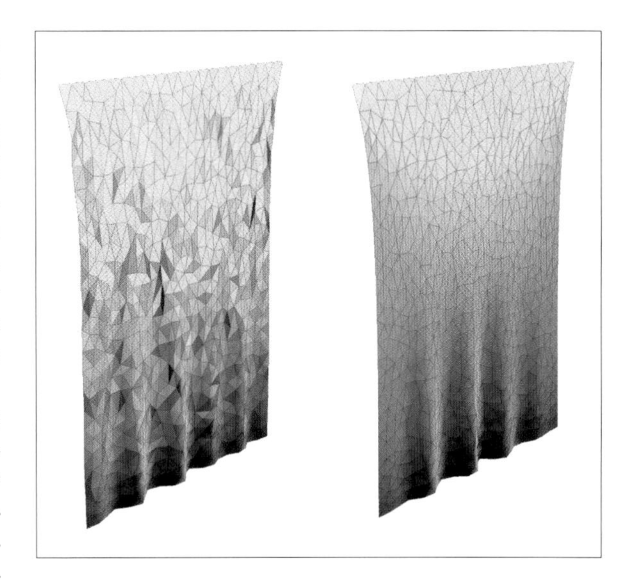

FIGURE 4.
Draping accuracy of a piece of textile simulated using a spring-mass system (left) and an accurate particle system model (right) [9]

Geometric refinements

The mechanical accuracy of the textile simulation, however, is not the only requirement for obtaining the best experience in cloth visualisation. A very important factor which should allow the user to correlate the visual and the tactile perception is *the geometric perception of the surface's shape*. The geometric rendering quality can be increased, e.g., by smoothing the textile's contours and edges through an appropriate mesh subdivision scheme. Such a subdivision scheme can be applied on-the-fly to the mesh at rendering time, and allows subdividing each geometric polygon into several pieces fitting adequately the local surface curvature. This very fast scheme brings much more accurate display quality without any significant impact over the total computation time. Furthermore, it simplifies the visual transition between the global and the local mesh representations of the surface.

Haptic rendering

Little research has been done in the domain of haptic rendering of virtual textiles. In order to distinguish the multiple haptic sensations evoked by the act of feeling textiles, we differentiate between tactile and force rendering. Especially tactile rendering *is a new research frontier.* One of the main problems in this domain is the definition of appropriate skin excitation patterns, whose nature and origin are not well understood even in real situations. Their simulation in a virtual environment context needs, therefore, solutions for relating the complex mechanical properties of the object's surface to the local topology of the fingertip's skin. Moreover, many aspects of the exploratory movement (such as speed, contact pressure and direction) strongly influence tactile sensing and must be accurately taken into account.

Tactile perception and stimulation

Early attempts to encode tactile perception of fabrics [2] took advantage of surface measurements performed with the Kawabata System, whose probe and instrumentation accurately correlate the measured quantities with subjective assessment of the textile surface. These measurements, therefore, return the *perceived* textile surface, since they provide an approximation of the surface after it has been filtered through the surface/skin interface.

Among the pioneers investigating the tactile rendering of 3-D deformable surfaces are the researchers of the HAPTEX Project [12]. Their solution is composed of three layers:

1. a fabric surface model, describing physical and geometrical properties in high- and low resolution (small-scale pseudo-topology and large-scale non-uniformity description)
2. a localisation layer, mapping the fingertip's position to the fabric's model and sending tactile cues to a tactile renderer
3. a tactile renderer, converting filtered tactile stimuli into signals for the tactile array hardware

In the particular case of tactile manipulation of virtual 3-D fabrics, *the mechanical input to the skin's mechanoreceptors can be approximated from KES-f roughness and friction profiles of textile specimens.* This data is used to generate driving signals for the tactile stimulator array as follows: a spatial-frequency spectrum is computed considering the direction of the movement and the high-resolution surface model. Taking into account the speed of movement, the spatial frequency spectrum is converted into temporal-frequency components. By application of appropriate band pass filter functions, the temporal-frequency spectrum is reduced to match only two amplitudes which are then weighted according to the low-resolution surface model, generating a 40 Hz and a 320 Hz channel. This information is conveyed to the tactile actuators described in the next section.

Local textile deformations and force-feedback

Similarly to the tactile rendering, *information concerning the force rendering of virtual textiles also comes from KES-f measurements.* Specific small- and large-scale information is computed reflecting respectively the local and global behaviour of the textile. Large-scale data defines the reaction of the complete textile specimen to the performed interaction and is provided by the global simulation model (described in the previous section) at a visual simulation refresh rate of approximately 40 fps. *The small-scale behaviour, however, reflects the immediate surface changes arising at the contact points between the virtual fabric and the virtual finger.* This layer must directly feed the haptic device through a force renderer approximating forces arising from local deformations. Moreover, it should efficiently interpolate this local behaviour with the global motion through a reciprocal synchronisation. Due to this direct communication with the force-feedback device, the small-scale data must be computed at high frequencies (300–1,000 Hz). Figure 5 displays a piece of textile showing large- and small scale areas.

While the global behaviour of textiles requires a non-linear simulation model in order to describe the anisotropic characteristics of fabrics, the local geometry of the fabric at the contact points can be defined by a linear mechanical

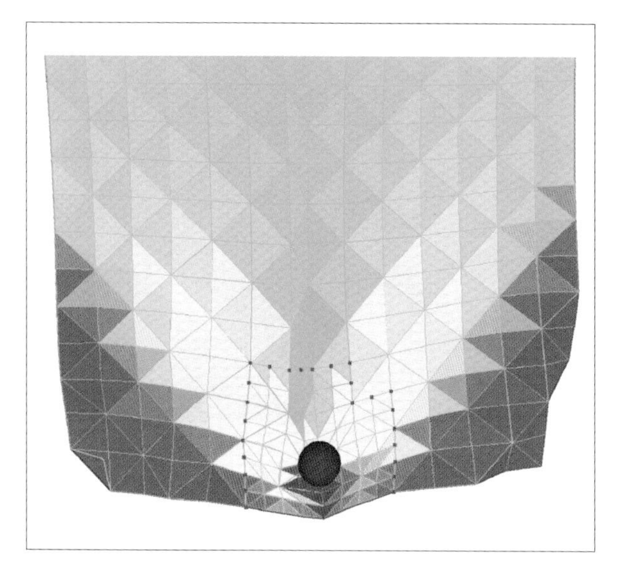

FIGURE 5.
Simulated textile displaying large-scale polygon mesh (over the whole area) and small-scale mesh (around the contact area) [13]

model. This can be explained by the entity of the deformations occurring during haptic interaction within a time frame of milliseconds, which is small enough to be described sufficiently well by the laws of elasticity, neglecting external forces like gravity. Therefore, *deformations of the local geometry can be described by a simple mass-spring model without a significant loss of accuracy* [13]. Such a mass-spring system can describe the forces occurring in the deformation of the textile by looking at changes in warp and weft directions for each triangular element of the mesh. The forces are then integrated over the triangle and distributed among the particles. To provide a comprehensive modelling of the forces acting in the contact region, the small-scale model should take into consideration:

- *Tensile forces* measured by the elongation in warp and weft direction compared to the rest state under the assumption that the stretch is constant over a triangle.
- *Shearing forces* generated by a movement parallel to a fixed axis, taking the KES-f measurements as a reference.
- *Bending forces* computed proportionally to the angle between two neighbouring triangles. Folds are displayed only over the edges of the triangles. In contrast to shearing and tensile forces, bending forces are not evenly distributed throughout the triangle.
- *Damping forces* modelling the energy dissipation during the deformation and ensuring simulation stability. These damping forces are opposing forces counteracting the stretch, shear and bend motions independently from each other.

To avoid instabilities at the contour of the contact area due to unsynchronised large- and small scale layers, constant velocity is assumed at the border of the local geometry, and no additional forces affect the corresponding particles.

For a discussion on the possible contact states allowed for the haptic interaction with textiles and possible implications of single and multiple contact points, see [13].

Tactile and force-feedback components

Efficient and accurate haptic rendering methods can only be useful in combination with appropriate hardware components, capable of returning the computed forces to the human operator. Therefore, tactile actuators and force-feedback devices must be integrated in order to provide a unified haptic interface delivering a comprehensive feeling of interacting with virtual textiles.

Tactile actuators

Different approaches may be used to provide information about the surface texture of the virtual object and about the contact between object and skin.

A variety of mechanisms applying normal forces to the skin through moving-coil technology or

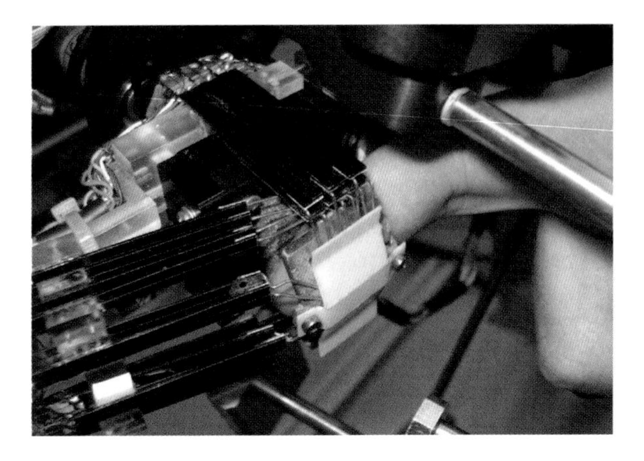

FIGURE 6.
Tactile array with dorsal encumbrance as used in [3]

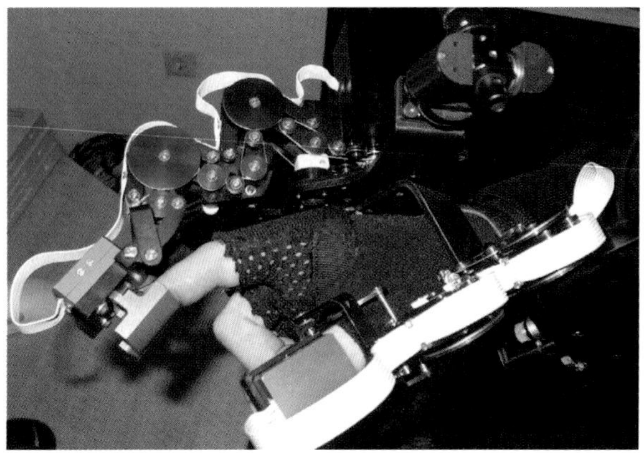

FIGURE 7.
A hand-exoskeleton especially conceived for manipulating light objects [16]

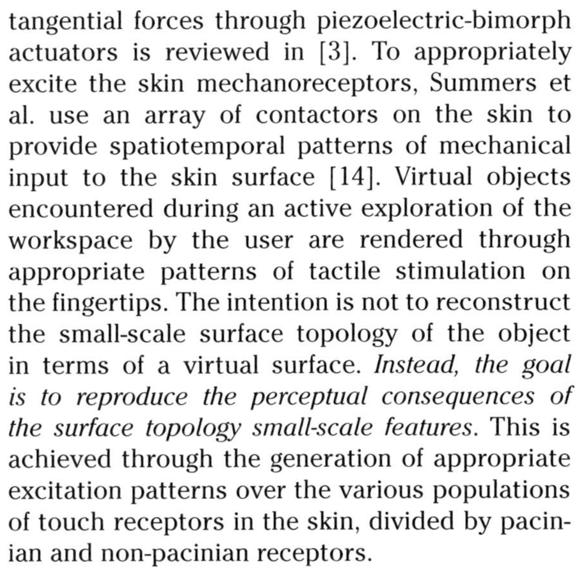

tangential forces through piezoelectric-bimorph actuators is reviewed in [3]. To appropriately excite the skin mechanoreceptors, Summers et al. use an array of contactors on the skin to provide spatiotemporal patterns of mechanical input to the skin surface [14]. Virtual objects encountered during an active exploration of the workspace by the user are rendered through appropriate patterns of tactile stimulation on the fingertips. The intention is not to reconstruct the small-scale surface topology of the object in terms of a virtual surface. *Instead, the goal is to reproduce the perceptual consequences of the surface topology small-scale features.* This is achieved through the generation of appropriate excitation patterns over the various populations of touch receptors in the skin, divided by pacinian and non-pacinian receptors.

Such a tactile stimulator can be used for evoking the sensation of stroking a textile surface. The tactile component shown in Figure 6 uses piezoelectric bimorphs to drive 24 contactors in a 6×4 array on the fingertip, with a spacing of 2 mm between contactor centres. The drive mechanism is placed to the side of the finger and ahead of the finger, rather than below the contactor surface. With one such array on the index finger and one on the thumb, this positioning of the drive mechanism allows the finger to move close to the thumb so that a virtual textile can be manipulated between the tips of the finger and the thumb.

Force-feedback device

The most intuitive force-feedback interface for manipulating thin objects is most probably a glove-like hand-exoskeleton due to its ergonomics. In order to be able to reproduce the forces arising during textile manipulation, *such a device must be expressly conceived for the accurate generation of light forces* [15].

Among the different types of hand-exoskeletons, *multi-phalanx devices* generate forces along each finger's phalanx in a fixed direction, while *fingertip devices* can generate forces only on the fingertips, but in any arbitrary direction in 3-D space.

As the second type of functionality promises the best results for assessing thin objects in 3-D space, researchers have realised an actuated and sensorised hand exoskeleton in form of a fingertip device [16], depicted in Figure 7.

The device is composed by two actuated and sensorised finger exoskeletons (one for the index and one for the thumb) and is body-grounded. It is connected to a sensorised mechanical tracker that provides the absolute position and orientation of the operator's hand. The whole mechanism has 4 degrees of freedom, even though it is actuated with only three motors, thanks to the coupling of the last joint with the previous one. The device is capable of exerting a continuous force on the fingertips of 5N with a resolution of 0.005 N.

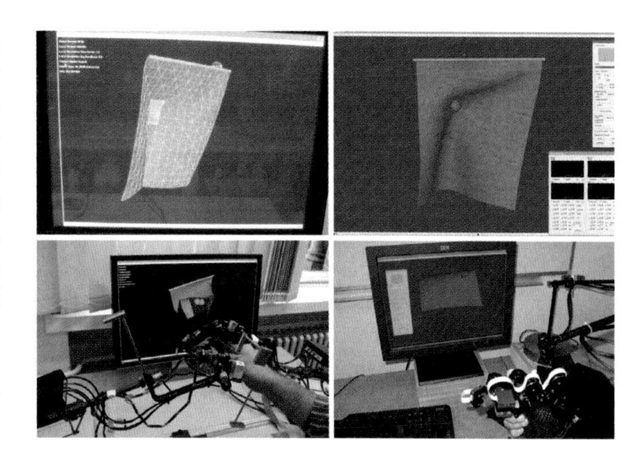

FIGURE 8.
The HAPTEX real-time textile simulation framework (top) and the final HAPTEX system in two different configurations (early version bottom left, final system bottom right) [16]

System integration and validation

The research and development of a haptic VR System featuring physically based simulation of textiles animated in real-time is a very challenging endeavour. It requires following a scientific roadmap and integrating a set of heterogeneous components described in the previous sections of this chapter.

Component integration

The communication between the various software components can be described according to the chronological succession of events taking place during the simulation. In the initial stage all threads concerning the global and local textile simulation, as well as tactile and force rendering, run separately at their dedicated update rate. This asynchronous behaviour must be coordinated through appropriate resampling and interpolation mechanisms. The HAPTEX approach for the synchronisation of multiple threads is described in [13].

On the hardware side, the integration of the force-feedback device with the tactile actuators should ensure the delivery of the separate mechanical stimulations to the fingertip without interferences between the devices. Two main problems should be avoided. Firstly, the global forces conveyed by the force-feedback device should not load the pins of the tactile array. Secondly, the mass of the tactile array drive mechanism should not load the force sensor. In the case of the HAPTEX System, the drive mechanism of the tactile actuators is directly attached to the final link of the force-feedback device. The tactile pins vibrate through holes in the force plate attached to the force sensor's load flange. An exhaustive description of this hardware integration is given by [17].

Figure 8 shows screenshots from the HAPTEX real-time textile simulation framework (top left and right) as well as the final HAPTEX system in two different configurations featuring an early force-feedback device (bottom left), and the hand-exoskeleton described in the previous section (bottom right) [16].

System validation

The evaluation of a system for visuo-haptic perception of virtual textile should assess all different aspects of the simulation. For example, the results of the physical simulation framework reproducing the virtual fabric should be compared with the motion of real fabrics, and likewise all different system components should be validated separately. Finally, the complete sys-

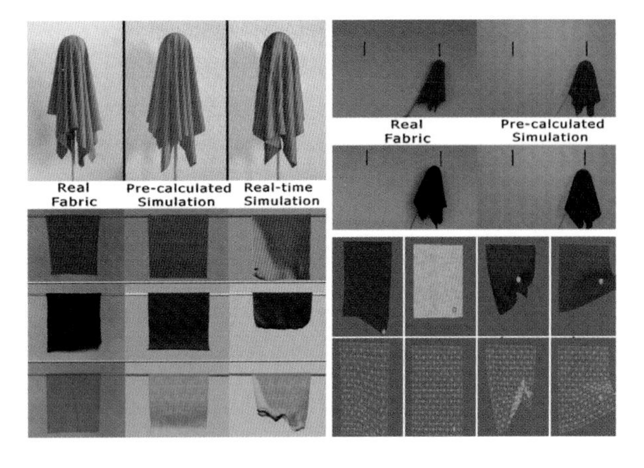

Figure 9.
Different validation tests: static simulation of fabric drape (top left), dynamic simulation of fabric falling down from a fixed stand (bottom left), and dynamic simulation of fabric on a moving sphere (top right). On the bottom right, manipulation procedures for assessing different physical properties

tem should be assessed in terms of functionality, i.e., *its validation should provide a comparison of results from a subjective evaluation of real fabrics and a subjective evaluation of virtual fabrics.* In order to analyse cross-modal interactions, experiments should be carried out in two variants: with and without vision.

A preliminary validation of the HAPTEX System assessed the physical and visual realism of simulation virtual textile simulation. In order to validate the software models derived from measurements on real textiles, three different evaluation methods were selected and performed on a variety of textile samples. Figure 9 shows the different tests: static simulation of fabric drape (top left), dynamic simulation of fabric falling down from a fixed stand (bottom left), and dynamic simulation of fabric on a moving sphere (top right).

Eventually, for the final system evaluation, virtual textiles were manipulated and their various properties felt during interaction were rated. These ratings were compared to those obtained in an equivalent assessment of the real textiles, and to physical data obtained using the Kawabata measurement system and other techniques. Assessed properties included tensile stiffness, surface roughness/friction, and bending stiffness/weight/drapeability. An evaluation procedure was defined and followed under different operating conditions (tests were performed in 'full vision' or 'wireframe' mode, as shown in Figure 9, bottom right). For the complex case of evaluating bending stiffness (difficult to assess due to the low forces arising during such interactions), the evaluator was asked to make a visual assessment – effectively an estimate of drapeability, which includes the effect of both bending stiffness and weight. *The performed evaluation provided a good correspondence between the ratings of the virtual textiles and the real textiles especially for the properties of tensile stiffness, surface roughness and surface friction* [18].

Summary

The presented results are mainly the output of the European Research Project HAPTEX – HAPtic Sensing of Virtual TEXtiles. Selecting the simulated textile from a broad range of samples, the final user of the HAPTEX System is able to interact with a virtual piece of fabric and see it moving according to his manipulation, as shown in Figure 8. During the interaction, the user is able to perceive the simulated fabric haptically, with the HAPTEX System providing the relevant tactile and kinaesthetic feedback to the operator.

The HAPTEX project has produced a workable system for the presentation of virtual textiles, providing evidence that the virtual system is delivering appropriate cues to the user. The first version of the interface (see Fig. 8, bottom left) had some non-ideal aspects which limited its functionality. For example, the system was not always easy to navigate and some manipulation movements were difficult to perform. However, with the final version of the complete system featuring the hand-exoskeleton (see Fig. 8, bottom right), many of these problems were solved.

Nonetheless, it is important to note that *the final goal of realising a VR System allowing the immediate identification of a variety of different fabrics has not been completely achieved yet.* The HAPTEX System allows discriminating between different fabrics – but a blindfold evaluation of fabric hand does not produce the same result as in the real world situation. The presented system, however, has significantly contributed to advancements beyond the state of the art in the domain of multimodal simulation in virtual reality environments, and will probably have a high impact on applications in the entertainment industry, in the design and training sector, and in the medical field.

Acknowledgements

The authors would like to thank Prof. Franz-Erich Wolter, Dr. Harriet Meinander, Dr. Ian Summers and Fabio Salsedo for their significant contribution to achieving the presented results. We are also grateful to the staff of MIRALab, University of Geneva, and particularly to Dr. Pascal Volino and Christiane Luible for their extensive commitment to perform high-quality research on physically based simulation of clothing materials, and to Ghislaine Wharton for improving the English text. Finally, special acknowledgements go to all members of the HAPTEX Project Consortium for their valuable contributions to the research on haptic sensing on virtual textiles.

Selected readings

Behery HM (2006) *Effects of mechanical and physical properties on fabric hand.* Cambridge: Woodhead Publishing

Kawabata S (1980) *The standardization and analysis of hand evaluation.* Osaka: The Textile Machinery Society of Japan

Volino P, Magnenat-Thalmann N (2000) *Virtual clothing.* Berlin/Heidelberg: Springer

House DH, Breen DE (eds) (2000) *Cloth modeling and animation.* Wellesley: AK Peters Ltd

Hayward V (2008) Physically-based haptic synthesis. In: Lin M, Otaduy M (eds): *Haptic rendering: foundations, algorithms and applications.* Wellesley: AK Peters Ltd, 297–309

Summers IR, Chanter CM (2002) A broadband tactile array on the fingertip. *J Acoust Soc Am* 112(5): 2118–2126

Barbagli F, Prattichizzo D, Salisbury K (eds) (2005) *Multi-point interaction with real and virtual objects.* Berlin/Heidelberg: Springer.

44

Haptic discrimination of paper

Ian R. Summers, Richard J. Irwin and Alan C. Brady

Introduction

This chapter describes a study on the haptic discrimination of different types of plain paper. The experiment is designed to replicate features of a 'banknote' scenario in which it may be possible to identify a counterfeit on the basis of only a few seconds contact. Multidimensional scaling (MDS) techniques are used to investigate the perceptual dimensions involved in the discrimination task. A related study of tactile perceptual space is summarised in Appendix 1.

It is an everyday experience to handle paper – turning the page of a book, opening mail, handling a banknote – and it takes only a short time to assess the paper in terms of its characteristic 'feel'. (See Lederman and Klatsky's investigation [1] of manipulation strategies for obtaining information about objects in general.) The present study investigates some of the perceptual processes which are involved in making such assessments, particularly in relation to features which are significant in the handling of banknotes. (In the United Kingdom, the distinctive feel of banknotes is officially recommended as an indicator for the detection of counterfeits, and there is anecdotal evidence that counterfeits may indeed be detected in this way.) Such features might include gross physical parameters such as paper thickness and stiffness, as well as parameters which relate to surface texture.

There have been few previous studies on the perception of thickness or stiffness for material in the form of thin sheets. Thickness discrimination might in principle be based either on direct perception of thickness (for example, when holding a sheet between finger and thumb, in terms of joint angle) or on perception of stiffness (which is determined both by the thickness of sheet and the mechanical properties of the material). Such discrimination has been investigated by John, Goodwin and Darian-Smith [2] and Ho [3]. The author of the latter study proposed an explanation of results from both investigations on the basis that, when sheets are sufficiently thin to deform under finger contact, thickness discrimination is based primarily on perception of the curvature of the deformed sheet.

The majority of published material on tactile perception of surface texture concerns the response to well-defined artificial surfaces such as gratings [4–6] or embossed patterns [7]. There are a few studies involving the surface texture of everyday objects, which is often difficult to describe objectively. In an early study, Katz [8] describes an experiment on the tactile discrimination of 14 types of paper (chosen with a wide range of properties and each intended to be clearly discriminable from the others) and develops the concept of *Modifikationen* (qualities), which provide scales on which surfaces may be rated. Hollins, Faldowski, Rao and Young [9] studied the dimensionality of 'natural' tactile stimuli such as wood, sandpaper, velvet, etc., and suggested that three perceptual dimensions were involved: one corresponding to roughness/smoothness, one to hardness/softness and a third tentatively ascribed to 'springiness'. In further studies [10–12], Hollins and co-workers have provided evidence for a sticky/slippery dimension, and demonstrated the role of Pacinian corpuscles in the tactile perception of surface roughness. In an experiment involving discrimination of a very wide range (124 types) of flat surface, Tiest and Kappers [13] identified four perceptual dimensions, none of which was well matched to measured values of surface roughness or surface compressibility.

In the case of paper, although it is possible to characterise the surface by means of a range of parameters such as mean height of surface features, typical separation of surface peaks, etc., it is not easy to predict how these parameters will contribute to aspects of perceived surface texture. Lyne, Whiteman and Donderi [14] found three main factors which influenced the perceived quality of paper towelling – rigidity, surface softness and embossment pattern. However, the tactile features of paper towels are rather different from those of the typing or photocopying paper used in the present study.

The literature also contains a variety of studies relating to 'fabric hand', i.e., the way in which a textile may be evaluated in terms of its characteristic feel. Kawabata [15] relates fabric hand to various measured properties of the textile: tensile properties, shearing and bending properties, thickness and compression properties, surface roughness and surface friction. Picard, Dacremont, Valentin and Giboreau [16] describe experiments which suggest a four-dimensional perceptual space for textiles stretched over a flat support.

The present study was designed to replicate some features of a 'banknote' scenario in which it is necessary, when handling a sequence of notes, to identify a counterfeit on the basis of only a few seconds contact (and generally without the advantage of comparing two notes directly). The intention was to investigate a set of relatively similar stimuli, with a view to establishing which aspects are important for discrimination of small differences between papers. Subjects were required to handle rectangles of various types of paper, each for a few seconds. The rectangles were presented in sequences of three, in an 'odd-one-out' task. Discrimination scores between the various papers were obtained, as the basis for constructing a multidimensional perceptual space for the papers. Broadly similar experimental strategies are described by Hollins, Bensmaia, Karlof and Young [10], involving tactile investigation of a wide range of everyday surfaces, and by Cooke, Steinke, Wallraven and Bülthoff [17], involving haptic exploration of object shape.

Method

Subjects were 12 unpaid volunteers: graduate students in the age range 22–27, nine male and three female. Two of the males were left-handed; the remainder of the subjects were right-handed.

Each stimulus was a rectangle of paper with dimensions 135 mm × 69 mm, corresponding to the size of a common UK banknote. Stimuli were produced from different types of plain white paper – one of these was rag paper of the type used for banknotes (in an unprinted state and consequently with somewhat different handling properties to actual banknotes) and the rest were intended for typing or photocopying, acquired from several stationery stores. 28 different types of paper were considered; ten were discarded after an initial inspection because of various anomalies, such as large surface features related to the watermark. From the remaining 18 papers, ten which appeared in an informal assessment to be perceptually similar were selected for use, including the rag paper. (Similar papers were selected in order to focus on the known acuity of discrimination in this context; this choice also facilitates the determination of discrimination indices d', as described below.) The selected papers varied in thickness from 98–131 μm and in area density from 73–102 gm m^{-2}. A large number of rectangles were prepared from each of the ten selected papers.

Procedure

The experiment used an 'odd one out from three' format with a three-alternative forced choice (3AFC), in which three samples of paper were presented in sequence to the subject, two being of the same type and the other of a different type. Subjects were instructed to pick up each sample with one hand (from a tray on which the experimenter had previously placed the required type of paper rectangle), pass it to the other hand, and then put it back down on the tray. No direct comparison between two papers was permitted and approximately one or two seconds was allowed for the complete operation. Subjects responded verbally to indicate which of the three papers in each trial was the 'odd one out'. It is clear that short-term memory of tactile stimuli may play an important part in such a task. This aspect of memory has been investigated by Bowers, Mollenhauer and Luxford [18], who found that accuracy for texture recollection was very high, suggesting that memory effects are not a major consideration in the present investigation.

Auditory and visual masking was present throughout testing, using white noise *via* headphones and opaque goggles, to ensure that only tactile cues were available. In a brief (around 10 min) familiarisation period in advance of the testing, subjects were allowed to perform the task without masking in order to become accustomed to the exact procedure, with feedback being given to correct any errors in procedure or timing.

Subjects were instructed to carefully wash and dry their hands prior to participating in the experiment in an attempt to minimise inter-subject variation in skin conditions. Air humidity and temperature in the test room were monitored to ensure that testing was not carried out under extreme conditions which might affect the paper characteristics (since paper is hydrophilic) or the subject's skin conditions (as a result of perspiration). Values recorded were typically on the order of 22°C for temperature, and 26% for humidity.

Each trial (i.e., each sequence of three papers) involved discrimination of a particular pair of papers. Each test block consisted of 45 trials – one for each of the 45 possible pairs available from the ten types of paper, with the sequence of pairs varied from subject to subject to avoid any order effects. Each subject completed four test blocks. Hence pooled data from all 12 subjects include results from 48 trials for each of the 45 pairs of papers. There are six possible patterns for a single trial with a given pair of papers (ABB, BAB, etc.), and the sequence of patterns within the test blocks was permuted so that all patterns were equally represented in the trials for a given pair of papers.

Results

Overall discrimination scores for the various paper pairs range from 15/48 to 47/48, i.e., from just below the chance score of 33% to just below the maximum score of 100%. Mean scores for each subject over all paper pairs range from 51–78%, with no obvious anomalies in terms of particularly good or particularly poor performance. The mean score for all subjects over all paper pairs is 67%, with a standard deviation of 8%, this moderate overall performance reflecting the initial selection of perceptually similar papers.

MDS procedure

Tables produced by Craven [19] were used to convert the percent-correct discrimination score for each paper pair to a corresponding value of discrimination index d'.

Perceptual spaces of various dimensions were constructed for the ten papers using multidimensional scaling (MDS) techniques [20]. For a given dimensionality, points corresponding to each paper were positioned within the space so that their Euclidean interpoint distances d_{ij} matched the corresponding inter-paper discrimination-index values d'_{ij} as closely as possible. The optimum configuration was obtained by minimising the stress [21], defined by:

$$\text{stress} = (\Sigma[d_{ij} - d'_{ij}]^2 \,/\, \Sigma\, d'^{\,2}_{ij})^{0.5} \quad (1)$$

where the summations are over all pairs of points. (The stress is the r.m.s. error for the configuration divided by the r.m.s. value of the 'target' interpoint distances d'_{ij}.)

It should be noted that this procedure, in contrast to many MDS techniques (e.g., [21]), produces a configuration whose interpoint distances are matched to the experimental data in terms of magnitude rather than in terms of rank order only, and hence a configuration whose interpoint distances d directly correspond to discrimination index d'.

Optimisation was achieved using a purpose-written iterative computer program which, starting from a random initial arrangement of points, moved each point in turn in a direction that reduced the stress. In this way the configuration homed in on a stress minimum – local minima were excluded by running the program from a wide range of initial conditions.

By repeating this procedure for spaces of different dimensions, it is possible to find the minimum number of dimensions which adequately fit the data. In this case, values of stress for one, two, three and four dimensions are 0.264, 0.124, 0.118, and 0.118, respectively, indicating that two dimensions are sufficient to fit the data (since the stress in three or more dimensions is not appreciably lower than in two dimensions). The optimised configuration of papers in the two-dimensional space is shown in Figure 1(a). (The plot is shown with an arbitrary displacement and rotation – neither the absolute displacement of the configuration nor its absolute angular position is constrained by the MDS procedure. A convenient choice for the absolute angular position is suggested below.)

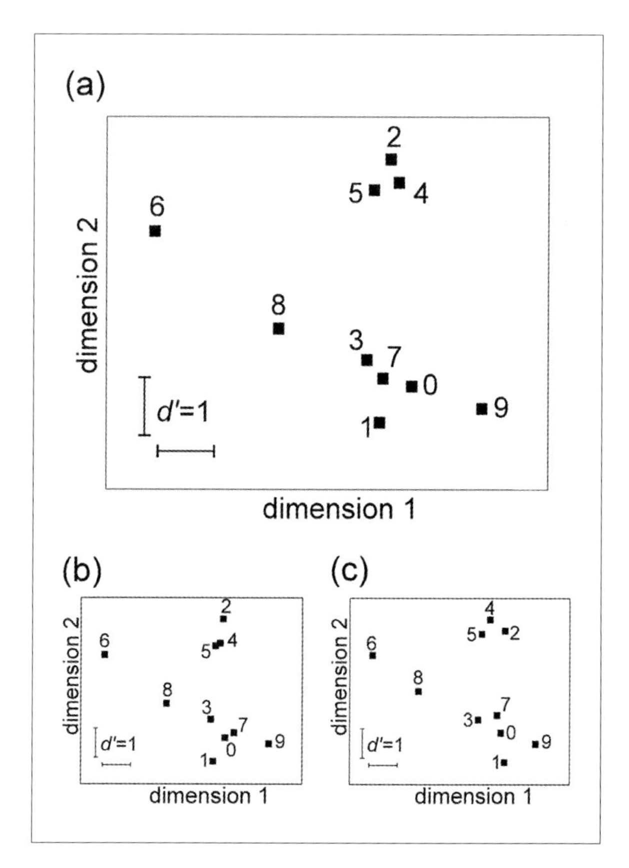

Figure 1.
Minimum-stress configurations for the ten papers in a 2-D perceptual space: (a) derived from data for all 12 subjects; (b) derived from first subgroup of six subjects; (c) derived from second subgroup of six subjects. The overall orientation of the configuration (i.e., with respect to displacement and rotation) is arbitrary.

The robustness of this configuration with respect to variability in the data was checked by randomly dividing the subject group into two subgroups of six subjects, and calculating two-dimensional configurations for each subgroup, shown in Figure 1(b) and Figure 1(c). It can be seen that there are no significant discrepancies between the two configurations obtained, and both are very similar to the Figure 1(a) configuration for the complete dataset. This suggests that the principal features of the configuration in Figure 1(a) do not derive from random variations in the data but are indeed 'real' features.

The fact that each of the two subgroups of subjects produces a similar perceptual space may indicate that all subjects used similar tactics to discriminate the papers, i.e., they made use of similar perceived aspects of the papers. To investigate this point further, the distribution pattern for correct responses over the 45 paper pairs was determined for each subject. Cross correlation between these patterns produces a measure of the similarity between individual subjects in respect of answer patterns, from which it is possible, using similar MDS techniques to those already applied to the d' data, to construct a two-dimensional 'subject space' (in which the interpoint distances correspond to inter-subject dissimilarity in respect of answer patterns). The configuration of subjects in this two-dimensional space shows a single cluster of points – there is no evidence of distinct subgroups of subjects within the configuration. This reinforces the above suggestion that all subjects used similar tactics for the discrimination.

It is implicit in the above analysis that d' values add vectorially within a perceptual space. Green and Swets [22] suggest that this should in general be the case, and data from Figure 1(a) are consistent with this – comparison of the interpoint distances d_{ij} with the 'target' distances d'_{ij} shows no obvious trend to suggest that a non-linear transformation of the d' data would produce a better-matched configuration in perceptual space.

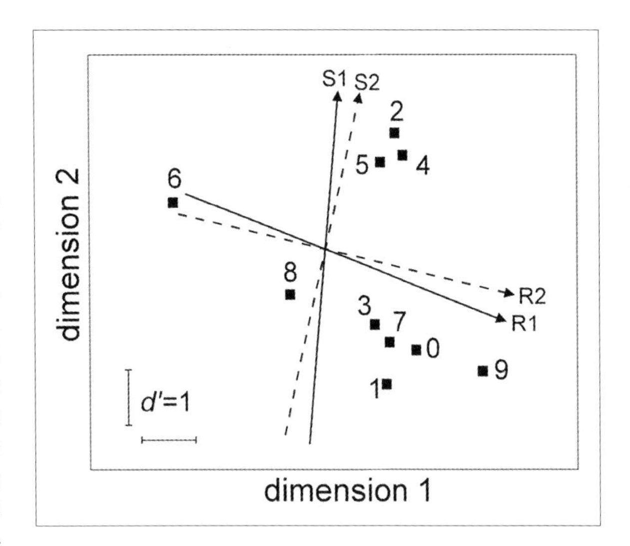

FIGURE 2.
The minimum-stress configuration for the ten papers, showing the directions R1 and S1 (full lines) onto which the configuration can be projected to give the best match to mean roughness rank and mean stiffness rank, respectively. Also shown are the directions R2 and S2 (dotted lines) which give the best match if orthogonal directions are specified. The overall orientation of the configuration (i.e., with respect to displacement and rotation) is arbitrary.

Interpretation of MDS findings

The MDS analysis suggests that discrimination of these papers is dominated by two perceptual dimensions. It was hypothesised that these dimensions were roughness and stiffness – informal comments offered by subjects after the testing often mentioned these attributes as giving important cues.

In order to test this hypothesis, a subsidiary experiment was carried out to establish subjects' estimates of roughness and stiffness for the ten papers, with a view to correlating these estimates with the perceptual space of Figure 1(a). Subjects were presented with samples of each of the ten papers and asked to arrange them in a row on a table, first in order of roughness and then in order of stiffness. (They were told that they could take as much time as they wanted to complete the task, and that they could compare the papers in any way that they thought necessary. They were also able to re-assess their initial orderings and make modifications.) In contrast to the main experiment, this subsidiary experiment was carried out without auditory or visual masking. Visual masking was removed because it would have seriously complicated the task of arranging the papers in rank order. Auditory cues were minimal in practice and so it was decided to also remove the auditory masking. A potential concern is that the multimodal assessments in the subsidiary experiment might not correlate well with the haptic-only assessments in the main experiment. However, comparison of the results from the two experiments (see below) shows a good correlation, suggesting that essentially the same aspects of roughness and stiffness are involved in the two experiments.

The participants in the subsidiary experiment were ten of the original 12 subjects (two were unavailable). For each paper, the average rank for roughness and the average rank for stiffness was calculated over the subject group. This procedure achieved a good separation of the papers giving mean values for roughness rank in the

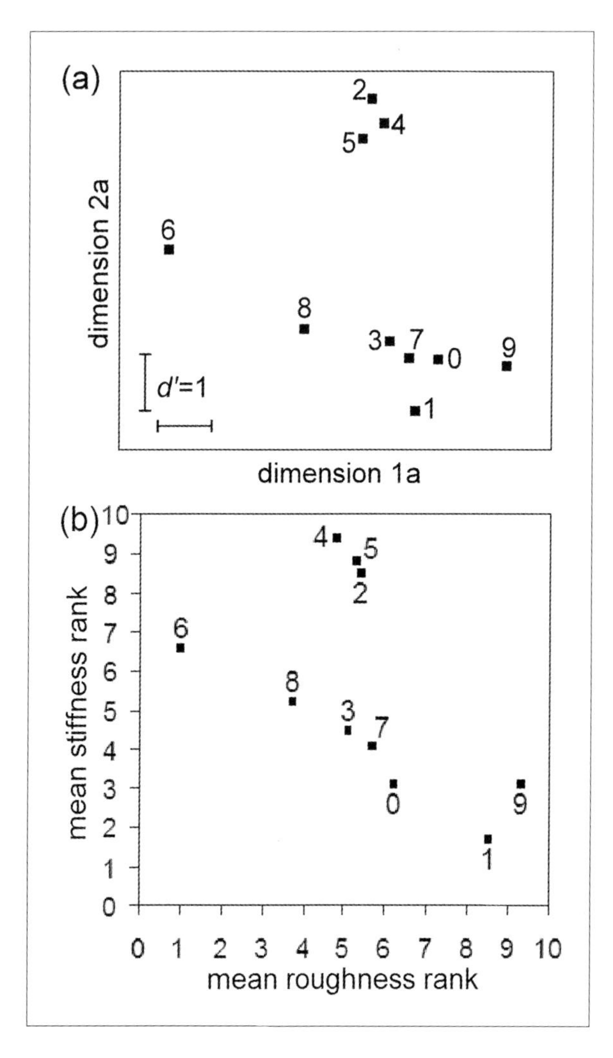

FIGURE 3.
(a) The minimum-stress configuration for the ten papers, as in Figure 1(a) but rotated through 13° counterclockwise (see text); (b) plot of mean (perceived) roughness rank versus mean (perceived) stiffness rank for the ten papers.

range 1.0–9.3 and mean values for stiffness rank in the range 1.7–9.4 (the available range in each case is 1.0–10.0, with higher numbers indicating higher roughness/stiffness).

In order to investigate the possible correspondence of perceived roughness or perceived stiffness to a dimension of the MDS space (Fig. 1(a)), one-dimensional projections of the points in the MDS space were obtained at various angles θ to the 'dimension 1' axis. For each value of θ, correlation coefficients r were calculated between the positions of the points in the one-dimensional projection and the mean roughness ranks from the subsidiary experiment, and the value of θ for the maximum correlation was determined. This procedure was repeated for the mean stiffness ranks. In each case the maximum correlation was high: $r=0.93$ for the roughness data and $r=0.98$ for the stiffness data, with an angle of 106° between the optimum values of θ for the two attributes (see Fig. 2, full lines labelled R1 and S1).

For both roughness and stiffness, the graph of correlation coefficient *versus* θ is somewhat flat-topped around the maximum, i.e., small changes in θ from the optimum value produce little decrease in the correlation. Hence, since the optimum values of θ for the two attributes differ by close to 90° it is possible to force an overall fit between *orthogonal* projections of the MDS space and perceived roughness and perceived stiffness, with little reduction in the individual correlations. Such a procedure, maximising the sum of the two correlation coefficients, gives $r=0.92$ for the roughness data and $r=0.97$ for the stiffness data (see Fig. 2, dotted lines labelled R2 and S2). Hence a rotation of the plot in Figure 2 (and also in Fig. 1(a)), so that the R2 and S2 directions become the principal axes, provides a convenient choice for the absolute angular position of the configuration. Such a rotation provides the optimum correspondence between the MDS space and a two-dimensional 'attribute space' whose orthogonal principal axes correspond to perceived roughness and perceived stiffness, as determined in the subsidiary experiment. This correspondence is shown in Figure 3 – panel (a) shows the MDS space of Figure 1(a), rotated so that the R2 and S2 directions become the principal axes but otherwise unchanged; panel (b) shows the 'attribute space'. It can be seen that, although there are small differences between the two configurations, the overall arrangements of the ten papers are very similar.

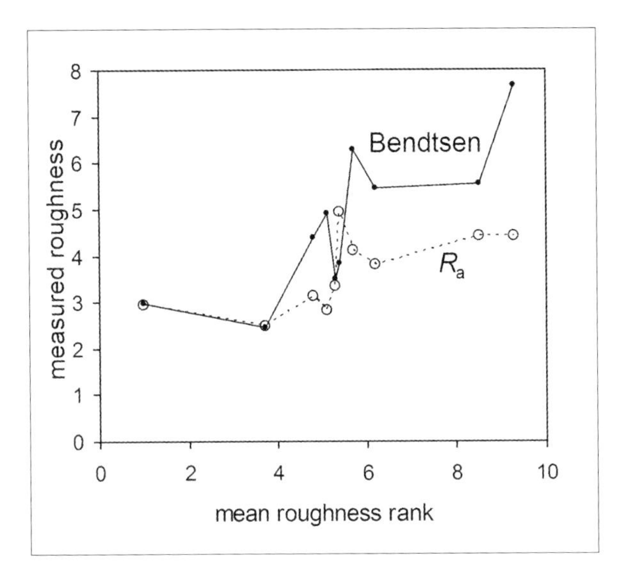

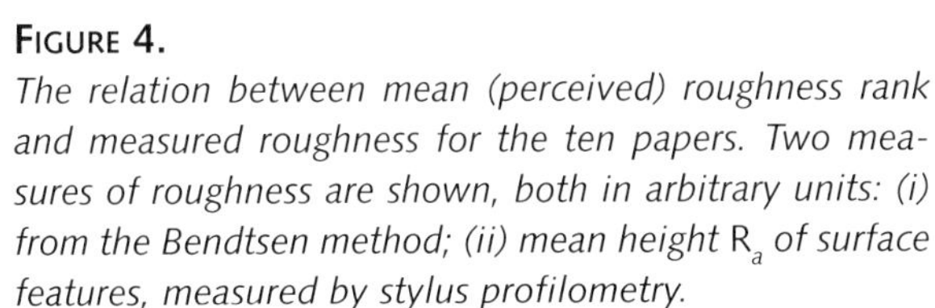

FIGURE 4.
The relation between mean (perceived) roughness rank and measured roughness for the ten papers. Two measures of roughness are shown, both in arbitrary units: (i) from the Bendtsen method; (ii) mean height R_a *of surface features, measured by stylus profilometry.*

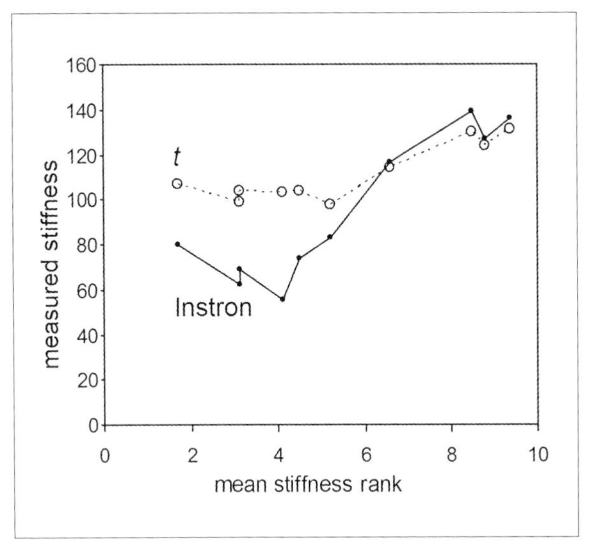

FIGURE 5.
The relationship between mean (perceived) stiffness rank and measured stiffness for the ten papers. Two measures relating to stiffness are shown: (i) measured by the Instron method, units of 2 gm wt: (ii) mean paper thickness t, *units of μm.*

Comparison with measured data

A range of parameters is used by paper manufacturers to characterise their products. Among these, surface roughness is described by a variety of measures relating to surface features, which may be derived from stylus profilometry [23]. Surface roughness may also be determined by the *Bendtsen method* [24]: an inverted metal cup, 3 cm in diameter, is placed on the horizontal paper surface; pressurised air at 1.5 kPa is introduced into the cup and leakage under the edge of the cup (machined to be a knife edge) is measured. Stiffness is described by *Instron Stiffness*, measured in terms of the force required to push a square paper sample, standard dimensions 67 mm × 67 mm, through a narrow slot placed under the midline of the sample (i.e., effectively, the force required to fold the paper in half).

Instron stiffness and a range of roughness measures were obtained for each of the papers used in these experiments. Measurements were also made of mean paper thickness – a principal determinant of stiffness, as mentioned above. Figure 4 shows two representative roughness measures for the ten papers, plotted against mean (perceived) roughness rank from the subsidiary experiment. Similarly, Figure 5 shows Instron stiffness and mean thickness, plotted against mean (perceived) stiffness rank. The data in Figure 4 and Figure 5 indicate the ranges of measured roughness and stiffness over the stimulus set which, as discussed above, were chosen to be relatively limited. From Figure 4, it appears that the data for Bendtsen roughness are more successful than the profilometry data as a predictor of perceived roughness: the larger difference in Bendtsen stiffness correspond quite well to the larger differences in perceived roughness, although this relation does not hold for the smaller differences. Similarly, from Figure 5 it appears that Instron stiffness is an approximate predictor of perceived stiffness: the Instron measures (and the closely related measures of paper thickness) suggest a group of six papers which are less stiff and a group of four papers which are more stiff, and this classification is consistent with the ranking of perceived stiffness.

Discussion

The results of this study demonstrate that subjects' discrimination of different types of paper can be successfully represented by a two-dimensional perceptual space – the MDS analysis produces a two-dimensional configuration with an acceptably low value of stress, and which is robust in terms of variability in the data. An alternative two-dimensional perceptual space which can be constructed for the ten papers from data for mean roughness rank and mean stiffness rank shows a close correspondence to the MDS space. This gives a further indication of the success of the MDS analysis, and provides persuasive evidence that the two dimensions of the MDS space correspond to perceived roughness and perceived stiffness.

The cumulative value of discrimination index d' across the range of the 'roughness' dimension 1a in Figure 3(a) is 5.9, and across the range of the 'stiffness' dimension 2a it is 4.8. Hence the two perceptual dimensions contribute in approximately equal measure to the separation of the papers in this experiment.

It can be seen from Figure 2 that the papers fall into two linear groups, one lying parallel to the R1 direction and one parallel to the S1 direction. This may indicate that, for a given paper, one or other of the two perceptual dimensions is dominant with information from the second dimension being masked.

As might be expected in view of the limited range of stimuli selected and the restricted handling procedure, the number of perceptual dimensions required to describe data from the present study is fewer than that suggested in some related studies. Only one perceptual dimension in the present study appears related to surface properties, in contrast to the three or four identified in other studies [9, 10, 13, 16]. It seems reasonable to conclude that the roughness dimension identified in the present study is related to the roughness dimension identified by Hollins, Faldowski, Rao and Young [9].

Most papers encountered in everyday situations have gross surface features deriving from watermarks or printing and it seems, on the basis of an informal investigation carried out in conjunction with this study, that such gross features (the perception of which may involve more than one additional dimension) can provide a strong additional cue for discrimination. As mentioned above, Lyne, Whiteman and Donderi [14] found that the three dimensions required to describe paper towelling included one related to emboss-ment pattern. Appendix 1 describes a study of the perceptual space for embossed patterns on a rigid sheet, also carried out in conjunction with the main study described in this chapter – in this case two perceptual dimensions are suggested.

Conventional measures used by paper manufacturers to characterise their products appear to be relatively poor predictors of the perceived attributes which are significant in this study. However, it must be remembered that the papers in this investigation were chosen to be perceptually similar, and so the measurement procedures are each being used over only a small part of their available range. Over a larger range, i.e., in terms of gross changes in roughness or stiffness, a better correspondence between measured and perceived aspects might be expected. (As outlined above, conventional measurement techniques for textiles [15] have been carefully designed to produce results which correlate with perceived attributes. However, conventional measurement techniques for paper have not been designed with this goal in mind, and so a good correlation with perceived attributes is perhaps not to be expected.)

There is no reason to believe that perceptual features of paper derive from anything other than large-scale or small-scale topological or mechanical features. Hence in principle it should be possible to establish measures of paper, or combinations of such measures, which correspond closely to the principal perceptual dimensions. For example, confocal laser scanning microscopy [25] can provide high-resolution topographic data in three dimensions, and can identify coherent surface structures (i.e., bundles of fibres) which are not apparent from a 1-D stylus profilometry scan. This is an area in which further study is required, and which should produce results of commercial as well as intrinsic interest.

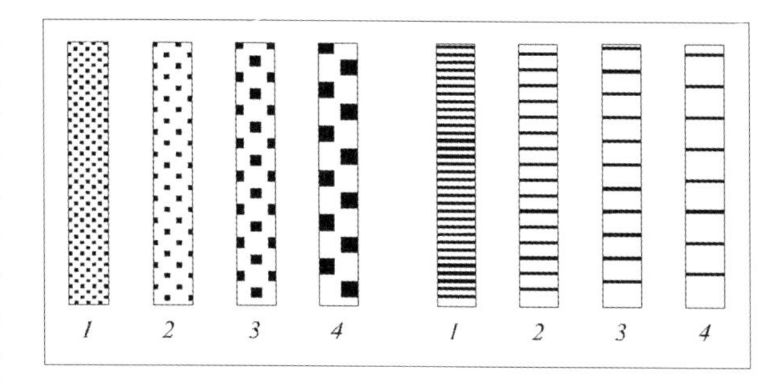

Figure 6.
The stimulus set for identification of embossed surface patterns (raised lines or chequerboards)

Summary

In a study on the discrimination of ten different types of plain paper, a three-alternative forced-choice procedure is used to obtain a measure of the dissimilarity of each of the possible pairs of papers. The experiment is designed to replicate features of a 'banknote' scenario in which it may be possible to identify a counterfeit on the basis of only a few seconds' contact. Using multidimensional scaling (MDS) techniques, two perceptual dimensions are found to satisfactorily represent the data. The nature of the data allows distances in MDS space to be measured in terms of the discrimination index d'. Ranking of the same papers in terms of perceived roughness and perceived stiffness produces a good correspondence to the perceptual dimensions derived from the discrimination experiment. Over the limited range of stimuli selected, conventional measurement methods for characterisation of paper are found to be relatively poor predictors of perceived roughness and perceived stiffness. Appendix 1 presents data for the discrimination of embossed patterns on a rigid surface – raised lines or chequerboards with a surface elevation of 35 µm.

Appendix 1: an example of tactile perceptual space

This study looks at the discrimination of embossed patterns on a rigid surface. It is assumed that the stimulus patterns occupy a multidimensional tactile space and are discriminated tactually on the basis of their different positions in this space.

Measurements were made on identification of embossed surface patterns (raised lines or chequerboards), with a surface elevation of 35 µm, on a rigid substrate of thickness 1 mm and area 10 mm by 70 mm (see Fig. 6). The stimulus tokens were made from commercially available printed-circuit board, using etching techniques to create the embossed patterns. After a period of training, subjects were tested on their ability to identify the different patterns. The test procedure involved the subject picking up the token with one hand, running the index finger of the other hand over the pattern, and then putting the token down again. Subjects were told to take between 1 and 2 s to feel each pattern. Visual masking was present throughout testing, using opaque goggles. Some of the test blocks included only bars, some only chequerboards, and some both bars and chequerboards. Results are presented here for the latter case (both bars and chequerboards), from three subjects.

Figure 7 shows a confusion matrix for all stimuli. Note the bar/chequerboard confusions (cells shaded grey) at both the shortest and longest length scales. (At the shortest scale the pattern features (~ 1 mm) are not so easily resolved by the sense of touch and the bar/chequerboard distinction is not always clear; at the longest scale the pattern features (~ 5 mm) are of similar

	bar response				cheq response			
	1	2	3	4	1	2	3	4
bar1	54				6			
bar2	2	56	2					
bar3		8	48	3				1
bar4			10	44			1	5
cheq1	8				52			
cheq2						52	8	
cheq3						9	43	8
cheq4				3			11	46

FIGURE 7.
Confusion matrix of results for identification of embossed surface patterns. Cells which indicate bar/chequerboard confusions are shaded grey.

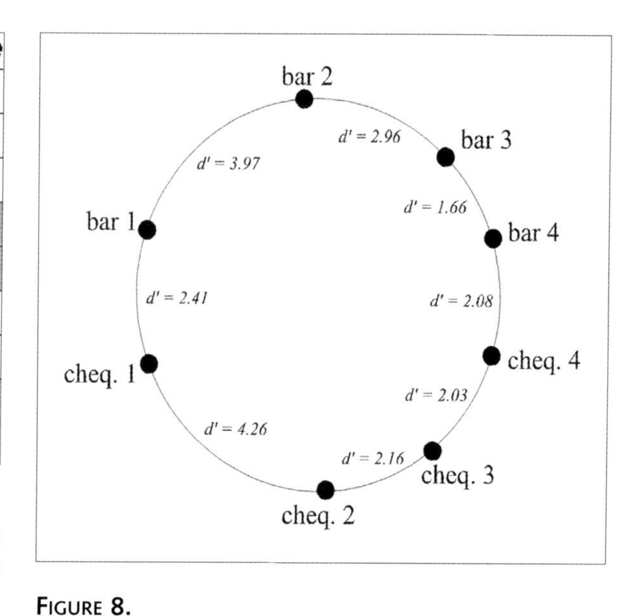

FIGURE 8.
A suggested perceptual space, constrained to a circle. Nearest-neighbour distances correspond to the inter-stimulus discrimination index d'*, with values as shown in the labels.*

size to the contact area of the fingertip and so the two-dimensional organisation of the pattern is not always apparent.)

The information transfer IT from the stimulus set to the subject may be calculated from the confusion matrix using the formula

$$\mathrm{IT} = \sum_i \sum_j p_{ij} \log_2 (p_{ij} / p_i p_j) \qquad (2)$$

where the index i indicates the various alternatives in the stimulus set, the index j indicates the various alternatives in the response set, p_i is the probability of stimulus i, p_j is the probability of response j, and p_{ij} is the joint probability of stimulus i and response j. The information transfer can be considered to indicate the number N_c of categories perceived by the subject, according to the relation $N_c = 2^{\mathrm{IT}}$. In this case, the information transfer is calculated to be 2.2 bits, corresponding to 4.6 discriminable categories (i.e., less than the eight categories presented, as may be inferred from subjects' errors).

A suggested perceptual space, constrained to a circle, is shown in Figure 8. This (non-rigorous) configuration is based on the observation that confusions within the stimulus set are observed between nearest neighbours in the subset of bars, between nearest neighbours in the subset of chequerboards, between bars and chequerboards at the shortest length scale, and between bars and chequerboards at the longest length scale. Nearest-neighbour distances in Figure 8 correspond to the inter-stimulus discrimination index d', which can be calculated from subjects' error patterns (as indicated by the confusion-matrix data) using the method proposed by Braida and Durlach [26]. (This method is designed for a one-dimensional perceptual space, but is here applied along the circumference of the circle.) It is possible to interpret the configuration as a two-dimensional space whose dimensions are related to spatial periodicity and the bar/chequerboard distinction – the rotational orientation of the figure has been chosen so that spatial periodicity runs horizontally and the bar/chequerboard distinction runs vertically.

Acknowledgements

This research was supported by the UK Engineering and Physical Sciences Research Council.

Selected readings

Hollins M, Bensmaia SJ (2007) The coding of roughness. *Can J Exp Psychol* 61: 184–195

Kennedy JM (2000) Recognizing outline pictures *via* touch: Alignment theory. In: MA Heller (ed): *Touch, representation and blindness*. Oxford University Press, Oxford, 67–98

Klatzky RL, Lederman SJ (2003) The haptic identification of everyday life objects. In: Y Hatwell, E Gentaz (eds): *Touching and knowing*. John Benjamins, Amsterdam, 123–159

Lederman SJ, Hamilton C (2002) Using tactile features to help functionally blind individuals denominate banknotes. *Hum Factors* 44: 413–428

Haptic banknote design

Hubert R. Dinse

Introduction

New developments in printing technologies have made it possible to create intaglio structures of enormous richness in fine spatial details. Nowadays relief height of printed intaglio structures can reach up to 50–100 microns thereby creating not only visual, but also strong and significant tactile and haptic sensations (Fig. 1).

FIGURE 1.
Example of an artificial banknote exemplifying the richness and visual details created by intaglio printing.

Measures to prevent counterfeiting: safety features

Since the introduction of banknotes centuries ago, banknotes were counterfeited. Particularly currencies with wide cross-country distributions are a target. According to "*The United States Treasury Bureau of Engraving and Printing*" (2005) counterfeited notes worth millions of US Dollars are detected year by year with a trend to increase in countries outside the US (Fig. 2).

Using current high-technology features, numerous security-features are integrated in the banknote design to prevent or to impede counterfeiting. Most widely used features are watermarks, security threads, ultraviolet fibres and ink, special foil elements, hologram patches or stripes, glossy stripes and iridescent stripe/shifting colours.

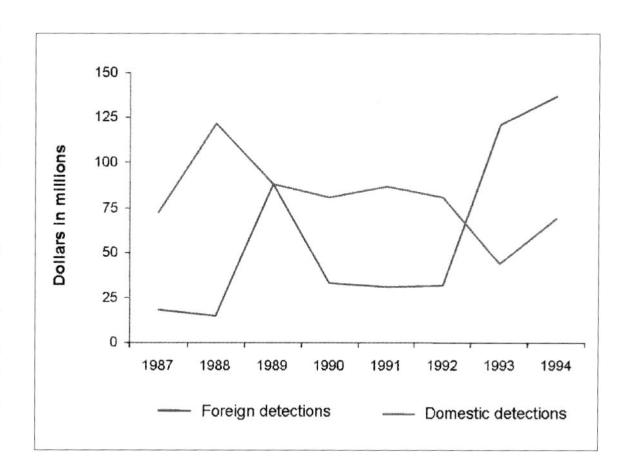

FIGURE 2.
Counterfeit detection of US$ according to data from Secret Service for Fiscal years 1987 to 1994. Source: The United States Treasury Bureau of Engraving and Printing © 2005.

Measures to prevent counterfeiting: tactility

Up to now, design options for banknotes were largely governed by aesthetical and graphical

Box 1

What governs perception in visuotactile stimulation, a situation relevant in banknote handling? Using transcranial magnetic stimulation as a means to induce a so-called 'virtual' lesion during a roughness and distance discrimination tasks in sighted and blind subjects revealed a double-dissociation effect in which roughness and distance are primarily processed in somatosensory and visual cortex, respectively. Low-frequency transcranial magnetic stimulation applied to somatosensory cortex disrupted roughness without affecting distance judgments, while transcranial magnetic stimulation to visual cortex disrupted distance but not roughness judgments indicating that in fact both modalities are involved in this tactile texture perception [1]. In another experiment visuotactile texture assessment consisting of the speeded discrimination of roughened textile samples, was tested in the presence of a congruent or an incongruent textile distractor. When discriminating between samples, visual assessment of textile roughness was modulated by incongruous tactile distractors, but not vice versa, even when visual distractors were more discriminable than tactile targets demonstrating a significant asymmetry in interference [2]. This view is supported by earlier experiments, in which subjects assembled puzzles under different conditions of visual-haptic perception which both separated these perceptual systems and required the simultaneous use of vision and touch. When vision and touch were used simultaneously to examine the same stimulus, and visual information was adequate for responding, vision dominated touch. However, in conditions of visual-haptic perception in which visual input was inadequate for responding or when haptic perception was independent of vision, subjects relied upon a form of information which was specifically tactual in nature and independent of visual or verbal recoding [3]. Using a visuo-haptic virtual reality environment, subjects had to judge the distance between two surfaces based on two perceptual cues: a visual stereo cue obtained when viewing the scene binocularly and a haptic cue obtained when subjects grasped the two surfaces between their thumb and index fingers. The results indicate that subjects recalibrated their interpretations of the stereo cue in a context-sensitive manner that depended on viewing distance, thereby making them more consistent with depth-from-haptic estimates at all viewing distances [4]. These findings suggest that observers' visual and haptic percepts are tightly coupled in the sense that haptic percepts provide a standard to which visual percepts can be recalibrated when the visual percepts are deemed to be erroneous. Combined, converging lines of evidence imply a guiding role of touch as compared to vision, possibly because the assessment of surfaces and textures is more ecologically suited to touch than to vision [2].

criterions. Owing to new printing technologies described above, tactile features based on intaglio structures are increasingly employed as a means to recognise authentic banknotes. Reasons for an increasing use of tactility is that safety features are often unknown to the user, and that in many cases technical devices are required to decode them such as ultraviolet ink.

The use of haptic features becomes immediately apparent in the following scenario: Imagine a person is receiving a banknote which is visually perfectly counterfeited as in case of high-quality colour prints. However, what is missing is tactility and haptic features. Once a person feels such a note with his fingers, the person will be immediately alarmed. The reason is that tactile/haptic information can override visual information (see Box 1).

How tactility is sensed: factors affecting tactility

Intaglio structures of the type described are detected by a set of highly specialised receptors residing in different layers of the skin. They signal variables such as force, pressure, velocity,

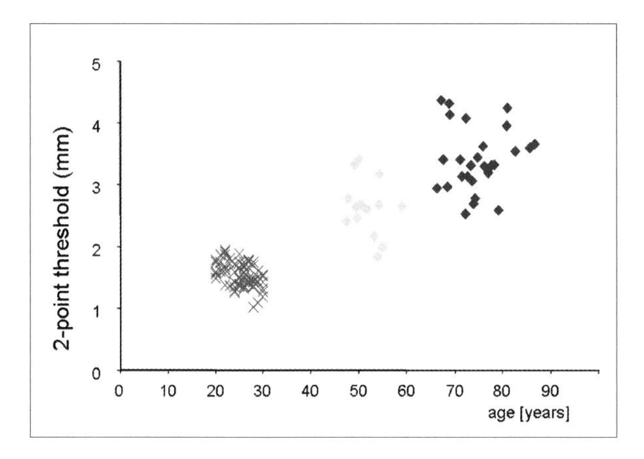

FIGURE 3.
Tactile two-point discrimination thresholds of the tip of the right index finger as a function of age (total of 120 subjects). Note almost linear decline over lifespan. Modified and reprinted with permission from Annals of Neurology *[7].*

acceleration exerted to the skin. The resulting information is coded into action potentials and transmitted to the brain for further processing in order to create conscious perception about texture properties.

A general scheme is that the distribution of the various receptor types is not uniform across the body, but high in those skin portions that are particularly sensitive, such as finger tips or lips. In addition, receptor distribution varies dramatically inter-individually making tactile and haptic sensation highly variable across a human population.

Among factors that have a major impact on tactile and haptic sensation is age. There is a tremendous age-related decline of performance over age, which starts already at an age of 30 to 35 and then almost linearly continuous to decline over the rest of the lifespan [5–7]. Figure 3 depicts an example of how tactile acuity, the ability to tell two closely spaced points in space apart, is worsening over age. Age-related deterioration of tactile and haptic sensation is not only critical from an individual point of view, but also from a socio-political view: We witness a unique

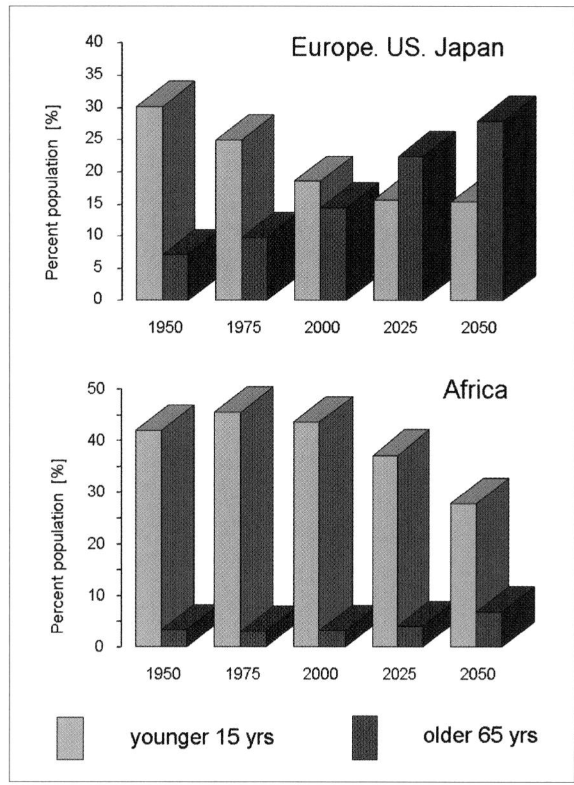

FIGURE 4.
Development of the ageing structure of the industrialised countries US, Europe and Japan, and for comparison from countries of Africa. Adopted from: World Population Prospects. The 2002 Revision. *UN, New York.*

restructuring of the ageing pattern in the societies of the industrial nations, characterised by an increasing probability to reach old age (Fig. 4). As a result, the constraints of everyday-living differ for a huge proportion of the population from that of the younger counterpart. In the context of designing haptic structures, ageing constitutes an important factor that has to be taken into account.

Besides ageing, factors such as deliberate practice, experience and use greatly influence haptic perception. Haptic performance of the fingers is significantly enhanced in subjects who are using their fingers much above average. Typical

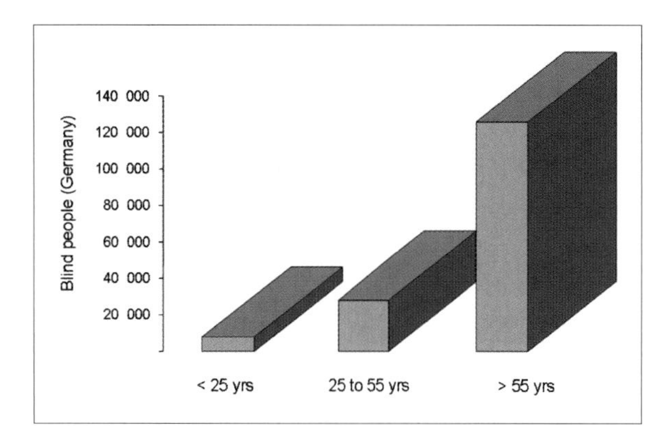

FIGURE 5.
Ageing structure of people in Germany suffering from blindness or severe vision impairment. Federal Statistical Office, Germany (2003). German Association of the Blind.

examples are musicians and those that are blind. Particularly, blind subjects who read Braille with their fingers develop an extraordinary level of tactile performance which is maintained into high age [8–10].

In Germany, about 0.05% of the total population suffers from blindness or severe vision impairments (Fig. 5). Despite the relatively small percentage, some countries have included Braille characters or specifically designed raised dot patterns to support blind people in recognising banknotes. For example, Lederman and Hamilton reported about the specific design of dot patterns that were tested and later incorporated into new banknotes issued by the Bank of Canada [11].

Constraints for banknote design

For the design of haptic features of banknotes, two aspects are important:

First, given the substantial technical possibilities to produce almost any types of extremely fine-grained intaglio structures, it is not clear, which tactile features are particularly suited to create strong haptic sensations in human users, and thus are optimal for being utilised in banknote authentication.

Second, in the context of haptic design and optimisation of tactile features for banknotes, a number of constraints have to be met which arise (arranged in the order of priority) from a number of factors that might not necessarily match those required for haptic authentication:

- politics
- technical (printing and paper quality)
- graphics and aesthetics
- tactility and haptics
- human factors

To consider human factors adequately, the following constraints have to be taken into account:

- A normal distribution of tactile and haptic performance in a 'idealised' population
- The effect of ageing, in particularly different age-groups
- the existence of specialised subgroups, such as blind individuals

Haptic banknote design: effective intaglio structures and their reliable recognition throughout the population

The following describes results from a study in cooperation with Giesecke & Devrient GmbH, Germany [12], in which tactile and haptic properties of banknote-like intaglio printing was investigated in three right-handed subgroups consisting of young adult subjects (aged 21–29 years), middle-aged to elderly subjects (aged 52–68 years), and middle-aged to elderly blind subjects who acquired blindness late in life and who were proficient Braille readers (aged 53–65 years).

Psychophysical tests

The task consisted of comparing in a two-alternative forced choice paradigm pairs of intaglio

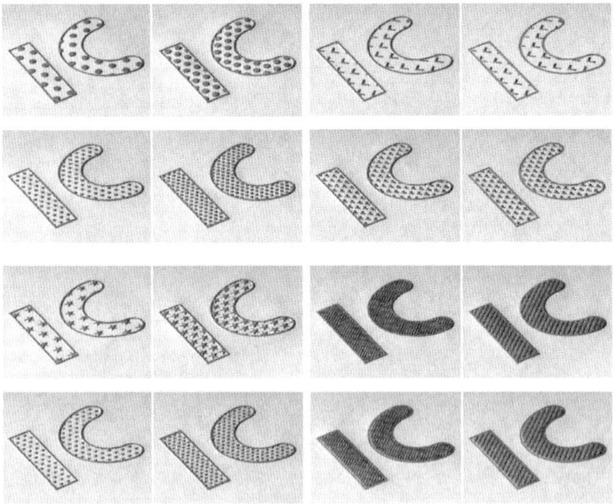

FIGURE 6.
Examples of the intaglio structures and value numerals used in the present investigation. Form elements used to fill the numerals were stars, dots, open triangles and line patterns that varied in size and spacing.

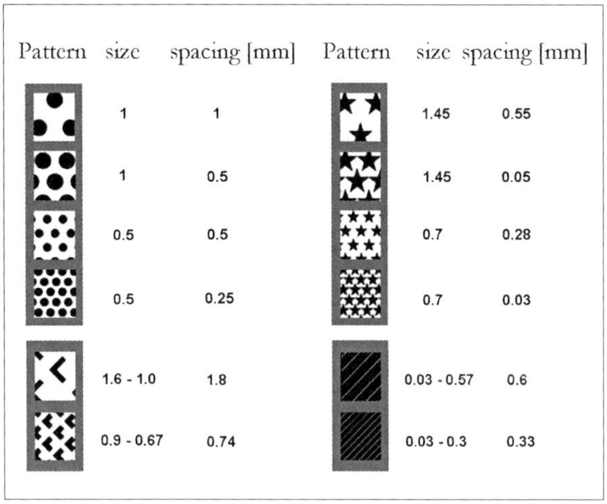

Pattern	size	spacing [mm]	Pattern	size	spacing [mm]
	1	1		1.45	0.55
	1	0.5		1.45	0.05
	0.5	0.5		0.7	0.28
	0.5	0.25		0.7	0.03
	1.6 - 1.0	1.8		0.03 - 0.57	0.6
	0.9 - 0.67	0.74		0.03 - 0.3	0.33

FIGURE 7.
Schematic illustration of the form elements and their dimensions (values given in millimetres).

structures on banknote-like sheets, which were printed on banknote-paper (105×50 mm); for illustrations of different specimens used see Figure 6. The papers contained intaglio pattern of different form elements of varying size and spacing embedded in two value numerals (width 3 mm, height 13 mm, separation 2 mm) approximately covering an area of 13×13 mm in size. According to optical interference measurement relief height was between 85 and 95 microns. For a schematic overview of the form elements tested see Figure 7. In addition, the value numerals could either have an outline with the same relief height as the form elements, or without an outline. Subjects were asked without sight to explore the value numeral with their thumb or index fingers and report whether pairs of notes were the same or different. To obtain information about the affective aspects of the pattern tested, subjects were also asked to rate each specimen according to pleasantness on a scale from 1 to 4 with 1 "very unpleasant" to 4 "very pleasant". Each pattern was compared to each other in one session (300 comparisons), once with the index finger and once with the thumb, according to discriminative and to affective properties. Each session was repeated eight times resulting to a total of more than 6,000 comparisons in each subject.

Discrimination/identification performance

Similarity rating depended on form, size, spacing and outlining. No general trend could be found accept that patterns made up from linings performed best. In Figure 8 the top 30th percentile of the pattern that yielded best and worst identification (correct rating) are shown. Among those performing best, all patterns investigated were represented except for open triangles, which performed worst. Outlining was found in most of the worst performing pattern indicating that this feature seems to prevent correct identification. No significant differences were observed for conditions when subjects used either the index finger or the thumb for exploration.

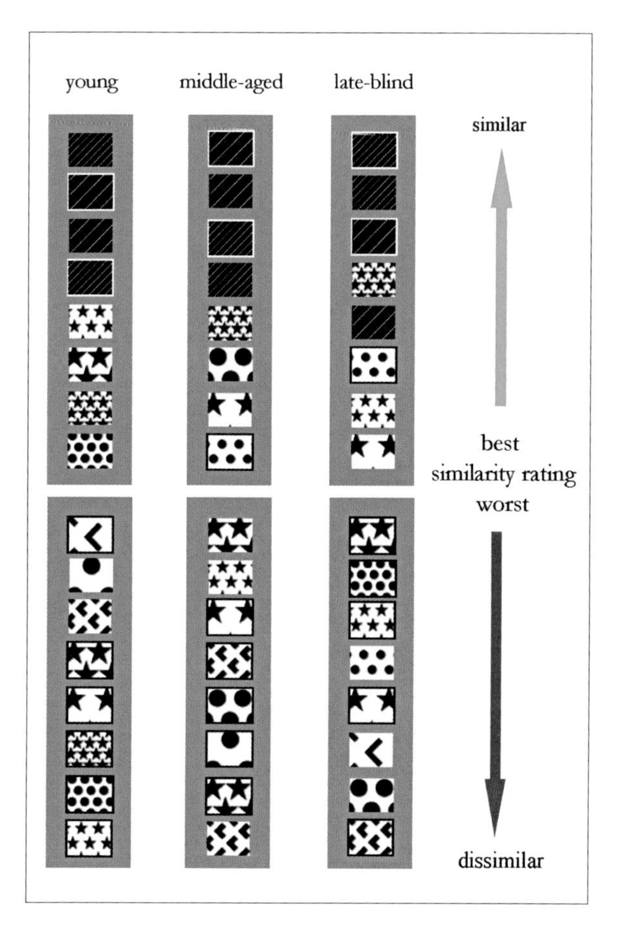

FIGURE 8.
Example of pattern of the 30th percentile yielding best (top) and worst (bottom) similarity rating. Shown are data for all subgroups tested: young adults (left), elderly (middle), and elderly, late blind people (right).

FIGURE 9.
Example of pattern of the 30th percentile yielding highest (top) and lowest (bottom) pleasantness rating. Shown are data for all subgroups tested: young adults (left), elderly (middle), and elderly, late blind people (right).

Affective performance

Subjects successfully rated patterns according to 'pleasant' and 'unpleasant'. In Figure 9 the top 30th percentile of the pattern that yielded highest or lowest pleasantness rating are shown. Again, there seemed to be no general trend. Surprisingly, there was a considerable match concerning pattern that yielded high similarity rating and those that yielded high pleasantness rating. Accordingly, a subgroup of patterns is similarly suited to provide effective discriminative as well as affective performance.

Dependence on pattern (form, size, spacing, outlining)

As noted above, there was no general trend in the sense that certain form elements, dimensions or spacing showed systematic superiority above others. The only form elements that yielded best performance were line patterns of different spacing. For the other form elements no general preference could be observed. The same was true for size and spacing of the form elements. Outlining had a crucial effect on the performance. In most cases, outlining impaired identification (outlining is present in only eight out of 24 pattern yielding high similarity rating, but in 16 out of 24 yielding poor similarity rating). For pleasantness rating the situation is slightly different, as young adults and elderly subjects prefer pattern without outlining, while in blind subjects the opposite was found (Fig. 9).

Local *versus* global features

Roughness is a conventional measurement of small-scale variations in the height of a physical surface. Usually, spacing, size and density of raised elements are crucial factors that determine texture perception. In the patterns studied here, four different components play a role in creating the perception of roughness and texture:

1. components resulting from the single form elements (i.e., edges in case of stars, smooth curves in case of dots, etc.)
2. components resulting from the spacing and size of the single form elements
3. components resulting from the outlining
4. components resulting from the overall size of the value numeral, which limits the number of form elements within

A recent multidimensional analysis of texture of raised-dot patterns revealed that three orthogonal perceptual dimensions can account for the judged tactile dissimilarities: blur, roughness, and clarity [13]. These authors also reported that dot-pattern roughness and individual-dot roughness corresponding to roughness produced by the single form elements and roughness produced by the spacing and size of the form elements affected perceived roughness differently.

The contributions of the single form elements were denoted as local, and those of the spacing and size as global. In order to illustrate the different texture properties evoked by the above mentioned components, the spatial frequencies were calculated by means of a spatially two-dimensional Fourier analysis for component 2 (spacing and size of the single form elements) and 4 (overall size of the value numeral including form elements). Figure 10 illustrates the spatial frequencies for all pattern combinations. This analysis demonstrated that the different components cover different frequency ranges with only a slight overlap. In addition, as depicted in Figure 10, the different components result in different similarities between patterns, which is one of the reasons for the poor predictability of pattern performance.

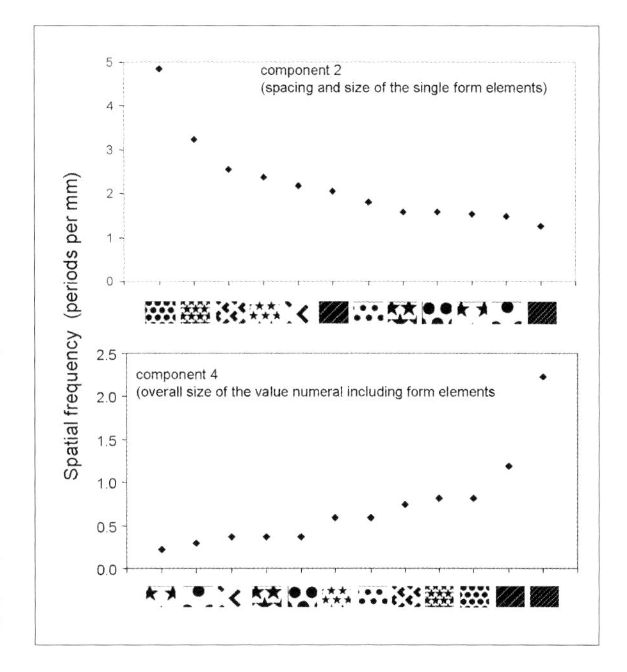

FIGURE 10.
Spatial frequencies as calculated by means of a spatially two-dimensional Fourier analysis for component 2 (spacing and size of the single form elements) and component 4 (overall size of the value numeral including form elements). According to this analysis, the different components cover different frequency ranges with only a slight overlap.

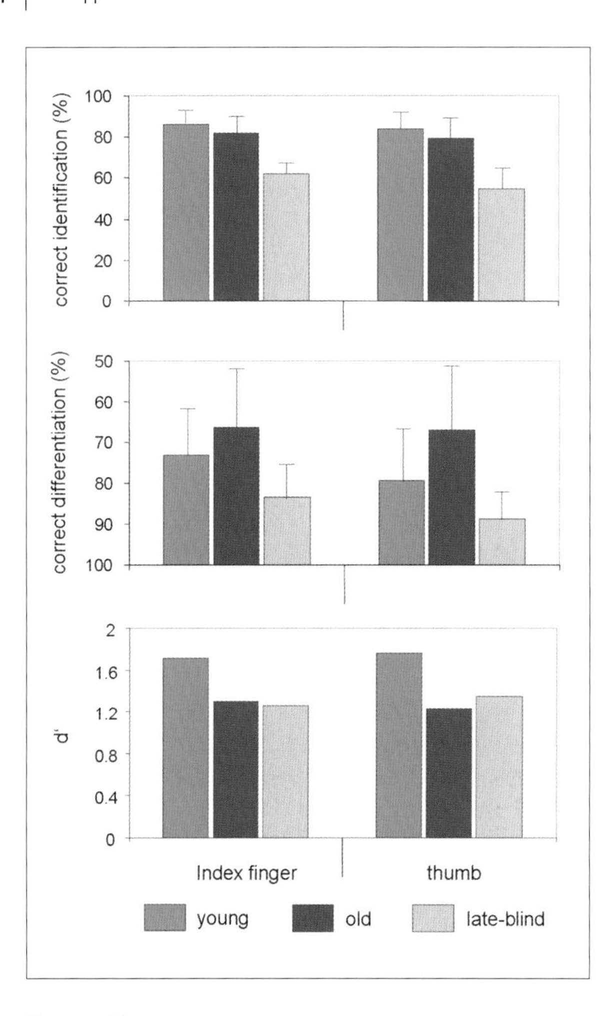

FIGURE 11.

Average performance scores for all three subgroups tested. a) similarity rating (% correct identification, 100% up) based on comparisons of 24 identical pattern. b) dissimilarity rating (% correct differentiation, 100% down) based on comparisons of 276 non-identical patterns. c) Overall detection performance (d-prime).

Comparison across subgroups: young adults – elderly – blind

Without exception, all participants were able to succeed the different tests. Figure 11 summarises the average performance measured across all subjects in all subgroups. To obtain a quantitative measure of correct performance, the so-called d-prime (or d') was calculated as a sensitivity index usually used in signal detection theory. An estimate of d' is obtained from measurements of the hit rate and false-alarm rate. A higher d' indicates that the signal can be more readily detected.

As expected, the group of elderly adults showed the poorest performance. Besides this observation, which is consistent with the age-related decline of tactile abilities, this group showed another characteristic feature in the pleasantness rating. It was found that the dynamic range was much smaller, presumably because elderly subjects were hesitant to assign extreme scores.

Surprisingly, the group of late blinds did not outperform the remaining participants, but showed a comparable performance as the age-matched sighted subjects (Fig. 11c). To understand this result, the performance obtained for identical and different pattern was more closely analysed. As can be seen in Figure 11b, blinds yielded best performance in a task of telling two different patterns apart. However, in a task consisting of identifying two prototypes with the same pattern, their performance dropped below that observed in the other groups. Questioning the participants after the end of the testing revealed an interesting feature: Blind subjects were able to detect even tiny differences of the prototypes. When two identical patterns had to be identified as 'same', because they were on physically different prototypes blinds were able to sense small differences in paper texture and surface properties, which led them more often report 'different' although the pattern were 'same', which resulted in an overall poor identification performance and thus in a high false alarm rate, which caused a small d' (Figs 11b and c). This outstanding ability explains their poor performance in this task, but highlights at the same time their unusual tactile abilities confirming other reports.

Comparing the outcome of the three subgroups revealed another important finding: While there were clear quantitative differences between subject groups, their overall performance was qualitatively comparable. As illustrated in Figure 8, the three subgroups shared five out of eight patterns that yielded best rating (young *versus* elderly: 5 of 8, young *versus* blind: 6 of 8, elderly *versus* blind: 7 of 8). For pleasantness rating a similar outcome was observed for young *versus* elderly (6 out of 8), while blind subjects showed a different preference (young *versus* blind: 3 of 8, elderly *versus* blind: 2 of 8). These findings demonstrate that despite the differences in absolute rating scores observed in the three subgroups, their preferred patterns share many similarities, both in terms of discriminative and affective performance. For making decisions about banknote design this feature is of particular implication as it facilitates haptic design by matching tactile sensing constrains found across different populations.

Non-predictability of pattern effectiveness: role of context

An unexpected finding was the poor predictability of pattern effectiveness. Decades of research on tactile texture and surface properties have revealed insight into crucial parameters determining texture and roughness perception. So one might have expected that pattern effectiveness can be more easily predicted from what is known about tactile texture and roughness perception. According to an elaborate study on texture discrimination abilities in humans, subjects seem to use spatial rather than intensive cues when discriminating patterns of similar density, suggesting that the similarity of form, i.e., the spatial arrangement of the closely spaced dots is more readily apparent than small differences in spacing along the axis of motion [14]. Besides spatial and intensity cues there is a discussion about the role of temporal cues such as the speed of scanning a texture with the finger, which corresponds to temporal frequency [15]. Remarkably, subjects appeared to be most successful in differentiating texture patterns when they were able to mentally picture the orientation and spacing of the pattern [14].

According to the experimental findings reported here, outlining is a very unpredictable source of disturbance and interference of tactile perception. In studies of tactile perception, interferences of this type have been intensively studied in the context of attempts to convey information to the skin by means of vibratory patterns such as letters of the alphabet [16]. For example, localisation of a tactile stimulus was affected by the temporal separation between the target and a tactile masker suggesting a role for temporal interferences [17]. In the visual domain, substantial modulation of target processing by introducing spatial context is well documented [18].

Another reason that outlining has such a severe modulatory effect comes from the fact that the form elements were confined to a small area within the value numeral. Dependent on the size and spacing of the form elements along the horizontal axis of the value numeral only one or two elements could be accommodated which increases the impact of the outlining in comparison to that of the form elements. As a consequence, designs of effective intaglio structures have to be individually tested and optimised in each case dependent on the overall graphical and other requirements.

Summary

Progress in printing technologies has made it possible to create intaglio structures of enormous spatial details. During the recent years, tactility and haptic features generated by intaglio structures are more and more used as additional security features to impede counterfeiting. Due to the strong visuotactile coupling the vision-touch combination offers unique advantages for spontaneous and almost unconscious banknote authentication of the end user. However, there is no *a priori* knowledge about which tactile features are particularly suited to create strong haptic sensations, and thus are optimal for banknote authentication. This is true because many human factors have to be considered such as the normal distribution of tactile and haptic performance in an 'idealised' population, the effect of ageing, because age deteriorates tactile perception, and the existence of specialised subgroups, such as blind subjects characterised by superior tactile abilities.

Psychophysical assessment of several form elements typically used in banknote design embedded in value numerals were tested according to their discriminative and affective properties in three groups of young adults, elderly subjects and age-matched late-blind people with profound Braille reading abilities. Size and spacing of the form elements were varied, and value numerals were presented with and without an outline. While there were clear quantitative differences of performance between groups, their overall performance was qualitatively comparable. These findings show that despite the differences in absolute performance observed in the three subgroups, their preferred patterns shared many similarities, both in terms of discriminative and affective properties. A number of patterns appear to match the different tactile sensing constrains found across different populations. This finding bears considerable implications for haptic banknote design.

One might have expected that pattern effectiveness can be more easily predicted from what is known from the literature about tactile perception. According to the data presented here this is not the case, as massive contextual factors such as an outline modulate perception making overall performance highly unpredictable. In addition, local (such as single form elements) and global features (such as the arrangement of form elements in terms of periodicity and density) contribute differentially to the overall tactile perception, i.e., the single form elements and the patterns created by assembling them through adjustment of size and spacing have different perceptual properties. As a consequence, the design of effective intaglio structures has to be individually tested and optimised in each case dependent on the overall graphical and other requirements.

Acknowledgements

Supported by Haptec – Haptic Research & Technology GmbH, Germany and Giesecke and Devrient GmbH, Germany. Thanks to Christine Heinisch and Stephanie Franzkowiak for the psychophysical testing, and Ingo Zaremba and Susanne Winter for valuable help in data analysis.

Suggested readings

Dinse HR, Kalisch T, Ragert P, Pleger B, Schwenkreis P, Tegenthoff M (2005) Improving human haptic performance in normal and impaired human populations through unattended activation-based learning. *Transaction Appl Perc* 2: 71–88

Kops CE, Gardner EP (1996) Discrimination of simulated texture patterns on the human hand. *J Neurophysiol* 76: 1145–1165

Lederman SJ, Hamilton C (2002) Using tactile features to help functionally blind individuals denominate banknotes. *Hum Factors* 44: 413–428

Important websites

Banknote2006. The definite forum on emerging trends and leading technologies in the banknote industry. http://www.banknote2005.com/banknote_home.html

Deutsche Bundesbank – Eurosystem: http://www.bundesbank.de/index.en.php

Giesecke and Devrient GmbH: http://www.gi-de.com/

Haptec Research and Technology GmbH: http://www.haptec.de

Haptics-L: The electronic mailing list for the international haptics community: http://www.roblesdelatorre.com/gabriel/hapticsl/

The Bureau of Engraving and Printing – United States Department of the Treasury: http://www.moneyfactory.gov/

The Haptics Community Web Site:
http://haptic.mech.northwestern.edu/
The Perceptual Learning mailing list:
http://cogmod.osu.edu/mailman/listinfo/plearn
The wiki of the international community of scholars studying perceptual learning:
http://www.perceptuallearning.org/
World Blind Union:
http://www.worldblindunion.org/home.cfm?id=91&nivel=3&orden=9

Get touched – bodycare as a design of media for self-perception

Eva Kristin Stein

Mirror, mirror on the wall, why are you just an aesthetic vision at all?

Well, here I am, a modern-day product designer. And besides just me: a part of my work. This is before you in the following, to me somewhat monotonous haptical performance; finished upon this keyboard designed for the purpose of the creation of text. As too often occurs in the creative design of objects, the resulting achievement and the process of creation must be presented in a reproduced form; that is, with text and photographs. Just like a printed rendering, this is a replacement for objects – unprinted, expressionless, a visual manifestation of meaning – printed out, it becomes simply an effigy. It remains problematic that I am unable to convey the subject matter in its comprehensible touchable form, because this is nothing more and nothing less than a pictorial narrative. Thus, I begin in traditional form:

Once upon a time men lived in an era that arose between a digital and a material culture. It was a time of change that occurred after the invention of the steam engine. As always in times of change, men must reorient themselves. Like in earlier centuries, a world of visual and acoustic communication dominates. The difference is that a new standard of interactive presentation has been introduced, and personal contact is being increasingly replaced by digital networking. Yet no platforms for multi-sensory perception and communication exist on this plane. The sensations of touch and smell are excluded. The 'in between' of which personal communication are constituted, social and cultural idiosyncrasies, facial expressions and gestures are becoming increasingly irrelevant. The means of communication remain fixed and firmly set within their setting. Notwithstanding this fact, communication on the digital level can happen in an authentic manner – like how I'm brilliantly smiling at you at this moment – but of course, you thought that already. What is new is firstly, that you can admire my smile through a webcam, and secondly, that the settings in which we spatially and physically find ourselves are becoming ever more distanced from the experiences that we obtain in digital spheres.

The monitor functions as a means of manifesting a superficial, arbitrary surface. It is a communicative projection screen. However, a large part of our modern, assumed reality functions no differently – look around!

For is it not the spitting image in a magazine or when logging into an email account so much as it is the stickers sold in chewing gum vending machines, with the girl in front or her silicon-infused mother next to her, clothed in pink from *Woolworth* or *Prada*, with a *Barbie* or 'Petruschka' under her arm and Paris Hilton in her eyes. But *Men's Health* is no problem, dressed in a black sports jacket and *Sony Vaio*, sitting at a poker table with a martini in hand, who isn't James Bond.

In the age of 'Copy & Paste', the imitation of surfaces and superficial approaches in product design have sparked the same debate over authenticity that has occupied photography [1] criticism and the Simulation culture á la *Baudrillard* [2]. How skilful are we with materials, and what effects will be achieved through an accelerated interchangeability of superficial surfaces? Intoxicating illusions and irritant disillusion from most immaterial material and pseudo-suitable, formal-material imitation are the borders of plausibility. Their common origin is the medium of the reproduction. The world of superficiality is the home of self-representation, real and digital. Therefore, the following holds true:

Scene it! So here I am. I shoot the breeze, ramble on, type emails, play a diverting computer game, work all day long like a good citizen and pine over the subject matter of my lifestyle – because free time is indeed very important for a well-balanced, aware state of being. Unfortunately, free time is far too seldom directed by desire than it is by social constraints and the pressure to perform. In this environment the manifestation of personal identity is indeed a difficult undertaking! My body, my clothing and the objects around me place me in the rank of a social personality. My capabilities, possibilities and their effectiveness create my societal status – which first must be situated [3].

Already through the course of industrialisation, the quality of the objects around us has steadily increased. Now, within the transition to an information culture, the abundance of references is also increasing. Fetishisation and overabundance form the boundaries of subjective materiality [3] – with which we are also digitally flooded – and the almost complete repression of the objective nearness to the sensual subject, in this case you.

For, who are you as 'I'? In your everyday lethargy, in your placating 'point-and-click' monotony, in your phlegmatic consumption you disappear within your artificial role, within which you confine yourself with shame, conventions and imaginations, in order to contrive yourself and live out your passion for social integration, the exchange of knowledge and personal progression.

Yet the interior of this facade, the place of self-perception, exposes and completes your being. Here you can experience desire and aversion and perceive this with all of your senses: your exterior and your interior. This is the individual retreat that should remain private for everyone, as a cocoon for your freedom as a living being.

In itself the vicissitude of the sense of wellbeing is of no continuity; it oscillates much more through a continual play of mental states, among which lust is the most positive form of motivation. However, this can also evaporate from one moment to the next. The manner in which we experience it is depending upon satiety, which in turn produces aversion. For it is the desire for newness and the new that drives us the most. We seek a personal pleasure garden in which we can roam, and there tarry and revel in our emotions. Man wishes to celebrate his zest for life in an unconstrained period within the depths of his mind, in order to again be capable of coming gracefully to terms with the weight of mundane routine.

Let us turn now to a field that offers like almost no other means towards self-adjustment and to self-perception. We deal now with the cosmetic market. More than almost any other branch of commerce, perfect corporeality is superficially staged; charm, communicated through visual perfection. This so-called perfection is created by the concealment and optimisation of the body. The tempting and flawless apparitions in the advertising spaces all around us create ideals, ideals that call forth in us the feeling that our own beauty and our personal desires are incomplete and boring. In the context of optimised archetypes, seemingly scientific findings within advertising texts and the wilfully indolent commercials that propagate the application of hi-tech cosmetics in order to prepare one's appearance for the 'Sex and the City' marathon, we come across as mundane in comparison.

Thus the formation of cosmetics serves as an exemplary subject of my following study. Is it possible to create in the products of self-presentation a metamorphosis, to free them from the purely visual-aesthetic consumption of satisfaction and open them to physical contact and multi-sensual self-experience?

Recreating 'cosmetics' as 'body care' means replacing the self-obsessed make-up mirror with an individual reflection surface. This should be created in an interpersonal, multi-sensual dialogue on pleasurable physical states and feelings, which exist in this form not in public, but in private.

The bathroom offers an especially intimate location for the private self. In the bathroom one frees oneself of public shamefacedness and conventions, in order to experience relaxation. In this way a space is created which offers the possibility of freedom from the everyday world, from jobs and other exertions. The success-driven lifestyle that leaves individuals constantly searching for stimulation and authenticity could here open itself to a diverse and multilayered responsiveness to the sensuality of the body. This is becoming ever more essential in a time of sensually monotonous overstraining. While our minds and intellects must increasingly think and interact at greater speeds and levels of specialisation, our bodies and the world of sensual experiences to which they are tied are ever more disregarded [4]. Consequently, one must turn especially to free time to find opportunities to repair this imbalance.

Sports, massage, creative and active pursuits which aren't goal-oriented are therefore ever more important. The longing to relax, to care for and to stimulate the body possesses meaning. The aspect of a symbiotic exchange forms the foundation of the experience of being touched, which accordingly leads to self-perception. Interactive and multi-sensual communication with oneself or with a partner can awaken a gentle empathy. The tactile stimulus acts as a means to break through the shell of the self-image, making self-realisation possible. Body care as enduring therapy that provides physical relaxation and spiritual balance also encourages a sexual attractiveness. Satisfaction is sustained through the process of stimulation. The brief thrills of a spectacle-based society (à la Guy Debord) are pit against reflective time for intimate solitude. Under good circumstances, the latter can even result in the reflective dissolution of a *Fließerlebnis*[1], in which one is so involved in an activity that the self as well as the world around oneself is completely forgotten. The needs of the imagination, fantasy-filled suspension and communication should inspire a thought-play that takes place in between the physical realities of space, objects and the developing ritual. Body care becomes an erotic moment which does not submit itself to a mass pornographised sex culture [5], but rather creates a personal staging that forms the backdrop for an individual conceptualisation of sensuality. Body care products can in this way act as media. With their formal language and idioms they should communicate a ritual that bestows security and shelter.

Here originates the desire to design a product that not only quiets lavish and superficial consumerist pleasure, but rather builds a long-lasting and satisfying connection to the individual. The product should evoke positive memories of togetherness as 'something to share'. At the same time, the individual in this situation should be able to give up oneself as material to be taken in its entirety, allowing oneself to be multi-sensually and sensitively stimulated.

The value of such body care products should be measured through these sensual experiences. One should be able to identify oneself with them, and be able to project personal desires upon them. For the outer surface is truly beautiful only when it radiates from within.

Care for the body has played many roles in human cultural traditions, and encompasses more than the theories of hygiene [6, 7]. Hot water and stimulating aromas relax the muscles and skin. Water surrounds the body and lets one submerge, to be enclosed within a flexible membrane. A feeling of security can thus propagate itself. Beauty care as emotional and tactile perception in this manner centres itself on you as a sensitive being. Grooming [8] as the maintenance of both bodies and relationships, cultivates a delicate responsiveness to outside influences, through which an emergence and unfolding of emotions can occur. Body contact and pleasurable surrender are thereby an alternative to the currently prevalent designs of pop-culture wellness, which in turn promotes the increase of

1 I think the feeling is not absolutely comparable with the 'Flow-Theory' of Mihaly Csikszentmihalyi, therefore I use the German term.

Figure 1.
Some documentary photos of the sensibility tests.

consumption within recreational time, often only in order to broaden its own inane banality with more monotonous aloe-vera-containing stimulations. The quick and cheap thrill is mostly conceived for specifically this utilisation of time. Thus, I seek not the tired thrill, but rather something refreshingly revitalising.

We come to the product. It is clear: in the creation of such products aren't only samples of cultural archetypes of importance, but also personally measured experimental values towards specific effects. The old methods of animal testing do not promise success; instead, people will be used to conduct test series. Of interest is not experimentation with novel chemical mixtures, but tactile reactions. What does an object feel like? How does the individual react to active and passive stimulation? With which kinds of stimulation is cold or warm contact preferred, or is this an arbitrary matter? How does one handle different materials in different ways, and how influential is the processing or workmanship of surfaces to the positive attributes of the perceiver? And how does the individual react when his eyes are opened and he sees what it is that feels so good [9]?

As a basic principle it is granted that feelings are above all very personal and unique for every individual. However, basic tendencies of sensual perception may be generalised: so may a self-administered caress be thoroughly more frigid than one performed by another. Squeaky and crinkly materials quickly produce an inferior effect. Lacquered wood was mostly perceived as objectionably greasy, while oiled wood possessed a pleasant velvet-like feeling when it was smoothly sanded. Porcelain and glass felt of higher quality, although no-one trusted themselves to handle them unreservedly. Things were quite different with plastic. The moulded corners were irritating, as were the greasy surfaces.

Further inquiries on dispositions, emotions and feelings in relation to situations yield similar results. What feels skilfully refined is pleasing, features an authentic surface and is suitable to its usage. Acoustic, olfactory and visual stimulations must stand in contextual agreement, or else irritations arise in the synesthetic perception.

With the intent of finding a practical application, I produce from this a scenario in which designed media work towards creating personal rituals, with the bathroom as region and the preciousness of free, shapable time. A combination and interplay of known methods and newly developed utilisations should offer an incentive to explore, care for and stimulate one's own body and that of a partner. Cosmetics offer the purification and care for the body, without propagating utopian panaceas. I appropriate cultural crossovers of traditions in hygiene and eroticism as basis elements of conceptual development and combine them with my findings from the sensibility tests.

Being clean is a positive starting point for physical caressing and offers the incentive to initially clean up, care and keep the skin soft and supple, in order to afterwards proceed with massage and corporeal exploration. Such approaches are already established and will quickly be recognised as luxury goods. The innovativeness of the concept guarantees a designed scenery by my evaluation. From this, individual rituals will be derived. Every action is linked to a topic. Subsequently, topics as well as actions will be placed within an emotional graph, which approximates a calculated emotional increase within a personal, expandable spectrum of experiences. From this arise specific requirements: the individual should undress, accustom himself to the presence of water, then sink in and enjoy, followed by scraping of the skin, aeration, swathing, enclosing, cleansing, purification and care, to finish with massage, stimulation, relaxation, rejuvenation and revitalisation.

The development of these individual physical rituals results from the complex interplay of the effects of outer forms, cosmetics, aromas, colours and the attributes of the materials used.

On this account, amenable prototypes uniting ergonomics and functionality were modelled. They were also tested on human bodies. The broader effects, consistency and individual tastes as well as general aesthetic irritations within the cosmetic contents were also examined and optimised.

Cosmetics as soap, shampoo and skin care compounds provide cleansing, care for and rejuvenation of the skin. Emotional associations should be opened by aromas, creating in combination with the colour scheme an inviting ambience. The flacons offer the incentive of active as well as passive tactile stimulation. With the means of the formal language they should provide the incentive of physical stimulation and create a pleasant and

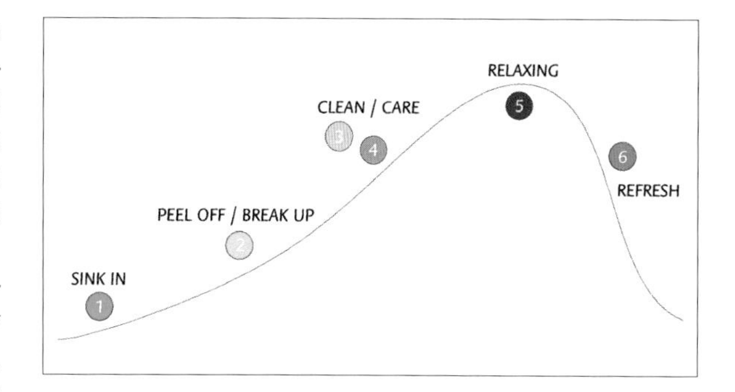

Figure 2.
Constructed curve for scenic formation to arrange a general set up and integrate emotional parameter

likable mediality with their materiality and outer surface. The combination of ABS and TPE plastics, as well as *Thermowood*, provides the opportunity to comply with the requisites of a moist environment. Inherently stable materials make easy an unhesitating manageability, because one need not worry that they might break. In addition, thermoplastic elastomers make it especially possible to create tactilely interesting textures and contrasts, which in model making are already approachingly imitable.

The formal language visualises the process of flowering, spanning from the bulb to the emergence of the flower until the last leaf, which bestows moisture and freshness. Finally, the packaging and graphics help to communicate the ritual, or more specifically the instructions for use. The graphics visualise the process of unfolding. The product's packaging should distinguish it from the image of popular wares, in order to present itself as a privately disclosed thing, which thenceforth will be treated as 'something to share' in private. The eroticism of the covering and the uniqueness of opening it should work to construct a tactile relationship to the 'Thing in itself' from the first visual contact on.

At the end of this journey through personal sensuality, a product series is created that contains media towards sensual self-perception, and whose purpose lies in the attainment of sensual enjoyment. The necessity of its use has in this

Figure 3.
Three examples of the created body care line: you can see the different materials in utilisation. Photography by Matthias Ritzmann

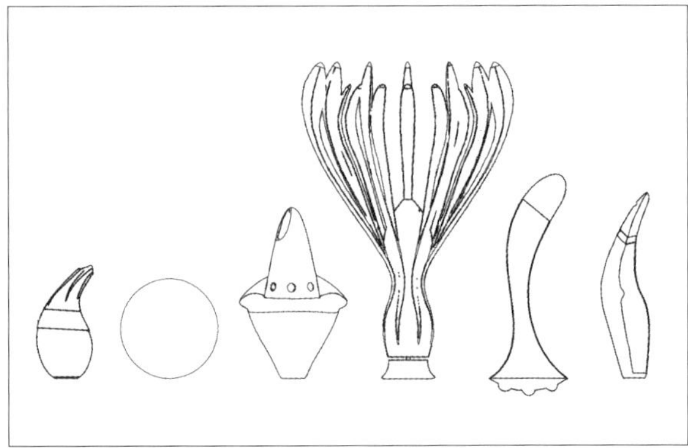

Figure 4.
The formal language of the line: you see the process of evolving shape – beginning with the bulb, to the blossom, ending with the last refreshing leaf.

connection only a secondary function. Much greater priority lies in the allure of luxury, which through the desire to possess acquires a preciousness that reflects a fineness of the senses and aesthetic taste. Because I am, if you appreciate me as a being, where I can be, how I like to conceive myself. A completely new mirror opens itself to me.

You have just had the possibility to rudimentarily immerse yourself in my offer of self-experience. Time, space and rituality, far from cults and mystical witchery, can be an alternative to relaxation and stress relief. Reflection as a possibility to radiate upon or against another. Here my conviction is affirmed, that the individual is not stimulated by curiosity and excess, but rather by the identification with sensual and spiritual experiences. With regard to sensual perception, the chaos of individual feelings, the tremors of the body, breath, exhalation and the magic of the moment are of enormous importance and cannot be artificially imitated, reproduced and copied, as long as the feeling itself is not yet emotionally and physically experienced.

Perhaps it is not yet, perhaps never, that cyborgs and androids [10] will be able to satisfy us as well as human personalities. Nevertheless my wish will continue to persist that reproduced experiences of 'good vibrations' also contain real sensations. For real people continue to carry the ability to be surprised, which machines can only grasp with difficulty. Interpersonally there exists not only restricted, if also noteworthy mechanical interaction, but also a direct and intuitive interchange. The requirements upon people as partners as a result of technical possibilities will decrease as well as increase. In many, perhaps most cases, a machine will represent an inferior partner. Real existence: endangered by technical improvement? Provoked by the widening of the scope of experience?

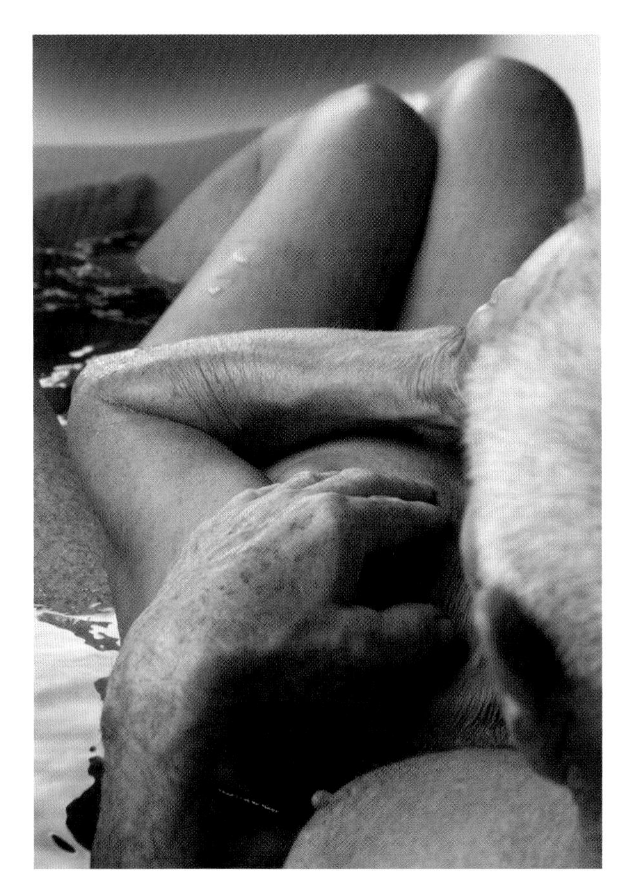

FIGURE 5.
Ambience of one way to celebrate the situation. Photography by Matthias Ritzmann

Yet you remain a subject! And you will be able to receive more stimulation in the future! But how will we react and to which contact? Will we become desensitised? Here as well arises the question of the competence of usage. As usual, abuse will take place. But then we will again take up the discussion of whether we will have become tactilely just as 'pornographised' and overstimulated as is now visually the case. We must be able to sense what is good for us and what overburdens us. If we survive the metamorphosis, each individual will have to decide for himself and in relation to his milieu what he wants or doesn't want. We will have to experiment and find forms in which the respective desired stimulations can be fashioned.

For as long as cyber-visions continue to be conceived, we can also address the cooperative networks of which they are formed, and mould these networks with connections and stimulations carried over from non-digital relationships. To accomplish this we require the symbolic form of the 'thing that was'. A given object rises up from its deterioration and is spared from the trash heap. If an object is simply 'a thing' with which one represents oneself, it gratifies its possessor only with marginal, fleeting satisfaction. The consumed article is simply too frigid to arouse responsiveness and emotionalism.

Because we now *via* digital realms possess enormous possibilities for representing and embodying ourselves, possibilities which surely will only continue to develop, for me it is a necessity to use material objects for self-perception. Where we can be touched, we should be, intensively, exquisitely and diversely. For only so can we affirm ourselves as beings within the illusory facade of everyday life – GET TOUCHED!

Suggested readings

Benjamin W (2006) *Das Kunstwerk im Zeitalter seiner technischen Reproduzierbarkeit*. Suhrkamp Verlag, Frankfurt/M.

De la Croix A (1999) *Liebeskunst und Lebenslust*. Jan Thorbecke Verlag, Ostfildern

Flusser V (1997) *Medienkultur*. Fischer Taschenbuchverlag, Frankfurt/M.

GEO (2004) *Das Verlangen nach Berührung*. Ausgabe Juli

Haug WF (1971) *Kritik der Warenästhetik*. Suhrkamp Verlag, Frankfurt/M.

Nagel L, Sandbothe M (2005) *Systematische Medienphilosophie*. Akademie Verlag, Berlin

Stein EK (February 2007a) *[Lust_Wandel] – Schriftlicher Teil der Diplomarbeit*. University of Art and Design, Halle/Saale

Stein EK (June 2007b) *[Lust_Wandel] – Praktischer Teil der Diplomarbeit*. University of Art and Design, Halle/Saale

Appendix

References

1 Haptic perception: an historic approach

1 Müller RFG (1951) *Grundsätze altindischer Medizin*. Munksgaard, Copenhagen
2 Hsu E (2001) Figurately speaking of 'danger of death' in Chinese pulse diagnostics. In: R Jütte, M Eklöf, MC Nelson (eds): *Historical aspects of unconventional medicine. Approaches, concepts, case studies*. EAHMH-Publications, Sheffield, 193–210
3 Stratton GM (1964) *Theophrastus and the Greek physiological psychology before Aristotle*. Bonset, Amsterdam
4 Neuhäuser J (1878) *Aristoteles Lehre von dem sinnlichen Erkenntnisvermögen und seinen Organen*. Koschny, Leipzig; Beare JI (1906) *Greek theories of elementary cognition from Alcmaeon to Aristotle*. Clarendon, Oxford
5 Theiß P (1997) *Die Wahrnehmungspsychologie und Sinnesphysiologie des Albertus Magnus. Ein Modell der Sinnes- und Hirnfunktion aus der Zeit des Mittelalters. Mit einer Übersetzung aus De anima*. Peter Lang, Frankfurt/M, 78–79
6 Tellkamp JA (1999) *Sinne, Gegenstände, Sensibilia. Zur Wahrnehmungslehre des Thomas von Aquin*. Brill, Leiden
7 Aquinas (1994) *Commentary on Aristotle's 'De anima'*, transl. Foster K, Humphries S (rev eds) Dumb Ox Books, Notre Dame, IN, 187
8 Aquinas (1994) *Commentary*, 187
9 Landauer S (1876) Die Psychologie des Ibn Sînâ. In: *Zeitschrift der Deutschen Morgenländischen Gesellschaft* 29: 335–418
10 Aquinas (1994) *Commentary*, 152
11 Aquinas (1994) *Commentary*, 152
12 Kaufmann D (1884) *Die Sinne. Beiträge zur Geschichte der Physiologie und Psychologie im Mittelalter aus hebräischen und arabischen Quellen*. Brockhaus, Leipzig, 188f
13 Josipovici G (1996) *Touch*. Yale Univ. Press, New Haven, London. Zeuch U (2000) *Umkehr der Sinneshierarchie. Herder und die Aufwertung des Tastsinns seit der frühen Neuzeit*. Niemeyer, Tübingen. Naumann-Beyer W (2003) *Anatomie der Sinne im Spiegel von Philosophie, Ästhetik, Literatur*. Böhlau, Cologne, Weimar, Vienna
14 Gilman SL (1993) Touch, sexuality and disease. In: WF Bynum, R Porter (eds): *Medicine and the five senses*. Cambridge Univ. Press, Cambridge, 198–224
15 Kauffmann H (1943) Die Fünfsinne in der niederländischen Malerei des 17. Jahrhunderts. In: H Tintelnot (ed): *Kunstgeschichtliche Studien*. Gauverlag, Breslau, 133–157; O'Rourke Boyle M (1998) *Senses of touch. Human dignity and deformity from Michelangelo to Calvin*. Brill, Leiden
16 Jütte R (2005) *A history of the senses. From Antiquity to Cyberspace*, translated by James Lynn. Polity Press, Cambridge, 175ff
17 Porter R (1993) The rise of physical examination. In: WF Bynum, R Porter (eds): *Medicine and the five senses*. Cambridge Univ. Press, Cambridge, 179–197
18 Honselmann D (1992) Die Geschichte des Eßbestecks. In: M Andritzky (ed): *Oikos. Von der Feuerstelle zur Mikrowelle. Haushalt und Wohnen im Wandel*. Anabas, Gießen, 378–381
19 Spitzer R (1928) Geographische Verteilung der Hautkrankheiten. In: J Jadassohn (ed): *Handbuch der Haut- und Geschlechtskrankheiten*, vol. 14: 2. Springer, Berlin, 253–328
20 Weber EH (1905) *Tastsinn und Gemeingefühl*. E Hering (ed), Engelmann, Leipzig, 1; Bueck-Rich U (1970) *Ernst Heinrich Weber (1795–1878) und der Anfang einer Physiologie der Hautsinne*. Juris Druck, Zürich
21 Boring EG (1977) *Sensation and perception in the*

history of experimental psychology. Appleton-Century, New York
22 Becker P (1998) Die Rezeption der Physiologie in der Kriminalistik und Kriminologie: Variationen über Norm und Ausgrenzung. In: J Tanner, P Sarasin (eds): *Physiologie und industrielle Gesellschaft. Studien zur Verwissenschaftlichung des Körpers im 19. und 20. Jahrhundert*. Suhrkamp, Frankfurt/M, 453–490
23 Friedmann H (1930) *Die Welt der Formen. System eines morphologischen Idealismus*. 2nd ed. C. H. Beck, Munich, 33
24 Révész G (1944) *Die menschliche Hand. Eine psychologische Studie*. Karger, Basel, New York
25 Baudrillard J (1996) *The System of objects*, transl. James Benedict. Verso, London, New York, 53
26 Naisbitt J (1999) *High tech/high touch. Technology and our accelerated search for meaning*. Broadway Books, New York
27 Kerckhove D (1993) Touch *versus* vision: Ästhetik Neuer Technologien. In: W Welsch (ed): *Die Aktualität des Ästhetischen*. Fink, Munich, 135–168
28 Heitmüller E (1994) Kybernetische Sinnlichkeit. SM-Körper – Authenzität – Digitalisierung. *Ästhetik & Kommunikation* 23: 14–22
29 Geary J (2002) *The body electric. An anatomy of the new bionic senses*. Weidenfeld & Nicolson, London, 134

2 German pioneers of research into human haptic perception

1 Weber EH (1834) *De pulsu, resortione, auditu et tactu annotatines anatomicae et physiologicae*. Köhler, Leipzig
2 Weber EH (1835) Über den Tastsinn. *Archiv für Anatomie, Physiologie und wissenschaftliche Medicin*, 152–160
3 Weber EH (1851) *Die Lehre vom Tastsinn und Gemeingefühl – auf Versuche gegründet*. Verlag Friedrich Vieweg und Sohn, Braunschweig: translated (1978) The sense of touch. Academic Press, London
4 Hoffmann C (2001) Haut und Zirkel. Ein Entstehungsherd. Ernst Heinrich Webers Untersuchung über den Tastsinn. In: M Hagner (ed): *Ansichten der Wissenschaftsgeschichte*. Fischer, Frankfurt/M, 191–226
5 Frey M von (1911) Allgemeine Muskelmechanik. In: R Tigerstedt (Ed): *Handbuch der physiologischen Methodik*. Hirzel Verlag, Bd. 2, Hälfte 1, Leipzig
6 Frey M von (1911) *Vorlesungen über Physiologie*. 2.Aufl., Springer, Berlin
7 Frey M von (1929) Physiologie der Haut. In: J Jadassohn (Ed): *Handbuch der Haut- und Geschlechtskrankheiten*. Bd. 1, Springer, Berlin
8 Frey M von (1896) Untersuchungen über die Sinnesfunctionen der menschlichen Haut. *Abhandl. d. Kgl. Sächs. Ges. d. Wiss. Math.-phys. Cl.*, Hirzel, Leipzig
9 Frey M von (1912) Die Wirkung einfacher Druckempfindungen aufeinander. *Verhandlungen der physik.-med. Gesellschaft zu Würzburg*; N.F., 41,3, S. 39–43, Kabitzsch, Würzburg
10 Hausmann T (1922) Berührungsempfindung und Druckempfindung, insbesondere die tiefe Druckempfindung. *Pflügers Archiv European Journal of Physiology*. Springer, Berlin, 611–628
11 Frey M von (1918) *Die Bedeutung des Drucksinns für die Wahrnehmung von Bewegung und Lage der Glieder*. Bayerische Akademie der Wissenschaften/München: Mathematisch-Physikalische Klasse; Sitzungsberichte 93–106
12 Dessoir M (1892) Über den Hautsinn. *Arch. f. Anat. u. Physiol., Physiol. Abt.*, 175–339
13 Hill R, Watanabe R (1894) 'Sensorial' and 'Muscular' Reaction. *Am J Psychol* VI, 239–246
14 Dessoir M (1902) *Geschichte der neueren deutschen Psychologie*. Dunker Verlag, Berlin
15 Dessor M (1931) *Vom Jenseits der Seele*. Enke Verlag, Stuttgart
16 Dessoir M (1946) *Buch der Erinnerung*. Enke, Stuttgart
17 Révész G (1946) *Einführung in die Musikpsychologie*. Francke, Bern
18 Révész G (1913) *Zur Grundlegung der Tonpsychologie*. Veit, Leipzig
19 Révész G (1946) *Ursprung und Vorgeschichte der Sprache*. Francke, Bern
20 Révész G (1952) *Talent und Genie: Grundzüge einer Begabungspsychologie*. Lehnen, München
21 Réveész G (1950) *Psychology and art of the blind*. Longmans, Green, London, New York
22 Révész G (1958) *The human hand; a psychological study*. Routledge & Paul, London

23 Révész G (1925) *The psychology of a musical prodigy*. K. Paul, Trench, Trubner & Co Ltd, London; Harcourt, Brace & Company, Inc., New York

24 Révész G (1938) *Die Formenwelt des Tastsinnes: Grundlegung der Haptik und der Blindenpsychologie*, Bd. 1 & 2 Haag, Martinus Nijhoff

25 Gelb A, Goldstein K (1920) Über den Einfluss des vollständigen Verlustes des optischen Vorstellungsvermögens auf das taktile Erkennen. *Psychologische Analysen hirnpathologischer Fälle*, Barth, Leipzig 157–250

26 Hippius R (1934) Erkennendes Tasten. *Neue Psychologische Studien* 10

27 Katz D (1925) *Der Aufbau der Tastwelt*. Verlag von J. A. Barth, Leipzig [Katz D (1989) *The World of Touch*. Edited and translated by Lester E. Krueger, Lawrence Erlbaum Associates Inc, US

28 Helmholtz H von (1879) *Die Tatsachen in der Wahrnehmung*. Hirschwald, Berlin

29 Skramlik E von (1937) *Psychophysiologie der Tastsinne*. Bd. 1 u. II. *Archiv für die gesamte Psychologie*: 4. Ergänzungsband, Akad. Verlags-Gesellschaft, Leipzig

30 Skramlik E von (1928) *Anleitung zum Physiologischen Praktikum*. G. Fischer, Jena

31 Skramlik E von (1926) *Handbuch der Physiologie der niederen Sinne*. 1. Auflage, Band 1 Thieme, Leipzig

32 Uexküll J von (1928) *Theoretische Biologie*. Springer, Berlin

33 Turro R (1920) Der Tastraum. *Arch de neurobiol* 1/148

3 British pioneers of research into human haptic perception

1 Bell C (1833) *The Hand; Its mechanism and vital endowments as evincing design*. William Pickering, London. Reprinted by Pilgrims' Press, Brentwood, UK (1979)

2 Cole J (1991, 1995) *Pride and a Daily Marathon*. Duckworth, London, UK and The MIT Press, Cambridge, MA

3 Olausson H, Lamarre Y, Backlund H, Morin C, Wallin BG, Starck G, Ekholm S, Strigo I, Worsley K, Vallbo AB et al (2002) Unmyelinated tactile afferents signal touch and project to insular cortex. *Nat Neurosci* 5: 900–904

4 Mott FW, Sherrington CS (1895) Experiments upon the influence of sensory nerves upon movement and nutrition of the limbs. *Proc Roy Soc* 57: 481–488

5 Sacks O (1984) A *Leg to Stand On*. Duckworth, London

6 Cole J (2004) On the relation between sensory input and action. *J Motor Behaviour* 36-3: 243–244

7 Taub E, Berman AJ (1968) Movement and learning in the absence of sensory feedback. In: SJ Freeman (ed): *The Neurophysiology of spatially orientated behaviour*. Dorsey, Homewood, Illinois, 173–191

8 Bossom J, Ommaya AK (1968) Visuo-motor adaptation, (to prismatic transformation of the retinal image), in monkeys with bilateral dorsal rhizotomy. *Brain* 91: 161–172

9 Head H (1920) *Studies in Neurology*, Volumes 1 and 2. Oxford University Press, Oxford and Hodder and Stoughton, London

10 Cole J (2004) *Still Lives*. The MIT Press, Cambridge, MA and London, UK

11 Wittgenstein L (1980) *Remarks on the Philosophy of Psychology*. Blackwell, Oxford

12 Rivers WH, Head Sir H (1908) A human experiment in nerve division. *Brain* 31: 323–450

13 Gallagher S (2005) *How the Body Shapes the Mind*. Oxford University Press, Oxford

14 Adrian ED (1932) *The Mechanism of Nervous Action*. Milford, London and Oxford University Press, Oxford

15 Adrian ED (1928) *The Basis of Sensation*. Reprinted: Hafner, New York and London (1964)

16 Gordon-Taylor Sir G, Walks EW (1958) *Sir Charles Bell: His Life and Times*. Livingstone, Edinburgh and London

17 Cole J (2007) Wittgenstein's neurophenomenology. Medical Humanities. *Medical Humanities* 33: 59–64

18 Sherrington CS (1900) The Muscular Sense. From Schafer, *Textbook of Physiology*, Pentland, Edinburgh

19 Haggard P, Magno E (1999) Localising awareness of action with transcranial magnetic stimulation. *Exp Brain Res* 127 (1): 102–107

20 Haggard P, Clark S, Kalogeras J (2002) Voluntary action and conscious awareness. *Nature Neurosci* 5 (4): 382–385

21 Haggard P, Clark S (2003) Intentional action: Conscious experience and neural prediction. *Consciousness and Cognition* 12: 695–707

22 Wolpert DM (1997) Computational approaches to motor control. *Trends in Cognitive Science* 1: 209–216

23 Marcel A (2002) The Sense of Agency: Awareness and Ownership of Action. Chapter 2. In: J Roessler, N Eilan (eds): *Agency and Self-awareness*. Oxford University Press, Oxford, 48–93

24 Sherrington CS (1900) Cutaneous Sensation. From Schafer, *Textbook of Physiology*. Pentland, Edinburgh

25 Sherrington Sir C (1906) *Integrative action of the Nervous System*. Scribner, Yale (reprinted 1947 Cambridge University Press, Cambridge)

4 Early psychological studies on touch in France

1 Hatwell Y (1966) *Privation Sensorielle et Intelligence*. Presses Universitaires de France, Paris

2 Hatwell Y (1985) *Piagetian reasoning and the blind* (translated from: *Privation sensorielle et intelligence* (1966) Presses Universitaires de France, Paris). American Foundation for the Blind, New York

3 Diderot D (1749/1972) *Lettre sur les aveugles à l'usage de ceux qui voient*. Garnier-Flammarion, Paris

4 Locke J (1689/1975) *An essay concerning human understanding*. Oxford University Press, Oxford

5 Jeannerod M (1975) Déficit visuel persistant chez les aveugles-nés: données cliniques et expérimentales. *L'Année Psychologique* 75: 169–196

6 Senden von M (1960) *Space and Sight*. The Free Press, Glencoe, Illinois

7 Weygand Z (2003) *Vivre sans voir. Les aveugles dans la société française du Moyen-Age au siècle de Louis Braille*. CREAPHIS, Paris

8 Villey P (1914) *Le Monde des aveugles*. Flammarion, Paris (2nd edition (1954) Corti, Paris)

9 Villey P (1927) *L'aveugle dans le monde des voyants*. Flammarion, Paris

10 Villey P (1930) *Pédagogie des aveugles*. Félix Alcan, Paris

11 Dolanski W (1930) Les aveugles possèdent-ils le "sens des obstacles"? *L'Année Psychologique* 30: 1–50

12 Supa M, Cotzin M, Dallenbach KM (1944) "Facial vision": The perception of obstacles by the blind. *Am J Psychology* 57: 133–183

13 Itard J (1801/1964) Mémoire et rapport sur Victor de l'Aveyron. In: L Malson (ed): *Les enfants sauvages*, 127–189, 10/18, Paris

14 Itard J (1807/1964) Rapport sur les nouveaux développements de Victor de L'Aveyron. In: L Malson (ed): *Les enfants sauvages*, 190–247, 10/18, Paris

15 Seguin E (ed) (1846) *Traitement moral, hygiène et éducation des idiots*. Baillères, Paris

16 Montessori M (1926/1958) *Pédagogie scientifique*. Desclée de Brower, Alençon

17 Bloch H (2006) *La psychologie scientifique en France*. Armand Colin, Paris

18 Nicolas S (2002) *Histoire de la psychologie française*. In Press Editions, Paris

19 Ribot T (1870) *La psychologie anglaise contemporaine*. Félix Alcan, Paris

20 Ribot T (1879) *La psychologie allemande contemporaine*. Félix Alcan, Paris

21 Binet A (1894) *Introduction à la psychologie expérimentale*. Alcan, Paris

22 Bernard C (1865/1966). *Introduction à l'Etude de la Médecine Expérimentale*. Gallimard, Paris

23 Bernard C (1878) *La science expérimentale*. Baillères, Paris

24 Magendie F (1822) Expériences sur les fonctions des racines des nerfs rachidiens. *J Physiologie Expérimentale et Pathologique* 2: 276–279

25 Magendie F (1822) Expériences sur les fonctions des racines des nerfs qui naissent dans la moëlle épinière. *J Physiologie Expérimentale et Pathologique* 2: 366–372

26 Skramlik von E (1937) *Psychophysiologie der Tastsinne. Archiv für die gesamte Psychologie*; vierter Ergänzungsband: Akademische Verlagsgesellschaft

27 Beaunis H (1888) Recherche sur la mémoire des sensations musculaires. *Revue Philosophique* 569 (quoted by Skramlik, 1937)

28 Brown-Sequard C (1858) Sur la sensation tactile et sur un moyen de mesurer l'anesthésie et

l'hyperesthésie. *J Physiologie* 1: 344 (quoted by Skramlik, 1937)

29 Heinrich W (1932) Sur la disparition périodique de certaines sensibilités. *J Psychologie* 29: 576 (quoted by Skramlik, 1937)

30 Lalanne X (1876) Note sur la durée de la sensation tactile. *J d'Anatomie et de Physiologie* 449

31 Charpentier A (1890) Influence des efforts musculaires sur les sensations de poids. *Compte Rendu des Séances de la Conférence de la Société de Biologie*, Paris (quoted by Skramlik, 1937)

32 Pallot G (1934) *Recherches cytologiques sur les fuseaux neuromusculaires.* C.R. de l'Académie des Sciences de Paris 198 (973)

33 Delage Y (1886) Etude expérimentale sur les illusions statiques et dynamiques de direction pour servir à déterminer les fonctions des canaux semicirculaires de l'oreille interne. *Archives de Zoologie Expérimentale* 2: 523 (quoted by Skramlik, 1937)

34 Bizet A (1861) Sur une hallucination du toucher propre aux amputés. *Gazette Médicale de Paris*

35 Pitres L (1897) Etude sur les sensations illusoires des amputés. *Annales Médico-Psychologiques* 5: 177

36 Müller J (1833) *Handbuch der Physiologie des Menschen*

37 Piéron H (1935/1945) *La Sensation, guide de vie.* Gallimard, Paris

38 Binet A (1900) Technique de l'esthésiomètre. *L'Année Psychologique* 7: 240–248

39 Piéron HM (1933) Le Gravimètre de Piéron. *Bulletin de l'Institut National d'Orientation Professionnelle* 5: 145–149

40 Piéron H, Piéron HM (1935) Le Pacho-esthésiomètre et son étalonnage. *Bulletin de l'Institut National d'Orientation Professionnelle* 7: 33–42

41 Piéron H (1932) Le toucher. In: G Dumas (ed): *Nouveau Traité de Psychologie.* Masson, Paris, 90–218

42 Piéron H (1935) Le toucher. In GH Roger, L Binet (eds): *Nouveau Traité de Physiologie normale et pathologique*. Masson, Paris, 1055–1219

43 Fessard A, Piéron H (1930) *Du minimum d'énergie nécessaire pour l'excitation tactile.* Paper presented at the Compte Rendu des Séances de la Conférence de la Société de Biologie, Paris

44 Piéron H (1922) Nouvelles recherches sur le temps de latence sensorielle et la la loi qui relie ce temps au temps de l'excitation. *L'Année Psychologique* 22: 58–60

45 Piéron H, Segal J (1938) Recherches sur la sensibilité tactile digitale pat stimulation électrique du nerf cutané. *L'Année Psychologique* 38: 89–135

46 Piéron H (1935) Recherches expérimentales sur la sensation vibratoire cutanée. *L'Année Psychologique* 35: 82–102

47 Toltchinsky A (1920) Recherches topographiques sur la discrimination tactile. *L'Année Psychologique* 20: 160–181

48 Gault RH (1933) Les sens vibro-tactiles. Enquête sur leur rôle dans ses rapports avec le langage parlé. *L'Année Psychologique* 34: 1–22

49 Binet A (1900) Un nouvel esthésiomètre. *L'Année Psychologique* 7: 231–239

50 Laureys J (1900) Comment l'oeil et la main nous renseignent différemment sur le volume des corps. *L'Année Psychologique* 7: 264–274

51 Streri A (1994) L'étude des relations entre les modalités sensorielles au début du siècle. In: P Fraisse, J Segui (eds): *Les origines de la psychologie scientifique française: le centenaire de L'Année Psychologique.* Presses Universitaires de France, Paris, 205–222

52 Philippe J (1922) A la recherche d'une sensation tactile pure. *L'Année Psychologique* 22: 167–183

53 Tastevin J (1937) En partant de l'expérience d'Aristote. Les déplacements artificiels d'une partie du corps ne sont pas suivis par le sentiment de ces parties. *L'Encéphale* 32: 5–84

54 Tastevin J (1937) En partant de l'expérience d'Aristote (suite). *L'Encéphale* 32: 140–158

55 Welch RB (1978) *Perceptual modification.* Academic Press, New York

56 Hatwell Y (2003) Intermodal coordination in children and adults. In: Y Hatwell, A Streri, E Gentaz (eds): *Touching for knowing. Cognitive psychology of haptic manual perception* (Vol. 207–219). John Benjamin Publishers, Amsterdam and Philadelphia

57 Piéron H (1923) Contribution expérimentale à l'étude du phénomène de transfert sensoriel. *L'Année Psychologique* 23: 76–124

58 Hatwell Y, Streri A,Gentaz E (eds) (2003) *Touching for knowing. Cognitive psychology of haptic manual*

perception. John Benjamins Publishing Company, Amsterdam and Philadelphia
59 Streri A (1993) *Seeing, reaching, touching. The relations between vision and touch in infancy*. Harvester Wheatsheaf, London
60 Streri A (2003) Intermodal relations in infancy. In: Y Hatwell, A Streri, E Gentaz (eds): *Touching for knowing. Cognitive psychology of haptic manual perception*. Benjamin Publishers, Amsterdam and Philadelphia, 191–206
61 Strelow ER, Brabyn, JA (1982) Locomotion of the blind controlled by natural sound cues. *Perception* 11: 635–640
62 Ashmead DH, Hill EW, Talor CR (1989) Obstacle perception by congenitally blind children. *Perception and Psychophysics* 46: 425–433
63 Fernald G (1943) *Remedial Techniques in Basic School Subjects*. McGraw-Hill, New York
64 Bryant P, Bradley L (1985) *Children's reading problems: psychology and education*. Blackwell, Oxford and New York
65 Hulme C (1981) *Reading retardation and multisensory teaching*. Routledge and Kegan Paul, London
66 Ofman W, Shaevitz M (1963) The kinesthetic method in remedial reading. *J Exp Education* 3: 317–320
67 Bara F, Gentaz E, Colé P (2004) The visuo-haptic and haptic exploration of letters increases the kindergarten-children's reading acquisition. *Cognitive Development* 19: 433–449
68 Bara F, Gentaz E, Colé P (2007) The visuo-haptic and haptic exploration increases the decoding level of children coming from low-socioeconomic status families. *Br J Dev Psychology 25*: 643–663
69 Gentaz E, Colé P, Bara F (2003) Evaluation d'entraînements multisensoriels de préparation à la lecture pour les enfants de grande section de maternelle. *L'Année Pschologique* 104(4): 561–584
70 Worchel P (1951) Space perception and orientation in the blind. *Psychological Monograph* 65 (15)
71 Hatwell Y (1986) *Toucher l'espace*. Presses Universitaires de Lille, Lille
72 Gentaz E, Hatwell Y (1998) The haptic oblique effect in the perception of rod orientation by blind adults. *Perception & Psychophysics* 60: 157–167
73 Hatwell Y (1960) Etude de quelques ilusions géométriques tactiles chez les aveugles. *L'Année Psychologique* 1: 11–27

5 Haptics in the United States before 1940

1 Hall GS (1879) Laura Bridgeman. *Mind* No. 14 April: 149–172
2 Gitter E (2001) *Imprisoned guest: Samuel Howe and Laura Bridgeman, the original deaf-blind girl*. Farrar, Straus and Giroux, New York
3 Schwartz H (1952) Samuel Gridley Howe as phrenologist. *American Historical Review* 63: 644–651
4 Freeberg E (2001) *The education of Laura Bridgeman: The first deaf and blind person to learn language*. Harvard University Press
5 Howe SG (1909) *The letters and journals of Samuel Gridley Howe*. Vol. 2 L. E. Richards (ed) Dana Estes & Company, Boston
6 Jastrow J (1894) Psychological notes on Hellen Kellar (sic). *Psychological Review*: 356–362
7 Ross D (1972) *G. Stanley Hall: The psychologist as prophet*. University of Chicago Press, Chicago
8 Donaldson HH (1885) On the temperature sense. *Mind* 10: 399–416
9 Blix M (1882) Experimentela bidrag till lösning af fragan om hudnerveras specifika energies. *Uppsala Läkfören Förhandlingar* 18: 87–102
10 Blix M (1883) Experimentela bidrag till lösning af fragan om hudnerveras specifika energies. *Uppsala Läkfören Förhandlingar* 18: 427–440
11 Goldscheider A (1884) Die specifische Energie der Temperaturnerven I. *Monatshefte für Praktische Dermatologie* 3: 198–208. Die specifische Energie der Temperaturnerven II. *Monatshefte für Praktische Dermatologie* 3: 325–341
12 Blix M (1884) Experimentelle Beiträge zur Lösung der Frage über die specifische Energie der Hautnerven. *Zeitschrift für Biologie* 20: 141–156
13 Blix M (1885) Experimentelle Beiträge zur Lösung der Frage über die specifische Energie der Hautnerven. *Zeitschrift für Biologie* 21: 145–160
14 Altruz S (1897) On the temperature-senses I. *Mind* NS 6: 445–448
15 Altruz S (1898) On the temperature-senses II. *Mind* NS 7: 141–145
16 Donaldson HH (1885) Action of the muriate of cocaine on the temperature nerves. *Maryland Medical Journal* 12: 475
17 Hall GS, Donaldson HH (1885) Motor sensations on the skin. *Mind* 10: 557–572

18 Jastrow J (1886) The perception of space by disparate senses. *Mind* NS 11 No. 44: 539–554
19 Hall GS, Motora Y (1887) Dermal sensitiveness to gradual changes. *Am J Psychol* 1: 72–98
20 Jastrow J (1893) Section of psychology. In: Hardy MP (ed): *Official catalogue of the world's Columbian exposition*. W. B. Conkey, Chicago, Part vii, 50–60
21 Münsterberg H (1893) The new psychology and Harvard's equipment for studying it. *Harvard Graduate Magazine* 1: 201–209
22 Titchener EB (1898) A psychological laboratory. *Mind* NS 7: 311–331
23 Titchener EB (1900) The equipment of a psychological laboratory. *Am J Psychol* 11: 251–265
24 Dessoir M (1892) Über den Hautsinn. *Archiv für Physiologie* 3: 175–339
25 Baldwin JM (1905) *Dictionary of philosophy and psychology*. Vol. 1. Macmillan, New York
26 Revesz G (1950) *The psychology and art of the blind*. Longmans Green, London
27 Titchener EB (1896) *An outline of psychology*. Macmillan, New York
28 Titchener EB (1899) *An outline of psychology*. 2nd Ed. Macmillan, New York, 50
29 Bentley M (1900) The synthetic experiment. *Am J Psychol* 11: 405–425
30 Titchener EB (1909) *A textbook of psychology*. Part I. Macmillan, New York
31 Bastian HC (1880) *The brain as an organ of mind*. D. Appleton and Company, Boston
32 Czermak JN (1852) Über den Raumsinn. *Berichte über die Verhandlungen der Königlich-Sächsischen Gesellschaft der Wissenschaften zu Leipzig, Mathematisch-Physische Klasse*.
33 Pillsbury WB (1895) Some questions of the cutaneous sensibility. *Am J Psychol* 7: 42–57
34 Washburn M (1895) Über den Einfluss den Gesichtsassociationen auf die Raumwahrnehmungen der Haut. *Philosophishe Studien* 11: 190–225
35 Parish C (1897) Localization of cutaneous impressions by arm movement without pressure upon the skin. *Am J Psychol* 8: 250–267
36 Boring EG (1942) *Sensation and perception in the history of psychology*. Appleton Century, New York
37 Friedline C (1918) The discrimination of cutaneous patterns below the two point threshold. *Am J Psychol* 29: 400–419
38 Thunberg T (1896) Förnimmelserne vid till samma ställe lokaliserad, samtidigt pågående köld- och värmereting. *Uppsala Läkfören Förhandlingar* 2nd Ser. 489–485
39 Geldard FA (1972) *The Human Senses*. John Wiley, New York, 2nd Ed., 359–362
40 Craig AD, Bushnell MC (1994) The thermal grill illusion: Unmasking the burn of cold pain. *Science* NS 265: 252–255
41 Murray E (1908) A qualitative analysis of tickling: Its relation to cutaneous sensation. *Am J Psychol* 19: 289–344
42 Murrray E (1909) Organic sensation. *Am J Psychol* 20: 386–446
43 Cutolo F (1918) A preliminary study of the psychology of heat. *Am J Psychol* 29: 442–448
44 Evans R (1972) E.B. Titchener and his lost system. *J History of the Behav Sci* 8: 168–180
45 Titchener EB (1920) Notes from the psychological laboratory of Cornell University. *Am J Psychol* 31: 212–214
46 Cobbey LW, Sullivan AH (1922) An experimental study of the perception of "oiliness". *Am J Psychol* 33: 121–127
47 Zigler MJ (1923) An experimental study of the perception of stickiness. *Am J Psychol* 34: 73–84
48 Meenes M, Zigler M (1923) An experimental study of the perceptions roughness and smoothness. *Am J Psychol* 34: 542–549
49 Zigler MJ (1923) An experimental study of the perception of clamminess. *Am J Psychol* 34: 550–561
50 Sullivan AH (1923) Perceptions of liquidity, semiliquidity and solidity. *Am J Psychol* 34: 531–541
51 Nafe JP (1927) Dermal sensitivity with special reference to the qualities of tickle and itch. *Pedagogical Seminary* 34: 14–27
52 Sullivan AH (1927) The cutaneous perceptions of softness and hardness. *J Exp Psychol* 10: 447–462
53 Alston JH (1920) The spatial condition of the fusion of warmth and cold in heat. *Am J Psychol* 31: 303–312
54 Burnet NC, Dallenbach KM (1927) The experience of heat. *Am J Psychol* 38: 418–431
55 Ferrall SC, Dallenbach KM (1930) The analysis and synthesis of burning heat. *Am J Psychol* 42: 72–82

56 Burnett NC, Dallenbach KM (1928) Heat intensity. *Am J Psychol* 40: 484–494
57 Gritman WB, Dallenbach KM (1929) The formula for the intensive graduation of heat. *Am J Psychol* 41: 460–464
58 Lowenstein E, Dallenbach KM (1937) The limen of heat and some conditions affecting it. *Am J Psychol* 49: 302–307
59 Ladd GT (1887) *Principles of physiological psychology*. Charles Scribner's Sons, New York
60 James W (1890) *Principles of psychology*. Vol. 2. Henry Holt & Co, New York
61 Hoisington LB (1920) On the non-visual perception of the length of lifted rods. *Am J Psychol* 31: 114–146
62 Schultz E (1922) On the non-visual perception of the length of vertically whipped rods. *Am J Psychol* 33: 135–139
63 Baker AS (1922) On the non-visual perception of the length of horizontally whipped rods. *Am J Psychol* 33: 139–144
64 Supa M, Cotzin M, Dallenbach KM (1944) "Facial vision": The perception of obstacles by the blind. *Am J Psychol* 57: 133–183
65 Levy WH (1872) *Blindness and the blind or A treatise on the science of typhlology*. Chapman and Hall, London
66 Hayes SP (1935) Facial vision or the sense of obstacles. *Perkins Publications* 12: 1–45
67 Dresslar FB (1893) On the pressure sense of the drum of the ear and "facial-vision". *Am J Psychol* 5: 313–368
68 Worchel P, Dallenbach KM (1947) Facial vision: The perception of obstacles by the deaf-blind. *Am J Psychol* 60: 502–553
69 Cotzin M, Dallenbach KM (1950) "Facial Vision": The role of pitch and loudness in the perception of obstacles by the blind. *Am J Psychol* 63: 485–515
70 Ammons CH, Worchel P, Dallenbach KM (1953) "Facial Vision": The perception of obstacles out of doors by blindfolded and blindfolded-deafened subjects. *Am J Psychol* 66: 519–553
71 Wertheimer M (1912) Über das Sehen von Bewegung. *Zeitschrift für Psychologie und Physiologie der Sinnesorgane* 61: 161–163
72 Benussi V (1913) Kinematohapische Erscheinungen. *Archiv für die gesamte Psychologie* 29: 385–388
73 Benusi V (1914) Kinematohatische Scheinbewegung und Auffassungsumformung. *Berichte VI Kongress für experimentelle Psychologie Göttingen*: 31–35
74 Benussi V (1916) Versuche zur Analyse taktil erweckter Scheinbewegungen. *Archiv für die gesamte Psychologie* 36: 59
75 Burt HE (1917) Auditory illusions of movement. *J Exp Psychol* 2: 63–75
76 Burt HE (1917) Tactual illusions of movement. *J Exp Psychol* 2: 371–385
77 Whitchurch AK (1921) The illusory perception of movement on the skin. *Am J Psychol* 32: 472–489
78 Geldard F, Sherrick CE (1972) The cutaneous "rabbit": A perceptual illusion. *Science* NS 178: 178–179
79 Geldard F (1982) Saltation in somesthesis. *Psychological Bulletin* 92: 136–175
80 Katz D (1925/1989) Der Aufbau der Tastwelt. *Zeitschrift für Psychologie und Physiologie der Sinnesorgane* Ergaenzungsband 11: 1–270; *The world of touch* (1989) tr. L. E. Krueger, Hillsdale, N.J. Earlbaum
81 Zigler M (1926) David Katz. Der Aufbau der Tastwelt. *Psychological Bulletin*: 326–336
82 Gibson JJ (1962) Observations on active touch. *Psychological Review* 69: 477–491
83 Nafe JP (1924) An experimental study of the affective qualities. *Am J Psychol* 35: 507–544
84 Nafe JP (1927) The psychology of felt experience. *Am J Psychol* 39: 367–389
85 Geldard F (1940) The perception of mechanical vibration: I. History of a controversy. *J General Psychol* 22: 243– 269
86 Alcorn S (1932) The Tadoma method. *Volta Review* 34: 195–198
87 Vivian RM (1966) Tadoma method: A tactual approach to speech and speech reading. *Volta Review* 68: 733–737
88 Goodfellow LD (1934) Vibratory sensitivity: Its present status. *Psychological Bulletin* 31: 560–571
89 Gault RH (1936) Recent developments in vibro-tactile research. *J Franklin Institute* 33: 776
90 Gault RH (1924) Progress in experiments on tactual interpretation of oral speech. *J Abnormal & Social Psychol* 19: 155–159
91 Katz D (1930) *The vibratory sense and other lectures*. The University of Maine Press, Orano

92 Anonymous (1932) The teletactor. *J Criminal Law and Criminology* 22: 765–766
93 Kirman JH (1982) Tactile communication of speech. In: W Schiff, E Foulke (eds): *Tactual perception: A sourcebook.* Cambridge University Press, New York, 240
94 Geldard FA (1940) The perception of mechanical vibration: II. The response of pressure receptors. *J General Psychol* 22: 271–280
95 Geldard FA (1940) The perception of mechanical vibration: III. The frequency function. *J General Psychol* 22: 281–289
96 Geldard FA (1940) The perception of mechanical vibration: IV. Is there a separate vibratory sense? *J General Psychol* 22: 291–308
97 Jenkins WL (1951) Somesthesis. In: SS Stevens (ed): *Handbook of Experimental Psychology.* John Wiley & Sons, New York, 1172–1190
98 Calne DB, Pallis CA (1966) The vibratory sense: A critical review. *Brain* 89: 723–746
99 Krueger LE (1982) Tactual perception in historical perspective: David Katz's world of touch. In: W Schiff, E Foulke (eds): *Tactual perception: A sourcebook.* Cambridge University Press, New York, 1–54
100 Helson H (1925) The psychology of "Gestalt". *Am J Psychol* 36: 342–370
101 Helson H (1925) The psychology of "Gestalt". *Am J Psychol* 36: 494–526
102 Koffka K (1922) Perception: An introduction to the Gestalt-theorie. *Psychological Bulletin* 19: 531–585
103 Gelb A (1914) Versuche auf dem Gebiete der Zeit- und Raumanschauung. In: F Schaumann (ed): *Berichte über die VI Kongress für Experimentelle Psychologie in Göttingen.* Leipzig, Barth, 36–42
104 Helson H (1930) The tau effect – An example of psychological relativity. *Science* NS 71: 536–537
105 Helson H, King S (1931) The tau effect; An examination of psychological relativity. *J Exp Psychol* 14: 202–217

6 Anatomy of receptors

1 Chambers MR, Andres KH, Duering Mv, Iggo A (1972) The structure and function of the slowly adapting type II mechanoreceptor in hairy skin. *Quart J Exp Physiol* 57: 417–445
2 Brisben AJ, Hsiao SS, Johnson KO (1999) Detection of vibration transmitted through an object grasped in the hand. *J Neurophysiol* 81: 1548–1558
3 Bridgman CF (1968) The structure of tendon organs in the cat: a proposed mechanism for responding muscle tension. *Anat Rec* 162: 209–220
4 Zelena J (ed) (1994) *Nerves and mechanoreceptors.* Chapmann & Hall, London, New York, Tokyo
5 Fukami Y, Wilkinson RS (1977) Responses of isolated Golgi tendon organs of the cat. *J Physiol* 265: 673–689
6 Stuart DG, Goslow GE, Mosher CG, Reinking RM (1970) Stretch responsiveness of Golgi tendon organs. *Exp Brain Res* 10: 463–476
7 Robertson JD (1960) Electron microscopy of the motor end-plate and the neuromuscular spindle. *Am J Phys Med* 39: 1–43
8 Steer JC, Horney FD (1968) Evidence for passage of cerebrospinal fluid among spinal nerves. *Can Med Assoc J* 98: 71–74
9 Cazzato G, Walton JN (1968) The pathology of the muscle spindle. A study of biopsy material in various muscular and neuromuscular diseases. *J Neurol Sci* 7: 15–70
10 Düring von M, Andres KH (1969) Zur Feinstruktur der Muskelspindel von Mammalia. *Anat Anz* 124: 566–573
11 Proske U, Wise AK, Gregory JE (2000) The role of muscle receptors in the detection of movements. *Prog Neurobiol* 60: 85–96
12 Vallbo AB, Johansson RS (1984) Properties of cutaneous mechanoreceptors in the human hand related to touch sensation. *Hum Neurobiol* 3: 3–14
13 Schimrigk K, Ruttinger H (1980) The touch corpuscles of the plantar surface of the big toe. *Eur Neurol* 19: 49–60
14 Bruce MF (1980) The relation of tactile thresholds to histology in the fingers of erderly people. *J Neurol Neurosurg Psychiat* 43: 730–734
15 Halata Z (1993) Sensory innervation of the hairy skin (light- and electronmicroscopic study. *J Invest Dermatol* 101: 75S–81S
16 Ide C, Kumagai K, Hayashi S (1985) Freeze-fracture study of mechanoreceptive digital corpuscle of mice. *J Neurocytol* 14: 1037–1052

17 Johansson RS, Landstrom U, Lundstrom R (1982) Responses of mechanoreceptive afferent units in the glabrous skin of the human hand to sinusoidal skin displacements. *Brain* Res 244: 17–25

18 Goodwin AW, Youl BD, Zimmerman NP (1981) Single quickly adapting mechanoreceptive afferents innervating monkey glabrous skin: response to two vibrating probes. *J Neurophysiol* 45: 227–242

19 Halata Z, Grim M, Baumann KI (2003) Friedrich Sigmund Merkel and his "Merkel cell", morphology, development, and physiology: Review and new results. *Anat Rec* A 271: 225–239

20 Iggo A, Muir AR (1969) The structure and function of a slowly adapting touch corpuscle in hairy skin. *J Physiol* 200: 763–796

21 Cahusac PM, Senok SS, Hitchcock IS, Genever PG, Baumann KI (2005) Are unconventional NMDA receptors involved in slowly adapting type I mechanoreceptor responses? *Neurosci* 133: 763–773

22 Cahusac PM, Senok SS (2006) Metabotropic glutamate receptor antagonists selectively enhance responses of slowly adapting type I mechanoreceptors. *Synapse* 59: 235–242

23 Fagan BM, Cahusac PMB (2001) Evidence for glutamate receptor mediated transmission at mechanoreceptors in the skin. *Neuroreport* 12: 341–347

24 Haeberle H, Fujiwara M, Chuang J, Medina MM, Panditrao MV, Bechstedt S, Howard J, Lumpkin EA (2004) Molecular profiling reveals synaptic release machinery in Merkel cells. *Proc Natl Acad Sci* 101: 14503–14508

25 Ogawa H (1996) The Merkel cell as a possible mechanoreceptor cell. *Prog Neurobiol* 49: 317–334

26 Senok SS, Baumann KI (1997) Functional evidence for calcium-induced calcium release in isolated rat vibrissal Merkel cell mechanoreceptors. *J Physiol* 500: 29–37

7 Physiological mechanisms of the receptor system

1 Christel M (1993) Grasping techniques and hand preferences in Hominoidea. In: H Preuschoft, DJ Chivers (eds): *Hands of Primates*. Springer-Verlag, Berlin

2 Vallbo AB, Johansson RS (1984) Properties of cutaneous mechanoreceptors in the human hand related to touch sensation. *Hum Neurobiol* 3: 3–14

3 Darian-Smith I (1984) The sense of touch: performance and peripheral neural processes. In: JM Brookhart, VB Mountcastle, I Darian-Smith, SR Geiger (eds): *Handbook of Physiology – The Nervous System* III. American Physiological Society, Bethesda

4 French AS (1992) Mechanotransduction. *Ann Rev Physiol* 54: 135–152

5 Johansson RS, Vallbo AB (1979) Tactile sensibility in the human hand: relative and absolute densities of four types of mechanoreceptive units in glabrous skin. *J Physiol* 286: 283–300

6 Goodwin AW, Wheat HE (2004) Sensory signals in neural populations underlying tactile perception and manipulation. *Ann Rev Neurosci* 27: 53–77

7 Johnson KO, Hsiao SS (1992) Neural mechanisms of tactual form and texture perception. *Ann Rev Neurosci* 15: 227–250

8 Srinivasan MA, LaMotte RH (1995) Tactual discrimination of softness. *J Neurophysiol* 73: 88–101

9 Monzee J, Lamarre Y, Smith AM (2003) The effects of digital anesthesia on force control using a precision grip. *J Neurophysiol* 89: 672–683

10 Pawluk DTV, Howe RD (1999) Dynamic lumped element response of the human fingerpad. *J Biomech Eng* 121: 178–183

11 Birznieks I, Jenmalm P, Goodwin AW, Johansson RS (2001) Encoding of direction of fingertip forces by human tactile afferents. *J Neurosci* 21: 8222–8237

12 LaMotte RH, Srinivasan MA (1996) Neural encoding of shape: Responses of cutaneous mechanoreceptors to a wavy surface stroked across the monkey fingerpad. *J Neurophysiol* 76: 3787–3797

13 Goodwin AW, Browning AS, Wheat HE (1995) Representation of curved surfaces in responses of mechanoreceptive afferent fibers innervating the monkey's fingerpad. *J Neurosci* 15: 798–810

14 Phillips JR, Johnson KO (1981) Tactile spatial resolution. II. Neural representation of bars, edges, and gratings in monkey primary afferents. *J Neurophysiol* 46: 1192–1203

15 Johansson RS, Westling G (1987) Signals in tactile afferents from the fingers eliciting adaptive motor responses during precision grip. *Exp Brain Res* 66: 141–154

16 Srinivasan MA, Whitehouse JM, LaMotte RH (1990) Tactile detection of slip: surface microgeometry and peripheral neural codes. *J Neurophysiol* 63: 1323–1332

17 Goodwin AW, Wheat HE (1999) Effects of nonuniform fiber sensitivity, innervation geometry, and noise on information relayed by a population of slowly adapting type I primary afferents from the fingerpad. *J Neurosci* 19: 8057–8070

18 LaMotte RH, Friedman RM, Lu C, Khalsa PS, Srinivasan MA (1998) Raised object on a planar surface stroked across the fingerpad: responses of cutaneous mechanoreceptors to shape and orientation. *J Neurophysiol* 80: 2446–2466

19 Blake DT, Hsiao SS, Johnson KO (1997) Neural coding mechanisms in tactile pattern recognition: The relative contributions of slowly and rapidly adapting mechanoreceptors to perceived roughness. *J Neurosci* 17: 7480–7489

20 Collins DF, Prochazka A (1996) Movement illusions evoked by ensemble cutaneous input from the dorsum of the human hand. *J Physiol* 496: 857–871

21 Edin BB, Abbs JH (1991) Finger movement responses of cutaneous mechanoreceptors in the dorsal skin of the human hand. *J Neurophysiol* 65: 657–670

22 Gandevia SC, McCloskey DI, Burke D (1992) Kinaesthetic signals and muscle contraction. *Trends Neurosci* 15: 62–65

8 Neural basis of haptic perception

1 Mountcastle VB, LaMotte RH, Carli G (1972) Detection thresholds for stimuli in humans and monkeys: comparison with threshold events in mechanoreceptive afferent nerve fibers innervating the monkey hand. *J Neurophysiol* 35: 122–136

2 Talbot WH, Darian-Smith I, Kornhuber HH, Mountcastle VB (1968) The sense of flutter-vibration: comparison of the human capacity with response patterns of mechanoreceptive afferents from the monkey hand. *J Neurophysiol* 31: 301–334

3 Hsiao SS, Johnson KO, Yoshioka T (2003) Processing of tactile information in the primate brain. In: M Gallagher, RJ Nelson (eds): *Comprehensive handbook of psychology*, Volume 3: *Biological psychology*. Wiley, New York, 211–236

4 Coleman GT, Zhang HQ, Rowe MJ (2003) Transmission security for single kinesthetic afferent fibers of joint origin and their target cuneate neurons in the cat. *J Neurosci* 23: 2980–2992

5 Wang X, Merzenich MM, Sameshima K, Jenkins WM (1995) Remodelling of hand representation in adult cortex determined by timing of tactile stimulation [see comments]. *Nature* 378: 71–75

6 Krubitzer LA, Clarey J, Tweedale R, Elston G, Calford MB (1995) A redefinition of somatosensory areas in the lateral sulcus of macaque monekeys. *J Neurosci* 15: 3821–3839

7 Fitzgerald PJ, Lane JW, Thakur PH, Hsiao SS (2004) Receptive field properties of the macaque second somatosensory cortex: evidence for multiple functional representations. *J Neurosci* 24: 11193–11204

8 Eickhoff SB, Schleicher A, Zilles K, Amunts K (2006) The human parietal operculum. I. Cytoarchitectonic mapping of subdivisions. *Cereb Cortex* 16: 254–267

9 Murray EA, Mishkin M (1984) Relative contributions of SII and area 5 to tactile discrimination in monkeys. *Behav Brain Res* 11: 67–85

10 Romo R, Hernandez A, Zainos A, Brody C, Salinas E (2002) Exploring the cortical evidence of a sensory-discrimination process. *Philos Trans R Soc Lond.B Biol Sci* 357: 1039–1051

11 Merabet L, Thut G, Murray B, Andrews J, Hsiao S, Pascual-Leone A (2004) Feeling by sight or seeing by touch? *Neuron* 42: 173–179

12 Zangaladze A, Epstein CM, Grafton ST, Sathian K (1999) Involvement of visual cortex in tactile discrimination of orientation. *Nature* 401: 587–590

13 Craig JC, Johnson KO (2000) The two-point threshold: not a measure of tactile spatial resolution. *Curr Dir Psychol Sci* 9: 29–32

14 Johnson KO, Phillips JR (1981) Tactile spatial resolution: I. Two-point discrimination, gap detection, grating resolution, and letter recognition. *J Neurophysiol* 46: 1177–1191

15 Johnson KO, Van Boven RW, Hsiao SS (1994) The perception of two points is not the spatial resolution threshold. In: J Boivie, P Hansson, U Lindblom (eds): *Touch, temperature, and pain in health and disease: Mechanisms and assessments*. IASP Press, Seattle, 389–404

16 Phillips JR, Johnson KO, Browne HM (1983) A comparison of visual and two modes of tactual letter resolution. *Percept Psychophys* 34: 243–249
17 Hsiao SS (1998) Similarities between touch and vision. In: JW Morley (ed): *Neural aspects of tactile sensation*. Elsevier, Amsterdam, 131–165
18 Johnson KO, Hsiao SS, Yoshioka T (2002) Neural coding and the basic law of psychophysics. *Neuroscientist* 8: 111–121
19 Hsiao SS, Lane JW, Fitzgerald P (2002) Representation of orientation in the somatosensory system. *Behav Brain Res* 135: 93–103
20 Fitzgerald PJ, Lane JW, Thakur PH, Hsiao SS (2006a) Receptive field (RF) properties of the macaque second somatosensory cortex: RF size, shape, and somatotopic organization. *J Neurosci* 26: 6485–6495
21 Fitzgerald PJ, Lane JW, Thakur PH, Hsiao SS (2006b) Receptive field properties of the macaque second somatosensory cortex: representation of orientation on different finger pads. *J Neurosci* 26: 6473–6484
22 Hsiao SS, Vega-Bermudez F (2002) Attention in the somatosensory system. In: RJ Nelson (ed): *The somatosensory system: deciphering the brain's own body image*. CRC Press, Boca Raton, 197–217

9 The neural bases of haptic working memory

1 Goldman-Rakic PS (1995) Architecture of the prefrontal cortex and the central executive. *Ann NY Acad Sci* 769: 71–83
2 Goldman-Rakic PS (1996) The prefrontal landscape: implications of functional architecture for understanding human mentation and the central executive. *Philos Trans R Soc Lond B Biol Sci* 351(1346): 1445–1453
3 Loomis JM, Lederman SJ (1986) Tactual perception. In: K Boff, L Kaufman, J Thomas (eds): *Handbook of perception and human performance*. Wiley, New York, chapter 31
4 Gibson JJ (1962) Observations on active touch. *Psychol Rev* 69: 477–491
5 Gibson JJ (1966) *The senses considered as perceptual systems*. Houghton-Mifflin, Boston, Mass.
6 Baddeley A (2003) Working memory: looking back and looking forward. *Nat Rev Neurosci* 4(10): 829–839
7 Beaumont JG (1981) Activation and interference in tactile perception. *Neuropsychologia* 19(1): 151–154
8 Gilson EQ, Baddeley AD (1969) Tactile short-term memory. *Q J Exp Psychol* 21(2): 180–184
9 Sullivan EV, Turvey MT (1972) Short-term retention of tactile stimulation. *Q J Exp Psychol* 24(3): 253–261
10 Murray DJ, Ward R, Hockley WE (1975) Tactile short-term memory in relation to the two-point threshold. *Q J Exp Psychol* 27(2): 303–312
11 Miles C, Borthwick M (1996) Tactile short-term memory revisited. *Memory* 4(6): 655–668
12 Sullivan EV (1989) Hemispheric asymmetry in tactile forgetting induced by tactually-guided movement. *Cortex* 25(1): 83–92
13 Sinclair RJ, Kuo JJ, Burton H (2000) Effects on discrimination performance of selective attention to tactile features. *Somatosens Mot Res* 17(2): 145–157
14 Harris JA, Harris IM, Diamond ME (2001) The topography of tactile working memory. *J Neurosci* 21(20): 8262–8269
15 Harris JA, Miniussi C, Harris IM, Diamond ME (2002) Transient storage of a tactile memory trace in primary somatosensory cortex. *J Neurosci* 22(19): 8720–8725
16 Fiehler K, Burke M, Engel A, Bien S, Roesler F (2007) Kinesthetic working memory and action control within the dorsal stream. *Cereb Cortex* 18(2): 243–253
17 Millar S (1977) Early stages of tactual matching. *Perception* 6(3): 333–343
18 Millar S (1978) Aspects of memory for information from touch and movement. In: G Gordon (ed): *Active touch*. Pergamon Press, Oxford, 215–227
19 Millar S (1974) Tactile short-term memory by blind and sighted children. *Br J Psychol* 65(2): 253–263
20 Kiphart MJ, Auday BC, Cross HA (1988) Short-term haptic memory for three-dimensional objects. *Percept Mot Skills* 66(1): 79–91
21 Bowers RL, Mollenhauer MS, Luxford J (1990) Short-term memory for tactile and temporal stim-

uli in a shared-attention recall task. *Percept Mot Skills* 70(3 Pt 1): 903–913

22 Zuidhoek S, Kappers AM, van der Lubbe RH, Postma A (2003) Delay improves performance on a haptic spatial matching task. *Exp Brain Res* 149(3): 320–330

23 Zuidhoek S (2005) Representations of space based on haptic input. Doctoral dissertation, Utrecht University

24 Bliss I, Hamalainen H (2005) Different working memory capacity in normal young adults for visual and tactile letter recognition task. *Scand J Psychol* 46(3): 247–251

25 Bliss I, Kujala T, Hamalainen H (2004) Comparison of blind and sighted participants' performance in a letter recognition working memory task. *Brain Res Cogn Brain Res* 18(3): 273–277

26 Marmor GS, Zaback LA (1976) Mental rotation by the blind: does mental rotation depend on visual imagery? *J Exp Psychol Hum Percept Perform* 2(4): 515–521

27 Carpenter PA, Eisenberg P (1978) Mental rotation and the frame of reference in blind and sighted individuals. *Percept Psychophys* 23(2): 117–124

28 Fuster JM, Alexander GE (1971) Neuron activity related to short-term memory. *Science* 173(997): 652–654

29 Smith EE, Jonides J (1999) Storage and executive processes in the frontal lobes. *Science* 283(5408): 1657–1661

30 Fletcher PC, Henson RN (2001) Frontal lobes and human memory: insights from functional neuroimaging. *Brain* 124: 849–881

31 Lebedev MA, Messinger A, Kralik JD, Wise SP (2004) Representation of attended *versus* remembered locations in prefrontal cortex. *PLoS Biol* 2(11): e365

32 Zhou YD, Ardestani A, Fuster JM (2007) Distributed and associative working memory. *Cereb Cortex* Suppl 1: i77–87

33 Rowe JB, Toni I, Josephs O, Frackowiak RS, Passingham RE (2000) The prefrontal cortex: response selection or maintenance within working memory? *Science* 288(5471): 1656–1660

34 Postle BR (2006) Working memory as an emergent property of the mind and brain. *Neuroscience* 139(1): 23–38

35 Pasternak T, Greenlee MW (2005) Working memory in primate sensory systems. *Nat Rev Neurosci* 6(2): 97–107

36 Linden DE (2007) The working memory networks of the human brain. *Neuroscientist* 13(3): 257–267

37 Romo R, Salinas E (2003) Flutter discrimination: neural codes, perception, memory and decision making. *Nat Rev Neurosci* 4(3): 203–218

38 Zhou YD, Fuster JM (1996) Mnemonic neuronal activity in somatosensory cortex. *Proc Natl Acad Sci USA* 93(19): 10533–10537

39 Bodner M, Shafi M, Zhou YD, Fuster JM (2005) Patterned firing of parietal cells in a haptic working memory task. *Eur J Neurosci* 21(9): 2538–2546

40 Salinas E, Hernandez A, Zainos A, Romo R (2000) Periodicity and firing rate as candidate neural codes for the frequency of vibrotactile stimuli. *J Neurosci* 20(14): 5503–5515

41 Romo R, Hernandez A, Zainos A, Lemus L, Brody CD (2002) Neuronal correlates of decision-making in secondary somatosensory cortex. *Nat Neurosci* 5(11): 1217–1225

42 Romo R, Brody CD, Hernandez A, Lemus L (1999) Neuronal correlates of parametric working memory in the prefrontal cortex. *Nature* 399(6735): 470–473

43 Klingberg T, Kawashima R, Roland PE (1996) Activation of multi-modal cortical areas underlies short-term memory. *Eur J Neurosci* 8(9): 1965–1971

44 Liu LC, Fenwick PB, Laskaris NA, Schellens M, Poghosya NV, Shibata T, Ioannides AA (2003) The human primary somatosensory cortex response contains components related to stimulus frequency and perception in a frequency discrimination task. *Neuroscience* 121(1): 141–154

45 Grunwald M, Weiss T, Krause W, Beyer L, Rost R, Gutberlet I, Gertz HJ (2001) Theta power in the EEG of humans during ongoing processing in a haptic object recognition task. *Brain Res Cogn Brain Res* 11(1): 33–37

46 Grunwald M, Weiss T, Krause W, Beyer L, Rost R, Gutberlet I, Gertz HJ (1999) Power of theta waves in the EEG of human subjects increases during recall of haptic information. *Neurosci Lett* 260(3): 189–192

47 Peltier S, Stilla R, Mariola E, LaConte S, Hu X, Sathian K (2007) Activity and effective connectivity

of parietal and occipital cortical regions during haptic shape perception. *Neuropsychologia* 45(3): 476–483
48 Numminen J, Schurmann M, Hiltunen J, Joensuu R, Jousmaki V, Koskinen SK, Salmelin R, Hari R (2004) Cortical activation during a spatiotemporal tactile comparison task. *Neuroimage* 22(2): 815–821
49 Stoeckel MC, Weder B, Binkofski F, Buccino G, Shah NJ, Seitz RJ (2003) A fronto-parietal circuit for tactile object discrimination: an event-related fMRI study. *Neuroimage* 19(3): 1103–1114
50 Stoeckel MC, Weder B, Binkofski F, Choi HJ, Amunts K, Pieperhoff P, Shah NJ, Seitz RJ (2004) Left and right superior parietal lobule in tactile object discrimination. *Eur J Neurosci* 19(4): 1067–1072
51 Kaas AL, van MH, Goebel R (2007) The neural correlates of human working memory for haptically explored object orientations. *Cereb Cortex* 17(7): 1637–1649
52 Pietrini P, Furey ML, Ricciardi E, Gobbini MI, Wu WH, Cohen L, Guazzelli M, Haxby JV (2004) Beyond sensory images: Object-based representation in the human ventral pathway. *Proc Natl Acad Sci USA* 101(15): 5658–5663
53 Ricciardi E, Bonino D, Gentili C, Sani L, Pietrini P, Vecchi T (2006) Neural correlates of spatial working memory in humans: a functional magnetic resonance imaging study comparing visual and tactile processes. *Neuroscience* 139(1): 339–349
54 Kostopoulos P, Albanese MC, Petrides M (2007) Ventrolateral prefrontal cortex and tactile memory disambiguation in the human brain. *Proc Natl Acad Sci USA* 104(24): 10223–10228
55 Preuschhof C, Heekeren HR, Taskin B, Schubert T, Villringer A (2006) Neural correlates of vibrotactile working memory in the human brain. *J Neurosci* 26(51): 13231–13239
56 Soros P, Marmurek J, Tam F, Baker N, Staines WR, Graham SJ (2007) Functional MRI of working memory and selective attention in vibrotactile frequency discrimination. *BMC Neurosci* 8: 48
57 Sakata H, Taira M (1994) Parietal control of hand action. *Curr Opin Neurobiol* 4(6): 847–856
58 Binkofski F, Buccino G, Posse S, Seitz RJ, Rizzolatti G, Freund H (1999) A fronto-parietal circuit for object manipulation in man: evidence from an fMRI-study. *Eur J Neurosci* 11(9): 3276–3286
59 Ehrsson HH, Fagergren A, Jonsson T, Westling G, Johansson RS, Forssberg H (2000) Cortical activity in precision *versus* power-grip tasks: an fMRI study. *J Neurophysiol* 83(1): 528–536
60 Grefkes C, Weiss PH, Zilles K, Fink GR (2002) Crossmodal processing of object features in human anterior intraparietal cortex: an fMRI study implies equivalencies between humans and monkeys. *Neuron* 35(1): 173–184
61 Culham JC, Danckert SL, DeSouza JF, Gati JS, Menon RS, Goodale MA (2003) Visually guided grasping produces fMRI activation in dorsal but not ventral stream brain areas. *Exp Brain Res* 153(2): 180–189
62 Frey SH, Vinton D, Norlund R, Grafton ST (2005) Cortical topography of human anterior intraparietal cortex active during visually guided grasping. *Brain Res Cogn Brain Res* 23(2–3): 397–405
63 Binkofski F, Kunesch E, Classen J, Seitz RJ, Freund HJ (2001) Tactile apraxia: unimodal apractic disorder of tactile object exploration associated with parietal lobe lesions. *Brain* 124(Pt 1): 132–144
64 Petrides M, Pandya DN (2002) Comparative cytoarchitectonic analysis of the human and the macaque ventrolateral prefrontal cortex and corticocortical connection patterns in the monkey. *Eur J Neurosci* 16(2): 291–310
65 Jones EG, Powell TP (1970) Connexions of the somatic sensory cortex of the rhesus monkey. 3. Thalamic connexions. *Brain* 93(1): 37–56
66 Jones EG, Powell TP (1969) Connexions of the somatic sensory cortex of the rhesus monkey. I. Ipsilateral cortical connexions. *Brain* 92(3): 477–502
67 Binkofski F, Amunts K, Stephan KM, Posse S, Schormann T, Freund HJ, Zilles K, Seitz RJ (2000) Broca's region subserves imagery of motion: a combined cytoarchitectonic and fMRI study. *Hum Brain Mapp* 11(4): 273–285
68 Koechlin E, Jubault T (2006) Broca's area and the hierarchical organization of human behavior. *Neuron* 50(6): 963–974
69 Amedi A, Malach R, Hendler T, Peled S, Zohary E (2001) Visuo-haptic object-related activation in the ventral visual pathway. *Nat Neurosci* 4(3): 324–330

70 Baddeley A (1986) *Working memory*. Clarendon Press, Oxford
71 Melzack R, Eisenberg H (1968) Skin sensory afterglows. *Science* 159(813): 445–447
72 Baddeley AD, Hitch GJ (1974) Working memory. In: GA Bower (ed): *Recent advances in learning and motivation*. Academic Press, New York, 47–89
73 Baddeley AD (1990) *Human memory*. Lawrence Erlbaum, HOve
74 Kandel ER, Kupfermann I, Iversen S (2000) Learning and memory. In: ER Kandel, JH Schwartz, TM Jessel (eds): *Principles of neural science*. McGraw-Hill, New York, 1227–1246
75 Brown J (1958) Some tests of the decay theory of immediate memory. *Quart J Exp Psych* 10: 12–21
76 Peterson LR, Peterson MJ (1959) Short-term retention of individual verbal items. *J Exp Psychol* 58: 193–198
77 Gevins A, Cutillo B (1993) Spatiotemporal dynamics of component processes in human working memory. *Electroencephalogr Clin Neurophysiol* 87(3): 128–143
78 Shepard R, Metzler J (1971) Mental rotation of three dimensional objects. *Science* 171: 701–703
79 Zarahn E, Aguirre G, D'Esposito M (1997) A trial-based experimental design for fMRI. *Neuroimage* 6(2): 122–138
80 Lederman SJ, Klatzky RL (1987) Hand movements: a window into haptic object recognition. *Cognit Psychol* 19(3): 342–368
81 Curtis CE, D'Esposito M (2003) Persistent activity in the prefrontal cortex during working memory. *Trends Cogn Sci* 7(9): 415–423

10 Neuronal plasticity of the haptic system

1 Penfield W, Boldrey E (1937) Somatic motor and sensory representation in the cerebral cortex of man as studied by electrical stimulation. *Brain* 60: 389–443
2 Sur M, Merzenich MM, Kaas JH (1980) Magnification, receptive-field area, and "hypercolumn" size in areas 3b and 1 of somatosensory cortex in owl monkeys. *J Neurophysiol* 44: 297–311
3 Merzenich M, Nelson RJ, Stryker MP, Cynader MS, Schoppmann A, Zook JM (1984) Somatosensory cortical map changes following digit amputation in adult monkeys. *J Comp Neurol* 224: 591–605
4 Yang T, Gallen CC, Ramachandran VS, Cobb S, Schwarz BJ, Bloom FE (1994) Noninvasive detection of cerebral plasticity in adult human somatosensory cortex. *Neuroreport* 5: 701–704
5 Flor H, Elbert T, Knecht S, Wienbruch C, Pantev C, Birbaumer N, Larbig W, Taub E (1995) Phantom-limb pain as a perceptual correlate of cortical reorganization following arm amputation. *Nature* 375: 482–484
6 Flor H (2002) Phantom-limb pain: characteristics, causes and treatment. *Lancet Neurol* 19: 345–355
7 Giummarra M, Gibson SJ, Georgiou-Karistianis N, Bradshaw JL (2007) Central mechanisms in phantom limb perception: the past, present and future. *Brain Res Rev* 54: 219–232
8 Lotze M, Grodd W, Birbaumer N, Erb M, Huse E, Flor H (1999) Does use of a myoelectric prosthesis prevent cortical reorganization and phantom limb pain? *Nature Neurosci* 2: 501–502
9 Maihöfner C, Handwerker HO, Neundörfer B, Birklein F (2003) Patterns of cortical reorganization in complex regional pain syndrome. *Neurol* 61: 1707–1715
10 Pleger B, Ragert P, Schwenkreis P, Förster AF, Wilimzig C, Dinse H, Nicolas V, Maier C, Tegenthoff M (2006) Patterns of cortical reorganization parallel impaired tactile discrimination and pain intensity in complex regional pain syndrome. *Neuroimage* 15: 503–510
11 Ramachandran V, Rogers-Ramachandran D, Stewart M (1992) Perceptual correlates of massive cortical reorganization. *Science* 258: 1159–1160
12 Knecht S, Henningsen H, Elbert T, Flor H, Höhling C, Pantev C, Birbaumer N, Taub E (1995) Cortical reorganization in human amputees and mislocalization of painful stimuli to the phantom limb. *Neurosci Lett* 201: 262–264
13 Maihöfner C, Neundörfner B, Birklein F, Handwerker HO (2006) Mislocalization of tactile stimulation in patients with complex regional pain syndrome. *J Neurol* 253: 1432–1459
14 Jenkins W, Merzenich MM, Ochs MT, Allard T, Guíc-Robles E (1990) Functional reorganization of primary somatosensory cortex in adult owl monkeys

after behaviorally controlled tactile stimulation. *J Neurophysiol* 63: 82–104

15 Recanzone G, Schreiner CE, Merzenich MM (1993) Plasticity in the frequency representation of primary auditory cortex following discrimination training in adult owl monkeys. *J Neurosci* 13: 87–103

16 Coq J, Xerri C (1998) Environmental enrichment alters organizational features of the forepaw representation in the primary somatosensory cortex of adult rats. *Exp Brain Res* 121: 191–204

17 Elbert T, Stewart M, Flor H, Rockstroh B, Knecht S, Pantev C, Wienbruch C, Taub E (1997) Input-increase and input-decrease types of cortical reorganization after upper extremity amputation in humans. *Exp Brain Res* 117: 161–164

18 Recanzone G, Jenkins WM, Hradek GT, Merzenich MM (1992) Progressive improvement in discriminative abilities in adult owl monkeys performing a tactile frequency discrimination task. *J Neurophysiol* 67: 1015–1030

19 Recanzone G, Merzenich MM, Dinse HR (1992) Expansion of the cortical representation of a specific skin field in primary somatosensory cortex by intracortical microstimulation. *Cereb Cortex* 2: 181–196

20 Recanzone G, Merzenich MM, Jenkins WM, Grajski KA, Dinse HR (1992) Topographic reorganization of the hand representation in cortical area 3b owl monkeys trained in a frequency-discrimination task. *J Neurophysiol* 67: 1031–1056

21 Recanzone G, Merzenich MM, Schreiner CE (1992) Changes in the distributed temporal response properties of SI cortical neurons reflect improvements in performance on a temporally based tactile discrimination task. *J Neurophysiol* 67: 1071–1091

22 Recanzone G, Merzenich MM, Jenkins WM (1992) Frequency discrimination training engaging a restricted skin surface results in an emergence of a cutaneous response zone in cortical area 3a. *J Neurophysiol* 67: 1057–1070

23 Elbert T, Pantev C, Wienbruch C, Rockstroh B, Taub E (1995) Increased cortical representation of the fingers of the left hand in string players. *Science* 270: 305–307

24 Hashimoto I, Suzuki A, Kimura T, Igouchi Y, Tanosaki M, Takino R, Harauta Y, Taira M (2004) Is there training-dependent reorganization of digit representations in area 3b of string players? *Clin Neurophysiol* 115: 435–447

25 Hebb D (1949) *The organization of behavior.* Wiley, New York

26 Godde B, Stauffenberg B, Spengler F, Dinse HR (2000) Tactile coactivation-induced changes in spatial discrimination performance. *J Neurosci* 20: 1597–1604

27 Joublin F, Spengler F, Wacquant S, Dinse HR (1996) A columnar model of somatosensory reorganizational plasticity based on Hebbian and non-Hebbian learning rules. *Biol Cybern* 74: 275–286

28 Dinse H, Recanzone GH, Merzenich MM (1993) Alterations in correlated activity parallel ICMS-induced representational plasticity. *Neuroreport* 5: 173–176

29 Hodzic A, Veit R, Karim AA, Erb M, Godde B (2007) Improvement and decline in tactile discrimination behavior after cortical plasticity induced by passive tactile coactivation. *J Neurosci* 24: 442–446

30 Allard T, Clark SA, Jenkins WM, Merzenich MM (1991) Reorganization of somatosensory area 3b representations in adult owl monkeys after digital syndactyly. *J Neurophysiol* 66: 1048–1058

31 Wang X, Merzenich M, Sameshima K, Jenkins WM (1995) Remodelling of hand representation in adult cortex determined by timing of tactile stimulation. *Nature* 378: 71–75

32 Mogilner A, Grossman JA, Ribary U, Joliot M, Volkmann J, Rapaport D, Beasley RW, Llinás RR (1993) Somatosensory cortical plasticity in adult humans revealed by magnetoencephalography. *Proc Natl Acad Sci USA* 90: 3593–3597

33 Sterr A, Müller MM, Elbert T, Rockstroh B, Pantev C, Taub E (1998) Perceptual correlates of changes in cortical representation of fingers in blind multifinger Braille readers. *J Neurosci* 18: 4417–4423

34 Sterr A, Müller MM, Elbert T, Rockstroh B, Pantev C, Taub E (1998) Changed perceptions in Braille readers. *Nature* 391: 134–135

35 Pelled G, Chuang KH, Dodd SJ, Koretsky AP (2007) Functional MRI detection of bilateral cortical reorganization in the rodent brain following peripheral nerve deafferentation. *Neuroimage* 37: 262–273

36 Rossini P, Martino G, Narici L, Pasquarelli A, Peres-

son M, Pizzella V, Tecchio F, Rorrioli G, Romani GL (1994) Short-term brain “plasticity“ in humans: transient finger representation changes in sensory cortex somatotopy following ischemic anesthesia. *Brain Topogr* 642: 169–177

37 Weiss T, Miltner WH, Hounker R, Friedel R, Schmidt I, Taub E (2000) Rapid functional plasticity of the somatosensory cortex after finger amputation. *Exp Brain Res* 134: 199–203

38 Birbaumer N, Lutzenberger W, Montoya P, Larbig W, Unertl K, Töpfner S, Grodd W, Taub E, Flor H (1997) Effects of regional anesthesia on phantom limb pain are mirrored in changes in cortical reorganization. *J Neurosci* 17: 5503–5508

39 Braun C, Wilms, Schweizer R, Godde B, Preissl H, Birbaumer N (2000) Activity patterns of human somatosensory cortex adapt dynamically to stimulus properties. *Neuroreport* 11: 2977–2980

40 Stavrinou M, Della Penna S, Pizzella V, Torquati K, Cianflone F, Franciotti R, Bezerianos A, Romani GL, Rossini PM (2007) Temporal dynamics of plastic changes in human primary somatosensory cortex after finger webbing. *Cereb Cortex* 17: 2134–2142

41 Wühle A, Fahlbusch JJ, Braun C (2006) Effects of motor activity on the organization of primary somatosensory cortex. *Neuroreport* 17: 39–43

42 Braun C, Heinz U, Schweizer R, Wiech K, Birbaumer N, Topka H (2001) Dynamic organization of the somatosensory cortex induced by motor activity. *Brain* 124: 2259–2267

43 Lim V, Bradshaw JL, Nicholls ME, Altenmüller E (2004) Abnormal sensorimotor processing in pianists with focal dystonia. *Adv Neurol* 94: 267–273

44 Schenk T, Mai N (2001) Is writer’s cramp caused by a deficit of sensorimotor integration? *Exp Brain Res* 136: 321–330

45 Bara-Jimenez W, Catalan MJ, Hallett M, Gerloff C (1998) Abnormal somatosensory homunculus in dystonia of the hand. *Ann Neurol* 44: 828–831

46 Byl N, Merzenich MM, Jenkins WM (1996) A primate genesis model of focal dystonia and repetitive strain injury: I. Learning-induced dedifferentiation of the representation of the hand in the primary somatosensory cortex in adult monkeys. *Neurol* 47: 508–520

47 Hallett M (2006) Pathophysiology of writer’s cramp. *Hum Mov Sci* 25: 454–463

48 Marsden C, Sheehy MP (1990) Writer’s cramp. *Trends Neurosci* 13: 148–153

49 Newmark J (1999) Musicians’ dystonia: the case of Gary Graffman. *Semin Neurol* 19 (Suppl 1): 41–45

50 Elbert T, Candia V, Altenmüller E, Rau H, Sterr A, Rockstroh B, Pantev C, Taub E (1998) Alteration of digital representations in somatosensory cortex in focal hand dystonia. *Neuroreport* 9: 3571–3575

51 Lederman R (1999) Robert Schumann. *Semin Neurol* 19 Supplement 1: 17–24

52 Ibáñez V, Sadato N, Karp B, Deiber MP, Hallett M (1999) Deficient activation of the motor cortical network in patients with writer’s cramp. *Neurol* 53: 96–105

53 Byl N, Merzenich MM, Cheung S, Bedenbaugh P, Nagarajan SS, Jenkins WM (1997) A primate model for studying focal dystonia and repetitive strain injury: effects on the primary somatosensory cortex. *Phys Ther* 77: 269–284

54 Byl N (2004) Focal hand dystonia may result from aberrant neuroplasticity. *Adv Neurol* 94: 19–28

55 Byl N (2007) Learning-based animal models: task-specific focal hand dystonia. *In Lab Anim Res* J 48: 411–431

56 Braun C, Schweizer R, Heinz U, Wiech K, Birbaumer N, Topka H (2003) Task-specific plasticity of somatosensory cortex in patients with writer’s cramp. *Neuroimage* 20: 1329–1338

57 Braun C, Schweizer R, Elbert T, Birbaumer N, Taub E (2000) Differential activation in somatosensory cortex for different discrimination tasks. *J Neurosci* 20: 446–450

58 Schweizer R, Maier M, Braun C, Birbaumer N (2000) Distribution of mislocalizations of tactile stimuli on the fingers of the human hand. *Somatosens Mot Res* 17: 309–316

59 Buchner H, Richrath P, Grunholz J, Noppeney U, Waberski TD, Gobbele R, Willmes K, Treede RD (2000) Differential effects of pain and spatial attention on digit representation in the human primary somatosensory cortex. *Neuroreport* 11: 1289–1293

60 Iguchi Y, Hoshi Y, Hashimoto I (2001) Selective spatial attention induces short-term plasticity in human somatosensory cortex. *Neuroreport* 12: 3133–3136

61 Braun C, Haug M, Wiech K, Birbaumer N, Elbert T, Roberts LE (2002) Functional organization of pri-

mary somatosensory cortex depends on the focus of attention. *Neuroimage* 17: 1451–1458
62 Schaefer M, Heinze HJ, Rotte M (2005) Task-relevant modulation of primary somatosensory cortex suggests a prefrontal-cortical sensory gating system. *Neuroimage* 27: 130–135
63 Holmes N, Spence C (2007) The body schema and multisensory representation(s) of peripersonal space. *Cog Process* 5: 94–105
64 Schaefer M, Rothemund Y, Heinze HJ, Rotte M (2004) Short-term plasticity of the primary somatosensory cortex during tool use. *Neuroreport* 15: 1293–1297
65 Pons T (1996) Novel sensations in the congenitally blind. *Nature* 380: 479–480
66 Sadato N, Pascual-Leone A, Grafman J, Ibáñez V, Deiber MP, Dold G, Hallett M (1996) Activation of the primary visual cortx by Braille reading in blind subjects. *Nature* 380: 526–528
67 Uhl F, Franzen P, Podreka I, Steiner M, Deecke L (1993) Increased regional cerebral blood flow in inferior occipital cortex in early blind persons. *Neurosci Lett* 150: 162–164
68 Uhl F, Franzen P, Lindinger G, Lang W, Deecke L (1991) On the functionality of the visually deprived occipital cortex in early blind persons. *Neurosci Lett* 124: 256–259
69 Rauschecker J (1995) Compensatory plasticity and sensory substitution in the cerebral cortex. *Trends Neurosci* 18: 36–43
70 Cohen L et al (1997) Functional relevance of cross-modal plasticity in blind humans. *Nature* 389: 180–183
71 Ptito M, Fumual A, de Noordhout AM, Schoenen J, Gjedde A, Kupers R (2007) TMS of the occipital cortex induces tactile sensations in the fingers of blind Braille readers. *Exp Brain Res* Epub ahead of print
72 Gerloff C, Braun C, Staudt M, Li Hegner Y, Dichgans J, Krägeloh-Mann I (2006) Coherent corticomuscular oscillations originate from primary motor cortex: evidence from patients with early brain lesions. *Hum Brain Map* 27: 789–798
73 Staudt M, Braun C, Gerloff C, Erb M, Grodd W, Krägeloh-Mann I (2006) Developing somatosensory projections bypass periventricular brain lesions. *Neurol* 67: 522–525
74 Hickmott PW (2005) Changes in intrinsic properties of pyramidal neurons in adult rat S1 during cortical reorganization. *J Neurophysiol* 94: 501–511
75 Carew T, Hawkins RD, Kandel ER (1983) Differential classical conditioning of a defensive withdrawal reflex in Aplysia californica. *Science* 219: 397–400
76 Hawkins R, Abrams TW, Carew TJ, Kandel ER (1983) A cellular mechanism of classical conditioning in Aplysia: activity-dependent amplification of presynaptic facilitation. *Science* 219: 400–405
77 Moore C, Nelson SB (1998) Spatio-temporal subthreshold receptive fields in the vibrissa representation of rat primary somatosensory cortex. *J Neurophysiol* 80: 2882–2892
78 Sheth B, Moore CI, Sur M (1998) Temporal modulation of spatial borders in rat barrel cortex. *J Neurophysiol* 79: 464–470
79 Dykes R (1997) Mechanisms controlling neuronal plasticity in somatosensory cortex. *Can J Physiol Pharm* 75: 535–545
80 Calford M, Tweedale R (1988) Immediate and chronic changes in responses of somatosensory cortex in adult flying-fox after digit amputation. *Nature* 332: 446–448
81 Jones E (1993) GABAergic neurons and their role in cortical plasticity in primates. *Cereb Cortex* 3: 361–372
82 Garraghty P, LaChica EA, Kaas JH (1991) Injury-induced reorganization of somatosensory cortex is accompanied by reductions in GABA staining. *Somatosens Mot Res* 8: 347–354
83 Dykes R, Landry P, Metherate R, Hicks TP (1984) Functional role of GABA in cat primary somatosensory cortex: shaping receptive fields. *J Neurophysiol* 52: 1066–1093
84 Calford M (2002) Mechanisms for acute changes in sensory maps. *Adv Exp Med Biol* 508: 451–460
85 Lømo T (1966) Frequency potentiation of excitatory synaptic activity in the dentate area of the hippocampal formation. *Acta Physiol Scand* 68 (Suppl 277): 128

11 Haptic perception in the human foetus

1 Hepper PG (1992) Fetal psychology: an embryonic science. In: JG Nijhuis (ed) *Fetal behaviour. Devel-*

opmental and perinatal aspects. Oxford University, Oxford, 129–156

2 Kurjak A, Carrera JM, Medic M, Azumendi G, Andonotopo W, Stanojevic M (2005) The antenatal development of fetal behavioral patterns assessed by four-dimensional sonography. *J Maternal-Fet Neonat Med* 17: 401–416

3 de Vries JIP, Visser GHA, Prechtl HFR (1985) The emergence of fetal behaviour. II. quantitative aspects. *Ear Hum Dev* 12: 99–120

4 Allister L, Lester BEM, Carr S, Liu J (2001) The effects of maternal depression on fetal heart rate response to vibroacoustic stimulation. *Dev Neuropsych* 20: 639–651

5 Groome LJ, Mooney DM, Holland SB, Bentz LS, Atterbury JL, Dykman RA (1997) The heart rate deceleratory response in low-risk human fetuses: Effect of stimulus intensity on response topography. *Dev Psychobiol* 30: 103–113

6 Hepper PG, Shahidullah S (1994) The development of fetal hearing. *Fet Mat Med Rev* 6: 167–179

7 Hepper PG, Scott D, Shahidullah S (1993) Newborn and fetal response to maternal voice. *J Reprod Inf Psych* 11: 147–153

8 Lecanuet J-P, Granier-Deferre C, Jacquet A-Y, Busnel M-C (1992) Decelerative cardiac responsiveness to acoustical stimulation in the near term foetus. *Q J of Exp Psych* 44B: 279–303

9 Moore RJ, Vadeyar S, Fulford J, Tyler DJ, Gribben C, Baker PN, James D, Gowland PA (2001) Antenatal determination of fetal brain activity in response to an acoustic stimulus using functional magnetic resonance imaging. *Hum Brain Map* 12: 94–99

10 Hepper PG, Shahidullah S (1994) *Noise and the fetus: A critical review of the literature*. HSE Books, Sudbury, Suffolk

11 Hepper PG, Shahidullah S (1994) Development of fetal hearing. *Arch Dis Child* 71: F81–F87

12 Hepper PG (1991) An examination of fetal learning before and after birth. *Irish J Psych* 12: 95–107

13 Moon CM, Fifer WP (2000) The fetus: evidence of transnatal auditory learning. *J Perinatol* 20: S37–S44

14 Querleu D, Renard X, Versyp F, Paris-Delrue L, Crepin G (1988) Fetal hearing. *Eur J Obstet Gynecol Reprod Biol* 29: 191–212

15 Maeda K, Tatsumura M (1993) Antepartum development of fetal behavior and fetal sensitivity to acoustic and photic stimuli: actocardiographic studies. *Asian Med J* 36: 277–288

16 Peleg D, Goldman JA (1980) Fetal heart rate acceleration in response to light stimulation as a clinical measure of fetal well-being. A preliminary report. *J Perinat Med* 8: 38–41

17 Polishuk WZ, Laufer N, Sadovsky E (1975) Fetal reaction to external light. *Harefuah* 89: 395

18 Schaal B, Hummel T, Soussignan R (2004) Olfaction in the fetal and premature infant: functional status and clinical implications. *Clin Perinatol* 31: 261–285

19 De Snoo K (1937) Das trinkende Kind im Uterus. *Monatsschr Geburtsh Gynaekol* 105: 88–97

20 Liley AW (1972) The foetus as a personality. *Aust NZ J Psychiat* 6: 99–105

21 Hepper PG (1995) Human fetal "olfactory" learning. *Int J Pre- Perinatal Psychol Med* 7: 147–151

22 Mennella JA, Jagnow CP, Beauchamp GK (2001) Prenatal and postnatal flavor learning by human infants. *Pediat* 107: e88

23 Schaal B, Marlier L, Soussignan R (2000) Human foetuses learn odours from their pregnant mother's diet. *Chemical Senses* 25: 729–737

24 Mennella JA, Johnson A, Beauchamp GK (1995) Garlic ingestion by pregnant women alters the odor of amniotic fluid. *Chem Sens* 20: 207–209

25 Schaal B, Orgeur P, Rognon C (1995) Odor sensing in the human fetus: Anatomical, functional, and chemoecological bases. In: J-P Lecanuet, WP Fifer, NA Krasnegor, WP Smotherman (eds): *Fetal development. A psychobiological perspective*. LEA, Hillsdale, NJ, 205–237

26 Hooker D (1942) Fetal reflexes and instinctual processes. *Psychosom Med* 4: 199–205

27 Baxi LV, Randolph P, Miller K (1988) Fetal heart rate response to intrauterine saline solution flush. *Am J Obstet Gynecol* 159: 547–549

28 Ververs IAP, de Vries JIP, van Geijn HP, Hopkins B (1994) Prenatal head position from 12–38 weeks. I. Developmental aspects. *Ear Hum Dev* 39: 83–91

29 Ververs IAP, de Vries JIP, van Geijn HP, Hopkins B (1994) Prenatal head position from 12–38 weeks. II. The effects of fetal orientation and placental localization. *Ear Hum Dev* 39: 93–100

30 Ververs IAP, van Gelder-Hasker MR, de Vries JIP,

Hopkins B, van Geijn HP (1998) Prenatal development of arm posture. *Ear Hum Dev* 51: 61–70
31 Hooker D (1952) *The prenatal origin of behavior.* University of Kansas Press, Kansas
32 Reynolds SRM (1962) Nature of fetal adaptation to the uterine environment: a problem of sensory deprivation. *Am J Obstet Gynecol* 83: 800–808
33 Glover V, Fisk NM (1999) Fetal pain: Implications for research and practice. *Brit J Obstet Gynaecol* 106: 881–889
34 Hill LM, Platt LD, Manning FA (1979) Immediate effect of amniocentesis on fetal breathing and gross body movements. *Am J Obstet Gynecol* 135: 689–690
35 Ron M, Yaffe H, Polishuk WZ (1976) Fetal heart rate response to amniocentesis in cases of decreased fetal movements. *Obstet Gynecol* 48: 456–459
36 Giannakoulopoulos X, Sepulveda W, Kourtis P, Glover V, Fisk NM (1994) Fetal plasma cortisol and beta-endorphin response to intrauterine needling. *Lancet* 344: 77–81
37 Kurjak A, Azumendi G, Vecek N, Kupesic S, Solak M, Varga D, Chervenak F (2003) Fetal hand movements and facial expression in normal pregnancy studied by four-dimensional sonography. *J Perinat Med* 31: 496–508
38 De Vries JIP, Wimmers RH, Ververs IAP, Hopkins B, Savelsbergh GJP, van Geijn HP (2001) Fetal handedness and head position preference: a developmental study. *Dev Psychobiol* 39: 171–178
39 Arabin B, Gembruch U, van Eyck J (1993) Registration of fetal behaviour in multiple pregnancy. *J Perinat Med* 21: 285–294
40 Gallagher MW, Costigan K, Johnson TRB (1992) Fetal heart rate accelerations, fetal movement, and fetal behavior patterns in twin gestations. *Am J Obstet Gynecol* 167: 1140–1144
41 Sadovsky E, Ohel G, Simon A (1987) Ultrasonographical evaluation of the incidence of simultaneous and independent movements in twin fetuses. *Gynecol Obstet Invest* 23: 5–9
42 Samueloff A, Younis JS, Strauss N, Baras M, Sadovsky E (1991) Incidence of spontaneous and evoked fetal movements in the first half of a twin pregnancy. *Gynecol Obstet Invest* 31: 200–203
43 Zimmer EZ, Goldstein I, Alglay S (1988) Simultaneous recording of fetal breathing movements and body movements in twin pregnancy. *J Perinat Med* 16: 109–112
44 Piontelli A, Bocconi L, Kustermann A, Tassis B, Zoppini C, Nicolini U (1997) Patterns of evoked behaviour in twin pregnancies during the first 22 weeks of gestation. *Ear Hum Dev* 50: 39–45
45 Hepper PG, Shahidullah S, White RG (1991) Handedness in the human fetus. *Neuropsycholog* 29: 1107–1111
46 Hepper PG, McCartney GR, Shannon EA (1998) Lateralised behaviour in first trimester human foetuses. *Neuropsycholog* 36: 531–534
47 Hepper PG, Wells DL, Lynch C (2005) Prenatal thumb sucking is related to postnatal handedness. *Neuropsychol* 43: 313–315
48 McCartney GR, Hepper PG (1999) Development of lateralized behaviour in the human fetus from 12 to 27 weeks' gestation. *Dev Med Child Neurol* 41: 83–86
49 McManus C (2002) *Right hand, left hand. The origins of asymmetry in brains, bodies atoms and cultures*. Weidenfeld & Nicolson, London
50 Fox K, Wong ROL (2005) A comparison of experience-dependent plasticity in the visual and somatosensory systems. *Neuron* 48: 465–477

12 Haptic behavior in social interaction

1 Sachs F (1988) The intimate sense. *Sciences* 28(1): 28–34
2 Andersen PA, Guerrero LK, Jones SM (2006) Nonverbal intimacy. In: V Manusov, ML Patterson (eds): *The Sage handbook of nonverbal communication.* Sage, Thousand Oaks, CA, 259–277
3 Floyd K (2006) *Communicating affection: Interpersonal behavior and social context.* Cambridge University Press, Cambridge, UK
4 Guerrero LK, Floyd K (2006) *Nonverbal Communication in Close Relationships*. Erlbaum, Mahwah, NJ
5 Fisher JD, Rytting M, Heslin R (1976) Hands touching hands: Affective and evaluative effects of an interpersonal touch. *Sociometry* 39: 416–421
6 Frank LK (1971) Tactile communication. *Genetic Psychology Monographs* 56: 204–255

7 Montagu A (1978) *Touching: The human significance of the skin*. Harper & Row, New York
8 Reite M (1990) Touch, attachment, and health: Is there a relationship? In: KE Barnard, TB Brazelton (eds): *Touch: The foundation of experience*. International Universities Press, Madison, CT, 195–225
9 Moszkowski RJ, Stacks DM (2007) Infant touching behaviour during mother-infant face-to-face interactions. *Infant & Child Dev* 16: 307–319
10 Harlow HF (1958) The nature of love. *Am Psychologist* 13: 673–685
11 Harlow HF, Harlow MK, Hansen EW (1963) The maternal affectional system of rhesus monkeys. In HL Rhinegold (ed): *Maternal behaviors in mammals.* Wiley, New York, 254 –281
12 Harlow HF, Zimmerman RR (1958) The development of affectional responses in infant monkeys. *Proc Am Phil Soc* 102: 501–509
13 Spitz RA (1945) Hospitalism: An inquiry into the genesis of psychiatric conditions in early childhood. *Psychoanalytic Study of the Child* 1: 53–74
14 Spitz RA (1946) Hospitalism: A follow-up report on investigation described in Volume 1, 1945. *Psychoanalytic Study of the Child* 1: 113–117
15 Money J, Annecillo C (1976) IQ changes following change in domicile in the syndrome of reversible hyposomatotropinism (psychosocial dwarfism): Pilot study. *Psychoneuroendocrinology* 1: 427–429
16 Rutter M (1998) Developmental catch-up and deficit following adoption after severe global early privation. *J Child Psychology and Psychiatry* 39: 465–476
17 Rutter M, Andersen-Wood L, Beckett C, Bredenkamp D, Castel J, Groothenus C, Deppner J, Keaveny L, Lord C, O'Connor TG (1999) Quasi-autistic patterns following severe early global privation. *J Child Psychology and Psychiatry* 40: 537–549
18 Dennis W (1973) *Chldren of the creche.* Appleton-Century-Croft, New York
19 Gerhart S (2004) *Why love matters: How affection shapes a baby's brain.* Brunner-Routledge, New York
20 Perry BD (2002) Childhood experience and the expression of genetic potential: What childhood neglect tells us about nature and nurture. *Brain and Mind* 3: 79–100
21 Honig AS (2005) Take time to touch. *Scholastic Parent & Child* 12(5): 32–34
22 Rymer R (1994) *Genie: A scientific tragedy*. Harper Collins, New York
23 Weiss SJ, Wilson P, Morrison D (2004) Maternal tactile stimulation and the neurodevelopment of low birth weight infants. *Infancy* 5: 85–107
24 Caulfield R (2000) Beneficial effects of tactile stimulation on early development. *Early Childhood Education Journal* 27: 255–257
25 Bowlby J (1969) *Attachment and loss:* Vol. 1. *Attachment.* Basic Books, New York
26 Bowlby J (1973) *Attachment and loss*: Vol 2. *Separation*. Basic Books, New York
27 Ainsworth MDS, Bowlby J (1991) An ethological approach to personality development. *Am Psychologist* 46: 333–341
28 Ainsworth MDS, Blehar MC, Waters E, Wall S (1978) *Patterns of attachment: A psychological study of the strange situation*. Lawrence Erlbaum, Hillsdale, NJ
29 Bartholomew K, Horowitz LM (1991) Attachment styles among young adults: A test of a four-category model. *J Personality and Social Psychol* 61: 226–244
30 Carlson EA (1998) A prospective longitudinal study of attachment disorganization/ disorientation. *Child Dev* 69: 1107–1128
31 Perry BD (2002) Childhood experience and the expression of genetic potential: What childhood neglect tells us about nature and nurture. *Brain and Mind* 3: 79–100
32 Davila J, Karney BR, Bradbury TN (1999) Attachment change processes in the early years of marriage. *J Personality and Social Psychol* 76: 783–802
33 Guerrero LK (1996) Attachment-style difference in intimacy and involvement: A test of the four-category model. *Communication Monographs* 63: 269–292
34 Tucker JS, Anders SL (1998) Adult attachment style and nonverbal closeness in dating couples. *J Nonverbal Behavior* 22: 109–124
35 Goleman D (7 May 1991) *Kids who got hugs found to be happy adults*. Tucson: Arizona Daily Star, 1C, 3C
36 Jones SE, Brown BC (1996) Touch attitudes and behaviors, recollections of early childhood touch,

and social self-confidence. *J Nonverbal Behavior* 20: 147–163

37 Jones SE, Yarbrough AE (1985) A naturalistic study of the meanings of touch. *Communication Monographs* 52: 19–56

38 McEwan B, Johnson SL (2008) Relational violence: The darkest side of haptic communication. In: LK Guerrero, ML Hecht (eds): *The nonverbal communication reader.* Waveland Press, Prospect Heights, IL, 232–241

39 Andersen PA (1998) The cognitive valence theory of intimate communication. In: MT Palmer, GA Barnett (eds): *Progress In Communication Sciences, Volume XIV: Mutual Influence in Interpersonal Communication: Theory and Research In Cognition, Affect, and Behavior.* Ablex, Stamford, CT, 39–72

40 Stafford L (2003) Maintaining romantic relationships: Summary and analysis of one research program. In: DJ Canary, M Dainton (eds): *Maintaining relationships through communication: Relational, contextual, and cultural variations.* Lawrence Erlbaum Associates, Mahwah, NJ, 51–77

41 Floyd K (2002) Human affection exchange V: Attributes of the highly affectionate. *Communication Quarterly* 50: 135–152

42 Floyd K, Hess JA, Mizco LA, Halone KK, Mikkelson AC, Tusing KJ (2005) Human affective exchange: VIII. Further evidence of the benefits of expressed affection. *Communication Quarterly* 53: 285–303

43 Weiss SJ (1990) Effects of differential touch on nervous system arousal of patients recovering from cardiac disease. *Heart and Lung* 19: 474–480

44 Whitcher SJ, Fisher JD (1979) Multidimensional reaction to therapeutic touch in a hospital setting. *J Personality and Social Psychol* 37: 87–96

45 Dolin D, Booth-Butterfield M (1993) Reach out and touch someone: Analysis of nonverbal comforting responses. *Communication Quarterly* 41: 383–393

46 Christopher FS, Lloyd SA (2001) Physical and sexual aggression in relationships. In: C Hendrick, SS Hendrick (eds): *Close relationships.* Sage, Thousand Oaks, CA, 331–343

47 Marshall LL (1994) Physical and psychological abuse. In: WR Cupach, BH Spitzberg (eds): *The dark side of interpersonal communication.* Lawrence Erlbaum Associates, Hillsdale, NJ, 281–311

48 Johnson MP (1995) Patriarchal terrorism and common couple violence: Two forms of violence against women. *J Marriage and Family* 57: 283–294

49 Johnson MP, Ferraro KJ (2000) Research on domestic violence in the 1990s: Making distinctions. *J Marriage and Family* 62: 948–963

50 Henley NM (1973) Status and sex: Some touching observations. *Bulletin of the Psychonomic Society* 2: 91–93

51 Major B, Schmidlin AM, Williams L (1990) Gender patterns in social touch: The impact of setting and age. *J Personality and Social Psychol* 58: 634–643

52 Burgoon JK, Bacue AE (2003) Nonverbal communication skills. In: J Greene, BR Burleson (eds): *Handbook of communication and social interaction skills.* Lawrence Erlbaum Associates, Mahwah, NJ, 179–219

53 Guerrero LK, Andersen PA (1994) Patterns of matching and initiation: Touch behavior and avoidance across romantic relationship stages. *J Nonverbal Behavior* 18: 137–153

54 Willis EN, Briggs LE (1992) Relationship and touch in public settings. *J Nonverbal Behavior* 16: 55–62

55 Willis FN, Dodds RA (1998) Age, relationship, and touch initiation. *J Social Psychol* 138: 115–123

56 Hall JA, Veccia EM (1990) More "touching" observations: New insights on men, women, and interpersonal touch. *J Personality and Social Psychol* 59: 1155–1162

57 Guerrero LK, Andersen PA (1991) The waxing and waning of relational intimacy: Touch as a function of relational stage, gender and touch avoidance. *J Social and Personal Relationships* 8: 147–165

58 McDaniel ER, Andersen PA (1998) Intercultural variations in tactile communication. *J Nonverbal Communication* 22: 59–75

59 Emmers TM, Dindia K (1995) The effect of relational stage and intimacy on touch: An extension of Guerrero and Andersen. *Personal Relationships* 2: 225–236

60 Andersen PA, Leibowitz K (1978) The development and nature of the construct touch avoidance. *Env Psychol and Nonverbal Behavior* 3: 89–106

61 Hall ET (1966) *The hidden dimension* (2nd ed). Anchor/Doubleday, Garden City, NY

62 Miller PM, Commons ML, Gutheil TG (2006) Clinicians' perceptions of boundaries in Brazil and the

United States. *J American Academy of Psychiatric Law* 34: 33–41
63 Samovar LA, Porter RE (2001) *Communication between cultures*. Wadsworth, Belmont, CA
64 Albert RD, Ah Ha I (2004) Latino/Anglo-American differences in attributions to situations involving touch and silence. *Int J Intercultural Relations* 28: 353–280
65 Jones SE (1994) *The right touch: Understanding and using the language of physical contact*. Hampton Press, Cresshill, NJ
66 Heslin R, Alper T (1983) Touch: A bonding gesture. In: JM Wiemann, R Harrison (eds): *Non-verbal Interaction*. Sage, Beverly Hills, CA, 47–75
67 Remland MS, Jones TS (1988) Cultural and sex differences in touch avoidance. *Perceptual and Motor Skills* 67: 544–546
68 Andersen PA, Hecht ML, Hoobler GD, Smallwood M (2002) Nonverbal communication across culture. In: B Gudykunst, B Mody (eds): *Handbook of International and Intercultural Communication*. Sage, Thousand Oaks, CA, 89–106
69 Andersen PA, Wang H (2006) Unraveling cultural cues: Dimensions of nonverbal communication across cultures. In: LA Samovar, RE Porter, ER McDaniel (eds): *Intercultural Communication: A reader*. Wadsworth, Belmont, CA, 250–266
70 Condon JC, Yousef F (1983) *An introduction to intercultural communication*. Bobbs-Merrill, Indianapolis, IN
71 Jones SE, Remland MS (1982) *Cross cultural differences in self-reported touch avoidance*. Paper presented at the annual convention of the Eastern Communication Association, Hartford, CT
72 Hofstede G (1982) *Culture's consequences*. (Abridged Edition). Sage, Beverly Hills, CA
73 Andersen PA, Lustig MW, Andersen JF (1990) Changes in latitude, changes in attitude: The relationship between climate and interpersonal communication predispositions. *Communication Quarterly* 38: 291–311
74 Pennebaker JW, Rimé B, Blankenship J (1996) Stereotypes of emotional expressiveness of Northerners and Southerners: A cross-cultural test of Montesquieu's hypotheses. *J Personality and Social Psychol* 70: 372–380
75 Jourard SM (1966) An exploratory study of body-accessibility. *Br J Social and Clin Psychol* 5: 221–231
76 Jourard SM, Rubin JE (1968) Self-disclosure and touching: A study of two modes of interpersonal encounter and their inter-relation. *J Humanistic Psychol* 8: 39–48
77 Leibowitz K, Andersen PA (December 1976) *The development and nature of the construct touch avoidance*. Paper presented at the annual meeting of the Speech Communication Association, San Francisco, CA
78 Sorensen G, Beatty MJ (1988) The interactive effects of touch and touch avoidance on interpersonal perceptions. *Communication Research Reports* 5: 84–90
79 Andersen PA, Sull KK (1985) Out of touch, out of reach: Tactile predispositions as predictors of interpersonal distance. *The Western J Speech Communication* 49: 57–72
80 Andersen JF, Andersen PA, Jensen AD (1979) The measurement of nonverbal immediacy. *J Applied Communication Res* 7: 153–180
81 Andersen PA (1985) Nonverbal immediacy in interpersonal communication. In: AW Siegman, S Feldstein (eds): *Multichanneled Integrations of Nonverbal Behavior*. Erlbaum, Hillsdale, NJ, 1–36
82 Fromme DK, Jaynes WE, Taylor DK, Harold EG, Daniell J, Rountree JR, Fromme ML (1989) Nonverbal behavior and attitudes toward touch. *J Nonverbal Behavior* 13: 3–14
83 Mehrabian A (1971) *Silent messages*. Wadsworth, Belmont, CA
84 Andersen JF, Andersen PA, Lustig MW (1987) Opposite-sex touch avoidance: A national replication and extension. *J Nonverbal Behavior* 11: 89–109
85 Ciceraro LDL, Andersen PA (February 2008) *The Effect of Touch Avoidance on Relational Satisfaction in Romantic Dyads*. Paper presented at the Western States Communication Association Convention, Denver, CO
86 Lower HM (1980) Fear of touching as a form of communication apprehension in professional nursing students. *Australian Scan: Journal of Human Communication* 7–8: 71–78
87 Andersen PA (2005) The Touch Avoidance Measure. In: V Manusov (ed): *The Sourcebook of Non-*

verbal Measures: Going Beyond Words. Erlbaum, Englewood Cliffs, NJ, 57–65
88 Floyd K (2000) Affectionate same-sex touch: Understanding influences on observers' perceptions. *J Social Psychol* 140: 774–788
89 Guerrero LK, Andersen PA, Afifi W (2007) *Close Encounters: Communication in Relationships*. Sage, Thousand Oaks, CA
90 Crawford CB (1994) Effects of sex and sex roles on avoidance of same- and opposite-sex touch. *Perceptual and Motor Skills* 79: 107–112
91 Eman VA, Dierks-Stewart KJ, Tucker RK (November 1978) *Implications of sexual identity and sexually identified situations on nonverbal touch*. Paper presented at the annual meeting of the Speech Communication Association, Minneapolis, MN
92 Andersen PA (2004) *The complete idiot's guide to body language*. Alpha Books, Indianapolis, Indiana
93 Andersen PA (2008) *Nonverbal communication: Forms and functions* (Second edition). Waveland Press, Long Grove, Illinois

13 Learning effects in haptic perception

1 Gibson EJ (1953) Improvement in perceptual judgements as a function of controlled practice or training. *Psychol B* 50: 401–431
2 Gibson JJ (1966) *The Senses Considered As Perceptual Systems*. Houghton Mifflin Co
3 Novak CB, Patterson JM, Mackinnon SE (1999) Evaluation of hand sensibility with single and double latex gloves. *Plast Reconstr Surg* 103: 128–131
4 Rabin E, Gordon AM (2004) Tactile feedback contributes to consistency of finger movements during typing. *Exp Brain Res* 155: 362–369
5 Goldstone RL (1998) Perceptual learning. *Ann Rev Psychol* 49: 585–612
6 Fahle M, Poggio T (2002) *Perceptual Learning*. MIT Press
7 Doane SM, Alderton DL, Sohn YW, Pellegrino JW (1996) Acquisition and transfer of skilled performance: Are visual discrimination skills stimulus specific? *J Exp Psychol* Gen 22: 1218–1248
8 Sagi D, Tanne D (1994) Perceptual learning: learning to see. *Curr Opin Neurobiol* 4: 195–199
9 Fahle M (2005) Perceptual learning: specificity *versus* generalization. *Curr Opin Neurobiol* 15: 154–160
10 Karni A, Sagi D (1993) The time course of learning a visual skill. *Nature* 365: 250–252
11 Eckstein MP, Abbey CK, Pham BT, Shimozaki SS (2004) Perceptual learning through optimization of attentional weighting: human *versus* optimal Bayesian learner. *J Vis* 4: 1006–1019
12 Recanzone GH, Jenkins WM, Hradek GT, Merzenich MM (1992) Progressive improvement in discriminative abilities in adult owl monkeys performing a tactile frequency discrimination task. *J Neurophysiol* 67: 1015–1030
13 Ahissar M, Hochstein S (2004) The reverse hierarchy theory of visual perceptual learning. *Trends Cogn Sci* 8: 457–464
14 Wright BA, Sabin AT (2007) Perceptual learning: how much daily training is enough? *Exp Brain Res* [Epub ahead of print]
15 Kalisch T, Dinse HR (submitted) The age-related decrease of cross-modal visuo-haptic performance reveals gender-specific differences.
16 Karni A, Sagi D (1991) Where practice makes perfect in texture discrimination: evidence for primary visual cortex plasticity. *Proc Natl Acad Sci USA* 88: 4966–4970
17 Schoups AA, Vogels R, Orban GA (1995) Human perceptual learning in identifying the oblique orientation: retinotopy, orientation specificity and monocularity. *J Physiol* 483: 797–810
18 Crist RE, Kapadia MK, Westheimer G, Gilbert CD (1997) Perceptual learning of spatial localization: specificity for orientation, position, and context. *J Neurophysiol* 78: 2889–2894
19 Fahle M (1997) Specificity of learning curvature, orientation, and vernier discriminations. *Vision Res* 37: 1885–1895
20 Herzog MH, Fahle M (1997) The role of feedback in learning a vernier discrimination task. *Vision Res* 37: 2133–2141
21 Fahle M, Edelman S, Poggio T (1995) Fast perceptual learning in hyperacuity. *Vision Res* 35: 3003–3013
22 Seitz AR, Nanez JE Sr, Holloway S, Tsushima Y, Watanabe T (2006) Two cases requiring external reinforcement in perceptual learning. *J Vis* 6: 966–973

23 Tricomi E, Delgado MR, McCandliss BD, McClelland JL, Fiez JA (2006) Performance feedback drives caudate activation in a phonological learning task. *J Cogn Neurosci* 18: 1029–1043

24 Polley DB, Steinberg EE, Merzenich MM (2006) Perceptual learning directs auditory cortical map reorganization through top-down influences. *J Neurosci* 26: 4970–4982

25 Desimone R, Duncan J (1995) Neural mechanisms of selective visual attention. *Ann Rev Neurosci* 18: 193–222

26 LaBerge D (2002) Attentional control: Brief and prolonged. *Psychol Res* 66: 220–233

27 Shipp S (2004) The brain circuitry of attention. *Trends Cog Sci* 8: 223–230

28 Sarter M, Bruno JP (1997) Cognitive functions of cortical acetylcholine: toward a unifying hypothesis. *Brain Res Rev* 23: 28–46

29 Sarter M, Gehring WJ, Kozak R (2006) More attention must be paid: The neurobiology of attentional effort. *Brain Res Rev* 51: 145–160

30 Seitz A, Watanabe T (2005) An unified model for perceptual learning. *Trends Cogn Sci* 9: 329–334

31 Seitz AR, Dinse HR (2007) A common framework for perceptual learning. *Curr Opin Neurobiol* 17: 148–153

32 Dosher BA, Lu ZL (1998) Perceptual learning reflects external noise filtering and internal noise reduction through channel reweighting. *Proc Natl Acad Sci USA* 95: 13988–13993

33 Gold J, Bennett PJ, Sekuler AB (1999) Signal but not noise changes with perceptual learning. *Nature* 402: 176–178

34 Sathian K, Zangaladze A (1997) Tactile learning is task specific but transfers between fingers. *Percept Psychophys* 59: 119–128

35 Sathian K, Zangaladze A (1998) Perceptual learning in tactile hyperacuity: complete intermanual transfer but limited retention. *Exp Brain Res* 118: 131–234

36 Nagarajan SS, Blake DT, Wright BA, Byl N, Merzenich MM (1998) Practice-related improvements in somatosensory interval discrimination are temporally specific but generalize across skin location, hemisphere, and modality. *J Neurosci* 18: 1559–1570

37 Simmons RW, Locher PJ (1979) Role of extended perceptual experience upon haptic perception of nonrepresentational shapes. *Percept Mot Skills* 48: 987–991

38 Wagman JB, Shockley K, Riley MA, Turvey MT (2001) Attunement, calibration, and exploration in fast haptic perceptual learning. *J Mot Behav* 33: 323–327

39 Lederman SJ, Klatzky RL (1990) Haptic classification of common objects: knowledge-driven exploration. *Cognit Psychol* 22: 421–459

40 Newell FN, Ernst MO, Tjan BS, Bülthoff HH (2001) Viewpoint dependence in visual and haptic object recognition. *Psychol Sci* 12: 37–42

41 Recanzone GH, Merzenich MM, Jenkins WM, Grajski K, Dinse HR (1992) Topographic reorganization of the hand representation in cortical area 3b of owl monkeys trained in a frequency discrimination task. *J Neurophysiol* 67: 1031–1056

42 Pleger B, Dinse HR, Ragert P, Schwenkreis P, Malin JP, Tegenthoff M (2001) Shifts in cortical representations predict human discrimination improvement. *Proc Natl Acad Sci USA* 98: 12255–12260

43 Pleger B, Foerster AF, Ragert P, Dinse HR, Schwenkreis P, Nicolas V, Tegenthoff M (2003) Functional imaging of perceptual learning in human primary and secondary somatosensory cortex. *Neuron* 40: 643–653

44 Dinse HR, Ragert P, Pleger B, Schwenkreis P, Tegenthoff M (2003) Pharmacological modulation of perceptual learning and associated cortical reorganization. *Science* 301: 91–94

45 Goldberg RF, Perfetti CA, Schneider W (2006) Perceptual knowledge retrieval activates sensory brain regions. *J Neurosci* 26: 4917–4921

46 Kennedy KM, Raz N (2005) Age, sex and regional brain volumes predict perceptual-motor skill acquisition. *Cortex* 41: 560–569

47 Pascual-Leone A, Torres F (1993) Plasticity of the sensorimotor cortex representation of the reading finger in Braille readers. *Brain* 116: 39–52

48 Elbert T, Pantev C, Wienbruch C, Rockstroh B, Taub E (1995) Increased cortical representation of the fingers of the left hand in string players. *Science* 270: 305–307

49 Pantev C, Oostenveld R, Engelien A, Ross B, Roberts LE, Hoke M (1998) Increased auditory cortical representation in musicians. *Nature* 392: 811–814

50 Sterr A, Muller MM, Elbert T, Rockstroh B, Pantev C, Taub E (1998) Changed perceptions in Braille readers. *Nature* 391: 134–135
51 Rauschecker JP (2001) Cortical plasticity and music. *Ann NY Acad Sci* 930: 330–336
52 Schlaug G (2001) The brain of musicians: A model for functional and structural plasticity. *Ann NY Acad Sci* 930: 281–299
53 Münte TF, Altenmüller E, Jäncke L (2002) The musician's brain as a model of neuroplasticity. *Nature* Rev *Neurosci* 3: 473–478
54 Ragert P, Pleger B, Völker B, Meier C, Schwenkreis P, Tegenthoff M, Dinse HR (2003) Impaired tactile performance in patients with hand immobilisation. *Soc Neurosci Abstr* 29: 379.13
55 Watanabe T, Nanez JE, Sasaki Y (2001) Perceptual learning without perception. *Nature* 413: 844–848
56 Seitz AR, Watanabe T (2003) Psychophysics: Is subliminal learning really passive? *Nature* 422: 36
57 Seitz A, Lefebvre C, Watanabe T, Jolicoeur P (2005) Requirement for high-level processing in subliminal learning. *Curr Biol* 15: 753–755
58 Seitz AR, Nanez JE, Holloway SR, Koyama S, Watanabe T (2005) Seeing what is not there shows the costs of perceptual learning. *Proc Natl Acad Sci USA* 102: 9080–9085
59 Godde B, Spengler G, Dinse HR (1996) Associative pairing of tactile stimulation induces somatosensory cortical reorganization in rats and humans. *Neuroreport* 8: 281–285
60 Godde B, Stauffenberg B, Spengler F, Dinse HR (2000) Tactile coactivation induced changes in spatial discrimination performance. *J Neurosci* 20: 1597–1604
61 Dinse HR, Merzenich MM (2002) Adaptation of inputs in the somatosensory system. In M Fahle, T Poggio (eds): *Perceptual Learning*. MIT Press, 19–42
62 Godde B, Ehrhardt J, Braun C (2003) Behavioral significance of input-dependent plasticity of human somatosensory cortex. *Neuroreport* 14: 543–546
63 Dinse HR, Ragert P, Pleger B, Schwenkreis P, Tegenthoff M (2003) GABAergic mechanisms gate tactile discrimination learning. *Neuroreport* 14: 1747–1751
64 Hodzic A, Veit R, Karim AA, Erb M, Godde B (2004) Improvement and decline in tactile discrimination behavior after cortical plasticity induced by passive tactile coactivation. *J Neurosci* 24: 442–446
65 Dinse HR, Kalisch T, Ragert P, Pleger B, Schwenkreis P, Tegenthoff M (2005) Improving human haptic performance in normal and impaired human populations through unattended activation-based learning. *Transaction Appl Perc* 2: 71–88
66 Dinse HR, Kleibel N, Kalisch T, Ragert P, Wilimzig C, Tegenthoff M (2006) Tactile coactivation resets age-related decline of human tactile discrimination. *Ann Neurol* 60: 88–94
67 Ragert P, Dinse HR, Pleger B, Wilimzig C, Frombach E, Schwenkreis P, Tegenthoff M (2003) Combination of 5 Hz repetitive transcranial magnetic stimulation (rTMS) and tactile coactivation boosts tactile discrimination in humans. *Neurosci Letters* 348: 105–108
68 Ragert P, Schmid A, Altenmueller E, Dinse HR (2004) Superior tactile performance and learning in professional pianists: Evidence for meta-plasticity in musicians. *Europ J Neurosci* 19: 473–478
69 Ragert P, Kalisch T, Dinse HR (2005) Perceptual changes in human tactile discrimination behavior induced by coactivation using LTP- and LTD-protocols. *Soc Neuroscience Abstr.* 173.6
70 Ragert P, Kalisch T, Bliem B, Franzkowiak S, Dinse HR (2008) Differential effects in human tactile discrimination behavior evoked by tactile high- and low-frequency stimulation.
71 Bliem B, Frombach E, Ragert P, Knossalla F, Woitalla D, Tegenthoff M, Dinse HR (2007) Dopaminergic influences on changes in human tactile acuity induced by tactile coactivation. *Exp Brain Res* 181: 131–137
72 Kalisch T, Tegenthoff M, Dinse HR (2007) Differential effects of synchronous and asynchronous multifinger coactivation on human tactile performance. *BMC Neurosci* 8: 58
73 Kilgard MP, Merzenich MM (1998) Cortical map reorganization enabled by nucleus basalis activity. *Science* 279: 1714–1718
74 Buonomano DV, Merzenich MM (1998) Cortical plasticity: from synapses to maps. *Annu Rev Neurosci* 21: 149–186
75 Hebb DO (1949) *The Organization Of Behaviour.* Wiley

76 Ragert P, Becker M, Tegenthoff M, Pleger B, Dinse HR (2004) Sustained increase of somatosensory cortex (SI) excitability by 5Hz repetitive transcranial magnetic stimulation (rTMS) studied by paired median nerve stimulation. *Neurosci Letters* 356: 91–94
77 Tegenthoff M, Ragert P, Pleger B, Schwenkreis P, Förster AF, Nicolas V, Dinse HR (2005) Persistent improvement of tactile discrimination performance and enlargement of cortical somatosensory maps after 5 Hz rTMS. *PloS Biol* 3: e362
78 Sterr A, Green L, Elbert T (2003) Blind Braille readers mislocate tactile stimuli. *Biol Psychol* 63: 117–127
79 Herron CE, Lester RA, Coan EJ, Collingridge GL (1986) Frequency-dependent involvement of NMDA receptors in the hippocampus: a novel synaptic mechanism. *Nature* 322: 265–268
80 Cotman CW, Monaghan DT, Ganong AH (1988) Excitatory Amino Acid neurotransmission: NMDA receptors and Hebb-type synaptic plasticity. *Annu Rev Neurosci* 11: 61–80
81 Lynch MA (2004) Long-term potentiation and memory. *Physiol Rev* 84: 87–136
82 Parsons CG, Danysz W, Quack G (1999) Memantine is a clinically well tolerated N-methyl-D-aspartate (NMDA) receptor antagonist – a review of preclinical data. *Neuropharm* 38: 735–767
83 Rosenkranz K, Williamon A, Rothwell JC (2007) Motorcortical excitability and synaptic plasticity is enhanced in professional musicians. *J Neurosci* 27: 5200–5206
84 Bock O, Schneider S (2002) Sensorimotor adaptation in young and elderly humans. *Neurosci Biobehav Rev* 26: 761–776
85 Sawaki L, Yaseen Z, Kopylev L, Cohen LG (2003) Age-dependent changes in the ability to encode a novel elementary motor memory. *Ann Neurol* 53: 521–524
86 Kornatz KW, Christou EA, Enoka RM (2005) Practice reduces motor unit discharge variability in a hand muscle and improves manual dexterity in old adults. *J Appl Physiol* 98: 2072–2080
87 Smith CD, Walton A, Loveland AD, Umberger GH, Kryscio RJ, Gash DM (2005) Memories that last in old age: motor skill learning and memory preservation. *Neurobiol Aging* 26: 883–890
88 Celnik P, Stefan K, Hummel F, Duque J, Classen J, Cohen LG (2006) Encoding a motor memory in the older adult by action observation. *Neuroimage* 29: 677–684
89 Dinse HR (2007) Cortical reorganization in the aging brain. *Prog Brain Res* 157: 57–80
90 Heinisch C, Kalisch T, Dinse HR (2006) Tactile and learning abilities in early and late-blind subjects. *Soc Neurosci Abstr* 52.4/N2
91 Dhruv NT, Niemi JB, Harry JD, Lipsitz LA, Collins JJ (2002) Enhancing tactile sensation in older adults with electrical noise stimulation. *Neuroreport* 13: 597–600
92 Priplata AA, Niemi JB, Harry JD, Lipsitz LA, Collins JJ (2003) Vibrating insoles and balance control in elderly people. *Lancet* 362: 1123–1124
93 Floel A, Breitenstein C, Hummel F, Celnik P, Gingert C, Sawaki L, Knecht S, Cohen LG (2005) Dopaminergic influences on formation of a motor memory. *Ann Neurol* 58: 121–130
94 Simon SA, de Araujo IE, Gutierrez R, Nicolelis MA (2006) The neural mechanisms of gustation: a distributed processing code. *Nat Rev Neurosci* 7: 890–901
95 Carmena JM, Lebedev MA, Crist RE, O'Doherty JE, Santucci DM, Dimitrov DF, Patil PG, Henriquez CS, Nicolelis MA (2003) Learning to control a brain-machine interface for reaching and grasping by primates. *PLoS Biol* 1: E42
96 Nicolelis MA, Ghazanfar AA, Stambaugh CR, Oliveira LM, Laubach M, Chapin JK, Nelson RJ, Kaas JH (1998) Simultaneous encoding of tactile information by three primate cortical areas. *Nat Neurosci* 1: 621–630
97 Wilimzig C, Dinse HR (2007) The role of cortical interaction for spatial discrimination, localization and its learning-induced changes. COSYN2007. *Computational and Systems Neuroscience Abstracts* 248
98 Amari S (1977) Dynamics of pattern formation in lateral-inhibition type neural fields. *Biol Cybern* 27: 77–87
99 Wilson HR, Cowan JD (1973) A mathematical theory of the functional dynamics of cortical and thalamic nervous tissue. *Kybernetik* 13: 55–80
100 Dinse HR, Jancke D (2001) Comparative population analysis of cortical representations in parametric

spaces of visual field and skin. *Prog Brain Res* 130: 155–173

101 Pouget A, Dayan P, Zemel RS (2003) Inference and computation with population codes. *Ann Rev Neurosci* 26: 381–410

102 Höffken O, Veit M, Knossalla F, Lissek S, Bliem B, Ragert R, Dinse HR, Tegenthoff M (2007) Sustained increase of somatosensory cortex excitability by tactile coactivation studied by paired median nerve stimulation in humans correlates with perceptual gain. *J Physiol* 584: 463–471

14 Implicit and explicit memory effects in haptic perception

1 Tulving E (1972) Episodic and semantic memory. In: E Tulving, W Donaldson (eds): *Organization and memory.* Academic Press, New York, 381–411

2 Graf P, Schacter DL (1985) Implicit and explicit memory for new associations in normal and amnesic patients. *J Exp Psych: Lear, Mem, and Cogn* 11: 501–518

3 Heller MA (1991) Introduction. In: MA Heller, W Schiff (eds): *The psychology of touch.* Lawrence Erlbaum Associates, Hillsdale, NJ, 1–19

4 Millar S (1999) Memory in touch. *Psicothema* 11: 747–767

5 Ballesteros S, Heller MA (2006) Conclusions: Touch and blindness, psychology and neuroscience. In: MA Heller, S Ballesteros (eds): *Touch and Blindness. Psychology and Neuroscience.* Lawrence Erlbaum Associates, Hillsdale NJ, 197–218

6 Jones LA, Lederman SJ (2006) *Human hand function.* Oxford University Press, New York

7 Klatzky RL, Lederman SJ, Metzger VA (1985) Identifying objects by touch: An "expert system". *Perception & Psychophysics* 37: 299–302

8 Amedi A, Kriegstein K, van Atteveldt NM, Beauchamp MS, Naumer MJ (2005) Functional imaging of human crossmodal identification and object recognition. *Exp Brain Res* 166: 559–571

9 Amedi A, Malach R, Hendler T, Peled S, Zohary E (2001) Visuo-haptic object-related activation in the ventral visual pathway. *Nature Neuros* 4: 324–330

10 James TW, Humphrey GK, Gati JS, Servos P, Menon RS, Goodale MA (2002) Haptic study of three-dimensional objects activates extrastriate visual areas. *Neuropsychologia* 40: 1706–1714

11 Pascual-Leone A, Hamilton R (2001) The metamodal organization of the brain. In: C Casanova, M Ptito (eds): *Progress in Brain Research,* Vol. 134: 1–19

12 Sathian K, Prather SC (2006) Cerebral cortical processing of tactile form: Evidence from functional imaging. In: MA Heller, S Ballesteros (eds): *Touch and Blindness: Psychology and Neuroscience.* Lawrence Erlbaum Associates, Hillsdale, NJ, 157–170

13 Heller MA (1987) Improving the passive tactile digit span. *Bulletin of the Psychonomic Society* 25: 257–258

14 Paz S, Mayas J, Ballesteros S (2007) Haptic and visual working memory in young adults, older healthy adults and mild cognitive impairment adults. WorldHaptics Proceedings (WHC'07) Tsukuba, Japan. *IEEE Computer Society* 553–554

15 Tulving E, Schacter DL (1990) Priming and human memory systems. *Science* 247: 301–306

16 Cooper LA, Schacter DE, Ballesteros S, Moore C (1992) Priming and recognition of transformed three-dimensional objects: Effects of size and reflection. *J Exp Psych: Lear, Mem, Cogn* 18: 43–57

17 Ballesteros S, Reales JM, Mayas J (2007) Picture priming in aging and dementia. *Psicothema* 19: 239–244

18 Biederman I, Cooper EE (1991) Evidence for complete translational and reflectional invariance in visual object priming. *Perception* 20: 585–593

19 Srinivas K (1993) Perceptual specificity in nonverbal priming. *J Exp Psych: Lear, Mem, Cogn* 19: 582–602

20 Kappers AML, Koenderink JJ, Lichtenegger I (1994) Haptic identification of curved surfaces. *Perc & Psych* 5: 53–61

21 Kappers AML, Koenderink JJ, The Pas SF (1994) Haptic discrimination of doubly curved surfaces. *Perception* 23: 1483–1490

22 Ballesteros S, Manga D, Reales JM (1997) Haptic discrimination of bilateral symmetry in two-dimensional and three-dimensional unfamiliar displays. *Perception & Psychophysics* 59: 37–50

23 Ballesteros S, Millar S, Reales JM (1998) Symmetry in haptic and in visual shape perception. *Perception & Psychophysics* 60: 389–404

24 Ballesteros S, Reales JM (2004) Visual and haptic discrimination of symmetry in unfamiliar displays extended in the z-axis. *Perception* 33: 315–327
25 Locher PJ, Simmons RW (1978) Influence of stimulus symmetry and complexity upon haptic scanning strategies during detection, learning, and recognition tasks. *Perception & Psychophysics* 23: 110–116
26 Simmons RW, Locher PJ (1979) The role of extended perceptual experience upon haptic perception of nonrepresentational shapes. *Perceptual & Psychophysics* 48: 887–991
27 Loomis J, Lederman SJ (1986) Tactual perception. In: KR Boff, L Kaufman, JP Thomas (eds): *Handbook of perception and human performance.* Wiley, New York, vol. 2, 31-1–31-44
28 Millar S (1994) *Understanding and representing space: Theory and evidence from studies with blind and sighted children.* Oxford University Press, Oxford
29 Ballesteros S, Reales JM, Manga D (1999) Implicit and explicit memory for familiar and novel objects presented to touch. *Psicothema* 11: 785–800
30 Srinivas K, Greene AJ, Easton RD (1997) Implicit and explicit memory for haptically experienced two-dimensional patters. *Psychological Science* 8: 243–246
31 Mussen G, Treisman A (1990) Implicit memory for visual patterns. *J Exp Psych: Learning, Memory, and Cognition* 16: 127–137
32 Easton RD, Srinivas J, Greene AJ (1997) Do vision and haptic shared common representations? Implicit and explicit memory between and within modalities. *J Exp Psych: Learning, Memory, and Cognition* 23: 153–193
33 Hamann SB (1996) Implicit memory in the tactile modality. Evidence from Braille stem completion in the blind. *Psychological Science* 7: 284–288
34 Wippich W, Warner V (1989) Auch Hände haben ein Gedächtnis: Implizite und explizite Erinnerungen and haptische Erfahrungen. (Hands have a memory too. Implicit and explicit memory for haptic object exploration). *Sprache and Kognition* 8: 166–177
35 Wippich W (1994) Unbewußte Effekte und Voreingenommenheiten bei Ur-teilen zu Per-sonennamen. *Zeitschrift für experimentelle und ange-wandte Psychologie* 41: 154–172
36 Ballesteros S, Reales JM (2004) Intact haptic priming in normal aging and Alzheimer's disease: evidence for dissociable memory systems. *Neuropsychologia* 42: 1063–1070
37 Easton RD, Greene AJ, Srinivas K (1997) Transfer between vision and touch: Memory for 2-D patterns and 3-D objects. *Psych Bull & Rev* 4: 403–410
38 Reales JM, Ballesteros S (1999) Implicit and explicit memory for visual and haptic objects: Cross-modal priming depends on structural descriptions. *J Exp Psych: Learning, Memory, and Cognition* 118: 219–235
39 Ballesteros S, Reales JM, Mayas J, González M, García B (2005) Priming between modalities in normal aging. Paper presented at the Symposium Experiencing objects through vision and touch. 6th International Multisensory Research Forum, June 2005. Roveretto, Italy
40 Schacter DL, Chiu CYP, Ochsner KN (1993) Implicit memory: A selective review. *Ann Rev Neurosci* 16: 159–182
41 Sathian K, Zangaladze A, Hoffman JM, Grafton ST (1997) Feeling with the mind's eye. *Neuroreport* 8: 3877–3881
42 James TW, Harman James K, Humphrey GK, Goodale MA (2006) Do visual and tactile object representations share the same neural substrate? In: MA Heller, S Ballesteros (eds): *Touch and Blindness: Psychology and Neuroscience.* Lawrence Erlbaum Associates, Hillsdale, NJ, 139–155
43 Zangaladze A, Epstein CM, Grafton ST, Sathian K (1999) Involvement of visual cortex in tactile discrimination of orientation. *Nature* 401: 587–590
44 Sadato N, Pascual-Leone A, Grafman J, Ibáñez V, Deiber, MP, Dold G, Hallett M (1996) Activation of the primary visual cortex by Braille reading in blind subjects. *Nature* 380: 526–528
45 Pascual-Leone A, Theoret H, Merabet L, Kaufman, Schlaug G (2006) The role of the visual cortex in tactile processing: A metamodal brain. In: MA Heller, S Ballesteros (eds): *Touch and Blindness: Psychology and Neuroscience.* Lawrence Erlbaum Associates, Hillsdale, NJ, 171–195
46 Smith AD (1977) Adult age differences in cued recall. *Developmental Psychology* 13: 326–331

47 Bartlett JC, Leslie JE, Tubb A, Fulton A (1989) Aging and memory for pictures of faces. *Psychology and Aging* 4: 276–283
48 Frieske DA, Park DC (1999) Memory for news in young and old adults. *Psychology and Aging* 14: 90–98
49 Nilsson LG (2003) Memory function in normal aging. *Acta Neurolo Scand* 107: 7–13
50 Park DC, Lautenschlager G, Hedden T, Davidson NS, Smith AD, Smith PK (2002) Models of visuospatial and verbal memory across the adult life span. *Psychology and Aging* 17: 299–320
51 Fleischman DA, Gabrieli JDE (1998) Repetition priming in normal aging and in Alzheimer's desease. A review of findings and theories. *Psychology and Aging* 13: 88–119
52 Standtlander LM, Murdock LD, Heiser SM (1998) Visual and haptic influences on memory: Age differences in recall. *Exp Aging Res* 24l: 257–272
53 Wippich W (1991) Haptic information processing in direct and indirect memory tests. *Psych Res* 53: 162–166
54 Gabrieli JDE, Keane MM, Stanger BZ, Kjelgaard M (1994) Dissociations among structural-perceptual, lexical-semantic, and event-fact memory systems in Alzheimer's, amnesic, and normal subjects. *Cortex* 30: 75–103
55 Amedi A, Jacobson G, Hendler T, Malach R, Zohary E (2002) Convergence of visual and tactile shape processing in the human lateral occipital cortex. *Cerebral Cortex* 12: 1202–1212
56 Prull MW, Gabrieli JDE, Bunge SA (2000) Age-related changes in memory: A cognitive neuroscience perspective. In: FIM Craik, TA Salthouse (eds): *Handbook of aging and cognition.* Lawrence Erlbaum Associates, Mahwah, NJ, 91–153
57 Squire LR (1992) Memory and the hippocampus: A synthesis from findings with rats, monkeys and humans. *Psychological Review* 99: 195–213
58 Cohen NJ, Squire LR (1980) Preserved learning and retention of pattern-analysing skill in amnesia: Dissociations of knowing how and knowing that. *Science* 210: 207–210
59 deToledo-Morrell L, Sullivan MP, Morrell F, Wilson RS, Bennett DA, Spencer S (1997) Alzheimer's disease: *Iin vivo* detection of differential vulnerability of brain regions. *Neurobiology of Aging* 18: 463–468

15 Attention in sense of touch

1 Hillyard SA, Mangun GR, Woldorff MG, Luck SJ (1995) Neural systems mediating selective attention. In: MS Gazzaniga (eds): *The cognitive neurosciences*. MIT Press, Cambridge
2 Posner MI (1980) Orienting of attention. Quarterly *J Exp Psychol*ogy 32: 3–25
3 Martin JH, Jessel TM (1991) Modality coding in the somatic sensory system. In: TM Jessell (eds): *Principles of neural science*. Prentice Hall International, London
4 Kandel ER, Jessel TM (1991) Touch. In: TM Jessell (ed): *Principles of neural science*. Prentice Hall International, London
5 Zimmermann M (1990) Das somatoviscerale sensorische System. In: G Thews (ed): *Physiologie des Menschen*. Springer-Verlag, Berlin
6 James W (1890) *Principles of Psychology*. Henry Holt, New York
7 Driver J (2001) A selective review of selective attention research from the past century. *Br J Psychol*ogy 92: 53–78
8 Hsiao SS, Vega-Bermudez F (2002) Attention in the somatosensory system. In: RJ Nelson (ed): *The somatosensory system: Deciphering the brain's own body image.* CRC Press, Boca Raton
9 Forster B, Eimer M (2004) The attentional selection of spatial and non-spatial attributes in touch: ERP evidence for parallel and independent processes. *Biological Psychology* 66: 1–20
10 Harrington GS, Hunter Downs III J (2001) fMRI mapping of the somatosensory cortex with vibratory stimuli. Is there a dependency on stimulus frequency? *Brain Research* 897: 188–192
11 Iguchi Y, Hoshi Y, Hashimoto I (2001) Selective spatial attention induces short-term plasticity in human somatosensory cortex. *Neuroreport* 12: 3133–3136
12 Hillyard SA, Anllo-Vento L (1998) Event-related brain potentials in the study of visual selective attention. *PNAS* 95: 781–787
13 Hansen JC, Hillyard SA (1980) Endogenous brain

potentials associated with selective auditory attention. *Electroencephalography and Clinical Neurophysiology* 49: 277–290

14 Johannsen-Berg H, Llyoyd D (2000) The physiology and psychology of selective attention to touch. *Frontiers in Bioscience* 5: 894–904

15 Hsiao SS, O'Shaughnessy DM, Johnson KO (1993) Effects of selective attention on spatial form processing in monkey primary and secondary cortex. *J Neurophysiol*ogy 70: 444–447

16 Hansson T, Brismar T (1999) Tactile stimulation of the hand causes bilateral cortical activation: a functional magnetic resonance study in humans. *Neurosci Lett* 271: 29–32

17 Buchner H, Reinartz U, Waberski TD, Gobbele R, Noppeney U, Scherg M (1999) Sustained attention modulates the immediate effect of de-afferentation on the cortical representation of the digits: Source localization of somatosensory evoked potentials in humans. *Neurosci Lett* 260: 57–60

18 Burton H, Abend NS, MacLeod AMK, Sinclair RJ, Snyder AZ, Raichle ME (1999) Tactile attention tasks enhance activation in somatosensory regions of parietal cortex: a positron emission tomography study. *Cerebral Cortex* 9: 662–674

19 Macaluso E, Frith CD, Driver J (2000) Selective spatial attention in vision and touch: unimodal and multimodal mechanisms revealed by PET. *J Neurophysiol*ogy 83: 3062–3075

20 Hämäläinen H, Kekoni J, Sams M, Reinikainen K, Näätänen R (1990) Human somatosensory evoked potentials to mechanical pulses and vibration: contributions of SI and SII somatosensory cortices to P50 and P100 components. *Electroencephalography and Clin Neurophysiol* 75(2): 13–21

21 Ishiko N, Hanamori T, Murayama N (1980) Spatial distribution of somatosensory responses evoked by tapping the tongue and finger in man. *Electroencephalography and Clin Neurophysiol* 50: 1–10

22 Taylor-Clarke M, Kennett S, Haggard P (2002) Vision modulates somatosensory cortical processing. *Current Biology* 12(3): 233–236

23 Zopf R, Giabbiconi CM, Gruber T, Müller MM (2004) Attentional modulation of the human somatosensory evoked potential in a trial-by-trial spatial cueing and sustained spatial attention task measured with high density 128 channels EEG. *Cognitive Brain Research* 20: 491–509

24 Mima T, Nagamine T, Nakamura K, Shibasaki H (1998) Attention modulates both primary and secondary somatosensory cortical activites in humans: A magnetoencephalogrpahic study. *J Neurophysiol*ogy 80: 2215–2221

25 Eimer M, Forster B (2003) Modulations of early somatosensory ERP components by transient and sustained attention. *Exp Brain Res* 151: 24–31

26 Chapman CE, Meftah EM (2005) Independent controls of attentional influences in primary and secondary somatosensory cortex. *J Neurophysiol*ogy 94: 4094–4107

27 Driver J, Grossenbacher PinG (1996) Multimodal spatial constraints on tactile selective attention. In: JL McClelland (ed): *Attention and performance* XVI. MIT Press, Cambridge, MA

28 Kida T, Wasaka T, Nakata H, Akatsuka K, Kakigi R (2006) Active attention modulates passive attention-related neural responses to sudden somatosensory input against a silent background. *Exp Brain Research* 175: 609–617

29 Kida T, Nishihira Y, Wasaka T, Nakata H, Sakamoto M (2004) Passive enhancement of the somatosensory P100 and N140 in an active attention task using deviant alone condition. *Clin Neurophysiol* 115: 871–879

30 Kida T, Nishihira Y, Wasaka T, Nakata H, Sakamoto M (2004) Differential modulation of temporal and frontal components of the somatosensory N410 and the effect of interstimulus interval in a selective attention task. *Cognitive Brain Research* 19: 33–39

31 Sperling G, Reeves A (1980) Measuring the reaction time of a shift of visual attention. In: RS Nickerson (ed): *Attention and performance.* VIII. Erlbaum, Hillsdale

32 Weichselgartner E, Sperling G (1987) Dynamics of automatic controlled visual attention. *Science* 238: 778–780

33 Müller MM, Teder-Sälejärvi W, Hillyard SA (1998) The time course of cortical facilitation during cued shifts of spatial attention. *Nature Neuroscience* 1: 631–634

34 Duncan J, Ward R, Shapiro K (1994) Direct mea-

surement of attentional dwell time in human vision. *Nature* 369: 313–315

35 Lakatos S, Shepard R (1997) Time-distance relations in shifting attention between locations on one's body. *Perception & Psychophysics* 59: 557–566

36 Nakajima Y, Imamura N (2000) Probability and interstimulus interval effects on the N140 and the P300 components of somatosensory ERPs. *Int J Neurosci* 104: 75–91

37 Yantis S, Johnson DN (1990) Mechanisms of attentional priority. *J Exper Psychol Hum Percept Perform* 16: 812–825

38 Kelly EF, Folger SE (1999) EEG evidence of stimulus-directed response dynamics in human somatosensory cortex. *Brain Research* 815: 326–336

39 Kelly EF, Trulsson M, Folger SE (1997) Periodic microstimulation of single mechanoreceptive afferents produces frequency-following responses in human EEG. *J Neurophysiology* 77: 137–144

40 Snyder AZ (1992) Steady-state vibration evoked potentials: description of technique and characterization of responses. *Electroenceph Clin Neurophysiol* 84: 257–268

41 Tobimatsu S, Zhang YM, Kato M (1999) Steady-state vibration somatosensory evoked potentials: physiological characteristics and tuning function. *Clin Neurophysiol* 110: 1953–1958

42 Tobimatsu S, Zhang YM, Suga R, Kato M (2000) Differential temporal coding in vibratory sense in the hand and foot in man. *Clin Neurophysiol* 111: 398–404

43 Goff GD (1967) Differential discrimination of frequency of cutaneous mechanical vibration. *J Exp Psychology* 74: 294–299

44 Hollins M, Goble AK, Whitsel BL, Tommerdahl M (1990) Time course and actionspectrum of vibrotactile adaptation. *Somatosensory and Motor Research* 2: 205–221

45 LaMotte RH, Mountcastle VB (1975) Capacities of humans and monkeys to discriminate between vibratory stimuli of different frequency and amplitude: a correlation between neural events and psychophysical measurements. *J Neurophysiology* 38: 539–559

46 Mountcastle VB, Talbot WH, Sakata H, Hyvarinen J (1969) Cortical mechanisms in flutter vibration studied in unanesthetized monkeys. *J Neurophysiology* 32: 453–484

47 Mountcastle VB, Steinmetz MA, Romo R (1990) Frequency discrimination in the sense of flutter: psychophysical measurements correlated with postcentral events in behaving monkeys. *J Neuroscience* 10: 3032–3044

48 Tommerdahl M, Delemos KA, Whitsel BL, Favorov OV, Metz CB (1999) Response of anterior parietal cortex to cutaneous flutter *versus* vibration. *J Neurophysiology* 82: 16–33

49 Salinas E, Hernandez A, Zainos A, Romo R (2000) Periodicity and firing rate as candidate neural codes for the frequency of vibrotactile stimuli. *J Neuroscience* 20(14): 5503–5515

50 Tommerdahl M, Whitsel BL, Favorov OV, Metz CB, O'Quinn BL (1999) Response of contralateral SI and SII in cat to same-site cutaneous flutter *versus* vibration. *J Neurophysiology* 82: 1982–1992

51 Giabbiconi CM, Dancer C, Zopf R, Gruber T, Müller MM (2004) Selective spatial attention to left or right hand flutter sensation modulates the steady-state somatosensory evoked potential. *Cognitive Brain Research* 20: 58–66

52 Müller MM, Gruber T (2001) Induced gamma-band responses in the human EEG are related to attentional information processing. *Visual Cognition* 8: 579–592

53 Niebur E, Krieger Z (2002) Electrophysiological correlates of synchronous neural activity and attention: a short review. *Biosystems* 67: 157–166

54 Niebur E, Koch C, Rosin C (1993) An oscillation-based model for the neuronal basis of attention. *Vision Research* 33: 2789–2802

55 Niebur E, Hsiao SS, Johnson KO (2002) Synchrony: a neural mechanism for attentional selection? *Curr Opin Neurobiol* 12: 190–194

56 Giabbiconi CM, Trujillo-Barreto NJ, Gruber T, Müller MM (2007) Sustained spatial attention to vibration is mediated in primary somatosensory cortex. *NeuroImage* 35: 255–262

57 Burton H, Sinclair RJ (2000) Tactile-spatial and cross-modal attention effects in the primary somatosensory cortical areas 3b and 1-2 of rhesus monkeys. *Somatosensory and Motor Research* 17: 213–228

58 Staines WR, Graham SJ, Black SE, McIlroy WE

(2002) Task-relevant modulation of contralateral and ipsilateral primary somatosensory cortex and the role of a prefrontal-cortical sensory gating system. *NeuroImage* 15: 190–199

59 Burton H, Sinclair RJ, Hong SY, Pruett JR, Whang KC (1997) Tactile-spatial and cross-modal attention effects in the second somatosensory and 7b cortical areas of rhesus monkeys. *Somatosensory and Motor Research* 14: 237–267

60 von der Malsburg C, Schneider W (1986) A neural cocktail-party processor. *Biological Cybernetics* 54: 29–40

61 Gruber T, Müller MM, Keil A, Elbert T (1999) Selective visual-spatial attention alters induced gamma band responses in the human EEG. *Clin Neurophysiol* 110: 2074–2085

62 Fries P, Reynolds JH, Rorie AE, Desimone R (2001) Modulation of oscillatory neural synchronization by selective visual attention. *Science* 291: 1560–1563

63 Steinmetz PN, Roy A, Fitzgerald PJ, Hsiao SS, Johnson KO, Niebur E (2000) Attention modulates synchronized neuronal firing in primate somatosensory cortex. *Nature* 404(6774): 187–190

64 Bauer M, Oostenveld R, Peeters M, Fries P (2006) Tactile spatial attention enhances gamma-band activity in somatosensory cortex and reduces low-frequency activity in parieto-occipital areas. *J Neuroscience* 26(2): 490–501

16 Haptic object identification

1 Heller MA, Ballesteros S (2006) Introduction: Approaches to touch and blindness. In: MA Heller, S Ballesteros (eds): *Touch and Blindness: Psychology and Neuroscience*. Lawrence Erlbaum Associates, Hillsdale, NJ, 1–24

2 Gibson JJ (1962) Observations on active touch. *Psych Rev* 69: 477–491

3 Gibson JJ (1966) *The senses considered as perceptual systems*. Houghton-Mifflin, Boston

4 Symmons MA, Richardson BL, Wuillemin DB (2004) Active *versus* passive touch: Superiority depends more on the task than the mode. In: S Ballesteros, MA Heller (eds): *Touch, Blindness and Neuroscience.* UNED, Madrid, Spain, 179–185

5 Heller MA, Myers DS (1983) Active and passive tactual recognition of form. *J Gen Psych* 108: 225–229

6 Heller MA (1984) Active and passive touch: The influence of exploration time on form recognition. *J Gen Psych* 110: 243–249

7 Heller MA, Rogers GJ, Perry CL (1990) Tactile pattern recognition with the Optacon: Superior performance with active touch and the left hand. *Neuropsychologia* 28: 1003–1006

8 Heller MA (1987) Improving the passive tactile digit span. *Bull Psych Soc* 25: 257–258

9 Critchley M (1953) The parietal lobes. Hafner, New York

10 Lederman SJ, Klatzky RL (1987) Hand-movements: A window into haptic object recognition. *Cog Psych* 19: 342–368

11 Lederman SJ, Klatzky RL (1990) Haptic classification of common objects: Knowledge-driven exploration. *Cog Psych* 22: 421–459

12 Klatzky RL, Lederman SJ (2003) The haptic identification of everyday life objects. In: Y Hatwell, A Streri, E Gentaz (eds): *Touching for knowing: Cognitive psychology of haptic manual perception.* Benjamins, Amsterdam, 105–121

13 Klatzky RL, Lederman SJ, Metzger V (1985) Identifying objects by touch: An expert system. *Perc & Psych* 37: 299–302

14 Jansson G, Monaci L (2004) Haptic identification of objects with different numbers of fingers. In: S Ballesteros, MA Heller (eds): *Touch, blindness and neuroscience*. UNED Press, Madrid, Spain, 209–219

15 Lederman SJ, Klatzky RL, Chataway C, Summers C (1990) Visual mediation and the haptic recognition of two-dimensional pictures of common objects. *Perc & Psych* 47: 54–64

16 Jones LA, Lederman SJ (2006) *Human hand function.* Oxford University Press, New York

17 Heller MA, Calcaterra JA, Burson LL, Tyler LA (1996) Tactual picture identification by blind and sighted people: Effects of providing categorical information. *Perc & Psych* 58: 310–323

18 Heller MA (1989) Picture and pattern perception in the sighted and blind: The advantage of the late blind. *Perception* 18: 379–389

19 Heller MA (2006) Picture perception and spatial cognition in visually impaired people. In: S Ballesteros, MA Heller (eds): *Touch and blindness:*

Psychology and neuroscience. Lawrence Erlbaum Associates, Mahwah, New Jersey, 49–71
20 Heller MA, Brackett DD, Scroggs E (2002) Tangible picture matching in people who are visually impaired. *J Visual Imp and Blindness* 96: 349–353
21 Heller MA, Hasselbring K, Wilson K, Shanley M, Yoneyama K (2004) Haptic illusions in the sighted and blind. In: S Ballesteros, MA Heller (eds): *Touch, blindness and neuroscience*. UNED Press, Madrid, Spain, 135–144
22 Heller MA, Kennedy JM, Clark A, McCarthy M, Borgert A, Fulkerson E, Wemple LA, Kaffel, N, Duncan A, Riddle T (2006) Viewpoint and orientation influence picture recognition in the blind. *Perception* 35: 1397–1420
23 Carello C, Turvey MT (2000) Rotational invariants and dynamic touch. In: MA Heller (ed): *Touch, representation and blindness*. Oxford University Press, Oxford, UK, 27–66
24 Heller MA (2000) Conclusions: The San Marino discussion. In: MA Heller (ed): *Touch, representation and blindness*. Oxford University Press, Oxford, UK, 183–214
25 Michaels CF, Weier Z, Harrison SJ (2007) Using vision and dynamic touch to perceive the affordances of tools. *Perception* 36: 750–772
26 Klatzky RL, Lederman SJ, Reed C (1987) There's more to touch than meets the eye: Relative salience of object dimensions for touch with and without vision. *J Exp Psych: General* 116: 356–369
27 Katz D (1925/1989) *The world of touch* (LE Krueger Trans.). Erlbaum, Hillsdale, NJ (Original published in 1925)
28 Hollins M, Faldowsky R, Rao S, Young F (1993) Perceptual dimensions of tactile surface texture: A multidimensional scaling analysis. *Perc & Psych* 54: 697–705
29 Hollins M, Bensmaïa S, Karlof K, Young F (2000) Individual differences in perceptual space for tactile textures: Evidence from multidimensional scaling. *Perc & Psych* 62: 1534–1544
30 Picard D, Dacremont C, Valintin D, Giboreau A (2003) Perceptual dimensions of tactile textures. *Acta Psychologica* 114: 165–184
31 Johnson KO, Hsiao SS (1992) Neural mechanisms of tactual form and texture perception. *Ann Rev Neurosc* 15: 227–250
32 Johnson KO (2004) Neural mechanisms of tactile sensation. In: S Ballesteros, MA Heller (eds): *Touch, Blindness, and Neuroscience*. UNED, Madrid, Spain, 37–38
33 Heller MA (1989) Texture perception in sighted and blind observers. *Perc & Psych* 45: 49–54
34 Hollins M, Bensmaïa SJ, Washburn S (2001) Vibrotactile adaptation impairs discrimination of fine, but not coarse, textures. *Somatosensory & Motor Research* 18: 253–262
35 Bensmaïa SJ, Hollins M (2005) Pacinian representations of fine surface texture. *Perc & Psych* 67: 842–854
36 Bensmaïa SJ, Hollins M, Yau J (2005) Vibrotactile information in the Pacinian system: a psychophysical model. *Perc & Psych* 67: 828–841
37 Ballesteros S, Reales JM, Ponce de León L, García B (2005) The perception of ecological textures by touch: Does the perceptual space change under bimodal visual and haptic exploration. *WorldHaptics Proceedings (WHC'05) Pisa. IEEE Computer Society*, Los Alamitos, CA, 635–638
38 Goldstone R (1994) An efficient method for obtaining similarity data. *Behavior Research Methods, Instruments, & Computers* 26: 381–386
39 Heller MA (1982) Visual and tactual texture perception: Intersensory cooperation. *Perc & Psych* 31: 339–344
40 Ballesteros S, Reales JM, García B, Ponce de León L (2006) Perceptual dimensions of haptic textures space by young and older adults. In: S Ballesteros (ed): *Aging, cognition, and neuroscience*. UNED, Madrid, Spain, 127–147
41 Ellis RR, Lederman SJ (1993) The role of haptic *versus* visual volume cues in the size-weight illusion. *Perc & Psych* 53: 315–324
42 Kawai S (2002) Heaviness perception. I. Constant involvement of haptically perceived size in weight discrimination. *Exp Brain Res* 147: 16–22
43 Kawai S (2002) Heaviness perception. II. Contributions of object weight, haptic size, and density to the accurate perception of heaviness or lightness. *Exp Brain Res* 147: 23–28
44 Ellis RR, Lederman SJ (1998) The golf-ball illusion: Evidence for top-down processing in weight perception. *Perception* 27: 193–201

45 Srinivasan MA, LaMotte RH (1995) Tactual discrimination of softness. *J Neurophysiol* 73: 88–101
46 LaMotte RH (2000) Softness discrimination with a tool. *J Neurophysiol* 83: 1777–1786
47 Davidson PW (1972) Haptic judgments of curvature by blind and sighted human. *J Exp Psych* 9: 43–55
48 Henriques DY, Soechting JF (2003) Bias and sensitivity in the haptic perception of geometry. *Exp Brain Res* 150: 95–108
49 Kappers AML, Koenderink JJ, Lichtenegger I (1994) Haptic identification of curved surfaces. *Perc & Psych* 56: 53–61
50 Louw S, Kappers AML, Koenderink JJ (2000) Haptic detection thresholds of Gaussian profiles over the whole range of spatial scales. *Exp Brain Res* 132: 369–374
51 Gibson JJ (1933) Adaptation, after-effect and contrast in the perception of curved lines. *J Exp Psych* 16: 1–31
52 Vogels I (1996) Haptic after-effect of curved surfaces. *Perception* 25: 109–119
53 Faineteau H, Gentaz E, Viviani P (2005) Factors affecting the size of the detour effect in the kinesthetic perception of Euclidean distance. *Exp Brain Res* 163: 503–514
54 Heller MA, Kappers AML, McCarthy M, Clark A, Riddle T, Fulkerson E, Wemple L (submitted) The effects of curvature on haptic judgments of extent in sighted and blind people.
55 Appelle S (1972) Perception and discrimination as a function of stimulus orientation: The “oblique effect” in man and animals. *Psych Bull* 78: 266–278
56 Heller MA, Calcaterra J, Green S, Lima F (1999) The effect of orientation on braille recognition in persons who are sighted and blind. *J Visual Impairment and Blindness* 93: 416–419
57 Appelle S (1991) Haptic perception of form. In: MA Heller, W Schiff (eds): *The psychology of touch*. Lawrence Erlbaum Associates, Hillsdale, NJ, 169–188
58 Heller MA, Brackett DD, Salik SS, Scroggs E, Green S (2003) Objects, raised-lines and the haptic horizontal-vertical Illusion. *Quarterly J Exp Psych A*: 56: 891–907
59 Julesz B (1971) *Foundations of cyclopean perception*. University of Chicago Press, Chicago
60 Locher PJ, Nodine C (1973) Influence of stimulus symmetry on visual scanning patterns. *Perc & Psych* 13: 408–412
61 Palmer SE (1989) Reference frames in the perception of shape and orientation. In: BS Shepp, S Ballesteros (eds): *Object perception: Structure and processes*. Lawrence Erlbaum Associates, Hillsdale, NJ, 121–163
62 Wagemans J (1995) Detection of visual symmetries. *Spatial Vision* 9: 9–32
63 Millar S (1978) Aspects of memory for information from touch and movement. In: G Gordon (ed): *Active touch: The mechanism of recognition of objects by manipulation. A multidisciplinary approach*. Pergamon Press, Oxford, 215–227
64 Locher PJ, Simmons RW (1978) Influence of stimulus symmetry and complexity upon haptic scanning strategies during detection, learning, and recognition tasks. *Perc & Psych* 23: 110–116
65 Simmons RW, Locher PJ (1979) The role of extended perceptual experience upon haptic perception of non-representational shapes. *Perc & Psych* 48: 987–991
66 Ballesteros S, Manga D, Reales JM (1997) Haptic discrimination of bilateral symmetry in two-dimensional and three-dimensional unfamiliar displays. *Perc & Psych* 59: 37–50
67 Ballesteros S, Millar S, Reales JM (1998) Symmetry in haptic and in visual shape perception. *Perc & Psych* 60: 389–404
68 Ballesteros S, Reales JM (2004) Visual and haptic discrimination of symmetry in unfamiliar displays extended in the z-axis. *Perception* 33: 315–327
69 Pascual-Leone A, Theoret H, Merabet L, Kauffmann T, Schlaug G (2006) The role of visual cortex in tactile processing: A metamodal brain. In: MA Heller, S Ballesteros (eds): *Touch and blindness: Psychology and Neuroscience*. Lawrence Erlbaum Associates, Mahwah, NJ, 171–195
70 Heller MA, Brackett DD, Scroggs E, Steffen H, Heatherly K, Salik S (2002) Tangible pictures: Viewpoint effects and linear perspective in visually impaired people. *Perception* 31: 747–769
71 Kennett S, Taylor-Clarke M, Haggard P (2001) Non-informative vision improves the spatial resolution of touch. *Curr Biol* 11: 1188–1191
72 Millar S (2006) Processing spatial information

from touch and movement: Implications from and for neuroscience. In: MA Heller, S Ballesteros (eds): *Touch and blindness: psychology and neuroscience.* Lawrence Erlbaum Associates, Mahwah, NJ, 25–48
73 Heller MA, Calcaterra JA, Green SL, Brown L (1999) Intersensory conflict between vision and touch: The response modality dominates when precise, attention-riveting judgments are required. *Perc & Psych* 61: 1384–1398
74 Graziano MSA, Gross CG, Taylor CSR, Moore T (2004) A system of multimodal areas in the primate brain. In: C Spence, J Driver (eds): *Crossmodal space and crossmodal attention.* Oxford University Press, Oxford, UK 51–67
75 Heller MA, Calcaterra JA, Tyler LA, Burson LL (1996) Production and interpretation of perspective drawings by blind and sighted people. *Perception* 25: 321–334

17 Haptic perceptual illusions

1 Revesz G (1934) System der optischen und haptishen Raumtäuschungen. *Zeitschrift für Physiologie* 131: 296–375
2 Bean CH (1938) The blind have "optical" illusions. *J Exp Psychol* 22: 283–289
3 Gentaz E, Hatwell Y (2004) Geometrical haptic illusion: Role of exploratory movements in the Muller-Lyer, Vertical-Horizontal and Delboeuf illusions. *Psychonomic Bulletin and Review* 11: 31–40
4 Hatwell Y, Streri A, Gentaz E (2000) *Toucher pour connaître. Psychologie cognitive de la perception tactile manuelle.* PUF, Paris
5 Hatwell Y, Streri A, Gentaz E (2003) *Touching for knowing.* John Benjamins Publishing Company, Amsterdam/Philadelphia
6 Heller M (ed) (2000) *Touch, representation and blindness.* Oxford University Press, Oxford
7 Millar S (1994) *Understanding and representing space. Theory and evidence from studies with blind and sighted children.* Clarendon Press, Oxford
8 Cornoldi C, Bertuccelli B, Rocchi P, Sbrana B (1993) Processing capacity limitations in pictorial and spatial representations in the totally congenitally blind. *Cortex* 29: 675–689
9 Hatwell Y (1978) Form perception and related issues in blind humans. In: R Held, HW Leibowitz, HL Teuber (eds): *Handbook of sensory physiology, VII: Perception.* Springer Verlag, New York and Berlin
10 Revesz G (1950) *Psychology and art of the blind.* Longmans Green, London
11 Worchel P (1951) Space perception and orientation in the blind. *Psychological Monograph* 65 (15)
12 Hatwell Y (1966) *Privation sensorielle et intelligence.* PUF, Paris
13 Hatwell Y (1985) *Piagetian reasoning and the blind* (translated from *Privation sensorielle et intelligence*, 1966, Paris: Presses Universitaires de France). American Foundation for the Blind, New York
14 Hatwell Y (1986) *Toucher l'espace.* Presses Universitaires de Lille, Lille
15 Thinus-Blanc C, Gaunet F (1997) Representation of space in the blind: Vision as a spatial sense? *Psychological Bulletin* 121: 20–42
16 Faineteau H, Gentaz E, Viviani P (2003) The kinaesthetic estimation of Euclidean distances in blindfolded adults: A study on the detour path effect. *Exp Brain Res* 153: 166–172
17 Faineteau H, Gentaz E, Viviani P (2005) Factors affecting the size of the detour effect in the kinaesthetic perception of Euclidian distance. *Exp Brain Res* 163: 503–514
18 Gentaz E, Gaunet F (2006) L'inférence haptique d'une localisation spatiale chez les adultes et les enfants: étude de l'effet du trajet et de l'effet du délai dans une tâche de complétion de triangle. *L'Année Psychologique* 2: 167–190
19 Lederman SJ, Klatzky RL, Barber PO (1985) Spatial and movement-based heuristics for encoding pattern information through touch. *J Exp Psychology: General* 114: 33–49
20 Gentaz E, Hatwell Y (1996) Role of gravitational cues in the haptic perception of orientation. *Perception & Psychophysics* 58: 1278–1292
21 Gentaz E, Hatwell Y (1998) The haptic oblique effect in the perception of rod orientation by blind adults. *Perception & Psychophysics* 60: 157–167
22 Gentaz E, Hatwell Y (1999) Role of memorisation conditions in the haptic processing of orientations and the 'oblique effect'. *Br J Psychol* 90: 377–388

23 Heller M, Calcaterra J, Burson L, Green S (1997) The tactual horizontal-vertical illusion depends on radial motion of the entire arm. *Perception & Psychophysics* 59: 1297–1311
24 Millar S, Al-Attar Z (2002) The Müller-Lyer illusion in touch and vision: implications for multisensory processes. *Perception & Psychophysics* 64 (3): 353–365
25 Casla M, Blanco F, Traviesco D (1999) Haptic perception of geometric illusions by persons who are totally congenitally blind. *J Visual Impairment* 93: 583–588
26 Hatwell Y (1960) Etude de quelques ilusions géométriques tactiles chez les aveugles. *L'Année Psychologique* 60: 11–27
27 Heller M, Brackett D, Wilson K, Yoneyama K, Boyer A, Steffen H (2002) The haptic Müller-Lyer illusion in sighted and blind people. *Perception* 31: 1263–1274
28 Heller M, McCarthy M, Schultz J, Greene J, Shanley M, Clark A, Skoczylas S, Procjuk I (2005) The influence of exploration mode, orientation, and configuration on the haptic Muller-Lyer illusion. *Perception* 34: 1475–1500
29 Over R (1966) Context and movement as factors influencing haptic illusions. *Australian J Psychol* 3: 262–265
30 Over R (1966) A comparison of haptic and visual judgments of some illusions. *Am J Psychol* 79: 590–595
31 Patterson J, Deffenbacher K (1972) Haptic perception of the Müller-Lyer illusion by the blind. *Perceptual and Motor Skills* 35: 819–824
32 Rudel RG, Teuber HL (1963) Discrimination of direction of line in children. *J Comparative & Physiological Psychol* 56: 892–898
33 Suzuki K, Arashida R (1992) Geometrical haptic illusions revisited: haptic illusions compared with visual illusions. *Perception & Psychophysics* 52 (3): 329–335
34 Tsai LS (1967) Mueller-Lyer illusion by the blind. *Perceptual and Motor Skills* 25: 641–644
35 Wong TS (1975) The respective role of limb and eye movements in the haptic and visual Müller-Lyer illusion. Quarterly *J Exp Psychol* 27: 659–666
36 Rudel RG, Teuber HL (1963) Decrement of visual and haptic Müller-Lyer illusion on repeated trials: A study of crossmodal transfer. *Quarterly J Exp Psychol* 15: 125–131
37 Over R (1967) Haptic illusions and inappropriate constancy scaling. *Nature* 214: 629
38 Coren S, Girgus JS (1978) *Seeing is deceiving: The psychology of visual illusions*. Lawrence Erlbaum Associates, Hillsdale, NJ
39 Gentaz E, Camos V, Hatwell Y, Jacquet AY (2004) The haptic and visual Müller-Lyer illusions: correlation studies. *Curr Psychol Letts (revue électronique)*
40 Gregory R (1998) *Eye and brain. The psychology of seeing* (5th edition). Oxford University Press, London
41 Day RH, Wong TS (1971) Radial and tangential movement directions as determinants of the haptic illusion in an L figure. *J Exp Psychol* 87: 19–22
42 Deregowski J, Ellis HD (1972) Effect of stimulus orientation upon haptic perception of the horizontal-vertical illusion. *J Exp Psychol* 95: 14–19
43 Heller MA, Joyner TD (1993) Mechanisms in the haptic horizontal-vertical illusion: Evidence from sighted and blind subjects. *Perception & Psychophysics* 53: 422–428
44 Heller M, Brakett D, Wilson K, Yoneyame K, Boyer A (2002) Visual experience and the haptic horizontal-vertical illusion. *Br J Visual Impairment* 20: 105–109
45 McFarland J, Soechting JF (2007) Factors influencing the radial-tangential illusion in haptic perception. *Exp Brain Res* 178: 99–114
46 Millar S (2000) Vertical and bisection bias in active touch. *Perception* 29: 481–500
47 Von Collani G (1979) An analysis of illusion components with L and T (inversed) figures in active touch. Quarterly *J Exp Psychol* 31: 241–248
48 Wong TS (1977) Dynamic properties of radial and tangential movements as determinants of the haptic horizontal-vertical illusion with an L figure. *J Exp Psychol: Human Perception & Performance* 3: 151–164
49 Künnapas TM (1955) An analysis of the "vertical-horizontal illusion". *J Exp Psychol* 49: 134–140
50 Künnapas T (1957) Interocular differences in the vertical-horizontal illusion. *Acta Psychologica* 13: 253–259

51 Künnapas T (1957) Vertical-horizontal and the surrounding field. *Acta Psychologica* 13: 35–42
52 Künnapas T (1957) The vertical-horizontal illusion and the visual field. *J Exp Psychol* 53: 405–407
53 Prinzmetal W, Gettleman L (1993) Vertical-horizontal illusion: One eye is better than two. *Perception & Psychophysics* 53: 81–88
54 Day RH, Avery GC (1970) Absence of the horizontal-vertical illusion in haptic space. *J Exp Psychol* 83: 172–173
55 Cheng MF (1968) Tactile-kinesthetic perception of length. *Am J Psychol* 81: 74–82
56 Wong TS (1979) Developmental study of a haptic illusion in relation to Piaget's centration theory. *J Exp Child Psychol* 27: 489–500
57 Piaget J (1957) *Les mécanismes perceptifs*. PUF, Paris
58 Wenderoth P, Alais D (1990) Lack of evidence for a tactual Poggendorff illusion. *Perception & Psychophysics* 48: 243–242
59 Lucca A, Dellantonio A, Riggio L (1986) Some observations on the Poggendorf and the Müller-Lyer tactual illusions. *Perception & Psychophysics* 39: 374–380
60 Hatwell Y, Orliaguet JP, Brouty G (1990) Effects of object properties, attentional constraints and manual exploratory procedures on haptic perceptual organization: A developmental study. In: H Bloch, B Bertenthal (eds): *Sensory-motor organization and development in infancy and early childhood*. Klumer Academic Publishers, Dordrecht 315–335
61 Kappers A (2002) Haptic perception of parallelity in the midsagittal plane. *Acta Psychologica* 109: 25–40
62 Kappers AM (1999) Large systematic deviations in the haptic perception of parallelity. *Perception* 28: 1001–1012
63 Kappers AM, Koenderink JJ (1999) Haptic perception of spatial relations. *Perception* 28: 781–795
64 Cuijpers RH, Kappers AM, Koenderink JJ (2000) Large systematic deviations in visual parallelism. *Perception* 29: 1467–1482
65 Cuijpers RH, Kappers AM, Koenderink JJ (2002) Visual perception of collinearity. *Perception & Psychophysics* 64: 1467–1482
66 Blumenfeld W (1937) The relationship between the optical and haptic construction of space. *Acta Psychologica* 2: 133–157
67 Hillebrand F (1902) Theorie der scheinbaren Grösse bei binocularem Sehen [Theory of apparent size in binocular vision]. *Denkschriften der Wiener Akademie, math.-nat.* 72: 255–307
68 Sanders D, Kappers AM (2007) Haptically straight lines. *Perception* 36: 1688–1697
69 Walkers JT (1971) Visual capture in illusions. *Perception and Psychophysics* 10: 71–73
70 Gallace A, Spence C (2005) Examining the cross-modal consequences of viewing the Muller-Lyer illusion. *Exp Brain Res* 162: 490–496
71 Rock I, Victor J (1964) Vision and touch: An experimentally induced conflict between the two senses. *Science* 143: 594–596
72 Misceo GF, Hershberger WA, Mancini RL (1999) Haptic estimates of discordant visual-haptic size vary developmentally. *Perception & Psychophysics* 61 (4): 608–614
73 Power RP, Graham A (1976) Dominance of touch by vision: Generalization of the hypothesis to a tactually experienced population. *Perception* 5: 161–166
74 Congedo M, Lecuyer A, Gentaz E (2006) The influence of spatial de-location on perceptual integration of vision and touch. *Presence: Teleoperators & Virtual Environments* 15: 353–357
75 Hershberger W, Misceo G (1996) Touch dominates haptic estimates of discordant visual-haptic size. *Perception & Psychophysics* 58: 1124–1132
76 Lederman SJ, Thorne G, Jones B (1986) Perception of texture by vision and touch: Multidimensionality and intersensory integration. *J Exp Psychol: Human Perception & Performance* 12: 169–180
77 Ernst MO, Banks MS (2002) Humans integrates visual and haptic information in a statistically optimal fashion. *Nature* 415: 429–433
78 Heller MA (1983) Haptic dominance in form perception with blurred vision. *Perception* 12: 607–613
79 Pavani F, Spence C, Driver J (2000) Visual capture of touch: out-of-the-body experiences with rubber gloves. *Psychological Science* 11: 353–359
80 Spence C, Walton M (2005) On the inability to ignore touch when responding to vision in the

crossmodal congruency task. *Acta Psychologica* 118: 47–70

81 Caclin A, Soto-Faraco S, Kingstone A, Spence C (2002) Tactile capture of audition. *Perception & Psychophysics* 64: 616–630

82 Bruno N, Jacomuzzi A, Bertamini M, Meyer G (2007) A visual-haptic Necker cube reveals temporal constraints on intersensory merging during perceptual exploration. *Neuropsychologia* 45: 469–475

83 Robles-de-la-Torre G, Hayward V (2001) Force can overcome object geometry in the perception of shape through active touch. *Nature* 412: 445–448

84 Drewing K, Ernst MO (2003) Integration of force and position cues in haptic curvature perception. *Abstracts of the Psychonomic Society 44th Annual Meeting:* 115

85 Wydoodt P, Gentaz E, Streri A (2006) Role of force cues in the haptic estimations of a virtual length. *Exp Brain Res* 171: 489–489

86 Millar S (2001) Illusions in reading maps by touch: Reducing distance errors. *Br J Psychol* 92: 643–657

87 Hayward V (2008) A brief taxonomy of tactile illusions and demonstrations that can be done in a hardware store. *Brain Res Bulletin* 75: 742–752

18 Haptic perception in interaction with other senses

1 O'Hare JJ (1991) Perceptual integration. *J Washington Acad Sci* 81: 44–59

2 Alais D, Burr D (2004) The ventriloquist effect results from near-optimal bimodal integration. *Curr Biol* 14: 257–262

3 Bernstein IH, Clark MH, Edelstein BA (1969) Effects of an auditory signal on visual reaction time. *J Exp Psychol* 80(3): 567–569

4 Ernst MO, Banks MS (2002) Humans integrate visual and haptic information in a statistically optimal fashion. *Nature* 415(6870): 429–433

5 Gepshtein S, Banks MS (2003) Viewing geometry determines how vision and haptics combine in size perception. *Curr Biol* 13(6): 483–488

6 Gielen SC, Schmidt RA, den Heuvel PJV (1983) On the nature of intersensory facilitation of reaction time. *Perception & Psychophysics* 34(2): 161–168

7 Hershenson M (1962) Reaction time as a measure of intersensory facilitation. *J Exp Psychology* 63: 289–293

8 Morrell LK (1968) Temporal characteristics of sensory interaction in choice reaction times. *J Exp Psychology* 77(1): 14–18

9 Nickerson RS (1973) Intersensory facilitation of reaction time: energy summation or preparation enhancement? *Psychol Rev* 80(6): 489–509

10 Welch R, Warren D (1986) Intersensory interactions. In: K Boff, L Kaufman, JP Thomas (eds): *Handbook of Perception and Human Performance.* Wiley, New York, 25.1–25.36

11 Wu W-C, Basdogan C, Srinivasan MA (1999) Visual, haptic, and bimodal perception of size and stiffness in virtual environments. *ASME Dynamic Systems and Control Division*, DSC-Vol. 67: 19–26

12 Luria SM, Kinney JA (1970) Underwater vision. *Science* 167(924): 1454–1461

13 Ross HE, Nawaz S (2003) Why do objects appear enlarged under water? *Arquivos Brasileiros de Oftalmologia* 66: 69–76

14 Hay JC, Pick HL, Ikeda K (1965) Visual capture produced by prism spectacles. *Psychonomic Sci* 2: 215–216

15 Howard IP, Templeton WB (1966) *Human spatial orientation*. Wiley, New York

16 Kinney JAS, Luria SM (1970) Conflicting visual and tactual-kinesthetic stimulation. *Perception and Psychophysics* 8(3): 189–192

17 Power RP, Grahman A (1976) Dominance of touch by vision: Generalization of the hypothesis to a tactually experienced population. *Perception* 5: 161–166

18 Rock I, Victor J (1964) Vision and touch: An experimentally created conflict between the two senses. *Science* 143: 594–596

19 Singer G, Day RN (1969) Visual capture of haptually judged depth. *Perception and Psychophysics* 5: 315–316

20 Welch RB, Warren DH (1980) Immediate perceptual response to intersensory discrepancy. *Psychological Bulletin* 88(3): 638–667

21 Shipley T (1964) Auditory flutter-driving of visual flicker. *Science* 145: 1328–1330

22 Welch RB, DuttonHurt LD, Warren DH (1986) Contributions of audition and vision to temporal

rate perception. *Perception & Psychophysics* 39(4): 294–300

23 Backus BT, Banks MS (1999) Estimator reliability and distance scaling in stereoscopic slant perception. *Perception* 28(2): 217–242

24 Banks MS, Hooge IT, Backus BT (2001) Perceiving slant about a horizontal axis from stereopsis. *J Vision* 1(2): 55–79

25 Battaglia PW, Jacobs RA, Aslin RN (2003) Bayesian integration of visual and auditory signals for spatial localization. *J Optical Society of America. A, Optics, image science, and vision* 20(7): 1391–1397

26 Buckley D, Frisby JP (1993) Interaction of stereo, texture and outline cues in the shape perception of three-dimensional ridges. *Vision Res* 33: 919–933

27 Drewing K, Ernst MO (2006) Integration of force and position cues for shape perception through active touch. *Brain* Res 1078(1): 92–100

28 Frisby JP, Buckley D, Horsman JM (1995) Integration of stereo, texture, and outline cues during pinhole viewing of real ridge-shaped objects and stereograms of ridges. *Perception* 24(2): 181–198

29 Jacobs RA (1999) Optimal integration of texture and motion cues to depth. *Vision Res* 39(21): 3621–3629

30 Johnston EB, Cumming BG, Landy MS (1994) Integration of stereopsis and motion shape cues. *Vision Res* 34(17): 2259–2275

31 Johnston EB, Cumming BG, Parker AJ (1993) Integration of depth modules: stereopsis and texture. *Vision Res* 33(5–6): 813–826

32 Knill DC (1998) Ideal observer perturbation analysis reveals human strategies for inferring surface orientation from texture. *Vision Res* 38(17): 2635–2656

33 Körding KP, Wolpert DM (2004) Bayesian integration in sensorimotor learning. *Nature* 427(6971): 244–247

34 Landy MS, Maloney LT, Johnston EB, Young M (1995) Measurement and modeling of depth cue combination: in defense of weak fusion. *Vision Res* 35(3): 389–412

35 Rogers BJ, Bradshaw MF (1995) Disparity scaling and the perception of frontoparallel surfaces. *Perception* 24(2): 155–179

36 van Beers RJ, Sittig AC, van der Gon JJD (1998) The precision of proprioceptive position sense. *Exp Brain Res* 122(4): 367–377

37 van Beers RJ, Sittig AC, van der Gon JJD (1999) Integration of proprioceptive and visual position-information: An experimentally supported model. *J Neurophysiol*ogy 81(3): 1355–1364

38 Young MJ, Landy MS, Maloney LT (1993) A perturbation analysis of depth perception from combinations of texture and motion cues. *Vision Res* 33(18): 2685–2696

39 Bresciani J-P, Dammeier F, Ernst MO (2006) Vision and touch are automatically integrated for the perception of sequences of events. *J Vision* 6(5): 554–564

40 Helbig HB, Ernst MO (2007) Optimal integration of shape information from vision and touch. *Exp Brain Res* 179(4): 595–606

41 Hillis JM, Watt SJ, Landy MS, Banks MS (2004) Slant from texture and disparity cues: optimal cue combination. *J Vision* 4(12): 967–992

42 Knill DC, Saunders JA (2003) Do humans optimally integrate stereo and texture information for judgments of surface slant? *Vision Res* 43(24): 2539–2558

43 Landy MS, Kojima H (2001) Ideal cue combination for localizing texture-defined edges. *J Optical Society of America. A, Optics, image science, and vision* 18(9): 2307–2320

44 Shams L, Ma WJ, Beierholm U (2005) Sound-induced flash illusion as an optimal percept. *Neuroreport* 16(17): 1923–1927

45 van Beers RJ, Wolpert DM, Haggard P (2002) When feeling is more important than seeing in sensorimotor adaptation. *Curr Biol* 12(20): 834–837

46 Bülthoff HH, Mallot HA (1988) Integration of depth modules: stereo and shading. *J Optical Society of America. A, Optics and image science* 5(10): 1749–1758

47 Banks MS, Backus BT (1998) Extra-retinal and perspective cues cause the small range of the induced effect. *Vision Res* 38(2): 187–194

48 Knill DC, Kersten D, Yuille AL (1996) Introduction: A Bayesian formulation of visual perception. In: DC Knill, W Richards (eds): *Perception as Bayesian Inference.* Cambridge University Press, Cambridge, UK 1–21

49 Maloney LT (2001) Statistical decision theory and

biological vision. In: D Heyer, R Mausfeld (eds): *Perception and the Physical World*. Wiley, Chichester, UK

50 Mamassian P, Landy MS, Maloney LT (2002) Bayesian modeling of visual perception. In: RPN Rao, BA Olshausen, MS Lewicki (eds): *Probabilistic Models of the Brain: Perception and Neural Function*. MIT Press, Cambridge, MA, 13–36

51 Yuille AL, Bülthoff HH (1996) Bayesian theory and psychophysics. In D Knill, W Richards (eds): *Perception as Bayesian Inference*. Cambridge University Press, Cambridge, MA

52 Ernst MO, Bülthoff HH (2004) Merging the senses into a robust percept. *Trends in Cognitive Sci* 8(4): 162–169

53 Bayes T (1783) An essay towards solving a problem in the doctrine of chances. *Philosophical Transactions of the Royal Society* 53: 370–418

54 Ramachandran VS (1988) Perception of shape from shading. *Nature* 331: 163–166

55 Adams WJ, Graf EW, Ernst MO (2004). Experience can change the 'light-from-above' prior. *Nature Neuroscience* 7(10): 1057–1058

56 Trommershäuser J, Maloney LT, Landy MS (2003) Statistical decision theory and the selection of rapid, goal-directed movements. *J Opt Soc Am A Opt Image Sci Vis* 20(7): 1419–1433

57 Trommershäuser J, Landy MS, Maloney LT (2006) Humans rapidly estimate expected gain in movement planning. *Psychol Sci* 17(11): 981–988

58 Clark JJ, Yuille AL (1990) *Data Fusion for Sensory Information Processing Systems*. Kluwer Academic Publishers

59 Ernst MO (2005) A bayesian view on multimodal cue integration. In: G Knoblich, M Grosjean, I Thornton, M Shiffrar (eds): *Human body perception from the inside out*, chapter 6, (105–131). Oxford University Press, New York, NY

60 Ernst MO, Banks MS, Bülthoff HH (2000) Touch can change visual slant perception. *Nature Neuroscience* 3(1): 69–73

61 Howard IP, Rogers BJ (2002) *Seeing in Depth. Volume II Depth Perception*. I. Porteous, Toronto

62 Atkins JE, Fiser J, Jacobs RA (2001) Experience-dependent visual cue integration based on consistencies between visual and haptic percepts. *Vision Res* 41(4): 449–461

63 Knill DC (1998b) Surface orientation from texture: ideal observers, generic observers and the information content of texture cues. *Vision Res* 38(11): 1655–1682

64 Knill DC (2001) Contour into texture: information content of surface contours and texture flow. *J Opt Soc Am A Opt Image Sci Vis* 18(1): 12–35

65 Ernst MO (2007) Learning to integrate arbitrary signals from vision and touch. *J Vision* 7(5): 1–14

66 Choe C, Welch R, Gilford R, Juola J (1975). The "ventriloquist effect": Visual dominance or response bias? *Perception & Psychophysics* 18: 55–60

67 Jack CE, Thurlow WR (1973). Effects of degree of visual association and angle of displacement on the "ventriloquism" effect. *Perceptual and Motor Skills* 37(3): 967–979

68 Klemm O (1909) Lokalisation von Sinneseindrücken bei disparaten Nebenreizen (localization of sense impressions with discordant additional stimulation). *Psychologische Studien* 5: 73–161

69 Radeau M, Bertelson P (1977) Adaptation to auditory-visual discordance and ventriloquism in semi-realistic situations. *Perception & Psychophysics* 22: 137–146

70 Radeau M, Bertelson P (1987) Auditory-visual interaction and the timing of inputs. *Psychological Res* 49: 17–22

71 Thomas GJ (1941) Experimental study of the influence of vision on sound localization. *J Exp Psychology* 28: 163–177

72 Bertelson P, Radeau M (1981) Crossmodal bias and perceptual fusion with auditory-visual discordance. *Perception & Psychophysics* 29: 578–584

73 Colin C, Radeau M, Deltenre P, Morais J (2001) Rules of intersensory integration in spatial scene analysis and speech reading. *Psychologica Belgica* 41: 131–144

74 Jackson CV (1953) Visual factors in auditory localization. *Quarterly J Exp Psycho*logy 5: 52–65

75 Meredith MA, Stein BE (1986) Spatial factors determine the activity of multisensory neurons in cat superior colliculus. *Brain Res* 365(2): 350–354

76 Thurlow WR, Jack CE (1973) Certain determinants of the "ventriloquism effect". *Perceptual and Motor Skills* 36(3): 1171–1184

77 Bresciani J-P, Ernst MO, Drewing K, Bouyer G, Maury V, Kheddar A (2005) Feeling what you hear:

auditory signals can modulate tactile tap perception. *Exp Brain Res* 162(2): 172–180

78 Shams L, Kamitani Y, Shimojo S (2002) Visual illusion induced by sound. *Cognitive Brain Res* 14(1): 147–152

79 Slutsky DA, Recanzone GH (2001) Temporal and spatial dependency of the ventriloquism effect. *Neuroreport* 12(1): 7–10

80 Gepshtein S, Burge J, Ernst MO, Banks MS (2005) The combination of vision and touch depends on spatial proximity. *J Vision* 5(11): 1013–1023

81 Warren DH, Cleaves WT (1971) Visual-proprioceptive interaction under large amounts of conflict. *J Exp Psychol*ogy 90(2): 206–214

82 Witkin HA, Wapner S, Leventhal T (1952) Sound localization with conflicting visual and auditory cues. *J Exp Psychol*ogy 43(1): 58–67

83 Bedford FL (2001) Towards a general law of numerical/object identity. *Cahiers de Psychologie Cognitives/Current Psychology of Cognition* 20: 113–175

84 Miller EA (1972) Interaction of vision and touch in conflict and nonconflict form perception tasks. *J Exp Psychol*ogy 96(1): 114–123

85 Welch RB (1972) The effect of experienced limb identity upon adaptation to simulated displacement of the visual field. *Perception & Psychophysics* 12: 453–456

86 Warren DH, Welch RB, McCarthy TJ (1981) The role of visual-auditory “compellingness” in the ventriloquism effect: implications for transitivity among the spatial senses. *Perception & Psychophysics* 30(6): 557–564

87 Radeau M, Bertelson P (1974) The after-effects of ventriloquism. *The Quarterly J Exp Psychol*ogy 26(1): 63–71

88 Radeau M, Bertelson P (1978) Cognitive factors and adaptation to auditory-visual discordance. *Perception & Psychophysics* 23(4): 341–343

89 Radeau M, Bertelson P (1976) The effect of a textured visual field on modality dominance in a ventriloquism situation. *Perception & Psychophysics* 20: 227–235

90 Helbig HB, Ernst MO (2007a) Knowledge about a common source can promote visual haptic integration. *Perception* 36: 1523–1533

91 Körding KP, Beierholm U, Ma WJ, Quartz S, Tenenbaum JB, Shams L (2007) Causal inference in multisensory perception. *PLoS ONE* 2(9): e943

19 Haptically evoked activation of visual cortex

1 Pascual-Leone A, Hamilton RH (2001) The metamodal organization of the brain. *Prog Brain Res* 134: 427–445

2 Sathian K, Zangaladze A, Hoffman JM, Grafton ST (1997) Feeling with the mind’s eye. *Neuroreport* 8: 3877–3881

3 Sergent J, Ohta S, MacDonald B (1992) Functional neuroanatomy of face and object processing. A positron emission tomography study. *Brain* 115: 15–36

4 Mellet E, Tzourio N, Crivello F, Joliot M, Denis M, Mazoyer B (1996) Functional anatomy of spatial imagery generated from verbal instructions. *J Neurosci* 16: 6504–6512

5 Zhang M, Mariola E, Stilla R, Stoesz M, Mao H, Hu X, Sathian K (2005) Tactile discrimination of grating orientation: fMRI activation patterns. *Hum Brain Mapp* 25: 370–377

6 van Boven RW, Ingeholm JE, Beauchamp MS, Bikle PC, Ungerleider LG (2005) Tactile form and location processing in the human brain. *Proc Natl Acad Sci USA* 102: 12601–12605

7 Kitada R, Kito T, Saito DN, Kochiyama T, Matsumura M, Sadato N et al (2006) Multisensory activation of the intraparietal area when classifying grating orientation: A functional magnetic resonance imaging study. *J Neurosci* 26: 7491–7501

8 Walsh V, Rushworth M (1999) A primer of magnetic stimulation as a tool for neuropsychology. *Neuropsychologia* 37: 125–135

9 Sack AT (2006) Transcranial magnetic stimulation, causal structure-function mapping and networks of functional relevance. *Curr Opin Neurobiol* 16: 593–599

10 Zangaladze A, Epstein CM, Grafton ST, Sathian K (1999) Involvement of visual cortex in tactile discrimination of orientation. *Nature* 401: 587–590

11 Merabet L, Thut G, Murray B, Andrews J, Hsiao S, Pascual-Leone A (2004) Feeling by sight or seeing by touch? *Neuron* 42: 173–179

12 Merabet L, Swisher JD, McMains SA, Halko MA, Amedi A, Pascual-Leone A et al (2007) Combined activation and deactivation of visual cortex during tactile sensory processing. *J Neurophysiol* 97: 1633–1641

13 Malach R, Reppas JB, Benson RR, Kwong KK, Jiang H, Kennedy WA et al (1995) Object-related activity revealed by functional magnetic resonance imaging in human occipital cortex. *Proc Natl Acad Sci USA* 92: 8135–8139

14 Amedi A, Malach R, Hendler T, Peled S, Zohary E (2001) Visuo-haptic object-related activation in the ventral visual pathway. *Nature Neurosci* 4: 324–330

15 Stilla R, Sathian K (2007) Selective visuo-haptic processing of shape and texture. *Hum Brain Mapp* Advance online doi: 10.1002/hbm.20456

16 Zhang M, Weisser VD, Stilla R, Prather SC, Sathian K (2004) Multisensory cortical processing of object shape and its relation to mental imagery. *Cogn Affect Behav Ne* 4: 251–259

17 Amedi A, Jacobson G, Hendler T, Malach R, Zohary E (2002) Convergence of visual and tactile shape processing in the human lateral occipital complex. *Cereb Cortex* 12: 1202–1212

18 Stoesz M, Zhang M, Weisser VD, Prather SC, Mao H, Sathian K (2003) Neural networks active during tactile form perception: common and differential activity during macrospatial and microspatial tasks. *Int J Psychophysiol* 50: 41–49

19 Prather SC, Votaw JR, Sathian K (2004) Task-specific recruitment of dorsal and ventral visual areas during tactile perception. *Neuropsychologia* 42: 1079–1087

20 Amedi A, Stern WM, Camprodon JA, Bermpohl F, Merabet L, Rotman S et al (2007) Shape conveyed by visual-to-auditory sensory substitution activates the lateral occipital complex. *Nature Neurosci* 10: 687–689

21 Feinberg TE, Rothi LJ, Heilman KM (1986) Multimodal agnosia after unilateral left hemisphere lesion. *Neurology* 36: 864–867

22 James TW, James KH, Humphrey GK, Goodale MA (2006) Do visual and tactile object representations share the same neural substrate? In: MA Heller, S Ballesteros (eds): *Touch and blindness: psychology and neuroscience*. Lawrence Erlbaum Associates, Mahwah, NJ, 139–155

23 Grefkes C, Geyer S, Schormann T, Roland P, Zilles K (2001) Human somatosensory area 2: observer-independent cytoarchitectonic mapping, interindividual variability, and population map. *NeuroImage* 14: 617–631

24 Grefkes C, Weiss PH, Zilles K, Fink GR (2002) Crossmodal processing of object features in human anterior intraparietal cortex: an fMRI study implies equivalencies between humans and monkeys. *Neuron* 35: 173–184

25 Saito DN, Okada T, Morita Y, Yonekura Y, Sadato N (2003) Tactile-visual cross-modal shape matching: a functional MRI study. *Cognitive Brain Res* 17: 14–25

26 Mechelli A, Price CJ, Friston KJ, Ishai A (2004) Where bottom-up meets top-down: neuronal interactions during perception and imagery. *Cereb Cortex* 14: 1256–1265

27 Stilla R, Deshpande G, LaConte S, Hu X, Sathian K (2007) Posteromedial parietal cortical activity and inputs predict tactile spatial acuity. *J Neurosci* 27: 11091–11102

28 Klatzky RL, Lederman SJ, Reed C (1987) There's more to touch than meets the eye: the salience of object attributes for haptics with and without vision. *J Exp Psychol Gen* 116: 356–369

29 Heller MA (1989) Texture perception in sighted and blind observers. *Percept Psychophys* 45: 49–54

30 Blake R, Sobel KV, James TW (2004) Neural synergy between kinetic vision and touch. *Psychol Sci* 15: 397–402

31 Hagen MC, Franzen O, McGlone F, Essick G, Dancer C, Pardo JV (2002) Tactile motion activates the human middle temporal/V5 (MT/V5) complex. *Eur J Neurosci* 16: 957–964

32 Ricciardi E, Vanello N, Sani L, Gentili C, Scilingo EP, Landini L et al (2007) The effect of visual experience on the development of functional architecture in hMT+. *Cereb Cortex* 17: 2933–2939

33 James TW, Blake R (2004) Perceiving object motion using vision and touch. *Cogn Affect Behav Ne* 4: 201–207

34 Craig JC (2006) Visual motion interferes with tactile motion perception. *Perception* 35: 351–367

35 Poirier C, Collignon O, De Volder AG, Renier L, Vanlierde A, Tranduy D et al (2005) Specific activation of the V5 brain area by auditory motion

processing: An fMRI study. *Cognitive Brain Res* 25: 650–658
36 Beauchamp MS, Yasar NE, Kishan N, Ro T (2007) Human MST but not MT responds to tactile stimulation. *J Neurosci* 27: 8261–8267
37 Freides D (1974) Human information processing and sensory modality: Cross-modal functions, information complexity, memory and deficit. *Psychol Bull* 81: 284–310
38 Newman SD, Klatzky RL, Lederman SJ, Just MA (2005) Imagining material *versus* geometric properties of objects: an fMRI study. *Cognitive Brain Res* 23: 235–246
39 De Volder AG, Toyama H, Kimura Y, Kiyosawa M, Nakano H, Vanlierde A et al (2001) Auditory triggered mental imagery of shape involves visual association areas in early blind humans. *NeuroImage* 14: 129–139
40 Lacey S, Campbell C (2006) Mental representation in visual/haptic crossmodal memory: evidence from interference effects. *Q J Exp Psychol* 59: 361–376
41 Sathian K (2004) Modality, quo vadis?: Comment. *Behav Brain Sci* 27: 413–414
42 Lewkowicz DJ (1994) Development of intersensory perception in human infants. In: DJ Lewkowicz, R Lickliter (eds): *The development of intersensory perception: comparative perspectives*. Lawrence Erlbaum Associates, Hove, UK, 165–203
43 Easton RD, Greene AJ, Srinivas K (1997) Transfer between vision and haptics: memory for 2-D patterns and 3-D objects. *Psychon Bull Rev* 4: 403–410
44 Easton RD, Srinivas K, Greene AJ (1997) Do vision and haptics share common representations? Implicit and explicit memory within and between modalities. *J Exp Psychol Learn* 23: 153–163
45 Reales JM, Ballesteros S (1999) Implicit and explicit memory for visual and haptic objects: cross-modal priming depends on structural descriptions. *J Exp Psychol Learn* 25: 644–663
46 James TW, Humphrey GK, Gati JS, Servos P, Menon RS, Goodale MA (2002) Haptic study of three-dimensional objects activates extrastriate visual areas. *Neuropsychologia* 40: 1706–1714
47 Pietrini P, Furey ML, Ricciardi E, Gobbini MI, Wu W-HC, Cohen L et al (2004) Beyond sensory images: Object-based representation in the human ventral pathway. *Proc Natl Acad Sci USA* 101: 5658–5663
48 Blajenkova O, Kozhevnikov M, Motes MA (2006) Object-spatial imagery: a new self-report imagery questionnaire. *Appl Cognitive Psych* 20: 239–263
49 Kozhevnikov M, Kosslyn SM, Shephard J (2005) Spatial *versus* object visualisers: a new characterisation of cognitive style. *Mem Cognition* 33: 710–726
50 Kozhevnikov M, Hegarty M, Mayer RE (2002) Revising the visualiser-verbaliser dimension: evidence for two types of visualisers. *Cognition Instruct* 20: 47–77
51 Lacey S, Peters A, Sathian K (2007) Cross-modal object representation is viewpoint-independent. *PLoS ONE* 2: e890. doi: 10.1371/journal.pone0000890
52 Newell FN, Ernst MO, Tjan BS, Bulthoff HH (2001) Viewpoint dependence in visual and haptic object recognition. *Psychol Sci* 12: 37–42
53 James TW, Humphrey GK, Gati JS, Menon RS, Goodale MA (2002) Differential effects of viewpoint on object-driven activation in dorsal and ventral streams. *Neuron* 35: 793–801
54 Eger E, Henson RNA, Driver J, Dolan RJ (2004) BOLD repetition decreases in object-responsive ventral visual areas depend on spatial attention. *J Neurophysiol* 92: 1241–1247
55 Valyear KF, Culham JC, Sharif N, Westwood D, Goodale MA (2006) A double dissociation between sensitivity to changes in object identity and object orientation in the ventral and dorsal visual streams: A human fMRI study. *Neuropsychologia* 44: 218–228
56 Grill-Spector K, Kushnir T, Edelman S, Avidan G, Itzchak Y, Malach R (1999) Differential processing of objects under various viewing conditions in the human lateral occipital complex. *Neuron* 24: 187–203
57 Kosslyn SM (1973) Scanning visual images: some structural implications. *Percept Psychophys* 14: 90–94
58 Kosslyn SM, Ball TM, Reiser BJ (1978) Visual images preserve metric spatial information: evidence from studies of image scanning. *J Exp Psychol Human* 4: 47–60
59 Röder B, Rösler F (1998) Visual input does not

facilitate the scanning of spatial images. *J Mental Imagery* 22: 165–181

60 Shepard RN, Metzler J (1971) Mental rotation of three-dimensional objects. *Science* 171: 701–703

61 Marmor GS, Zaback LA (1976) Mental rotation by the blind: does mental rotation depend on visual imagery? *J Exp Psychol Human* 2: 515–521

62 Carpenter PA, Eisenberg P (1978) Mental rotation and the frame of reference in blind and sighted individuals. *Percept Psychophys* 23: 117–124

63 Hollins M (1986) Haptic mental rotation: more consistent in blind subjects? *J Visual Impair Blin* 80: 950–952

64 Dellantonio A, Spagnolo F (1990) Mental rotation of tactual stimuli. *Acta Psychol* 73: 245–257

65 Prather SC, Sathian K (2002) Mental rotation of tactile stimuli. *Cognitive Brain Res* 14: 91–98

66 Alivisatos B, Petrides M (1997) Functional activation of the human brain during mental rotation. *Neuropsychologia* 36: 11–118

67 Peltier S, Stilla R, Mariola E, LaConte S, Hu X, Sathian K (2007) Activity and effective connectivity of parietal and occipital cortical regions during haptic shape perception. *Neuropsychologia* 45: 476–483

68 Falchier A, Clavagnier S, Barone P, Kennedy H (2002) Anatomical evidence of multimodal integration in primate striate cortex. *J Neurosci* 22: 5749–5759

69 Rockland KS, Ojima H (2003) Multisensory convergence in calcarine visual areas in macaque monkey. *Int J Psychophysiol* 50: 19–26

70 Lakatos P, Chen C-M, O'Connell MN, Mills A, Schroeder CE (2007) Neuronal oscillations and multisensory interactions in primary auditory cortex. *Neuron* 53: 279–292

71 Schroeder CE, Smiley J, Fu KG, McGinnis T, O'Connell MN, Hackett TA (2003) Anatomical mechanisms and functional implications of multisensory convergence in early cortical processing. *Int J Psychophysiol* 50: 5–17

72 Haenny PE, Maunsell JHR, Schiller PH (1988) State dependent activity in monkey visual cortex. II. Retinal and extraretinal factors in V4. *Exp Brain Res* 69: 245–259

73 Zhou Y-D, Fuster JM (1997) Neuronal activity of somatosensory cortex in a cross-modal (visuohaptic) task. *Exp Brain Res* 116: 551–555

20 Haptic perception and synaesthesia

1 Day S (2005) Some demographic and socio-cultural aspects of synesthesia. In: LC Robertson, N Sagiv (eds): *Synesthesia: Perspectives from cognitive neuroscience*. Oxford University Press, Oxford

2 Baron-Cohen S, Burt L, Smith-Laittan F, Harrison J, Bolton P (1996) Synaesthesia: Prevalence and familiality. *Perception* 25: 1073–1079

3 Ward J, Mattingley JB (2006) Synaesthesia: An overview of contemporary findings and controversies. *Cortex* 42: 129–136

4 Robertson LC (2003) Binding, spatial attention and perceptual awareness. *Nature Reviews Neuroscience* 4: 93–102

5 Ward J, Simner J (2005) Is synaesthesia an X-linked dominant trait with lethality in males? *Perception* 34: 611–623

6 Simner J, Mulvenna C, Sagiv N, Tsakanikos E, Witherby SA, Fraser C, Scott K, Ward J (2006) Synaesthesia: The prevalence of atypical cross-modal experiences. *Perception* 35: 1024–1033

7 Hollister LE (1968) *Chemical psychoses: LSD and related drugs*. Charles C. Thomas, Springfield, Illinois

8 Rao AL, Nobre AC, Alexander I, Cowey A (2007) Auditory evoked visual awareness following sudden ocular blindness: an EEG and TMS investigation. *Exp Brain Res* 176: 288–298

9 Baron-Cohen S, Harrison J, Goldstein LH, Wyke M (1993) Coloured speech perception: Is synaesthesia what happens when modularity breaks down? *Perception* 22: 419–426

10 Ramachandran VS, Hubbard EM (2001) Psychophysical investigations into the neural basis of synaesthesia. *Proceedings of the Royal Society of London B* 268: 979–983

11 Rouw R, Scholte HS (2007) Increased structural connectivity in grapheme-color synesthesia. *Nature Neuroscience* 10: 792–797

12 Falchier A, Clavagnier S, Barone P, Kennedy H (2002) Anatomical evidence of multimodal inte-

gration in primate striate cortex. *J Neurosci* 22: 5749–5759

13 Ward J (2008) *The Frog who Croaked Blue: Synesthesia and the Mixing of the Senses*. Routledge, Oxford

14 Ward J, Huckstep B, Tsakanikos E (2006) Sound-colour synaesthesia: To what extent does it use cross-modal mechanisms common to us all? *Cortex* 42: 264–280

15 Marks LE (2004) Cross-modal interactions in speeded classification. In: G Calvert, C Spence, BE Stein (eds): *The handbook of multisensory processes*. MIT Press, Cambridge, MA

16 Kennett S, Taylor-Clarke M, Haggard P (2001) Non-informative vision improves the spatial resolution of touch in humans. *Current Biology* 11: 1188–1191

17 Pavani F, Spence C, Driver J (2000) Visual capture of touch: Out-of-the-body experiences with rubber gloves. *Psychological Science* 11: 353–359

18 Ramachandran VS, Hirstein W (1998) The perception of phantom limbs. *Brain* 121: 1603–1630

19 Ramachandran VS, Rogers-Ramachandran D (1996) Synaesthesia in phantom limbs induced with mirrors. *Proceedings of the Royal Society of London B* 263: 377–386

20 Mon-Williams M, Wann JP, Jenkinson M, Rushton K (1997) Synaesthesia in the normal limb. *Proceedings of the Royal Society of London B* 264: 1007–1010

21 Halligan PW, Hunt M, Marshall JC, Wade DT (1996) When seeing is feeling: Acquired synaesthesia or phantom touch? *Neurocase* 2: 21–29

22 Bradshaw JL, Mattingley JM (2001) Allodynia: A sensory analogue of motor mirror neurons in a hyperaesthetic patient reporting instantaneous discomfort to another's perceived sudden minor injury? *J Neurology, Neurosurgery and Psychiatry* 70: 135–136

23 Blakemore SJ, Bristow D, Bird G, Frith C, Ward J (2005) Somatosensory activations during the observation of touch and a case of vision-touch synesthesia. *Brain* 128: 1571–1583

24 Banissy M, Ward J (2007) Mirror touch synaesthesia enhances empathy. *Nature Neuroscience* 10: 815–816:

25 Keysers C, Wicker B, Gazzola V, Anton JL, Fogassi L, Gallese V(2004) A touching sight: SII/PV activation during the observation and experience of touch. *Neuron* 42: 335–346

26 Armel KC, Ramachandran VS (1999) Acquired synaesthesia in retinitis pigmentosa. *Neurocase* 5: 293–296

27 Macaluso E, Frith CD, Driver J (2002) Crossmodal spatial influences of touch on extrastriate visual areas take current gaze direction into account. *Neuron* 34: 647–658

28 Smith HL (1905) Synesthesia. *John Hopkins Hospital Bulletin* 16: 258–263

29 Dudycha GJ, Dudycha MM (1935) A case of synesthesia: Visual pain and visual audition. J *Abnormal Psychology* 30: 57–69

30 Whipple GM (1900) Two cases of synaesthesia. *Am J Psycho*logy 11: 377–404

31 Wheeler RH, Cutsforth TD (1921) The role of synaesthesia in learning. *J Exp Psycho*logy 4: 448–468

32 Steven MS, Blakemore C (2004) Visual synaesthesia in the blind. *Perception* 33: 855–868

33 Dixon MJ, Smilek D, Merikle PM (2004) Not all synaesthetes are created equal: Projector vs. associator synaesthetes. *Cognitive, Affective and Behavioral Neuroscience* 4: 335–343

34 Martino G, Marks LE (2000) Cross-modal interaction between vision and touch: The role of synesthetic correspondence. *Perception* 29: 745–754

35 Mattingley JB, Rich AN, Bradshaw JL (2001) Unconscious priming eliminates automatic binding of colour and alphanumeric form in synaesthesia. *Nature* 410: 580–582

21 Haptic perception in sexuality

1 Anzieu D (1996) *Das Haut-Ich*. Suhrkamp, Frankfurt am Main

2 Harlow HF, Harlow MK, Hansen EW (1963) The maternal affectional systems of rhesus monkeys. In: HL Rheingold (ed): *Maternal behavior in mammals*. Wiley, New York

3 Harlow HF, Zimmerman R (1959) Affectional responses in the infant monkey. *Science* 130: 421–422

4 Gieler U (1988) Skin and body experience. In: E Brähler (ed): *Body experience. The subjective dimension of psyche and soma*. Springer, Berlin,

Heidelberg, New York, London, Paris, Tokyo, Hong Kong, Barcelona, Budapest, 62–73

5 Montagu A (1986) *Touching. The human significance of the skin.* 3rd Edition. Harper Perennial, London

6 Seikowski K (1997) Psychologische Aspekte der erektilen Dysfunktion. *WMW*: 105–108

7 Gödtel R (1996) Sexualität schon im Uterus? *Sexualmed* 18: 311–312

9 Spitz RA (1957) *Die Entstehung der ersten Objektbeziehungen.* Klett, Stuttgart

9 Schmidt-Sibeth F (1989) Glückliche Eltern haben liebesfähige Kinder. *Sexualmed* 18: 441–447

10 Worm G (1996) Berührung als Abstinenzverletzung – Berührung als Heilungsweg. In: H Richter-Appelt (ed) V*erführung, Trauma, Missbrauch (1896–1996).* Psychosozial-Verlag, Giessen, 51–67

11 Egle UT, Hoffmann SO, Joraschky P (eds) (1997) *Sexueller Missbrauch, Misshandlung, Vernachlässigung: Erkennung und Behandlung psychischer und psychosomatischer Folgen früher Traumatisierung.* Schattauer, Stuttgart, New York

12 Ferenczi S (1933) Sprachverwirrungen zwischen den Erwachsenen und dem Kind. Die Sprache der Zärtlichkeit und der Leidenschaft. In: *Bausteine zur Psychoanalyse*, Bd. II, Fischer, Frankfurt am Main

13 Bornemann E (1980) Puberale Amnesie. Die Sexualität des Kindes und ihre erkenntnistheoretischen Folgen. *Psychoanalyse* 1: 62–76

14 Baurmann MC (1996) *Sexualität, Gewalt und psychische Folgen.* Bundeskriminalamt 2, Wiesbaden Aufl

15 Julius H, Boehme U (1997) *Sexuelle Gewalt gegen Jungen.* Verlag für Angewandte Psychologie, Göttingen

16 Lange C (1998) *Sexuelle Gewalt gegen Mädchen.* Ergebnisse einer Studie zur Jugendsexualität. Enke, Stuttgart

17 Bauserman R, Rind B (1997) Psychological correlates of male child and adolescent sexual experiences with adults: review of the non-clinical literature. *Arch Sex Behav* 26: 105–141

18 Rind B, Tromovitch P, Bauserman R (1998) A metaanalytic examination of assumed properties of child sexual abuse using college samples. *Psychol Bulletin* 124: 22–53

19 Struckman-Johnson C, Struckmann-Johnson D (1994) Men pressured and forced into sexual experience. *Arch Sex Behav* 23: 93–114

20 Niemeier V, Winckelsesser Th, Gieler U (1997) Hautkrankheit und Sexualität. Eine empirische Studie zum Sexualverhalten von Patienten mit Psoriasis vulgaris und Neurodermitis im Vergleich mit Hautgesunden. *Hautarzt* 48: 629–633

21 Bogaerts F, Boeckx W (1992) Burns and sexuality. *J Burn Care Rehabil* 13: 39–43

22 Sachsse U (1997) *Selbstverletzendes Verhalten.* Vandenhoeck & Ruprecht, Göttingen

23 Bornemann E (1995) Sexualität heute (I). *Sexualmed* 17: 309–312

24 Bornemann E (1995) Sexualität heute (II). *Sexualmed* 17: 329–332

25 Fröhlich HH (1997) Sexualität heute und künftig. Von partnerschaftlicher Sexualität zu Sexualität ohne Partnerschaft. *Sexualmed* 19: 268–276

26 Fröhlich HH (1996) Sexualität im Alter. Eine sexualpsychologische und –soziologische Studie. *Sexualmed* 18: 334–340

27 Kockott G (1997) Sexualität kennt keine Altersgrenze. *Sexualmed* 19: 10–14

28 Haeberle EJ (1998) Was ist sexuelle Gesundheit? Eine kritische Würdigung der WHO-Definition. *Sexualmed* 20: 143–148

29 Tiefer L (1993) Über die fortschreitende Medikalisierung männlicher Sexualität. *Z Sexualforsch* 6: 119–131

30 Seikowski K (Ed.) (2005) *Sexualität und Neue Medien.* Pabst Science Publishers, Lengerich, Berlin, Bremen, Riga, Rom, Viernheim, Wien, Zagreb

31 Voigt H (1991) Enriching the sexual experience of couples: the Asian traditions and sexual counseling. *J Sex Marital Ther* 17: 214–219

32 Masters WH, Johnson VE (1966) *Human Sexual Response*. Churchill, Livingstone

33 Kochenstein P (1999) Der Weg zur gemeinsamen Lust. Verhaltenstherapeutische Behandlung funktioneller Sexualstörungen. *Sexualmed* 21: 70–74

34 Derbolowsky Y (1989) Wenn tabuisierte Körperzonen Probleme bereiten. *Sexualmed* 18: 392–401

35 Borelli S (1967) Psyche und Haut. In: J Jadassohn (Ed.) *Handbuch der Haut- und Geschlechtskrankheiten.* Ergänzungswerk Bd 8, Springer-Verlag, Berlin, Heidelberg, New York, 264–568

36 Seikowski K, Haustein UF (1996) Chronische Erkrankungen der Haut – Vorbeugung und Verlaufsbeeinflussung. In: H Schröder, K Reschke (Eds.) *Intervention zur Gesundheitsförderung für Klinik und Alltag*. S. Roderer-Verlag, Regensburg: 121–143

22 Haptic perception in space travel

1 Fowler B, Comfort D, Bock O (2000) A review of cognitive and perceptual-motor performance in space. *Aviation, Space, and Environmental Medicine* 71, No. 9, Section 11: A66–68
2 Space Studies Board (1998) Sensorimotor integration. Ch.5. In: *A strategy for research in space biology and medicine in the new century*. Washington, D.C., National Research Council, 63–79
3 Reschke MF, Bloomberg JJ, Harm DL, Paloski WH, Layne C, McDonald V (1998) Posture, locomotion, spatial orientation, and motion sickness as a function of space flight. *Brain Research Reviews* 28: 102–117
4 Kornilova LN (1997) Orientation illusions in spaceflight. *J Vestibular Res* 7: 429–439
5 Jeka JJ, Lackner JR (1995) Fingertip touch as an orientation reference for human postural control. In: T Mergner, F Hlavacka (eds): *Multisensory control of posture*. Plenum Press, New York, 213–221
6 Gentez E, Hatwell Y (1996) Role of gravitational cues in the haptic perception of orientation. *Perception and Psychophysics* 58: 1278–1292
7 Ross HE (1989) Perceptual and motor skills of divers under water. *Int Rev Ergonomics* 2: 155–181
8 Young LR, Shelhamer M (1990) Microgravity enhances the relative contribution of visually-induced motion sensation. *Aviation, Space, and Environmental Medicine* 61: 525–530
9 Arnesen NT, Olsen MH, Sylvestre B, Clément G (2005) Perception of body vertical in microgravity during parabolic flight. *J Gravitational Physiology* 12: P17–18
10 Young LR, Oman CM, Watt DGD, Money KE, Lichtenberg BK (1984) Spatial orientation in weightlessness and readaptation to Earth's gravity. *Science* 225: 205–208
11 Fisk J, Lackner JR, DiZio P (1993) Gravitoinertial force level influences arm movement control. *J Neurophysiology* 69: 504–511
12 Veringa F (1987) Arm positioning in microgravity during D1 Challenger flight. In: PR Sahm, R Jansen, MH Keller (eds): *Proc Norderney Symposium on Scientific Results of the German Spacelab Mission D1*. DFVLR, Cologne, 537–540
13 Ross HE, Farkin B (1991) Knowledge of arm position under varied gravitoinertial force in parabolic flight. In: V Pletser, JF Couffy (eds): *Microgravity Experiments during Parabolic Flights with Caravelle*. ESTEC, Netherlands, ESA WPP-021, 147–152
14 Bock O (1994) Joint position sense in simulated changed-gravity environments. *Aviation, Space, and Environmental Medicine* 65: 621–626
15 Watt DGD, Money KE, Bondar RL, Thirsk RB, Garneau M, Scully-Power P (1985) Canadian medical experiments on shuttle flight 41-G. *Canadian Aeronautical and Space Journal* 31: 215–226
16 Young LR, Oman CM, Merfeld D et al (1993) Spatial orientation and posture during and following weightlessness: human experiments on Spacelab Life Sciences1. *J Vestibular Res* 3: 231–239
17 Watt DGD (1997) Pointing at memorized targets during prolonged microgravity. *Aviation, Space and Environmental Medicine* 68: 99–103
18 Whiteside TCD (1961) Hand-eye coordination in weightlessness. *Aerospace Medicine* 32: 318–322
19 Bock O, Howard IP, Money KE, Arnold KE (1992) Accuracy of aimed arm movements in changed gravity. *Aviation, Space, and Environmental Medicine* 63: 994–998
20 Bock O, Fowler B, Comfort D (2001) Human sensorimotor coordination during spaceflight: An analysis of pointing and tracking responses during the "Neurolab" space shuttle mission. *Aviation, Space, and Environmental Medicine* 72: 877–883
21 Ross HE (1991) Motor skills under varied gravitoinertial force in parabolic flight. *Acta Astronautica* 23: 85–89
22 Bock O (1998) Problems of sensorimotor coordination in weightlessness. *Brain Res Reviews* 28: 155–160
23 Sangals J, Heuer H, Manzey D, Lorenz B (1999) Changed visuomotor transformations during and after prolonged microgravity. *Exp Brain Res* 129: 378–390

24 Flanagan JR, Vetter P, Johansson RS, Wolpert DM (2003) Prediction precedes control in motor learning. *Current Biology* 13: 146–150

25 Hermsdörfer J, Marquardt C, Philipp J, Zierdt A, Nowak D, Glasauer S, Mai N (2000) Moving weightless objects. Grip force control during microgravity. *Exp Brain Res* 132: 52–64

26 Ross HE, Reschke MF (1982) Mass estimation and discrimination during brief periods of zero gravity. *Perception and Psychophysics* 31: 429–436

27 Ross HE, Brodie EE (1987) Weber fractions for weight and mass as a function of stimulus intensity. *Quarterly J Exp Psychol*ogy 39A: 77–78

28 Ross HE, Brodie EE, Benson AJ (1986) Mass-discrimination in weightlessness and readaptation to earth's gravity. *Exp Brain Res* 64: 358–366

29 Ross HE, Schwartz E, Emmerson P (1987) The nature of sensorimotor adaptation to G-levels: Evidence from mass-discrimination. *Aviation, Space, and Environmental Medicine* 58 (9 Suppl): A148–152

30 Bock O, Arnold KE, Cheung BSK (1996) Performance of a simple aiming task in hypergravity. II. Detailed response characteristics. *Aviation, Space, and Environmental Medicine* 67: 133–138

31 Durlach NI, Mavor AS (1995) *Virtual reality – Scientific and technological challenges*. National Academy Press, Washington D.C.

32 Tan HZ, Srinivasan MA, Eberman B, Cheng B (1994) Human factors for the design of force-reflecting haptic interfaces. In: CJ Radcliffe (ed): *Dynamic systems and control*, Vol. 55–1. The American Society of Mechanical Engineers

33 Bock O (1996) Grasping of virtual objects in changed gravity. *Aviation, Space, and Environmental Medicine* 67: 1185–1189

23 Phantom sensations

1 Shukla GD, Sahu SC, Tripathi RP, Gupta DK (1982) A psychiatric study of amputees. *Br J Psychiatry* 141: 50–53

2 Shukla GD, Sahu SC, Tripathi RP, Gupta DK (1982) Phantom limb: A phenomenological study. *Br J Psychiatry* 141: 54–58

3 Hrbek V (1976) New pathophysiological interpretation of the so-called phantom limb and phantom pain syndromes. *Act Univ Palack Olomuc* 80: 79–90

4 Sherman RA, Sherman CJ, Parker L (1984) Chronic phantom and stump pain among American veterans: Results of a survey. *Pain* 18: 83–95

5 Sherman RA (1997) *Phantom pain*. Plenum, New York

6 Flor H, Nikolajsen L, Staehelin Jensen T (2006) Phantom limb pain: a case of maladaptive CNS plasticity? *Nat Rev Neurosci* 7: 873–881

7 Rothemund Y, Grusser SM, Liebeskind U, Schlag PM, Flor H (2004) Phantom phenomena in mastectomized patients and their relation to chronic and acute pre-mastectomy pain. *Pain* 107: 140–146

8 Waxman SG, Hains BC (2006) Fire and phantoms after spinal cord injury: Na^+ channels and central pain. *Trends Neurosci* 29: 207–215

9 Saadah ES, Melzack R (1994) Phantom limb experiences in congenital limb-deficient adults. *Cortex* 30: 479–485

10 Wilkins KL, McGrath PJ, Finley GA, Katz J (1998) Phantom limb sensations and phantom limb pain in child and adolescent amputees. *Pain* 78: 7–12

11 Krane EJ, Heller LB (1995) The prevalence of phantom sensation and pain in pediatric amputees. *J Pain Symptom Manage* 10: 21–29

12 Riese W, Bruck G (1950) Le membre fantôme chez l'entant. Rev *Neurol* 83: 221–222

13 Weinstein S, Sersen EA, Vetter RJ (1964) Phantoms and somatic sensation in cases of congenital aplasia. *Cortex* 1: 276–290

14 Melzack R, Israel R, Lacroix R, Schultz G (1997) Phantom limbs in people with congenital limb deficiency or amputation in early childhood. *Brain* 9: 1603–1620

15 Melzack R (1990) Phantom limbs and the concept of a neuromatrix. *Trends Neurosci* 13: 88–92

16 Melzack R (2005) Evolution of the neuromatrix theory of pain. *Pain Pract* 5: 85–94

17 Dettmers C, Liepert J, Adler T, Rzanny R, Rijntjes M, van Schayck R, Kaiser W, Brückner L, Weiller C (1999) Abnormal motor cortex organization contralateral to early upper limb amputation in humans. *Neurosci Lett* 263: 41–44

18 Weiss T, Miltner WHR, Dillmann J, Meissner W, Huonker R, Nowak H (1998) Reorganization of the

somatosensory cortex after amputation of the index finger. *Neuroreport* 9: 213–216

19 Ramachandran VS, Stewart M, Rogers RD (1992) Perceptual correlates of massive cortical reorganization. *Neuroreport* 3: 583–586

20 Ramachandran VS, Rogers Ramachandran D, Stewart M (1992) Perceptual correlates of massive cortical reorganization. *Science* 258: 1159–1160

21 Grüsser SM, Winter C, Mühlnickel W, Denke C, Karl A, Villringer K, Flor H (2001) The relationship of perceptual phenomena and cortical reorganization in upper extremity amputees. *Neuroscience* 102: 263–272

22 Knecht S, Henningsen H, Elbert T, Flor H, Hohling C, Pantev C, Taub E (1996) Reorganizational and perceptional changes after amputation. *Brain* 119: 1213–1219

23 Grüsser SM, Mühlnickel W, Schaefer M, Villringer K, Christmann C, Koeppe C, Flor H (2004) Remote activation of referred phantom sensation and cortical reorganization in human upper extremity amputees. *Exp Brain Res* 154: 97–102

24 Knecht S, Henningsen H, Hohling C, Elbert T, Flor H, Pantev C, Taub E (1998) Plasticity of plasticity? Changes in the pattern of perceptual correlates of reorganization after amputation. *Brain* 121: 717–724

25 Nyström B, Hagbarth KE (1981) Microelectrode recordings from transected nerves in amputees with phantom limb pain. *Neurosci Lett* 27: 211–216

26 Jänig W, McLachlan E (1984) On the fate of sympathetic and sensory neurons projecting into a neuroma of the superficial peroneal nerve in the cat. *J Comp Neurol* 225: 302–311

27 Fried K, Govrin-Lippman R, Rosenthal F, Ellisman MH, Devor M (1991) Ultrastructure of afferent axon endings in a neuroma. *J Neurocytol* 20: 682–701

28 Wall PD, Gutnick M (1974) Ongoing activity in peripheral nerves: the physiology and pharmacology of impulses originating from a neuroma. *Exp Neurol* 43: 580–593

29 Devor M (2005) Response of injured nerve. In: M Koltzenburg, SB McMahon (eds): *Wall and Melzack's textbook of pain*. Elsevier, Amsterdam, 905–927

30 Devor M (2006) Sodium channels and mechanisms of neuropathic pain. *J Pain* 7: S3–S12

31 Gorodetskaya N, Constantin C, Janig W (2003) Ectopic activity in cutaneous regenerating afferent nerve fibers following nerve lesion in the rat. *Eur J Neurosci* 18: 2487–2497

32 Blumberg H, Jänig W (1982) Activation of fibers via experimentally produced stump neuromas of skin nerves: ephaptic transmission or retrograde sprouting? *Exp Neurol* 76: 468–482

33 Blumberg H, Jänig W (1982) Changes in unmyelinated fibers including sympathetic postganglionic fibers of a skin nerve after peripheral neuroma formation. *J Auton Nerv Syst* 6: 173–183

34 Jensen TS, Nikolajsen LJ (2005) Phantom limb. In: M Koltzenburg, SB McMahon (eds): *Wall and Melzack's textbook of pain*. Elsevier, Amsterdam, 961–971

35 Chen Y, Michaelis M, Jänig W, Devor M (1996) Adrenoreceptor subtype mediating sympathetic-sensory coupling in injured sensory neurons. *J Neurophysiol* 76: 3721–3730

36 Katz J (1992) Psychophysiological contributions to phantom limbs. *Canadian J Psychiatry* 37: 282–298

37 Chabal C, Jacobson L, Russell LC, Burchiel KJ (1992) Pain response to perineuromal injection of normal saline, epinephrine, and lidocaine in humans. *Pain* 49: 9–12

38 Birbaumer N, Lutzenberger W, Montoya P, Larbig W, Unertl K, Topfner S, Grodd W, Taub E, Flor H (1997) Effects of regional anesthesia on phantom limb pain are mirrored in changes in cortical reorganization. *J Neurosci* 17: 5503–5508

39 Baron R, Maier C (1995) Phantom limb pain: are cutaneous nociceptors and spinothalamic neurons involved in the signaling and maintenance of spontaneous and touch-evoked pain? A case report. *Pain* 60: 223–228

Schmidt AP, Takahashi ME, de Paula Posso I (2005) Phantom limb pain induced by spinal anesthesia. *Clinics* 60: 263–264

40 Schmidt AP, Takahashi ME, de Paula Posso I (2005) Phantom limb pain induced by spinal anesthesia. *Clinics* 60: 263–264

41 Woolf CJ (2004) Dissecting out mechanisms responsible for peripheral neuropathic pain: implications for diagnosis and therapy. *Life Sci* 74: 2605–2610

42 Moore KA, Kohno T, Karchewski LA, Scholz J, Baba

H, Woolf CJ (2002) Partial peripheral nerve injury promotes a selective loss of GABAergic inhibition in the superficial dorsal horn of the spinal cord. *J Neurosci* 22: 6724–6731

43 Coull JA, Beggs S, Boudreau D, Boivin D, Tsuda M, Inoue K, Gravel C, Salter MW, De Koninck Y (2005) BDNF from microglia causes the shift in neuronal anion gradient underlying neuropathic pain. *Nature* 438: 1017–1021

44 Wang S, Lim G, Yang L, Zeng Q, Sung B, Martyn JA, Mao J (2005) A rat model of unilateral hindpaw burn injury: slowly developing rightwards shift of the morphine dose-response curve. *Pain* 116: 87–95

45 Wiesenfeld-Hallin Z, Xu XJ, Hökfelt T (2002) The role of spinal cholecystokinin in chronic pain states. *Pharmacol Toxicol* 91: 398–403

46 Squire LR, Kandel ER (1999) *Memory. From mind to molecules*. Scientific American Library, New York

47 Ueda H (2006) Molecular mechanisms of neuropathic pain-phenotypic switch and initiation mechanisms. *Pharmacol Ther* 109: 57–77

48 Devor M, Wall PD (1978) Reorganisation of spinal cord sensory map after peripheral nerve injury. *Nature* 276: 75–76

49 Melzack R, Loeser JD (1978) Phantom body pain in paraplegics: evidence for a central “pattern generating mechanism” for pain. *Pain* 4: 195–210

50 Davis KD, Kiss ZHT, Luo L, Tasker RR, Lozano AM, Dostrovsky JO (1998) Phantom sensations generated by thalamic microstimulation. *Nature* 391: 385–387

51 Florence SL, Taub HB, Kaas JH (1998) Large-scale sprouting of cortical connections after peripheral injury in adult macaque monkeys. *Science* 282: 1117–1121

52 Jones EG, Pons TP (1998) Thalamic and brainstem contributions to large-scale plasticity of primate somatosensory cortex. *Science* 282: 1121–1125

53 Kaas JH, Florence SL, Jain N (1999) Subcortical contributions to massive cortical reorganizations. Neuron 22: 657–660

54 Jain N, Florence SL, Qi HX, Kaas JH (2000) Growth of new brainstem connections in adult monkeys with massive sensory loss. *Proc Natl Acad Sci USA* 97: 5546–5550

55 Ergenzinger ER, Glasier MM, Hahm JO, Pons TP (1998) Cortically induced thalamic plasticity in the primate somatosensory system. *Nat Neurosci* 1: 226–229

56 Kaas JH (1999) Is most of neural plasticity in the thalamus cortical? *Proc Natl Acad Sci USA* 96: 7622–7623

57 Merzenich M, Nelson RJ, Stryker MP, Cynader MS, Schoppmann A, Zook JM (1984) Somatosensory cortical map changes following digit amputation in adult monkeys. *J Comp Neurol* 224: 591–605

58 Pons TP, Garraghty PE, Ommaya AK, Kaas JH, Taub E, Mishkin M (1991) Massive cortical reorganization after sensory deafferentation in adult macaques. *Science* 252: 1857–1860

59 Yang TT, Gallen CC, Schwartz B, Bloom FE, Ramachandran VS, Cobb S (1994) Sensory maps in the human brain. *Nature* 368: 592–593

60 Elbert T, Flor H, Birbaumer N, Knecht S, Hampson S, Larbig W, Taub E (1994) Extensive reorganization of the somatosensory cortex in adult humans after nervous system injury. *Neuroreport* 5: 2593–2597

61 Flor H, Elbert T, Knecht S, Wienbruch C, Pantev C, Birbaumer N, Larbig W, Taub E (1995) Phantom-limb pain as a perceptual correlate of cortical reorganization following arm amputation. *Nature* 375: 482–484

62 Flor H, Elbert T, Muhlnickel W, Pantev C, Wienbruch C, Taub E (1998) Cortical reorganization and phantom phenomena in congenital and traumatic upper-extremity amputees. *Exp Brain Res* 119: 205–212

63 Wu CWH, Kaas JH (1999) Reorganization in primary motor cortex of primates with long-standing therapeutic amputations. *J Neurosci* 19: 7679–7697

64 Cohen LG, Bandinelli S, Findley TW, Hallett M (1991) Motor reorganization after upper limb amputation in man. *Brain* 114: 615–627

65 Karl A, Birbaumer N, Lutzenberger W, Cohen LG, Flor H (2001) Reorganization of motor and somatosensory cortex in upper extremity amputees with phantom limb pain. *J Neurosci* 21: 3609–3618

66 Karl A, Muhlnickel W, Kurth R, Flor H (2004) Neuroelectric source imaging of steady-state movement-related cortical potentials in human upper extremity amputees with and without phantom limb pain. *Pain* 110: 90–102

67 Lotze M, Flor H, Grodd W, Larbig W, Birbaumer N

(2001) Phantom movements and pain – An MRI study in upper limb amputees. *Brain* 124: 2268–2277
68 Calford MB, Tweedale R (1991) C-fibres provide a source of masking inhibition to primary somatosensory cortex. *Proc R Soc Lond B* 243: 269–275
69 Weiss T, Miltner WHR (2006) Selektive C-Faser-Stimulation durch Stimulation winziger Hautareale. *Schmerz* 20: 238–244
70 Draganski B, Moser T, Lummel N, Ganssbauer S, Bogdahn U, Haas F, May A (2006) Decrease of thalamic gray matter following limb amputation. *NeuroImage* 31: 951–957
71 Bowlus TH, Lane RD, Stojic AS, Johnston M, Pluto CP, Chan M, Chiaia NL, Rhoades RW (2003) Comparison of reorganization of the somatosensory system in rats that sustained forelimb removal as neonates and as adults. *J Comp Neurol* 465: 335–348
72 Churchill JD, Muja N, Myers WA, Besheer J, Garraghty PE (1998) Somatotopic consolidation: a third phase of reorganization after peripheral nerve injury in adult squirrel monkeys. *Exp Brain Res* 118: 189–196
73 Weiss T, Miltner WHR, Huonker R, Friedel R, Schmidt I, Taub E (2000) Rapid functional plasticity of the somatosensory cortex after finger amputation. *Exp Brain Res* 134: 199–203
74 Churchill JD, Tharp JA, Wellman CL, Sengelaub DR, Garraghty PE (2004) Morphological correlates of injury-induced reorganization in primate somatosensory cortex. *BMC Neurosci* 5: 43
75 Weiss T, Miltner WHR, Liepert J, Meissner W, Taub E (2004) Rapid functional plasticity in the primary somatomotor cortex and perceptual changes after nerve block. *Eur J Neurosci* 20: 3413–3423
76 Levy LM, Ziemann U, Chen R, Cohen LG (2002) Rapid modulation of GABA in sensorimotor cortex induced by acute deafferentation. *Ann Neurol* 52: 755–761
77 Canu M-H, Treffort N, Picquet F, Dubreucq G, Guerardel Y, Falempin M (2006) Concentration of amino acid neurotransmitters in the somatosensory cortex of the rat after surgical or functional deafferentation. *Exp Brain Res* 173: 623–628
78 Larbig W, Montoya P, Flor H, Bilow H, Weller S, Birbaumer N (1996) Evidence for a change in neural processing in phantom limb pain patients. *Pain* 67: 275–283
79 Kaas JH, Florence SL, Jain N (1997) Reorganization of sensory systems of primates after injury. *Neuroscientist* 3: 123–130
80 Elbert T, Sterr A, Flor H, Rockstroh B, Knecht S, Pantev C, Wienbruch C, Taub E (1997) Input-increase and input-decrease types of cortical reorganization after upper extremity amputation in humans. *Exp Brain Res* 117: 161–164
Taub E, Uswatte G, Elbert T (2002) New treatments in neurorehabilitation founded on basic research. *Nat Rev Neurosci* 3: 228–236
81 Taub E, Uswatte G, Elbert T (2002) New treatments in neurorehabilitation founded on basic research. *Nat Rev Neurosci* 3: 228–236
82 Harris AJ (1999) Cortical origin of pathological pain. *Lancet* 354: 1464–1466
83 McCabe CS, Haigh RC, Ring EFJ, Halligan PW, Wall PD, Blake DR (2003) A controlled pilot study of the utility of mirror visual feedback in the treatment of complex regional pain syndrome (type 1). *Rheumatology* 42: 97–101
84 Ramachandran VS, Rogers Ramachandran D (1996) Synaesthesia in phantom limbs induced with mirrors. *Proc R Soc Lond B Biol Sci* 263: 377–386
85 Giraux P, Sirigu A (2003) Illusory movements of the paralyzed limb restore motor cortex activity. *NeuroImage* 20: S107–S111
86 MacLachlan M, McDonald D, Waloch J (2004) Mirror treatment of lower limb phantom pain: a case study. *Disabil Rehabil* 26: 901–904
87 McCabe CS, Haigh RC, Halligan PW, Blake DR (2005) Simulating sensory-motor incongruence in healthy volunteers: implications for a cortical model of pain. *Rheumatology* 44: 509–516
88 Treede RD, Kenshalo DR, Gracely RH, Jones AKP (1999) The cortical representation of pain. *Pain* 79: 105–111
89 Apkarian AV, Bushnell MC, Treede RD, Zubieta JK (2005) Human brain mechanisms of pain perception and regulation in health and disease. *Eur J Pain* 9: 463–484
90 Weiss T, Miltner WHR (2007) Zentralnervöse nozizeptive Verarbeitung: Netzwerke, Schmerz und Reorganisation. *Manuelle Medizin* 45: 38–44
91 Appenzeller O, Bicknell JM (1969) Effects of ner-

vous system lesions on phantom experience in amputees. *Neurology* 19: 141–146

92 Doetsch GS (1998) Perceptual significance of somatosensory cortical reorganization following peripheral denervation. *Neuroreport* 9: R29–R35

93 Katz J, Melzack R (1990) Pain "memories" in phantom limbs: Review and clinical observations. *Pain* 43: 319–336

94 Nikolajsen L, Ilkjaer S, Kroner K, Christensen JH, Jensen TS (1997) The influence of preamputation pain on postamputation stump and phantom pain. *Pain* 72: 393–405

95 Hanley MA, Jensen MP, Smith DG, Ehde DM, Edwards WT, Robinson LR (2007) Preamputation pain and acute pain predict chronic pain after lower extremity amputation. *J Pain* 8: 102–109

96 Nikolajsen L, Ilkjaer S, Jensen TS (2000) Relationship between mechanical sensitivity and postamputation pain: a prospective study. *Eur J Pain* 4: 327–334

97 Flor H, Braun C, Elbert T, Birbaumer N (1997) Extensive reorganization of primary somatosensory cortex in chronic back pain patients. *Neurosci Lett* 224: 5–8

98 Maihöfner C, Handwerker HO, Neundörfer B, Birklein F (2003) Patterns of cortical reorganization in complex regional pain syndrome. *Neurology* 61: 1707–1715

99 Maihöfner C, Handwerker HO, Neundörfer B, Birklein F (2004) Cortical reorganization during recovery from complex regional pain syndrome. *Neurology* 63: 693–701

100 Woolf CJ, Chong MS (1993) Preemptive analgesia – treating postoperative pain by preventing the establishment of central sensitization. *Anesth Analg* 77: 362–379

101 Nikolajsen L, Ilkjaer S, Christensen JH, Kroner K, Jensen TS (1997) Randomized trial of epidural bupivacaine and morphine in prevention of stump and phantom pain in lower-limb amputation. *Lancet* 350: 1353–1537

102 Wiech K, Kiefer RT, Topfner S, Preissl H, Braun C, Unertl K, Flor H, Birbaumer N (2004) A placebo-controlled randomized crossover trial of the N-methyl-D-aspartic acid receptor antagonist, memantine, in patients with chronic phantom limb pain. *Anesth Analg* 98: 408–413

103 Wilder-Smith OH, Arendt-Nielsen L (2006) Postoperative hyperalgesia: its clinical importance and relevance. *Anesthesiology* 104: 601–607

104 Bornhövd K, Quante M, Glauche V, Bromm B, Weiller C, Büchel C (2002) Painful stimuli evoke different stimulus-response functions in the amygdala, prefrontal, insula and somatosensory cortex: a single-trial fMRI study. *Brain* 125: 1326–1336

105 Büchel C, Bornhövd K, Quante M, Glauche V, Bromm B, Weiller C (2002) Dissociable neural responses related to pain intensity, stimulus intensity, and stimulus awareness within the anterior cingulate cortex: A parametric single-trial laser functional magnetic resonance imaging study. *J Neurosci* 22: 970–976

106 Rosen G, Hugdahl K, Ersland L, Lundervold A, Smievoll AI, Barndon R, Sundberg H, Thomsen T, Roscher BE, Tjolsen A, Engelsen B (2001) Different brain areas activated during imagery of painful and non-painful 'finger movements' in a subject with an amputated arm. *Neurocase* 7: 255–260

107 Willoch F, Rosen G, Tolle TR, Oye I, Wester HJ, Berner N, Schwaiger M, Bartenstein P (2000) Phantom limb pain in the human brain: Unraveling neural circuitries of phantom limb sensations using positron emission tomography. *Ann Neurol* 48: 842– 849

108 Wei F, Zhuo M (2001) Potentiation of sensory responses in the anterior cingulate cortex following digit amputation in the anaesthetised rat. *J Physiol-London* 532: 823–833

109 Flor H (2002) Phantom-limb pain: characteristics, causes, and treatment. *Lancet Neurol* 1: 182–189

110 Recanzone GH, Merzenich MM, Jenkins WM, Grajski KA, Dinse HR (1992) Topographic reorganization of the hand representation in cortical area 3b owl monkeys trained in a frequency-discrimination task. *J Neurophysiol* 67: 1031–1056

111 Elbert T, Pantev C, Wienbruch C, Rockstroh B, Taub E (1995) Increased cortical representation of the fingers of the left hand in string players. *Science* 270: 305–307

112 Sterr A, Muller MM, Elbert T, Rockstroh B, Pantev C, Taub E (1998) Changed perceptions in Braille readers. *Nature* 391: 134–135

113 Sterr A, Muller MM, Elbert T, Rockstroh B, Pantev C, Taub E (1998) Perceptual correlates of changes

in cortical representation of fingers in blind multifinger Braille readers. *J Neurosci* 18: 4417–4423

114 Liepert J, Miltner WHR, Bauder H, Sommer M, Dettmers C, Taub E, Weiller C (1998) Motor cortex plasticity during constraint-induced movement therapy in stroke patients. *Neurosci Lett* 250: 5–8

115 Liepert J, Bauder H, Miltner WHR, Taub E, Weiller C (2000) Treatment-induced cortical reorganization after stroke in humans. *Stroke* 31: 1210–1216

116 Weiss T, Miltner WHR, Adler T, Bruckner L, Taub E (1999) Decrease in phantom limb pain associated with prosthesis-induced increased use of an amputation stump in humans. *Neurosci Lett* 272: 131–134

117 Lotze M, Grodd W, Birbaumer N, Erb M, Huse E, Flor H (1999) Does use of a myoelectric prosthesis prevent cortical reorganization and phantom limb pain? *Nat Neurosci* 2: 501–502

118 Flor H, Denke C, Schaefer M, Grüsser S (2001) Effect of sensory discrimination training on cortical reorganisation and phantom limb pain. *Lancet* 357: 1763–1764

24 The neuroscience and phenomenology of sensory loss

1 Sterman AB, Schaumburg HH, Asbury AK (1980) The acute sensory neuronopathy syndrome; a distinct clinical entity. *Ann Neurol* 7(4): 354–358

2 Rolke R, Magerl W, Andrews Campbell K, Schalber C, Caspari S, Birklein F, Treede RD (2006) Quantitative sensory testing: a comprehensive protocol for clinical trials. *Eur J Pain* 10: 77–86

3 Shy ME, Frohman EM, So YT, Arezzo JC, Cornblath DR, Giuliani MJ, Kincald JC, Ochoa JL, Parry GJ, Weimer LH (2003) Quantitative sensory testing. Report of the Therapeutics and Technology Assessment Subcommittee of the American Academy of Neurology. *Neurology* 60: 898–904

4 Koenig M (2003) Rare forms of autosomal recessive neurodegenerative ataxia. *Semin Pediatr Neurol* 10(3): 183–192

5 Schaumburg HH, Kaplan J, Windebank A, Vick N, Rasmus S, Pleasure Brown MJ (1983) Sensory neuropathy from pyridoxine abuse. A new megavitamin syndrome. *N Engl J Med* 309(8): 445–448

6 Gregg RW, Molepo JM, Monpetit VJ, Mikael NZ, Redmond D, Gadia M, Stewart DJ (1992) Cisplatin neurotoxicity: the relationship between dosage, time, and platinum concentration in neurologic tissues, and morphologic evidence of toxicity. *J Clin Oncol* 10 (5): 795–803

7 van der Bent MJ, van Putten WL, Hilkens PH, de Wit R, van der Burg ME (2002) Re-treatment with dose-dense weekly cisplatin after previous cisplatin chemotherapy is not complicated by significant neuro-toxicity. *Eur J Cancer* 38(3): 387–391

8 O'Leary CP, Willison HJ (1997) Autoimmune ataxic neuropathies (sensory ganglionopathies). *Curr Opin Neurol* 10(5): 366–370

9 Comi G, Roveri L, Swan A et al (2002) for the Inflammatory Neuropathy Cause and Treatment Group. A randomised controlled trial of intravenous immunoglobulin in IgM paraprotein associated demyelinating neuropathy. *J Neurol* 249(10): 1370–1377

10 Camdessanche JP, Antoine JC, Honnorat J, Vial C, Petiot P, Convers P, Michel D (2002) Paraneoplastic peripheral neuropathy associated with anti-Hu antibodies. A clinical and electrophysiological study of 20 patients. *Brain* 125(1): 166–175

11 Lucchinetti CF, Kimmel DW, Lennon VA (1998) Paraneoplastic and oncologic profiles of patients seropositive for type 1 antineuronal nuclear autoantibodies. *Neurology* 50: 652–657

12 Sillevis Smitt P, Grefkens J, de Leeuw B, van den Bent M, van Putten W, Hooijkaas H, Vecht C (2002) Survival and outcome in 73 anti-Hu positive patients with paraneoplastic encephalomyelitis/sensory neuronopathy. *J Neurol* 249(6): 745–753

13 Font J, Ramos-Casals M, de la Red G, Pou A, Casanova A, Garcia-Carrasco M, Cervera R, Molina JA, Valls J, Bove A et al (2003) Pure sensory neuropathy in primary Sjogren's syndrome. Long-term prospective follow-up and review of the literature. *J Rheumatol* 30(7): 1552–1557

14 Rothwell JC, Traub MM, Day BL, Obeso JA, Marsden CD, Thomas PK (1982) Manual motor performance in a deafferented man. *Brain* 105: 515–542

15 Forget R, Lamarre Y (1987) Rapid elbow flexion in the absence of proprioceptive and cutaneous feedback. *Human Neurobiol* 6: 27–37

16 Cole J (1991) *Pride and a Daily Marathon*. Duck-

worth, London (reprinted by The MIT Press, London and Cambridge, MA, 1995)
17 Cole JD, Sedgwick EM (1992) The perceptions of force and of movement in a man without large myelinated sensory afferents below the neck. *J Physiol* 449: 503–515
18 Fleury M, Bard C, Teasdale N, Paillard J, Cole J, Lajoie Y, Lamarre Y (1995) Weight judgment. The discrimination capacity of a deafferented subject. *Brain* 118: 1149–1156
19 Cole J, Gallagher S, McNeill D (2002) Gesture following deafferentation: A phenomenologically informed experimental study. *Phenomenology and the Cognitive Sciences* 1(1): 49–67
20 McNeill D (2005) *Gesture and Thought*. University of Chicago Press, Chicago
21 Nowak D, Glasauer S, Hermsdorfer J (2004) How predictive is grip force control in the complete absence of somatosensory feedback? *Brain* 127: 1–11
22 Phillips C (1985) *Movements of the hand*. Liverpool University Press, Liverpool
23 Alper G, Narayanan V (2003) Friedreich's ataxia. *Pediatr Neurol* 28 (5): 335–341
24 Miall RC, Cole JD (2007) Evidence for stronger visuo-motor than visuo-proprioceptive conflict during mirror drawing performed by a deafferented subject and control subjects. *Exp Brain Res* 176(3): 432–439

25 Focal dystonia: diagnostic, therapy, rehabilitation

1 Fahn S, Bressman SB, Marsden CD (1998) Classification of dystonia. *Adv Neurol* 78: 1–10
2 Jankovic J (ed) (2005) *Dystonia*. Seminars in Clinical Neurology, Vol 3. Demos Medical Publishing LLC, New York
3 Fahn S (1989) Assessment of primary dystonias. In: TL Munsat (ed): *Quantification of neurologic deficit*. Butterworths, Boston
4 Rosenbaum F, Jankovic J (1988) Focal task-specific tremor and dystonia; categorization of occupational movement disorders. *Neurology* 38: 522–527
5 Deuschl G, Bain P (2002) Deep brain stimulation for tremor [correction of trauma]: patient selection and evaluation. *Mov Disord* 17 (Suppl 3): S102–11
6 Jabusch HC, Altenmüller E (2006) Epidemiology, phenomenology and therapy of musician's cramp. In: E Altenmüller, J Kesselring, M Wiesendanger (eds): *Music, motor control and the brain*. Oxford University Press, Oxford
7 Bara-Jimenez W, Shelton P, Hallett M (2000) Spatial discrimination is abnormal in focal hand dystonia. *Neurology* 55 (12): 1869–1873
8 Sanger TD, Tarsy DM, Pascual-Leone A (2001) Abnormalities of spatial and temporal sensory discrimination in writer's cramp. *Mov Disord* 16: 94–99
9 Fiorio M, Tinazzi M, Bertolasi L, Aglioti SM (2003) Temporal processing of visuotactile and tactile stimuli in writer's cramp. *Ann Neurol* 53(5): 630–635
10 Naumann M, Magyar-Lehmann S, Reiners K, Erbguth F, Leenders KL (2000) Sensory tricks in cervical dystonia: perceptual dysbalance of parietal cortex modulates frontal motor programming. *Ann Neurol* 47(3): 322–328
11 Altenmüller E (2003) Focal dystonia: advances in brain imaging and understanding of fine motor control in musicians. *Hand Clin* 19(3): 523–538
12 Nutt JG, Muenter MD, Melton LJ 3rd, Aronson A, Kurland LT (1988) Epidemiology of dystonia in Rochester, Minnesota. *Adv Neurol* 50: 361–365
13 Nakashima K, Kusumi M, Inoue Y, Takahashi K (1995) Prevalence of focal dystonias in the western area of Tottori Prefecture in Japan. *Mov Disord* 10(4): 440–443
14 Lim V, Altenmüller E (2003) Musicians Cramp: Instrumental and gender differences. *Med Probl Perf Artists* 18: 21–27
15 Schmidt A, Jabusch HC, Altenmüller E, Hagenah J, Brüggemann N, Hedrich K, Saunders-Pullman R, Bressman SB, Kramer PL, Klein C (2006) Dominantly transmitted focal dystonia in families of patients with musician's cramp. *Neurology* 67(4): 691–693
16 Naumann M, Pirker W, Reiners K, Lange KW, Becker G, Brucke T (1998) Imaging the pre- and postsynaptic side of striatal dopaminergic synapses in idiopathic cervical dystonia: a SPECT study using [123I] epidepride and [123I] beta-CIT. *Mov Disord* 13 (2): 319–323

17 Preibisch C, Berg D, Hofmann E, Solymosi L, Naumann M (2001) Cerebral activation patterns in patients with writer's cramp: a functional magnetic resonance imaging study. *J Neurol* 248(1): 10–17
18 Lenz FA, Byl NN (1999) Reorganization in the cutaneous core of the human thalamic principal somatic sensory nucleus (Ventral caudal) in patients with dystonia. *J Neurophysiol* 82(6): 3204–3212
19 Delmaire C, Krainik A, Tézenas du Montcel S, Gerardin E, Meunier S, Mangin JF, Sangla S, Garnero L, Vidailhet M, Lehéricy S (2005) Disorganized somatotopy in the putamen of patients with focal hand dystonia. *Neurology* 64(8): 1391–1396
20 Byl NN, Merzenich MM, Jenkins WM (1996) A primate genesis model of focal dystonia and repetitive strain injury: Learning-induced dedifferentiation of the representation of the hand in the primary somatosensory cortex in adult monkeys. *Neurology* 47(2): 508–520
21 Molloy FM, Carr TD, Zeuner KE, Dambrosia JM, Hallett M (2003)Abnormalities of spatial discrimination in focal and generalized dystonia. *Brain* 126: 2175–2182
22 Bara-Jiminez W, Catalan MJ, Hallett M, Gerloff C (1998) Abnormal sensory homunculus in dystonia of the hand. *Ann Neurol* 44(5): 828–831
23 McKenzie AL, Nagarajan SS, Roberts TP, Merzenich MM, Byl NN (2003) Somatosensory representation of the digits and clinical performance in patients with focal hand dystonia. *Am J Phys Med Rehabil* 82: 737–749
24 Elbert T, Candia V, Altenmüller E, Rau H, Rockstroh B, Pantev C, Taub E (1998) Alteration of digital representations in somatosensory cortex in focal hand dystonia. *Neuroreport* 16: 3571–3575
25 Elbert T, Pantev C, Wienbruch C, Rockstroh B, Taub E (1995) Increased cortical representation of the fingers of the left hand in string players. *Science* 270(5234): 305–307
26 Altenmüller E (2007) From the Neanderthal to the concert hall: development of sensory motor skills and brain plasticity in music performance. In: A Williamon, D Coimbra (eds): *Proceedings of the International Symposium on Performance Science 2007.* European Association of Conservatoires (AEC), Utrecht
27 Ragert P, Schmid A, Altenmüller E, Dinse R (2004) Meta-plasticity in musicians. *Eur J Neurosci* 19: 473–478
28 Braun C, Schweizer R, Heinz U, Wiech K, Birbaumer N, Topka H (2003) Task-specific plasticity of somatosensory cortex in patients with writer's cramp. *Neuroimage* 20: 1329–1338
29 Lederman RJ (1991) Focal dystonia in instrumentalists: clinical features. *Med Probl Perform Art* 6(4): 132–136
30 Charness ME, Ross MH, Shefner JM (1996) Ulnar neuropathy and dystonic flexion of the fourth and fifth digits: clinical correlation in musicians. *Muscle Nerve* 19(4): 431–437
31 Flor H, Braun C, Elbert T, Birbaumer N (1997) Extensive reorganization of primary somatosensory cortex in chronic back pain patients. *Neurosci Lett* 224(1): 5–8
32 Tinazzi M, Fiaschi A, Rosso T, Faccioli F, Grosslercher J, Aglioti SM (2000) Neuroplastic changes related to pain occur at multiple levels of the human somatosensory system: A somatosensory-evoked potentials study in patients with cervical radicular pain. *J Neurosci* 20(24): 9277–9283
33 Rosenkranz K, Altenmüller E, Siggelkow S, Dengler R (2000) Alteration of sensorimotor integration in musician's cramp: Impaired focusing of proprioception. *Clin Neurophysiol* 111: 2040–2045
34 Rosenkranz K, Williamon A, Butler K, Cordivari C, Lees AJ, Rothwell JC (2005) Pathophysiological differences between musician's dystonia and writer's cramp. *Brain* 128: 918–931
35 Rosenkranz K, Butler K, Williamon A, Cordivari C, Lees AJ, Rothwell JC (2008) Sensorimotor reorganization by proprioceptive training in musician's dystonia and writer's cramp. *Neurology* 70(4): 304–315
36 Jabusch HC, Müller SV, Altenmüller E (2004) Anxiety in musicians with focal dystonia and those with chronic pain. *Mov Disord* 19(10): 1169–1175
37 Nygaard TG, Marsden CD, Fahn S (1991) Dopa-responsive dystonia: long-term treatment response and prognosis. *Neurology* 41(2 (Pt 1)): 174–181
38 Jabusch HC, Zschucke D, Schmidt A, Schuele S, Altenmüller E (2005) Focal dystonia in musicians:

treatment strategies and long-term outcome in 144 patients. *Mov Disord* 20(12): 1623–1626
39 Jabusch HC, Schneider U, Altenmüller E (2004) Delta-9-Tetrahydrocannabinol (THC) improves motor control in a patient with musician's dystonia. *Mov Disord* 19(8): 990–991
40 Topka H, Jankovic J, Dichgans J (2002) Dyskinesias. In: T Brandt, L Caplan, J Dichgans, HC Diener, C Kennard (eds): *Neurological disorders: course and treatment*. Academic Press, New York
41 Schuele SU, Jabusch HC, Lederman RJ, Altenmüller E (2005) Botulinum toxin injections of musician's dystonia. *Neurology* 64: 341–343
42 Fukaya C, Katayama Y, Kano T, Nagaoka T, Kobayashi K, Oshima H, Yamamoto T (2007) Thalamic deep brain stimulation for writer's cramp. *J Neurosurg* 107(5): 977–982
43 Candia V, Schafer T, Taub E, Rau H, Altenmüller E, Rockstroh B, Elbert T (2002) Sensory motor retuning: a behavioral treatment for focal hand dystonia of pianists and guitarists. *Arch Phys Med Rehabil* 83(10): 1342–1348
44 Candia V, Wienbruch C, Elbert T, Rockstroh B, Ray W (2003) Effective behavioral treatment of focal hand dystonia in musicians alters somatosensory cortical organization. *Proc Natl Acad Sci USA* 100(13): 7942–7946
45 Liepert J, Tegenthoff M, Malin JP (1995) Changes in cortical motor area size during immobilization. *Clin Neurophysiol* 97: 382–386
46 Pesenti A, Barbieri S, Priori A (2004) Limb immobilization for occupational dystonia: a possible alternative treatment for selected patients. *Adv Neurol* 94: 247–254
47 Byl NN, Priori A (2006) The development of focal dystonia in musicians as a consequence of maladaptive plasticity: implications for intervention. In: E Altenmüller, J Kesselring, M Wiesendanger (eds): *Music motor control and the brain*. Oxford University Press, Oxford
48 Byl N, Nagarajan SS, Newton N, McKenzie AL (2000) Effect of sensory discrimination training on structure and function in a musician with focal hand dystonia. *Phys Therapy Case Reports* 3: 94–113
49 Byl NN, McKenzie A (2002) The effect of sensory discrimination training on structure and function in patients with focal hand dystonia: 3 case series. *Arch of Phys Med and Rehab* 84: 1505–1514
50 Byl NN, McKenzie A (2000) Treatment effectiveness for patients with a history of repetitive hand use and focal hand dystonia: a planned prospective follow-up study. *J Hand Ther* 13(4): 289–301
51 Zeuner KE, Bara-Jimenez W, Noguchi PS, Goldstein SR, Dambrosia JM, Hallett M (2002) Sensory training for patients with focal hand dystonia. *Ann Neurol* 51(5): 593–598
52 Zeuner KE, Hallett M (2003) Sensory training as treatment for focal hand dystonia: A 1-year follow up. *Mov Disord* 18: 1044–1047
53 Boullet L (2003) Treating focal dystonia – A new retraining therapy for pianists. In: R Kopiez, AC Lehmann, I Wolther, C Wolf (eds): *Abstracts of the 5th Triennial Conference of the European Society for the Cognitive Sciences of Music (ESCOM)*. Hanover University of Music and Drama, Hanover
54 Tubiana R, Chamagne P (2000) Prolonged rehabilitation treatment of musician's focal dystonia. In: R Tubiana, PC Amadio (eds): *Medical problems of the instrumentalist musician*. Martin Dunitz Ltd, London
55 Münte TF, Altenmüller E, Jäncke L (2002) The musician's brain as a model of neuroplasticity. *Nature Reviews* 3: 473–478

26 Self-injurious behavior

1 Frances A (1987) Introduction (to the section on self-mutilation). *J Personality Disorders* 1: 316
2 Favazza A (1996) *Bodies under siege: self-mutilation and body modification in culture and psychiatry*. Johns Hopkins University Press, Baltimore
3 Walsh BW, Rosen P (1987) Self-mutilation. Guilford, New York
4 Favazza A, Conterio K (1989) Female habitual self-mutilators. *Acta Psychiat Scandinavica* 79: 283–289

27 Language development disturbances

1 Goldin-Meadow S, Seligman ME, Gelman R (1976) Language in the two-year-old. *Cognition* 4: 189–202

2 Benedict H (1979) Early lexical development: Comprehension and production. *J Child Lang* 6: 183–200
3 Gentner D (1982) Why nouns are learned before verbs: Linguistic relativity *versus* natural partitioning. In: S Kuczaj II (ed): *Language development. Vol. 2: Language, thought, and culture*. Erlbaum, Hillsdale/NJ, 301–334
4 Goldfield BA, Reznick JS (1990) Early lexical acquisition: Rate, content, and the vocabulary spurt. *J Child Lang* 17: 171–183
5 Bates E, Dale P, Thal S (1995) Individual differences and their implications for theories of language development. In: P Fletcher, B MacWhinney (eds): *The handbook of child language*. Blackwell, Oxford, 96–151
6 Smith LB, Jones SS, Landau B, Gershkoff-Stowe L, Samuelson L (2002) Object name learning provides on the-job-training for attention. *Psychol Sci* 13: 13–19
7 Gibson JJ (1966) *The senses considered as perceptual systems*. Houghton Mifflin, Boston
8 Gibson JJ (1979) *The ecological approach to visual perception*. Hougthon Mifflin, Boston
9 Gibson EJ (1988) Exploratory behavior in the development of perceiving, acting, and the acquiring of knowledge. *Ann Rev Psychol*, 39: 1–41
10 Sinclair H (1970) The transition from sensory-motor behaviour to symbolic activity. *Interchange* 1: 119–126
11 Martin A (2007) The representations of objects in the brain. *Annu Rev Psychol* 58: 25-45
12 Spelke ES, Breinlinger K, Macomber L, Jacobson K (1992) Origins of knowledge. *Psychol Rev* 99: 605–632
13 Baillargeon R (1993) The object concept revisited: New directions in the investigations of infants' physical knowledge. In: CE Grangrud (ed): *Visual perception and cognition in infancy*. Erlbaum, Hillsdale, NJ, 265–315
14 Gopnik A, Meltzoff A (1987) The development of categorization in the second year and its relation to other cognitive and linguistic developments. *Child Dev* 58: 1523–1531
15 Pauen S, Schulz P, Waidhas J (2004) Vorsprachliche Objektkategorisierung und Sprachverstehen. Welche Beziehung besteht zwischen beiden Kompetenzen? (unpublished manuscript); Poster: How preverbal categorization performance relates to early word acquisition. XIVth Biennial International Conference on Infant Studies (ICIS), Chicago, USA
16 Streri A, Lhote M, Dutilleul S (2000) Haptic perception in newborns. *Dev Sci* 3: 319–327
17 Révész G (1944) *Die menschliche Hand. Eine psychologische Studie*. Karger, Basel
18 Lederman SJ, Klatzky RL (1987) Hand movements: A window into haptic object recognition. *Cogn Psychol* 19: 342–368
19 Bushnell EW, Boudreau JP (1991) The development of haptic perception during infancy. In: MA Heller, W Schiff (eds): *The psychology of touch*. Erlbaum, Hillsdale, NJ, 139–161
20 Klatzky RL, Lederman SJ, Metzger VA (1985) Identifying objects by touch: An "expert system". *Perception & Psychophysics* 37: 299–302
21 Klatzky RL, Lederman SJ, Reed C (1989) Haptic integration of object properties: Texture, hardness, and planar contour. *J Exp Psychol: Human Perception & Perform* 15: 45–57
22 Bushnell EW, Boudreau JP (1998) Exploring and exploiting objects with the hands during infancy. In: KJ Conolly (ed): *The psychobiology of the hand*. Mac Keith Press, London, 144–161
23 Jouen F, Molina M (2005) Exploration of the newborn's manual activity: A window onto early cognitive processes. *Infant Behav Dev* 28: 227–239
24 Molina M, Jouen F (1998) Modulation of palm grasp behavior in neonates according to texture property. *Infant Behav Dev* 21: 659–666
25 Molina M, Jouen F (2004) Manual cyclical activity as an exploratory tool in neonates. *Infant Behav Dev* 27: 42–53
26 Striano T, Bushnell EW (2005) Haptic perception of material properties by 3-month-old infants. *Infant Behav Dev* 28: 266–289
27 Adamson-Macedo EN, Barnes CR (2004) Grasping and fingering (active or haptic touch) in healthy newborns. *Neuro Endocrinol Lett* 25 (Suppl 1): 157–168
28 Stack DM, Tsonis M (1999) Infants' haptic perception of texture in the presence and absence of visual cues. *Br J Dev Psychol* 17: 97–110
29 Schellingerhout R, Smitsman AW, van Galen GP

(1997) Exploration of surface texture in congenitally blind infants. *Child Care Health Dev* 23: 247–264
30 Pauen S (2000) Beeinflußt sprachlicher Input die Objektkategorisierung im Säuglingsalter? *Sprache & Kognition* 19: 39–50
31 Landau B, Smith LB, Jones SS (1988) The importance of shape in early lexical learning. *Cogn Dev* 3: 299–321
32 Samuelson LK, Smith LB (1999) Early noun vocabularies. Do ontology, category structure and syntax correspond? *Cognition* 73: 1–33
33 Sandhofer CM, Smith LB, Luo J (2000) Counting nouns and verbs in the input. Differential frequencies, different kinds of learning? *J Child Lang* 27: 561–585
34 Ruff HA (1984) Infants' manipulative exploration of objects: effects of age and object characteristics. *Dev Psychol* 20: 9–20
35 Rescorla L, Mirak J (1997) Normal language acquisition. *Semin Pediatr Neurol* 4: 70–76
36 Mandler JM (1992) How to build a baby: II. Conceptual primitives. *Psychol Rev* 99 : 587–604
37 Mandler JM (1993) On concepts. *Cogn Dev* 8: 141–148
38 Mandler JM, McDonough L (1993) Concept information in infancy. *Cogn Dev* 8: 291–318
39 Markman EM (1989) *Categorization and naming in children. Problems of induction.* MIT Press, Cambridge/MA
40 Markman E (1990) Constraints children place on word meanings. *Cogn Sci* 14: 57–78
41 Markman E (1994) Constraints on word meaning in early language acquisition. In: I Gleitman, B Landau (eds): *The acquisition of the lexicon.* MIT Press, Cambridge/MA, 199–227
42 Ryde Brandt B (1996) Impaired tactual perception in children with Down's syndrome. *Scand J Psychol* 37: 312–316
43 Rogers SJ, Hephurn S, Wehner E (2003) Parent reports of sensory symptoms in toddlers with autisms and those with other developmental disorders. *J Autism Dev Disord* 33: 631–642
44 Baranek GT, Boyd BA, Poe MD, David FJ, Watson, LR (2007) Hyperresponsive sensory patterns in young children with autism, developmental delay, and typical development. *Am J Ment Retard* 112: 233-245
45 Blakemore SJ, Tavassoli T, Caló S, Thomas RM, Catmur C, Frith U Haggard P (2006) Tactile sensitivity in Asperger syndrome. *Brain Cogn* 61: 5-13
46 Güçlü B, Tanidir C, Motavalli Mukaddes N, Ünal F (2007) Tactile sensitivity of normal and autistic children. *Somatosens Mot Res* 24: 21-33
47 Gernsbacher MA, Sauer EA, Geye HM, Schweigert EK, Goldsmith HH (2007) Infant and toddler oral- and manual-motor skills predict later speech fluency in autism. *J Child Psychol Psychiatry* 49: 43–50
48 Johnston RB, Stark RE, Mellits E, Tallal P (1981) Neurological status of language-impaired and normal children. *Ann Neurol* 10: 159–163
49 Stark RE, Tallal P (1981) Perceptual and motor deficits in language-impaired children. In: RW Keith (ed): *Central auditory and language disorders.* College Hill Press, Houston/Texas, 121–144
50 Kamhi AG (1981) Nonlinguistic symbolic and conceptual abilities of language-impaired and normally developing children. *J Speech Hear Res* 24: 446–453
51 Kamhi AG, Catts HW, Koenig LA, Lewis BA (1984) Hypothesis-testing and nonlinguistic symbolic abilities in language-impaired children. *J Speech Hear Disord* 49: 169–176
52 Kiese-Himmel C, Schiebusch-Reiter U (1995) Taktil-kinästhetisches Erkennen bei sprachentwicklungsgestörten Kindern – erste empirische Ergebnisse. *Sprache & Kognition* 14: 126–137
53 Kiese-Himmel C, Kruse E (1998) Höhere taktile und kinästhetische Funktionen bei ehemals sprech-/sprachentwicklungsgestörten Kindern: eine neuropsychologische Studie. *Fol Phoniatr Logop* 50: 195–204
54 Kiese-Himmel C, Schiebusch-Reiter U (1999) Haptische Formdiskrimination. Gruppenvergleich von sprachunauffälligen und ehemals sprachentwicklungsgestörten Kindern. *HNO* 47: 45–50
55 De la Peña A, Hirsh R, Eisenson J (1973) Form discrimination performance of linguistically retarded children. *Percept Mot Skills* 36: 187–194
56 Accardi DW (1997) *Linguistic and kinaesthetic ability in preschool children with a language impair-*

ment. Dissertation Abstracts International: Section B: The Sciences and Engineering, 58 (4–B): 2150

57 Hill EL (1998) A dyspraxic deficit in specific language impairment and developmental coordination disorder? Evidence from hand and arm movements. *Dev Med Child Neurol* 40: 388–395

58 Fabbro F, Libera L, Tavano A (2002) A callosal transfer deficit in children with developmental language disorder. *Neuropsychologica* 40: 1541–1546

59 Broesterhuizen MLHM (1997) Psychological assessment of deaf children. *Scand Audiol* 26: Suppl 46: 43–49

60 Colleti EA, Geffner D, Schlanger P (1976) Oral sterognostic ability among tongue thrusters with interdental lisp, tongue thrusters without interdental lisp and normal children. *Percept Mot Skills* 42: 259–268

61 McNutt JC (1977) Oral sensory and motor behaviours of children with /s/ or /r/ misarticulations. *J Speech Hear Res* 20: 694–703

62 Ruscello DM, Lass NJ (1977) Articulation improvement and oral tactile changes in children. *Percept Mot Skills* 44: 155–159

63 Hetrick RD, Sommers RK (1988) Unisensory and bisensory processing skills of children having misarticulations and normally speaking peers. *J Speech Hear Res* 31: 575–581

64 Fucci D, Petrosino L, Underwood G, Clark KC (1992) Differences in lingual vibrotactile threshold shifts during magnitude-estimation scaling between normal-speaking children and children with articulation problems. *Percept Mot Skills* 75: 495–504

65 Schliesser HF, Cary MH (1973) Oral sterognosis in predicting speech performance: Preliminary report. *Percept Mot Skills* 36: 707–711

66 Laasonen M, Haapanen ML, Mäenpää P, Pulkkinen J, Ranta R, Virsu V (2004) Visual, auditory, and tactile temporal processing in children with oral clefts. *J Craniofac Surg* 15; 510-518

67 Morrongiello BA, Humphrey GK, Timney B, Choi J, Rocca PT (1994) Tactual object exploration and recognition in blind and sighted children. *Perception* 23: 833–848

68 Kiese-Himmel C unter Mitarbeit von Kiefer S (2000) *Diagnostischer Elternfragebogen zur taktil-kinästhetischen Responsivität bei Kleinkindern.* Beltz, Göttingen

69 Kiese–Himmel C (2003) *Göttinger Entwicklungstest der Taktil–Kinästhetischen Wahrnehmung (TAKI-WA).* Beltz, Göttingen

70 Ayres AJ (1989) *Sensory integration and praxis tests.* Western Psychological Service, Los Angeles

71 Leonard LB (1979) Language impairment in children. *Merrill–Palmer Quarterly* 25: 205–232

28 Haptic perception in Anorexia nervosa

1 Grunwald M, Weiss T, Krause W, Beyer L, Rost R, Gutberlet I, Gertz HJ (1999) Power of the theta waves in the EEG of human subjects increases during recall of haptic information. *Neurosci Lett* 260(3): 189–192

2 Grunwald M, Weiss T, Krause W, Beyer L, Rost R, Gutberlet I, Gertz HJ (2001) Theta power in the EEG of humans during ongoing processing in a haptic object recognition tasks. *Cognitive Brain Res* 11: 33–37

3 Fairburn CG, Harrison PJ (2003) Eating disorders. *Lancet* 361: 407–416

4 Deter HC, Herzog W (1994 Anorexia nervosa in a long-term perspective: results of the Heidelberg-Mannheim Study. *Psychosom Med* 56(1): 20–27

5 Birmingham CL, Su J, Hlynsky JA, Goldner EM, Gao M (2005) The mortality rate from anorexia nervosa. *Int J Eating Disorders* 38 (2): 143–146

6 Skrzypek S, Wehmeier PM, Remschmidt H (2001) Body image assessment using body size estimation in recent studies on anorexia nervosa. A brief review. *Eur Child Adolesc Psychiatry* 10(4): 215–221

7 Cash TF, Deagle EA (1997) The nature and extent of body-image disturbances in anorexia nervosa and bulimia nervosa: A meta-analysis. *Int J Eating Disorders* 22: 107–125

8 Hsu LKG, Sobkiewicz TA (1991) Body image disturbance: Time to abandon the concept for eating disorders? *Int J Eating Disorders* 14: 427–431

9 Tovim B, Walker DI, Gilchrist P, Freeman R, Kalucy R, Esterman A (2001) Outcome in patients with eating disorders: a 5-year study. *Lancet* 357: 1254–1257

10 Fernandez-Aranda F, Dahme B, Meermann R (1999) Body image in eating disorders and analysis of its relevance: A preliminary study. *J Psychosomatic Res* 47: 419–428

11 Gallagher S (1995) Body schema and intentionality. In: JL Bermudez, A Marcel, N Eilan (eds): *The body and the self.* MIT Press, Cambridge, MA, 275–243

12 Röhricht F, Seidler KP, Joraschky P, Borkenhagen A, Lausberg H, Lemche E, Loew T, Porsch U, Schreiber-Willnow K, Tritt K (2005) Konsensuspapier zur terminologsichen Abgrenzung von Teilaspekten des Körpererlebens in Forschung und Praxis. *Psychother Psychosom Med Psychol* 55(3–4):183–190

13 Munk H (1890) *Über die Functionen der Grosshirnrinde*, 2 Aufl. Berlin: Aug. Hirschwald

14 Pick A (1908) Über Störungen der Orientierung am eigenen Körper. In: A Pich (Hrsg): *Arbeiten aus der deutschen psychiatrischen Klinik in Prag*. Karger, Berlin, 1–19

15 Pick A (1922) Störung der Orientierung am eigenen Körper. *Psychologische Forschung* 1: 303–318

16 Head H, Holmes G (1911–1912) Sensory disturbances from cerebral lesions. *Brain* 34: 102–254

17 Schilder PF (1923) *Das Körperschema*. Springer, Berlin [The Image and the Appearance of the Human Body; Studies in Constructive Energies of the Psyche. London, 1935]

18 Coslett HB (1998) Evidence for a disturbance of the body schema in neglect. *Brain Cogn* 37(3): 527–544

19 Buxbaum LJ, Giovannetti T, Libon D (2000) The role of the dynamic body schema in praxis: evidence from primary progressive apraxia. *Brain Cogn* 44(2):166–191

20 Kammers MP, van der Ham IJ, Dijkerman HC (2006) Dissociating body representations in healthy individuals: differential effects of a kinaesthetic illusion on perception and action. *Neuropsychologia* 44(12): 2430–2436

21 Thombs BD, Haines JM, Bresnick MG, Magyar-Russell G, Faucrbach JA, Spencc RJ (2007) Depression in burn reconstruction patients: symptom prevalence and association with body image dissatisfaction and physical function. *Gen Hosp Psychiatry* 29(1): 14–20

22 Himelein MJ, Thatcher SS (2006) Depression and body image among women with polycystic ovary syndrome. *J Health Psychol* 11(4): 613–625

23 Kittler JE, Menard W, Phillips KA (2007) Weight concerns in individuals with body dysmorphic disorder. *Eat Behav* 8(1): 115–120

24 Gualdi-Russo E, Albertini A, Argnani L, Celenza F, Nicolucci M, Toselli S (2008) Weight status and body image perception in Italian children. *J Hum Nutr Diet* 21(1): 39–45

25 Ivarsson T, Svalander P, Litlere O, Nevonen L (2005) Weight concerns, body image, depression and anxiety in Swedish adolescents. *Eat Behav* 7(2): 161–175

26 Huang JS, Lee D, Becerra K, Santos R, Barber E, Mathews WC (2006) Body image in men with HIV. *AIDS Patient Care STDS* 20(10): 668–677

27 Williams LK, Ricciardelli LA, McCabe MP, Waqa GG, Bavadra K (2006) Body image attitudes and concerns among indigenous Fijian and European Australian adolescent girls. *Body Image* 3(3): 275–287

28 Holmqvist K, Lunde C, Frisén A (2007) Dieting behaviors, body shape perceptions, and body satisfaction: cross-cultural differences in Argentinean and Swedish 13-year-olds. *Body Image* 4(2): 191–200

29 Swami V, Knight D, Tovée MJ, Davies P, Furnham A (2007) Preferences for female body size in Britain and the South Pacific. *Body Image* 4(2): 219–223

30 Haggard P, Wolpert DM (2005) Disorders of body schema. In: Freund, Jeannerod, Hallett, Leiguarda (eds) *Higher-order motor disorders*. Oxford University Press, Oxford

31 Kolb B, Whishaw IQ (1993) *Fundamentals of human neuropsychology*. Oxford, New York

32 Haggard P, Taylor-Clarke M, Kennett S (2003) Tactile perception, cortical representation and the bodily self. *Curr Biol* 4: 13(5) R170–R173

33 Bruch H (1980) *Der Goldene Käfig – Das Rätsel der Magersucht*. G Fischer Verlag [Golden cage: The enigma of anorexia nervosa. Harvard University Press, 2001]

34 Gruber AJ, Pope HG, Borowiecki JJ, Cohane G (2000) The development of the somatomorphic matrix: a bi-axial instrument for measuring body image in men and women. In: K Norton, T Olds, J Kollman (eds): *Kinanthropometry*. VI. International

Society for the Advancement of Kinanthropometry, Adelaide, 217–231
35 Penner LA, Thompson JK, Coovert DL (1991) Size overestimation among anorexics: much ado about very little? *J Abnorm Psychol* 100: 90–93
36 Fernández F, Probst M, Meermann R, Vandereycken W (1994) Body size estimation and body dissatisfaction in eating disorder patients and normal controls. I*nt J Eat Disord* 16(3): 307–310
37 Probst M, Vandereycken W, van Coppenolle H (1997) Body size estimation in eating disorders using video distortion on an life-size screen. *Psychother Psychosom* 66: 87–91
38 Garner DM (2002) Body image and anorexia nervosa. In: TF Cash, T Pruzinsky (eds): *Body image: a handbook of theory, research, and clinical practice.* Guilford, New York, 295–303
39 Hennighausen K, Enkelmann D, Wewetzer C, Remschmidt H (2003). Body image distortion in anorexia nervosa – is there really a perceptual deficit? *Eur Child Adolesc Psychiatry* 8: 200–206
40 Bisiach E, Capitani E, Luzzatti C, Perani D (1981) Brain and conscious representation of outside reality. *Neuropsychologia* 19: 543–552
41 Renzi E de, Faglioni P, Scotti G (1968) Tactiles, spatial impairment and unilateral cerebral damage. *J Nervous and Mental Disorders* 146: 468–475
42 Renzi E de, Scotti G (1969) The influence of spatial disorders in impairing tactual discrimination of shapes. *Cortex* 5: 53–62
43 Corkin S (1978) The role of different cerebral structures in somesthetic perception. In: EC Carterette, MP Friedman (eds): *Handbook of perception.* Academic Press, Bd. 6, New York
44 Kolb B, Sutherland RJ, Whishaw IQ (1983) A comparison of the contributions of the frontal and parietal association cortex to spatial localization in rats. *Behavioral Neuroscience* 97: 13–27
45 Barnard KE, Brazelton TB (eds) (1990) *Touch: the foundation of experience.* National Center for Clinical Infant Programs, Arlington
46 Heller MA, Schiff W (eds) (1991) *The Psychology of touch.* Lawrence Erlbaum Associates, New Jersey
47 Florin I (1987) Untersuchungen zur Körperwahrnehmung von Probandinnen mit Anorexia nervosa und Bulimia nervosa. In: WD Geber (ed): *Verhaltensmedizin: Ergebnisse und Perspektiven interdisziplinärer Forschung.* Edition Medizin, Weinheim, 473–480
48 Lautenbacher S, Pauls AM, Strian F, Pirke K, Krieg JC (1991) Pain sensitivity in anorexia nervosa and bulimia nervosa. *Biological Psychiatry* 29: 1073–1078
49 Kinsbourne M, Bemporad B (1984) Lateralization of emotion: A model and the evidence. In: NA Fox, RJ Davidson (eds): *The psychobiology of affective development.* Lawrence Erlbaum, Hillsdale, NJ, 259–291
50 Rovet J, Bradley S, Goldberg E, Wachsmuth R (1988) Hemispheric lateralization in anorexia nervosa. A Pilot Study. *J Clin Exp Neuropsychology* 10: 24
51 Pendleton Jones B, Duncan CC, Brouwers P, Mirsky AF (1991) Cognition in eating disorders. *J Clin Exp Neuropsychology* 13: 711–728
52 Bradley SJ, Taylor MJ, Rovet JF, Goldberg E, Hood J (1997). Assessment of Brain Function in adolescent anorexia nervosa before and after weight gain. *J Clin Exp Neuropsychology* 19: 20–33
53 Brouwers P, Duncan CC, Mirsky AF (1986) Cognitive and personality concomitants of eating disorders. *J Clin Exp Neuropsychology* 8: 135
54 Laessle RG, Fischer M, Fichter MM, Pirke KM, Krieg JC (1992) Cortisol levels and vigilance in eating disorder patients. *Psychoneuroendocrinology* 17: 475–484
55 Szmukler GI, Anrewes D, Kingston K, Chen L, Stargatt R, Stanley R (1992) Neuropsychological impairment in anorexia nervosa: before and after refeeding. *J Clin Exp Neuropsychology* 14: 347–352
56 Gordon DP, Halmi KA, Ippolito PM (1984) A comparison of the psychological evaluation of adolescents with anorexia nervosa and of adolescents with conduct disorders. *J Adolescence* 7: 245–266
57 Schmidt MH, Lay B, Blanz B (1997) Verändern sich kognitive Leistungen Jugendlicher mit Anorexia nervosa unter der Behandlung? *Zeitschrift für Kinder-und Jugendpsychiatrie* 25: 17–26
58 Rescher B, Rappelsberger P (1999) Gender dependent EEG-changes during a mental rotation task. *Int J Psychophysiol* 33(3): 209–222
59 Gevins AS, Zeitlin GM, Doyle JC, Yingling CD, Schaffer RE, Callaway E, Yeager CL (1979) Electro-

encephalogram correlates of higher cortical functions. *Science* 203: 665–667

60 Gevins AS, Schaffer RE (1980) A critical review of electroencephalographic (EEG) correlates of higher cortical functions. *CRC Critical Review in Bioengineering* 4: 113–164

61 Klimesch W, Schimke H, Schwaiger J (1994) Episodic and semantic memory: an analysis in the EEG theta and alpha band. *Electroencephalography and Clinical Neurophysiology* 91: 428–441

62 Tewes U (1983) *HAWIK-R: Hamburg-Wechsel Intelligenztest für Kinder; Handbuch und Testanweisung.* 2. korr. Auflage, Bern; Huber, Stuttgart

63 Rappelsberger P, Petsche H (1988) Probability mapping: Power and coherence analyses of cognitive processes. *Brain Topog*raphy Vol 1, 1: 46–53

64 Rappelsberger P, Mayerweg M, Kriegelsteiner S, Petsche H (1988) EEG-mapping: Application to spatial imagination studies. *J Psychophysiology* 2: 153–154

65 Grunwald M, Ettrich C, Assmann B, Dähne A, Krause W, Beyer L, Rost R, Gertz HJ (1999) Haptische Wahrnehmung und EEG Veränderungen bei Anorexia nervosa. *Zeitschrift für Kinder- und Jugendpsychiatrie* 27(4): 241–250

66 Grunwald M, Ettrich C, Krause W, Assmann B, Dähne A, Weiss T, Gertz HJ (2001) Haptic perception in anorexia nervosa before and after weight gain. *J Clin Exp Neuropsychology* 23: 520–529

67 Grunwald M, Ettrich C, Assmann B, Dähne A, Krause W, Busse F, Gertz HJ (2001) Deficits in haptic perception and right parietal theta-power changes in patients with anorexia nervosa before and after weight gain. *Int J Eating Disorders* 29: 417–428

68 Grunwald M, Weiss T, Assmann B, Ettrich C (2004) Stable asymmetric interhemispheric theta power in patients with anorexia nervosa during haptic perception even after weight gain: A longitudinal study. *J Clin Exp Neuropsychology* 26(5): 608–620

69 Springer SP, Deutsch G (1993) *Left brain, right brain*. W.H. Freeman and Company, New York

70 Grunwald M, Ettrich C, Busse F, Assmann B, Dähne A, Gertz HJ (2002) Angle paradigm: a new method to measure right parietal dysfunction in anorexia nervosa. *Archives of Clinical Neuropsychology* 17: 485–496

71 Baraldi P, Porro CA, Serafini M, Pagnoni G, Murari C, Corazza R, Nichelli P (1999) Bilateral representation of sequential finger movements in human cortical areas. *Neurosci Lett* 269: 95–98

72 Salmelin R, Forss N, Knuutila J, Hari R (1995) Bilateral activation of the human somatomotor cortex by distal hand movements. *Electroencephalography Clin Neurophysiol* 95: 444–452

73 Schnitzler A, Salmelin R, Salenius S, Jousmaki V, Hari R (1995) Tactile information from the human hand reaches the ipsilateral primary somatosensory cortex. *Neurosci Lett* 200: 25–28

74 Singh LN, Higano S, Takahashi S, Abe Y, Sakamoto M, Kurihara N, Furuta S, Tamura H, Yanagawa I, Fujii T et al (1998) Functional MR imaging of cortical activation of the cerebral hemispheres during motor tasks. *Am J Neuroradiology* 19: 275–280

75 Marsden JF, Werhahn KJ, Ashby P, Rothwell J, Noachtar S, Brown P (2000) Organization of cortical activities related to movement in humans. *J Neurosci* 20: 2307–2314

76 Critchley M (1953) *The parietal lobes*. Arnold, London

77 Birbaumer N, Schmidt RF (1996) *Biologische Psychologie* (3rd ed.) Springer, Berlin, Heidelberg, New York

78 Harris IM, Egan GF, Sonkkila C, Tochon-Danguy HJ, Paxinos G, Watson JD (2000) Selective right parietal lobe activation during mental rotation: a parametric PET study. *Brain* 123: 65–73

79 Karnath HO (1997) Spatial orientation and the representation of space with parietal lobe lesions. *Philos Trans R Soc Lond B Biol Sci* 352: 1411–1419

80 Knecht S, Kunesch E, Schnitzler A (1996) Parallel and serial processing of haptic information in man: effects of parietal lesions on sensorimotor hand function. *Neuropsychologia* 34: 669–687

81 Hansson T, Brismar T (1999) Tactile stimulation of the hand causes bilateral cortical activation: a functional magnetic resonance study in humans. *Neurosci Lett* 271: 29–32

82 Oliveri M, Rossini PM, Pasqualetti P, Traversa R, Cicinelli P, Palmieri MG, Tomaiuolo F, Caltagirone C (1999) Interhemispheric asymmetries in the perception of unimanual and bimanual cutaneous stimuli. A study using transcranial magnetic stimulation. *Brain* 122: 1721–1729

83 Maravita A, Iriki A (2004) Tools for the body (schema). *Trends Cogn Sci* 8(2):79–86
84 Grunwald M, Weiss T (2005) Inducing sensory stimulation in treatment of anorexia nervosa. *Quarterly J Medicine* 98: 379–380

29 History of haptic interfaces

1 Kontarinis DA, Howe RD (1995) Tactile display of vibratory information in teleoperation and virtual environment. *Presence* 4 (4): 387–402
2 Minsky M, Lederman SJ (1997) Simulated haptic textures: roughness. Symposium on haptic interfaces for virtual environment and teleoperator systems. *Proceedings of the ASME Dynamic Systems and Control Division*, DSC, Vol 58
3 Kawai Y, Tomita F (2000) A support system for the visually impaired to recognize three-dimensional objects. IOS Press, *Technology and Disability* 12, 1: 13–20
4 Moy G, Wagner C, Fearing RS (2000) A compliant tactile display for teletaction. *IEEE Int. Conf. on Robotics and Automation*
5 Burdea GC (1996) *Force and touch feedback for virtual reality.* Wiley-Interscience Publication
6 Asanuma N, Yokoyama N, Shinoda H (1999) A method of selective stimulation to epidermal skin receptors for realistic touch feedback. *Proc of IEEE Virtual Reality* 274–281
7 Kajimoto H, Kawakami N, Maeda T, Tachi S (1999) Tactile feeling display using functional electrical stimulation. *Proc of ICAT'99* 107–114
8 Brooks FP Jr (1986) A dynamic graphics system for simulating virtual buildings. *Proceedings of the 1986 Workshop on Interactive 3D Graphics (Chapel Hill, NC)*, ACM, New York, 9–21
9 Brooks FP, Ouh-Young M, Batter JJ, Jerome O (1990) Project GROPE – Haptic displays for scientific visualization. *Computer Graphics* 24: 4
10 Minsky M, Ouh-Young M, Steele O, Brooks FP, Behensky M (1990) Feeling and seeing: issues in force display. *Computer Graphics* 24: 2
11 Iwata H (1990) Artificial reality with force-feedback: Development of desktop virtual space with compact master manipulator, ACM SIGGRAPH. *Computer Graphics* 24: 4
12 Iwata H (1990) Artificial reality for walking about large scale virtual space. *Human Interface News and Report* 5,1: 49–52 (in Japanese)
13 Burdea G, Zhuang J, Roskos E, Silver D, Langlana L (1992) A portable dextrous master with force feedback. *Presence* 1: 1
14 Iwata H (1993) Pen-based haptic virtual environment. *Proc of IEEE VRAIS'93*
15 Massie T, Salisbury K (1994) The PHANToM Haptic Interface: A device for probing virtual objects. *ASME Winter Anual Meeting*, DSC-Vol 55–1
16 Tachi S, Maeda T, Hirata R, Hoshino H (1994) A construction method of virtual haptic space. *Proc of ICAT'94*
17 McNeely W (1993) Robotic graphics: A new approach to force feedback for virtual reality. *Proc of IEEE VRAIS'93*
18 Hirota K, Hirose M (1996) Simulation and presentation of curved surface in virtual reality environment through surface display. *Proc of IEEE VRAIS'96*

30 Principles of haptic perception in virtual environments

1 Hayward V, Astley OR, Cruz-Hernandez M, Grant D, Robles-De-La-Torre G (2004) Haptic interfaces and devices. *Sensor Rev* 24: 16–29
2 Biggs J, Srinivasan MA (2002) Haptic interfaces. In: K Stanney (ed): *Handbook of virtual environments.* Lawrence Erlbaum, London, 93–116
3 Ogata K (2004) *System dynamics*. Prentice Hall, New Jersey
4 Lederman SJ, Taylor MM (1972) Fingertip force, surface geometry and the perception of roughness by active touch. *Percept Psychophys* 12: 401–408
5 Basdogan C, Srinivasan MA (2002) Haptic rendering in virtual environments. In: K Stanney (ed): *Handbook of virtual environments*. Lawrence Erlbaum, London, 117–134
6 Iwamoto T, Shinoda H (2005) Ultrasound tactile display for stress field reproduction: Examination of non-vibratory tactile apparent movement. *Proc World Haptics* 220–228
7 Murayama J, Bouguila L, Luo Y, Akahane K, Hasegawa S, Hirsbrunner B, Sato M (2004) SPIDAR G&G:

A two-handed haptic interface for bimanual VR interaction. *Proc Eurohaptics* 138–146
8 Campion G, Hayward V (2005) Fundamental limits in the rendering of virtual haptic textures. *Proc World Haptics* 263–270
9 Klatzky RL, Lederman SJ (2003) Touch. In: AF Healy, RW Proctor (eds): *Experimental psychology*. Volume 4, IB Weiner (Ed) *Handbook of Psychology*. John Wiley & Sons, New York, 147–176
10 Hayward V, Astley OR (1996) Performance measures for haptic interfaces. In: G Giralt, G Hirzinger (eds): *Robotics research: The 7th International Symposium*. Springer Verlag, 195–207
11 Cavusoglu MC, Feygin D, Tendick F (2002) A critical study of the mechanical and electrical properties of the PHANToM (TM) haptic interface and improvements for high performance control. *Presence* 11: 555–568
12 Milgram P, Kishino AF (1994) Taxonomy of mixed reality visual displays. *IEICE Trans on Inform and Systems* E77-D (12): 1321–1329
13 Robles-De-La-Torre G, Hayward V (2001) Force can overcome object geometry in the perception of shape through active touch. *Nature* 412: 445–448
14 Robles-De-La-Torre G, Hayward V (2000) Virtual surfaces and haptic shape perception. Proc Haptics Symposium, Orlando, Florida, USA. *Proc ASME* DSC-69-2: 1081–1087
15 Hollerbach JM, Mills R, Tristano D, Christensen RR, Thompson WB, Xu Y (2001) Torso force feedback realistically simulates slope on treadmill-style locomotion interfaces. *Int J of Robotics Res* 20: 939–952
16 Robles-De-La-Torre G (2002) Comparing the role of lateral force during active and passive touch: lateral force and its correlates are inherently ambiguous cues for shape perception under passive touch conditions. *Proc Eurohaptics* 159–164
17 Portillo-Rodriguez O, Avizzano CA, Bergamasco M, Robles-De-La-Torre G (2006) Haptic rendering of sharp objects using lateral forces. *Proc IEEE RO-MAN* 431–436
18 Morgenbesser HB, Srinivasan MA (1996) Force shading for haptic shape perception. *Proc ASME* DSC-58: 407–412
19 Minsky M (1995) Computational haptics: The sandpaper system for synthesizing texture for a force-feedback display. Ph.D. dissertation, Massachusetts Institute of Technology
20 Robles-De-La-Torre G (2006) The importance of the sense of touch in virtual and real environments. *IEEE Multimedia* 13: 24–30

31 Haptic shape cues, invariants, priors and interface design

1 Ernst MO, Bulthoff HH (2004) Merging the senses into a robust percept. *Trends Cogn Sci* 8(4): 162–169
2 Coren S, Ward LM, Enns JT (2003) *Sensation and perception*. J. Wiley & Sons, New York
3 Goldstein EB (2001) *Sensation and perception*. Wadsworth Publishing Company, Belmont
4 Bregman AS (1990) *Auditory scene analysis*. MIT Press, Cambridge
5 Krueger LE (1982) Tactual perception in historical perspective: David Katz's world of touch. In: W Schiff, E Foulke (eds): *Tactual perception; a source-book*. Cambridge University Press, 1–55
6 Loomis JM, Lederman SJ (1986) Tactual perception. In: JTK Boff, L Kaufman (eds): *Handbook of human perception and performance*. J. Wiley & Sons, New York, 1–41
7 Yao HY, Hayward V, Ellis RE (2005) A tactile enhancement instrument for minimally invasive surgery. *Comput Aided Surg* 10(4): 233–239
8 Hayward V, Yi D (2003) Change of height: An approach to the haptic display of shape and texture without surface normal. In: B Siciliano, P Dario (eds): *Experimental robotics VIII*. Springer Tracts in Advanced Robotics 5. Springer Verlag, Heidelberg, 570–579
9 Loomis JM (1990) Distal attribution and presence. *Presence* 1(1): 113–119
10 Askenfelt A, Jansson EV (1991) From touch to string vibrations. II: The motion of the key and hammer. *J Acoust Soc Am* 90(5): 2383–2393
11 Greenish S, Hayward V, Chial V, Okamura A, Steffen T (2002) Measurement, analysis and display of haptic signals during surgical cutting. *Presence* 6(11): 626–651
12 Hayward V, Astley OR, Cruz-Hernandez M, Grant D,

Robles-De-La-Torre G (2004) Haptic interfaces and devices. *Sensor Review* 24(1): 16–29
13 Goldstein H (1950) *Classical mechanics*. Addison-Wesley, New York
14 Kuchenbecker KJ, Fiener J, Niemeyer G (2006) Improving contact realism through event-based haptic feedback. *IEEE Trans Vis Comput Graph* 12(2): 219–230
15 Goldstein EB (1981) The ecology of J. J. Gibson's perception. *Leonardo* 14(3): 191–195
16 Marr D (1982) *Vision*. Freeman, New York
17 Turvey MT (1996) Dynamic touch. *Am Psychol* 51: 1134–1152
18 O'Regan JK, Noe A (2001) A sensorimotor account of vision and visual consciousness. *Behav Brain Sci* 24(5)
19 McIntyre J, Senot P, Prevost P, Zago M, Lacquaniti F, Berthoz A (2003) The use of online perceptual invariants *versus* cognitive internal models for the predictive control of movement and action. In: *Proc of the First International IEEE EMBS Conference on Neural Engineering* 438–441
20 Berthoz A, Melvill Jones G (eds) (1985) *Adaptive mechanisms in gaze control*. Elsevier, Amsterdam
21 Johnson KL (1985) *Contact mechanics*. Cambridge University Press
22 Serina ER, Mockensturm E, Mote Jr. CD, Rempel D (1998) A structural model of the forced compression of the fingertip pulp. *J Biomech* 31: 639–646
23 Pawluk DTV, Howe RD (1999) Dynamic contact of the human fingerpad against a flat surface. *J Biomech Eng* 121: 605–611
24 Srinivasan MA, LaMotte RH (1991) Encoding of shape in the response of cutaneous mechanoreceptors. In: O Franzen, J Westman (eds): *Information processing in the somatosensory system*. MacMillan, London, 59–69
25 Goodwin AW, John KT, Marceglia AH (1991) Tactile discrimination of curvature by humans using only cutaneous information from the fingerpads. *Exp Brain Res* 86: 663–672
26 Bicchi A, Scilingo EP, Dente D, Sgambelluri N (2005) Tactile flow and haptic discrimination of softness. In: D Prattichizzo, K Salisbury (eds): *Multi-point interaction with real and virtual objects*. Springer tracts in advanced robotics 18, Springer Verlag, Heidelberg, 165–176
27 Rabinowicz E (1956) Stick and slip. *Sci Am* 194(5)
28 Levesque V, Hayward V (2003) Experimental evidence of lateral skin strain during tactile exploration. In: *Proc of Eurohaptics* 261–275
29 Tada M, Kanade T (2004) An imaging system of incipient slip for modelling how human perceives slip of a fingertip. In: *Proc of the 26th Annual International Conference of the Engineering in Medicine and Biology Society*, 2045–2048
30 Essick GK, Franzen O, Whitsel BL (1988) Discrimination and scaling of velocity of stimulus motion across the skin. *Somatosens Mot Res* 6: 21–40
31 Dostmohamed H, Hayward V (2005) Trajectory of contact region on the fingerpad gives the illusion of haptic shape. *Exp Brain Res* 164: 387–394
32 Robles-De-La-Torre G, Hayward V (2001) Force can overcome object geometry in the perception of shape through active touch. *Nature* 412: 445–448
33 Hayward V (2004) Display of haptic shape at different scales. In: *Proc of Eurohaptics* 20–27
34 Hayward V (2007) A brief taxonomy of tactile illusions and demonstrations that can be done in a hardware store. *Brain Res Bull* 75: 742–752
35 Drewing K, Ernst MO (2006) Integration of force and position cues for shape perception through active touch. *Brain Res* 1078: 92–100
36 Voisin J, Lamarre Y, Chapman CE (2002) Haptic discrimination of object shape in humans: Contribution of cutaneous and proprioceptive inputs. *Exp Brain Res* 145(2): 251–260
37 Johansson RS, Westling G (1987) Signals in tactile afferents from the fingers eliciting adaptive motor responses during precision grip. *Exp Brain Res* 66: 141–154
38 Lederman SJ, Klatzky RL (1993) Extracting object properties through haptic exploration. *Acta Psychol* 84(1): 29–40
39 Louw S, Kappers AML, Koenderink JJ (2000) Haptic detection thresholds of Gaussian profiles over the whole range of spatial scales. *Exp Brain Res* 132: 369–374
40 Phillips J (1994) *Freedom in machinery*. Cambridge University Press
41 Salada MA, Colgate JE, Lee MV, Vishton PM (2002) Validating a novel approach to rendering fingertip contact sensations. In: *Proc 10th Symposium on*

Symposium on Haptic Interfaces for Virtual Environment and Teleoperator Systems 217–224
42 Dépeault A, Meftah EM, Chapman CE (2008) Tactile speed scaling: contributions of time and space. *J Neurophysiol* 99: 1422–1434
43 Keyson DV, Houtsma AJM (1995) Directional sensitivity to a tactile point stimulus moving across the fingerpad. *Percept Psychophys* 57(5): 738–744
44 Webster III RJ, Murphy TE, Verner LN, Okamura AM (2005) A novel two-dimensional tactile slip display: design, kinematics and perceptual experiments. *ACM Trans Appl Percept* 2(2): 150–165
45 Hirota K, Hirose M (1993) Development of surface display. In: *Proc oIEEE Virtual Reality Annual International Symposium* 256–262
46 Yokokohji Y, Nuramori N, Sato Y, Yoshikawa T (2005) Designing an encountered-type haptic display for multiple fingertip contacts based on the observation of human grasping behaviors. *Int J Rob Res* 24(9): 717–729
47 Provancher WR, Cutkosky MK, Kuchenbecker KJ, Niemeyer G (2005) Contact location display for haptic perception of curvature and object motion. *Int J Rob Res* 24(9): 1–11
48 Solazzi M, Frisoli A, Salsedo F, Bergamasco M (2007) A fingertip haptic display for improving local perception of shape cues. In: *Proc Second Joint Eurohaptics Conference and Symposium on Haptic Interfaces For Virtual Environment And Teleoperator Systems* 409–414

32 Design guidelines for generating force feedback on fingertips using haptic interfaces

1 Douglas SA, Kirkpatrick AE (2002) Application based evaluation of haptic interfaces. *Proceedings of the Tenth Symposium on haptic interfaces for virtual environment and teleoperator systems*
2 Lederman S, Klatzky R (1990) Procedures for haptics exploration vs. manipulation. In: *Vision and Action: The Control of Grasping*. Canadian Institute for Advanced Research Series in Artificial Intelligence and Robotics, 110–127
3 Jansson G, Monaci L (2003) Haptic identification of objects with different numbers of fingers. In: Ballesteros S, Heller MA (eds) *Touch, Blindness and Neurosciences*. UNED Press, Madrid
4 Frisoli A, Barbagli F, Wu SL, Ruffaldi E, Bergamasco M (2004) Comparison of multipoint contact interfaces in haptic perception. *Workshop on Multipoint Interaction in Robotics and Virtual Reality, International Conference on Robotics and Automation*
5 Jansson G (2000) Effects of number of fingers involved in exploration on haptic identification of objects. Excerpt from pure-form: The museum of pure form; haptic exploration for perception of the shape of virtual objects. Technical report, EU-PURE FORM
6 Jansson G, Bergamasco M, Frisoli A (2003) A new option for the visually impaired to experience 3d art at museums: manual exploration of virtual copies. *Visual Impairment Res* 5: 1
7 Jansson G, Monaci L (2006) Identification of real objects under conditions similar to those in haptic displays: providing spatially distributed information at the contact areas is more important than increasing the number of areas. *Virtual Reality* 9: 243–249
8 Frisoli A, Bergamasco M, Wu S, Ruffaldi E (2005) Multi-point interaction with real and virtual objects. Springer tracts in advanced robotics, Vol. 18 Evaluation of multipoint contact interfaces in haptic perception of shapes. *Advanced Robotics* 177–188
9 Barbagli F, Frisoli A, Salisbury K, Bergamasco M (2004) Simulating human fingers: a soft finger proxy model and algorithm. *12th International Symposium on Haptic Interfaces for Virtual Environment and Teleoperator Systems* 9–17
10 Barbagli F, Salisbury K, Devengenzo R (2004) Toward virtual manipulation: from one point of contact to four. *Sensor Review* 24(1)
11 McNeely W, Puterbaugh K, Troy J (1999) Six degree-of freedom haptic rendering using voxel sampling. *ACM SIGGRAPH* 401–408
12 Wan M, McNeely WA (2003) Quasi-static approximation for 6 degrees-of-freedom haptic rendering. *IEEE Visualization*
13 Kuroda Y, Nakao M, Kuroda T, Oyama H, Yoshihara H (2005) Shape perception with friction model for indirect touch. *IEEE WorldHaptics*
14 Klatzky R, Lederman S, Hamilton C, Ramsay G

(1999) Perceiving roughness *via* a rigid probe: effects of exploration speed. *Proc Int Mech Eng Congress, Dynamic Systems and Control Division (Haptic Interfaces for Virtual Environments and Teleoperator Systems)* 67: 27–33

15 Westling G, Johansson R (1984) Factors influencing the force control during precision grip. *Exp Brain Res* 53: 277–284

16 Johansson R, Westling G (1984) Roles of glabrous skin receptors and sensorimotor memory in automatic control of precision grip when lifting rougher or more slippery objects. *Exp Brain Res* 56: 550–564

17 Goodwin AW, Johansson R (1998) Control of grip force when tilting objects: Effect of curvature of grasped surfaces and applied tangential torque. *J Neuroscience* 18(24): 10724–10734

18 Avizzano CA, Marcheschi S, Angerilli M, Fontana M, Bergamasco M, Gutierrez T, Mannegeis M (2003) A multi-finger haptic interface for visually impaired people. *Robot and Human Interactive Communication* 165–170

19 Melder N, Harwin W (2004) Extending the friction cone algorithm for arbitrary polygon based haptic objects. *12th International Symposium on Haptic Interfaces for Virtual Environment and Teleoperator Systems (HAPTICS04)* 234–241

20 Barbagli F, DeVengenzo R, Salisbury K (2003) Dual-handed virtual grasping. *Int Conference on Robotics and Automation, ICRA* 1: 1259–1263

21 Johannes V, Green M (1973) Role of the rate of application of the tangential force in determining the static friction coefficient. *WEAR* 24(3): 381–385

22 Kinoshita H, Backstrom L, Flanagan J, Johansson R (1997) Tangential torque effects on the control of grip forces when holding objects with a precision grip. *Am Physiological Soc* 78(3): 1619–1630

23 Kuchenbecker K, Provancher W, Niemeyer G, Cutkosky M (2004) Haptic display of contact location. Haptic Interfaces for Virtual Environment and Teleoperator Systems. *HAPTICS'04* 40–47

24 Salada M, Colgate J, Lee M, Vishton P (2002) Validating a novel approach to rendering fingertip contact sensations. *Haptic Interfaces for Virtual Environment and Teleoperator Systems, 2002. HAPTICS 2002. Proceedings: 10th Symposium* 217–224

25 Dostmohamed H, Hayward V (2005) Trajectory of contact region on the finger-pad gives the illusion of haptic shape. *Exp Brain Res* 164: 387–394

26 Yokokohji Y, Muramori N, Sato Y, Yoshikawa T (2005) Designing an encountered-type haptic display for multiple fingertip contacts based on the observation of human grasping behaviors. *Int J Robotics Res* 24: 717

27 Cini G, Frisoli A, Marcheschi S, Salsedo F, Bergamasco M (2005) A novel fingertip haptic device for display of local contact geometry. *Haptic Interfaces for Virtual Environment and Teleoperator Systems, 2005. WHC 2005. First Joint Eurohaptics Conference and Symposium* 602–605

28 Basdogan C, Srinivasan M (2002) *Haptic rendering in virtual environments*. Lawrence Erlbaum Associates

29 Hayward V, Astley O, Cruz-Hernandez M, Grant D, Robles-De-La-Torre G (2004) Haptic interfaces and devices. *Sensor Review* 24: 16–29

30 Minsky M (1995) *Computational haptics: the Sandpaper System for synthesizing texture for a force-feedback display.* Massachusetts Institute of Technology. Program in Media Arts and Sciences, School of Architecture and Planning

31 Morgenbesser H, Srinivasan M (1996) Force shading for haptic shape perception. *Proceedings of the ASME Dynamics Systems and Control Division* 58: 407–412

32 Ho P, Adelstein B, Kazerooni H (2004) Judging 2d *versus* 3d square-wave virtual gratings. *Haptic Interfaces for Virtual Environment and Teleoperator Systems, 2004. HAPTICS'04. Proceedings. 12th International Symposium* 176–183

33 Robles-De-La-Torre G, Hayward V (2000) Virtual surfaces and haptic shape perception. ASME (ed): *Haptic interfaces for virtual environments and teleoperator systems symposium, International Mechanical Engineering Congress & Exposition*

34 Robles-De-La-Torre G, Hayward V (2001) Force can overcome object geometry in the perception of shape through active touch. *Nature* 412: 389–391

35 Portillo O, Avizzano CA, Raspolli M, Bergamasco M (2005) Haptic desktop for assisted handwriting and drawing. *Robot and Human Interactive Communication, 2005. ROMAN 2005. IEEE International Workshop* 512–517

36 Okamura A, Dennerlein J, Howe R (1998) Vibration feedback models for virtual environments. *Robotics and Automation, 1998. Proceedings. 1998 IEEE International Conference* Vol. 1

37 Haans A, IJsselsteijn W (2006) Mediated social touch: a review of current research and future directions. *Virtual Reality* 9: 149–159, Springer

38 Sama M, Pacella V, Farella E, Benini L, Ricco B (2006) 3did: a low-power, low-cost hand motion capture device. *Proceedings of the Conference on Design, Automation and Test in Europe: Designers' Forum* 136–141, European Design and Automation Association. 3001 Leuven, Belgium

39 Kessler G, Hodges L, Walker N (1995) Evaluation of the cyberglove as a whole-hand input device. *ACM Transactions on Computer-Human Interaction (TOCHI)* 2: 263–283

40 Tarasewich P (2002) Wireless devices for mobile commerce: User interface design and usability. In: Mennecke BE, Strader TK (eds) *Mobile Commerce: Technology.* Idea Group Publishing, Hershey, PA, 25–50

41 Murray A, Klatzky R, Khosla P (2003) Psychophysical characterization and testbed validation of a wearable vibrotactile glove for telemanipulation. *Presence: Tele-operators and Virtual Environments* 12: 156–182

42 http://www.immersion.com/3d/products/ (1993) Cybertouch. Tech. rep., Immersion

43 Simone L, Kamper D (2005) Design considerations for a wearable monitor to measure finger posture. *J NeuroEngineering and Rehabilitation* 2: 5

44 Wise S, Gardner W, Sabelman E, Valainis E, Wong Y, Glass K, Drace J, Rosen J (1990) Evaluation of a fiber optic glove for semi-automated goniometric measurements. *J Rehabil Res Dev* 27: 411–424

45 Salsedo F, Ullrich G, Bergamasco M, Villella P (2004), Goniometric sensor. Patent WO/2004/059,249

46 Johansson R (1978) Tactile sensibility in the human hand: receptive field characteristics of mechanoreceptive units in the glabrous skin area. *J Physio*logy 281: 101–125

47 Carrozzino M, Tecchia F, Bacinelli S, Cappelletti C, Bergamasco M (2005) Lowering the development time of multimodal interactive application: the real-life experience of the xvr project. *Proceedings of the 2005 ACM SIGCHI International Conference on Advances in Computer Entertainment Technology* 270–273. ACM Press, New York

48 Tecchia F (2006) Building a complete virtual reality application. *Proceedings of the ACM Symposium on Virtual Reality Software and Technology* 383. ACM Press, New York

33 Haptic rendering and control

1 Colgate JE, Schenkel G (1997) Passivity of a class of sampled data systems: Application to haptic interfaces. *Journal of Robotic Systems* 14: 37–47

2 Hannaford B, Ryu JH (2001) Time-domain passivity control of haptic interfaces. *Proc. of the IEEE Int. Conf. on Robotics and Automation (ICRA)*, 1863–1869

3 Trenkel S, Weller R, Zachmann G (2007) A benchmarking suite for static collision detection algorithms. In: V Skala (ed.) In: *International Conference in Central Europe on Computer Graphics, Visualization and Computer Vision (WSCG)*, Plzen, Czech Republic. Union Agency

4 Zilles CB, Salisbury JK (1995) A constraint-based god-object method for haptic display. *Proc. IEEE/RSJ Int. Conf. on Intelligent Robots and Systems (IROS)*, Vol. 3: 146–151. IEEE Computer Society

5 Ho C-H, Basdogan C, Srinivasan MA (1999) Efficient point-based rendering techniques for haptic display of virtual objects. *Presence: Teleoper. Virtual Environ* 8(5): 477–491

6 Lin MC, Gottschalk S (1998) Collision detection between geometric models: a survey. *Proc. of IMA Conference on Mathematics of Surfaces* 37–56

7 Cameron S (1997) Enhancing GJK: computing minimum and penetration distances between convex polyhedra. *Int. Conf. Robotics & Automation*

8 Gregory A, Mascarenhas A, Ehmann S, Lin M, Manocha D (2000) Six degree-of-freedom haptic display of polygonal models. In: T Ertl, B Hamann, A Varshney (eds) In: *Proceedings Visualization* 2000, 139–146

9 Lin MC, Canny JF (1991) A fast algorithm for incremental distance calculation. *IEEE Int. Conf. on Robotics and Automation* 1008–1014

10 Jung D, Gupta KK (1997) Octree-based hierarchical

distance maps for collision detection. *Journal of Robotic Systems* 14(11): 789–806

11 Colgate JE, Stanley MC, Brown JM (1995) Issues in the haptic display of tool use. *IEEE/RSJ Int. Conf. on Intelligent Robots and Systems*, Vol. 3: 140–145. Pittsburgh, PA

12 Otaduy MA, Lin MC (2005) Stable and responsive six-degree-of freedom haptic manipulation using implicit integration. *Worldhaptics Conf.* 247–256

13 Iwata H, Noma H (1993) Volume haptization. *IEEE Symposium on Research Frontiers in Virtual Reality*, 16–23

14 Hasegawa S, Fujii N (2003) Real-time rigid body simulation based on volumetric penalty method. In: *11th Symposium on Haptic Interfaces for Virtual Environment and Teleoperator Systems* 326–332

15 Thompson TV, Johnson DE, Cohen E (1997) Direct haptic rendering of sculptured models. *Symposium on Interactive 3D Graphics* 167–176

16 Chang B, Colgate J (1997) Real-time impulse-based simulation of rigid body systems for haptic display. *Proceedings of the ASME International Mechanical Engineering Congress and Exhibition*. Dallas, Texas, 1–8

17 Constantinescu D, Salcudean SE, Croft EA (2005) Haptic rendering of rigid contacts using impulsive and penalty forces. *IEEE Transactions on Robotics* 21(3): 309–323

18 McNeely WA, Puterbaugh KD, Troy JJ (1999) Six degree-of-freedom haptic rendering using voxel sampling. *Proc. of SIGGRAPH* 401–408

19 Kuchenbecker KJ, Fiene JP, Niemeyer G (2006) Improving contact realism through event-based haptic feedback. *IEEE Transactions on Visualization and Computer Graphics* 12(2): 219–230

20 Ruspini DC, Kolarov K, Khatib O (1997) The haptic display of complex graphical environments. *Computer Graphics (SIGGRAPH 97 Conference Proceedings)*. ACM SIGGRAPH, Los Angeles, CA, 345–352

21 Corso J, Chhugani J, Okamura A (2002) Interactive haptic rendering of deformable surfaces based on the medial axis transform. *Eurohaptics* 92–98

22 Pauly M, Pai DK, Guibas LJ (2004) Quasi-rigid objects in contact. *Eurographics/ACM SIGGRAPH Symposium on Computer Animation*

23 Barbič J, James DL (2007) Time-critical distributed contact for 6-dof haptic rendering of adaptively sampled reduced deformable models. *2007 ACM SIGGRAPH/Eurographics Symposium on Computer Animation*

24 Hayward V (2006) Haptic synthesis. SYROCO 2006, 8th Int. *IFAC Symposium on Robot Control*. Keynote paper, Bologna, Italy, 19–24

25 Pai DK, Doel Kvd, James DL, Lang J, Lloyd JE, Richmond JL, Yau SH (2001) Scanning physical interaction behavior of 3D objects. In: E Fiume (ed): *SIGGRAPH 2001, Computer Graphics Proceedings*. ACM Press/ACM SIGGRAPH

26 Otaduy MA, Jain N, Sud A, Lin MC (2004) Haptic display of interaction between textured models. *Visualization* 297–304

27 Renz M, Preusche C, Pätke M, Kriegel H-P, Hirzinger G (2001) Stable haptic interaction with virtual environments using an adapted Voxmap-Pointshell Algorithm. In: *Proc. of Eurohaptics 2001*, Birmingham, UK

28 McNeely WA, Puterbaugh KD, Troy JJ (2006) Voxel-based 6-dof haptic rendering improvements. H*aptics-e* 3(7)

29 Ruffaldi E, Morris D, Barbagli F, Salisbury K, Bergamasco M (2008) Voxel-based haptic rendering using implicit sphere trees. *Proc. of the 2008 IEEE Haptics Symposium*

30 Basdogan C, Srinivasan MA (2001) Haptic rendering in virtual environments. In: K Stanney (ed): *Virtual Environments HandBook*. Lawrence Erlbaum Associates Inc, 117–134

31 Massie TH, Salisbury JK (1994) The phantom haptic interface: A device for probing virtual objects. *Proc. of the ASME Winter Annual Meeting, Symposium on Haptic Interfaces for Virtual Environment and Teleoperator Systems*. Chicago, 295–302

32 Burdea GC (1996) *Force and touch feedback for virtual reality*. Wiley Interscience

33 Ryu JH, Preusche C, Hannaford B, Hirzinger G (2005) Time domain passivity control with reference energy behavior. *IEEE Transactions on Control Systems Technology* 13(5): 737–742

34 Gosline AH, Hayward V (2007) Time-domain passivity control of haptic interfaces with tunable damping hardware. *World Haptics 2007* 164–169

35 Stramigioli S, Secchi C, van der Schaft AJ, Fantuzzi C (2005) Sampled data systems passivity and dis-

crete port-Hamiltonian systems. *IEEE Transactions on Robotics* 21(4): 574–587

36 Adams RJ, Hannaford B (1999) Stable haptic interaction with virtual environments. *IEEE Trans. on Robotics and Automation* 15(3): 465–474

37 Artigas J, Preusche C, Hirzinger G (2004) Wave variables based bilateral control with a time delay model for space robot applications. *Robotik 2004, VDI-Bericht 1841*. Munich, Germany, 101–108

38 Diolaiti N, Niemeyer G, Barbagli F, Salisbury JK (2006) Stability of haptic rendering: Discretization, quantization, time delay, and coulomb effects. *IEEE Transactions on Robotics* 22(2): 256–268

39 Iskakov R, Alin AS, Schedl M, Hirzinger G, Lopota V (2007) Influence of sensor quantization on the control performance of robotics actuators. *Proc. IEEE/RSJ Int. Conf. on Intelligent Robots and Systems (IROS)*. San Diego, CA, 1085–1092

40 Salcudean SE, Vlaar TD (1997) On the emulation of stiff walls and static friction with a magnetically levitated input/output device. *ASME Journal of Dynamic Systems, Measurement and Control*

41 Gil JJ, Avello A, Rubio A, Flórez J (2004) Stability analysis of a 1 dof haptic interface using the routh-hurwitz criterion. *IEEE Transactions on Control Systems Technology* 583–588

42 Hulin T, Preusche C, Hirzinger G (2006) Stability boundary and design criteria for haptic rendering of virtual walls. *Proc. of the 8th Int. IFAC Symposium on Robot Control*. Bologna, Italy

43 Hulin T, Preusche C, Hirzinger G (2006) Stability boundary for haptic rendering: Influence of physical damping. Proc. *IEEE/RSJ Int. Conf. on Intelligent Robots and Systems (IROS)*. Beijing, China, 1570–1575

44 Colgate JE, Brown JM (1994) Factors affecting the z-width of a haptic display. *Proc. of the IEEE Int. Conf. on Robotics and Automation* 3205–3210

45 Mehling J, Colgate JE, Peshkin MA (2005) Increasing the impedance range of a haptic display by adding electrical damping. *Proc. of the IEEE World Haptics Conference (WHC)*. Pisa, Italy, 257–262

46 Gil JJ, Sánchez E, Hulin T, Preusche C, Hirzinger G (2007) Stability boundary for haptic rendering: Influence of damping and delay. In: *Proc. of the IEEE Int. Conf. on Robotics and Automation (ICRA)*. Rome, Italy

47 Hulin T, Gil JJ, Sánchez E, Preusche C, Hirzinger G (2006) Experimental stability analysis of a haptic system. *Proc. of 3rd Int. Conf. on Enactive Interfaces*. Montpellier, France

48 Preusche C, Rettig A, Hirzinger G (2002) Assembly verification in digital mock-ups using force feedback. *12th Int. Symposium on Measurement and Control in Robotics Towards Advanced Robot Systems and Virtual Reality*. Bourges, France

49 Preusche C, Koeppe R, Albu-Schäffer A, Hähnle M, Sporer N, Hirzinger G (2001) Design and haptic control of a 6 DoF force- feedback device. *Workshop on Advances in Interactive Multimodal Telepresence Systems*. Munich, Germany

50 Hayward V, MacLean KE (2007) Do it yourself haptics, part-i. *IEEE Robotics and Automation Magazine* 14(4): 88–104

51 MacLean KE, Hayward V (2008) Do it yourself haptics, part-ii. *IEEE Robotics and Automation Magazine* 15(1)

52 Laycock SD, Day AM (2007) A survey of haptic rendering techniques. *Computer Graphics Forum* 26(1): 50–65

53 Gil JJ, Sánchez E, Hulin T, Preusche C, Hirzinger G (2008) Stability boundary for haptic rendering: Influence of damping and delay. *J Computing and Information Science in Eng* 8(3)

54 Hirzinger G, Sporer N, Albu-Schäffer A, Hähnle M, Krenn R, Pascucci A, Schedl M (2002) DLR's torque-controlled light weight robot III – are we reaching the technological limits now? In: *IEEE Int. Conf. Robot. Autom.*. Washington D.C., 1710–1716

34 Haptic perception in human robotic systems

1 Reimer EM (1999) Cavity sensor technology for low cost automotive safety & control devices, Airbag technology'99, Detroit

2 Osumi H, Ishii N, Takahashi K, Umeda K, Kinoshita G (1999) Optimal grasping for a parallel two-fingered hand with compliant tactile sensors. *IEEE/RSJ International Conference on Intelligent Robots and Systems*, Korea, vol. 2: 799–804

3 Pressure Profile, Inc., Los Angeles, CA, Homepage. http://www.pressureprofile.com

4 du Moncel T (1878) *Le téléphone, le microphone et le phonographe*. Bibliothèque des merveilles, Hachette, Paris

5 Shimojo M, Ishikawa M, Kanaya K (1991) A flexible high resolution tactile imager with video signal output. *IEEE International Conference on Robotics and Automation*, Sacramento, CA

6 Russell RA (1991) A tactile sensory skin for measuring surface contours. *IEEE Region 10 Conference TENCON*, Melbourne, Australia

7 Shimojo M, Namiki A, Ishikawa M, Makino R, Mabuchi K (2004) A tactile sensor sheet using pressure conductive rubber with electrical-wires stitched method. *IEEE Sensors journal* Vol. 4, No. 5: 589–596

8 Kerpa O, Weiss K, Wörn H (2003) Development of a flexible tactile sensor system for a humanoid robot. *IEEE/RSJ International Conference on Intelligent Robots and Systems* Vol. 1, 1–6

9 Krivopal B (1999) Pressure sensitive ink means, and methods of use. Patent No. US005989700A, date of patent: 23 November 1999

10 Foster JK (1991) Effects of carbon black properties on conductive coatings. *2nd International Exhibition of Paint Industry Suppliers*, São Paolo, Brazil

11 Weiss K, Woern H (2005) The working principle of resistive tactile sensor cells. *IEEE International Conference on Mechatronics and Automation*, Niagara Falls, Canada

12 Ferry J (1980) *Viscoelastic properties of polymers*. John Wiley and Sons

13 Woicke N (2006) *Viskoelastizität von Polypropylen im Glasübergang*. Dissertation, Fakultaet Maschinenbau der Universitaet Stuttgart

14 Dillmann R, Becher R, Steinhaus P (2004) ARMAR II – a learning and cooperative multimodal humanoid robot system. *Int J Humanoid Robotics* 1(1): 143–155

15 Asfour T, Regenstein K, Azad P, Schroeder J, Dillmann R (2006) ARMAR-III: A humanoid platform for perception-action integration. *International Workshop on Human-Centered Robotic Systems (HCRS)*, Munich, 51–56

16 Cutkosky MR, Wright PK (1986) Modelling manufacturing grips and correlations with the design of robotic hands. *IEEE Int Conf Robotics and Automation* 1533–1539

35 Haptic design of vehicle interiors at AUDI

1 J. Meyer: Lastenheft Sitzkomfort

2 ECIA: Firmenpräsentation

36 Visual-haptic interfaces in car design by BMW

Garbin CP (1988) Visual-haptic perceptual nonequivalence for shape information and its impact upon cross-modal performance. *J Exp Psychology: Human Perception and Performance* 14 (4): 547–553

37 Haptics in reasearch at Daimler AG

1 Hacker W (1998) *Allgemeine Arbeitspsychologie. Psychische Regulation von Arbeitstätigkeiten*. Verlag Hans Huber, Bern

2 Osgood CE, Suci GJ, Tannenbaum DH (1957) *The Measurement of Meaning*. University of Illinois Press, Urbana, Illinois

3 Sprung L, Sprung H (1987) *Grundlagen der Methodologie und Methodik der Psychologie*. VEB Deutscher Verlag der Wissenschaften, Berlin

4 Enigk H (2003) *Ein psychologisches Vorgehensmodell zur Entwicklung von Unterstützungssystemen für Kraftfahrzeuge*. dissertation.de – Verlag im Internet GmbH, Berlin

5 Ericsson KA, Simon HA (1978) *Retrospective Verbal Reports as Data*. CIP Working Paper No. 388, Carnegie-Mellon University

6 Kriz J, Lisch R (1988) *Methoden-Lexikon für Mediziner, Psychologen, Soziologen*. Psychologie Verlags Union, München

7 Millward LJ (1995) Focus Groups. In: GM Breakwell, S Hammond, C Fife-Shaw (eds): *Research methods in psychology*. SAGE Publications, London

8 Thurstone LL (1927) A Law of Comparative Judgement. *Psychological Review* 34: 273 – 286

38 Haptic design of handles

1 Klingel J (2003) *Analyses of individual rubber-*

coated handles of a power tool. IMK (now IKTD), Universitaet Stuttgart
2 Seeger H (2005) *Design technischer Produkte, Produktprogramme und -systeme*. 2. Auflage: Springer-Verlag, Berlin, Heidelberg
3 Kunsch K (2000) *Der Mensch in Zahlen. Datensammlung in Tabellen*. 2. Auflage: Spektrum Akademischer Verlag, Heidelberg, 174
4 FLEX-Elektrowerkzeuge GmbH (2003) *Katalog Elektrowerkzeuge*. Steinheim an der Murr
5 DIN 33402-2. Supplement 1 (2006–08) *Human body dimensions – Part 2: Values; Application of body dimensions in practice*. Beuth Verlag, Berlin
6 Flügel B, Greil H, Sommer K (1986) *Anthropologischer Atlas. Grundlagen und Daten*. Minerva – Edition, Wissen, Berlin
7 Cacha CA (1999) *Ergonomics and safety in hand tool design*. CRC Press LLC, Boco Raton, Florida 52–55
8 Bullinger HJ (1994) *Ergonomie. Produkt- und Arbeitsplatzgestaltung*. B. G. Teubner, Stuttgart, 322

39 Vestibular sensory substitution using tongue electrotactile display

1 Reed CM, Rabinowitz WM, Durlach NI, Delhorne LA, Bradia LD, Pemberton JC, Mulcahey BD, Washington DL (1992) Analytic study of the Tadoma method: Improving performance through the use of supplementary tactile displays. *J Speech Hearing Res* 35(April): 450–465
2 Collins CC, Saunders FA (1970) Pictorial display by direct electrical stimulation of the skin. *J Biomed Sys* 1(2): 3–16
3 Collins CC (1985) On mobility aids for the blind, In DH Warren, ER Strelow (eds): *Electronic spatial sensing for the blind.* Matinus Nijhoff, Dordrecht, The Netherlands
4 Machts L (1920) Device for converting light effects into effects perceptible by blind persons. German Patent 326283
5 Loomis JM (1992) Distal attribution and presence. *Presence: Teleoperators and virtual environments* 1(1): 113–119
6 Bach-y-Rita P (1971) A tactile vision substitution system based on sensory plasticity. In: TD Sterling, EA Bering Jr, SV Pollack, HG Vaughn Jr (eds): *Visual prosthesis: the interdisciplinary dialogue.* Academic Press, New York, 281–290
7 Unitech Research Inc (1993) SBIR Phase I Final Report: A new electrotactile prosthesis for the blind: Unitech Research Inc
8 Kaczmarek KA, Bach-y-Rita P (1995) Tactile displays. In: W Barfield, T Furness (eds): *Virtual environments and advanced interface design*. Oxford University Press, New York, 349–414
9 Szeto AYJ, Riso RR (1990) Sensory feedback using electrical stimulation of the tactile sense. In: RV Smith, JH Leslie Jr (eds): *Rehabilitation engineering.* CRC Press, Boca Raton, FL, 29–78
10 Haase SJ, Kaczmarek KA (2005) Electrotactile perception of scatterplots on the fingertips and abdomen. *Med Biol Eng Comput* 43: 283–289
11 Bach-y-Rita P, Kaczmarek K, Tyler M, Garcia-Lara J (1998) Form perception with a 49-point electrotactile stimulus array on the tongue. *J Rehab Res Develop* 35: 427–430
12 Bach-y-Rita P, Kaczmarek KA (2002) *Tongue placed tactile output device*. US Patent 6,430,450
13 Ptito M, Moesgaard S, Gjedde A, Kupers R (2005) Cross-modal plasticity revealed by electrotactile stimulation of the tongue in the congenitally blind. *Brain* 128(3): 606–614
14 Kupers R, Ptito M (2004) "Seeing“ through the tongue: cross-modal plasticity in the congenitally blind. *International Congress Series* 1270: 79–84
15 Bach-y-Rita P (1995) Rehabilitation following brain damage: conceptual and practical issues. In: O Kahyan, Z Guven, N Ozaras (eds): *Proceedings of the International Congress of Rehabilitation*, Istanbul Turkey. Monduzzi, Milan 13–19
16 Doidge N (2007) *The Brain that Changes Itself.* Viking, New York
17 Saunders FA, Hill WA, Franklin B (1981) A wearable tactile sensory aid for profoundly deaf children. *J Med Sys* 5(4): 265–270
18 Weisenberger JM, Broadstone SM, Saunders FA (1989) Evaluation of two multichannel tactile aids for the hearing impaired. *J Acoust Soc Am* 86(5): 1764–1175
19 Sparks DW, Ardell LA, Bourgeois M, Wiedmer B, Kuhl PK (1979) Investigating the MESA (Multipoint Electrotactile Speech Aid): The transmission of

connected discourse. *J Acoust Soc Am* 65(3): 810–815

20 Reed CM, Durlach NI, Bradia LD (1982) Research on tactile communication of speech: A review. *AHSA Monographs* 20: 1–23

21 Weisenberger JM, Broadstone SM, Kozma-Spytek L (1991) Relative performance of single-channel and multichannel tactile aids for speech perception. *J Rehab Res Dev* 28(2): 45–56

22 Hannaford B, Venema S (1995) Kinesthetic displays for remote and virtual environments. In: W Barfield (ed): *Virtual environments and advanced interface design.* Oxford, New York, 415 –436

23 Droessler N, Hall D, Tyler ME, Ferrier NJ (2001) Tongue-based electrotactile feedback to perceive objects grasped by a robotic manipulator. In: *Proc of 23rd Annual Int Conf of the IEEE Eng Med Biol Soc.* IEEE, Istanbul

24 Loomis JM, Lederman SJ (1986) Tactual perception. In: KR Boff, L Kaufman, JP Thomas (eds): *Handbook of perception and human performance: vol. II, cognitive processes and performance.* Wiley, New York, 31.1–31.41

25 Burdea G (1996) *Force and touch feedback for virtual reality.* Wiley, New York

26 Tyler M, Danilov Y, Bach-y-Rita P (2003) Closing an open-loop control system: Vestibular substitution through the tongue. *J Integrat Neurosci* 2(2): 159–164

27 Rupert AH, Guedry FE, Reschke MF (1994) The use of tactile interface to convey position and motion perceptions. *AGARD-CP* 541: 20.1–20.7

28 Mergner T, Anastasopoulos D, Becker W, Deecke L (1981) Discrimination between trunk and head rotation; a study comparing neuronal data from the cat with human psychophysics. *Acta Psychol* (Amst) 48(1–3): 291–301

29 Berthoz A, Israel I, Georges-Francois P, Grasso R, Tsuzuku T (1995) Spatial memory of body linear displacement: what is being stored? *Science* 269(5220): 95–98

30 Berthoz A (1996) The role of inhibition in the hierarchical gating of executed and imagined movements. *Brain Res Cogn Brain Res* 3(2): 101–113

31 Bloomberg J, Jones GM, Segal B, McFarlane S, Soul J (1988) Vestibular-contingent voluntary saccades based on cognitive estimates of remembered vestibular information. *Adv Otorhinolaryngol* 41: 71–75

32 Nakamura T, Bronstein AM (1995) The perception of head and neck angular displacement in normal and labyrinthine-defective subjects. A quantitative study using a 'remembered saccade' technique. *Brain* 118 (Pt 5): 1157–1168

33 Andersen GJ (1989) Perception of three-dimensional structure from optic flow without locally smooth velocity. *J Exp Psychol Hum Percept Perform* 15(2): 363–371

34 Bishop PO (1974) Stereopsis and fusion. *Trans Ophthalmol Soc NZ* 26(0): 17–27

35 Mergner T, Rottler G, Kimmig H, Becker W (1992) Role of vestibular and neck inputs for the perception of object motion in space. *Exp Brain Res* 89(3): 655–668

36 Mesland BS, Finlay AL, Wertheim AH, Barnes GR, Morland AB, Bronstein AM, Gresty MA (1996) Object motion perception during ego-motion: patients with a complete loss of vestibular function vs. normals. *Brain Res Bull* 40(5–6): 459–465

37 Wiener SI (1993) Spatial and behavioral correlates of striatal neurons in rats performing a self-initiated navigation task. *J Neurosci* 13(9): 3802–3817

38 Honrubia V, Greenfield A (1998) A novel psychophysical illusion resulting from interaction between horizontal vestibular and vertical pursuit stimulation. *Am J Otol* 19(4): 513–520

39 Demer JL, Oas JG, Baloh RW (1992) Visual-vestibular interaction during high-frequency, active head movements in pitch and yaw. *Ann NY Acad Sci* 656: 832–835

40 Demer JL, Oas JG, Baloh RW (1993) Visual-vestibular interaction in humans during active and passive, vertical head movement. *J Vestib Res* 3(2): 101–114

41 Foster CA, Demer JL, Morrow MJ, Baloh RW (1997) Deficits of gaze stability in multiple axes following unilateral vestibular lesions. *Exp Brain Res* 116(3): 501–509

42 Brandt T (1991) Man in motion. Historical and clinical aspects of vestibular function. A review. *Brain* 114 (Pt 5): 2159–2174

43 Brandt T (2002) Visual acuity, visual field and visual scene characteristics affect postural balance. In: M Igarashi, F Black (eds): *Vestibular and visual con-*

trol on posture and locomotion equilibrium. Karger, Basel, 93

44 Strupp M, Glasauer S, Jahn K, Schneider E, Krafczyk S, Brandt T (2003) Eye movements and balance. *Ann NY Acad Sci* 1004: 352–358

45 Wall C 3rd, Merfeld DM, Rauch SD, Black FO (2002) Vestibular prostheses: the engineering and biomedical issues. *J Vestib Res* 12(2–3): 95–113

46 Wall C 3rd, Oddsson LE, Horak FB, Wrisley DW, Dozza M (2004) Applications of vibrotactile display of body tilt for rehabilitation. *Conf Proc IEEE Eng Med Biol Soc* 7: 4763–4765

47 Wall C 3rd, Kentala E (2005) Control of sway using vibrotactile feedback of body tilt in patients with moderate and severe postural control deficits. *J Vestib Res* 15(5–6): 313–325

48 Cholewiak RW, Sherrick CE (1981) A computer-controlled matrix system for presentation to the skin of complex spatiotemporal patterns. *Behav Res Method Instr* 13(5): 667–673

49 Monkman GJ (1992) An electrorheological tactile display. *Presence: Teleoperators and virtual environments* 1(2): 219–228

50 Hegeman J, Honegger F, Kupper M, Allum JH (2005) The balance control of bilateral peripheral vestibular loss subjects and its improvement with auditory prosthetic feedback. *J Vestib Res* 15(2): 109–117

51 Chiari L, Dozza M, Cappello A, Horak FB, Macellari V, Giansanti D (2005) Audio-biofeedback for balance improvement: an accelerometry-based system. *IEEE Trans Biomed Eng* 52(12): 2108–2111

52 Dozza M, Chiari L, Hlavacka F, Cappello A, Horak FB (2006) Effects of linear *versus* sigmoid coding of visual or audio biofeedback for the control of upright stance. *IEEE Trans Neural Syst Rehabil Eng* 14(4): 505–512

53 Dozza M, Horak FB, Chiari L (2007) Auditory biofeedback substitutes for loss of sensory information in maintaining stance. *Exp Brain Res* 178(1): 37–48

54 Bliss JC (1970) Dynamic tactile displays in man-machine systems: Editorial. *IEEE Trans Man-Mach Sys* MMS-11 (1): 1

55 Dursun E, Hamamci N, Donmez S, Tuzunalp O, Cakci A (1996) Angular biofeedback device for sitting balance of stroke patients. *Stroke* 27(8): 1354–1357

56 Hirvonen TP, Aalto H, Pyykko I (1997) Stability limits for visual feedback posturography in vestibular rehabilitation. *Acta Otolaryngol Suppl* 529: 104–107

57 Cholewiak RW, Collins AA (1995) Vibrotactile pattern discrimination and communality at several body sites. *Percept & Psychophys* 57(5): 724–737

58 Lamore PJ, Keemink CJ (1988) Evidence for different types of mechanoreceptors from measurements of the psychophysical threshold for vibrations under different stimulation conditions. *J Acoust Soc Am* 83(6): 2339–2351

59 Lamore PJ, Keemink CJ (1990) Conditions and possibilities for the detection of speech-signal elements by means of vibrotaction. *Acta Otolaryngol Suppl* 469: 47–54

60 Rupert AH (2000) An instrumentation solution for reducing spatial disorientation mishaps. *IEEE Eng Med Biol Mag* 19(2): 71–80

61 Rupert AH (2000) Tactile situation awareness system: proprioceptive prostheses for sensory deficiencies. *Aviat Space Environ Med* 71(9 Suppl): A92–99

62 Pavlik AE, Inglis JT, Lauk M, Oddsson L, Collins JJ (1999) The effects of stochastic galvanic vestibular stimulation on human postural sway. *Exp Brain Res* 124(3): 273–280

63 Scinicariello AP, Eaton K, Inglis JT, Collins JJ (2001) Enhancing human balance control with galvanic vestibular stimulation. *Biol Cybern* 84: 475–480

64 Scinicariello AP, Inglis JT, Collins JJ (2002) The effects of stochastic monopolar galvanic vestibular stimulation on human postural sway. *J Vestib Res* 12(2–3): 77–85

65 Merfeld DM, Gong W, Morrissey J, Saginaw M, Haburcakova C, Lewis RF (2006) Acclimation to chronic constant-rate peripheral stimulation provided by a vestibular prosthesis. *IEEE Trans Biomed Eng* 53(11): 2362–2372

66 Merfeld DM, Haburcakova C, Gong W, Lewis RF (2007) Chronic vestibulo-ocular reflexes evoked by a vestibular prosthesis. *IEEE Trans Biomed Eng* 54(6 Pt 1): 1005–1015

67 Della Santina C, Migliaccio A, Patel A (2005) Electrical stimulation to restore vestibular function

development of a 3-d vestibular prosthesis. *Conf Proc IEEE Eng Med Biol Soc* 7: 7380–7385
68 Della Santina CC, Migliaccio AA, Patel AH (2007) A multichannel semicircular canal neural prosthesis using electrical stimulation to restore 3-d vestibular sensation. *IEEE Trans Biomed Eng* 54(6 Pt 1): 1016–1030
69 Bach-y-Rita P, Kaczmarek KA, Tyler ME, Garcia-Lara M (1998) Form perception with a 49-point electrotactile stimulus array on the tongue: A technical note. *J Rehab Res Dev* 35(4): 427–430
70 Van Boven RW, Johnson KO (1994) The limit of tactile spatial resolution in humans: Grating orientation discrimination at the lip, tongue, and finger. *Neurol* 44: 2361–2366
71 Kaczmarek KA, Bach-y-Rita P, Tyler ME (1998) *Electrotactile pattern perception on the tongue.* BMES, Cleveland, OH, 5–131
72 Danilov Y, Tyler M (2005) Brainport: an alternative input to the brain. *J Integr Neurosci* 4(4): 537–550
73 Darian-Smith I (1973) The trigeminal system. In: A Iggo (ed): *Handbook of sensory physiology*, Vol. 2. Springer-Verlag, New York, 271–314
74 Care UoIH (2002) Comprehensive management of vestibular disorders. *Currents* 3(2)
75 Disorders NIoDaOC (1989) A Report of the Task Force on the National Strategic Research Plan. National Institutes of Health, Bethesda, Maryland, 74
76 University of Virginia Health System DoOHNS, Vestibular & Balance Center (June 2003); cited, available from: http://www.healthsystem.Virginia.edu/internet/otolaryngology/patients_vbc.cfm.
77 Brown KE, Whitney SL, Wrisley DM, Furman JM (2001) Physical therapy outcomes for persons with bilateral vestibular loss. *Laryngoscope* 111(10): 1812–1817
78 Telian SA, Shepard NT, Smith-Wheelock M, Kemink JL (1990) Habituation therapy for chronic vestibular dysfunction: preliminary results. *Otolaryngol Head Neck Surg* 103(1): 89–95
79 Whitney SL, Wrisley DM, Marchetti GF, Furman JM (2002) The effect of age on vestibular rehabilitation outcomes. Laryngoscope 112(10): 1785–1790
80 Whitney SL, Rossi MM (2000) Efficacy of vestibular rehabilitation. *Otolaryngol Clin North Am* 33(3): 659–672
81 Furman JM, Whitney SL (2000) Central causes of dizziness. *Phys Ther* 80(2): 179–187
82 Gillespie MB, Minor LB (1999) Prognosis in bilateral vestibular hypofunction. *Laryngoscope* 109(1): 35–41

40 The blind get a taste of vision

1 Ptito M, Desgent S (2006) Sensory input-based adaptation and brain architecture. In: P Baltes, P Reuter-Lorenz, F Rosler (eds): *Lifespan development and the brain.* Cambridge University Press, 111–133
2 Lyckman AW, Sur M (2002) Role of afferent activity in the development of cortical specification. *Results Probl Cell Differ* 39: 139–156
3 Ptito M, Giguère JF, Boire D, Frost DO, Casanova C (2001) When the auditory cortex turns visual. *Prog Brain Res* 134: 447–458
4 Frost DO, Boire D, Gingras G, Ptito M (2000) Surgically created neural pathways mediate visual pattern discrimination. *Proc Natl Acad Sci* 97: 11068–110733
5 Merabet LB, Rizzo JF, Amedi A, Somers DC, Pascual-Leone A (2005) What blindness can tell us about seeing again: merging neuroplasticity and neuroprostheses. *Nat Rev Neurosci* 6: 71–77
6 Weeks R, Horwitz B, Aziz-Sultan A, Tian B, Wessinger CM, Cohen LG, Hallett M, Rauschecker JP (2000) A positron emission tomographic study of auditory localization in the congenitally blind. *J Neurosci* 20: 2664–2672
7 Burton H, Sinclair RJ, McLaren DG (2004) Cortical activity to vibrotactile stimulation: an fMRI study in blind and sighted individuals. *Hum Brain Mapp* 23: 210–228
8 Bach-Y-Rita P, Collins CC, Scadden LA, Holmlund GW, Hart BK (1970) Display techniques in a tactile vision-substitution system. *Med Biol Illus* 20: 6–12
9 Collins CC, Bach-y-Rita P (1973) Transmission of pictorial information through the skin. *Adv Biol Med Phys* 14: 285–315
10 Jansson G (1983) Tactile guidance of movement. *Int J Neurosci* 19: 37–46
11 Bach-y-Rita P, Kaczmarek KA, Tyler ME, Garcia-Lara J (1998) Form perception with a 49-point elec-

trotactile stimulus array on the tongue: a technical note. *J Rehabil Res Dev* 35: 427–430

12 Chan JS, Maucher T, Schemmel J, Kilroy D, Newell FN, Meier K (2007) The virtual haptic display: a device for exploring 2-D virtual shapes in the tactile modality. *Behav Res Methods* 39: 802–810

13 Colwell C, Petrie H, Kornbrot D, Hardwick A, Furner S (1998) Haptic virtual reality for blind computer users. *Proceedings of ASSETS '98*, April, Los Angeles

14 Shim JW, Liu W, Tang H (2006) System development for multichannel electrotactile stimulation on the lips. *Med Eng Phys* 28: 734–739

15 Liu W, Tang H (2005) An initial study on lip perception of electrotactile array stimulation. *J Rehabil Res Dev* 42: 705–714

16 Simpson R, LoPresti E, Hayashi S, Guo S, Ding D, Ammer W, Sharma V, Cooper R (2005) A prototype power assist wheelchair that provides for obstacle detection and avoidance for those with visual impairments. *J Neuroeng Rehabil* 2: 30

17 Chebat DR, Rainville C, Kupers R, Ptito M (2007) Tactile-'visual' acuity of the tongue in early blind individuals. *Neuroreport* 18: 1901–1904

18 Ptito M, Moesgaard S, Gjedde A, Kupers R (2005) Cross-modal plasticity revealed by electrotactile stimulation in the congenitally blind. *Brain* 128: 606–614

19 Sathian K (2005) Visual cortical activity during tactile perception in the sighted and the visually deprived. *Dev Psychobiol* 46: 279–286

20 De Volder AG, Catalan-Ahumada M, Robert A, Bol A, Labar D, Coppens A, Michel C, Veraart C (1999) Changes in occipital cortex activity in early blind humans using a sensory substitution device. *Brain Res* 826: 128–134

21 Ptito M, Kupers R (2005) Cross-modal plasticity in early blindness. *J Integr Neurosci* 4: 479–488

22 Nolte J (2007) *Integrated neuroscience*. Mosby (Elsevier), Philadelphia

23 Ptito M, Matteau I, Kupers R (2005) Activation of the dorsal pathway in congenital blindness. *NeuroImage* 30 (Suppl 1): 173

24 Matteau I, Kupers R, Ptito M (2006) Tactile motion discrimination through the tongue in blindness. *NeuroImage* 31 (Suppl 1): 132

25 Matteau I, Kupers R, Ptito M (2008) Activation of the ventral visual pathway in congenital blindness. *Vision Vis Imp Res Suppl* (in press)

26 Armstrong JD (1977) Mobility aids and the limitations of technological solutions. *New Beacon* 61: 113–115

27 Jansson G (2000) Verbal and tactile map information for travelling without sight. In: *Proceedings of the AAATE Conference 1999*, November 1–4, Düsseldorf, Germany

28 Chebat D-R, Madsen K, Paulson OB, Kupers R, Ptito M (2007) Navigational skills in the blind using a sensory substitution device. Soc Neurosci Abst, San Diego, CA

29 Maguire EA, Spiers HJ, Good CD, Frackowiak RS, Burgess N (2003) Navigation expertise in the human hippocampus: a structural brain imaging analysis. *Hippocampus* 13: 250–259

30 Chebat D-R, Chen JK, Schneider F, Ptito A, Kupers R, Ptito M (2007) Alterations in right posterior hippocampus in early blind individuals. *Neuroreport* 18: 329–333

31 Loomis JM, Klatzky RL, Golledge RG, Cicinelli JG, Pellegrino JW, Fry PA (1993) Nonvisual navigation by blind and sighted: assessment of path integration ability. *J Exp Psychol* Gen 122: 73–91

32 Passini R, Proulx G, Rainville C (1990) The spatio-cognitive abilities of the visually impaired population. *Environ & Behav* 22: 91–118

33 Kupers R, Fumal A, de Noordhout AM, Gjedde A, Schoenen J, Ptito M (2006) Transcranial magnetic stimulation of the visual cortex induces somatotopically organized qualia in blind subjects. *Proc Natl Acad Sci* 103: 13256–13260

34 Ptito M, Fumal A, de Noordhout AM, Schoenen J, Gjedde A, Kupers R (2008a) TMS of the occipital cortex induces tactile sensations in the fingers of blind Braille readers. *Exp Brain Res* 184: 193–200

35 Ptito M, Schneider FCG, Paulson OB, Kupers R (2008b) Alterations of the visual pathways in congenital blindness. *Exp Brain Res*. On Line Access 1273–1284

36 Ungar S, Blades M, Spencer C (1995) Mental rotation of a tactile layout by young visually impaired children. *Perception* 24(8): 891–900

41 Tactile ground surface indicators in public places

1 Hurt HH, Ouellet JV, Thom DR (1981) *Motorcycle accident cause factors and identification of countermeasures*. Volume 1: Technical Report, Traffic Safety Center, University of Southern California, Los Angeles, California 90007, Contract No. DOT HS-5-01160

2 American Foundation for the Blinds. *Blindness statistics*. <http://www.afb.org/Section.asp?SectionID=15#num>

3 US Department of Justice. *Americans with Disabilities Act*. <http://www.usdoj.gov/crt/ada/adahom1.htm>

4 Der Beauftragte der Bundesrepublik für die Belange behinderter Menschen. *Das Behindertengleichstellungsgesetz* <http://www.behindertenbeauftragte.de/index.php5?nid=20&Action=home>

5 United Nations. 56/168 (2001) *Comprehensive and integral international convention to promote and protect the rights and dignity of persons with disabilities*. Resolution adopted by the General Assembly. <http://www.un.org/esa/socdev/enable/disA56168e1.htm>

6 The Center for Universal Design (1997) *Principles of Universal Design*. <http://www.design.ncsu.edu/cud/about_ud/udprinciples.htm>

7 European Conference of Ministers of Transport (ECMT). *Tactile surfaces and audible signals* (2000) <http://www.cemt.org/topics/handicaps/pdf/TPH0010Fe.pdf>

8 Department for Transport. *Guidance on the use of tactile paving surfaces*. <http://www.dft.gov.uk/transportforyou/access/tipws/guidanceontheuseoftactilepav6167>

9 European Conference of Ministers of Transport (ECMT). *Improving the transport for people with mobility handicaps*. OECD Publications Service. Paris. 2000 <http://www.cemt.org/pub/pubpdf/TPHguideE.pdf>

10 *Tactile footway surfaces for the blind* (1991) TRL Contractor Report 257, TRL Crowthorne, UK

11 <http://www.pr-o.info/makeframe.asp?url=/bc/uvv/1/inhalt.HTM>

12 Böhringer D (2004) *Wertlos – brauchbar – sehr gut: Über Sinn und Unsinn von Bodenindikatoren*. <http://www.dbsv.org/dbsv/download/GFUV/Wertlos-brauchbar-sehr%20gut%20-%20Bodenindikatoren.rtf>

13 Kooi FL (year?) *The perceptibility of tactile pathways and warnings*. Translation of the TNO report TM-98-C072. TNO Human Factors Research Institute, Soesterberg, NL

14 Ståhl A, Almén M, Wemme M. *Orientation using guidance surfaces – Blind tests of tactility in surfaces with different materials and structures* [Att orientera med hjälp av ledytor – Blinda testar taktiliteten i ytor med olika material och struktur], Swedish Road Administration, Borlänge, 2004 <http://www.vv.se/templates/page3____11372.aspx>

15 Grahmbeek AJ (2005) *Standardising provisions on substrates for the visually impaired*. Translation of a lecture in Dutch for the European Commission, Sept. 2005, Brussels

16 Bundesministerium für Verkehr (Hrsg.) (1997) *Bürgerfreundliche und behindertengerechte Gestaltung von Haltestellen des öffentlichen Personennahverkehrs*. Direkt 51. Fach Media Service, Bad Homburg

17 Mizukami N (2005) Installation of tactile ground surface indicators for blind persons on railway platforms. *Rail Technology Avalanche* 7: 43

18 Bentzen BL, Barlow JM, Tabor LS (2000) *Detectable Warnings: Synthesis of US and International Practice* <http://kiewit.oregonstate.edu/pdf/acc_DW-synthesis.pdf>

19 *Taktile Bodenelemente für sehbehinderte und blinde Fahrgäste in U-Bahn-Haltestellen und Eisenbahnbetriebsanlagen im Bereich des Hamburger Verkehrsverbundes*, Ergebnisbericht, Hamburg, October 2001

20 *DIN 18024 – Teil 1: Barrierefreies Bauen. Straßen, Plätze, Wege, öffentliche Verkehrs- und Grünanlagen sowie Spielplätze – Planungsgrundlagen. Teil 2: Barrierefreies Bauen. Öffentlich zugängliche Gebäude und Arbeitsstätten – Planungsgrundlagen*. Beuth Verlag, Berlin

21 Kobayashi Y, Takashima T, Hayashi M, Fujimoto H (2005) Gait analysis of people walking on tactile ground surface indicators. *IEEE Trans Neural Syst Rehabil Eng* 13(1): 53–59

22 O'Neil K (2001) *Galileo – European Satellite Navi-*

gation System. Advanced Aviation Technology Ltd <http://www.aatl.net/publications/galileo.htm>

42 HapticWalker – haptic foot device for gait rehabilitation

1 Kolominsky-Rabas PL, Sarti C, Heuschmann PU, Graf C, Siemonsen S, Neundoerfer B, Katalinic A, Lang E, Gassman KG, von Stockert TR (1998) A prospective community-based study of stroke in Germany: the Erlangen Stroke Project (ESPro): incidence and case fatality at 1, 3, and 12 months. *Stroke* 2501–2506
2 Carr JH, Shepherd RB (1998) *Neurological rehabilitation: Optimizing motor performance*. Butterworth-Heinemann
3 Asanuma H, Keller A (1991) Neurobiological basis of motor learning and memory. *Concepts Neuroscience* 2: 1–30
4 Hesse S, Bertelt C, Schaffrin A, Malezic M, Mauritz KM (1994) Restoration of gait in non-ambulatory hemiparetic patients by treadmill training with partial body weight support. *Arch Phys Med Rehabil* 75: 1087–1093
5 Colombo G, Joerg M, Schreier R, Dietz V (2000) Treadmill training of paraplegic patients using a robotic orthosis. *J Rehab Res Dev* 37 (6): 313–319
6 Reinkensmeyer DJ, Wynne JH, Harkema SJ (2002) A robotic tool for studying locomotor adaptation and rehabilitation. In: *Proceedings of the IEEE Engineering in Medicine and Biology Conference (EMBC)*. Houston, TX
7 Boian RF, Kourtev H, Erickson K, Deutsch JE, Lewis JA, Burdea GC (2003) Dual stewart platform gait rehabilitation system for individuals post stroke. In: *Proceedings of 2nd International Workshop on Virtual Rehabilitation (IWVR)*, Piscataway, NJ
8 Hollerbach JM (2002) Locomotion interfaces. In: Stanney KM (eds): *Handbook of virtual environments: design, implementation, and applications*. Lawrence Erlbaum, Mahwah, NJ, 239–254
9 Roston GP, Peurach T (1997) A whole body kinesthetic display device for virtual reality applications. In: *Proceedings of IEEE International Conference on Robotics & Automation (ICRA)*, Albuquerque, NM, 3006–3011
10 Shiozawa N, Arima S, Makikawa M (2004) Virtual walkway system and prediction of gait mode transition for the control of the gait simulator. In: *Proceedings of the IEEE Engineering in Medicine and Biology Conference (EMBC)*, San Francisco, CA
11 Yano H, Noma H, Iwata H, Miyasato T (2000) Shared walk environment using locomotion interfaces. In: *Proceedings of ACM Conference on Computer Supported Cooperative Work (CSCW)*, Philadelphia, PA, 163–170
12 Yano H, Kasai K, Saitou H, Iwata H (2003) Development of a gait rehabilitation system using a locomotion interface. *J Visualization and Computer Animation* 14 (5): 243–252
13 Yoon J, Ryu J, Burdea G (2003) Design and analysis of a novel virtual walking machine. In: *Proceedings of the 11th Symposium on Haptic Interfaces for Virtual Environment and Teleoperator Systems (Haptics'03)*, Los Angeles, CA
14 Schmidt H, Sorowka D, Hesse S, Bernhardt R (2002) Design of a robotic walking simulator for neurological rehabilitation. In: *Proceedings of IEEE International Conference on Intelligent Robots and Systems (IROS)*, Lausanne, Switzerland
15 Woltring HJ (1986) A FORTRAN package for generalized cross-validatory spline smoothing and differentiation. *Adv Eng Software* 8 (2): 104–113
16 Perry J (1992) *Gait analysis*. Slack, Thorofare, NJ
17 Winter DA (1990) *Biomechanics and motor control of human movement*, 2nd ed. Wiley, New York
18 Schmidt H, Hesse S, Bernhardt R (2004) Safety concept for robotic gait trainers. In: *Proceedings of the IEEE Engineering in Medicine and Biology Conference (EMBC)*, San Francisco, CA
19 Albus JS, McCain HG, Lumia R (1987) NASA/NBS standard reference model for telerobot control system architecture (NASREM). *Tech Rep* 1253, National Institute of Standards and Technology (NIST), Gaithersburg, MD
20 Schmidt H, Sorowka D, Piorko F, Marhoul N, Hesse S, Bernhardt R (2004) Control system for a robotic walking simulator. In: *Proceedings of IEEE International Conference on Robotics and Automation (ICRA)*, New Orleans, LA
21 Schmidt H, Sorowka D, Piorko F, Marhoul N, Bernhardt R (2004) Generation of cyclic trajectories

for a robotic walking simulator. In: *Proceedings of 'Robotik 2004'*, Munich, Germany

22 Schmidt H, Sorowka D, Piorko F, Marhoul N, Hesse S, Bernhardt R (2003) Trajectory generation for gait rehabilitation. In: *Proceedings of the 8th International Conference on Rehabilitation Robotics (ICORR)*. Daejeon, Korea, 1: 62–63

43 Haptic sensing of virtual textiles

1 Magnenat-Thalmann N, Volino P (2005) From early draping to haute couture models: 20 years of research. *The Visual Computer* 21(8): 506–519

2 Govindaraj M, Garg A, Raheja A, Huang G, Metaxas D (2003) Haptic simulation of fabric hand. *Proc of Eurohaptics Conference 2003* 1: 253–260

3 Magnenat-Thalmann N, Volino P, Bonanni U, Summers IR, Bergamasco M, Salsedo F, Wolter FE (2007) From physics-based simulation to the touching of textiles: The HAPTEX Project. *Int J Virtual Reality* 6(3): 35–44

4 Magnenat-Thalmann N, Volino P, Bonanni U, Summers IR, Brady AC, Qu J, Allerkamp D, Fontana M, Tarri F, Salsedo F et al (2007) Haptic simulation, perception and manipulation of deformable objects. In: K Myszkowski, V Havran (eds): *Eurographics 2007 – Tutorials*. Eurographics Association, Prague, 1–22

5 Hatch KL (1993) *Textile Science*. West Publishing Company, Minneapolis

6 AATCC Committee RR89 (2001) Fabric hand: guidelines for the subjective evaluation of. *AATCC Evaluation Procedure* 5

7 Luible C, Varheenmaa M, Magnenat-Thalmann N, Meinander H (2007) Subjective fabric evaluation. *Proc Int Conference on Cyberworlds* 285–291

8 Luible C, Magnenat-Thalmann N (2008) The simulation of cloth using accurate physical parameters. *Computer Graphics and Imaging* 123–128

9 Volino P, Davy P, Bonanni U, Luible C, Magnenat-Thalmann N, Mäkinen M, Meinander M (2007) From measured physical parameters to the haptic feeling of fabric. *The Visual Computer* 23(2): 133–142

10 Etzmuss O, Gross J, Strasser W (2003) Deriving a particle system from continuum mechanics for the animation of deformable objects. *IEEE Transactions on Visualization and Computer Graphics* 9(4): 538–550

11 Volino P, Magnenat-Thalmann N (2005) Accurate garment prototyping and simulation. *Computer-Aided Design & Applications* 2(5): 645–654

12 Allerkamp D, Böttcher G, Wolter FE, Brady AC, Qu J, Summers IR (2007) A vibrotactile approach to tactile rendering. *The Visual Computer* 23(2): 97–108

13 Böttcher G, Allerkamp D, Wolter FE (2007) Virtual reality systems modelling haptic two-finger contact with deformable physical surfaces. *Proc Int Conference on Cyberworlds 2007*, 292–299

14 Summers IR, Brady AC, Syed M, Chanter CM (2005) Design of array stimulators for synthetic tactile sensations. *Proc World Haptics '05*, 586–587

15 Bergamasco M, Salsedo F, Fontana M, Tarri F, Avizzano C, Frisoli A, Ruffaldi E, Marcheschi S (2007) High performance haptic device for force rendering in textile exploration. *The Visual Computer* 23(4): 247–256

16 *The HAPTEX Project Website* (2008) Deliverable 4.3: Whole Haptic Interface Hardware. 18 February 2008 <http://haptex.miralab.unige.ch/public/deliverables/HAPTEX-D4.3.pdf>

17 Fontana M, Marcheschi S, Tarri F, Salsedo F, Bergamasco M, Allerkamp D, Böttcher G, Wolter FE, Brady AC, Qu J et al (2007) Integrating force and tactile rendering into a single VR system. *Proc Int Conference on Cyberworlds 2007*, 277–284

18 *The HAPTEX Project Website* (2008) Deliverable 5.2: Final Demonstrator. 18 February 2008 <http://haptex.miralab.unige.ch/public/deliverables/HAPTEX-D5.2.pdf>

44 Haptic discrimination of paper

1 Lederman SJ, Klatsky RL (1993) Extracting object properties through haptic exploration. *Acta Psychol* 84: 29–40

2 John KT, Goodwin AW, Darian-Smith I (1989) Tactual discrimination of thickness. *Exp Brain Res* 78: 62–68

3 Ho C-H (1991) *Human haptic discrimination of*

thickness. MSc Thesis, Massachusetts Institute of Technology

4 Lederman SJ, Loomis JM, Williams DA (1982) The role of vibration in the tactual perception of roughness. *Percept Psychophys* 32: 109–116

5 Taylor MM, Lederman SJ (1975) Tactile roughness of grooved surfaces: a model and the effect of friction. *Percept Psychophys* 17: 23–36

6 Sathian K (1989) Tactile sensing of surface features. *Trends Neurosci* 12: 513–519

7 Culbert SS, Stellwagen WT (1963) Tactual discrimination of textures. *Percept Motor Skill* 16: 545–552

8 Katz D (1925) *Der Aufbau der Tastwelt*, translated by Lester E Krueger (1989) as *The World of Touch*. Lawrence Erlbaum Associates, New Jersey

9 Hollins M, Faldowski SR, Rao S, Young F (1993) Perceptual dimensions of tactile surface texture: A multidimensional scaling analysis. *Percept Psychophys* 54: 697–705

10 Hollins M, Bensmaia S, Karlof K, Young F (2000) Individual differences in perceptual space for tactile textures: Evidence from multidimensional scaling. *Percept Psychophys* 62: 1534–1544

11 Hollins M, Bensmaia SJ, Roy EA (2002) Vibrotaction and texture perception. *Behav Brain Res* 135: 51–56

12 Bensmaia S, Hollins M (2005) Pacinian representations of fine surface texture. *Percept Psychophys* 67: 842–854

13 Tiest WMB, Kappers AML (2006) Analysis of haptic perception of materials by multidimensional scaling and physical measurements of roughness and compressibility. *Acta Psychol* 121: 1–20

14 Lyne MB, Whiteman A, Donderi DC (1984) Multidimensional scaling of tissue quality. *Pulp Pap-Canada* 85: 43–50

15 Kawabata S (1980) *The standardization and analysis of hand evaluation* (2nd edition). The Textile Machinery Society of Japan, Osaka

16 Picard D, Dacremont C, Valentin D, Giboreau A (2003) Perceptual dimensions of tactile textures. *Acta Psychol* 114: 165–184

17 Cooke T, Steinke F, Wallraven C, Bülthoff HH (2005) A similarity-based approach to perceptual feature validation. *Proc 2nd Symposium on Applied Perception in Graphics and Visualization*, New York, 59–66

18 Bowers RL, Mollenhauer MS, Luxford J (1990) Short-term memory for tactile and temporal stimuli in a shared-attention recall task. *Percept Motor Skill* 70: 903–913

19 Craven BJ (1992) A table for d' for M-alternative odd-man-out forced-choice procedures. *Percept Psychophys* 51: 379–385

20 Schiffman SS, Reynolds ML, Young FW (1981) *Introduction to multidimensional scaling – theory, methods, and applications*. Academic Press, New York

21 Kruskal JB (1964) Nonmetric multidimensional scaling: a numerical method. *Psychometrika* 29: 115–129

22 Green DM, Swets JA (1966) *Signal detection theory and psychophysics*. Wiley, New York

23 Rust JP, Keadle TL, Allen DB, Shalev I, Barker RL (1994) Tissue softness evaluation by mechanical stylus scanning. *Text Res J* 63: 163–168

24 Heinemann S (1996) Estimation of surface-roughness and air permeability of paper according to Bendtsen. *Papier* 50: 233–241

25 Moss PA, Retulainen E, Paulapuro H, Aaltonen P (1993) Taking a new look at pulp and paper – Applications of Confocal Laser Scanning Microscopy (CLSM) to Pulp and Paper Research. *Pap Puu-Pap Timb* 75: 74–79

26 Braida LD, Durlach NI (1972) Intensity perception: II. Resolution in one-interval paradigms. *J Acoust Soc Am* 51: 483–502

45 Haptic banknote design

1 Merabet L, Thut G, Murray B, Andrews J, Hsiao S, Pascual-Leone A (2004) Feeling by sight or seeing by touch? *Neuron* 42: 173–179

2 Guest S, Spence C (2003) Tactile dominance in speeded discrimination of textures. *Exp Brain Res* 150: 201–207

3 Locher PJ (1982) Influence of vision on haptic encoding processes. *Percept Mot Skills* 55: 59–74

4 Atkins JE, Jacobs RA, Knill DC (2003) Experience-dependent visual cue recalibration based on discrepancies between visual and haptic percepts. *Vision Res* 43: 2603–2613

5 Stevens JC, Choo KK (1996) Spatial acuity of the

body surface over the life span. *Somatosens Motor Res* 13: 153–166
6 Sathian K, Zangaladze A, Green J, Vitek JL, DeLong MR (1997) Tactile spatial acuity and roughness discrimination: impairments due to aging and Parkinson's disease. *Neurology* 49: 168–177
7 Dinse HR, Kleibel N, Kalisch T, Ragert P, Wilimzig C, Tegenthoff M (2006) Tactile coactivation resets age-related decline of human tactile discrimination. *Ann Neurol* 60: 88–94
8 Van Boven RW, Hamilton RH, Kauffman T, Keenan JP, Pascual-Leone A (2000) Tactile spatial resolution in blind braille readers. *Neurology* 54: 2230–2236
9 Goldreich D, Kanics IM (2003) Tactile acuity is enhanced in blindness. *J Neurosci* 23: 3439–3445
10 Heinisch C, Kalisch T, Dinse HR (2006) Tactile and learning abilities in early and late-blind subjects. *Soc Neurosci Abstr* 52.4/N2
11 Lederman SJ, Hamilton C (2002) Using tactile features to help functionally blind individuals denominate banknotes. *Hum Factors* 44: 413–428
12 Dinse HR (2005) Tactile Banknote Authentication: Effective intaglio structures and their reliable recognition throughout the population. Panel Discussion: The Advantages of Two-Sided Intaglio Printing. Banknote 2005. Washington DC, 20–23 February 2005. A Banknote Conference® in association with Kelly, Anderson & Associates, Inc.
13 Gescheider GA, Bolanowski SJ, Greenfield TC, Brunette KE (2005) Perception of the tactile texture of raised-dot patterns: a multidimensional analysis. *Somatosens Mot Res* 22: 127–140
14 Kops CE, Gardner EP (1996) Discrimination of simulated texture patterns on the human hand. *J Neurophysiol* 76: 1145–1165
15 Cascio CJ, Sathian K (2001) Temporal cues contribute to tactile perception of roughness. *J Neurosci* 21: 5289–5296
16 Craig JC (1985) Tactile pattern perception and its perturbations. *J Acoust Soc Am* 77: 238–246
17 Craig JC (1989) Interference in localizing tactile stimuli. *Percept Psychophys* 45: 343–355
18 Gilbert CD, Wiesel TN (1990) The influence of contextual stimuli on the orientation selectivity of cells in primary visual cortex of the cat. *Vision Res* 30: 1689–1701

46 Get touched – bodycare as a design of media for self-perception

1 Sontag S (1997) *Über Fotografie*. Fischer Taschenbuchverlag, Frankfurt/M.
2 Baudrillard J Hrsg (1989) *Ars Electronica. Philosophie der neuen Technologien*. Merve Verlag GmbH, Berlin
3 Böhme H (2006) *Fetischismus und Kultur – eine andere Theorie der Moderne*. Rowohlt, Reinbek
4 Jütte R (2000) *Geschichte der Sinne – Von der Antike bis zum Cyperspace*. C.H. Beck Verlag, München
5 Metelmann J Hrsg (2005) *Porno-Pop – Sex in der Oberflächenwelt*. Verlag Könighausen & Neumann, Würzburg
6 Corbin A (1984) *Pesthauch und Blütenduft*. Wagenbach Verlag, Berlin
7 Vigarello G (1988) *Wasser und Seife, Puder und Parfüm – Geschichte der Körperhygiene seit dem Mittelalter*. Campus, F/M
8 http://de.wikipedia.org/wiki/Grooming
9 Grunwald M, Beyer L (Hrsg) (2001) *Der bewegte Sinn*. Birkhäuser Verlag, Basel, Boston, Berlin
10 http://service.spiegel.de/digas/find?DID=54230942

List of contributors

Eckart Altenmüller, University for Music and Drama, Hannover, Institute for Music Physiology and Musicians' Medicine, Hohenzollernstr. 47, 30161 Hannover; Germany *e-mail: altenmueller@hmt-hannover.de*

Peter A. Andersen, School of Communication, San Diego State University, San Diego, CA 92182-4561, USA; *e-mail: peterand@mail.sdsu.edu*

Carlo Alberto Avizzano, PERCRO Laboratory, Scuola Superiore Sant'Anna, 56127 Pisa, Italy; *e-mail: carlo@sssup.it; website: www.percro.org*

Bernd Bader, BMW Group, Entwicklung Anzeige- und Bedienkonzept, Knorrstr. 14780788 München, Germany; *e-mail: bernd.bader@bmw.de*

Soledad Ballesteros, Universidad Nacional de Educación a Distancia, Department of Basic Psychology II, Juan del Rosal, 10, 28040 Madrid, Spain; *e-mail: mballesteros@psi.uned.es*

Michael J. Banissy, Department of Psychology, University College London, London WC1E 6BT, UK

Klaus I. Baumann, Institute of Experimental Morphology, University of Hamburg, UKE, Martinistr. 52, 20246 Hamburg, Germany; *e-mail: mail@klaus-baumann.de*

Klaus Bengler, BMW Group Forschung und Technik, Connected Drive, Projects human machine interaction, Hanauerstr. 46, 80992 München, Germany; *e-mail: Klaus-josef.Bengler@bmw.de*

Massimo Bergamasco, PERCRO Laboratory, Scuola Superiore Sant'Anna, 56127 Pisa, Italy; *e-mail: bergamasco@sssup.it; website: www.percro.org*

Alec Bernstein, Strategy, Research & Strategic Partnering, BMW Group, Designworks USA, 2201 Corporate Center Drive, Newbury Park, CA 91320, USA; *e-mail: Alec.Bernstein@designworksusa.com*

Ugo Bonanni, MIRALab, University of Geneva, Battelle, Bâtiment A, route de Drize 7, 1227 Carouge/Genève, Switzerland; *e-mail: ugo.bonanni@miralab.unige.ch*

Alan C. Brady, Biomedical Physics Group, School of Physics, University of Exeter, Exeter, EX4 4QL, UK; *e-mail: Alan.C.Brady@exeter.ac.uk; website: http://newton.ex.ac.uk/research*

Christoph Braun, Institute of Medical Psychology and Behavioural Neurobiology, University of Tübingen, Germany, MEG-Centre, University of Tübingen, Germany, and Center for Mind/Brain Sciences, University of Trento, Italy; *e-mail: christoph.braun@uni-tuebingen.de*

Catherina R. Burghart, Institute of Process Control and Robotics, University of Karlsruhe, Building 40.28, Kaiserstrasse 12, 76128 Karlsruhe, Germany; *e-mail: burghart@ira.uka.de*

Daniel-Robert Chebat, Department of Psychology, Université de Montréal, Montréal, Qc, H3C 3J7 Canada; *e-mail : dchebat00@hotmail.com*

Jonathan Cole, Department of Clinical Neurophysiology, Poole Hospital, Longfleet Road, Poole, BH15 2JB, UK; *e-mail: jonathan.cole@poole.nhs.uk*

Yuri P. Danilov, Department of Orthopedics and Rehabilitation Medicine University of Wisconsin – Madison, 1300 University Avenue, Madison, WI 53706, USA, *e-mail: ydanilov@wisc.edu; website: http://kaz.med.wisc.edu*

Hubert R. Dinse, Institut für Neuroinformatik, Dept. Theoretical Biology, Neural Plasticity Lab, Ruhr University, Building ND04, 44780 Bochum, and Haptec – Haptic Research & Technology GmbH, 58740 Fröndenberg, Germany; *e-mails: hubert.dinse@neuroinformatik.rub.de, info@haptec.de; website: www.haptec.de*

Holger Enigk, Daimler AG, Customer Research Center (GR/VER), HPC: 059/H602, Leibnizstr. 2, 71032 Boeblingen, Germany; *e-mail: holger.enigk@daimler.com*

Marc O. Ernst, Max Planck Institute for Biological Cybernetics, Spemannstr. 41, 72076 Tübingen, Germany; *e-mail: marc.ernst@tuebingen.mpg.de; website: www.kyb.mpg.de/de/ernstgroup/index.html*

Rand B. Evans, Department of Psychology, East Carolina University, Greenville, NC, USA; *e-mail: evansr@ecu.edu*

Armando R. Favazza, Department of Psychiatry, University of Missouri-Columbia, 3 Hospital Drive, Columbia, MO 65201, USA; *e-mail: favazzaa@health.missouri.edu*

Antonio Frisoli, PERCRO Laboratory, Scuola Superiore Sant'Anna, 56127 Pisa, Italy; *e-mail: a.frisoli@sssup.it; website: www.percro.org*

Uli Foehl, Daimler AG, Customer Research Center (GR/VER), HPC: 059/H602, Leibnizstr. 2, 71032 Boeblingen, Germany; *e-mail: uli.foehl@daimler.com*

Edouard Gentaz, Laboratoire de Psychologie et Neurocognition (UMR 5105 CNRS), Domaine Universitaire, 1251 Avenue Centrale, P.O. Box 47, Université Pierre Mendès-France, 38040 Grenoble Cedex 9, France ; *e-mail: Edouard.Gentaz@upmf-grenoble.fr*

Claire-Marie Giabbiconi, Universität Leipzig, Institut für Psychologie I, Seeburgstr. 14-20, 04103 Leipzig, Germany; *e-mail: giabbico@rz.uni-leipzig.de*

Rainer Goebel, Department of Cognitive Neuroscience, Faculty of Psychology, Maastricht University, Universiteitssingel 40, Postbus 616, 6200 MD Maastricht, The Netherlands; *e-mail: r.goebel@psychology.unimaas.nl; website: www.psychology.unimaas.nl*

Dirk Göger, Institute of Process Control and Robotics, University of Karlsruhe, Building 40.28, Kaiserstrasse 12, 76128 Karlsruhe, Germany; *e-mail: goeger@ira.uka.de*

Sabine Gollek, University of Leipzig, Department of Psychiatry, Johannisallee 20, 04317 Leipzig, Germany; *e-mail: sabine.gollek@medizin.uni-leipzig.de*

Antony W. Goodwin, Department of Anatomy and Cell Biology, University of Melbourne, Parkville, Victoria 3010, Australia; *e-mail: a.goodwin@unimelb.edu.au*

Martin Grunwald, University of Leipzig, Paul Flechsig Institute for Brain Research, Haptic and EEG-Research Laboratory, Johannisallee 34, 04103 Leipzig, Germany; *e-mail: Martin.Grunwald@medizin.uni-leipzig.de; website: www.haptik-labor.de*

Laura K. Guerrero, The Hugh Downs School of Human Communication, Arizona State University, Tempe, AZ 85287-1205, USA; *e-mail: aura.guerrero@asu.edu*

Zdenek Halata, Institute of Experimental Morphology, University of Hamburg, UKE, Martinistr. 52, 20246 Hamburg, Germany, and Department of Anatomy, First Faculty of Medicine, Charles University Prague, Czech Republic; *e-mail: halata@uke.uni-hamburg.de*

Yvette Hatwell, Laboratoire de Psychologie et Neurocognition (UMR 5105 CNRS), Domaine Universitaire, 1251 Avenue Centrale, P.O. Box 47, Université Pierre Mendès-France, 38040 Grenoble Cedex 9, France, *e-mail: Yvette.Hatwell@upmf-grenoble.fr*

Vincent Hayward, Haptics Laboratory, Center for Intelligent Machines, McGill University, Montréal, H3A 2A7 Canada; *e-mail: Hayward@cim.mcgill.ca*

Hannah B. Helbig, Max Planck Institute for Biological Cybernetics, Spemannstr. 41, 72076 Tübingen, Germany; *e-mail: helbig@tuebingen.mpg.de; website: www.kyb.mpg.de/de/ernstgroup/index.html*

Morton A. Heller, Eastern Illinois University, 600 Lincoln Ave., Physical Science Building, Charleston, IL 61920, USA; *e-mail: maheller@eiu.edu*

Peter G. Hepper, School of Psychology, Queens University Belfast, Belfast BT7 1NN, N. Ireland, UK; *e-mail: p.hepper@qub.ac.uk*

Stefan Hesse, Charité University Hospital/Klinik Berlin, Department of Neurological, Rehabilitation, Kladower Damm 223, 14089 Berlin, Germany; *e-mail: bhesse@zedat.fu-berlin.de*

Gerd Hirzinger, Institute of Robotics and Mechatronics, German Aerospace Center (DLR) Oberpfaffenhofen, 82234 Wessling, Germany; *e-mail: Gerd.Hirzinger@dlr.de*

Steven Hsiao, 338 Krieger Hall, Krieger Mind/Brain institute, The Johns Hopkins University, Baltimore, Maryland 21218, USA; *e-mail: Steven.Hsiao@jhu.edu; website: http://neuroscience.jhu.edu/StevenHsiao.php*

Thomas Hulin, Institute of Robotics and Mechatronics, German Aerospace Center (DLR) Oberpfaffenhofen, 82234 Wessling, Germany; *e-mail: Thomas.Hulin@dlr.de*

Richard J. Irwin, Biomedical Physics Group, School of Physics, University of Exeter, Exeter, EX4 4QL, UK; *e-mail: richard.irwin@ballinger.co.uk; website: http://newton.ex.ac.uk/research*

Hiroo Iwata, University of Tsukuba, Department of Intelligent Interaction Technologies, Tennoudai 1-1-1, Tsukuba, 305-8573, Japan; *e-mail: iwata@kz.tsukuba.ac.jp*

Hans-Christian Jabusch, University for Music and Drama, Hannover, Institute for Music Physiology and Musicians' Medicine, Hohenzollernstr. 47, 30161 Hannover; Germany; *e-mail: jabusch@hmt-hannover.de*

Matthias John, University of Leipzig, Haptic Research Laboratory, Johannisallee 34, 04103 Leipzig, Germany; *e-mail: kraf@medizin.uni-leipzig.de*

Clare N. Jonas, Department of Psychology, University of Sussex, Falmer, Brighton, BN1 9QH, UK

Robert Jütte, Institut für Geschichte der Medizin der Robert Bosch Stiftung, Straussweg 17, 70184 Stuttgart, Germany: *e-mail: Robert.juette@igm-bosch.de*

Amanda L. Kaas, Maastricht University, Department of Cognitive Neuroscience, Faculty of Psychology, Universiteitssingel 40, Postbus 616, 6200 MD, Maastricht, The Netherlands; *e-mail: a.kaas@psychology.unimaas.nl; website: www.psychology.unimaas.nl*

Kurt A. Kaczmarek, Department of Orthopedics and Rehabilitation , University of Wisconsin-Madison, 1300 University Ave., Madison, WI 53706, USA; *e-mail: kakaczma@wisc.edu*

Tobias Kalisch, Institut für Neuroinformatik, Dept. Theoretical Biology, Neural Plasticity Lab Ruhr University, Building ND04, 44780 Bochum, Germany; *e-mail: tobias.kalisch@rub.de*

Christiane Kiese-Himmel, Department of Otorhinolaryngology – Head and Neck Surgery: Phoniatrics and Pedaudiology, University of Göttingen, Robert-Koch-Str. 40, 37075 Göttingen, Germany; *e-mail: ckiese@med.uni-goettingen.de*

Jörg Krüger, Fraunhofer Institute for Production Systems and Design Technology (IPK), Automation Technology Division, Pascalstrasse 8-9, 10587 Berlin, Germany; *e-mail: joerg.krueger@ipk.fraunhofer.de*

Hermann Künzner, BMW Group, Entwicklung Anzeige- und Bedienkonzept, Knorrstr. 147, 80788 München, Germany; *e-mail: Hermann.Kuenzner@bmw.de*

Ronald E. Kupers, Department of Surgical Pathophysiology and Positron Emission Tomography Unit, Rigshosptitalet, 2100 Copenhagen, Denmark; *e-mail: ron@pet.rh.dk*

Simon Lacey, Department of Neurology, Emory University, 101 Woodruff Circle, WMB 6000, Atlanta, GA 30322, USA

Nadia Magnenat-Thalmann, MIRALab, University of Geneva, Battelle, Bâtiment A, route de Drize 7, 1227 Carouge/Genève, Switzerland; *e-mail: thalmann@miralab.unige.ch*

Thomas Maier, Institute for Engineering Design and Industrial Design (IKTD), Universität Stuttgart, Pfaffenwaldring 9, 70569 Stuttgart, Germany; *e-mail: thomas.maier@iktd.uni-stuttgart.de; website: www.iktd.uni-stuttgart.de/design*

Matthias M. Müller, Universität Leipzig, Institut für Psychologie I, Seeburgstr. 14-20, 04103 Leipzig; *e-mail: m.mueller@rz.uni-leipzig.de*

Carsten Preusche, Institute of Robotics and Mechatronics, German Aerospace Center (DLR) Oberpfaffenhofen, 82234 Wessling, Germany; *e-mail: Carsten.Preusche@dlr.de*

Maurice Ptito, Chaire de recherche Harland Sanders, Ecole d'Optométrie, Université de Montréal, 3744 Jean-Brillant, Montréal, Qc, H3T 1P1, Canada; *e-mail: maurice.ptito@umontreal.ca*

Gabriel Robles-De-La-Torre, International Society for Haptics, MDM Coyoacán, Apdo. Postal 21-058, c.p. 04021 Coyoacán, México D.F., México; *e-mail: Gabriel@RoblesDeLaTorre.com*

Timm Rosburg, Department of Epileptology, University of Bonn, Sigmund-Freud-Str. 25, 53105 Bonn, Germany; *e-mail: timm.rosburg@ukb.uni-bonn.de*

Helen E. Ross, Department of Psychology, University of Stirling, Scotland FK9 4LA, UK; *e-mail: h.e.ross@stir.ac.uk*

Krish Sathian, Departments of Neurology, Rehabilitation Medicine, and Psychology, Emory University, 101 Woodruff Circle, WMB 6000, Atlanta, GA 30322, and Rehabilitation R&D Center of Excellence, Atlanta VAMC, Decatur, GA, USA; *e-mail: krish.sathian@emory.edu*

Henning Schmidt, Fraunhofer Institute for Production Systems and Design Technology (IPK), Automation Technology Division, Rehabilitation Robotics Group, Pascalstrasse 8-9, 10587 Berlin, Germany; *e-mail: henning.schmidt@ipk.fraunhofer.de*

Kurt Seikowski, University of Leipzig, Department of Dermatology, Training Centre of the European Academy of Andrology, Philipp-Rosenthal-Str. 23-25, 04103 Leipzig, Germany; *e-mail: kurt.seikowski@medizin.uni-leipzig.de*

Ian R. Summers, Biomedical Physics Group, School of Physics, University of Exeter, Exeter, EX4 4QL, UK; *e-mail: i.r.summers@exeter.ac.uk; website: http://newton.ex.ac.uk/research*

Eva Kristin Stein, Burg Giebichenstein - University of Art and Design Halle, Hegelstraße 3, 06114 Halle/Saale, Germany; *e-mail: Eva@stign.de; website: www.stign.de*

M. Cornelia Stoeckel, University of Oxford, J R Hospital, Centre for Functional MRI of the Brain (FMRIB), Headley Way, Headington, Oxford, OX3 9DU, UK; *e-mail: cornelia@fmrib.ox.ac.uk; website: www.fmrib.ox.ac.uk*

Werner Tietz, AUDI AG, I/EK-3, 85045 Ingolstadt, Germany; *e-mail: werner.tietz@audi.de; website: www.audi.com*

Mitchell E. Tyler, Senior Lecturer, Biomedical Engineering, Researcher, Ortho-Rehabilitation Medicine, University of Wisconsin, Madison, WI 53706, USA; *e-mail: metyler1@wisc.edu*

Verena Wagner, Daimler AG, Customer Research Center (GR/VER), HPC: 059/H602, Leibnizstr. 2, 71032 Boeblingen, Germany; *e-mail: verena.wagner@daimler.com*

Jamie Ward, Department of Psychology, University of Sussex, Falmer, Brighton, BN1 9QH, UK; *e-mail: jamiew@sussex.ac.uk*

Thomas Weiss, Friedrich-Schiller-University Jena, Department of Biological and Clinical Psychology, Am Steiger 3, H. 1, 07743 Jena, Germany; *e-mail: thomas.weiss@uni-jena.de*

Karsten Weiß, Institute of Process Control and Robotics, University of Karlsruhe, Building 40.28, Kaiserstrasse 12, 76128 Karlsruhe, Germany; *e-mail: kweiss@ira.uka.de*

Heather E. Wheat, Department of Anatomy and Cell Biology, University of Melbourne, Parkville, Victoria 3010, Australia; *e-mail: h.wheat@unimelb.edu.au*

Claudia Wilimzig, California Institute of Technology, Division of Biology, Pasadena, CA 91125, USA; *e-mail: claudia@klab.caltech.edu*

Heinz Wörn, Institute of Process Control and Robotics, University of Karlsruhe, Building 40.28, Kaiserstrasse 12, 76128 Karlsruhe, Germany; *e-mail: woern@ira.uka.de*

Jeffrey Yau, 338 Krieger Hall, Krieger Mind/Brain institute, The Johns Hopkins University, Baltimore, Maryland 21218, USA; *e-mail: yau@mbi.mb.jhu.edu*

Glossary

ACTIVE TOUCH: Touch in which, say the hand, moves to manipulate and explore an object to determine its properties, so using both touch on the skin but touch informed by and as a result of active exploratory movement too. As such active touch uses feedback of movement and position sense and comparison of this with motor commands as well as peripheral sensory feedback to discriminate. In passive touch, in contrast, the skin is touched by an object and does not move to explore.

ACTUATION POINT: The position at which the jump mechanism is activated on the force-travel characteristic of a push-push switch.

ADAPTATION: The systematic change in the response of a neuron to constant stimuli. Slowly adapting neurons show a mild decline in firing rate to an indented bar. Rapidly adapting neurons respond only with brief bursts of action potentials to the indentation and retraction of the bar.

AFFECTION TOUCH: A cornerstone of personal relationship, this haptic behavior communicates messages of warmth, comfort, liking and affection.

AFFECTIVE TOUCH: Touch may be divided into discriminatory functions (how large, soft, hot etc an object is, for instance), but also how pleasant it is. This pleasant, emotional or affective dimension is often encountered within a more social affiliative context.

AFFERENT NERVES: Nerves whose impulses travel towards the central nervous system, i.e. sensory nerves.

ALLODYNIA (MECHANICAL): A pathological condition in which light touch on the skin is perceived as being painful. A stimulus which is normally painless becomes felt as painful.

ALTERNOBARIC VERTIGO: Dizziness affecting ascending divers with a blocked eustachian tube, causing unequal middle ear pressure in the two ears. There is a sensation of spinning towards the blocked ear, which retains a higher pressure and stimulates the semicircular canals.

ALZHEIMER'S DISEASE (AD): It is a progressive neurodegenerative disorder, the most common type of senile dementia and the fourth cause of death in older adults. The neuropathological characteristics of ad are the amyloid plaques and neurofibrillary tangles that occur very early in the course of the disease (even before the clinical diagnostic), starting in the mesial-temporal lobe (entorhinal cortex and hippocampus. The disease progresses to other cortical association neocortical areas in an anterior-to-posterior projection.

AMERICANS WITH DISABILITIES ACT (ADA): Civil right law in the USA that prohibits discrimination based on disability, signed 1990.

AMPA: α-amino-3-hydroxy-5-methyl-4-isoxazolepropionic acid is an agonist for a subtype of synaptic glutamate receptors. It mediates fast synaptic transmission.

ANTHROPOMORPHIC ARM: A robotic arm using the same number and configuration of joints as a human arm including joint constraints.

APLYSIA: A large sea slug. Because of its big and few neurons it serves a model to study the cellular mechanisms of learning.

ARTIFICIAL SKIN: Mat/suit sensors used to envelop a robot. The artificial skin comprises tactile sensors and others, i.e. sensors for temperature, distance etc.

ATTACHMENT STYLE: A relational predisposition in infants and adults developed through one's early interaction with caregivers.

AUTISM: A neurodevelopmental disorder accompanied by severe deficiencies of speech/language development and nonverbal communication.

BARREL CORTEX: Cortical representation area of whiskers in rodents. Since the representations of individual whiskers are quite large and somatotopically organised, barrel cortex serves as model for the study of plastic changes in cortical representations.

BARRIER-FREE: Allowing the accessibility and usability of buildings and facilities for physically disadvantaged or disabled persons by architectural or technical means; example: Ramp for wheel chairs.

BAYESIAN DECISION THEORY: Statistical approach to model decision making based on the probability that expresses knowledge and uncertainty of states (likelihood function, prior knowledge) and a utility function that expresses the preferences for consequences of the decision or action.

BENCHMARK: Comparative analysis of products, processes and services.

BILATERAL SYMMETRY: Symmetry in which similar anatomical parts are arranged on opposite sides of a median axis so that only one plane can divide the individual into essentially identical halves.

BILATERAL VESTIBULAR LOSS (BVL): A condition in which the vestibular apparatus in the inner ear does not provide adequate accurate information to the brain concerning head acceleration. BVL is sometimes referred to as bilateral vestibular disorder or dysfunction (BVD).

BLINDNESS: Loss of vision due physiological or neurological reasons; often also used to describe severe visual impairment with some residual vision; total blindness describes the complete loss of form and light perception.

BOTTOM-UP: Bottom-up perceptually-driven processing that deals with a physical stimulus as opposed to top-down processing which refers to higher-level processing about the nature of the stimulus.

BOTULINUM TOXIN: A neurotoxic protein produced by the bacterium Clostridium botulinum. In very small dosages, botulinum toxin is used for chemical denervation for therapeutic purposes, e.g. for the treatment of dystonias and spastic disorders.

BRAILLE: The braille system, devised in 1821 by Louis Braille, is a method that enables blind people to read and write. Each braille character or 'cell' is made up of six dot positions, arranged in a rectangle containing two columns of three dots each. A dot may be raised at any of the six positions to form 64 combinations (including the combination in which no dots are raised). These dot patterns code for letters and numbers.

BRAIN ORGANISATION: Different brain areas can be distinguished with respect to anatomy, histology and function. Moreover even within a histologically homogenous brain area functional specialisation can be observed.

BRAINSTEM: The anatomically lowest part of the brain, and serves to exchange information between the brain and the spinal cord and cranial nerves, as well a providing a variety of subconscious information processing and regulatory functions.

BURGER MODEL: A mechanical model describing linear visco-elastic behaviour using four parameters: Two springs and two dashpots.

CAPACITIVE SENSORS: Utilise the change of capacitance between two electrodes covering a deformable clielectric in order to detect a contact between sensor and object/human being.

CELL RECORDING: An invasive technique in which a microelectrode is inserted in a neuron to measure changes in voltage or current related to action potentials.

CEREBELLUM: Also called 'little brain'. A part of the brain anatomically behind the brainstem that provides a variety of sensory integration and both gross and fine motor control functions.

CLOTH SIMULATION: Digital reproduction of the appearance and dynamic behaviour of clothes and textile materials.

COACTIVATION: Tactile stimulation protocol to induce learning processes. Coactivation utilizes hebbian learning rules to drive synaptic changes in the cortical maps representing the stimulated skin sites. As a consequence, tactile processing is altered which alters the entire haptic perception.

COLLISION DETECTION: Uses tactile sensors to detect an unintended collision between a robot and an object or human being. Spatial resolution of the detected contact is not necessarily required.

CONCEPTUALIZATION (CONCEPT FORMATION): Perceptual level at which stimuli are responded to as a class (or concept) based on some similarity or common features (e.g. object properties).

CONNECTIVITY: Neuronal connectivity refers to a pattern of anatomical links ('anatomical connectivity'), of statistical dependencies ('functional connectivity') or of causal interactions ('effective connectivity') between distinct units within a nervous system. Connectivity studies investigate the network(s) of different brain areas that are involved in performing a particular task.

CONTACT CULTURES: Societies where interpersonal touch is more normative, typically closer to the equator and non-asian.

CONTACT MECHANICS: Analysis of the deformation of solids when they come into contact.

CONTROL CONTACTS: Are initiated by persons in the environment of the robot wishing the robot to perform a definite action i.e. pushing the robot out of the way, manually guiding the robot etc.

CO-OPERATION: Is a team temporally working together to achieve a common aim.

COROLLARY DISCHARGE: When a movement command is made within the brain initiated at the same time is a signal within the brain, a corollary discharge, which allows comparison between the command and any peripherally originating feedback, so allowing the animal to distinguish between self-generated, internal commands and external feedback.

CORTEX SEGREGATION: Segregation with respect to the functional organization of cortex means that representational zones for different functions are clearly separated and do not show any overlap.

CORTICAL REPRESENTATION: Input at sensory epithelia is forwarded to cortex forming a representation of the sensory input.

COUPLING DEGREE: Here: The force-fit or form-fit connection between the interface of hand and handle.

CROSS-MODAL (PRIMING) IMPLICIT MEMORY: It refers to the facilitation showed (in accuracy or response time) when a stimulus (e.g., an object) is presented to a perceptual modality (e.g., vision) at encoding and the same stimulus is again presented but now to other perceptual modality (e.g., touch).

CROSSMODAL PLASTICITY: The brain activity associated with a given function can move to a different location as a result of both normal and abnormal experience.

CUTANEOUS INNERVATION DENSITY: The number of primary afferent fibres innervating a unit area of skin.

CUTANEOUS RECEPTIVE FIELD: The region of skin from which a primary afferent fibre's response can be modulated.

DAMPING: An effect that dissipates energy and slows down motion. Damping elements are used in control theory to stabilize a system.

DATA GLOVE: Digitally equipped device that allow to track the posture of the user's hand.

DEAFFERENTATION: Loss of sensory or afferent nerves. This may occur after sectioning of the sensory nerve roots or following disease of the sensory peripheral nerves, or from some central nervous system lesions.

DELAYED MATCHING-TO-SAMPLE TASK: An experimental task in which the participant is presented with a sample of one or more stimuli. After a delay, a probe stimulus is presented and the participant decides whether this was part of the original sample.

DISCRIMINATION INDEX (d′): A measure of the perceived difference between two objects, normalised according to the perceptual uncertainty of the relevant object attributes. In the context of object discrimination, the discrimination index is the perceptual equivalent of signal-to-noise ratio.

DISPLAYS (HAPTIC DISPLAYS): Devices designed to mediate a physical signal intended to be perceived.

DOCTRINE OF SPECIFIC NERVE ENERGIES: A physiological doctrine devised by johannes mueller in the 1830s that held that every nerve fiber communicates one and only one quality to the "sensorium." He held that it was the nerve and not the stimulating object that was responsible for the sensation.

DORSAL COLUMNS: Large posterior part of the spinal cord through which most sensory nerve fibres run up to the brain.

DORSAL ROOTS: Posterior nerve roots which convey sensory nerve axons to the spinal cord. The anterior or ventral roots are efferent or motor.

DORSAL VISUAL PATHWAY: Begins with V1, goes through visual area V2, then to the dorsomedial area and visual area mt (also known as V5) and to the posterior parietal cortex. The dorsal stream is associated with motion.

D-PRIME: d-prime (or d′) is a so-called sensitivity index used in signal detection theory. An estimate of d′ is obtained from measurements of the hit rate and false-alarm rate. A higher d′ indicates that the signal can be more readily detected. Statistically, d′ measures the separation between the means of the signal and noise distributions in units of the standard deviation of the noise distribution.

DRIVING MODE: Mode of the driver that is based on tasks and conditions such as traffic, weather or passengers. The driver's window of comprehension and ability to handle information complexity is highly dependent on the actual driving mode.

DYSTONIA: A syndrome characterised by involuntary sustained muscle contractions, frequently causing twisting and repetitive movements, or abnormal postures. Dystonia localised to a single body part such as the hand or neck is referred to as focal dystonia.

ELECTROENCEPHALOGRAPHY: Electroencephalography (EEG) is a non-invasive diagnostic tool for the examination of brain functions. Electrodes placed on the scalp measure changes in the electrical potential caused by neuronal activity. The recordings are analyzed with a number of different techniques to obtain either brain responses triggered by an external or internal event. Using multiple electrodes, eeg can also be used to localise brain functions and to estimate their time course with a millisecond temporal resolution.

ELECTROTACTILE STIMULATION: Electrotactile stimulation produces localized tactile sensations on the skin by passing a small, controlled electric current into the skin using surface electrodes.

ENCODING: The process of converting sensations into (working) memory representations/mental representations.

ENCOUNTERED HAPTICS: A contactless haptic interface that only encounters human body when a touch sensation is required by exploration or interaction with dynamic or movable objects.

EPIPHENOMENAL: Epiphenomenal brain activity that occurs during task performance but which is not causally related to it.

ESTHESIOMETER: Instrument for measuring the cutaneous sensory discrimination. The two ends of a kind of compass are pressed simultaneously on the skin and their distance is varied until the participant perceives two separate points and not a unique point.

EXOSKELETON: A set of actuators attached to a hand or a body.

EXPLICIT (HAPTIC) MEMORY: A type of long-term memory for haptically explored stimuli that refer to conscious recollection of previous experience with the stimuli. It is usually assessed by free recall, cued recall, and recognition memory tests.

EXPLORATION CONTACTS: Contacts occurring /measured when exploring the shape and structure of an object by active touching.

EXTERME VIRTUAL REALITY (XVR): A 3D graphical software development environment used for fast development and deploy of interactive virtual environments. With respect to existing 3D tools, XVR allows realtime interaction, haptic control and an open interface for 3rd part libraries.

EXTRASTRIATE VISUAL CORTEX: A belt of visual association areas surrounding the primary visual cortex and involved in higher visual processing.

FABRIC HAND: The sum of the sensations experienced when a fabric is touched or otherwise handled with the fingers.

FAST ADAPTING AFFERENT: A primary afferent fibre that responds to indentation of the skin with a dynamic (when the stimulus is changing) response only.

FIELD EXPERIMENT: Scientific method to investigate experimental interventions in real world situations.

FMRI (FUNCTIONAL MAGNETIC RESONANCE IMAGING): Functional magnetic resonance imaging is a non-invasive method to visualise the concentration of oxygenated haemoglobin. Neuronally active brain regions show increased oxygen consumption which is satisfied after a delay of some seconds by an overshoot of oxygen. Although the blood oxygen level dependent (BOLD) response depends on haemodynamical and vasodynamical a high correlation between BOLD and the local field potentials (LF potentials) have been found. LF potentials are summated electrical potentials that can be measured by extracellular recordings from cortical tissue.

FOCAL DYSTONIA: Focal dystonia is a neuromuscular disease affecting the control of muscles in specific parts of the body. The loss of control can be seen in involuntary muscular contractions or twistings.

FOETUS: The developing individual from eight weeks gestational age until term (delivery).

FORCE FEEDBACK: Technique employed to realize the simulation of haptic interaction either remotely or in a virtual reality.

FORCE-TRAVEL: Describes the force which has to be applied along the characteristic distance to be covered between touching the switch and triggering the pressure function. The force-travel characteristic is crucial to the overall haptic impression and experienced appeal.

GABA: Gamma-aminobutyric acid is an inhibitory neurotransmitter in the central nervous system.

GESTATIONAL AGE: The age of the fetus determined from the first day of the mother last menstrual period.

GLABROUS SKIN: Skin, lacking hair, covering the palms of the hands and soles of the feet.

GOLGI TENDON ORGANS: Tension receptors at the juncture between skeletal muscle and tendon.

GRAB, GRAPHICAL ACCESS FOR BLIND: A two arms haptic system developed to deploy haptic applications to visual impaired people.

GRAVIMETER: Instrument for measuring the sensory discrimination of weights (or more generally of gravities). In its initial and most simple form, it consisted in a horizontal bar which could rotate around its centre and was ended by a small weight. The participant pressed on the other extremity of the bar in order to compare successively two weights.

GRIPPING: For example, the way different handles are activated or held.

HALL EFFECT SENSOR: A magnetic field revealer that detects the amount of field along a given direction. Coupled with magnets allows to determine the field and the associated distance of the magnet.

HAND MOVEMENTS: Stereotyped exploratory procedures (EPs) executed by the active human explorer intends with the hands to extract different object attributes such as texture, weight, shape, hardness, etc. Lederman and Klatzky (1987) described different types of exploratory stereotyped hand movements executed by the haptic explorer when intend to extract a particular type of information.

HAPTIC: The science of all biological and psychological aspects of human touch processes. This term was first used in 1892 by the german psychologist max dessoir.

HAPTIC DEVICES: Devices designed to mediate haptic interactions involving bilateral mechanical signals between a subject and the world outside.

HAPTIC FEEDBACK: Information for a system or user i.e. whether an item is firmly grasped or sliding or of the elasticity of an object touched in a virtual environment.

HAPTIC ILLUSION: Break of physical coherence between real motion and feedback forces, used to create illusion of non-existing feature or compensate by illusion the sensation of an unwanted detail.

HAPTIC INTERACTION POINT (HIP): The point in the simulated space representing the position of the haptic interface, i.e. The actual user's hand position.

HAPTIC MEMORY: Type of sensory memory for stimuli received through the sense of touch. It allows the user of visual-haptic interfaces to bulid a mental construct of the information space and use the system without even having to look.

HAPTIC PERCEPTION: Touch perception in which both the cutaneous sense and kinesthesis convey significant information about distal objects and events and body parts.

HAPTIC SIGNAL: A haptic variable that supplies important information for haptically perceiving one or more characteristics of a real or virtual environment.

HAPTIC SURFACE: A spatial profile used as reference to analyze proxy motion space and consequently generate force feedback.

HAPTIC VARIABLE: An entity playing a role in the physics of haptic interaction with real or virtual environments (e.g., contact forces during object palpation, motion and mass of limbs, mass of objects in the environment, weight of objects, stiffness of objects, etc.). A haptic variable may or may not supply important information for haptically perceiving specific characteristics of the environment. See also haptic signal.

HAPTIC VIRTUAL OBJECT: A virtual object that can be manipulated or touched with one's hands or body. It is a computational entity that is physically realized through haptic technology. Such virtual objects may reproduce mechanical characteristics (weight, shape, stiffness, surface texture, etc.) Of real objects in everyday environments, or have mechanical characteristics that do not normally exist in nature (see paradoxical haptic object).

HAPTIC-VISUAL ALIGNMENT: The process of aligning the visual model and the physical haptic action.

Hebb's rule: Theory introduced by donald hebb in 1949 stating that an increase in synaptic efficacy arises from the presynaptic cell's repeated and persistent stimulation of the postsynaptic cell.

Hemispheric specialisation: The fact that the two halves of the brain have different specialised functional centres in the cortical regions, for example, broca's and wernicke's areas in the left hemisphere are specialised for language and speech skills.

Homunculus: The term homunculus (latin: "Little man") is used to illustrate the somatotopy of sensory and motor functions in primary somatosensory and motor cortex, reflecting a complete image of the body.

Human factors: Human factors is a term for several areas of research that include human performance, technology, design, and human-computer interaction. It describes how people interact with products, tools, procedures, and any processes likely to be encountered in the modern world.

Humanoid robot: A robotic system having human-like kinematics, being able of human-like movements, communicating in a human-like manner and showing aspects of human-like intelligence.

Implicit (haptic) memory: A type of long-term memory that refers to previous experience with haptically explored stimuli (e.g., raised-line shapes, three-dimensional familiar or unfamiliar objects) that does not require intentional, conscious retrieval. It is assessed by indirect or implicit tests with no reference to previous experience. It is demonstrated by showing repetition priming effects (better performance with repeated compared to non-repeated stimuli).

Impossible haptic object: See paradoxical haptic object.

Infant touch: Haptic behavior between adult caregivers and infants that is vital for physical and psychological health of the infant.

Information transfer (IT): For a stimulus which is presented to an observer, a quantity which represents the number of distinct categories of stimulus which can be distinguished by the observer. If the number of distinct categories is N, the information transfer is log2(N). Information transfer is measured in dimensionless units of bits.

Intaglio: Intaglio is a family of printmaking techniques in which the image is incised into a surface, known as the matrix or plate. Usually, copper or zinc plates are used as a surface, and the incisions are created by etching, engraving, drypoint, aquatint or mezzotint. New technologies allow high-precision, computer-based incisions.

Integration: Representational zones for different functions overlap strongly.

Interface effect: Between metal electrode and structured sensor material results in a change in resistance which is a function of the applied load and time.

Interpersonal touch: Haptic behavior between individuals that communication warmth, affection, intimacy, immediacy, or conversely, threat or hostility.

Intimate terrorism: An unfortunately widespread pattern of abuse in one's close relationships often characterized by violent touch.

Inverse kinematics: Mathematical relationship among the position of a robotic limb and the joint coordinates required to get that position.

ISO: International organization for standardization; an international standard-setting body Composed of representatives from various national standard-setting organizations, head quarter located in geneva (www.iso.org).

Isoinertial balls: Balls of constant volume that vary in mass, but have equal moments of inertia due to the spacing arrangement of mass within the balls.

Isomorphic representation: A pattern of neural activity which has the same form or shape as the peripheral input. For example, when seeing the letter "a" there is an representation of the letter "a" across the receptor sheet of the retina that is in the same shape as the letter.

Just noticable difference (JND): Difference between two stimuli that can be detected some percentage of time. Often the JND is defined as 75% or 84% correct discrimination performance. JND is usually coded in arbitrary parameter conditioned by the virtual environment. The closer "JND in VE" is to human perception limit, the better is the realism of the environment.

Kinaesthesis: The ability to sense static and dynamic body posture, i.e. Relative positioning of the head, torso, limbs, and end effectors.

Language disorder: Language behaviours that exhibit delayed and/or impaired than expected normal language development.

LATERAL FORCE (IN HAPTIC ILLUSION): Haptic effect to provide the sensation of a vertical bump by laterally stretching the human skin.

LATERALISED BEHAVIOUR: Refers to behaviour that is distributed unevenly between the sides of the body, e.g. Handedness where 90% of the population use their right hand, for example, for writing. It is argued to be linkd to hemispheric specialisation.

LEAD TRAVEL: In the force-travel characteristic of a push-push switch this is the distance up to the actuation point.

LEARNING: Learning is the ability of humans, animals and artificial cognitive systems to acquire new skills, values, understanding, and preferences due to perceived information. It allows individuals to adapt to changes in their environment. Learning has to be distinguished from maturation, fatigue and changes in behaviour directly elicited by injuries.

LIGHT-WEIGHT ROBOT: Is a serially-linked robot arm with seven joints and an outstanding load to weight ratio of 1:1. It has been developed at the German Aerospace Center (DLR). Electronics, including torque and position sensors, are integrated in the robot.

LONG CANE: Cane used by blind pedestrians for maintaining orientation; used for scanning the ground in walking direction.

LONG-TERM DEPRESSION (LTD): Following long lasting deprivation of stimulation synaptic transmittion is decreased.

LONG-TERM POTENTIATION (LTP): Following strong, long lasting repetitive stimulation synaptic transmittion is increased.

MAGNETOENCEPHALOGRAPHY (MEG): Magnetoencephalography (MEG) is an imaging technique based on the measurement of magnetic fields generated by the summed neuroelectrical activity in the brain. The magnetic brain activity is usually measured at up to 300 sensors with a temporal resolution in the millisecond range. From the topographical distribution of the magnetic activtiy the underlying brain areas being activated can be inferred.

MAGNIFICATION: The extent of cortical representations of different skin areas varies considerably. Larger cortical representations for finger tips and mouth than for other parts of the body reflect strong magnification.

MAINTENANCE: The process of keeping information available in working memory.

MALADAPTIVE PLASTICITY: Reorganization of neuronal pathways and / or topographical representations with pathological consequences. Maladaptive plasticity is, for example, hypothesized to be involved in the pathophysiology of task specific focal dystonia.

MANIPULATION: The performance of mental operations on information in working memory.

MAP: Functions of the brain are localised and organised in specific regions. Thus the topographical representation of functions can be regarded as a map.

MAXIMUM LIKELIHOOD ESTIMATE (MLE): Perceptual estimate based on the maximum of the likelihood distribution that is used to model noisy sensory signals.

MEAN LENGTH OF UTTERANCES (MLU): Measure of language development based on average number of morphemes per utterance.

MECHANICAL SIGNAL: See Haptic signal.

MECHANICAL VARIABLE: See Haptic variable.

MEISSNER CORPUSCLES: Rapidly adapting mechanoreceptors (RA or FA) in the dermal papillae of glabrous skin responding to changes in pressure or indentation.

MENTAL MODELS: Representations of external reality in the mind of the user.

MERKEL CELL NERVE ENDINGS: Slowly adapting mechanoreceptors (SA I) in the epidermis of glabrous and hairy skin responding to pressure or indentation.

MI: Primary motor cortex. In humans MI is located in the pre-central gyrus. Basically it sends long axons to the spinal cord and is thus the brain area controlling muscular acitivty. Primary motor cortex reveals a somatotopic organisation, i.e. different muscles are controlled from different areas within MI.

MICRONEUROGRAPHY: A method of recording from single nerve fibres in the peripheral nerve by inserting a sharp microelectrode through the skin.

MILD COGNITIVE IMPAIRMENT (MCI): Is a condition in which an individual has problems with memory, language, or another mental function in a degree that is noticeable and to show up on tests, but not too serious to interfere with daily life activities. The best studied mci involves the deterioration of memory.

Mirror neuron: A neuron that responds equivalently depending on whether self or other is an agent.

Mirror-touch synaesthesia: A type of synaesthesia in which observed touch to another person is experienced as a tactile sensation on the observer's body.

Mislocalisation: If tactile stimuli with intensities close to the sensory threshold are presented subjects make errors. However, the errors don't follow a chance distribution. Instead subjects localise the erroneously localised stimuli closer to the actual stimulation site than more far apart.

Monotonic: Monotonic relationship between two variables means that each value for one corresponds to a unique value for the other such that they increase/decrease smoothly together when graphed with no 'steps'.

Motor schema: An implicit (non-conscious) motor programme stored within the brain and which enables performance of a movement.

Multimodal interface: Multimodal interface is a man machine interface that interacts with more than one human sense (modality), but typically visual, acoustic and haptic.

Multimodal simulation of virtual textiles: Computer-based simulation aiming at reproducing the interaction with textiles in a virtual environment. The multimodal characteristic requires the simulation to provide adequate physical stimuli for multiple perceptual modalities (e.g. visual, audio, tactile, or haptic) in a synchronised way.

Multiple pregnancies: Pregnancies with more than one fetus. Where fetuses are in the same sac (chorion) they are termed as monochorionic, when in different chorions they are termed dichorionic.

Multisensory integration: Integration of information derived from different sensory modalities, such as vision audition or touch.

Muscle spindles: Mechanoreceptors embedded between muscle fibres in skeletal muscle monitoring muscle length.

Musician's dystonia or musician's cramp: A type of task specific focal dystonia that may affect the control of hand and finger movements in instrumentalists of all instrument groups or the control of the embouchure in brass and wind players.

Near infrared spectroscopy (NIRS): A neuroimaging method visualising blood supply in the brain. Oxygenated and deoxygenated blood can be differentiated.

Neural code: The systematic relationship between the neural activity of a population of neurons and behavior. An example of a neural code is the firing rate of a population of neurons each with a different orientation tuning function for the perceived orientation of a bar indented on the skin.

Neuroplasticity: The ability of brains o adapt continuously to changing conditions in the internal and external environment. Neuroplasticity processes are the basis of learning, and are not limited to early developmental phases but present through the whole life-span into high-age.

Nitric oxide (NO): Nitric oxide is a neurotransmitter. Because of its high capacity to diffuse widely it can act on several nearby neurons. It is also assumed to play a role in retrograd signalling, transmitting information from the post- to the presysnaptic neuron.

N-methyl-D-aspartat (NMDA): N-methyl-D-aspartat is an agonist of a subtype of synaptic glutamate receptors. Due to the voltage dependent conductivity for cations NMDA receptors are assumed to play a role in associative learning.

Normal ageing: All human beings are involved in the ageing process and none can escape it. Many human abilities peak before age 30, while others continue to grow through life. Normal aging or "healthy aging" is a term that is widely used to describe the natural changes that occur in the absence of any disease. The great majority of people over age 65 are healthy and independent but some individuals experience signs of cognitive decline. A healthy lifestyle may slow negative effects of ageing.

Nystagmus: The set of involuntary eye movements that help the eyes to fixate on a target when the head is turning.

Object-oriented-type haptic interface: A device that physically simulate surface of virtual objects.

Oculoagravic illusion: Movement illusion caused by vertical linear acceleration, which adds or subtracts an accelerative force to that of earth's gravitational force. The effective force increases when an elevator accelerates up or decelerates down, and it decreases in the opposite cases.

Under increasing force the passenger feels himself pressed down and the framework of the elevator appears to rise for a moment, and under decreasing force the opposite occurs.

Oculogravic illusion: Movement illusion caused by horizontal linear acceleration, which combines with earth's gravitational force to form the gravitoinertial vertical. This causes an upright passenger to feel tilted and to see the visual world as tilted. Apparent movement of the visual world occurs when the accelerative force changes.

Orbital flight: In orbital spaceflight a vehicle such as the international space station orbits around the earth. It is in constant free fall, producing a state of weightlessness.

Oscillopsia: A visual disturbance, caused by impaired eye movement control, in which objects appear to blur or jump around. Oscilloscopia is common in individuals with vestibular disorders.

Parabolic flight: A parabolic flight path is the same as that of an object in free fall, such as a cannonball fired into the air. Weightlessness is achieved inside an aircraft as it travels "up-and-over the hump".

Paradoxical haptic object: A haptic virtual object (or a combination of haptic virtual objects with real objects) having mechanical characteristics that are not normally found in everyday objects. Paradoxical objects typically provide conflicting haptic sensory nformation. For example, during haptic exploration, paradoxical objects may supply contact forces that do not correspond to the object's actual geometry. Such objects provide powerful stimuli for studying haptic perception in new ways.

Parahippocampal place area (PPA): A subregion of the parahippocampal cortex that plays an important role in the encoding and recognition of scenes. This region of the brain becomes highly active when human subjects view topographical scene stimuli such as images of landscapes, cityscapes, or rooms (i.e. images of "places").

Parasthesia: Commonly referred to as pins and needles in an area of the skin. It can be a sensation of tingling, pricking, or numbness of a person's skin with no apparent long-term physical effect.

Passive stimulation: Comprises approaches to induce learning that do not require training or practicing. Instead, pure exposure to sensory stimulation, electrical stimulation of the brain or stimulation with transcranial magnetic stimulation drives changes of cortical processing thereby altering perception. In order to be effective, similar to coactivation certain stimulation requirements have to be met such as high-frequency, burst-like or prolonged stimulation.

Passive system: A system that does not generate energy, but only stores or dissipates the energy, which is supplied to the system.

Passive touch: It refers to a non-moving observer that obtains tactile information passively from cutaneous sources only

Peg-in-hole task: Performance test to assess goodness of a control system for robots. Peg in hole requires at once both position and force accuracy.

Perception: The process of acquiring, interpreting, selecting, and organizing sensory information.

Perceptual constancy: Perceptual phenomenon where the conscious experience tends to remain constant despite changes in the stimulus.

Perceptual learning: The ability of an organism to improve perception through training and practice.

Phantom limb: The feeling that an amputated limb still exists (e.g. can be moved; is painful; can sense touch).

Phantom pain: In case of limb or organ loss patients might still feel the part of the body that is not physically present. Often patients report painful sensations. The effect can be explained by uncontrolled activation of the still existing representations in cortex.

Phosphene: An entoptic phenomenon characterized by the experience of seeing light without light actually entering the eye.

Phrenology: A popular 19th century movement originated by franz josef gall in the late 18th century. It emphasized the brain as the organ of mind and that different faculties located in the brain served different human abilities. Important in phrenological teaching was that through exercising the faculties, a person could realize a greater portion of their maximum capacity than someone who does not.

Plasticity: Neuronal plasticity describes the potential

of the nervous system to react with adaptive changes to intrinsic or extrinsic inputs. Neuronal plasticity is a property of the neurons or rather neuronal networks to change, temporarily or permanently their biochemical, physiological and morphological characteristics. The term captures the ability of the nervous system to adapt its structural organization to new situations emerging from changes of developmental and environmental situations, as well as injuries.

Plate (in encoutered haptics): The planar shape used as end effector dish for showing the contact position and orientation during object touching.

Point of subjective equality (PSE): Different stimuli that are perceived to be equal.

Pointshell: A cloud of points describing the surface of an object. The points are approximately equidistant and represent a discretization of the object. A pointshell is used for the detection and computation of collisions.

Population response: The responses of a collection of neurons that respond to the stimulus.

Positron emission tomography (PET): A nuclear medicine imaging technique which produces a three-dimensional image or map of functional processes in the brain or body.

Pressure profile: Tactile sensors designed to measure a spatial resolution of the applied force as well as its amplitude deliver tactile data in form of images. Within these images texels, to which a force has been applied a coloured. The amplitude of the force at the specific texel is coded by different colours. This image is also called pressure profile.

Primary afferent fibre: A nerve fibre arising from a sensory receptor in the body and travelling in a peripheral neve to convey information to the central nervous system.

Proprioception: Usually considered to be the perception of movement and position sense of the body and limbs.

Proxy: A virtual point coupled with haptic device end effectors and used as reference to generate force feedback. Usually a proxy point tracks the haptic end effector with the exception that it cannot enter the surface of virtual objects. The distance vector among end effector and proxy, associated with the stiffness of contact defines the reflection force to be displayed to the user.

Pulse width modulation (PWM): A time based waveform used to control electrical motors with maximum energy savings with respect to linear driving systems. Side effect of pwm is residual em noise at typical frequencies of 20–30 KHz.

Push-push switch: Pressable switches which return to their original position (e.g. air-conditioning system, radio).

Reactive control: Uses a direct translation of acquired sensor data into a robotic action i.e. A force applied to the tool of a robot by a person results in a movement of the robotic arm.

Receptive field: The receptiv field of a somatosensory neuron comprises the region of the skin that when stimulated activates this neuron.

Receptive field profile: The spatial distribution of response modulation across a receptive field.

Relational stage: The degree of closeness or intimacy in a relationship that is crucial in determining the appropriateness of tactile behavior.

Reliability: Inverse of the variance of a perceptual estimate.

Reorganisation: Functions of the brain area localised and organised in specific regions of the brain. Changes in the organisation due to injuries, lesions, and due to altered sensory input and behavioural output as well as due to modified cerebral processing are referred to reorganisation.

(Haptic) repetition priming: It is the facilitation in cognitive processing that occurs for repeated exposures to stimuli. It is a nonconscious and indirect manifestation of the memory trace as is used as a way to assessing implicit memory. Here, it refers to the better performance in terms of accuracy and/or response time for haptically explored repeated stimuli compared to nonrepeated haptic stimuli.

Repetitive transcranial magnetic stimulation (RTMS): Repetitive tms application of a sequence of tms pulses producing a relatively long-lasting cortical deactivation, generally used when it is impractical for TMS to be applied during the experimental task, so that participants can be tested ‘offline’.

Requirements: Describe the exact needs regarding, for example, components, products, processes, systems in terms of fixed and sector requirements and preferences.

RESISTIVE TACTILE SENSORS: Use conductive polymers covering an electrode matrix. Dependent on load resistivity changes when pressure is exercised on this type of tactile sensors.

RETRIEVAL: The process of recalling information from (working) memory in response to some cue for use in a task.

RIGID-BODY DISPLACEMENT: Average displacement of all the particles of a solid expressed with six quantities.

ROBOT'S MODES OF OPERATION: Describe the actual state of the robotic system, whether it is active or passive (idle).

ROBOTIC ENVIRONMENT: Is the world a robot moves or acts in. This comprises static objects, rooms / houses as well as moving objects or persons and animals.

ROUGHNESS: Roughness is a measurement of small-scale variations in the height of a physical surface. Usually, spacing, size and density of raised elements are crucial factors that determine texture perception.

RUFFINI CORPUSCLES: Slowly adapting mechanoreceptors (SA II) in the connective tissue of the skin, ligaments and joint capsules responding to tissue stretch.

SAFETY MARGIN (IN GRASPING): The percentage of force used in excess with respect the minimum force required to hold an object.

SAME DIFFERENCE PROCEDURE: Experimental analysis to discriminate perceptual sensitivity by means of discrete thresholds. A subject is shown with two examples and has to assert if they look the same or not.

SECURE ATTACHMENT: A pattern of trust, exploration, and positive relational experiences that creates optimal closeness with other people.

SECURITY FEATURES (ON BANKNOTES): Security or safety features on banknotes are used to prevent or to impede counterfeiting. Widely used features are watermarks, security threads, ultraviolet fibres and ink, special foil elements, hologram patches or stripes, glossy stripes and iridescent stripe/shifting colours.

SENSOR MODEL: Mechanical and mathematical description/function modelling the behaviour of the sensor.

SENSORY ATAXIA: Loss of controlled movement and still posture at rest seen after profound loss of movement and position sensory input. This ataxia can occur after loss of this input without the person realising its cause. In others words one can have a sensory ataxia without perceived loss of proprioception.

SENSORY SUBSTITUTION: The use of one human sensory channel to receive information normally received by another sense, for example, to substitute touch for vision.

SHAPE: Geometrical attributes of a solid object that are invariant under any translation or rotation.

SI: Primary somatosensory cortex. In humans si brain area is located in the post-central gyrus. It receives input from various nuclei of the thalamus and is thus the first cortical relay station for somatosensory information. In different subregions the whole surface of the body is mapped multiply.

SINGLE-PULSE TMS: Single-pulse tms application of a single, precisely timed tms pulse during task performance when participants can be tested 'online'.

SLOWLY ADAPTING AFFERENT: A primary afferent fibre that responds to indentation of the skin with both a dynamic (when the stimulus is changing) and a static (when the stimulus is constant) response.

SNAP: Section of the force-travel characteristic of a push-push switch which manifests a drop in force between the actuation point and the snap point.

SNAP POINT: The point on the force-travel characteristic of a push-push switch as of which the force characteristic rises again.

SNAP TRAVEL: On the force-travel characteristic of a push-push switch this represents the section between the actuation point and the snap point.

SOMATOSENSORY EVOKED POTENTIAL (SEP): Neural response to transient stimuli. The sep consits of typical positive and negative going peaks. According to their latency (occurrence after stimulus onset) and polarity (negative or positive) they are labeled P50, N80, P100, N140 and late component (LC).

SOMATOSENSORY: A term referring to body senses. These include inputs related to the sense of temperature, pain, proprioception, mechanoreception and itch.

SOMATOTOPIC MAP: A term referring to the orderly representation of the body surface across a set of neurons or fiber tract. Thus, the head is represented next to the hands, which is next to the body trunk which is next to a representation of the legs.

SOMATOTOPY: The cortical representation of haptic sensation from the body surface in the cortex is organised such that adjacent areas of the body are represented in adjacent cortical regions.

SPATIAL ACUITY: The minimum distance at which two points can be discriminated from each other (see also two-point discrimination).

SPECIFIC LANGUAGE IMPAIRMENT (SLI): Term for the most frequent developmental disorder in childhood. A language disorder which is unrelated to sensory, motor, emotional causes, to general intellectual deficits or to environmental factors.

STABILITY IN TERMS OF CONTROL: Stability in terms of control theory means that the output of a system is bounded, if the input is bounded (also bibo-stability).

STEADY-STATE EVOKED POTENTIAL (SSEP): Sinusoidal continuous brain response that is elicited with repetitive stimuli (such as flicker light or vibrotactile stimuli). The ssep has the same frequency as the driving stimulus and can be measured with eeg and/or meg.

STEREOGNOSIS: The facility of knowing the size, shape, weight of an external object through exploratory, active touch.

STIFFNESS: Stiffness of a contact describes the rigidity of a contact in terms of a relation between the penetration depth and the reflecting force.

SURFACE TEXTURE: Visual and tactile aspects of a surface, particularly the small-scale visual and tactile aspects.

SYNAESTHESIA: A condition in which a stimulus triggers an 'extra' perceptual quality not experienced by most members of the population.

SYNDACTYLY: Pathological fusion of fingers or toes.

SYNTHETIC EXPERIMENT: An experimental situation in introspective psychology in which the analyzed components of a perceptual whole are recombined artificially to test the introspective analysis. It was also used to create an artificial illusion of a naturally occurring experience.

TACTILE DISPLAYS: Displays that present information to the human tactile sense that may be perceived as spatial and temporal patterns on the skin. Tactile displays may use mechanical, electrical, thermal, magnetic, or chemical means to stimulate the tactile receptors and nerves.

TACTILE FEEDBACK: The provision of sensory feedback in the form of vibration or pressure in response to the user's choices on the interface.

TACTILE FLATNESS: The experience of touching something flat.

TACTILE GROUND SURFACE INDICATOR: Any floor element which is directly or indirectly perceived by tactile or haptic sensation; tactile ground surface indicators serve as warning elements for potential hazards and for orientation.

TACTILE/TACTUAL PERCEPTION: All perceptions mediated by variations of passive cutaneous stimulations (e.g. pressure, vibrations). In contrast to haptic perception processes the perceived subject does not generate active finger/hand/ body movements.

TASK CONTACTS: Contacts that are related to the actual task of the robot like handling an object.

TASK SPECIFIC FOCAL DYSTONIA: A neurological disorder characterised by the loss of voluntary motor control of highly overlearned complex and skilled movement patterns in a specific sensory-motor task.

TAU EFFECT: A tactual illusion of apparent movement. Similar to the phi phenomenon of visual apparent movement, when a series of short and discrete pressure sensations are produced on the skin, they are perceived as though they were produced by a single stimulus drawn continuously between the points.

TEXEL: A discrete sensor cell of a tactile sensor system. Texels are arranged in homogeneous matrices, detecting an applied load profile.

TEXTURE: A regular (haptic) pattern shown with tactile and/or haptic information. Textures allow discrimination of surface properties different from friction and stiffness. Haptic textures are generated as zero means force feedback during haptic exploration.

THE POSTERIOR PARIETAL CORTEX: A portion of the parietal lobe which manipulates mental images, and integrates sensory and motor portions of the brain. It is involved in formation of plans.

TONGUE DISPLAY UNIT: A sensory substitution device developed by paul bach-y-rita that converts an image into electro-tactile pulses applied to the tongue.

Tool-handling-type haptic interface: A device that applies force through a tool such as pen or stylus.

Top-down: See bottom-up.

Touch avoidance: A predispostion to avoid or approach touch that varies individually or culturally.

Touch taboos: Haptic behaviors that are inappropriate in a given culture or relationship.

Transcranial magnetic stimulation (TMS): A noninvasive method to excite neurons in the brain. The excitation is caused by weak electric currents induced in the tissue by rapidly changing magnetic fields (electromagnetic induction). As a result, brain activity can be triggered or modulated without the need for invasive approaches.

Two-point discrimination: Two-point discrimination (E.H. Weber) is the ability to discern that two nearby objects touching the skin are truly two distinct points, and not one. It is often tested with two sharp points. The two-point discrimination varies from 2–4 mm (finger tips, lip) to 30–40 mms (back) across the surface of the body.

Unity assumption: Belief that sensory signals arise from the same object.

Usability: Denotes the ease and efficiency with which people can employ a particular tool or other human-made objects for their potential use; closely associated to user friendliness.

Use-dependent plasticity: A particular form of neuroplasticity, which can be induced by manipulating the amount of use of a body part. Typical examples are blind braille readers or musicians. Brain reorganization following use-dependent plasticity can be monitored using non-invasive imaging methods and is paralleled by alterations in perception and behaviour.

User friendliness: See usability.

Vater-pacini corpuscles or pacinian corpuscles: Vibration receptors in the deeper layers of the skin and connective tissue around joints.

Ventral visual pathway: Begins with V1, goes through visual area V2, then through visual area V4, and to the inferior temporal cortex. The ventral stream, is associated with form recognition and object representation.

Ventriloquism: Derived from the ventriloquist effect in which a person (the ventriloquist) manipulates his or her voice so that it appears that the voice is coming from elsewhere (a puppet). It is used as a synonym for the dominance of vision over audition for spatial information.

Vestibular sense: The vestibular sense provides information on linear and angular head acceleration to the balance, body and eye movement control centers of the brain. The gravity vector is also sensed as alinear acceleration.

Vestibular system: A: System of the inner ear that contributes to balance and the sense of orientation, along with visual and tactile-haptic informations. The semicircular canals indicate rotational acceleration, and the otoliths linear acceleration. The neural signals cause reflex eye movements and postural adjustments.

Virtual coupling: A simulated coupling between a virtual object position and the haptic interaction point (user's hand). Virtual coupling is mostly composed of a spring-damper system and produces forces that the user can feel during a haptic interaction.

Virtual prototyping: The use of haptics and virtual reality to integrate cad/cam process in industrial design.

Visual capture: Dominance of the visual information over information from other sensory modalities in creating a percept.

Visual overload: The volume of visual information being presented to the driver in the cockpit is too large to be processed entirely on time.

Visual ventral pathway: Visual ventral pathway dealing with perception for object recognition as opposed to the dorsal pathway specialized for spatial localization and action.

Visual-haptic interface: Car-driver interface that aligns the two senses – sight and touch – rather than treaing them as two separate systems (for example turn-push knob input devices).

Visual-haptic matching: The appropriate alignment of the visual model and the physical haptic action, so that there is a direct relationship between what you are feeling or doing and what you are seeing in response.

Voxel: A cubical volume element, representing a value on a regular grid in the three dimensional space, comparable with a pixel in 2D. A voxel stores the status of this volume in terms of occupancy.

WORK LOAD CASES: There are different work load cases which can physically and psychologically stress the user.

WORKING MEMORY: Concept from cognitive psychology. It refers to the structures and processes used for temporarily storing and manipulating information for upcoming tasks.

WORKING POSITION: The posture and the position a user takes to perform a task. Here a difference is made between static and dynamic postures.

Index